The Elements

	Symbol	Atomic No.	Atomic Mass		Symbol	Atomic No.	Atomic Mass
Actinium	Ac	89	[227]†	Mendelevium	Md	101	[258]†
Aluminum	Al	13	26.98154	Mercury	Hg	80	200.59
Americium	Am	95	[243]†	Molybdenum	Mo	42	95.94
Antimony	Sb	51	121.75	Neodymium	Nd	60	144.24
Argon	Ar	18	39.948	Neon	Ne	10	20.180
Arsenic	As	33	74.9216	Neptunium	Np	93	[237]†
Astatine	At	85	[210]†	Nickel	Ni	28	58.69
Barium	Ba	56	137.33	Niobium	Nb	41	92.9064
Berkelium	Bk	97	[247]†	Nitrogen	N	7	14.0067
Beryllium	Be	4	9.0122	Nobelium	No	102	[259]†
Bismuth	Bi	83	208.9804	Osmium	Os	76	190.2
Bohrium	Bh	107	[262]†	Oxygen	O	8	15.9994
Boron	B	5	10.811	Palladium	Pd	46	106.4
Bromine	Br	35	79.904	Phosphorus	P	15	30.97376
Cadmium	Cd	48	112.411	Platinum	Pt	78	195.08
Calcium	Ca	20	40.08	Plutonium	Pu	94	[244]†
Californium	Cf	98	[251]†	Polonium	Po	84	[209]†
Carbon	C	6	12.011	Potassium	K	19	39.098
Cerium	Ce	58	140.12	Praseodymium	Pr	59	140.9077
Cesium	Cs	55	132.9054	Promethium	Pm	61	[145]†
Chlorine	Cl	17	35.453	Protactinium	Pa	91	[231]†
Chromium	Cr	24	51.996	Radium	Ra	88	[226]†
Cobalt	Co	27	58.9332	Radon	Rn	86	[222]†
Copper	Cu	29	63.546	Rhenium	Re	75	186.21
Curium	Cm	96	[247]†	Rhodium	Rh	45	102.9055
Dubnium	Db	105	[262]†	Rubidium	Rb	37	85.4678
Dysprosium	Dy	66	162.50	Ruthenium	Ru	44	101.07
Einsteinium	Es	99	[254]†	Rutherfordium	Rf	104	[261]†
Erbium	Er	68	167.26	Samarium	Sm	62	150.36
Europium	Eu	63	151.96	Scandium	Sc	21	44.9559
Fermium	Fm	100	[253]†	Seaborgium	Sg	106	[263]†
Fluorine	F	9	18.99840	Selenium	Se	34	78.96
Francium	Fr	87	[223]	Silicon	Si	14	28.086
Gadolinium	Gd	64	157.25	Silver	Ag	47	107.868
Gallium	Ga	31	69.723	Sodium	Na	11	22.98977
Germanium	Ge	32	72.61	Strontium	Sr	38	87.62
Gold	Au	79	196.9665	Sulfur	S	16	32.066
Hafnium	Hf	72	178.49	Tantalum	Ta	73	180.9479
Hassium	Hs	108	[265]†	Technetium	Tc	43	[98]†
Helium	He	2	4.00260	Tellurium	Te	52	127.60
Holmium	Ho	67	164.9303	Terbium	Tb	65	158.9253
Hydrogen	H	1	1.0079	Thallium	Tl	81	204.383
Indium	In	49	114.82	Thorium	Th	90	232.038
Iodine	I	53	126.9045	Thulium	Tm	69	168.9342
Iridium	Ir	77	192.22	Tin	Sn	50	118.71
Iron	Fe	26	55.847	Titanium	Ti	22	47.87
Krypton	Kr	36	83.80	Tungsten	W	74	183.85
Lanthanum	La	57	138.9055	Uranium	U	92	238.0289
Lawrencium	Lr	103	[260]†	Vanadium	V	23	50.9415
Lead	Pb	82	207.2	Xenon	Xe	54	131.29
Lithium	Li	3	6.941	Ytterbium	Yb	70	173.04
Lutetium	Lu	71	174.97	Yttrium	Y	39	88.9059
Magnesium	Mg	12	24.305	Zinc	Zn	30	65.39
Manganese	Mn	25	54.9380	Zirconium	Zr	40	91.22
Meitnerium	Mt	109	[266]†				

Only 109 elements are listed. Elements 110–112 have not yet been named.

†Mass number of most stable or best-known isotope.

Source: American Chemical Society, *Chemistry in Context: Applying Chemistry to Society,* Third Edition, © 2000 McGraw-Hill Companies, New York, NY.

WATER CHEMISTRY

McGraw-Hill Series in Water Resources and Environmental Engineering

Consulting Editor

George Tchobanoglous, University of California, Davis

The Series

Bailey and Ollis: *Biochemical Engineering Fundamentals*

Benjamin: *Water Chemistry*

Bishop: *Pollution Prevention: Fundamentals and Practice*

Canter: *Environmental Impact Assessment*

Chanlett: *Environmental Protection*

Chapra: *Surface Water-Quality Modeling*

Chow, Maidment, and Mays: *Applied Hydrology*

Crites and Tchobanoglous: *Small and Decentralized Wastewater Management Systems*

Davis and Cornwell: *Introduction to Environmental Engineering*

deNevers: *Air Pollution Control Engineering*

Eckenfelder: *Industrial Water Pollution Control*

Eweis, Ergas, Chang, and Schroeder: *Bioremediation Principles*

LaGrega, Buckingham, and Evans: *Hazardous Waste Management*

Linsley, Franzini, Freyberg, and Tchobanoglous: *Water Resources Engineering*

McGhee: *Water Supply and Sewage*

Metcalf & Eddy, Inc.: *Wastewater Engineering: Collection and Pumping of Wastewater*

Metcalf & Eddy, Inc.: *Wastewater Engineering: Treatment, Disposal, Reuse*

Peavy, Rowe, and Tchobanoglous: *Environmental Engineering*

Rittmann: *Environmental Biotechnology: Principles and Applications*

Rubin: *Introduction to Engineering and the Environment*

Sawyer, McCarty, and Parkin: *Chemistry for Environmental Engineering*

Tchobanoglous, Theisen, and Vigil: *Integrated Solid Waste Management: Engineering Principles and Management Issues*

Wentz: *Safety, Health, and Environmental Protection*

WATER CHEMISTRY

MARK M. BENJAMIN
Department of Civil and Environmental Engineering
University of Washington

Boston   Burr Ridge, IL   Dubuque, IA   Madison, WI   New York   San Francisco   St. Louis
Bangkok   Bogotá   Caracas   Kuala Lumpur   Lisbon   London   Madrid   Mexico City
Milan   Montreal   New Delhi   Santiago   Seoul   Singapore   Sydney   Taipei   Toronto

McGraw-Hill Higher Education

*A Division of The **McGraw-Hill** Companies*

WATER CHEMISTRY

Published by McGraw-Hill, a business unit of The McGraw-Hill Companies, Inc., 1221 Avenue of the Americas, New York, NY 10020. Copyright © 2002 by The McGraw-Hill Companies, Inc. All rights reserved. No part of this publication may be reproduced or distributed in any form or by any means, or stored in a database or retrieval system, without the prior written consent of The McGraw-Hill Companies, Inc., including, but not limited to, in any network or other electronic storage or transmission, or broadcast for distance learning.

Some ancillaries, including electronic and print components, may not be available to customers outside the United States.

This book is printed on acid-free paper.

1 2 3 4 5 6 7 8 9 0 QPF/QPF 0 9 8 7 6 5 4 3 2 1

ISBN 0–07–238390–9

Publisher: *Thomas E. Casson*
Executive editor: *Eric M. Munson*
Editorial coordinator: *Zuzanna Borciuch*
Senior marketing manager: *John Wannemacher*
Project manager: *Vicki Krug*
Media technology senior producer: *Phillip Meek*
Production supervisor: *Enboge Chong*
Coordinator of freelance design: *Michelle D. Whitaker*
Cover designer: *Pam Verros*
Cover image: *© CSIRO Land and Water/William van Aken 1992*
Senior photo research coordinator: *Carrie K. Burger*
Photo research: *Cheryl DuBois, Feldman & Associates, Inc.*
Supplement producer: *Jodi K. Banowetz*
Compositor: *Interactive Composition Corporation*
Typeface: *10.5/12 Times Roman*
Printer: *Quebecor World Fairfield, Inc.*

Library of Congress Cataloging-in-Publication Data

Benjamin, Mark M.
    Water chemistry / Mark M. Benjamin. — 1st ed.
      p.   cm. — (McGraw-Hill series in water resources and environmental engineering.)
    Includes index.
    ISBN 0–07–238390–9
    1. Water chemistry. I. Title. II. Series.

GB855 .B46 2002
551.46'01'54—dc21   00–048986
                                                     CIP

www.mhhe.com

DEDICATION

To those who have nurtured me

Arthur and Hannah Benjamin

Doc and Hune Smith

Judith and Mara Benjamin

TABLE OF CONTENTS

PREFACE

OBJECTIVES AND TEXT ORGANIZATION

The past three decades have witnessed a virtual explosion in the range of topics gathered under the umbrella of environmental chemistry. Throughout this period, the basic principles of equilibrium chemistry, particularly as they apply to aquatic systems, have served as indispensable tools for understanding the composition of, and direction of change in, environmental systems. This fact, it seems to me, owes as much to the seminal book that established our current paradigm for studying and interpreting the chemistry of aquatic systems as to the centrality of the equilibrium principles themselves. That book, of course, is *Aquatic Chemistry,* by Stumm and Morgan, first published in 1970 and currently in its third edition. Since the publication of that text, the tools available for solving the equations that define and constrain the equilibrium composition of aquatic systems have been improved significantly, and those tools have been applied to an ever-expanding range of systems, but the basic approach for analyzing the systems has remained largely unaltered.

Though it is unarguably the definitive text in the field, *Aquatic Chemistry* is widely perceived as too advanced for students taking their first course in the subject area, particularly those with little background beyond an introductory course in general chemistry. As a result, over the years, a number of texts have emerged that attempt to convey the key concepts of equilibrium chemistry in a more accessible format. This text follows in that line, covering much of the same material but diverging in a few ways both substantive and stylistic. A brief outline of the text highlights both the similarities and differences.

The text starts with an overview of a few simple, well-known physical/chemical concepts: conservation of mass and energy, and the tendency for any system to change toward a more stable (less reactive) condition. In Chapter 1, a good deal of the vocabulary of equilibrium aquatic chemistry is defined, links between chemical parameters and reactivity are introduced, and the kinetic model for chemical equilibrium is developed.

Chapter 2 provides a more formalized approach to understanding and predicting chemical change, via the concepts of chemical thermodynamics. The presentation and level of coverage in this chapter, particularly the first half, differ substantially from those in most other texts in this field, and a case can be made that the presentation is beyond what is necessary or appropriate in an introductory course. Frankly, at times I have persuaded myself that this assessment is accurate. However, after deliberation, I always returned to the opinion that if I wanted students to understand how thermodynamics applies to aquatic systems, as opposed to simply understanding how to carry out useful thermodynamic calculations, I had to devote substantial space to the topic.

This decision reflects, in large part, my own frustration at having studied thermodynamics so often without quite seeing the connections among the various pieces. For instance, the relationships $\Delta \overline{G} = \Delta \overline{H} - T \, \Delta \overline{S}$ and $\Delta \overline{G} = \Delta \overline{G}^{\circ} + RT \ln a$ make it clear that enthalpy and entropy must be related to chemical activity a, but until recently, I did not understand how. Similarly, I could calculate the redox potential of a solution (E_H) and the surface potential on a suspended colloid (Ψ), but I was never quite sure if, or how, these two electrical potentials were related.

Once I sorted out those issues, I found the insights they provided immensely satisfying, and found that several ideas and principles that I had previously thought disconnected could be interpreted coherently. This cohesiveness is the essential beauty of thermodynamics, and in Chapter 2 I have attempted to convey some of that cohesiveness to students.

In Chapters 3 through 5 and 7 through 10, applications of the chemical principles introduced in Chapters 1 and 2 are presented in the context of specific types of chemical reactions. The first such reactions described focus on acid/base chemistry, in a section that comprises Chapters 3 through 5. This section differs from the discussion of acid/base equilibria in other texts in two ways that are significant. First, in presenting an algorithm for solving for the equilibrium pH of solutions prepared with known inputs, I have chosen to introduce both the proton condition and the *TOT*H equation. In my experience, although students can rapidly master the use of the *TOT*H equation to solve an acid/base problem, they gain a firmer grasp of the qualitative chemistry and the quantitative analysis of equilibrium solutions by writing out the proton condition table. On the other hand, the *TOT*H equation provides an excellent introduction to the development of the tableau that is at the core of numerical solutions to such problems. The essential identity of these two equations for characterizing the proton mass balance is emphasized, so that students understand that both equations provide the same information.

A second skill that is developed in this section is the ability to predict a priori the dominant acid/base species expected to be present at equilibrium, even when a complicated mixture of acids and bases has been used to prepare a solution. I derived this algorithm almost two decades ago, with a good deal of assistance from Dimitri Spyridakis. It has been very gratifying and more than a little surprising to see how enthusiastically other instructors have adopted the algorithm, now that I have begun to publicize what I had assumed was a widespread approach.

Chapter 6 diverges from the preceding and subsequent chapters, being devoted to a presentation of the most common features of some currently available software for solving chemical equilibrium problems. I have emphasized the solution approach taken in the MINEQL family of programs, without tying the discussion to any particular software package. While some instructors may choose to skip this chapter or to have students start using the software packages without going into the solution algorithms, I believe that understanding the basics of those algorithms is valuable, both pedagogically and to ease the learning of the program mechanics when applied to some important systems that are not covered in the manuals.

Chapters 7 through 10 describe, respectively, equilibrium between solutions and a gas phase, reactions of metals in aqueous systems (both complexation and precipitation/dissolution), equilibrium in systems where oxidation-reduction reactions are occurring, and equilibrium between solutions and solid surfaces (adsorption). In each of these chapters, the presentation includes both a formal mathematical analysis of the reactions of interest and a discussion of how those reactions are analyzed by using chemical equilibrium software. All these chapters also refer to the thermodynamic developments in Chapter 2, and the last two chapters rely heavily on that development in the analysis of how the local electrical potential can affect chemical behavior. It is particularly in these latter chapters that, I hope, the detailed discussion of electrical potential and activity coefficients in Chapter 2 pays dividends.

A Comment on the Text Length and a Philosophy of Instruction

One of the most difficult parts of writing this text has been finding the right balance between attention to fundamental concepts and problem-solving techniques. In striking that balance, I have been guided by my experience teaching water chemistry courses over the past 20 years, which has convinced me that students want and can handle more fundamentals than most instructors (including myself) have been providing. Ironically, in my opinion, what frustrates these students and sometimes leads them to believe that water chemistry is overwhelmingly difficult is that, as instructors, we have tried too hard to *simplify* the concepts. Too often, the simplifications we offer provide students the tools to derive correct answers to numerical problems, but only by following algorithms that they do not fully understand. Then, they feel intimidated and lost when faced with a problem for which the algorithm is inapplicable (or worse, they fail to realize that the algorithm is inapplicable and so apply it inappropriately).

I have therefore chosen to write longer and more detailed explanations of both the relevant chemistry and mathematics than are found in most other texts. Undeniably, this decision has lengthened the text, perhaps to the dismay of those who are perusing it for the first time. However, the range of topics covered is no greater than in other water chemistry texts, so the added length does not represent an increase in the conceptual material that readers are asked to master. To the contrary, my belief is that the extra explanatory material will actually *reduce* the time that students need to devote to learning the course content, while simultaneously facilitating a deeper understanding of the subject matter.

Having said that, I recognize that water chemistry courses are taught in many different formats—as semester courses, quarter courses, with and without laboratory components, etc.—and that many instructors will choose to cover only a portion of the text in their courses. In such cases, I believe that a successful course could be taught by omitting coverage of Chapters 10, 9, and 2, in that order, depending on the severity of the time constraints. While an understanding of chemical thermodynamics (Chapter 2) is certainly helpful for interpreting all types of chemical reactions, it is more central to discussions of redox and adsorption reactions (Chapters 9 and 10, respectively) than the reactions covered in earlier chapters. Alternatively, an instructor might consider omitting coverage of chemical equilibrium software packages (Chapter 6 and easily identifiable sections of Chapters 7 through 10).

One of my goals throughout the writing of the text has been to integrate the material within each chapter and between chapters as seamlessly as possible. The benefits of such integration are self-evident, but the integration does impair any effort to fashion a course based on reading of disparate sections. Therefore, my personal preference is not to respond to time constraints by eliminating coverage of selected, isolated sections of the text. I believe that, in the end, students are better served by reading and mastering Chapters 1 and 3 through 8 in their entirety, than by being exposed to all 10 chapters but feeling unsure about their mastery of any of them. However, I realize that different courses have different objectives, and I hope that instructors will experiment freely with various ways to use the text and provide feedback to me on how well those approaches work.

ACKNOWLEDGMENTS

In the end, this book was written because of the encouragement I received from students who flattered me into believing that I could write about water chemistry in a way that made sense to them. The faculty and students who use the book will be the ultimate judges of whether that flattery was merited. But regardless of the verdict, I owe a debt of gratitude to all the students over the years who have suffered through this process with me and who have challenged and rewarded me so.

At the risk of offending the many, I would be remiss if I did not acknowledge by name the few whose support has been so very far above and beyond the call. Paul Anderson has been, first and foremost, a friend for lo these 20 years. That he has been such while simultaneously playing the role of student and later colleague, and always that of gentle but firm critic, surely qualifies him for some sort of award. John Ferguson, Bruce Honeyman, Gregory Korshin, Jim Morgan, Mickey Schurr, John van Benschoten, Ray Simons, and David Waite all contributed generously of their time to help me understand bits of water chemistry that had me confused, and to point out to me portions of the text that needed revision. Desmond Lawler contributed portions of Chapter 7 as part of our joint efforts to write a textbook on physical and chemical water treatment processes. Jill Nordstrom provided student feedback at a level of detail that no author of a textbook deserves, but every author must dream of.

My wife, Judith, has been a source of support and encouragement throughout the years that I devoted to this project. When it seemed that both of our lives were being dominated by the writing effort, I could be reenergized by my fascination with the subject matter and a sense of making progress toward a lifelong dream. Judith shared neither of those sources of inspiration, yet she has remained steadfast throughout, energized by her love and her willingness to share my dream as her own. For that, I will be forever grateful. Finally, I thank my daughter Mara for giving me the joy of fatherhood.

I also gratefully acknowledge the support provided by the University of Washington throughout my career and by the University of New South Wales in Sydney, Australia, during my sabbatical there in 2000.

ABOUT THE AUTHOR

Mark M. Benjamin is the Alan and Ingrid Osberg Professor of Civil and Environmental Engineering at the University of Washington, where he has served on the faculty since 1978. He received his undergraduate degree in chemical engineering at Carnegie-Mellon University and his Master's and Ph.D. degrees in environmental engineering from Stanford University.

Dr. Benjamin has directed and published research on various aspects of water chemistry and water treatment processes. His long-term interests have been in the behavior of metals and their interactions with mineral surfaces, and in the reactions of natural organic matter in water treatment systems. He and his students have received three patents for treatment processes that they have developed. Their publications have won several awards, and three of his students have won awards for best doctoral dissertation in environmental engineering.

Dr. Benjamin is a member of several professional societies and has served on the board of the Association of Environmental Engineering and Science Professors.

1

CONCEPTS IN AQUATIC CHEMISTRY

CHAPTER OUTLINE

1.1   INTRODUCTION

Over the past several decades, the field of environmental aquatic chemistry has expanded to encompass studies of the source, distribution, transport, reaction rates, and fate of chemicals in natural aquatic systems, while maintaining its historical links to the chemistry of water treatment processes. The subject matter of this text is a small but important subset of that field, specifically the prediction of the chemical composition of solutions that have reached chemical equilibrium. Such predictions are useful both because many systems closely approach their equilibrium state within the time frame of interest and because, even in systems that are far from this condition, the direction of chemical change and the expected ultimate composition are of interest.

In addition to describing approaches for computing the equilibrium composition of a system, the text provides an introduction to the types of chemicals that are present in natural waters, drinking water, and certain wastewaters, and to the reactions that those chemicals might undergo when the water is used, mixes with other waters, or interacts with soils and sediments. The techniques presented in the text form the basis for analyzing or designing facilities to bring about many desired changes in water quality via water or waste treatment.

The focus of the text is on aqueous solutions of environmental significance, which are almost always relatively dilute systems: typically, the reactants of interest comprise around 10^{-8} to 0.1 percent of all molecules in the system, with the remaining molecules being water. Such systems are typical of natural waters (lakes, rivers, oceans, groundwater) and also of domestic and industrial wastewaters, runoff from residential or agricultural areas, and many other solutions.

The text emphasizes the determination of the equilibrium chemical speciation, where speciation refers to the distribution of the chemicals among their various forms. For instance, in anaerobic (oxygen-free) environments, dissolved sulfur can be present as hydrogen sulfide (H_2S), but in aerobic environments it is usually converted to sulfate ion[1] (SO_4^{2-}). Hydrogen sulfide is a weak acid that reacts strongly with metals to form solid precipitates. It can also escape from solution into the gas phase, generating unpleasant odors at extremely low concentrations and creating significant health risks at higher (but still very low) concentrations. By contrast, sulfate ion is a very weak base that cannot exist as a gas, does not react significantly with most metals, and is generally innocuous as a health hazard. Thus, knowledge of the speciation of the sulfur is essential if we are to understand its behavior and its impact on water quality, or to design engineering processes to remove it from solution. Similar comparisons could be made regarding the properties of various forms of chlorine, chromium, arsenic, and many other chemicals of environmental importance. The calculations emphasized in the text allow us to determine the conditions that favor formation of one chemical form of an element or another.

Concentrations of some species of interest in a variety of solutions are presented in Table 1.1, and the range of solute concentrations in terrestrial waters is shown in Figure 1.1. In some situations, dissolved contaminant concentrations of a few micrograms per liter or less can be a cause of major concern. For example, some carcinogenic compounds can form when drinking water is chlorinated, and the regulatory limits for several of these compounds are in the range of one to a few tens of micrograms per liter ($\mu g/L$). Similar limits apply to the concentrations of some heavy metals, and even lower concentrations of these metals can have significant ecological effects in streams if the metals are in particularly toxic forms. While a concentration of such a species at a level of, say, 50 $\mu g/L$ might be considered very high with respect to some ecological or regulatory

---

[1]An ion is any charged chemical species. Ions carrying positive charges are called cations, and those that carry negative charges are called anions.

Table 1.1 Concentrations of major dissolved constituents in some natural water bodies and of some potentially toxic constituents in various solutions.

		Savannah River	Mississippi River at St. Paul	Colorado River near Phoenix	Typical Groundwater	Mean Seawater
Ca^{2+}	mg/L	4	23	77	135	408
Mg^{2+}	mg/L	1	5	29	60	128
Na^+	mg/L	12	10		325	10,800
Cl^-	mg/L	9	21	88	35	19,400
SO_4^{2-}	mg/L	7	23	250	650	2710
HCO_3^-	mg/L as $CaCO_3{}^a$	23	150	135	550	120
DOC	mg/L	3	8	3.0	1.0	1.0
pH	pH units	7.0	8.6	8.3	7.2	7.9

		Electroplating Rinse Water[b]	Pretreatment Req't for Discharge to POTW[c]	Water Quality Standards[d]	MCL[e] for Drinking Water
Cu	mg/L	3	3	0.009	1.3
Zn	mg/L	2	5	0.117	—
Cr	mg/L	3	2.75	0.027	0.1
As	mg/L	—	1	0.148	0.05
Pb	mg/L	0.7	2	0.0025	0.015
Ni	mg/L	3	2.5	0.052	0.1
Cd	mg/L	0.5	0.5	0.0021	0.005
pH	pH units	5–12	5.5–12.0	6.5–8.5	

[a]Alkalinity, expressed as milligrams per liter as $CaCO_3$. This way of expressing concentration is explained shortly in the text. As explained in Chapter 5, for many natural waters, the HCO_3^- concentration is approximately equal to the alkalinity, when both are expressed as equivalents per liter or as milligrams per liter as $CaCO_3$.

[b]Values shown are in the normal range, but note that these waters have highly variable composition, depending on the electroplating operation being carried out.

[c]A POTW is a publicly owned treatment works; values shown are for King Co., WA.

[d]Chronic water quality criteria for protection of aquatic life in ambient water, assuming hardness of 100 mg/L as $CaCO_3$. Values shown for arsenic and chromium are for these metals in the +3 oxidation state.

[e]MCL is the maximum contaminant level allowed by law in the United States.

limit, it is important to recognize how phenomenally dilute such a species truly is. For instance, a concentration of 10 μg/L (10 ppb) is equivalent to 60 people out of the world population of 6 billion. Imagine viewing the earth from a spaceship and being told that there are 60 very dangerous criminals scattered around the globe, and that you must find and capture 30 of them in a few minutes. This impossible-sounding task is equivalent to what water treatment engineers are asked to do, day in and day out, with respect to certain contaminants.

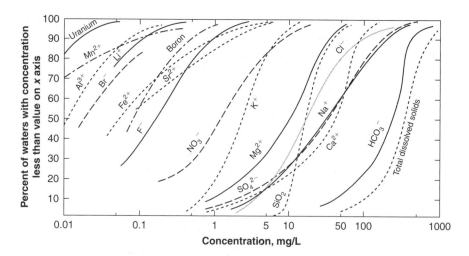

Figure 1.1 The cumulative frequency distribution of solute concentrations in terrestrial waters.

From S. Davis and R. deWiest, *Hydrogeology*, John Wiley & Sons, Inc. New York (1966). Reprinted by permission of the author.

Natural waters have a wide range of compositions. **(a)** Bottled drinking water. Bottled water often contains higher solute concentrations than the local tap water. **(b)** The Dead Sea, where the salt pillars form when the salt concentration exceeds the amount that can remain dissolved **(c)** The notice board reports the composition of a hot spring in northern Thailand.

1.2   THE STRUCTURE OF WATER MOLECULES AND INTERACTIONS AMONG THEM

The shape of a water molecule can be represented roughly by a tetrahedron (a four-sided, triangular pyramid), with the oxygen in the center and hydrogen ions at two corners.[2] The region between the oxygen and each of the hydrogen ions is occupied by a pair of electrons, one contributed by the hydrogen and one by the oxygen. Since the oxygen atom has six electrons in its outer shell, this arrangement leaves four unshared electrons, which occupy orbitals pointing toward the other two corners of the tetrahedron (two electrons in each orbital). The structure is shown schematically in Figure 1.2.

Although the electrons forming the hydrogen-oxygen bonds are shared by the two atoms, they are not shared equally: the oxygen attracts them more than the hydrogen does. As a result, although water molecules are electrically neutral overall (i.e., they are not ions), they do have local regions of finite charge. Specifically, a positive charge of about 0.24 esu (electrostatic unit, the charge on a single electron) resides near each hydrogen atom, and a negative charge of around 0.48 esu resides on the side of the oxygen opposite the hydrogen atoms. Because of this separation of charge, the molecule is referred to as being *polar*.

Hydrogen atoms have only one electron and one proton. Because that single electron usually resides between the H and O atoms in a water molecule, from the perspective of the rest of the solution, each hydrogen ion looks almost like a bare proton. The absence of other electrons associated with the hydrogen allows

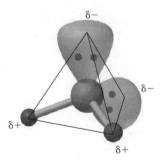

Figure 1.2   Schematic representation of the structure and charge distribution of a water molecule. The hydrogen ions and two sets of unshared electron pairs reside at the corners of a tetrahedron. Two other pairs of electrons are shared—one pair between the oxygen and each hydrogen ion.

From M. S. Silberberg, *Chemistry: The Molecular Nature of Matter and Change.* Copyright © 2000 The Mc-Graw-Hill Companies, New York, NY. Reproduced by permission of The McGraw-Hill Companies.

[2]The shape of the molecule actually corresponds to a slightly distorted tetrahedron, with the H—O—H angle being ~105° as opposed to 109° in an ideal tetrahedron.

other molecules to approach the "unprotected" proton quite closely, and in fact more closely than they could approach the nucleus of any other positively charged atom. This closeness of approach allows unusually strong electrostatic bonds to form between the proton and negatively charged portions of other molecules. Such bonds, called **hydrogen bonds,** are particularly important in aqueous solutions, because they can form between the electron-poor hydrogen ions of one water molecule and the electron-rich parts of another (i.e., the orbitals with unshared electrons). As a result, every water molecule can bond to other water molecules by as many as four hydrogen bonds (Figure 1.3). All four such bonds form in ice; experimental evidence suggests that in liquid water, each molecule is hydrogen-bonded to three others at any instant. These bonds cause water to be more cohesive than almost any other liquid and affect the physical and chemical behavior of water in many other important ways.

1.3   THE BEHAVIOR OF SOLUTES IN WATER; DISSOLUTION OF SALTS

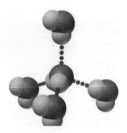

Figure 1.3
Hydrogen-bonded water molecules.

From M. S. Silberberg, *Chemistry: The Molecular Nature of Matter and Change.* Copyright © 2000 The McGraw-Hill Companies, New York, NY. Reproduced by permission of The McGraw-Hill Companies.

The transfer of any substance into or out of a solution changes the energy distribution in both environments. For instance, when a pure substance A enters solution, bonds among water molecules break as the molecules rearrange to create a "hole" for the solute to enter. Bonds that linked A molecules to one another prior to dissolution also break, and A-to-H_2O bonds form. In general, the solutions we will consider are sufficiently dilute that molecules of A are widely dispersed in solution, and A-to-A bonds can be ignored.[3]

The tendency for molecules of A to dissolve is enhanced by the formation of strong A-to-H_2O bonds and is diminished by the existence of strong A-to-A bonds in "undissolved" A (i.e., solid or liquid A). Therefore, other factors being equal, the relative strengths of A-to-A and A-to-H_2O bonds can give a good indication of how soluble molecules of A will be.[4] Molecules whose dissolution is favorable are called **hydrophilic** (water-loving), and those whose dissolution is unfavorable are called **hydrophobic** (water-hating). Because these terms are relative, though, the same molecule might be considered hydrophilic in one context and hydrophobic in another, depending on what it is being compared with and what is deemed a favorable amount of dissolution in the context of the discussion.

The electric charge on dissolved ions tends to orient the neighboring water molecules to maximize favorable (plus-to-minus) electrical interactions (Figure 1.4). The structured orientation encouraged by these electrical interac-

---

[3]If A-to-A bonds in solution are strong, molecules of the type A_n ($n = 2$ or larger) might form, but the A_n that forms is represented as a new type of molecule in the system rather than as a group of individual A molecules.

[4]The dissolution process also affects the spatial distribution of molecules in the system in a way that might either favor or oppose the reaction. As a result, one cannot conclude whether dissolution will be favorable based solely on bond formation and breakage. This phenomenon is quantified as part of the entropy change accompanying dissolution. Entropy is discussed in some detail in Chapter 2.

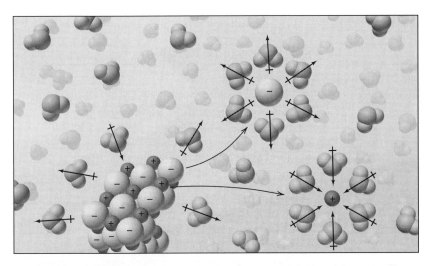

Figure 1.4   The tendency of water molecules to orient themselves around ions facilitates dissolution of salts.

From M. S. Silberberg, *Chemistry: The Molecular Nature of Matter and Change.* Copyright © 2000 The McGraw-Hill Companies, New York, NY. Reproduced by permission of The McGraw-Hill Companies.

tions is opposed by the randomizing influence of thermal motion, so the water molecules are not frozen in place, and some transient unfavorable (+ to + or − to −) interactions might occur. Nevertheless, at any instant, the favorable electrostatic interactions are likely to far outnumber the unfavorable ones. These electrostatic interactions stabilize dissolved ions significantly, making all ions hydrophilic, and making it possible for large concentrations of individual, isolated ions to be present in solution. No comparable electrostatic interactions are possible in a gas phase, so ions are extremely unstable in gases, and the concentration of ions present in a gas under normal environmental conditions is always exceedingly small. Therefore, throughout this text, the assumption is made that gases are made up entirely of neutral molecules. (Note that, overall, aqueous solutions are electrically neutral, just as gases are. However, gases are neutral because they contain only neutral molecules, whereas solutions are neutral because the total positive charge on dissolved cations is exactly balanced by the total negative charge on dissolved anions.)

Solutes that are neutral but polar can orient neighboring water molecules in much the same way that ions can, but the strength of the orienting force is lower. Nonpolar molecules, on the other hand, usually interact only weakly with water molecules. As a result, the hydrophilicity of neutral molecules generally increases with increasing polarity of the molecule.

Some ions are so stable when surrounded by water molecules that, upon contact with water, they are very likely to break off of any compound they are associated with and become surrounded by water molecules, and are then very unlikely to participate in any subsequent reactions. Sodium (Na^+) and chloride

Table 1.2   General solubility rules for inorganic compounds[a]

Ion	Characteristic Solubility of Compounds Containing Ion
Nitrate, NO_3^-	All nitrates are soluble.
Chloride, Cl^-	All chlorides are soluble except $AgCl$, $PbCl_2$, and Hg_2Cl_2.
Sulfate, SO_4^{2-}	Sulfates are soluble, except $BaSO_4$ and $PbSO_4$; Ag_2SO_4, $CaSO_4$, and Hg_2SO_4 are only slightly soluble.
Carbonate, CO_3^-; phosphate, PO_4^{3-}; silicate, SiO_4^{4-}	Carbonates, phosphates, and silicates are insoluble, except those of sodium, potassium, and ammonium.
Hydroxide, OH^-	Most hydroxides are insoluble. Exceptions include $LiOH$, $NaOH$, KOH, and NH_4OH (soluble); $Ba(OH)_2$ (moderately soluble); and $Ca(OH)_2$ and $Sr(OH)_2$ (slightly soluble).
Sulfide, S^{2-}	All sulfides are insoluble, with the exception of alkali metal sulfides (Na_2S, K_2S, etc.), $(NH_4)_2S$, MgS, CaS, and BaS.
Sodium, Na^+; potassium, K^+; ammonium, NH_4^+	All sodium, potassium, and ammonium compounds are soluble, with the exception of a few compounds that contain these ions along with a heavy metal (for example, K_2PtCl_6).

[a]Based on Dean, J. A., *Lange's Handbook of Chemistry,* 14th ed., McGraw-Hill, 1992.

(Cl^-) are examples of ions that behave in this way. Therefore, compounds of the form $Na_aX(s)$ or $ZCl_b(s)$ tend to dissociate completely to Na^+ plus X^{a-} or Z^{b+} plus Cl^-, respectively, whenever they are exposed to water.

$$Na_aX \xrightarrow{+H_2O} a\,Na^+ + X^{a-}$$

$$ZCl_b \xrightarrow{+H_2O} Z^{b+} + b\,Cl^-$$

The X^{a-} and Z^{b+} ions might react with other dissolved species, but the Na^+ and Cl^- rarely do. Such compounds, which dissociate completely when they dissolve, are called *salts.*[5]

General guidelines for the solubility of some environmentally important ions are provided in Table 1.2. Again, it is important to remember that the term *soluble* is qualitative and can have different quantitative meaning in different circumstances.

Logically enough, ions that are present in relative abundance in the geosphere and are highly stable as dissolved aqueous species tend to accumulate in natural waters and to dominate the ionic composition of such systems. This group of ions includes sodium (Na^+), potassium (K^+), calcium (Ca^{2+}), magnesium (Mg^{2+}), chloride (Cl^-), sulfate (SO_4^{2-}), and nitrate (NO_3^-). The other major ion in most natural waters is bicarbonate (HCO_3^-) which, although it is more reactive than the other ions listed, is present in relatively high concentrations because of continual

[5]Other definitions of salts are based on the crystal form and/or chemical bonding arrangements of the compound in the solid form. The definition given here is commonly used by water chemists.

inputs of CO_2 from the atmosphere and from biological activity. The CO_2 reacts with water to form bicarbonate by the reaction $CO_2 + H_2O \leftrightarrow HCO_3^- + H^+$; this reaction is discussed in detail in later chapters.[6]

1.4 COMMON WAYS OF EXPRESSING CONCENTRATIONS IN ENVIRONMENTAL CHEMISTRY

1.4.1 MASS/VOLUME AND MASS/MASS UNITS FOR SOLUTE CONCENTRATIONS

Concentrations of species in aqueous systems are reported in a wide variety of ways. Clearly, the behavior of a chemical does not depend on the units chosen to represent its concentration, any more than the velocity of an object depends on whether that velocity is given in miles per hour or meters per second. However, it is also clear that certain choices of units are more convenient than others (no one would want to have a speedometer read in millimeters per year).

The most common way to report the concentration of a chemical dissolved or suspended in water is as mass per volume, with units such as milligrams per liter (mg/L) or micrograms per liter (μg/L). Dimensions of moles per liter (mol/L) (often designated simply by an M) are also widely used for expressing concentrations of dissolved constituents; these dimensions are convenient to use in conjunction with chemical reactions, because ratios of stoichiometric coefficients in chemical reactions correspond to molar ratios of the reacting species.[7]

Because chemical concentrations can range over many orders of magnitude, it is convenient to express their values in logarithmic units. By convention, in chemical literature, a lowercase p preceding a symbol is used to designate the negative, base-10 logarithm ($\log_{10}$) of the value associated with the symbol. When this convention is applied to the concentration of a solute, the units of solute concentration are understood to be moles per liter. Thus, for example, if the concentration of Cu^{2+} ions in a solution ($c_{Cu^{2+}}$) is 2×10^{-5} mol/L, then $\log c_{Cu^{2+}} = -4.70$, so $p(c_{Cu^{2+}}) = 4.70$. The p convention is applied not only to solute concentrations, but also to chemical activities and equilibrium constants, both of which are described later in the chapter.

Sometimes, concentrations are normalized based on the mass of the phase in which they are found, rather than the volume. For instance, the concentration of a dissolved species i might be given as a *mass fraction*, i.e., the mass of i per unit mass of solution, with units such as milligrams per kilogram or micrograms per

---

[6]Many molecules of interest consist of a "core" anion to which one or more H^+ ions can attach. In most cases, when one H^+ is attached, the name of the molecule is the name of the core anion with the prefix *bi*. In this case, the core anion is carbonate, CO_3^{2-}.

[7]One mole is Avogadro's number (6.023×10^{23}) of items. One mole of a chemical is therefore 6.023×10^{23} molecules of that chemical.

kilogram. Concentrations reported using this convention are often given as dimensionless quantities [mg/kg = parts per million by mass (ppm_m), g/100 g = percent, etc.]. This approach is commonly used to express the concentrations of constituents in a solid phase; for example, a concentration of 1 ppm_m in a solid phase means that there is 1 mg of the given species per kilogram of solid. A similar approach can be used to express concentrations as *mole fractions,* i.e., the number of moles of a species as a fraction of the total number of moles of all substances in that phase.

Because most aqueous solutions of interest contain far more water than other species, they usually have a density very close to 1000 g/L. In such solutions, concentrations expressed as milligrams per liter or micrograms per liter (i.e., mass/volume) are nearly equivalent to those expressed as parts per million or parts per billion by mass, respectively, e.g.,

$$1 \frac{mg\ A}{L} \approx 1 \frac{mg\ A}{1000\ g\ solution} = \frac{1\ mg\ A}{10^6\ mg\ solution} = 1\ ppm_m$$

Solute concentrations are also sometimes reported as mass or moles of solute per unit mass of solvent (rather than per unit mass of the total solution). Concentrations computed using this approach are independent of the density of solution, making the approach particularly useful for describing the preparation of reagents for laboratory use. Concentrations given as moles of solute per 1000 g of solvent are referred to as *molal* concentrations.

Example 1.1

A solution contains 8 μg/L lead (Pb, atomic mass 207). Express this concentration as moles per liter, ppm_m, and the mole fraction of Pb. Assume the solution has a density of 1 g/mL.

Solution

The molar concentration of Pb is computed simply by dividing the mass concentration by the atomic mass:

$$\frac{8\ \mu g/L\ Pb}{207 \times 10^6\ \mu g\ Pb/mol\ Pb} = 3.86 \times 10^{-8} \frac{mol\ Pb}{L}$$

For a solution with density 1 g/mL, 1 μg/L = 1 ppb_m = 0.001 ppm_m. Therefore, 8 μg/L Pb corresponds to 0.008 ppm_m. This equality is derived formally as follows:

$$\frac{8 \times 10^{-6}\ g\ Pb/L\ of\ solution}{10^3\ g\ solution/L\ of\ solution} = 0.008 \times 10^{-6} \frac{g\ Pb}{g\ solution} = 0.008\ ppm_m$$

The molar concentration of Pb in the solution is computed above. The molar concentration of water can be computed by assuming the solution is essentially pure water, as follows:

$$\frac{1000\ g\ H_2O/L\ of\ solution}{18\ g\ H_2O/mol\ H_2O} = 55.6 \frac{mol\ H_2O}{L}$$

Given the molar concentrations of Pb and H_2O, the mole fraction of Pb can be computed as

$$\frac{3.86 \times 10^{-8} \text{ mol Pb/L of solution}}{55.6 \text{ mol } H_2O/L \text{ of solution} + 3.86 \times 10^{-8} \text{ mol Pb/L of solution}}$$

$$= 6.95 \times 10^{-10} \frac{\text{mol Pb}}{\text{total moles in solution}}$$

Thus, the mole fraction of Pb is 6.95×10^{-10}.

1.4.2   GAS-PHASE CONCENTRATIONS

Like concentrations in solution, concentrations in gas phases are often reported in mass-per-volume units. However, because gases are so much less dense than water, typical units for dilute constituents are micrograms per cubic meter, rather than milligrams per liter or micrograms per liter. In some cases, the concentration of a gaseous species is given as its *partial pressure* or its *volume fraction*. These units of concentration are a bit more obscure, but can be understood as follows.

Under environmental conditions, virtually all gases behave very nearly as ideal gases. For such gases, the pressure, volume, and temperature are interrelated by the ideal gas law, two forms of which are shown in Equations (1.1) and (1.2):

$$P_i = \frac{n_i}{V_{tot}} RT = c_{g,i} RT \qquad \textbf{(1.1)}$$

$$V_i = \frac{n_i}{P_{tot}} RT \qquad \textbf{(1.2)}$$

where P_i, V_i = partial pressure and partial volume of species i, respectively

n_i = number of moles of species i in the gas phase

P_{tot}, V_{tot} = total pressure and volume of the gas phase, respectively

$c_{g,i}$ = gas-phase concentration of i, moles per unit volume

R = universal gas constant

T = absolute temperature

The value of R in a few commonly encountered sets of units is 8.314 J/(mol-K), 0.0821 L-atm/(mol-K), or 1.987 cal/(mol-K).

The partial pressure of i can be defined as the pressure that i would exert if it were present in a hypothetical system that was at the same temperature and that contained the same concentration of gaseous i as the real system, but from which all other gases had been removed. Similarly, the partial volume of i is the volume that i would occupy if all other gases were removed from the system and the total system pressure were held constant. Partial pressure is expressed in the usual units of pressure, e.g., bars, atmospheres, kilopascals (kPa), or pounds per square inch.[8]

According to Equation (1.1), the partial pressure is *not* a direct measure of the concentration of i, since the same concentration of i is associated with different P_i values at different temperatures. However, if the temperature is specified, the partial pressure can be used as a surrogate for the concentration of i in the gas phase. Use of partial pressure in this way is common in environmental chemistry, often with the implicit assumption that the temperature is 25°C.

Equation (1.1) can be rearranged to yield an expression for the molar concentration of i (moles per unit volume) as follows:

$$c_{g,i} = \frac{P_i}{RT} \tag{1.3}$$

In an ideal gas phase, the sums of P_i, V_i, and n_i for all gases in the system equal P_{tot}, V_{tot}, and n_{tot}, respectively.

$$P_{tot} = \sum P_i \tag{1.4a}$$

$$V_{tot} = \sum V_i \tag{1.4b}$$

$$n_{tot} = \sum n_i \tag{1.4c}$$

Combining these identities with Equations (1.1) and (1.2) yields

$$\frac{P_i}{P_{tot}} = \frac{V_i}{V_{tot}} = \frac{n_i}{n_{tot}} \tag{1.5}$$

Equation (1.5) indicates that the volume fraction of i (i.e., V_i/V_{tot}) is the same as the mole fraction, which was identified above as one way of representing concentrations. Thus, the volume fraction can also be considered a measure of concentration. Volume fractions of dilute gaseous constituents are typically expressed in parts per million by volume (ppm_v).[9]

1.4.3 COMPOSITE PARAMETERS, CONCENTRATIONS REPORTED "AS X," AND OTHER UNUSUAL REPRESENTATIONS OF SOLUTE CONCENTRATION

Aquatic chemistry, more than other fields of chemistry, deals with complex, uncontrolled mixtures of chemicals in samples whose history is often poorly known. Under the circumstances, a detailed chemical characterization of the samples is often impossible. Furthermore, in many cases, our interest is in the

---

[8]The SI unit for pressure is bars: 1.0 bar = 0.987 atm = 1000 kPa.

[9]It should be clear that 1 ppm_m and 1 ppm_v can represent quite different concentrations. Although it would be convenient if subscripts were consistently used with these terms to avoid confusion, that is rarely done. Rather, both ppm_m and ppm_v are commonly designated simply ppm, with the implicit understanding that the units are mass-based (1 ppm = 1 mg/kg) when applied to a solid or liquid and are volume-based (1 ppm = 1 mL/m^3) when applied to a gas.

behavior or effects of diverse groups of chemicals, rather than a single species. Thus, for both fundamental and practical reasons, concentrations of environmentally significant chemicals are frequently reported in ways other than as concentrations of individual chemical species. In this section, a few of those ways of expressing concentration are presented.

SPECIES' CONCENTRATIONS REPRESENTED BY A SINGLE ELEMENT: NITROGEN SPECIATION TOC, AND TOX   One approach for describing the collective concentration of several different chemical species, or to compare the concentrations of several species that contribute to a "pool" of interest, is to report the concentration of a single element that is common to all the species. For instance, nitrogen species commonly found in aquatic environmental samples include molecular nitrogen (N_2), nitrate ion (NO_3^-), nitrite ion (NO_2^-), ammonium ion (NH_4^+), and nitrogen incorporated into unspecified organic compounds. These species can all be interconverted by microbial action. The concentrations of these species (especially the latter four) are commonly reported in terms of the amount of elemental N that each one contains, rather than in milligrams per liter or moles per liter of the entire species. This approach is useful for comparing the concentrations of the different species and for tracking conversions among them because 1.0 mg of N in any of the species can be converted to exactly 1.0 mg of N in another form. The same 1:1 ratio does not apply if the concentrations are computed for each species in its entirety because of the different molecular weights of the different species and because organic compounds might contain more than one N atom per molecule.

When this reporting convention is employed, the symbols NO_2-N, NO_3-N, NH_4-N, and org-N, respectively, are used to represent the concentrations of nitrite, nitrate, ammonium, and organic species containing N. For instance, the statement that a solution contains 1.4 mg/L of NO_2-N means that nitrite is present in the solution in an amount sufficient to include 1.4 mg/L of N. This concentration is sometimes conveyed in both writing and speaking by the statement that the concentration of nitrite is 1.4 mg/L *as N*. The molecular weight of nitrite ion is 46, so the molar and mass concentrations of nitrite ion in such a solution are

$$\left(1.4 \frac{\text{mg NO}_2\text{-N}}{\text{L}}\right)\left(\frac{1 \text{ mol NO}_2\text{-N}}{14{,}000 \text{ mg NO}_2\text{-N}}\right)\left(\frac{1 \text{ mol NO}_2^-}{\text{mol NO}_2\text{-N}}\right) = 10^{-4} \frac{\text{mol NO}_2^-}{\text{L}}$$

$$\left(10^{-4} \frac{\text{mol NO}_2^-}{\text{L}}\right)\left(\frac{46{,}000 \text{ mg NO}_2^-}{\text{mol NO}_2^-}\right) = 4.6 \frac{\text{mg NO}_2^-}{\text{L}}$$

If all the nitrite in this solution were converted to nitrate, the NO_3-N concentration in solution would increase by 1.4 mg/L; that is, the increase in NO_3-N is numerically identical to the decrease in NO_2-N. However, since the molecular weight of NO_3^- is 62, the increase in the mass concentration of nitrate would be

$$\left(1.4 \frac{\text{mg NO}_3\text{-N}}{\text{L}}\right)\left(\frac{62 \text{ mg NO}_3^-}{14 \text{ mg NO}_3\text{-N}}\right) = 6.2 \frac{\text{mg NO}_3^-}{\text{L}}$$

Thus, 1.4 mg/L NO_2-N can be converted to 1.4 mg/L NO_3-N, corresponding to conversion of 4.6 mg/L NO_2^- to 6.2 mg/L NO_3^-.

As noted above, the total concentration of a particular element in a group of different species is also of interest in cases where it is impractical to analyze for each of the species individually. For instance, the overall organic concentration in environmental samples is almost always reported in terms of the total concentration of carbon in all the organic compounds present, because the samples contain a large and disparate collection of organic molecules which would be impossible to identify individually. Such data are usually reported either as the total organic carbon (TOC) concentration of the sample or, if the sample is filtered and only the solution is analyzed, as the dissolved organic carbon (DOC) concentration.[10] By analogy, the concentration of total inorganic carbon in a solution [generally considered to include only carbonic acid (H_2CO_3), bicarbonate ion (HCO_3^-), and carbonate ion (CO_3^{2-})] is sometimes designated by the acronym TIC.

Example 1.2

A natural water has a DOC concentration of 3.5 mg/L. The average composition of the organic molecules in the solution is $C_{30}H_{33}O_{18}N$. Assuming that all the molecules actually have that composition, compute the molar concentration and mass concentration of the molecules in the sample. (Note that, in truth, the solution contains many molecules with disparate compositions, so the properties of an "average" molecule should not be interpreted too literally.)

Solution

The molecular weight of the hypothetical average molecule is

$$30 \times 12 + 33 \times 1 + 18 \times 16 + 1 \times 14 = 695$$

Carbon represents a fraction equal to 360/695, or 52 percent, of the mass of the molecule. Since the sample contains 3.5 mg/L DOC, the total mass concentration of the dissolved organics is

$$\frac{3.5 \text{ mg C/L}}{0.52 \text{ mg C/mg organic molecule}} = 6.8 \frac{\text{mg organic molecules}}{\text{L}}$$

The molar concentration of the hypothetical organic molecules is therefore

$$\frac{6.8 \text{ mg organic molecules/L}}{695{,}000 \text{ mg organic molecules/mol}} = 9.7 \times 10^{-6} \frac{\text{mol organics}}{\text{L}}$$

A similar situation applies to certain organic contaminants that can form when chlorine-based disinfectants are used to treat drinking water or wastewater.

---

[10]Another way of characterizing the composite organic content of a sample is to report the amount of oxygen that is consumed when the organic matter is oxidized, either chemically [in the chemical oxygen demand (COD) test] or by a combination of biological and chemical processes [in the biochemical oxygen demand (BOD) test].

In particular, when hypochlorous acid (HOCl) is added to water, it reacts not only with microbes, but also with some of the organic compounds in the water, generating products known as *disinfection by-products,* or *DBPs.* Some of these DBPs are carcinogenic, so regulations have been established limiting their concentrations in drinking water. The concentration limit established by law is referred to as the *maximum contaminant level (MCL).*

The DBPs thought to pose the greatest risk to public health are organic compounds into which chlorine and/or bromine has been incorporated. (Bromide ions are present in some source waters, and the addition of chlorine to the water converts them to a form that can react with the organic matter.) The dominant DBPs in most drinking water systems are the trihalomethanes (CHX_3) and haloacetic acids ($X_nH_{3-n}C$—COOH), where X can be Cl, Br, or a mixture of the two, and n can be 1, 2, or 3. The trihalomethanes and haloacetic acids are commonly designated by their respective acronyms, THMs and HAAs. The MCLs for total THMs and total HAAs are currently 80 and 60 $\mu g/L$, respectively.

The total concentration of chlorine and bromine incorporated in all the DBPs is also of interest. This parameter is commonly referred to as the *total organic halogen (TOX)* and is reported in units of micrograms per liter *as Cl*. If a DBP contains bromine, the contribution of the bromine atoms to the TOX is computed by treating them as chlorine atoms, as demonstrated in the following example.

Example 1.3

Consider a treated drinking water that contains 31 $\mu g/L$ chloroform ($CHCl_3$), 11 $\mu g/L$ bromoform ($CHBr_3$), 8 $\mu g/L$ dichloroacetic acid (Cl_2HC—COOH), and 19 $\mu g/L$ trichloroacetic acid (Cl_3C—COOH) as the major DBPs. What is the TOX concentration of the sample, expressed in milligrams per liter *as Cl*?

Solution

The TOX concentration is the total concentration of halogen in the four compounds, computed as if all the halogens were Cl atoms. The molecular weights of the compounds are 119.5, 252.7, 129.0, and 163.5, respectively, so the molar halogen concentrations in the compounds are

$$\left(\frac{31\ \mu g\ CHCl_3/L}{119.5\ \mu g\ CHCl_3/\mu mol}\right)\left(3\ \frac{mol\ halogen}{mol\ CHCl_3}\right) = 0.78\ \frac{\mu mol\ halogen}{L}$$

$$\left(\frac{11\ \mu g\ CHBr_3/L}{252.7\ \mu g\ CHBr_3/\mu mol}\right)\left(3\ \frac{mol\ halogen}{mol\ CHBr_3}\right) = 0.13\ \frac{\mu mol\ halogen}{L}$$

$$\left(\frac{8\ \mu g\ Cl_2HC-COOH/L}{129\ \mu g\ Cl_2HC-COOH/\mu mol}\right)\left(2\ \frac{mol\ halogen}{mol\ Cl_2HC-COOH}\right) = 0.12\ \frac{\mu mol\ halogen}{L}$$

$$\left(\frac{19\ \mu g\ Cl_3C-COOH/L}{163.5\ \mu g\ Cl_3C-COOH/\mu mol}\right)\left(3\ \frac{mol\ halogen}{mol\ Cl_3C-COOH}\right) = 0.35\ \frac{\mu mol\ halogen}{L}$$

The total halogen concentration in the four compounds is the sum of the above values, or 1.38 $\mu mol/L$. If all this halogen were Cl, the solution would contain 1.38 $\mu mol/L$

organic Cl, corresponding to a mass concentration of 49.0 μg Cl/L:

$$\left(1.38 \frac{\mu\text{mol Cl}}{\text{L}}\right)\left(\frac{35.5 \ \mu\text{g Cl}}{\mu\text{mol Cl}}\right) = 49.0 \frac{\mu\text{g Cl}}{\text{L}}$$

The concentration of TOX in the sample is therefore 49.0 μg/L *as Cl*.

SPECIES' CONCENTRATIONS REPRESENTED AS THE CONCENTRATION OF A DIFFERENT COMPOUND: HARDNESS AND ALKALINITY In some cases, the logic behind reporting the concentration of a species *as X* can be fairly obscure. For instance, calcium ion (Ca^{2+}) concentrations are often reported as though the calcium were present as calcium carbonate solid ($CaCO_3$), even if no calcium carbonate is actually present in the system. That is, a concentration of 40 mg/L Ca^{2+} might be reported as 100 mg/L *as $CaCO_3$*, meaning that if all the Ca^{2+} in solution were converted to $CaCO_3$, the concentration of $CaCO_3$ formed would be 100 mg/L. The history that led to this reporting approach is that calcium ion used to be analyzed by adding various reagents to a solution and causing calcium carbonate to precipitate. The mass of $CaCO_3$ solid was then measured, and the Ca^{2+} concentration in the original solution was inferred. Rather than report the Ca^{2+} concentration that was inferred, though, people simply reported the concentration of $CaCO_3$ that had precipitated.

The collective concentration of several different divalent (doubly charged) ions is also commonly reported *as $CaCO_3$*. Specifically, Ca^{2+}, Mg^{2+}, and Fe^{2+} all contribute to the hardness of a water (an indicator of its mineral content and its reactivity with soap and certain other chemicals). In applications where the hardness of the water is of concern, the relative contributions of Ca^{2+}, Mg^{2+}, and Fe^{2+} are usually not particularly important. As a result, the hardness is usually reported as a composite parameter that includes all three contributions. By convention, this value is computed as though all the hardness were contributed by Ca^{2+} ions. For instance, if a solution contains 10^{-3} mol/L Ca^{2+}, 5×10^{-4} mol/L Mg^{2+}, and 1×10^{-4} mol/L Fe^{2+}, the total hardness is 1.6×10^{-3} mol/L. The hardness of this solution might be expressed in units of milligrams per liter *as Ca^{2+}* or *as $CaCO_3$* as follows:

$$\left(1.6 \times 10^{-3} \frac{\text{mol hardness}}{\text{L}}\right)\left(\frac{1 \text{ mol } Ca^{2+}}{\text{mol hardness}}\right)\left(\frac{40,000 \text{ mg } Ca^{2+}}{\text{mol } Ca^{2+}}\right)$$

$$= 64 \frac{\text{mg hardness}}{\text{L}} \ as \ Ca^{2+}$$

$$\left(64 \frac{\text{mg hardness}}{\text{L}} \ as \ Ca^{2+}\right)\left(\frac{100 \text{ mg } CaCO_3}{40 \text{ mg } Ca^{2+}}\right) = 160 \frac{\text{mg hardness}}{\text{L}} \ as \ CaCO_3$$

Note that in both of the above expressions, the contributions of Mg^{2+} and Fe^{2+} are being reported as the amount of Ca^{2+} that would contribute an equivalent amount of hardness. In the latter case, the value of hardness being reported is even more obscure since, first, Mg^{2+} and Fe^{2+} are being reported as equivalent amounts of Ca^{2+} and then the total equivalent amount of Ca^{2+} is being reported as though it were present as $CaCO_3$.

One other important cumulative parameter that is commonly used to characterize water quality is the *alkalinity*, which is a measure of the capacity of the solution to neutralize acidic inputs; a corresponding, but slightly less commonly reported value (the *acidity*) is used to characterize a solution's capacity to neutralize inputs of bases. Natural aquatic systems as well as most other solutions of environmental interest contain a variety of species that can neutralize acids or bases, and it is often impractical to analyze for each of these species directly. The alkalinity and acidity of the solution are easily measured, composite parameters that characterize the collective effects of these species, at least with respect to one important aspect of water quality. Alkalinity is often reported *as $CaCO_3$*, meaning that the real solution contains the same amount of alkalinity as would a hypothetical solution containing the stated concentration of $CaCO_3$ and no other solutes. Alkalinity and acidity are discussed in detail in Chapter 5.

CONCENTRATIONS REPORTED AS EQUIVALENTS PER LITER   Sometimes (particularly in some of the older environmental engineering literature) concentrations of various solutes are reported in units of *equivalents per liter* (equiv/L), in which case the value is referred to as the *normality N* of the solute. Such an approach implies that the concentration of the substance of interest is "equivalent" in some way to the concentration of another chemical. Thus, this approach is similar to reporting concentrations *as X*. However, when normalities are reported, the *as X* label is always implicit. As a result, there is at least some risk of misunderstanding the basis for the comparison.

Currently, concentrations are widely reported in equivalents per liter in only three cases. When one is referring to salt ions, the implied unit defining one equivalent is one mole of electric charge (without regard to sign); when referring to acids and bases, it is one mole of H^+; and when referring to oxidation-reduction reactions, it is one mole of electrons (e^-). The meaning intended in a given application is usually clear from the context, although some ambiguity can arise (e.g., if a charged, acidic species is participating in an oxidation-reduction reaction). In general, it is best to avoid using these units or to identify the reference explicitly, e.g., by using the term *electron equivalents* to describe concentrations of species participating in oxidation-reduction reactions.

A number of chlorine-containing chemicals are used to disinfect water, each of which generates different types and amounts of DBPs. Chlorine dioxide (ClO_2, MW 67.5) has received a good deal of attention in recent years because it is a powerful disinfectant that generates much lower concentrations of trihalomethanes and haloacetic acids than does hypochlorous acid (HOCl, MW 52.5), which is the most commonly used disinfectant. It is common to express the dose of any chlorinated disinfectant in terms of the equivalent concentration of chlorine molecules (Cl_2, MW 71), where the equivalency is based on the

Example 1.4

number of electrons that can combine with the Cl atoms. The relevant reactions for ClO_2, HOCl, and Cl_2 are as follows:

$$HOCl + H^+ + 2e^- \rightarrow Cl^- + H_2O$$

$$ClO_2 + 4H^+ + 5e^- \rightarrow Cl^- + 2H_2O$$

$$Cl_2 + 2e^- \rightarrow 2Cl^-$$

Express doses of 5 mg/L HOCl and 5 mg/L ClO_2 in milligrams per liter *as Cl_2*.

Solution

The basis for the comparison is the number of electrons that each chemical can combine with. That is, because one mole of Cl_2 and one mole of HOCl can each combine with two moles of electrons, one mole of HOCl is "equivalent" to one mole of Cl_2. On the other hand, one mole ClO_2 is equivalent, in terms of its e^--consuming ability, to 2.5 moles of Cl_2. The calculations for expressing the doses of HOCl and ClO_2 as Cl_2 are therefore as follows:

$$\left(5\,\frac{\text{mg HOCl}}{\text{L}}\right)\left(\frac{1 \text{ mol HOCl}}{52.5 \text{ mg HOCl}}\right)\left(1\,\frac{\text{mol Cl}_2}{\text{mol HOCl}}\right)\left(71\,\frac{\text{mg Cl}_2}{\text{mol Cl}_2}\right) = 6.76\,\frac{\text{mg Cl}_2}{\text{L}}$$

$$\left(5\,\frac{\text{mg ClO}_2}{\text{L}}\right)\left(\frac{1 \text{ mol ClO}_2}{67.5 \text{ mg ClO}_2}\right)\left(2.5\,\frac{\text{mol Cl}_2}{\text{mol ClO}_2}\right)\left(71\,\frac{\text{mg Cl}_2}{\text{mol Cl}_2}\right) = 13.15\,\frac{\text{mg Cl}_2}{\text{L}}$$

COMPOSITE PARAMETERS CHARACTERIZING THE TOTAL SOLUTE CONCENTRATION: TDS AND IONIC STRENGTH   Two composite parameters are commonly reported as overall measures of the total solute or total ionic content of a sample. One of these, the **total dissolved solids (TDS)** concentration, includes the majority of both the ionic and nonionic dissolved species. It is determined by heating a sample of the water at a temperature slightly greater than 100°C until only a dry residue remains. During the drying step, essentially all the gases dissolved in the water [primarily oxygen (O_2), nitrogen (N_2), and carbon dioxide (CO_2)] escape. However, very few other substances that are likely to be dissolved form gases at 100°C, so they remain in the residue. These species include virtually all salt ions as well as any neutral inorganic molecules that are present (e.g., H_4SiO_4) and any neutral organic molecules that have been released into the water by plants or other organisms. Most of the water molecules also evaporate, although some might remain firmly attached to salts or other solids in the residue. In most natural waters, the majority of the residue is inorganic salts, so the TDS can be thought of as a rough measure of the ionic content of the water.

A second measure of the overall solute concentration in water takes into account the observation that the impact of an ion on certain chemical interactions in solution depends strongly on the ionic charge. Specifically, the strength of some interactions varies as the square of the ionic charge. To quantify the composite effect of all ions in solution, a term known as the *ionic strength* (commonly designated μ or I) has been defined as follows:

$$I = \frac{1}{2} \sum_{\text{all ions}} c_i z_i^2 \qquad \textbf{(1.6)}$$

where c_i is the concentration of i in moles per liter and z_i is the charge on species i. Generally z_i is taken to be dimensionless, so I formally has units of moles per liter. However, it is often reported without stating the units explicitly, i.e., as though it were dimensionless.

Compute the ionic strength of solutions containing 0.05 mol/L NaCl, 0.025 mol/L Na_2SO_4, or 0.025 mol/L $MgSO_4$ as the only salts. All these salts dissociate completely, so the resulting solutions contain only Na^+, Cl^-, Mg^{2+}, and SO_4^{2-} in addition to H_2O. The concentrations of the salts have been chosen such that each solution contains 0.05 mol/L of positive and negative charges.

Example 1.5

Solution

The ionic strengths of the various solutions can be computed by plugging appropriate values into Equation (1.6). Note that two moles of Na^+ enter solution per mole of Na_2SO_4 dissolving.

$$I_{NaCl} = \tfrac{1}{2}\left[c_{Na^+}z_{Na^+}^2 + c_{Cl^-}z_{Cl^-}^2\right] = \tfrac{1}{2}\left[(0.05\ M)(+1)^2 + (0.05\ M)(-1)^2\right] = 0.05\ M$$

$$I_{Na_2SO_4} = \tfrac{1}{2}\left[c_{Na^+}z_{Na^+}^2 + c_{SO_4^{2-}}z_{SO_4^{2-}}^2\right] = \tfrac{1}{2}\left[(0.05\ M)(+1)^2 + (0.025\ M)(-2)^2\right] = 0.075\ M$$

$$I_{MgSO_4} = \tfrac{1}{2}\left[c_{Mg^{2+}}z_{Mg^{2+}}^2 + c_{SO_4^{2-}}z_{SO_4^{2-}}^2\right] = \tfrac{1}{2}\left[(0.025\ M)(+2)^2 + (0.025\ M)(-2)^2\right] = 0.100\ M$$

Thus, the ionic strengths of the solutions vary by a factor of 2, even though all the solutions contain the same total concentrations of positive and negative charge.

Note that the ionic strength does not include any contribution from neutral dissolved species. As shown below, the ionic strength is a key parameter controlling the reactivity of dissolved ions (and, to a lesser extent, neutral solutes).

1.5   CHARACTERIZING CHEMICAL REACTIVITY

1.5.1   FACTORS AFFECTING THE REACTIVITY OF CHEMICALS

For any chemical reaction we wish to consider, we can arbitrarily define one group of compounds as reactants and another as products. At any instant, a system might contain a finite concentration of these relatively stable species, as well as a small concentration of molecules that are in transition between the two more stable states. Furthermore, at any instant, we might expect some molecules to be in the process of being converted from reactants to products and others to be undergoing the opposite reaction. If the system contains many chemicals, then many reactions are likely to be proceeding at once.

Based on the above scenario, a complete characterization of a chemical system requires knowledge about the various states in which chemicals can exist and the reactivity of chemicals in each of those states. In the sense it is being used here, the *reactivity* refers to the overall tendency for the chemical to participate in reactions, incorporating factors such as its concentration, the concentration of other constituents of the system, and temperature. In other words, reactivity characterizes the stability (or, more precisely, the instability) of a chemical.

Historically, chemical reactivity has been explained by focusing on either the energy stored in chemical species or the likelihood and consequences of molecular collisions. These two approaches for explaining reactivity are linked, in that both the energy of a species in a given system and the likelihood of collisions involving that species are directly related to the composition of the system. However, the approaches reflect slightly different perspectives and are couched in different terminology. The collision-based approach is presented in the following section, while the energy-based approach, which falls in the realm of chemical thermodynamics, is presented in Chapter 2.

Consider the reaction $A + B \rightarrow C$, where all three chemicals are dissolved in an aqueous solution. The reaction indicates that C can be generated by collisions of A and B molecules. Since the probability of an A molecule and a B molecule being in a given region of space is proportional to their respective concentrations (in mass or moles per unit volume), the rate of A-B collisions (and hence the rate of formation of C) is expected to be proportional to the concentrations of A and B, all other factors being equal. For instance, if the concentration of B is fixed and the concentration of A doubles, we expect twice as many A-B collisions in a given volume of solution per unit time; correspondingly, we expect the rate of formation of C to double. The same argument applies to collisions of any pair of molecules in a bulk fluid phase, i.e., to collisions between pairs of water molecules, between a water molecule and a solute, or between two molecules in a gas.

Often, we are interested in reactions in which one or more constituents must cross a phase boundary (solution to solid, gas to solution, etc.). In such a case, the likelihood that a given type of molecule will cross the boundary is, to a first approximation, proportional to that molecule's concentration in the bulk phase that it is leaving. That is, in a system with a gas/liquid interface, doubling the concentration of the constituent in solution doubles the frequency with which that species strikes the interface and therefore doubles its likelihood of crossing the phase boundary and entering the gas. [The likelihood of cross-boundary transfers also depends on the *amount* of interfacial area in the system. However, that dependency is generally considered to be a physical characteristic of the system (interfacial area per unit volume of the system) and is handled separately.] Thus, regardless of whether we are considering homogeneous (one-phase) or heterogeneous (multiphase) reactions, we expect the reactivity of any substance to be proportional to its concentration in the phase wherein it is found.

Factors other than concentration can also affect the reactivity of molecules. Among these are temperature, pressure, and the concentration of other species in the phase. For instance, increasing the temperature increases the likelihood of collisions in solution, because as temperature increases, the molecules themselves have more kinetic energy, and in addition the water becomes less viscous, providing less resistance to molecular movement. Furthermore, even if temperature, pressure, and the concentrations of A and B are the same in two solutions, the frequency of A-B collisions might not be identical in the two solutions if the other (non-reacting) components of the two solutions differ. For instance, if one solution contains species that surround molecules of A and shield them from molecules of B, the likelihood of a collision between A and B will be lower than in another solution lacking those shielding molecules. Thus, the reactivity of a given species is affected by system-specific chemical factors in addition to the species' concentration. We next consider ways to quantify this concept.

1.5.2   DEFINING CHEMICAL REACTIVITY (ACTIVITY) AND THE STANDARD STATE

The reactivity of a chemical A in a given system might be measured in a number of ways. For instance, it might be defined as the tendency of the substance to be consumed by a particular reaction. To quantify this tendency, an arbitrary, but well-characterized system containing A could be defined as a standard system and assigned a reactivity value of 1.0. The reactivity of A in any other system could then be quantified by comparison with the standard system: if A were more reactive in the system of interest than in the standard system, its reactivity would be >1.0. When quantified in this way, the reactivity of species A is called the **chemical activity** (or simply the **activity**) and is often designated by either a_A or $\{A\}$.[11] The complete set of conditions describing the standard system is called the **standard state.**

As noted above, reactivity is related to both the concentration of a species and its environment, i.e., the temperature, pressure, and overall composition of the phase. Therefore, all these factors must be clearly stated in the definition of the standard state. Historically, the concentration and environmental components of the standard state have been dealt with independently, with the "standard environment" being referred to as the **reference state.** Thus, the standard state can be described as the standard concentration of a species in its reference state.

---

[11]Different authors use different types of brackets to distinguish between the concentration and activity of A. For instance, [A], (A), and {A} might each be used to represent concentration in some contexts and activity in others. In this text, braces { } are used to designate activities, and square brackets [ ] are used to represent concentrations.

This relationship is summarized below.

$$\left\{ \begin{array}{c} \text{Standard} \\ \text{state} \end{array} \right\} = \left\{ \begin{array}{c} \text{Standard} \\ \text{concentration} \end{array} \right\} + \left\{ \begin{array}{c} \text{Reference state} \\ \text{environmental conditions} \end{array} \right\}$$

$$\left\{ \begin{array}{c} \text{Standard} \\ \text{state} \end{array} \right\} = \left\{ \begin{array}{c} \text{Standard} \\ \text{concentration} \end{array} \right\} + \left\{ \begin{array}{c} \text{Standard pressure} \\ \text{Standard temperature} \\ \text{Standard composition} \end{array} \right\}$$

Note that a species could be present in a system at its standard concentration but not be in its reference state, and it could be in its reference state but at a non-standard concentration; in neither case would it be in its standard state. Conventional assignments for standard state concentrations and reference state conditions are described next.

CONCENTRATION IN THE STANDARD STATE   In environmental systems, the most common choices for standard state concentrations are as follows:

- For dissolved solutes: a concentration of 1.0 mol/L.

- For bulk liquids and solids: the concentration of the species when it is present as a pure substance (mole fraction = 1.0) at 25°C and 1 bar total pressure.

- For gases: the concentration of a pure gas (mole fraction = 1.0) that is behaving in accord with the ideal gas law, at 25°C and 1 bar total pressure.

ENVIRONMENTAL CONDITIONS IN THE STANDARD STATE: THE REFERENCE STATE
Just as the standard state concentration establishes a baseline for the concentration of A, the reference state establishes a kind of baseline chemical environment in which those molecules reside. That is, it establishes a set of conditions for which chemical interactions are well defined, and to which other environments can be compared when we quantify a species' reactivity.

The temperature and pressure normally chosen for the reference state represent typical conditions at the surface of the earth, specifically 25°C and 1 bar. These choices are generally made regardless of whether the species of interest is a solid, liquid, or gas, and regardless of whether it is the major component of the system or just a trace component. Choices for the chemical composition of the system in the reference state are, however, generally different for different types of constituents.

For bulk solids and liquids, the reference state composition is usually defined as the pure substance. Thus, for example, the reference state for water corresponds to the environment in pure water at 25°C and 1 bar. If a solute were dissolved in the water, then the mole fraction of H_2O molecules would be <1.0; i.e., the solute would cause the *concentration* of H_2O to be less than its standard state concentration. In addition, the solute would change the *environment* in which the water molecules find themselves, and this change could make each individual water molecule either more or less reactive than it would be in the absence of the solute. Both of these factors would be expected to affect the overall reactivity of the water.

The reference state for gas molecules is usually chosen to be slightly different from the environment in pure, real gases. The ideal gas law represents molecules as hard spheres that occupy no space, that can engage in perfectly elastic collisions, and that have no attraction for one another. None of these assumptions is absolutely valid for real gases, although they are very good approximations for most gases under most environmental conditions. Because the characteristics of hypothetical, ideal gases are well established, the theoretical interactions in such a gas are generally chosen to define the reference state for gases. Therefore, for example, the standard state for oxygen gas is a *hypothetical* gas containing only O_2 molecules (i.e., a mole fraction of 1.0) at 25°C and 1.0 bar, in which each O_2 molecule behaves not as those molecules actually do in such a system, but rather as they would be expected to behave according to the ideal gas law.

The most common choice for the reference state of dissolved species is also based on a hypothetical, idealized situation. Specifically, the reference state is defined as a solution in which each molecule of solute interacts only with solvent (water) molecules, and not with other constituents. This hypothetical condition corresponds to what would actually occur in an infinitely dilute solution. Thus, for example, the standard state for dissolved zinc ions (Zn^{2+}) is a solution at 25°C and 1 bar total pressure, containing 1.0 mol/L Zn^{2+} (the standard state concentration), with each Zn^{2+} ion behaving as it would if it were the only ion in solution (the reference state environment). In a real solution containing 1.0 M Zn^{2+}, zinc ions might interact with one another and with the anions that must have entered the solution with the Zn^{2+} ions. As a result, that real solution might behave quite differently from the hypothetical standard state solution.

The most common definitions for the standard state for species in various phases are summarized in Table 1.3.

From the above description, it is apparent that standard states need not be real or attainable conditions. For instance, it is not possible to prepare a solution in which the Zn^{2+} concentration is 1.0 M and simultaneously to ensure that every Zn^{2+} ion behaves as it would if it were the only ion in solution. Nevertheless, despite the impossibility of preparing such a solution, we can imagine ways to infer what the reactivity of Zn^{2+} ions would be in such a standard state solution. For instance, we could measure the reactivity of successively more dilute real solutions containing Zn^{2+}, and then extrapolate to a hypothetical solution containing one molecule of Zn^{2+} per liter. If we multiplied the reactivity of Zn^{2+} in that (virtually) infinitely dilute solution by Avogadro's number (6.023×10^{23}),

Table 1.3   Common definitions for the standard state conditions in environmental systems

| | | Standard State | | |
| | | Reference State Conditions | | |
	Standard Concentration	**Temperature**	**Pressure**	**Other**
Solid	Concentration in pure solid	25°C	1 bar	—
Liquid	Concentration in pure liquid	25°C	1 bar	—
Gas	Concentration in pure gas	25°C	1 bar	Ideal gas behavior
Solute	1.0 M	25°C	1 bar	Infinite dilution

the result would be the hypothetical reactivity of a solution containing 1.0 mol/L Zn^{2+}, with each molecule of Zn^{2+} behaving as though the solution were infinitely dilute; i.e., it would be the reactivity of Zn^{2+} in its standard state.

The introduction to this section suggested that the chemical activity of any substance could be defined as its reactivity in the system of interest divided by its reactivity in a standard system, i.e., in the standard state. Thus, for example, if we determined the reactivity of Zn^{2+} in a solution to be 6.023×10^{20} times as great as that in a hypothetical solution containing one molecule of Zn^{2+} per liter and no other solutes, we could conclude that the activity of Zn^{2+} in the solution was 10^{-3} times as great as it would be in the standard state; hence its activity would be 10^{-3}. This means that although the real solution may not actually have a concentration of 6.023×10^{20} molecules of Zn^{2+} per liter, and the interactions of Zn^{2+} molecules with others might not be identical to those that would occur in an infinitely dilute solution, the Zn^{2+} behaves identically to how it would behave in a system in which those two conditions were met.

Note that, by definition, activities are always dimensionless, since they are ratios of the actual reactivity of a substance to the reactivity of that substance in the standard state. This result applies regardless of the phase in which the substance is found. For instance, the activity of a gas behaving the same as an ideal gas at a partial pressure of 0.1 bar and 25°C would be 0.1 (i.e., dimensionless) and would represent the ratio of the actual reactivity to that in the standard state.

SYSTEMS IN WHICH ENVIRONMENTAL CONDITIONS DIFFER FROM THE REFERENCE STATE; THE ACTIVITY COEFFICIENT   The ratio of the reactivity per molecule or per mole of A in a real system to the reactivity per molecule or per mole in the reference state is called the **activity coefficient** γ_A. The reactivity per molecule in the reference state is the same as in the standard state so, for a solute A, the activity coefficient can be represented as follows:

$$\gamma_A = \frac{\text{reactivity per molecule (or per mole) of } A \text{ in real system}}{\text{reactivity per molecule (or per mole) of } A \text{ in standard state}} \qquad \textbf{(1.7a)}$$

$$= \frac{\dfrac{\text{reactivity of } A \text{ in real system}}{\text{concentration of } A \text{ in real system}}}{\dfrac{\text{reactivity of } A \text{ in standard state}}{\text{concentration of } A \text{ in standard state}}} \qquad \textbf{(1.7b)}$$

$$= \frac{\dfrac{\text{reactivity of } A \text{ in real system}}{\text{reactivity of } A \text{ in standard state}}}{\dfrac{\text{concentration of } A \text{ in real system}}{\text{concentration of } A \text{ in standard state}}} \qquad \textbf{(1.7c)}$$

$$\gamma_A = \frac{a_A}{c_{A,\,\text{real}}/c_{A,\,\text{std. state}}} \qquad \textbf{(1.7d)}$$

Equations (1.7c) and (1.7d) indicate that the activity coefficient is dimensionless, being the ratio of two dimensionless terms. All the forms of Equation (1.7) apply to any constituent in any system, although for gases and solids it would be more conventional to write the denominator of Equation (1.7d) as the ratio of partial pressures and mole fractions, respectively.

Consider again a hypothetical solution in which the activity of dissolved Zn^{2+} is 10^{-3}, and assume that the concentration of Zn^{2+} in the system, designated $c_{Zn^{2+}}$ or $[Zn^{2+}]$, is 1.5×10^{-3} mol/L. If each molecule of Zn^{2+} in the solution behaved as it would in its reference state, the activity of Zn^{2+} in the system would be 1.5×10^{-3}. However, in the system of interest, the reactivity of the Zn^{2+} molecules is less than this; in fact, the data suggest that, on average, each molecule of Zn^{2+} in the real solution has a reactivity only two-thirds as great as that of a Zn^{2+} molecule in the reference state. We therefore conclude [based on Equation (1.7a) or (1.7d)] that the activity coefficient of Zn^{2+} in the system is 0.667:

$$\gamma_{Zn^{2+}} = \frac{10^{-3}}{1.5 \times 10^{-3}\ M/1.0\ M} = 0.667$$

The concentration and activity of dissolved glucose ($C_6H_{12}O_6$, MW 180) in a solution are determined to be 3.0 g/L and 0.015, respectively. What is the activity coefficient of glucose in the solution, and how can this value be interpreted in terms of the reactivity of glucose molecules compared to that in an infinitely dilute solution?

Example 1.6

Solution

The molar concentration of glucose in the solution of interest is 3.0/180, or 0.0167 M, and its concentration in the standard state is 1.0 M. Substituting these values and the known activity of glucose into Equation (1.7d), we obtain

$$\gamma_{C_6H_{12}O_6} = \frac{0.015}{0.0167/1.0} = 0.90$$

The result can be interpreted to mean that each molecule of glucose in the real solution is only 90 percent as reactive as it would be in an infinitely dilute solution.

Since the concentration of any species A in the standard state always has a numerical value of 1.0 (mol/L for dissolved species, bar for gases, and mole fraction for solids and liquids), Equation (1.7d) is frequently simplified as follows:

$$\gamma_A = \frac{a_A}{c_A/1.0} = \frac{a_A}{c_A} \tag{1.8a}$$

$$a_A = \gamma_A c_A \tag{1.8b}$$

where c_A is the concentration of A. (Note that c_A must be given in the same units as are used to define the standard state for the equation to be valid.) Based on Equation (1.8b), the activity of A can be thought of as the "equivalent concentration" of A in the reference state ($\gamma_A = 1.0$). That is, a system with a concentration of 0.2 mol/L A and $\gamma_A = 0.8$ is equivalent (in terms of the reactivity of A) to one that has a concentration of 0.16 mol/L A and $\gamma_A = 1.0$. By extending the analysis, dissolved A would be equally reactive in all the following solutions:

$$[A] = 0.20 \text{ mol/L} \qquad \gamma_A = 0.80 \qquad \{A\} = 0.16$$

$$[A] = 0.40 \text{ mol/L} \qquad \gamma_A = 0.40 \qquad \{A\} = 0.16$$

$$[A] = 0.16 \text{ mol/L} \qquad \gamma_A = 1.00 \qquad \{A\} = 0.16$$

$$[A] = 1.00 \text{ mol/L} \qquad \gamma_A = 0.16 \qquad \{A\} = 0.16$$

The wide range of activity coefficients of the hypothetical solutions listed above suggests that those solutions have very different compositions. Nevertheless, the equality of $\{A\}$ in the various solutions indicates that that species is equally reactive in all of them. Thus, the activity provides a consistent way to compare behavior of a given species in disparate systems.

Although activity coefficients are frequently represented as in Equation (1.8b), this representation can lead to some confusion about dimensions. As noted above, both activities and activity coefficients are always formally dimensionless. However, to make γ_A dimensionless when it is computed according to Equation (1.8a), aqueous phase activities are often represented as though they had dimensions of concentration. In reality, dimensions only *appear* to apply to the activity in Equation (1.8a) because the fact that the dimensions of c_A have been canceled (by normalizing the concentration in the real system to the concentration in the standard state) is hidden.

To repeat, the activity coefficient describes the reactivity of an average molecule in the actual system compared to the reactivity of such a molecule in the reference state. If the actual and reference states are quite similar chemically, then the activity coefficient will be close to 1.0, and the numerical values of the activity and concentration will be nearly equal. Correspondingly, to the extent that conditions in the real system differ from the reference condition, the activity coefficient will differ from a value of 1.0.

A species that behaves in the real system exactly as it would in the reference state is said to behave *ideally,* and a solution in which all species behave as they would in their respective reference states is called an *ideal solution*. The use of the term *ideal* in such a circumstance and the idea that activity coefficients "correct" for the non-ideality of real solutions are unfortunate. They can mislead one into thinking either that the activity coefficient is a fudge factor, or that activity coefficients far from 1.0 reflect extreme or highly unusual conditions in solution. This is not at all the case. Ideality as used in this context simply means conformity to the arbitrary reference state. To make the point in an exaggerated way, one could choose the reference state for a dissolved substance as a solution containing 10 mol/L of that substance at the critical point of water. In that case, dilute solutions at normal temperatures and pressures would have activity coefficients very far from 1.0, reflecting the fact that the ions in those solutions behave very differently from ions in the "ideal" reference state.

ALTERNATIVE CHOICES FOR THE REFERENCE STATE AND THE STANDARD STATE The reference state for solutes described above (infinite dilution) is the one that is most commonly used in dilute aquatic systems, but not the only one. For instance, one fairly common alternative choice for the reference state is a solution that contains, in addition to water molecules, the major species that are present in the real solution of interest. This reference state is frequently used by marine chemists. According to this convention, for instance, the reference state for Zn^{2+} ions in seawater would be a solution in which Zn^{2+} is infinitely dilute, but in which Na^+, Ca^{2+}, Mg^{2+}, Cl^-, SO_4^{2-}, and HCO_3^- are present at their normal concentrations in seawater. Infinite dilution of Zn^{2+} implies that zinc ions interact with H_2O and the six other ions listed, but not with other Zn^{2+} ions or other minor ions that might be present in the system. Because some of the major ions in seawater shield Zn^{2+} ions, each Zn^{2+} ion in seawater is less reactive than a Zn^{2+} ion in fresh water. As a result, Zn^{2+} ions in fresh water would have an activity coefficient greater than 1.0 when evaluated using the seawater reference state.

The activity coefficient of Sr^{2+} in seawater is approximately 0.26 for a reference state of infinite dilution and approximately 1.0 for a reference state of major ion seawater. What would $\gamma_{Sr^{2+}}$ be in infinitely dilute fresh water, if the reference state were defined as major ion seawater? Is Sr^{2+} an ideal solute in seawater? | **Example 1.7**

Solution

The reactivity of Sr^{2+} in any solution is determined by chemistry, not by arbitrary definitions. Therefore, the *relative* reactivities of Sr^{2+} in two solutions must be the same regardless of the scale we use to quantify activity. (The analogous statement for a non-chemical system is that if an object is twice as long as another, the ratio of their lengths is 2.0 regardless of whether we measure length in inches, centimeters, or leagues.) The problem statement indicates that the ratio of γ values in seawater (sw) and infinitely dilute fresh water (fw) is $\gamma_{sw}/\gamma_{fw} = 0.26$. If the reference state were major ion seawater, γ_{sw} would be 1.0, so γ_{fw} would be given by $\gamma_{fw} = 1.0/0.26 = 3.85$.

An ideal solute is one that behaves (in a real solution) exactly as it would in the reference solution, and for which γ is therefore 1.0. Thus, in seawater, Sr^{2+} behaves as an ideal solute if the reference state is major ion seawater, but it is quite non-ideal if the reference state is infinitely dilute fresh water.

Note that there is no need to define the standard state identically for all dissolved molecules in a system; all that is needed is for the standard state of each species to be defined unambiguously. For example, for solutes that can form a separate liquid phase (e.g., benzene), the pure compound in its liquid form is sometimes used as the reference state. In other words, even though the system of interest might be a very dilute aqueous solution of benzene, the behavior of benzene molecules in the aqueous solution is described by comparison with their behavior in pure liquid benzene. In this case, the activity coefficient of benzene might be very different from 1.0, reflecting the different chemical environment in the aqueous solution as opposed to the pure, non-aqueous liquid defining the reference state. Again, it is perfectly acceptable to use pure liquid as the reference state for dissolved benzene and infinite dilution as the reference state for Zn^{2+} in the same solution.

In sum, the chemical activity of a substance incorporates factors related to the substance's concentration and its physical and chemical environment into a single variable. Use of the term *activity* to describe this synthesis is apt; the value of the activity indicates how active the species is in the system, in terms of its tendency to participate in reactions. Either increasing the concentration of the species or changing the chemical environment in such a way that the species is more likely to collide with other substances increases its activity.

While at first the various standard and reference states used to quantify activity might seem confusing, the definitions are in fact nothing more than a convenient shorthand. They allow us to describe the reactivity of a solute in a real system simply by stating its activity, in lieu of stating all the details about the solution composition and environmental conditions.

1.5.3   PREDICTING ACTIVITY COEFFICIENTS FROM KNOWLEDGE OF THE SOLUTION COMPOSITION

Under some circumstances, activity coefficients of dissolved species can be estimated from theory. The ionic strength, which was introduced earlier in this chapter, turns out to be a key parameter in this calculation. The basis for the relationship between ionic strength and activity coefficients is that dissolved ions attract oppositely charged ions and repel like-charged ions. Therefore, each cation in a solution is surrounded by a diffuse cloud of other molecules consisting primarily of anions and water molecules, the latter oriented with the oxygen closest to the cation. Some cations are also in this cloud, but they are less concentrated than their average concentration in the whole solution, whereas the anions are more concentrated. An analogous but inverse description applies to dissolved anions. As noted earlier, the kinetic energy of the molecules prevents them from remaining in place, so ions and water molecules are constantly entering and exiting one another's "sphere of influence."

The activity coefficient characterizes the difference between the chemical environment in the system of interest and that in the reference state. Thus, if the reference state for the ion under consideration is infinite dilution, the activity coefficient of the ion reflects the fact that in the real system, the species surrounding the ion include solutes and other ions as well as water molecules. Most of the ions in this region are oppositely charged to the ion of interest and therefore have a greater tendency to remain close to the central ion than do water molecules. These relatively tightly held, oppositely charged ions shield the central ion from interactions with other dissolved species and reduce its activity coefficient. The more highly charged are the central ion and the other ions in solution, the greater is the attraction and the shielding effect, and the lower is the activity coefficient. Similarly, for a given central ion, the higher the concentration of other ions in solution, the greater the shielding and the lower the activity coefficient (Figure 1.5).

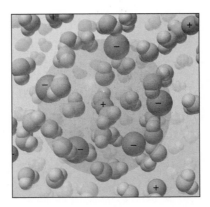

Figure 1.5   The arrangement of water molecules and ions around a dissolved ion. Although most of the molecules around a given ion are water, oppositely charged ions are selectively attracted to that environment.

From M. S. Silberberg, *Chemistry: The Molecular Nature of Matter and Change.* Copyright © 2000 The McGraw-Hill Companies, New York, NY. Reproduced by permission of The McGraw-Hill Companies.

In 1923, Debye and Huckel derived a mathematical model to predict the value of the activity coefficient of symmetric electrolytes (salts in which the magnitude of the charge is the same on the cation as on the anion) in aqueous solutions. The derivation can be developed fairly simply using principles of physical chemistry.[12] The resulting equation, known as the *Debye–Huckel limiting law,* proved to be quite accurate for predicting activity coefficients of individual ions in dilute solutions.[13]

One key assumption of the Debye–Huckel derivation is that ions can be treated as point charges and can therefore approach infinitely close to one another. Eliminating this assumption and taking ionic size into account yields another equation known as the *extended Debye–Huckel equation,* which is useful for predicting ion activity coefficients in significantly more concentrated solutions than can be modeled with the Debye–Huckel limiting law. However, the extended Debye–Huckel equation is strictly valid only if both the ion of interest and the shielding ions are all of the same ionic size.

Davies proposed adding a term to the extended Debye–Huckel equation to improve the fit between the equation and experimental observations. Later, Pitzer and coworkers proposed incorporating additional terms to account for interactions among specific pairs of ions, thereby effectively incorporating information about the size and properties of individual ions into the analysis. The additional terms have the same mathematical form for any pair of ions, but they have coefficients

---

[12]An excellent presentation is provided in *Modern Electrochemistry,* vol. 1, by Bockris and Reddy (Plenum Press, 1970).

[13]Since any salt must release both cations and anions to solution simultaneously, it is not possible to attribute the effect of the salt to one of the individual ions; only the overall effect of both ions on activity coefficients can be predicted or measured. Nevertheless, it is useful to assign activity coefficients to individual ions. The Debye–Huckel limiting law and others derived from it achieve this by making some arbitrary, but reasonable and consistent, assumptions.

Table 1.4a   Various equations for predicting activity coefficients in aqueous solutions

Name and Equation	Notes and Approximate Range of Applicability
Debye–Huckel limiting law $$\log \gamma_{D-H} = -Az^2 I^{1/2}$$	$A = 1.82 \times 10^6\, (\varepsilon T)^{-3/2}$, where ε is the dielectric constant of the medium. For water at 25°C, $A = 0.51$; z = ionic charge. Applicable at $I < 0.005\ M$
Extended Debye–Huckel $$\log \gamma_{\text{Ext.D-H}} = -Az^2 \frac{I^{1/2}}{1 + BaI^{1/2}}$$	$a \equiv$ ion size parameter (see Table 1.4b) $B = 50.3(\varepsilon T)^{1/2}$; for water at 25°C, $B = 0.33$. Appropriate in solutions where one salt dominates ionic strength. Applicable at $I < 0.1\ M$
Davies $$\log \gamma_{\text{Davies}} = -Az^2 \left(\frac{I^{1/2}}{1 + I^{1/2}} - 0.2I \right)$$	Applicable at $I < 0.5\ M$
Specific interaction model $$\log \gamma_{\text{Pitzer}} = \log \gamma_{\text{Ext.D-H}} + \sum_j B_{ij}\, Im_j$$	B_{ij} is specific interaction term between ions i and j; m_j is molality (mol/kg solution) of j. Applicable at $I < 1\ M$; additional terms can extend range to higher ionic strengths[a]

[a]See Pitzer (*J. Solution Chem.* 4, 249–265, 1975) or Millero (*Geochim. Cosmochim. Acta* 47, 2121–2129, 1983).

Table 1.4b   Values of ion size parameter (in Ångstroms) for use in the extended Debye–Huckel equation

	Cations	Anions
Singly charged ions	$a = 3$: Ag^+, K^+, NH_4^+ $a = 4$: Na^+ $a = 9$: H^+	$a = 3$: Cl^-, ClO_4^-, HS^-, I^-, NO_3^-, OH^- $a = 4$: CH_3COO^-, HCO_3^-, $H_2PO_4^-$, PO_4^{3-}
Doubly charged ions	$a = 5$: Ba^{2+}, Pb^{2+}, Sr^{2+} $a = 6$: Ca^{2+}, Cu^{2+}, Fe^{2+}, Mn^{2+}, Sn^{2+} $a = 8$: Be^{2+}, Mg^{2+}	$a = 4$: HPO_4^{2-}, SO_4^{2-} $a = 5$: CO_3^{2-}
Triply charged ions	$a = 9$: Al^{3+}, Ce^{3+}, Fe^{3+}, La^{3+}	$a = 4$: PO_4^{3-}

that have to be determined for each pair. Still more terms intended to account for three-way interactions among ions and for other factors have been proposed as well. The resulting equations have been used successfully to estimate activity coefficients in seawater and even more concentrated ionic solutions.[14]

[14]An excellent comparison of all these models for estimating γ is available in *Aqueous Environmental Geochemistry*, by Langmuir (Prentice-Hall, 1997).

The equations discussed above are collected in Table 1.4a, and predicted values of activity coefficients of a few ions are shown as a function of ionic strength in Figure 1.6. Two key points to note about the equations in Table 1.4a and the figure are that

1. In all the equations, the activity coefficient of an ion depends strongly on that ion's charge.
2. The variable that characterizes the ionic composition of the solution is the ionic strength.

One interesting phenomenon that has been observed experimentally and that can be modeled with some success by the specific interaction model is that the activity coefficients of cations increase dramatically with increasing ionic strength when the ionic strength exceeds approximately 1 mol/L. Part of the explanation for this phenomenon is the substantial reduction in the concentration of free water molecules in solution, because so many of them are held in the hydration sphere surrounding the ions, but other as yet unidentified factors must also be involved. The activity coefficient of OH^- follows a trend similar to that of cations, but for most anions γ either declines steadily or increases only slightly as ionic strength increases above 1 mol/L.

A consequence of the assumption that ionic interactions are the only factor that affects activity coefficients is that, according to the models discussed above, uncharged species are always predicted to behave ideally; i.e., they have activity coefficients of 1.0. Empirically, activity coefficients of neutral species usually increase with increasing ionic strength, but the dependence is quite weak (γ is usually in the range 1.00 ± 0.05 for natural fresh waters). This dependence is commonly modeled by the equation $\log \gamma_A = k_A I$, where k_A is positive for most species, but negative for a few. The error associated with assuming $\gamma_{neut} = 1.0$ is probably smaller than that associated with other uncertainties in analyzing most systems, and the assumption that neutral solutes behave ideally is made throughout this text.

Example 1.8

A model fresh water is prepared by adding $10^{-3}\ M$ NaCl, $10^{-3}\ M$ NaHCO$_3$, and $10^{-4}\ M$ CaCl$_2$ to water. Assume that when these salts dissolve, they dissociate to Na^+, Cl^-, Ca^{2+}, and HCO_3^- and then do not react further.

a. Compute the ionic strength of the solution.

b. What is the activity of each dissolved ion if the infinitely dilute solution approximation applies and the infinite dilution reference scale is used? What is the activity of each ion if the extended Debye–Huckel equation applies?

c. Calculate the activities of all the ions if the NaCl concentration is increased by two orders of magnitude to 0.1 M to model an estuarine water. Because the ionic strength is higher after the NaCl is added, use the Davies equation to calculate activity coefficients.

d. What would the activity coefficient of Ca^{2+} be in the model estuarine water if the reference state were defined as the conditions in the model estuarine water? What would the activity coefficient of Ca^{2+} in the model fresh water be on this newly defined scale?

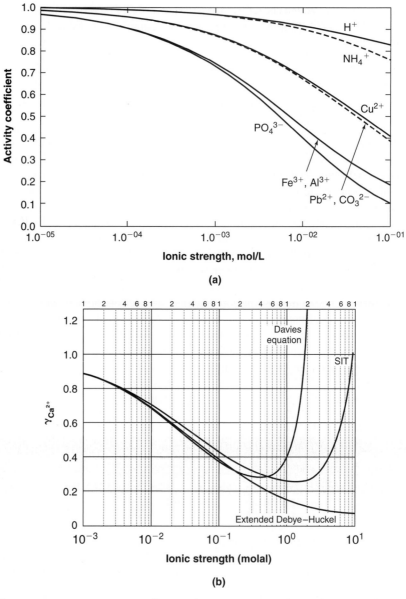

Figure 1.6 **(a)** Activity coefficients of various ions according to the extended Debye–Huckel law, based on the infinite dilution reference state. **(b)** The activity coefficient of Ca^{2+} in a solution prepared by dissolution of $CaCl_2$, according to three models. The specific ion interaction (SIT) model (of Pitzer) is the most complex of the three shown, and it fits the experimental data best. However, at low ionic strengths, the other two equations yield satisfactory results.

From *Aqueous Environmental Chemistry* by Donald Langmuir, © 1997. Reprinted by permission of Prentice-Hall, Inc., Saddle River, NJ.

Solution

a. After dissociation, the concentrations of the various ions and their contributions to the ionic strength are as follows:

Ion	Concentration (M)	$0.5\,c_i z_i^2$
Na^+	2.0×10^{-3}	1.0×10^{-3}
Cl^-	1.2×10^{-3}	6.0×10^{-4}
Ca^{2+}	1.0×10^{-4}	2.0×10^{-4}
HCO_3^-	1.0×10^{-3}	5.0×10^{-4}
	Sum	2.3×10^{-3}

As shown in the table, the ionic strength of the solution is 2.3×10^{-3}.

b. By definition, if the reference state is infinite dilution and the infinite dilution approximation applies, then $\gamma = 1.0$. In that case, the activity of each ion just equals its molar concentration; that is, $\{Na^+\} = 2.0 \times 10^{-3}$, $\{Cl^-\} = 1.2 \times 10^{-3}$, $\{Ca^{2+}\} = 1.0 \times 10^{-4}$, $\{HCO_3^-\} = 1.0 \times 10^{-3}$. The activity coefficients and ion activities computed according to the extended Debye–Huckel equation are summarized in the following table.

Ion	Size Parameter a	Log γ	γ	Activity
Na^+	4	-0.0226	0.95	0.00190
Cl^-	3	-0.0229	0.95	0.00114
Ca^{2+}	6	-0.0876	0.82	0.000082
HCO_3^-	4	-0.0226	0.95	0.00095

c. The ionic strength of the model estuarine water can be computed as in part (a), except that the new Na^+ and Cl^- concentrations are 0.101 and 0.1002 M, respectively, yielding a new ionic strength of 0.1013. Substituting this value into the Davies equation yields the values shown in the following table for activity coefficients and the corresponding activities of the ions. Note that the activities of Ca^{2+} and HCO_3^- decrease by 57 and 19%, respectively, when the NaCl concentration is increased, even though the concentrations of these two ions are not altered.

	Log γ	γ	Activity
Na^+	-0.113	0.77	7.79×10^{-2}
Cl^-	-0.113	0.77	7.73×10^{-2}
Ca^{2+}	-0.451	0.35	3.50×10^{-5}
HCO_3^-	-0.113	0.77	7.70×10^{-4}

d. Activity coefficients are assigned a value of 1.0 in the reference state and are determined experimentally or predicted by model equations if the system is not in the reference state. Therefore, if the reference state were defined as the conditions in the estuarine water, the activity coefficient of Ca^{2+} in that water would be 1.0 by definition.

According to the table in part (c), Ca^{2+} ions are 35% as reactive in the estuarine water as in an infinitely dilute solution, and according to part (b), they are 82% as reactive in the fresh water as in an infinitely dilute solution. Thus the ratio of Ca^{2+} activity coefficients in the estuarine and fresh waters is 0.35/0.82, or 0.43. This ratio, being chemically determined, is not altered by our choice of reference systems. Therefore, if we chose the estuarine water as the reference state and assigned Ca^{2+} ions an activity coefficient of 1.0 in that solution, their activity coefficient in fresh water would be 1.0/0.43, or 2.33.

1.5.4   THE ACTIVITY AS AN INTENSIVE PROPERTY AND A MEAN FIELD PARAMETER

Note that simply increasing the *amount* of a given phase present without changing its composition, temperature, or pressure does not change the activity of any chemical in that phase. Thus, the activity of Ca^{2+} in a $10^{-2} M$ solution of $CaCl_2$ is the same regardless of whether there is 1 L or 1000 L of solution in the system ($CaCl_2$ is a salt that dissociates to release Ca^{2+} and Cl^- ions to solution). Changing the concentration of $CaCl_2$ in the solution would change the activity of dissolved Ca^{2+}, and adding a different chemical to solution might do so as well, but changing the amount of solution present would not. Similarly, increasing the amount of a solid in a system does not change its activity. If, for instance, a pure solid is present in a system at 1.0 bar total pressure and 25°C, the activity of the solid will be 1.0 no matter how much of it is present. In thermodynamic jargon, properties that characterize a phase and that are independent of how much of the phase is present are called **intensive properties.**

The constancy of the activity of pure solids, regardless of the mass of solid present, can become confusing when one is dealing with systems in which a solid is dispersed in a solution. For instance, copper might be removed from an industrial wastewater by precipitation of solid copper hydroxide, $Cu(OH)_2(s)$. Assuming the solid that formed was pure $Cu(OH)_2(s)$, the activity of $Cu(OH)_2(s)$ would be 1.0 regardless of how much solid was present, since the mole fraction of $Cu(OH)_2(s)$ *in the phase in which it is found* would be 1.0. We may wish to quantify how much solid $Cu(OH)_2(s)$ is suspended per liter of water and refer to that value as the concentration of $Cu(OH)_2(s)$ in the water. However, such a concentration refers to the amount of one phase *dispersed* in another and is fundamentally different from the concentration of a substance *dissolved* in another. In the former case, we are just talking about two bulk substances sharing a region of space, and in the latter we are describing chemical mixing on the molecular level, with all molecules of the solute able, in theory, to interact with all molecules of the solvent. The use of the same term—*concentration*—to refer to both dispersed and dissolved substances often confuses students and leads to the error of treating the activity of the solid as being proportional to its concentration in the solid/solution mixture. Be careful to avoid this error!

A final point worth noting about chemical activity is that it is a continuous property that applies uniformly to a whole phase (like a gravitational or electrical field), and hence it has meaning even under conditions where the chemical concentration (a discrete parameter) does not. For instance, we are sometimes interested in the properties of solutions in which certain chemical species are fantastically dilute, so dilute in fact that their concentrations might be computed to be less than one molecule per liter. Discussing the concentration of such a species in the system is meaningless, in that a given liter of solution either does or does not contain a molecule of the substance; it cannot contain the average concentration of less than one molecule per liter. Furthermore, even if we knew (somehow) that one molecule of the substance were in a liter of solution, that molecule could not be viewed as uniformly distributed throughout the solution. Nevertheless, the activity of that species is considered to be well defined and uniform throughout the phase. A similar situation applies when we wish to characterize the environment very near a solid surface at a distance closer than a molecule could realistically approach. The concentration of molecules in that region is not definable, but their activity is.

One justification for treating chemical activities in this way is that the activity refers to the potential for a molecule to react, taking into account both the probability that the molecule will be in a given region of space and the likelihood that it will be in a given energy environment. Since the probability that a molecule will be in a location is finite, even if the molecule is not there at a given instant, the activity of the molecule at that location is nonzero. For most systems of interest, this subtle point is not critical to our understanding or our calculations. However, to avoid giving the impression that we are talking about "partial molecules" in systems with very low activities, and to justify talking about the activity of a species right up to the surface of a solid, it is useful to make the point.

1.6 CHEMICAL REACTIONS AND CHEMICAL EQUILIBRIUM

As noted in the introduction to this chapter, this text focuses on the analysis of **chemical equilibrium,** i.e., the state toward which a system tends as reactions proceed to reduce any chemical instabilities within it. Once equilibrium is attained, there is no net driving force for any further change in the chemical composition of the system. In the following section, the classical explanations of why reactions occur and the associated terminology are introduced.

1.6.1 HOW AND WHY CHEMICAL REACTIONS OCCUR: A MOLECULAR-LEVEL PICTURE OF AN ELEMENTARY REACTION AND THE FACTORS THAT AFFECT ITS RATE

Chemical reactions occur when the reactant molecules collide with sufficient energy to allow them to rearrange into a more stable configuration. Therefore, a complete description of the process requires knowledge about the frequency of collisions, the energy involved in the collisions, the energy required for molecular rearrangement, and the relative stability of the various configurations possible.

Often, the net observable reaction at a macroscopic scale is the result of several reactions proceeding in parallel and series. That is, when two reactants are mixed, they might form a new species, which then reacts with others in the system to form different species, which in turn might react with yet others. Eventually, a stable suite of products is formed. If all the intermediate steps occur very quickly, we might not even be aware of their existence.

Each reaction in the sequence, whether we can detect it or not, occurs as the result of individual, sufficiently energetic collisions between the reactants for that reaction step. Such reactions are called **elementary reactions.** Theory suggests that collisions among more than two molecules are rare, so that any overall reaction involving three or more reactants is virtually certain to be non-elementary. One example of a sequence of elementary reactions that leads to an overall, non-elementary reaction with four reactants is shown below. [The designation $2(Y + C \rightarrow F + Z)$ means that the elementary reaction is between one Y and one C molecule, but that two such elementary reactions occur each time the overall sequence proceeds.]

$$\begin{aligned}
\text{Sequence of elementary reactions:} \quad & A + B \rightarrow 2X \\
& 2(X + C \rightarrow Y + Z) \\
& \underline{2(Y + Z \rightarrow D)} \\
\text{Overall reaction:} \quad & A + B + 2C \rightarrow 2D
\end{aligned}$$

In the above example, species X, Y, and Z are intermediates that are first created and then destroyed in the elementary reactions. Therefore, the concentrations of these species are not affected at all by the overall reaction sequence.

Consider the progression of events as a single elementary reaction occurs. As the reactant molecules approach one another, interactions between their electric fields cause the bonds in each molecule to become strained, increasing the chemical potential energy stored in the molecules. The strain increases dramatically with decreasing separation. Since energy is conserved during the interaction, this increase in chemical potential energy must somehow be balanced by a corresponding decrease in energy elsewhere in the system. In this case, the dominant conversion process is from molecular kinetic energy into chemical potential energy: the molecules slow down as the distance between them decreases (some of the molecular energy associated with intramolecular vibrations and rotation may also be affected). In the absence of other factors, the molecules would eventually stop their mutual approach and then begin moving away from one another, thereby relieving the strain and converting the chemical energy back into kinetic energy. Conceptually, the process is identical to a ball rolling up an incline, stopping, and reversing itself.

Although the molecules might indeed separate without undergoing any long-term changes, it is also possible that, as the original molecular structures adjust in response to the strain, new bonds will begin to form. At some critical point, the original bonds may become sufficiently distorted, and the new bonds may form to a sufficient extent, that the strain can be more easily relieved by rearrangements that form product molecules rather than the original reactants (like a ball rolling over the crest of a hill and proceeding down the other side, rather than returning along the path of its climb). The amount of energy necessary to bring molecules from far apart (no interaction) to the critical point (the point where the strain is equally likely to be relieved by formation of products or re-formation of reactants) is called the **activation energy** E^*. Correspondingly, the process of reaching the critical condition is sometimes referred to as overcoming the *activation energy barrier*. At the critical point, the molecules are not identifiable as either the reactants or the products, but are an intermediate species of negligible stability (analogous to the ball being perched exactly at the crest of the hill).

In addition to the energy of the colliding molecules, the orientation of the molecules when they collide might be important in determining whether a reaction occurs. If the molecules are not spherically symmetric, only a fraction of all collisions can cause the bonds to distort in a way that leads to the formation of product, even if the collisions involve an amount of energy greater than the activation energy. Thus, the overall rate of reaction depends on the frequency with which reactant molecules collide, the likelihood that colliding molecules are properly oriented and have sufficient kinetic energy to overcome the activation barrier, and the rate at which the activated species is converted to products.

Catalysts operate by providing an alternative path by which a reaction can occur. Specifically, they allow reactants to be converted to products via a route that has a lower activation energy than the route that is taken in their absence. Continuing the analogy to a ball on a hill, a catalyst might be viewed as providing an alternative path for the ball to follow that does not require quite so much of a climb before arriving at a point on the downhill slope. In some cases, catalysts can increase the rate of an overall reaction by many orders of magnitude. However, catalysts cannot alter the overall energy change associated with the reaction or the ultimate mix of products that forms.

Two mathematical models for the rate at which chemical reactions proceed have been developed based on the conceptual picture described above. The models are called the *collision model* and the *activated complex* or *transition state model,* although the names may be somewhat misleading since both rely on the conceptual framework of collisions causing molecules to pass through an activated state before being transformed to products. The key difference between the models relates to how molecules are envisioned to approach and pass through the activated state.

In the collision model, the approach to the activated state is treated as an instantaneous process, like popcorn popping: the corn is unpopped one instant and popped the next, and at no point do we think of the kernel as being in between those two states (although obviously there must be a short time when it is).

In the transition state model, the passage from reactants to products is viewed more as a continuum, with the reactant molecules being converted to products by proceeding along a path on the "energy landscape" described above. While the concentration of molecules at the exact peak of the path is essentially zero at any instant, a finite concentration of molecules is envisioned to be on a flat part of the hill near the peak, and all these molecules are treated as being in the activated state. Such molecules are called **activated complexes.** For a generic elementary reaction between A and B, the activated complexes are commonly represented as AB^*; that is, the reaction is $A + B \rightarrow AB^* \rightarrow P$.

A schematic of the energy relationships during the transition from reactants through the high-energy state to products according to the two models is shown in Figure 1.7. A more thorough description of these models is provided in Appendix 1A.

While the above description provides a framework for understanding chemical reactions at the molecular level, in most cases we do not have the information necessary to describe reactions of interest at that level of detail. Furthermore, as explained above, most overall reactions of interest are more complicated than the description suggests, because they are in fact not elementary, but rather the net result of several elementary reactions. Nevertheless, even without knowing the details of a specific reaction at the molecular level, it is possible to make some general observations about the tendency for reactions to occur and the stable endpoint (i.e., the equilibrium condition) toward which they progress. The following section describes ways of thinking about and analyzing that condition.

1.6.2   MODELS OF CHEMICAL EQUILIBRIUM

Chemical equilibrium describes a state in which there is no chemical driving force favoring a change in the system's composition. Two conceptual models are widely employed to describe chemical equilibrium, corresponding closely to the two models mentioned earlier that are used to describe molecular reactivity. One model characterizes equilibrium as a dynamic steady state in which each type of molecule in the system is being formed by chemical reactions at the same rate that it is being destroyed; as a result, on a macroscopic level, the system behaves

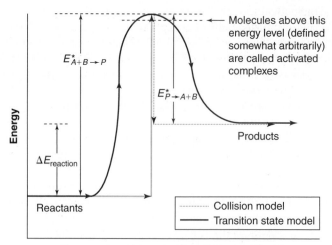

Progression of reaction (reaction coordinate)

Figure 1.7   Schematic representation of the transition from reactants to products according to the collision and activated complex (transition state) models. Here $\Delta E_{reaction}$ is the overall energy change accompanying the reaction, i.e., the energy of the products minus that of the reactants.

as though the molecules are not reacting at all. This representation is closely related to the model that explains the reactivity of a species A in terms of the likelihood of collisions between A and other molecules in the system.

The second model for equilibrium is more closely aligned with the energy-oriented perspective on chemical reactivity and can be described as the rock-on-a-hillside model. In the absence of other forces, a rock has the natural tendency to attain its condition of minimum gravitational potential energy; i.e., it rolls downhill. Similarly, chemicals can be thought of as having a chemical potential energy. In any system, in the absence of other factors affecting the chemical composition, reactions will occur in such a way that the total chemical potential energy approaches a minimum; once that condition is attained, they are at equilibrium, and no further (macroscopic) change will occur.

These two views, the dynamic molecular kinetic model of equilibrium and the "static" thermodynamic model, are equally valid. Neither is better than the other in an absolute sense, and both are useful in various circumstances. Either or both models can be used in a given situation, depending on which approach is more useful. If the available data are accurate, both models must predict the same final results, and these results must be experimentally valid.

Two important points that are basic to the study of equilibrium chemistry fit well with the rock-on-a-hill analogy. The first point allows us to generalize conclusions from single experiments to an infinite variety of other systems: the equilibrium condition is independent of previous history. The location of lowest potential energy for a rock is at the lowest altitude available, regardless of where the rock has been previously or where it is when we encounter it. Similarly, when chemicals are mixed to generate a system with a given set of initial concentrations, they react to form the same products (at equilibrium), regardless of how the reactants were formed, the order in which they were added to the system, how they were mixed, etc.

The second point limits the applicability of our calculations. Specifically, equilibrium considerations tell us the ultimate equilibrium conditions, but not how long it will take to attain them. As is the case with rocks, chemicals can sometimes remain in non-equilibrium states for geologic times. Indeed, all plants and animals are thermodynamically unstable in an atmosphere that contains oxygen; if they (we!) reached chemical equilibrium, all organisms exposed to the oxygen in the atmosphere would be oxidized to carbon dioxide and water.

Both chemical and physical processes require an energy input to destabilize the system before it can be rearranged into a more stable configuration. The required energy might be relatively small (e.g., for rocks, it might be provided by wind, a small flow of water, or a hiker's step) or very large (e.g., on the order of that provided by tectonic action). In the case of chemical reactions, the energy is required to destabilize existing bonds, so that atoms can rearrange to form different, more stable ones. This energy is defined above as the activation energy. Note that the magnitude of the activation energy is independent of the amount of potential energy that is released in the subsequent equilibration process.

How closely a particular system approaches chemical equilibrium depends on the relative rates of chemical processes (i.e., chemical reactions) and physical processes (bulk flow, mixing, diffusion, etc.) in the system. A simple example in which the reaction is the melting of ice can illustrate this point. If a mixture of ice at 0°C and water at 20°C is placed in an environment where the temperature of the overlying air is 25°C, some of the ice melts. Assume that if this system reached equilibrium, all the ice would melt and the temperature of the water and air would be 10°C. Whether the system actually reaches equilibrium depends on the rates of heat transfer and the amount of time that the various components are in contact. In the limit, if the three phases are in contact for a very short time and mixing is minimal, almost no ice melts, the water and air temperatures change negligibly, and the system remains far from equilibrium. On the other hand, after long periods of contact, the system reaches equilibrium. In addition to contact time, the mixing intensity in the system and the physical form of the constituents (whether the ice is present as a solid block or thin shavings, whether the water is present only as a bulk fluid or also as a mist) affect the rate of approach to equilibrium.

Like the physical changes in the form and temperature of water in the system described above, chemical changes (reactions) proceed at different rates in different environments. Also, the solutions in which those reactions occur might spend a short or long time in the environment of interest. If a chemical reaction in an aquatic system is rapid compared to the time that the reactants spend in the system, the reaction is expected to reach equilibrium by the time the water exits the system. On the other hand, if the reaction is slow compared to the time available, the materials exiting the system will not have progressed significantly toward equilibrium, and in fact will be minimally modified from their input conditions.

Because this text focuses almost exclusively on the determination of the equilibrium condition in aquatic systems, the calculations described are most directly applicable to systems in which the reactions are rapid compared to the time available. However, even in systems that do not attain equilibrium, the calculations are often useful as indicators of the ultimate equilibrium condition toward which the system tends. The remainder of this chapter focuses on the kinetic model of chemical equilibrium, describing the progress of individual (forward or reverse) reactions, the factors that control their rates, and the quantification of the equilibrium condition based on those rates. Chemical equilibrium is analyzed from a thermodynamic perspective in Chapter 2.

1.6.3 THE KINETIC MODEL FOR CHEMICAL EQUILIBRIUM AND THE DEFINITION OF THE EQUILIBRIUM CONSTANT

The kinetic model for chemical equilibrium uses the activity directly as a measure of the tendency for the molecule to participate in chemical reactions. That is, if a collision between A and B can lead directly to the formation of C and D ($A + B \leftrightarrow C + D$), the rate at which the reaction proceeds to the right is assumed to be

$$\text{Rate of conversion of } A \text{ and } B \text{ to } C \text{ and } D = k_f\{A\}\{B\} \qquad \textbf{(1.9)}$$

where k_f is called the forward **rate constant** and $\{A\}$ and $\{B\}$ are the activities of the respective species. The rate constant accounts for the inherent tendency for the reaction to proceed, including factors such as the activation energy. Another way of thinking of k_f is as the rate of conversion of A and B to C and D in a system containing A and B in their standard states (in which case $\{A\}$ and $\{B\}$ are 1.0). The dependence of the reaction rate on nonchemical environmental factors, such as temperature, is also incorporated into the value of k_f.

For a reaction requiring a collision between two A molecules, the probability that the reaction will occur is proportional to the activity of A, times itself; i.e., it is proportional to $\{A\}^2$. Extending this reasoning, for an elementary reaction, $aA + bB \leftrightarrow cC + dD$, the rate at which the forward reaction proceeds is proportional to the activities of A and B raised to powers equal to their stoichiometric coefficients:

$$\text{Rate of conversion of } A \text{ and } B \text{ to } C \text{ and } D = k_f\{A\}^a(B)^b \quad \textbf{(1.10)}$$

At the same time that some molecules of C and D are being generated by the forward reaction, others are being destroyed by the reverse reaction, at a rate $k_r\{C\}^c\{D\}^d$, where k_r is the reverse reaction rate constant. Therefore, the net production rate of C is

$$\left.\frac{d[C]}{dt}\right|_{net} = \text{rate of } C \text{ production} - \text{rate of } C \text{ destruction} \quad \textbf{(1.11a)}$$

$$= k_f\{A\}^a\{B\}^b - k_r\{C\}^c\{D\}^d \quad \textbf{(1.11b)}$$

Note that the net rate of production or destruction of C is measured as the rate of change of its *concentration,* but this rate is assumed to be proportional to the *activities* of the reacting species.[15]

The condition defining equilibrium in a batch system (one with no transfer of mass into or out of the system) is that there is no net change in the concentration of any of the reactants as a function of time. Applying this requirement to the concentration of C at equilibrium gives $d[C]/dt = 0$, so

$$\text{At equilibrium:} \quad k_f\{A\}^a\{B\}^b\big|_{eq} = k_r\{C\}^c\{D\}^d\big|_{eq} \quad \textbf{(1.12)}$$

$$\frac{k_f}{k_r} = \frac{\{C\}^c\{D\}^d}{\{A\}^a\{B\}^b}\bigg|_{eq} \quad \textbf{(1.13)}$$

Equation (1.13) indicates that if the reaction $aA + bB \leftrightarrow cC + dD$ is at equilibrium, the ratio $\{C\}^c\{D\}^d/\{A\}^a\{B\}^b$ must equal the fixed value given by the ratio k_f/k_r. If the actual ratio of the activities is different from k_f/k_r, the system is not at equilibrium, and the concentration of C will change over time. The absolute values of the concentrations and activities of A, B, C, and D might vary widely

---

[15]Some authors prefer to write Equation (1.11b) in such a way that the rates are proportional to the concentrations rather than the activities of the reacting species. In truth, the dependence of the reaction rate on the activity coefficients of the reacting species is somewhat more complicated than is indicated by Equation (1.11b) regardless of whether that equation is written in terms of activities or concentrations. The dependence can be interpreted in terms of activated complex theory, as described in Appendix 1A.

from one equilibrium system to the next, but the ratio given by Equation (1.13) is the same for the given reaction in all systems at a given temperature. The ratio k_f/k_r is given the symbol K_{eq} and is called the **equilibrium constant.** Note that it makes no sense to speak of a single species being in equilibrium or out of equilibrium; the concept of equilibrium applies only to a complete reaction.

The ratio $\{C\}^c\{D\}^d/(\{A\}^a\{B\}^b)$ appears in many important expressions that we will be developing and using. This ratio is called the **activity quotient** and is commonly represented by the symbol Q. When a reaction is at equilibrium, the activity quotient equals the equilibrium constant; that is, $Q = K_{eq}$, as shown above. If $Q \neq K_{eq}$, the reaction will proceed in a direction that causes Q to approach K_{eq}. That is, if $Q > K_{eq}$, products will be converted to reactants, decreasing the numerator and increasing the denominator of Q, so that its value decreases. Similarly, if $Q < K_{eq}$, reactants will be converted to products. A numerical example of this process is provided below.

Equating the equilibrium constant to the ratio of the forward and reverse rate constants is valid only for elementary reactions; for overall reactions that represent the net result of several steps, the rate of formation of the products is not necessarily related to the activities of the original reactants in any simple, predictable way. Rather, that rate depends on the forward and reverse rate constants of all the elementary reactions in the sequence. Nevertheless, combining the equilibrium constant expressions for all the elementary reactions leads to the conclusion that the overall reaction can always be characterized by an equilibrium constant of the form $K_{eq} = \{C\}^c\{D\}^d/(\{A\}^a\{B\}^b)$, even though that activity ratio does not correspond to a simple ratio of reaction rate constants. Thus, for a generic reaction of the form $aA + bB \rightarrow cC + dD$, the equilibrium constant is $K_{eq} = \{C\}^c\{D\}^d/(\{A\}^a\{B\}^b)$, regardless of whether the reaction occurs in a single step or via a sequence of steps.

Note that K and Q are ratios of dimensionless quantities (activities) and are therefore dimensionless as well. In some situations, it is useful to compute the ratio corresponding to Q or K_{eq}, but to use concentrations rather than activities in the calculations. This practice is most common when the reaction involves transfer of a molecule between solution and either a gas phase or the surface of a solid. Although this concentration ratio is sometimes referred to as an equilibrium constant, that designation is not really accurate: equilibrium constants must be ratios of activities and must be dimensionless. In this text, concentration ratios that characterize the equilibrium state of a system are referred to as **distribution coefficients** or **partition coefficients.**

The equilibrium constant expression can also be derived by using the thermodynamic model to interpret chemical processes. This approach is presented in Chapter 2 in the context of a broader discussion of energy relationships.

Example 1.9 | **OPTIMIZING DISINFECTION EFFICIENCY IN WATER TREATMENT**   Consider the following elementary reaction between H_3O^+, OCl^-, H_2O, and $HOCl$ in solution:

$$HOCl + H_2O \leftrightarrow H_3O^+ + OCl^-$$

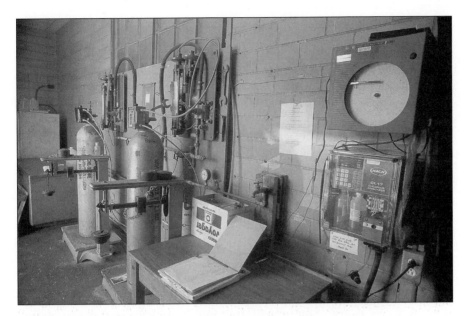

Chlorine is used as a disinfectant in the treatment of potable water and wastewater. The speciation of the Cl is critical. One mg Cl/L in the form of HOCl is toxic to most microorganisms. However, the oceans can support rich microbial life even though they contain 17,000 mg Cl/L, because it is present as the nontoxic ion Cl⁻.

| Tom Pantages.

This reaction is extremely important in water and wastewater treatment because, as noted previously, HOCl (hypochlorous acid) is the dominant disinfecting agent when water is chlorinated. Hypochlorite ion, OCl^-, is a much weaker disinfectant. Therefore, to optimize disinfection, it is important to adjust solution conditions so that most of the OCl in the system is bound to an H^+ ion and present as HOCl rather than as OCl^-.

The concentration of H_3O^+ (hydronium) ions is a fundamental property of any aqueous solution, since H_3O^+ (and OH^-) ions can be formed by the dissociation of water $(2H_2O \leftrightarrow H_3O^+ + OH^-)$.[16] Hydronium and hydroxide ions can also be added to solution as acids and bases, respectively. In a neutral solution at 25°C, $\{H_3O^+\} = \{OH^-\} = 10^{-7}$.

Using infinite dilution as the reference state for the various species and 1.0 mol/L as the standard state concentration, the equilibrium constant for the reaction of HOCl with H_2O shown above at 25°C is 2.5×10^{-8}. That is,

$$2.5 \times 10^{-8} = \frac{\{H_3O^+\}\{OCl^-\}}{\{HOCl\}\{H_2O\}} = \frac{k_f}{k_r}$$

a. Compute the equilibrium ratio of HOCl activity to OCl^- activity at H_3O^+ activities of 10^{-6}, 10^{-7}, and 10^{-8}. Assume that all the solutes behave ideally ($\gamma = 1.0$ in all

---

[16]In all likelihood, you have previously seen the products of the water dissociation reaction written as H^+ and OH^-, rather than as H_3O^+ and OH^-. As explained in Chapter 3, the symbol H^+ is really a shorthand way of representing a group of chemicals, of which H_3O^+ is the major one. For the purposes of this example, it is useful to represent this group of chemicals as H_3O^+.

cases), and that the water can be approximated as being in its standard state (pure water), so that its activity is 1.0.

b. At a particular instant, the activities of H_3O^+, HOCl, and OCl^- in a solution are 10^{-7}, 10^{-3}, and 10^{-4}, respectively. Determine whether the reaction discussed above is at equilibrium and, if not, whether the concentration of HOCl in solution will increase or decrease as the reaction proceeds.

Solution

a. From the equilibrium constant expression we can write

$$\left.\frac{\{H_3O^+\}\{OCl^-\}}{\{HOCl\}\{H_2O\}}\right|_{eq} = 2.5 \times 10^{-8}$$

$$\left.\frac{\{HOCl\}}{\{OCl^-\}}\right|_{eq} = \left.\frac{\{H_3O^+\}}{(2.5 \times 10^{-8})\{H_2O\}}\right|_{eq}$$

According to the problem statement, we can approximate $\{H_2O\}$ as 1.0, so that term drops out of the denominator. The resulting relationship indicates that, at equilibrium, the ratio $\{HOCl\}/\{OCl^-\}$ depends only on (H_3O^+). For instance, in a neutral solution (i.e., one with $\{H_3O^+\} = 10^{-7}$) at equilibrium

$$\frac{\{HOCl\}}{\{OCl^-\}} = \frac{10^{-7}}{(2.5 \times 10^{-8})(1.0)} = 4$$

Thus, in such a solution, if $\gamma_{HOCl} = \gamma_{OCl^-} = 1.0$, the total concentration of OCl is apportioned in the ratio of four HOCl molecules per OCl^- molecule, and 80 percent of the total OCl in the system is present in the desired form.

By adding a little acid or base to the solution we can increase $\{H_3O^+\}$ to 10^{-6} or decrease it to 10^{-8}, respectively. The total OCl (*TOTOCl*) would then be apportioned as follows:

Case 1: $\{H_3O^+\} = 10^{-6}$	**Case 2: $\{H_3O^+\} = 10^{-8}$**
$\dfrac{\{HOCl\}}{\{OCl^-\}} = \dfrac{10^{-6}}{(2.5 \times 10^{-8})(1.0)} = 40$	$\dfrac{\{HOCl\}}{\{OCl^-\}} = \dfrac{10^{-8}}{(2.5 \times 10^{-8})(1.0)} = 0.4$
$\dfrac{[HOCl]}{TOTOCl} = \dfrac{40}{41} = 97.5\%$ present as HOCl	$\dfrac{[HOCl]}{TOTOCl} = \dfrac{0.4}{1.4} = 29\%$ present as HOCl

Based on the distribution of OCl between the HOCl and OCl^- forms and the knowledge that HOCl is the more effective disinfectant, we conclude that, for a given amount of chlorine added, disinfection will be much more effective in a solution with $\{H_3O^+\} = 10^{-6}$ than in one with $\{H_3O^+\} = 10^{-8}$, and this is in fact observed.

b. At the given activities of H_3O^+, HOCl, and OCl^-, the activity quotient is

$$Q = \frac{\{H_3O^+\}\{OCl^-\}}{\{HOCl\}\{H_2O\}} = \frac{(10^{-7.0})(10^{-4.0})}{(10^{-3.0})(1.0)} = 10^{-8.0}$$

Since $Q < K_{eq}$, the reaction is not at equilibrium, and we expect the reaction to convert reactants to products. This process causes Q to increase and will continue until {HOCl} has decreased and {H_3O^+} and {OCl^-} have increased sufficiently that equilibrium is attained.

One gram of solid silver chloride, AgCl(s), is added to 1.0 L of estuarine water that contains 0.10 mol/L Cl^- and has an ionic strength of 0.12 mol/L. The solid begins to dissolve according to the following reaction, which has an equilibrium constant of $K_{eq} = 10^{-9.75}$: | **Example 1.10**

$$AgCl(s) \leftrightarrow Ag^+ + Cl^-$$

Estimate the mass of AgCl(s) that dissolves per liter of solution before the solution reaches equilibrium with the solid (assuming that the solid does not completely dissolve). Use the Davies equation to estimate activity coefficients.

Solution

The molecular weight of AgCl(s) is 143, so the initial molar concentration of the solid in the suspension is $1/143 = 7.0 \times 10^{-3}$ M. The solid will dissolve until the reaction reaches equilibrium. Designating the concentration of AgCl(s) that dissolves as x, the molar concentration of dissolved Ag^+ once equilibrium is attained will be x, that of Cl^- will be $0.1 + x$, and that of AgCl(s) will be $0.007 - x$.

The activity of each of the dissolved species (Ag^+ and Cl^-) will be given by the product $\gamma_i c_i$, but that of the solid will be 1.0 regardless of how much solid is present. Assuming that not enough AgCl(s) dissolves to change the ionic strength of the solution significantly, the activity coefficients for Ag^+ and Cl^- can be calculated from the Davies equation. In both cases, the value of γ is 0.76 (because each species carries a single charge). Therefore, the conditions at equilibrium can be represented by the following equation:

$$K_{eq} = 10^{-9.75} = \frac{\{Ag^+\}\{Cl^-\}}{\{AgCl(s)\}} = \frac{\gamma_{Ag^+}[Ag^+]\gamma_{Cl^-}[Cl^-]}{1.0}$$

$$10^{-9.75} = \frac{0.76x[0.76(0.1 + x)]}{1.0}$$

The above expression can be solved by the quadratic equation. Alternatively, we could make an assumption that the amount of solid that dissolves is much less than 0.1 M [since there is only 0.007 M AgCl(s) present initially, and only some of that dissolves]. If we make such an assumption, the expression simplifies to one that can be solved for x very easily:

$$10^{-9.75} \approx 0.76x(0.76)(0.1) = 0.058x$$

$$x = 3.05 \times 10^{-9} = 10^{-8.52}$$

Indeed $x \ll 0.1$, so the assumption is valid. The conclusion is that an exceedingly small amount of the solid dissolves. The activity of Ag^+ at equilibrium is $\gamma_{Ag^+}x$, or 2.32×10^{-9}

(equal to $10^{-8.64}$). [We will see in later chapters why the solubility of AgCl(s) in seawater is actually much greater than the value computed here.]

1.6.4   EFFECT OF TEMPERATURE ON REACTION RATE CONSTANTS AND THE EQUILIBRIUM CONSTANT

The preceding sections indicate that the rate of any elementary reaction can be computed as the product of a reaction rate constant and the activities of the reacting species, which, in a given solution, are proportional to their respective concentrations. The dependence of the reaction rate on the activities or concentrations of the reactants can be understood as a direct consequence of the relationship between concentration and collision frequency. The rate constant depends on the specific chemical reaction under study, the chemical conditions in the system, and the temperature. Seasonal changes in temperature can have a large effect on reaction rates in many natural and engineered aquatic systems, even if the bulk composition of the system remains relatively constant. We next consider the effect of temperature on the forward and reverse reaction rates, and thereby on the condition of chemical equilibrium.

The earliest successful attempt to describe the temperature dependence of reaction rate constants was made by Arrhenius, who derived the relationship

$$k = k_{Ar} \exp\left(\frac{-E_{Ar}}{RT}\right) \tag{1.14}$$

where k is the reaction rate constant and k_{Ar} is another constant with the same units as k (sometimes called the frequency factor); E_{Ar} is an empirical constant, unique to a particular reaction, with units of energy per mole; R is the universal gas constant, in appropriate units so that the argument of the exponential is dimensionless; and T is absolute temperature. Although E_{Ar} can be interpreted as the activation energy for elementary reactions, Arrhenius' result was empirical and was based on results from both elementary and non-elementary reactions.

The effect of temperature, according to the Arrhenius equation, is seen most easily by separating the temperature term from the others in the argument of the exponential and taking the logarithm of both sides of Equation (1.14):

$$\ln k = \ln k_{Ar} - \frac{E_{Ar}}{R}\frac{1}{T} \tag{1.15}$$

Equation (1.15) indicates that a plot of $\ln k$ versus $1/T$ should be a straight line with slope $-E_{Ar}/R$ and intercept $\ln k_{Ar}$, as shown in Figure 1.8. These values can then be used in conjunction with the equation to compute the rate constant at any temperature.

The effect of temperature on the equilibrium constant can be assessed by considering a reaction that is elementary in both the forward and reverse

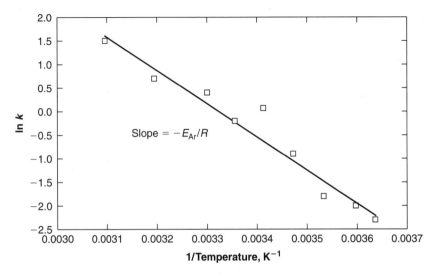

Figure 1.8   Characteristic plot of the rate constant versus inverse absolute temperature, from which the values of k_{Ar} and E_{Ar} can be computed; ln (k_{Ar}) is found by extrapolating the straight line to the hypothetical condition $1/T = 0$.

directions. According to Equation (1.14), the equilibrium constant for such a reaction can be written as follows:

$$K_{eq} = \frac{k_f}{k_r} = \frac{k_{Ar,f}\exp[-E_{Ar,f}/(RT)]}{k_{Ar,r}\exp[-E_{Ar,r}/(RT)]} = \frac{k_{Ar,f}}{k_{Ar,r}}\exp\left(-\frac{E_{Ar,f} - E_{Ar,r}}{RT}\right)$$

$$K_{eq} = \frac{k_{Ar,f}}{k_{Ar,r}}\exp\left(-\frac{\Delta E_{Ar}}{RT}\right) \tag{1.16}$$

$$\frac{K_{eq}|_{T_2}}{K_{eq}|_{T_1}} = \exp\left[\frac{\Delta E_{Ar}}{R}\left(\frac{1}{T_1} - \frac{1}{T_2}\right)\right]$$

$$\ln\frac{K_{eq}|_{T_2}}{K_{eq}|_{T_1}} = \frac{\Delta E_{Ar}}{R}\left(\frac{1}{T_1} - \frac{1}{T_2}\right) \tag{1.17}$$

where $\Delta E_{Ar} = E_{Ar,f} - E_{Ar,r}$.

Equation (1.17) provides a prediction of how the equilibrium constant is expected to change as a function of temperature. As noted above, for elementary reactions, E_{Ar} can be identified as the activation energy for a reaction E^*. Furthermore, as shown in Figure 1.7, the difference $E_f^* - E_r^*$ is the energy change for the overall conversion of reactants to products ΔE_{rxn}. Thus, Equation (1.17) indicates that the temperature dependence of the equilibrium constant is related to the net energy change accompanying the reaction. Although the relationship between E_{Ar} and E^* applies strictly only to elementary reactions, it turns out that the general form of Equation (1.17) applies to both elementary and non-elementary reactions, and that ΔE_{Ar} is the net change in heat content of the molecules accompanying the overall reaction. This net change in heat content is a thermodynamic quantity called the *molar enthalpy of reaction*, which is described in some detail in Chapter 2 and is represented as $\Delta \bar{H}_r$. Thus,

$$\ln \frac{K_{eq}\big|_{T_2}}{K_{eq}\big|_{T_1}} = \frac{\Delta \bar{H}_r}{R}\left(\frac{1}{T_1} - \frac{1}{T_2}\right)$$
(1.18)

Example 1.11 | The reaction for the dissociation of water is shown below as it is conventionally written. The equilibrium constant for this reaction at 25°C is 1.00×10^{-14}, and the enthalpy of the reaction is 55.81 kJ/mol. Estimate the activity of OH^- in a solution at 25°C and one at 4°C, if $\{H^+\} = 10^{-7.0}$ in both solutions.

$$H_2O \leftrightarrow H^+ + OH^-$$

Solution

Substituting the given values into Equation (1.18), we find

$$\ln \frac{K_{eq}\big|_{4°C}}{K_{eq}\big|_{25°C}} = \frac{55.81 \text{ kJ/mol}}{8.314 \times 10^{-3} \text{ kJ/(mol-K)}}\left(\frac{1}{298 \text{ K}} - \frac{1}{277 \text{ K}}\right) = -1.71$$

$$\frac{K_{eq}\big|_{4°C}}{K_{eq}\big|_{25°C}} = \exp(-1.71) = 0.18$$

$$K_{eq}\big|_{4°C} = 1.8 \times 10^{-15}$$

The equilibrium constant for the reaction of interest is $\{H^+\}\{OH^-\}/\{H_2O\}$. Since the activity of H_2O is ~1.0 (because its mole fraction is very nearly 100 percent), and the activity of H^+ is given as 10^{-7}, the activity of OH^- is simply $K_{eq}/10^{-7}$. Therefore, at 25°C, $\{OH^-\}$ is 1.0×10^{-7}, and at 4°C it is 1.8×10^{-8}. We conclude that lowering the temperature substantially decreases the tendency for water molecules to dissociate.

1.7   COMBINING CHEMICAL REACTIONS

Often we are interested in analyzing the equilibrium condition for chemical reactions for which we do not know the equilibrium constant. In some cases, we can derive that constant from related reactions. For instance, say we are interested in the following reaction:

$$H_3PO_4 + 3H_2O \leftrightarrow 3H_3O^+ + PO_4^{3-}$$

$$K_{eq} = \frac{\{H_3O^+\}^3\{PO_4^{3-}\}}{\{H_3PO_4\}} = ?$$
(1.19)

In the reaction of interest, three H_3O^+ ions are produced by transfer of H^+ ions from a phosphoric acid (H_3PO_4) molecule to water molecules. This reaction can be generated as the sum of the following three reactions, each of which is written for the transfer of one H^+ ion, and for each of which K_{eq} is known:

Reaction 1: $\qquad H_3PO_4 + H_2O \leftrightarrow H_3O^+ + H_2PO_4^- \qquad K_{eq} = 10^{-2.16}$

Reaction 2: $\qquad H_2PO_4^- + H_2O \leftrightarrow H_3O^+ + HPO_4^{2-} \qquad K_{eq} = 10^{-7.20}$

Reaction 3: $\qquad HPO_4^{2-} + H_2O \leftrightarrow H_3O^+ + PO_4^{3-} \qquad K_{eq} = 10^{-12.35}$

The sum of the above three reactions is

$$H_3PO_4 + \cancel{H_2PO_4^-} + \cancel{HPO_4^{2-}} + 3H_2O$$
$$\leftrightarrow 3H_3O^+ + \cancel{H_2PO_4^-} + \cancel{HPO_4^{2-}} + PO_4^{3-}$$

Net overall reaction: $\qquad H_3PO_4 + 3H_2O \leftrightarrow 3H_3O^+ + PO_4^{3-}$

Reactions 1 through 3 are called *acid dissociation reactions,* for which the equilibrium constants are commonly designated K_{a1}, K_{a2}, and K_{a3}, respectively. We will explore the significance of such reactions in Chapter 3; for now, our goal is just to evaluate the equilibrium constant for the net overall reaction. This constant can be generated by taking the product of the equilibrium constants for the corresponding three constituent reactions:

$$K_{overall} = K_{a1} K_{a2} K_{a3}$$
$$= \frac{\{H^+\}\{H_2PO_4^-\}}{\{H_3PO_4\}} \frac{\{H^+\}\{HPO_4^{2-}\}}{\{H_2PO_4^-\}} \frac{\{H^+\}\{PO_4^{3-}\}}{\{HPO_4^{2-}\}} = \frac{\{H^+\}^3\{PO_4^{3-}\}}{\{H_3PO_4\}}$$

By inserting values for K_{a1}, K_{a2}, and K_{a3} in the above expression, the equilibrium constant for the overall reaction, $K_{overall}$, is computed to be $10^{-21.71}$. In addition to demonstrating the computation of an equilibrium constant for an overall reaction from the K_{eq} values of related reactions, this calculation emphasizes that although chemical activities and equilibrium constants characterize chemical reactions, the values associated with those terms are simply numbers. Those numbers can be manipulated using any mathematical operations that we choose; i.e., adding, multiplying, and taking square roots of chemical activities are all mathematically valid operations. Thus, in the above example, the value of $K_{overall}$ is a certain *number* which can be determined from relationships among other *numbers* (K_{a1}, K_{a2}, K_{a3}, and the activities of the various solutes). The operations performed above are no more or less valid than if the activities were replaced by unspecified variables such as x, y, and z.

To generalize from the above example, the equilibrium constant for a reaction that is equivalent to the sum of several other reactions is simply the product of the equilibrium constants of the constituent reactions. Similar reasoning shows that if a reaction is reversed, the equilibrium constant is the inverse of the original equilibrium constant $(1/K)$.

Example 1.12 | Given the following equilibrium constants, find pK_{eq} for the reaction

$$HOCl + NH_3 \leftrightarrow NH_4^+ + OCl^-$$

(a) $HOCl + H_2O \leftrightarrow H_3O^+ + OCl^-$ $K_{a,HOCl} = 10^{-7.60}$

(b) $NH_4^+ + H_2O \leftrightarrow NH_3 + H_3O^+$ $K_{a,NH_4^+} = 10^{-9.25}$

Solution

The desired reaction can be generated by adding reaction (a) to the reverse of reaction (b):

$HOCl + H_2O \leftrightarrow H_3O^+ + OCl^-$ $K_{eq} = K_{a,HOCl} = 10^{-7.60}$

$+ [NH_3 + H_3O^+ \leftrightarrow NH_4^+ + H_2O]$ $K_{eq} = \dfrac{1}{K_{a,NH_4^+}} = 10^{+9.25}$

$HOCl + NH_3 + \cancel{H_3O^+} + \cancel{H_2O}$
$\leftrightarrow \cancel{H_3O^+} + NH_4^+ + OCl^- + \cancel{H_2O}$

Net reaction: $HOCl + NH_3 \leftrightarrow NH_4^+ + OCl^-$ $K_{eq} = \dfrac{K_{a,HOCl}}{K_{a,NH_4^+}} = \dfrac{10^{-7.60}}{10^{-9.25}} = 10^{+1.65}$

The equilibrium constant for the reaction of interest is $10^{1.65}$. Therefore pK_{eq} is $-\log 10^{+1.65} = -1.65$.

SUMMARY

Environmental aquatic chemistry comprises an enormous range of topics, many of which relate to the prediction or analysis of the speciation of chemicals in aquatic systems. In most systems of interest, the vast majority of the molecules are water molecules. Nevertheless, the relatively low concentrations of the other molecules determine the quality of the water and its usefulness for various potential applications.

This chapter has introduced the basic nomenclature and principles used to describe chemical concentrations, reactions, and equilibrium. One of the central concepts introduced is that of the chemical activity of a substance, which is a measure of a chemical's tendency to react. The chemical activity of a species in a given phase is a function of the species' concentration in that phase, the composition of the phase, temperature, and pressure.

The chemical activity of a solute is often represented as the product of two terms: the concentration of the solute relative to a standard concentration (c/c_{std}), and an activity coefficient (γ). Normally c_{std} is defined to be 1.0 mol/L. The activity coefficient is defined to have a value of 1.0 under certain, specified conditions (the reference state conditions); if the solution conditions conform to the reference state conditions, the (dimensionless) activity has the same value as the concentration in moles per liter. In such a case, the solute is said to behave ideally. In other solutions, the activity coefficient might be different from 1.0, indicating that the molecular interactions in the system of interest are different from those in the reference state. Although the standard state concentration and the reference state conditions can be chosen arbitrarily, once those are defined, the activity coefficient of any species A, denoted by γ_A, and the activity of A in any given solution are fixed.

Activity coefficients of ions and, to a much lesser extent, activity coefficients of neutral solutes depend on the ionic composition of the solution. Some simple equations have been developed to predict activity coefficients as a function of ionic strength. In general, the larger the charge on the ion of interest and the higher the ionic strength of the solution, the lower the activity coefficient. At ionic strengths greater than approximately 1.0 mol/L, the activity coefficients of many cations begin to increase with increasing ionic strength.

The chemical activities of solvents (e.g., water), gases, and solids can also be represented by a product of the form $\{A\} = \gamma_A c_A$. However, for these substances, c_A is often expressed as a mole fraction, and c_{std} corresponds to a mole fraction of 1.0, i.e., to the pure substance. Also, for these substances, the reference state is usually defined in such a way that γ_A is close to 1.0 under normal environmental conditions. When these conventions are used, $\{A\}$ is simply the mole fraction of A in the phase in which it is found if A is a solvent or a solid, and the magnitude of the partial pressure of A (when it is expressed in bars) if A is a gas.

This book focuses on the calculation of chemical speciation in systems that reach equilibrium, defined as a state in which there is no driving force for a change in the chemical composition of the system. Mechanistically, this situation can be interpreted as the result of a balance in which forward and reverse reactions proceed at equal overall rates, or as a condition of minimum chemical potential energy.

The kinetic interpretation of equilibrium can be understood by assuming that rates of elementary reactions are proportional to the chemical activities of the reactants. This model leads to the definition of an equilibrium constant K_{eq} that characterizes the activity ratio of products to reactants at equilibrium. The constant K_{eq} can be used to determine whether a system is at equilibrium and, if it is not, the direction in which the reaction will proceed in order for equilibrium to be established.

The effect of temperature on the rate of an elementary reaction is related to the activation energy for the reaction, and the effect on the equilibrium constant for any reaction is related to the enthalpy of the reaction.

Having covered these basic concepts, we can now move on to the thermodynamic analysis of equilibrium and to the study of specific types of environmentally important chemical reactions.

Problems

1. The following "total analysis" of a wastewater has been reported. Note that pH is not given.

NH_3	0.08 mg/L	as N
NO_2^-	0.008	as N
NO_3^-	2.0	as N
Na^+	227	as Na
K^+	18.3	as K
F^-	21.2	as F
Cl^-	24.1	as Cl
HCO_3^-	15	as C
Ca^{2+}	1.7	as $CaCO_3$
SO_4^{2-}	20	as SO_4

   a. Do a charge balance analysis on the data to see if such a solution would be electrically neutral.

   b. If H^+ and OH^- are the only ions missing from the analysis, what must the values of their activities be, given that K_{eq} for the dissociation of water ($H_2O \leftrightarrow H^+ + OH^-$) is $10^{-14.0}$ and that the activity of H_2O is approximately 1.0. Assume that activity coefficients can be modeled using the extended Debye–Huckel equation.

   c. Assuming your answer to part (b) is correct, what is the TDS of the water in milligrams per liter? Assume that during the drying of the sample, all the HCO_3^- is lost by the reaction $HCO_3^- + H^+ \leftrightarrow H_2O + CO_2$, and that all the NH_3 evaporates. The remaining ions are not volatilized. Keep in mind that the salt remaining after the drying step must be electrically neutral.

2. Express the concentration of Ca^{2+} in the Colorado River (composition shown in Table 1.1) in moles of Ca^{2+} per liter and in milligrams per liter as $CaCO_3$. The concentration of Na^+ is not given in the table. Assuming that Na^+ is the only significant species missing from the analysis, compute its value based on the electroneutrality requirement. Note that the HCO_3^- concentration is expressed in terms of alkalinity, which is normally quantified based on H^+-equivalents. The relevant conversions are: 1 mol Ca^{2+} = 40 g Ca^{2+} = 2 equiv, and 1 mol HCO_3^- = 61 g HCO_3^- = 1 equiv. Therefore, to compute the molar concentration of HCO_3^- in the river, you must convert the given alkalinity from mg/L $CaCO_3$ to meq/L, then assume that the concentration of HCO_3^- in meq/L is the same as the alkalinity in meq/L, then convert from meq/L of HCO_3^- to mmol/L of HCO_3^-. Then compute the ionic strength of the river water and the activities of Ca^{2+}, SO_4^{2-}, and Cl^-, using the Davies equation to estimate activity coefficients.

3. One gram of solid calcium sulfate, $CaSO_4(s)$, is added to 1.0 L of pure water. Immediately, the solid begins to dissolve according to the following reaction, which has an equilibrium constant of $K_{eq} = 10^{-4.85}$:

$$CaSO_4(s) \leftrightarrow Ca^{2+} + SO_4^{2-}$$

   a. What are the concentration (in moles per liter) and activity of $CaSO_4(s)$ before any dissolution occurs?

b. What is the activity of $CaSO_4(s)$ after enough dissolution has occurred that the reaction reaches equilibrium, assuming that not all the solid dissolves?

c. Assuming that $\gamma_{Ca^{2+}} = \gamma_{SO_4^{2-}} = \gamma_{CaSO_4(s)} = 1.0$, compute the concentrations of Ca^{2+} and SO_4^{2-} in moles per liter and in milligrams per liter at equilibrium. (*Note:* Use the information that, by stoichiometry, the molar concentrations of Ca^{2+} and SO_4^{2-} must be equal.)

d. Estimate the ionic strength of the solution based on the result of part (c). Then compute $\gamma_{Ca^{2+}}$ and $\gamma_{SO_4^{2-}}$, and make a new estimate of the amount of each ion that dissolves as equilibrium is approached. Iterate between calculations of the concentrations of Ca^{2+} and SO_4^{2-} and those of the corresponding ionic strength and activity coefficients until each calculation converges. What fraction of the original solid is dissolved at equilibrium?

4. A river contains 8 mg/L DOC in molecules whose average composition is $C_{10}H_{15}O_4N$.

a. What is the mass fraction of C in the organic molecules? What is its mole fraction?

b. What are the mass fraction and mole fraction of these molecules in the whole solution?

5. Having become thoroughly frustrated with the concept of moles, a reformer decides to defy convention and carry out all calculations using a new set of definitions for the standard state. Specifically, he chooses the standard state concentration to be 1 mg/L for all solutes and 1000 g/L for water. He chooses the reference state environment for solutes to be infinite dilution, and that for water to be pure water. The equilibrium constant for the following reaction using the conventional standard state definitions is $10^{11.32}$. Compute the equilibrium constant using the revised conventions. Include dimensions, if appropriate.

$$H_2CO_3 + 2OH^- \leftrightarrow CO_3^{2-} + 2H_2O$$

6. You wish to add 5 mg/L NaOCl *as* Cl_2 to a solution in a disinfection test, and you have a stock solution (household bleach) that contains 5% NaOCl by weight. Assuming that the density of the stock solution is 1.0 g/mL, how many milliliters of bleach should you add to each liter of test solution? Each mole of NaOCl can combine with two moles of electrons in a reaction that is essentially irreversible under normal environmental conditions, as follows:

$$NaOCl + 2H^+ + 2e^- \rightarrow Na^+ + H_2O + Cl^-$$

7. A river contains a relatively low concentration of dissolved salts, so that the ideal solution assumption applies. The river has the same pH (8.1) and total carbonate concentration (2×10^{-3} M) as bulk seawater. The activity coefficients of the major ions in seawater have been studied extensively, and when evaluated using infinite dilution as the reference state concentration, those of H^+, OH^-, and various carbonate-containing species are as follows:

Ion	γ
H_3O^+	0.590
OH^-	0.244
H_2CO_3	1.14
HCO_3^-	0.57
CO_3^{2-}	0.038

The activity coefficient of H_2O is assigned a value of 1.0 in both freshwater and seawater.

The equilibrium constants shown below are based on the use of the infinite dilution reference state. Compute the values of the same equilibrium constants using major ion seawater as the reference state.

$$2H_2O \leftrightarrow H_3O^+ + OH^- \qquad\qquad K_w = 10^{-14.0}$$

$$H_2CO_3 + H_2O \leftrightarrow H_3O^+ + HCO_3^- \qquad K_1 = 10^{-6.35}$$

$$HCO_3^- + H_2O \leftrightarrow H_3O^+ + CO_3^{2-} \qquad K_2 = 10^{-10.33}$$

8. In the early sanitary engineering literature, there was a great deal of interest in the effect of temperature on the rates of biodegradation and oxygen transfer in aquatic systems, where the range of temperatures is limited to around $20 \pm 20°C$ $(293 \pm 20\ K)$. In that work, an empirical model was developed suggesting that

$$\frac{k_{T_2}}{k_{T_1}} = \theta^{T_2 - T_1} \qquad\qquad\qquad \textbf{(1.20)}$$

$$\ln \frac{k_{T_2}}{k_{T_1}} = (T_2 - T_1) \ln \theta \qquad\qquad \textbf{(1.21)}$$

where θ is an empirical constant. Typically, the value of k at $T_1 = 20°C$ was determined in the laboratory, and its value at other temperatures was estimated from Equation (1.20). For T in degrees Celsius or kelvins, commonly cited values of θ were 1.047 for utilization of oxygen by microorganisms (BOD consumption) and 1.016 for oxygen transfer from the atmosphere to water. Compare the resulting expression for the temperature dependence of these two reactions to the Arrhenius equation and estimate E_{Ar} for each reaction.

APPENDIX 1A: MODELS FOR THE KINETICS OF ELEMENTARY REACTIONS

THE COLLISION MODEL

MOLECULAR INTERACTION (COLLISION) FREQUENCIES The simplest models for molecular interactions represent the individual molecules as hard spheres with energy content related to the temperature. In such models, the gradual slowing of the molecules and the rearrangement of the bonds as the molecules approach one another are ignored; i.e., the molecules are assumed to approach one another at a constant velocity and with no bond rearrangement until they get to the critical separation distance, at which point they collide and can react.

In 1917, Smoluchowski derived the following theoretical equation for the frequency of first encounters between pairs of dissolved, uncharged molecules:

$$\text{Rate of first encounters} = 4\pi N_A (r_A + r_B)(D_A + D_B)c_A c_B \qquad \textbf{(1.22)}$$

where N_A is Avogadro's number and r_i, D_i, and c_i are the radius, diffusion coefficient, and concentration of molecule i, respectively. The idea of a first encounter is that the calculation includes collisions that occur when molecules first approach one another, but does not count subsequent collisions that might occur if the molecules remain close to one another but do not react.

If the molecules are assumed to behave as hard spheres moving through a continuum, the diffusion coefficients D_A and D_B can be expressed in terms of the molecular radii and the solution viscosity μ,[17] in which case Equation (1.22) can be written as:

$$\text{Rate of first encounters} = \frac{2RT}{3\mu}\left(2 + \frac{r_A^2 + r_B^2}{r_A r_B}\right)c_A c_B \qquad \textbf{(1.23)}$$

The rate given by Equation (1.23) is the maximum possible rate of a second-order elementary reaction between uncharged molecules in aqueous solution. If the molecules of interest are ions, additional terms must be included that can increase the rate by up to about an order of magnitude if the ions have opposite charges, and can reduce it by up to about two orders of magnitude if they have like charges. Reactions that proceed at rates close to these maximum predicted values are said to be **diffusion-controlled.**

The concentration-independent terms in the equation can be grouped and represented as a rate constant k_{d-c} for diffusion-controlled reactions, as follows:

$$r_{d-c} = k_{d-c}c_A c_B \qquad \textbf{(1.24)}$$

where $k_{d-c} = \dfrac{2RT}{3\mu}\left(2 + \dfrac{r_A^2 + r_B^2}{r_A r_B}\right)$. For typical uncharged solutes, k_{d-c} in aqueous solutions at room temperature is on the order of 10^{10} (mol/L)$^{-1}$s^{-1}.

By substituting an expression to describe the approximate dependence of μ on temperature, k_{d-c} can be expressed as a function of temperature only. Comparison of such an equation with the Arrhenius expression [Equation (1.14)] allows one to estimate an apparent E_{Ar} of 12 to 21 kJ/mol for diffusion-controlled reactions between uncharged molecules. If there were no chemical activation barrier to overcome, the apparent activation energy of diffusion would be the main barrier restricting the rate of reaction.

In the more general case, not every collision leads to a reaction, because many collisions involve insufficient energy to overcome the activation energy barrier. The likelihood that a collision will actually cause a reaction can be equated with the likelihood that the energy of the collision exceeds the activation energy (E^*) for the reaction. This likelihood depends on temperature and is given by:[18]

$$\begin{array}{c}\text{Fraction of collisions that}\\ \text{involve an amount of energy} \geq E^*\end{array} = \exp\left(-\frac{E^*}{RT}\right) \qquad \textbf{(1.25)}$$

---

[17] The key relationship is known as the Stokes–Einstein equation, which is derived in Chapter 2.

[18] The given equation is an approximation that is applicable at $E^* \gg RT$, which is valid for virtually any reaction.

At 25°C, the numerical value of RT is 2.48 kJ/mol. For most reactions, E^* is at least several times that value, so, according to Equation (1.24), the likelihood that any single collision will involve enough energy to overcome the activation energy barrier is quite small. In other words, many collisions must occur, on average, for each successful conversion of reactants to products.

Although the measured reaction rates of some proton-transfer (acid-base) reactions approach those estimated based on diffusion control, the vast majority of reactions of interest in aquatic systems have chemical activation energy barriers that are much higher than the barrier imposed by diffusion; the corresponding rates are anywhere from a few orders of magnitude to tens of orders of magnitude slower than the diffusion-controlled rates. In such cases, the energy required for bond rearrangement provides the dominant impediment to the progress of the reaction, so the reaction rate is referred to as being **chemically controlled.**

THE ACTIVATED COMPLEX MODEL

As noted in the main body of the chapter, the activated complex model considers the whole continuum of energy conditions experienced by the molecules as they are converted from reactants through the activated state to products. This continuum can be represented as a topographical map, with the elevation corresponding to the energy of the species at that point. In such a representation, the reactants and products are located in two valleys separated by a ridge, and molecules on or above the ridge are activated complexes (Figure 1.9a). Although a fraction of the activated complexes in a system have enough energy to cross from reactants to products high on the ridge (e.g., as shown by the dashed line in Figure 1.9a), the vast majority of complexes that cross do so at or very near the lowest point. The lowest energy path leading from reactants to products (the dashed line in Figure 1.9b) is of particular interest. Progress along this path is quantified by a variable ε, referred to as the *reaction coordinate*. This coordinate is dimensionless and has a value of 0 for reactants and 1.0 for products.

The details of the derivation of the rate constant according to the activated complex model are based on the principles of statistical mechanics and are beyond the scope of this text. Formal derivations are provided in numerous physical chemistry and reaction kinetics texts.[19] However, some key assumptions and the model results can be summarized as follows.

In a system at constant temperature, molecules with a range of energies are present. The total kinetic energy of a group of identical molecules is approximately constant over time, but that energy is continuously redistributed within the group via collisions. If the group of molecules is at thermal equilibrium, the energy is distributed according to the Boltzmann distribution, which indicates that number of molecules with a given energy decreases exponentially as the amount of energy being considered increases.

[19]Two good references are *Physical Chemistry* by I. Levine (McGraw-Hill, 1988) and *Kinetics and Mechanism* by Moore and Pearson (Wiley, 1981).

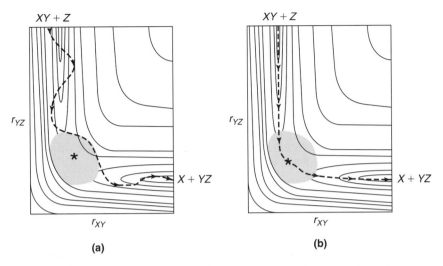

Figure 1.9   An energy contour map illustrating aspects of the activated complex model. The solid lines represent isopleths of constant energy for various separations of the reacting species. At small values of r_{XY} and large values of r_{YZ}, atoms X and Y are close to each other, forming compound XY, while atom Z is far away. At large r_{XY} and small r_{YZ}, compound YZ has formed and atom X is far away. Activated complexes are present in the stippled area, and the point marked with the asterisk is the lowest energy point at which species can "cross" from the reactant (species XY and Z) side of the barrier to the product (X and YZ) side. **(a)** A possible path for conversion of $XY + Z$ into $X + YZ$. In this path, molecules pass over the activation energy barrier with more than the minimum amount of energy required and then proceed to form the reaction products. **(b)** The lowest energy path for conversion of $XY + Z$ into $X + YZ$. This path defines the *reaction coordinate*.

    In a reactive system, the molecules with the highest energy are likely to be selectively eliminated (by conversion to different molecules). In that case, the Boltzmann function is no longer an accurate description of the molecular energy distribution of the system. However, if inter- and intramolecular collisions cause the system's energy to be redistributed rapidly compared to the rate of reaction, thermal equilibrium is quickly reestablished, and an assumption can be made that each chemical species in the system has an energy distribution that follows the Boltzmann distribution at all times. Such an assumption is applied to all species, including activated complexes, in the activated complex model.

    A second assumption of the model is that once reactants form activated complexes and cross the peak in the energy barrier, they continue in the direction leading to formation of products. The justification for this assumption is as follows. Recall that after passing over the activation barrier, chemical energy is converted into kinetic energy, and the products of the reaction begin to separate. If the reaction has proceeded along the lowest energy pathway, then as soon as this conversion of chemical potential energy to kinetic energy begins, the molecules have insufficient chemical potential energy to cross back over the ridge in the other direction. Because theoretical calculations show that, at the temperatures of interest in aquatic systems, almost all reactions proceed along or very near the lowest energy path, the assumption that all activated complexes formed from reactants generate products seems reasonable. Note that this assumption applies equally to the reverse reaction, i.e., it is also the case that all reactant molecules that form activated complexes proceed to form products. Thus, the model is consistent with the idea that the overall reaction is reversible; it simply treats the forward and reverse reactions independently.

The key result that emerges from an analysis based on the activated complex model is that the rate at which A and B molecules pass over the activation barrier for the reaction $A + B \xrightarrow{k_f} P$ is given by

$$\text{Rate at which } A \text{ and } B \text{ cross} \atop \text{activation barrier to form } P = \frac{k_B T}{h} K^* c_A c_B \qquad \textbf{(1.26)}$$

where k_B and h are the Boltzmann and Planck constants, respectively, and K^* is a constant that relates the energy distribution among activated complexes to the energy distribution of A and B molecules. Since the reaction $A + B \rightarrow P$ is elementary, the rate given in Equation (1.26) can also be written as $k_f c_A c_B$. Therefore, we can relate k_f, the empirical forward rate constant, to the theoretical parameters characterizing the reaction rate as follows:

$$k_f c_A c_B = \frac{k_B T}{h} K^* c_A c_B \qquad \textbf{(1.27)}$$

$$k_f = \frac{k_B T}{h} K^* \qquad \textbf{(1.28)}$$

The energy distributions of A, B, and AB^* that go into the term K^* are similar to those that lead to the calculation of equilibrium constants in statistical mechanics. That is, K^* bears some resemblance to an equilibrium constant for a hypothetical reversible reaction $A + B \leftrightarrow AB^*$. Since, in an ideal solution, the equilibrium constant for such a reaction could be written as $c_{AB}/(c_A c_B)$, the product $K^* c_A c_B$ is commonly designated c_{AB^*} and is called the *concentration of activated complexes*. This definition leads to the following form of Equation (1.26):

$$\text{Rate at which } A \text{ and } B \text{ cross} \atop \text{activation barrier to form } P = \frac{k_B T}{h} c_{AB^*} \qquad \textbf{(1.29)}$$

Equation (1.29) has the form of a simple reaction rate expression; i.e., it has the form expected for the rate of a first-order, irreversible, elementary reaction in which product P is formed from activated complexes AB^*. Furthermore, since this expression applies to *all* activated complexes, the equation seems to suggest that the rate constant for conversion of any activated complex to products ($k_B T/h$) is the same for all reactions, depending on temperature but not on the identity of the activated complex or the product.

Unfortunately, although the functional relationship expressed by Equation (1.29) is widely accepted as correct, the interpretations of c_{AB^*} as the true concentration of activated complexes and of K^* as a pseudo-equilibrium constant are overly simplistic, and have led to some misunderstandings about the activated complex model. The most common misconception is that the model assumes that activated complexes are in equilibrium with the reactants (based on the interpretation of K^* as a true equilibrium constant). In truth, the basic assumption of the model is that the reaction $A + B \rightarrow AB^*$ is not reversible and therefore cannot be at chemical equilibrium.[20] Furthermore, if an assumption of chemical equilibrium between the reactants and the activated complex were part of the model, it would have to hold for both forward and backward reactions. In that case, the reactants and products would both be in equilibrium with the activated complex, and thus with each other; no net reaction could occur under such conditions. The confusion surrounding this point is exacerbated by the fact that the assumption of *thermal* equilibrium (i.e., the Boltzmann distribution of molecular kinetic energy) *is* a key part of the model. Understanding the difference between thermal and chemical equilibrium can eliminate much of the confusion.

EFFECT OF IONIC STRENGTH ON REACTION RATE CONSTANTS   The strongest support for the activated complex model derives from its ability to account correctly for the effect of ionic strength on reaction rates. Empirically, if the ionic strength of a solution increases, the reaction rate increases if the reactants carry charge of the same sign (both positive or both negative) and decreases if the reactants are oppositely charged. Based on the collision model, one would predict that the increase in shielding of the ions that accompanies an increase in ionic strength would decrease reaction rates regardless of the charges on the two ions. However, the computed change in K^* and the corresponding change in c_{AB^*} as a function of ionic strength according to the activated complex model are qualitatively and quantitatively in agreement with the empirical results. A detailed discussion of these calculations is presented in many physical chemistry textbooks and is beyond the scope of the current discussion; however, the key result is that reaction rate constants can either increase or decrease with increasing ionic strength, and the activated complex model can account for these changes.

---

[20]Irreversible, in this context, means that none of the AB^* that is formed from collisions of A and B molecules reverts to A and B molecules; all the AB^* formed in this way proceeds to form P. As noted above, the reversibility of the overall reaction $A + B \leftrightarrow P$ derives from the fact that another irreversible reaction is forming AB^* from P, and these molecules do proceed to form $A + B$.

POTENTIALS, ENERGY, AND FORCES: WAYS TO INTERPRET CHANGES IN PHYSICAL/CHEMICAL SYSTEMS

CHAPTER OUTLINE

2.1 INTRODUCTION

This chapter introduces the thermodynamic concepts that underpin all physical and chemical changes in the world. The unifying principle that links all such changes is the second law of thermodynamics, which states that any process will occur spontaneously if it increases the total entropy of the universe.

Historically, the concept of entropy was first developed from the empirical evidence that, in systems isolated from their surroundings, certain types of changes always proceed in the same direction: hot objects transfer heat to cool ones, water flows downhill, odors distribute themselves throughout a gas phase rather than stay localized, etc. In a truly prodigious intellectual feat, using data primarily from thermomechanical systems (systems that can do work on, and can exchange thermal energy with, their surroundings), engineers of the 19th century inferred the form and significance of the entropy function. Their key insight is often expressed by some variation of the statement that isolated systems always change in a direction that decreases their ability to do useful work. Although "useful work" might sound imprecise and purely anthropocentric, it actually relates to the entropy and is a legitimate thermodynamic quantity.

Later, in an equally impressive effort, physical chemists demonstrated that entropy could be explained and quantified via a statistical analysis of the number of ways that energy and mass can be distributed in a system. The statistical analysis provided a satisfying explanation as to *why* universal entropy is always increasing and also extended the entropy concept to systems that involved no thermal energy transfer. It is from the statistical analysis that the popular characterization of entropy as a measure of the "disorderliness" of a system derives, and from that, the characterization of the second law as postulating that the disorderliness of the universe is always increasing.

Although the applicability of entropy analysis to any macroscopic change occurring in the world is what makes the concept so powerful, that very universality can be problematic at times. Specifically, the entropy principle is so broad that it can impede the recognition of simple but important insights applicable only to certain types of systems. To deal with such situations, it is common to recast the analysis in terms that are both more intuitive and more specific to the system of interest. Thus, for instance, we can use a simple analysis considering only gravity to explain why water flows downhill, recognizing that this analysis does not fully characterize the movement of water; for instance, a more extensive analysis (perhaps relying on the entropy principle in all its glory) would be required to explain why water can also be absorbed upward into a paper tissue.

In this chapter, the insights gained from both the thermomechanical and statistical analyses of entropy are exploited to explore the entropy concept and to derive the related function (the Gibbs energy function) that characterizes entropy changes in chemical systems at constant temperature and pressure.

The description of entropy and the exploration of the Gibbs energy function provided in the chapter are considerably more extensive than in comparable water chemistry texts. A more typical approach is to provide only a brief introduction to the terminology of thermodynamics and then present Equation (2.54) (showing the dependence of the molar Gibbs energy of a species on the species' activity) as the key equation linking thermodynamics to chemical processes. The topics following Equation (2.54) are then covered more or less as they are here.

The basis for the expanded discussion is that a good understanding of entropy allows linkages to be recognized among many topics that otherwise appear unconnected to one another, such as activity coefficients, the dependence of the molar Gibbs energy on concentration, the electrostatic contribution to adsorption reactions, and the effect of imposed electrical fields on redox reactions, to mention just a few. This chapter attempts to meet the challenge of providing some of the insights that an understanding of entropy allows, while remaining accessible and reasonably concise.

The chapter is organized as follows. Following this introduction, the entropy function is defined both as it was originally—for thermomechanical systems—and from a statistical perspective. The presentation then shifts to demonstrate the recasting of entropy analysis into terms applicable to specific types of systems, which leads to the development of the Gibbs energy function. The remainder of the chapter focuses on how the Gibbs function can be used to analyze systems of interest and how that function relates to the concepts of chemical activity and chemical equilibrium that were introduced in Chapter 1.

2.2   DEFINING AND QUANTIFYING ENTROPY

2.2.1   ENTROPY AND THE SECOND LAW: THE THERMOMECHANICAL INTERPRETATION

As noted above, the original proponents of the entropy concept did not start from any theoretical premise, but rather intuited the idea and its quantification from experimental observations. Here, we follow their lead by simply stating the definition of entropy that they used and describing the way that they measured entropy changes. Although our interest is not in thermomechanical systems *per se,* the brief discussion here will come in handy in the analysis of chemical systems.

Any object that can receive, store, and/or release thermal energy is called a *thermal reservoir.* In thermomechanical systems, entropy changes are defined with respect to energy exchanges with such reservoirs. Specifically, the change in entropy of a thermal reservoir ($\Delta S_{reservoir}$) is defined as the amount of thermal energy transferred into the reservoir ($\delta q_{reservoir}$) divided by the temperature of the reservoir ($T_{reservoir}$), i.e., $\Delta S_{reservoir} \equiv \delta q_{reservoir}/T_{reservoir}$.

Consider, for example, the setup shown schematically in Figure 2.1, consisting of a gas-filled cylinder-and-piston arrangement (the *system*) that is submerged in a water bath that serves as the thermal reservoir (the *surroundings*). The setup is isolated, meaning that neither mass nor energy crosses its exterior boundaries. Assume that the pressure inside the cylinder is different from that outside, with the piston being held in place by some external force. Also assume that the wall of the cylinder is infinitely thin, so that it transfers thermal energy efficiently but retains virtually none itself.

When the piston is released, the gas volume changes, causing the gas to either heat up (if it is compressed) or cool down (if it expands) transiently. Exchange of thermal energy between the gas and the thermal reservoir then re-equilibrates the gas to the original temperature. Conservation of energy (the first law of thermodynamics) requires that the change in total energy stored in the gas (ΔE_{gas}) during this process be related to the thermal energy transferred into the

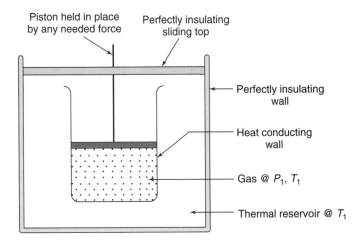

Piston held in place
by any needed force

Perfectly insulating
sliding top

Perfectly insulating
wall

Heat conducting
wall

Gas @ P_1, T_1

Thermal reservoir @ T_1

Figure 2.1   A thermomechanical system consisting of a cylinder and piston and a thermal reservoir. When the piston is released, thermal energy is exchanged with the reservoir. The increase in entropy of the reservoir equals the thermal energy entering it divided by its temperature.

gas (δq_{gas}, equal to $-\delta q_{reservoir}$) and the work done by the gas on its surroundings (δw_{gas}) (via movement of the piston). Specifically,

$$\Delta E_{gas} = \delta q_{gas} - \delta w_{gas} \qquad \textbf{(2.1)}$$

The signs in Equation (2.1) are based on the convention that δq is positive if thermal energy is transferred *into* the system, whereas δw is positive if work is done *by* the system. Since the energy of a system increases when work is done *on* it, the sign on the δw term is negative.

Because energy could be conserved no matter which way the piston moved, the first law as expressed by Equation (2.1) gives no clue as to the direction of change. The second law, however, as formulated in classical thermodynamics, establishes that such systems *always* change in the direction that increases the total entropy of the universe, i.e., of the gas and the thermal reservoir combined.

An important outcome of the early work on entropy was the recognition that entropy is a **state function,** meaning that the entropy change of the system when it shifts between any two states depends on the characteristics of those states, but not on the process by which the change is carried out.[1] For example, in the process described above, the change in entropy of the gas in the cylinder between

---

[1]A thermodynamic state is a condition in which all the intensive properties of a system are specified. An intensive property is one whose value is independent of the scale of the system (e.g., temperature, pressure, concentration). By contrast, the values of extensive properties (e.g., volume, mass) do depend on the scale of the system. For our purposes, specifying the temperature, pressure, and composition of all the phases in the system defines its state. More formal and detailed discussions of thermodynamic states can be found in many physical chemistry and chemical thermodynamics texts.

a highly energetic state 1 (high P) and a less energetic state 2 (lower P) would be the same regardless of how much friction opposes the movement of the piston. The larger the frictional resistance, the less work that the system would do on its surroundings as part of the process, and the more heat that would transfer to the reservoir. However, for a given state 1 and state 2, the entropy change *of the gas* would be the same, regardless of the details of how the change was accomplished. (The change in entropy of the reservoir would be greater in the case of higher friction, because the reservoir receives more thermal energy in that case.) The fact that entropy is a state function is critical in the analysis of changes in chemical systems, which we consider later in the chapter.

2.2.2   THE MICROSCOPIC VIEW OF ENTROPY AND THE SECOND LAW

RANDOM EVENTS AND NON-RANDOM OBSERVATIONS   Although the development of the second law provided a powerful tool for predicting changes in physical systems, the classical thermodynamicists who first recognized and elaborated the law could offer no explanation for its validity; they simply observed that it seemed always to be satisfied. It took the insights of subsequent workers, armed with the tools of statistical mechanics, to provide a logical, consistent explanation for the observations.

The basic idea underlying the statistical mechanical view of entropy can be understood by considering a very simplified picture of the way that bond energy is distributed in a molecule. The identity of a molecule is defined by its structure, i.e., the way that the various atoms are bonded to one another. Although we normally characterize a molecule as having a single bonding arrangement (e.g., H—O—H), the energy stored in the bonds of a given compound varies slightly from one molecule of that compound to the next, and within a given molecule over time. That is, a particular bond might be more energetic one moment and less so the next. Each of these arrangements of energy within the molecule is referred to as a *quantum state* of the molecule.[2]

When we observe a macroscopic quantity of some substance, we perceive the average behavior of all the molecules in it over a time span that is long compared to the frequency with which the molecules change quantum states. Given the huge number of water molecules in, say, a liter of seawater, the average quantum state of the whole collection of molecules is very stable, even though the quantum state of individual molecules is constantly changing. We are therefore unaware of the different quantum states and the shifts among them, but theoretical arguments and experimental evidence make it clear that such states exist.

---

[2]The jargon associated with the statistical mechanical interpretation of entropy can sound intimidating. Phrases such as *quantum state*, *ground state*, *accessible state*, *microstate*, and *macrostate*, all of which are introduced in the upcoming pages, are not part of our everyday lexicon, to say the least. However, in the end, these are just labels. If you keep in mind the relatively simple ideas that the phrases are meant to convey, this section should not be excessively difficult.

A certain minimum amount of energy is needed to form the bonds when the molecule is least energized, and the corresponding quantum state is referred to as the *ground state* of the molecule. At a temperature of absolute zero, all molecules are in their ground state, but at higher temperatures, molecules are distributed among various quantum states, with the number of such states "accessible" to the molecules increasing with increasing temperature. According to statistical mechanics, the entropy of a compound is directly related to the number of quantum states it can access in the given system.

Consider, for example, a hypothetical system consisting of just 100 molecules, some of which are the sugar glucose (G) and the rest of which are the closely related sugar fructose (F). (We will later expand the system to include a large amount of water, so that the model applies to a dilute solution.) Both of these sugars have the chemical formula $C_6H_{12}O_6$, but they differ in the details of how chemical energy is distributed among the molecular bonds. Assume that at the given temperature, glucose molecules have eight accessible quantum states and fructose molecules have two.

The fundamental assumption of the statistical mechanical view of entropy is that energy packets (quanta) move among the molecular bonds, both within a given molecule and from one molecule to another, with no preference. As they do so, the molecules shift among the various quantum states randomly, constrained only by the fact that, at the given temperature, the total amount of energy available (i.e., the number of quanta available for distribution) is fixed.

Since each $C_6H_{12}O_6$ molecule can exist in 10 different energy arrangements (quantum states), the whole collection of 100 molecules is characterized by a fairly astronomical number of possible permutations (10^{100}). If we had perfect information about the system, we would be able to distinguish among all these permutations. However, realistically, the only information about the mixture that can be accessed is its overall composition in terms of glucose and fructose molecules. Thus, at the level of resolution available, only 101 distinct possibilities can be recognized (0 G and 100 F; 1 G and 99 F; etc.; up to 100 G and 0 F).

Intuitively, if the energy distribution in the molecules shifts randomly among the possible arrangements so that each quantum state is equally likely, and if 8 of the 10 quantum states of any given molecule cause it to be identified as glucose, a given molecule would be expected to be identified as glucose 8/10 (80%) of the time and as fructose the other 2/10 (20%) of the time. Similarly, we expect the overall composition of the mixture to be approximately 80% glucose and 20% fructose.

At any instant, of course, the actual number of fructose molecules in the mixture might not be exactly 20% of the total, and there is some possibility that the number would be far from that value (for instance, in the extreme, the possibility that all 100 molecules would be fructose at some instant is 0.20^{100}, or $\sim 10^{-70}$). However, the more times that the mixture is analyzed, and the more molecules that are included in the analysis, the more likely it is that overall result will correspond closely to a ratio of four glucose molecules to one fructose molecule. As a result, if the test were conducted a very large number of times, we might well infer a "law" that the molecules have a strong and invariant tendency to be distributed in this ratio. Ironically, though, the key pillar supporting that law is a *lack* of information! That is, the 10 possible quantum states of a given molecule are in fact all different and all equally likely, but our inability to distinguish among the majority of those states leads us to perceive only two forms of the molecule, and to conclude that these forms have unequal likelihoods of occurring.

In statistical mechanics, each of the 10^{100} permutations of the suite of molecules is called a *microstate* of the system, and each of the 101 possible observable outcomes ($n_G'G + n_F'F$) is called a *macrostate*. Each macrostate has an entropy that depends only on the number of microstates associated with it. We next compute the number of microstates associated with two specific macrostates and then generalize the result to all other macrostates.

Consider first the macrostate consisting of 0 glucose and 100 fructose molecules (0G + 100F). For each molecule, there are two quantum states that cause it to be identified as fructose. To generate the desired macrostate, each of the 100 molecules must be in one of these two quantum states, so the macrostate can be generated in 2^{100} different ways; i.e., the macrostate comprises 2^{100} microstates.

Now consider the number of microstates that generate the macrostate of 1G + 99F. There are 8 quantum states corresponding to glucose, so the number of ways in which the first molecule can be glucose while all the rest are fructose is $(8)^1(2)^{99}$. However, there are another $(8)^1(2)^{99}$ microstates that cause the second molecule to be glucose while all the rest are fructose, and these microstates also correspond to the 1G + 99F macrostate. Extrapolating, we see that the 1G + 99F macrostate can be established by having any of the 100 molecules in the array be the only glucose molecule, and that each of these arrangements comprises $(8)^1(2)^{99}$ microstates. Therefore, the total number of microstates associated with the 1G + 99F macrostate is $(8)^1(2)^{99}(100)$.

Extending the above reasoning and applying some basic rules of probability theory lead to the following generic result for the number of microstates W associated with the macrostate $n_G'G + n_F'F$:

$$W = \begin{pmatrix} \text{number of ways that a} \\ \text{particular sequence of} \\ n_G' \text{ glucose and } n_F' \text{ fructose} \\ \text{molecules can be generated} \end{pmatrix} \begin{pmatrix} \text{number of different sequences} \\ \text{in which that mix of } n_G' \text{ glucose} \\ \text{and } n_F' \text{ fructose molecules can} \\ \text{be arranged} \end{pmatrix}$$

$$= \left[(8)^{n_G'}(2)^{n_F'} \right] \left[\frac{(n_G' + n_F')!}{n_G'! n_F'!} \right] \tag{2.2}$$

The first term in brackets in Equation (2.2) includes information about the number of quantum states that are accessible to glucose and fructose molecules in the system under consideration. As noted above, for each molecule, the number of such quantum states depends on the molecule's identity and the system temperature. By contrast, the second term in Equation (2.2) characterizes the number of ways that a given mixture of glucose and fructose molecules can be arranged in space. The value of this term is independent of how many quantum states correspond to each type of molecule. For example, the fact that there are 100 ways to arrange 1 glucose and 99 fructose molecules means that the second term in the equation has a value of 100 for the macrostate 1G + 99F regardless of the number of quantum states that characterize each type of molecule, and regardless of the system's temperature. For that matter, the second term would have a value of 100 for the 1A + 99B macrostate for *any* two molecules A and B, regardless of their identity.

Values of W become overwhelmingly large and difficult to compute directly when the number of molecules under consideration becomes sizable. In such cases, the calculation can be carried out by applying Sterling's approximation that, for large values of N, $\ln(N!) = N \ln N - N$. Applying this approximation to Equation (2.2) and carrying out some algebra yield

$$\ln W = n'_G \ln 8 + n'_F \ln 2 - n'_G \ln \frac{n'_G}{n'_G + n'_F} - n_F \ln \frac{n'_F}{n'_G + n'_F} \quad \textbf{(2.3)}$$

The relative values of W of each of the 101 macrostates in the example system are shown in Figure 2.2. The largest value is for a combination of 80 glucose and 20 fructose molecules. As a result, if we analyzed the system repeatedly and if all microstates were equally likely, we would find that it contained 80 glucose and 20 fructose molecules more often than any other composition. The corresponding data sets for systems with 20, 300, and 1000 molecules are also shown. In all cases, W is maximized when one-fifth of the molecules are fructose; however, the peak in W gets progressively more pronounced as the number of molecules increases. By extrapolating the results in the figure, it should be clear

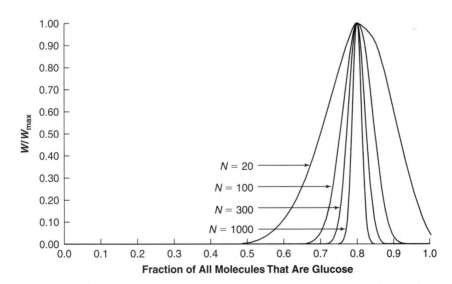

Figure 2.2   Relative values of W associated with various fractions of the molecules being glucose in the example system described in the text.

that in a system containing, say, 10^{-5} mol of sugar (6.02×10^{18} molecules), the peak would be virtually a spike.

Imagine now that all 100 molecules in the system were originally glucose. Over time, at a rate related to the frequency of the random transitions among quantum states, the system would shift toward the distribution of macrostates shown in Figure 2.2. In fact, the system would drift toward that distribution of macrostates no matter what the initial numbers of glucose and fructose molecules were, simply because the energy quanta move around randomly, and the figure reflects the outcome of a large number of those random changes.

Furthermore, once a system with a substantial number of molecules reached the 80/20 distribution, we would expect it to remain in or near that macrostate indefinitely, with excursions being undetectable at the macroscopic level. For instance, if a system containing 10^{-5} mol of sugar molecules were at the 80/20 distribution and 1 billion molecules suddenly shifted from a glucose to a fructose quantum state without any shifting in the other direction, the system composition would change by only $1.6 \times 10^{-8}\%$; i.e., the shift would not be noticed. We can therefore identify the 80/20 distribution as the equilibrium state of the system, i.e., the state toward which the system drifts and in which the system remains once the state has been attained. And, more generally, we can declare the steady change from any initial condition toward more likely arrangements, and ultimately to the equilibrium condition, a "law of nature," ignoring the fact that we fully expect the law occasionally to be broken in microscopic ways and for micro-durations.

Finally, consider how the value of W would change if one water molecule were present in the mixture and that water molecule had six accessible quantum states. Each of these quantum states could be combined with any of the combinations of glucose and fructose, so the number of microscopically distinct ways to generate n'_G glucose and n'_F fructose molecules in the three-component system [the first term on the right in Equation (2.2)] would increase to $(8)^{n'_G}(2)^{n'_F}(6)^1$. By extrapolating, if n'_W water molecules were present, this expression would become $(8)^{n'_G}(2)^{n'_F}(6)^{n'_W}$.

The addition of the water molecules would also increase the number of ways that a given set of n'_G glucose and n'_F fructose molecules could be distributed in space, i.e., the number of sequences that include $n'_G G + n'_F F + n'_W W$, so that the second term on the right in Equation (2.2) would be

$$\frac{(n'_G + n'_F + n'_W)!}{(n'_G!)(n'_F!)(n'_W!)}.$$

Applying Sterling's approximation, the expression corresponding to Equation (2.3) for this situation would be

$$\ln W = n'_G \ln 8 + n'_F \ln 2 + n'_W \ln 6 - n'_G \ln \frac{n'_G}{n'_G + n'_F + n'_W}$$

$$- n'_F \ln \frac{n'_F}{n'_G + n'_F + n'_W} - n'_W \ln \frac{n'_W}{n'_G + n'_F + n'_W} \qquad \textbf{(2.4)}$$

The ratios in the last three terms in Equation (2.4) equal the mole fractions X_i of glucose, fructose, and water in the system, respectively. By making that substitution and collecting terms, the equation can be rewritten as follows:

$$\ln W = n_G'(\ln 8 - \ln X_G) + n_F'(\ln 2 - \ln X_F) + n_W'(\ln 6 - \ln X_W) \quad \textbf{(2.5)}$$

$$\ln W = n_G' \ln W_G + n_F' \ln W_F + n_W' \ln W_W \quad \textbf{(2.6)}$$

where $\ln W_i = \ln Q_i - \ln X_i$, and Q_i is the number of quantum states in which species i can exist at the given temperature.

QUANTIFYING ENTROPY AND SEPARATING ITS THERMAL AND CONFIGURATIONAL COMPONENTS Statistical mechanics formulates the entropy of a given macrostate S_{tot} as being directly proportional to $\ln W$:

$$S_{tot} = k_B \ln W \quad \textbf{(2.7)}$$

where k_B is the Boltzmann constant (1.38×10^{-23} J/K). Therefore, Equation (2.6) can be rewritten to show the total entropy of the system in terms of three contributions, one from each of the constituents comprising the system:

$$k_B \ln W = n_G' k_B \ln W_G + n_F' k_B \ln W_F + n_W' k_B \ln W_W \quad \textbf{(2.8)}$$

$$S_{tot} = S_G + S_F + S_W \quad \textbf{(2.9)}$$

where $S_i \equiv n_i' k_B \ln W_i$.

If we write the expression for S_i as $(n_i')(k_B \ln W_i)$, we see that the contribution of species i to the entropy of the system can be expressed as the product of the number of molecules of i in the system and $k_B \ln W_i$. The term $k_B \ln W_i$ can therefore be interpreted as the entropy per molecule of i, denoted $\bar{S}_i'$. Frequently, the product $n_i' k_B \ln W_i$ is converted to a molar basis by writing it as $(n_i'/N_A)(N_A k_B \ln W_i)$, where N_A is Avogadro's number. When the expression is written in this way, the first term is the number of moles of i in the system (n_i), and the second term is the entropy per mole of i, commonly represented as $\bar{S}_i$. The product $N_A k_B$ is the universal gas constant R. So, summarizing and generalizing all the key results of the derivation, we have

$$\bar{S}_i' = k_B \ln W_i \quad \textbf{(2.10)}$$

$$\bar{S}_i = R \ln W_i \quad \textbf{(2.11)}$$

$$S_{tot} = \sum_i S_i = \sum_i n_i' \bar{S}_i' = \sum_i n_i \bar{S}_i \quad \textbf{(2.12)}$$

Because it is critical to parts of the discussion later in the chapter, the point made about W after Equation (2.2) is worth considering again in terms of the entropy per mole of a substance. Specifically, the molar entropy of any substance i can be conceptually divided into two terms. One of these terms depends on the identity of the species and is related to the number of quantum states it can access. This term depends on temperature (the higher the temperature, the larger the number of accessible states), but is independent of the concentration of i. By contrast, the second term relates to the number of spatial distributions that the molecules in the system can adopt. This term depends on the concentration of i (expressed as a mole fraction), but that dependence is the same regardless of the identity of i and regardless of the system temperature. By substituting $\ln Q_i - \ln X_i$ for W_i in Equation (2.11), we can quantify the two contributions to the molar entropy as follows:

$$\overline{S}_i = R \ln W_i = R \ln Q_i - R \ln X_i$$

$$\overline{S}_i = R \ln Q_i + R \ln \frac{1}{X_i} \tag{2.13}$$

Following the terminology used by Craig,[3] we refer to the entropy contribution from the different ways that energy can be distributed in a molecule as *thermal entropy* ($\overline{S}_{i,\text{therm}}$) and the contribution from the different ways that molecules can be arrayed in space as *configurational entropy* ($\overline{S}_{i,\text{config}}$). That is,

$$\overline{S}_{i,\text{therm}} = R \ln Q_i \tag{2.14a}$$

$$\overline{S}_{i,\text{config}} = R \ln \frac{1}{X_i} \tag{2.14b}$$

Thermal entropy $\overline{S}_{i,\text{therm}}$ can be calculated based on simulations of molecular and electronic structure, but the details of those calculations are beyond the scope of this text (which is why an arbitrary number of quantum states was postulated for the molecules in the example system). On the other hand, $\overline{S}_{i,\text{config}}$ can be calculated just by knowing the system composition, as indicated by Equation (2.14b).

In dilute aqueous solutions, the concentration of water is always close to 55.56 M,[4] so the mole fraction of a solute i is approximately $c_i/55.56$, where c_i is in moles per liter. In such cases, Equation (2.14) can be rewritten as

$$\overline{S}_{i,\text{config}} = -R \ln \frac{1}{c_i/55.56} = R \ln 55.56 + R \ln \frac{1}{c_i} \tag{2.15}$$

Since $R \ln 55.56$ is a constant, Equation (2.15) indicates that, in dilute solutions, $\overline{S}_{i,\text{config}}$ varies with $\ln (1/c_i)$. Thus, changing the concentration of i in a system from c_1 to c_2 changes the configurational entropy of i according to the following equation:

$$(\overline{S}_{i,\text{config}})_{c_2} - (\overline{S}_{i,\text{config}})_{c_1} = R \ln \frac{c_1}{c_2} \tag{2.16}$$

Note that, according to Equation (2.16), if an increment of i is added to a system, the concentration of i increases and the *molar* configurational entropy of i ($\overline{S}_{i,\text{config}}$) decreases. Nevertheless, the *total* configurational entropy of i ($= n_i \overline{S}_{i,\text{config}}$) increases.

---

[3]N. Craig, *Entropy Analysis: An Introduction to Chemical Thermodynamics*, VCH Publishers, New York, 1992.
[4]Assuming a solution density of 1000 g/L and a molecular weight of 18 for water, the concentration of water is (1000 g/L)/(18 g/mol) = 55.56 mol/L.

The molar entropy of cupric ion (Cu^{2+}) in a solution containing 1.0 M Cu^{2+} at 25°C is given in reference tables as -99.6 J/mol-K.

Example 2.1

a. Estimate the molar values of the thermal and configurational entropies under these conditions.

b. Estimate the values of the same two parameters in a solution containing 10^{-4} M Cu^{2+} at 25°C.

c. Compare the contribution of Cu^{2+} to the configurational entropy of the solution in parts (a) and (b).

Solution

a. The mole fraction of Cu^{2+} in the solution can be approximated as 1.0/55.56, or 0.018. Substituting this value into Equation (2.14b), we estimate the molar configurational entropy of the Cu^{2+} to be 33.4 J/mol-K. The molar thermal entropy can then be computed as $\overline{S}_{tot} - \overline{S}_{config}$, or -133.0 J/mol-K.

b. Since $\overline{S}_{therm}$ does not vary with concentration, its value is the same as in the 1.0 M solution, i.e., -133.0 J/mol-K. On the other hand, $\overline{S}_{config}$ increases upon dilution to 110.0 J/mol-K, so $\overline{S}_{tot}$ increases to -23.0 J/mol-K.

c. The contribution of Cu^{2+} to the configurational entropy of the solution equals $n_{Cu^{2+}}\overline{S}_{config,Cu^{2+}}$. The values of this product for parts (a) and (b) are 33.4 and 0.011 J/K, respectively. Thus, even though the configurational entropy per mole of Cu^{2+} increases when its concentration decreases, the total contribution of the Cu^{2+} to the configurational entropy of the system decreases.

As noted above, the thermal entropy of a species is not easily calculable. The one exception is that, at a temperature of absolute zero, there is no energy to distribute in the molecules beyond the minimum needed to form the bonds, so all the molecules are in their ground state, and $\overline{S}_{i,therm} = 0$. Furthermore, in a perfect crystal of any substance, there is only one way to arrange the molecules of i spatially, so $\overline{S}_{i,config} = 0$ as well. As a result, $\overline{S}_{i,tot} = 0$ for any perfect crystal at 0 K. Changes in $\overline{S}_i$ between any two chemical states can be determined either from theory or experimentally. Therefore, given the non-arbitrary baseline value of $\overline{S}_i$ at 0 K, the value of $\overline{S}_i$ under any other condition is unambiguously established.

Above, we concluded that any system tends to move toward the macrostate that is associated with the most microstates, and we identified that macrostate as the equilibrium state of the system. Now, having quantified the number of microstates that correspond to a given macrostate as W, and seeing that increasing W corresponds to increasing entropy, we can restate the conclusion as indicating that all systems tend to drift toward higher-entropy states. A corollary is that the equilibrium state, i.e., the state in which the system ultimately stabilizes, must be the state of maximum entropy. Note, however, that the changes in the system are not driven by any "invisible force," but are simply a consequence of the transfers of energy associated with random molecular collisions. The statement that entropy "tends to increase" in the universe is thus seen as no more than a statement that probability theory really works! When viewed in this

In the science museum exhibit pictured above, the balls are dropped onto the top of the pegboard above the middle peg. They fall through the pegboard and always accumulate at the bottom in a bell-shaped pattern. There is no true force directing the balls into that pattern, but the result is so consistent that we might infer the existence of a virtual force or a "law" that governs the outcome.
| Tom Pantages.

context, we can characterize entropy as a concise mathematical function that summarizes the results obtained from application of probability theory to energy shifts among molecules.

CLARIFYING SOME MISCONCEPTIONS ABOUT ENTROPY   The above discussion can help clarify a few common misconceptions about the second law. Specifically, based on Equations (2.14) through (2.16) and examples like an ink drop dispersing in water, the second law is sometimes misunderstood to imply that mass always disperses to the maximum extent possible. Configurational entropy does increase with dilution, so it is true that the configurational component of entropy always encourages mass dispersal. However, our everyday experience tells us that mass does not *always* disperse—sometimes liquids spontaneously coalesce rather than disperse, as when droplets of oil are placed in water.

The thermodynamic explanation for both spontaneous dispersal and spontaneous coalescence is that processes always occur in the direction that increases *total* entropy. Although shifts in configurational entropy always favor dispersal of mass, shifts in thermal entropy can favor either dispersal or coalescence. If, in a particular situation, coalescence leads to an increase in thermal entropy that is greater than the concomitant decrease in configurational entropy, it will be favored. The factors that might cause thermal entropy to increase when liquids coalesce are discussed later in the chapter. For now, the important point is that the second law does not state that mass always disperses, and in fact the law provides an explanation for why dispersal of mass does not always occur.

The preceding paragraph describes a misunderstanding that can arise from failure to consider thermal entropy properly when analyzing a potential change in a system's state. Ironically, a second misconception about the second law sometimes arises from a failure to recognize the contribution of configurational entropy. Specifically, when thermomechanical systems are operated in a cycle (such as the famous *Carnot cycle*), the chemicals in the system are all at the same concentration at the end of the cycle as they were in the beginning. As a result, the configurational entropy of the system does not change, and the change in total entropy can be equated exclusively with the change in thermal entropy.

In cyclic thermomechanical systems, the change in thermal entropy is simply the sum of the $\delta q_{reservoir}/T_{reservoir}$ terms for all the steps in the cycle. Therefore, in those systems, any spontaneous process must be characterized by the inequality $\Delta S_{tot} = \Delta S_{thermal} = \sum(\delta q_{reservoir}/T_{reservoir}) > 0$. However, the extrapolation of this result to other systems is not appropriate. When a system is not operated in a cycle (as is typical of chemically reactive systems), the total entropy change includes contributions from changes in both thermal and configurational entropy. In those cases, a process with $\Delta S_{thermal} \leq 0$ can be spontaneous, if ΔS_{config} is sufficiently positive that ΔS_{tot} is positive.

SOME COMPLICATIONS THAT HAVE BEEN IGNORED   The preceding discussion provides the basic introduction to entropy that will help us interpret the behavior of solutes in aquatic systems. The discussion has been simplified in two ways that are worth noting, both related to the transfer of energy in the system. First, the assumption that all molecules of a given species have the same number of accessible quantum states (eight, two, and six for glucose, fructose, and water in the example) implies that energy can shift among bonds within a molecule, but ignores the possible transfer of energy between molecules. If energy transfer between molecules were considered, then different molecules of the same species would have different numbers of accessible states. Incorporating that possibility into the analysis leads to the conclusion that energy is distributed among molecules according to a function known as the Boltzmann distribution. Including this energy distribution in the analysis provides a more realistic picture of chemical systems, but it also complicates the mathematics and is not essential for understanding the key points being made.

Second, we have ignored the fact that when molecules shift between quantum states, some bond energy might be converted to thermal energy. In particular, if the shift converts one species into a different one (e.g., glucose to fructose), an amount of energy is acquired from or released to the surroundings that equals, to a first approximation, the different amounts of energy needed to form the bonds in the ground states of the two molecules. This possibility can be accounted for in either of two ways. One option would be to assume that the thermal energy remains in the system, in which case each macrostate would be associated with a different temperature, and the number of accessible microstates available to each species would depend on the overall distribution of species in the system. Alternatively, we could assume that a thermal reservoir is available to provide or remove thermal energy as needed, so that the system temperature remains constant. In that case, the entropy change of the reservoir would have to be considered along with the entropy changes of the molecules in the analysis. Again, analysis of systems in which thermal energy is exchanged in concert with changes in speciation is more realistic than the situation described above, but it is not essential for our purposes.

Thus, by making the simplifying assumptions described above, we are able to obtain the same fundamental insights that can be gained by a more rigorous analysis, without getting bogged down in more complex mathematics. More detailed analyses are available in many texts on chemical thermodynamics.

SUMMARY   To summarize the key points from the above discussion, entropy is a property of a macroscopic system, characterizing the number of different microstates that cause the system to have a given set of average macroscopic properties, i.e., to be in a given thermodynamic state. Assuming that all the microstates of a system are equally likely and that the system can freely shift among all of them, probability theory requires that, over time, the system shift from lower-entropy to higher-entropy states. The endpoint of such shifts is, obviously, the state of maximum entropy, which is the equilibrium state.

In chemical systems, the microstates of interest relate to different energy distributions in and among molecules (generating thermal entropy) and different spatial arrangements of molecules among one another (generating configurational entropy). Each of these forms of entropy can be defined for individual species in the system, with the entropy of the system being simply the summation of the entropies associated with all the constituent species. The total entropy of a perfect crystal of any pure substance is zero at a temperature of absolute zero; at least in theory, the entropy of substances in any other condition and at any temperature can be determined based on this one value and the experimental or computed entropy change associated with the conversion to the conditions of interest.

Because the ability to distinguish different macrostates depends on the analytical tool being used, the entropy that we compute depends on the resolution of the measurements relative to the scale of the events being observed. For the chemical systems of interest in this text, the resolution needed and available is at the level of distinguishing one chemical species from another, so that we can distinguish among systems with different chemical compositions.

A complete analysis of entropy changes in a system can tell us the direction of spontaneous change in that system. However, such an analysis can get very complex in some cases. We next consider surrogate parameters for entropy that are convenient to use for analyzing a few specific types of systems.

2.3 ENTROPY ANALYSIS APPLIED TO SPECIFIC SYSTEMS: POTENTIAL ENERGY, POTENTIALS, AND FORCES

As most of us see it, the simplest explanation for the fact that water flows downhill, bicycles can move uphill, and carousels go in circles is not that those changes maximize entropy in the universe, but rather that *forces* are being applied to the objects to cause them to move in the observed way. Apparently, then, at least in some cases, force-based analyses can provide the same information as entropy-based analyses. In this section, a force-based analysis is developed for a simple and familiar physical system consisting of a ball rolling along an incline. The analysis is then extended to electrical systems and chemical systems, both of which are relevant for understanding chemical behavior in systems discussed in subsequent chapters.

2.3.1 POTENTIAL ENERGY, POTENTIAL, AND FORCE IN GRAVITATIONAL SYSTEMS

When a ball rolls down an incline, gravitational potential energy is converted to other forms of energy, primarily kinetic and thermal. If the ball moves upward, of course, the reverse conversion occurs. The amount of gravitational potential energy gained by the ball (ΔPE_{grav}) during any such change is related to the ball's mass m, the earth's gravitational constant g, and the elevation change Δh as follows:

$$\Delta PE_{grav} = mg\,\Delta h \qquad\qquad \textbf{(2.17a)}$$

$$= mg(h_2 - h_1) \qquad\qquad \textbf{(2.17b)}$$

The three parameters on the right-hand side of Equation (2.17a) are measurable, unambiguous quantities, so the change in potential energy when the ball moves between any two points can be computed. The two individual values of h in Equation (2.17b), however, are not uniquely defined; they can be evaluated only by comparison with some datum level. In geography, the datum level is commonly chosen to be mean sea level. In situations where the concern is more localized, e.g., the height of a building, the datum is commonly chosen as the local ground level. The point is that no absolute measure of elevation (or PE_{grav}) exists; nevertheless, one can measure differences in elevation or PE_{grav} unambiguously.

Equation (2.17), of course, applies to any object of mass m. In using the equation, it is important to recognize that $mg\,\Delta h$ represents not the total energy needed to raise the object by Δh, but only the change in the gravitational potential energy of the object. When the object is moved, some energy must be expended to overcome frictional resistance, and the amount of this energy depends on the details of the move; different "pathways" require different amounts of energy. However, for a given elevation change Δh, the change in the gravitational potential energy of the object is the same regardless of the path taken. For this reason, PE_{grav} is a state variable (as was noted earlier for entropy).

A slightly different perspective on the point made in the preceding paragraph is that $mg\,\Delta h$ represents the (fixed) increase in potential energy stored *in the system* (the object) when the object is moved, but a (variable) amount of energy in excess of $mg\,\Delta h$ must be expended to carry out the transfer; this "excess" energy accumulates *outside* the system as waste heat. In the limit of no frictional resistance, the total energy requirement approaches $mg\,\Delta h$; thus $mg\,\Delta h$ is the minimum amount of energy that must be expended to carry out the operation.

Return now to consideration of the ball on an incline. Because force can be equated with a change in energy per unit distance traveled, we can express the force acting on the ball in the direction of the incline as

$$F_{l,\mathrm{grav}} = -\frac{d\mathrm{PE}_{\mathrm{grav}}}{dl} = -mg\frac{dh}{dl} \qquad \textbf{(2.18)}$$

where l is the distance measured along the incline. (The minus signs are required because the force is in the direction of decreasing h and $\mathrm{PE}_{\mathrm{grav}}$.) If we consider a three-dimensional terrain over which the ball can move, the force in a given direction is the negative gradient of potential energy in that direction:

$$F_{l,\mathrm{grav}} = -mg\,\nabla h = -\nabla\mathrm{PE}_{\mathrm{grav}} \qquad \textbf{(2.19)}$$

Experience tells us that, absent other forces (buoyancy, friction, etc.), gravity always causes objects to move downhill. Thus, in systems where gravity is the only factor of importance, changes always occur in the direction that decreases the system's gravitational potential energy, and those changes can be attributed to the existence of a force that is proportional to the negative gradient in the potential energy.

According to the second law, changes occur in any system only if they increase the total entropy of the universe; so in the simple system under consideration, a change that decreases gravitational potential energy must also increase entropy. As a result, we could, in theory, derive the conclusion that objects spontaneously move downhill based on an analysis of the entropy of the universe for each possible location of the ball. That analysis would also show that the condition of minimum gravitational potential energy (and, therefore, maximum entropy) corresponds to *gravitational equilibrium,* i.e., a condition where there is no gravitational impetus for further change. Clearly, although both approaches lead to the same result, the analysis based on potential energy is more intuitive.

Before we leave this example, it is useful to define one other parameter that characterizes the system. The partial derivative of the gravitational potential energy with respect to mass is called the **gravitational potential,** which we will represent as $\overline{\mathrm{PE}}_{\mathrm{grav}}$:

$$\overline{\mathrm{PE}}_{\mathrm{grav}} \equiv \frac{\partial\mathrm{PE}_{\mathrm{grav}}}{\partial m} = \frac{\partial(mgh)}{\partial m} = gh \qquad \textbf{(2.20)}$$

Because gh also equals the ratio of the total gravitational potential energy of objects at height h (mgh) to the total mass at that height (m), an alternative definition of $\overline{\text{PE}}_{\text{grav}}$ is as the average potential energy of a unit of mass at height h:

$$\overline{\text{PE}}_{\text{grav}} = gh = \frac{mgh}{m} = \frac{\text{total PE}_{\text{grav}} \text{ at } h}{\text{total mass at } h} = \frac{\text{average PE}_{\text{grav}}}{\text{per unit mass at } h} \quad \textbf{(2.21)}$$

A subtle but significant distinction between gravitational potential energy and gravitational potential is that whereas the former relates to an object, the latter relates to a location where an object might reside. Therefore, the gravitational potential of a location can be defined even if no object is currently there.

2.3.2   POTENTIAL ENERGY, POTENTIAL, AND FORCE IN ELECTRICAL SYSTEMS

Consider next the case of a charged particle in an electrical field. The analogy between this system and the one discussed above is very close. Specifically, an electrical system can be fully characterized by choosing an arbitrary datum where the **electrical potential energy** is assigned a value of zero, and then determining the electrical potential energy at any other point in the system based on the minimum amount of electrical energy needed to move a unit of charge between the datum location and that point.

Using an approach that is essentially identical to that in the preceding section, we can define the electrical force on an object as the spatial gradient in its electrical potential energy. Furthermore, experience indicates that, in the absence of other forces, charged objects move in the direction that minimizes the electrical potential energy in the system, so such changes must also maximize universal entropy. As a result, a condition where the electrical potential energy is minimized can be defined as one of *electrical equilibrium*.

The main difference between electrical and gravitational systems is that whereas gravity acts on objects based on their mass, electricity acts on objects based on their charge (σ, typically expressed in coulombs). Therefore, it only makes sense that the **electrical potential** should be defined as the partial derivative of electrical potential energy of the system with respect to charge. One conventional way of representing this term is by the Greek letter psi (Ψ), with units of volts (volts do, in fact, have dimensions of energy per unit charge). Thus

$$\overline{\text{PE}}_{\text{elec}} \equiv \Psi = \frac{\partial \text{PE}_{\text{elec}}}{\partial \sigma} \quad \textbf{(2.22)}$$

Analogous to the gravitational potential, an alternative definition for the electrical potential is the average electrical potential energy per unit charge of an object at a given location:

$$\overline{\text{PE}}_{\text{elec}} = \frac{\text{total PE}_{\text{elec}} \text{ at } x}{\text{total charge at } x} = \frac{\Psi \sigma}{\sigma} = \Psi \quad \textbf{(2.23)}$$

Parallel to the case for PE_{grav} and $\overline{\text{PE}}_{\text{grav}}$, and PE_{elec} and Ψ relate to charged objects and to locations where such objects might reside, respectively. However, whereas the accumulation of mass at a location has a negligible effect on the local gravitational constant, the local value of Ψ can change significantly as charge accumulates at that location.

2.3.3 SYSTEMS WITH MORE THAN ONE TYPE OF POTENTIAL ENERGY

Because energy can be interconverted among various forms, and because the laws of thermodynamics apply regardless of the form of the energy, the above discussion also applies to systems where objects have potential energy contributions from more than one source. In the following example, the equilibrium condition is determined for a particle influenced by both gravitational and electrical fields.

| **Example 2.2** | A particle has mass of 1.0 mg and carries a positive electrical charge σ of 10^{-5} coulomb (C). Assume an electrical field exists in the region where the particle is located, with the electrical potential varying only in the vertical (x) direction according to the equation $\Psi = 30/x$ (Ψ in millivolts, x in centimeters, and x defined as increasing in the upward direction). Thus, the potential is 30 mV at $x = 1$ cm and 3 mV at $x = 10$ cm (and approaches zero as x gets very large). Since the potential increases in the downward direction and the particle is positively charged, the particle experiences increasing electrical repulsion as it moves downward. What is the equilibrium position of the particle if gravity and the electric field are the only two influences on its motion?

Solution A

One approach for solving this problem is to find the value of x at which the downward force due to gravity is exactly balanced by the upward electrical force, so the net force acting on the particle is zero. This equilibrium criterion can be expressed as

$$\text{Net force} = F_{grav} + F_{elec} = 0 \tag{2.24}$$

The gravitational and electrical forces are, respectively,

$$F_{grav} = -mg = -(1.0 \times 10^{-3} \text{g})(980 \text{ cm/s}^2) = -0.98 \text{ dyn} \tag{2.25}$$

$$F_{elec} = -\sigma \frac{d\Psi}{dx} \tag{2.26}$$

$$F_{elec} = -(10^{-5} \text{C})\left(-\frac{30}{x^2} \frac{\text{mV}}{\text{cm}}\right)\left(10^4 \frac{\text{dyn-cm}}{\text{mV-C}}\right) = \frac{3.00}{x^2} \quad \text{dyn}$$

[The term 10^4 dyn-cm/(mV-C) is simply a unit conversion identity.] Thus, at equilibrium,

$$F_{grav} + F_{elec} = 0 = -mg - \sigma \frac{d\Psi}{dx} \tag{2.27}$$

$$0 = -0.98 \text{ dyn} + \frac{3.00}{x^2} \text{ dyn}$$

$$x = 1.74 \tag{2.28}$$

The forces are equal and opposite at $x = 1.74$ cm (recall that the relationship $\Psi = 30/x$ specified that x was in centimeters). Thus, the particle will come to rest at that height.

Solution B

A second approach that is only slightly different is to recognize that systems change spontaneously in the direction that reduces their total potential energy, and that equilibrium therefore coincides with the condition of minimum potential energy. To determine the equilibrium position of the particle in this system, then, we can explore the total potential energy of the system (PE_{tot}) as a function of the particle's location (x). The total potential energy of the particle can be expressed as a function of x as

$$PE_{tot} = mgx + \sigma\Psi \tag{2.29}$$

$$= (1 \text{ mg})\left(980\frac{\text{cm}}{\text{s}^2}\right)(x)\left(10^{-3}\frac{\text{erg-s}^2}{\text{mg-cm}^2}\right) + (10^{-5}\text{ C})\left(\frac{30}{x}\text{ mV}\right)\left(10^4\frac{\text{erg}}{\text{mV-C}}\right)$$

$$= \left(0.98x + \frac{3.00}{x}\right)\text{ erg}$$

The total potential energy is plotted as a function of x in Figure 2.3, from which it is clear that the potential energy minimum is near the point identified by the force balance as the equilibrium point. To find the exact value of x at the minimum, the total potential energy function can be differentiated and set equal to zero:

$$\frac{dE_{tot}}{dx} = mg + \sigma\frac{d\Psi}{dx} = 0 \tag{2.30}$$

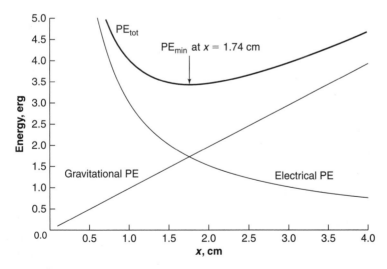

Figure 2.3   Total gravitational plus electrical potential energy of a particle as a function of distance above the datum plane, for the example system.

Equation (2.30) is identical to Equation (2.27), so the solution is once again $x = 1.74$ cm. Thus, the equilibrium condition can be identified either as a point where the net force on the particle is zero or as one where the particle's total potential energy is minimized.

2.4   POTENTIAL AND POTENTIAL ENERGY IN CHEMICAL SYSTEMS

Using the preceding discussion as a guide, it is possible to interpret virtually any change in a system as the result of a force corresponding to the downward gradient along some potential energy surface. This approach for characterizing a system can even be useful for systems in which there is no true force acting. For instance, consider the attraction of bees to a particularly fragrant flower. Without knowing the exact biochemical mechanisms at work, it is reasonable to assume that the fragrance of the flower is sensed by the bees, and that they find the source by following some path along which that fragrance gets more intense. From a strictly empirical point of view, it may appear that the flower is exerting a force on the bees, drawing them toward it, and one may speak of the bees being "attracted" by the flower. Although this attraction is clearly different from that of the moon to the earth, a conceptual or mathematical model representing the two phenomena in the same way may be useful. One could speak of a "virtual force" drawing the bees to the flower, meaning that the bees behave *as though* there were a real force pulling them toward it.

A somewhat analogous situation applies to the behavior of chemical species. Although it is certainly desirable to understand as much as possible about why chemical reactions occur, it is often sufficient to lump all these factors together into a parameter that we will characterize as the chemical potential energy of the system. The chemical driving force for a change in the system (e.g., for a chemical reaction) can then be represented as a gradient in that potential energy. If a small change in the system could reduce the potential energy dramatically, then the gradient is large, i.e., the force driving the system to change in that direction is large. Such a "force" may be more like the virtual force attracting bees to flowers than the forces that cause physical objects to move, but as long as the system behaves as predicted by the model, there is every reason to use the analogy.

In the next portion of the chapter, the tendency of a species to react is characterized in terms of its chemical potential energy. The discussion is presented in two stages. In the first stage, the function that represents chemical potential energy (the Gibbs energy function) is derived, and it is shown that, for conditions that are commonly of interest in aquatic systems, this function is a direct indicator of all the entropy changes accompanying a reaction. That result justifies the use of the Gibbs function in place of entropy as the basis for evaluating whether a reaction is thermodynamically favorable, in the same way that elevation can be used in place of entropy to determine whether a change is favorable in a system dominated by gravity.

We then switch gears, cease talking about entropy explicitly, and proceed with the analysis by defining the chemical potential as a partial derivative of the chemical potential energy (i.e., of the Gibbs energy). Using this parameter, and by analogy to gravitational and electrical systems, we calculate the condition of minimum chemical potential energy and interpret it as the condition of chemical equilibrium. Finally, we relate this thermodynamic interpretation of chemical change to the discussion of reactivity in Chapter 1, where the tendency to react was characterized in terms of chemical activity.

2.4.1 THE ENTROPY CHANGE ACCOMPANYING A CHEMICAL REACTION AT CONSTANT TEMPERATURE AND PRESSURE

Consider a system in which a chemical reaction can take place, causing some chemical species to disappear from the system and others to appear. Also, assume that the system is at the same temperature and pressure at the beginning and end of the change being investigated. These restrictions can be met by providing a thermal reservoir to acquire or release thermal energy as necessary and by allowing the system volume to change to ensure that the final pressure is the same as the initial pressure. The setup is essentially identical to the one shown schematically in Figure 2.1, except that now we assume that a chemical reaction is taking place inside the cylinder (the system), and we allow the piston to move freely to maintain a constant pressure in the system. Our goal is to determine the direction in which the reaction proceeds spontaneously and the composition of the system once it reaches equilibrium.

To determine whether the reaction is thermodynamically favorable, we need to evaluate all changes in entropy that occur when the reaction proceeds. Those changes can be expressed as follows:

$$\Delta S_{\text{tot}} = \Delta S_{\text{system}} + \Delta S_{\text{reservoir}} \qquad \textbf{(2.31)}$$

The right-hand side of Equation (2.31) can be evaluated most easily if $\Delta S_{\text{reservoir}}$ is expressed in terms of parameters that characterize the system, rather than those that characterize the reservoir. To accomplish this, we define another thermodynamic function, the **internal energy** U, as the total energy stored in the molecules in the system. The internal energy includes the bond energy holding all the "pieces" of each molecule together, as well as the molecules' kinetic, vibrational, and rotational energy. Like entropy, the internal energy is a state function, i.e., it depends on the pressure, temperature, and composition of the system, but not on how the system got to its present condition. This characteristic should make intuitive sense—we expect the total energy stored in a water molecule, for instance, to depend on the system temperature and to change slightly from one instant to the next in a solution, but not to depend on its history prior to entering the solution or on how it got into the solution in the first place.

Because both S and U are state functions, we can imagine a reaction proceeding in any way we choose, and the values of ΔS and ΔU for the imaginary process will be identical to those in the setup where the reaction is really occurring, as long as the starting and ending states are the same in the two processes. For the current analysis, we imagine that the reaction proceeds by a series of small steps, during each of which the temperature and pressure are maintained at the original values. Thus, in the hypothetical system, T_{system} and P_{system} are constant throughout the process, whereas in the real system they can vary, as long as the final values equal the initial ones. (Also $T_{reservoir}$ and $P_{reservoir}$ are constant throughout the process, but that is true in both the real and imaginary setups.)

Consider the change in internal energy ΔU accompanying the reaction. By the first law, ΔU must equal the difference between the energy transferred into the system from the thermal reservoir and the amount of energy lost from the system as work done on the surroundings ($= P \Delta V$). Because P is constant, $P \Delta V = \Delta(PV)$, so

$$\Delta U_{system} = \delta q_{system} - \Delta(PV)_{system} \tag{2.32}$$

Since $\delta q_{system} = -\delta q_{reservoir}$, we can rearrange the equation as follows:

$$\delta q_{reservoir} = -\Delta U_{system} - \Delta(PV)_{system} = -\Delta(U + PV)_{system} \tag{2.33}$$

The sum $U + PV$ appears frequently in thermodynamics and is called the **enthalpy** H. Being a combination of state functions, H is a state function as well. Equation (2.33) can therefore be written as follows:

$$\delta q_{reservoir} = -\Delta H_{system} \tag{2.34}$$

Since we have stipulated that the reaction occurs at a constant temperature equal to that of the reservoir, Equation (2.34) can be used to express the entropy change of the thermal reservoir in terms of the enthalpy change and temperature of the reactive part of the system:

$$\Delta S_{reservoir} \equiv \frac{\delta q_{reservoir}}{T_{reservoir}} = -\frac{\Delta H_{system}}{T_{system}} \tag{2.35}$$

Substituting Equation (2.35) into Equation (2.31), we obtain an expression for the total entropy change accompanying the reaction (i.e., the combined entropy changes of both the system and its surroundings) in terms of state variables that characterize only the reactive system

$$\Delta S_{tot} = \Delta S_{system} - \frac{\Delta H_{system}}{T_{system}} \tag{2.36}$$

2.4.2   DEFINITION OF THE GIBBS FUNCTION AND ITS USE AS A SURROGATE FOR ENTROPY CHANGES IN REACTIVE SYSTEMS

The Gibbs energy function G is defined by rearranging and combining the terms in Equation (2.36) as follows:

$$\Delta G_{system} \equiv \Delta H_{system} - T \Delta S_{system} = -T \Delta S_{tot} \tag{2.37}$$

Like S, U, and H, the Gibbs function is a state function. As a result, ΔG_{system} can be specified based on the conditions in the system at the beginning and end of the process of interest, without knowing how the process was carried out and without knowing what changes the process induced in the surroundings.

According to Equation (2.37), for systems that meet the specified restrictions on temperature and pressure, an increase in S_{tot} always corresponds to a decrease in G_{system}. Therefore, for the reaction to be thermodynamically favorable, it is necessary and sufficient that ΔG_{system} be negative. In other words, the Gibbs function tells us the same thing about the possibility of spontaneous change in these systems as an entropy analysis would.

The right-hand portion of Equation (2.37) is normally written without the subscripts, with the implicit understanding that each variable is referring to the material in the reactive system:

$$\Delta G \equiv \Delta H - T \Delta S \qquad \textbf{(2.38)}$$

As is the case for S and U, the enthalpy and Gibbs energy of the entire system can be written as the sum of the contributions of the species that make it up. Also, H and G of individual species are often normalized to the number of moles of those species present. That is,

$$H_{\text{system}} = \sum_i H_i = \sum_i n_i \overline{H}_i \qquad \textbf{(2.39)}$$

$$G_{\text{system}} = \sum_i G_i = \sum_i n_i \overline{G}_i \qquad \textbf{(2.40)}$$

where $\overline{H}_i$ and $\overline{G}_i$ are the **molar enthalpy** and **molar Gibbs energy** of i, respectively. Equations parallel to Equation (2.38) relate changes in G, H, and S for individual species, both in a system as a whole and when normalized to 1 mol of the species, i.e.,

$$\Delta G_i = \Delta H_i - T \Delta S_i \qquad \textbf{(2.41)}$$

$$\Delta \overline{G}_i = \Delta \overline{H}_i - T \Delta \overline{S}_i \qquad \textbf{(2.42)}$$

The advantages of carrying out the analysis of a chemical process in terms of the Gibbs function rather than entropy are essentially the same as those of analyzing the motion of an object in terms of gravitational potential energy. Specifically, (1) the Gibbs function describes energy, which most people can understand intuitively better than entropy; and (2) by using the Gibbs function, the criterion for a spontaneous change can be evaluated in terms of changes inside the system, without knowing or measuring anything in the surroundings.

Analogous to the criterion for change in systems where gravitational or electrical potential energy is of interest, the criterion for a spontaneous change in a chemical system is a decrease in chemical potential energy, i.e., in the Gibbs energy. Thus, the Gibbs function has all the essential characteristics of other types of potential energy, modified so that it is specifically applicable to chemically reactive systems in which the temperature and pressure are the same at the beginning and end of the process. Enough systems meet these restrictions (at least approximately) that the Gibbs function has tremendous utility.

2.5 DEFINING AND QUANTIFYING THE CHEMICAL POTENTIAL

2.5.1 DEFINITION OF THE CHEMICAL POTENTIAL

While the preceding paragraphs define the Gibbs energy function and make a case for its use instead of entropy in certain types of systems, we remain one step away from having a truly useful set of tools for analyzing reactivity. Specifically, we need a way to compute values of G at a given temperature and pressure based on the system composition, so that calculation of ΔG for a hypothetical reaction can be carried out without experimentation. In this section, we address that issue.

We begin by defining the **chemical potential** of a species i, denoted μ_i, as the partial differential of the chemical potential energy in the system with respect to the number of moles of i present (n_i):

$$\mu_i \equiv \left. \frac{\partial G_{\text{tot}}}{\partial n_i} \right|_{\text{constant } n_{j \neq i}, P, T} \tag{2.43}$$

where $n_{j \neq i}$ is the number of moles of all substances other than i. According to Equation (2.43), μ_i is the increment in the total chemical potential energy of the system when a differential amount of i is added to it, per unit amount of i added, if all other system conditions remain unchanged. The parallels between μ_i, on the one hand, and the gravitational potential $\left(= \dfrac{\partial \text{PE}_{\text{grav}}}{\partial m} \right)$ or the electrical potential $\left(= \dfrac{\partial \text{PE}_{\text{elec}}}{\partial \sigma} \right)$, on the other, should be apparent.

The constraint that "all other conditions remain unchanged" means that such a small amount of i is added that the concentration of i changes negligibly in the system. In other words, the equation defines μ_i for a hypothetical system in which a small amount of i is added to the system, but the composition of the system is not altered. Under such conditions, once it enters the system, the differential amount of i added would have the same Gibbs energy per mole ($\overline{G}_i$) as the molecules of i that were already present. The additional i would also not change the Gibbs energy of the other species already in the system. As a result, the change in G_{tot}, i.e., ∂G_{tot}, can be equated with $\overline{G}_i \, \partial n_i$. Substituting $\overline{G}_i \, \partial n_i$ for ∂G_{tot}, in Equation (2.43), we find that μ_i can be equated with $\overline{G}_i$:

$$\mu_i \equiv \frac{\overline{G}_i \, \partial n_i}{\partial n_i} = \overline{G}_i \tag{2.44}$$

2.5.2 Quantifying Molar Gibbs Energy (Chemical Potential) and Molar Enthalpy

As is true for other forms of potential energy, no absolute criteria exist for specifying conditions under which $\bar{G}_i$ is zero, so $\bar{G}_i$ values can only be assigned relative to an arbitrarily defined datum or baseline. The universal convention for this baseline is pure elements under standard state conditions. That is, $\bar{G}_i$ for any species i under any conditions is defined as the increment in Gibbs energy required to form i under those conditions, starting with only pure elements in their standard states. Obviously, it takes no energy to form a pure element in its standard state when that is the starting point, so the convention has the effect of assigning $\bar{G}_i = 0$ to any pure element under standard conditions. Thus, for instance,

$$\bar{G}_{\text{oxygen}} = \mu_{\text{oxygen}} \equiv 0 \text{ kJ/mol} \qquad \text{for pure } O_2 \text{ gas at 1 bar, 25°C}$$

$$\bar{G}_{\text{hydrogen}} = \mu_{\text{hydrogen}} \equiv 0 \text{ kJ/mol} \qquad \text{for pure } H_2 \text{ gas at 1 bar, 25°C}$$

$$\bar{G}_{\text{mercury}} = \mu_{\text{mercury}} \equiv 0 \text{ kJ/mol} \qquad \text{for liquid Hg at 25°C under } P_{\text{tot}} = 1 \text{ bar}$$

$$\bar{G}_{\text{iron}} = \mu_{\text{iron}} \equiv 0 \text{ kJ/mol} \qquad \text{for solid Fe at 25°C under } P_{\text{tot}} = 1 \text{ bar}$$

The value of $\bar{G}_i$ under standard state conditions is called the **standard molar Gibbs energy** of i (or, sometimes, the **standard Gibbs energy of formation** of i) and is designated $\bar{G}_i^o$ or $\bar{G}_{f,i}^o$. An alternative name is the **standard chemical potential** of i, designated μ_i^o.

Values of $\bar{G}_i^o$ for compounds (either pure compounds or solutes) are generally non-zero, since some chemical potential energy is either acquired or released when such compounds are formed from pure elements under standard state conditions. In theory, these $\bar{G}_i^o$ values can be determined by application of Equation (2.38) to a reaction in which the compound of interest is formed from its constituent elements, with all species in their standard states. In such a process, the enthalpy change (ΔH) can be measured based on the amount of energy exchanged with a thermal reservoir, and the entropy change (ΔS) can be determined from experiment or calculated from models, so ΔG can be computed. Because the Gibbs energy of the reactants in such a reaction is zero by definition, the computed value of ΔG can be equated with the Gibbs energy of the species i that was formed. Furthermore, because we have specified that the product i is in its standard state, the standard molar Gibbs energy of i ($\bar{G}_i^o$) can be determined by dividing ΔG by the number of moles of i that were formed:

$$\Delta G = G_{\text{product}} - G_{\text{reactants}}$$

$$\Delta G = G_{i,\text{std.state}} - \underset{\substack{\text{elements} \\ \text{in their} \\ \text{std.states}}}{\sum G}$$

$$\frac{\Delta G}{n_i} = \frac{G_{i,\text{std.state}}}{n_i} = \bar{G}_i^o$$

The same types of experiments can be used to determine $\overline{G}_i^\circ$ values of dissolved ions, except that one additional arbitrary baseline value must be established in that case because it is not possible to generate only one ion at a time. The convention that has been adopted is to assign $\overline{G}_{H^+}^\circ$ a value of 0 kJ/mol.

While the experiments described above for calculating $\overline{G}_i^\circ$ of compounds and ions are feasible conceptually, they are not always feasible in practice. In truth, $\overline{G}_i^\circ$ values are more often determined based on the enthalpy and entropy changes in reactions in which the reactants are other compounds (rather than pure elements) and in which neither the reactants nor the product is in the standard state. Examples of the corresponding calculations are provided later in the chapter. For now, the important point is that $\overline{G}_i^\circ$ values of many compounds can be and have been determined.

The same convention has been chosen for assigning $\overline{H}_i$ values as is used for $\overline{G}_i$; i.e., $\overline{H}_i$ is defined as the enthalpy input required to form 1 mol of species i from pure elements in their standard states. Again, this convention has the effect of assigning $\overline{H}_i$ a value of zero for pure elements in their standard state. Like the Gibbs function, the molar enthalpy (also molar entropy) under standard state conditions is commonly designated by a superscript o; i.e., it is represented as $\overline{H}_i^\circ$ (or, for entropy, $\overline{S}_i^\circ$). The $\overline{G}_i^\circ$, $\overline{H}_i^\circ$, and $\overline{S}_i^\circ$ values for some compounds of interest are provided in Table 2.1 and Appendix A at the back of the book.

It is important to recognize that the fundamental relationship among Gibbs energy, enthalpy, and entropy is given by Equation (2.42), which is based strictly on *changes* in those parameters:

$$\Delta\overline{G}_i = \Delta\overline{H}_i - T\,\Delta\overline{S}_i \qquad \textbf{(2.42)}$$

Because the baselines for $\overline{G}$ and $\overline{H}$ are different from that for $\overline{S}$, one cannot relate absolute values of $\overline{G}_i$, $\overline{H}_i$, and $\overline{S}_i$ by an equation analogous to Equation (2.42). That is, $\overline{G}_i \neq \overline{H}_i - T\overline{S}_i$. The simplest example of this inequality is for a pure element in its standard state, for which $\overline{G}_i$ and $\overline{H}_i$ are zero by definition, but $\overline{S}_i$ is non-zero. Rather, the relationship among $\overline{G}_i$, $\overline{H}_i$, and $\overline{S}_i$ is

$$\overline{H}_i = \overline{G}_i + T\overline{S}_i - T\overline{S}_i^\circ \qquad \textbf{(2.45)}$$

2.5.3 THE MOLAR GIBBS ENERGY UNDER NON-STANDARD STATE CONDITIONS

The preceding section indicates that the molar Gibbs energy of any species in its standard state can be determined. However, the vast majority of chemicals in systems of interest to us are not in their standard states, so values of $\overline{G}_i^\circ$ are not directly of value. Rather, what we need is an approach for calculating values of $\overline{G}_i$ under any non-standard conditions in which species i is found. To develop such an approach, we next consider the increment of Gibbs energy accompanying a change in which either the concentration of i or the environment around it is altered from standard state conditions to the conditions of interest.

According to Equation (2.42), the value of $\Delta\overline{G}_i$ accompanying any change in the system can be represented as $\Delta\overline{G}_i = \Delta\overline{H}_i - T\,\Delta\overline{S}_i$. Expanding the enthalpy

Table 2.1 Standard molar Gibbs energy of formation of some species of interest[1]

Species	$\bar{G}_f^o$, kJ/mol	Species	$\bar{G}_f^o$, kJ/mol
Ag^+	77.12	$HCrO_4^-$	764.8
$AgBr(s)$	−96.9	$Cr_2O_7^{2-}$	−1301
$AgCl(s)$	−109.8	$Cu(s)$	0
Al^{3+}	−489.4	Cu^{2+}	65.5
$Al(s)$	0	$Fe(s)$	0
Br^-	−104.0	H^+	0
$C(graphite)$	0	$H_2(g)$	0
$CO_2(g)$	−394.37	$H_2(aq)$	17.57
$H_2CO_3(aq)$	−623.2	$H_2O(l)$	−237.18
HCO_3^-	−586.8	$Hg(l)$	0
CO_3^{2-}	−527.9	$NH_3(aq)$	−26.57
$C_2H_4O_2$ (HAc)	−396.6	NH_4^+	−79.37
$C_2H_3O_2^-$ (Ac$^-$)	−369.4	Na^+	−240
Ca^{2+}	−553.54	$O_2(g)$	0
$CaO(s)$	−604.4	$O_2(aq)$	16.32
$CaCO_3(s,\ calcite)$	−1128.8	H_3PO_4	−1142.6
Cl^-	−131.3	$H_2PO_4^-$	−1130.4
$HOCl$	−79.9	HPO_4^{2-}	−1089.3
OCl^-	−36.8	PO_4^{3-}	−1018.8

[1]Values are provided in the table only for chemicals whose Gibbs energies are used in examples or problems in this chapter. Thermodynamic properties ($\Delta\bar{G}_f^o$, $\Delta\bar{H}_f^o$, and $\bar{S}^o$) for a much more extensive list of species are provided in Appendix A.

term, we can write that equation as

$$\Delta\bar{G}_i = \Delta\bar{U}_i + P\,\Delta\bar{V}_i - T\,\Delta\bar{S}_i \qquad (2.46)$$

For the purposes of this analysis, it is useful to consider the meaning of the individual terms on the right-hand side of Equation (2.46). These terms can be expressed in words as follows:

$$\Delta\bar{U}_i = \Delta\begin{pmatrix} \text{bonding energy} \\ \text{within } i, \text{ per} \\ \text{mole of } i \end{pmatrix} + \Delta\begin{pmatrix} \text{bonding energy between} \\ i \text{ and neighboring species,} \\ \text{per mole of } i \end{pmatrix} \qquad (2.47a)$$

$$P\,\Delta\bar{V}_i = P\Delta\begin{pmatrix} \text{volume occupied} \\ \text{by } i \text{ per mole of } i \end{pmatrix} \qquad (2.47b)$$

$$-T\,\Delta\bar{S}_i = -T\left[\Delta\begin{pmatrix} \text{thermal} \\ \text{entropy per} \\ \text{mole of } i \end{pmatrix} + \Delta\begin{pmatrix} \text{configurational} \\ \text{entropy per} \\ \text{mole of } i \end{pmatrix}\right] \qquad (2.47c)$$

Changes in the kinetic, rotational, and vibrational energies of molecules of i could, in theory, contribute to ΔU as well, but no changes are expected in those terms since we are restricting our analysis to systems at constant temperature.

In the following discussion, the effects of a change in the concentration of i and in its environment (specifically, the ionic composition of the solution) on each term in Equation (2.47) are considered. The discussion yields some important insights into the behavior of the Gibbs energy function and its relationship to measurable solution parameters.

EFFECT OF CONCENTRATION OF i ON $\overline{G}_i$ Consider first a change from the standard state concentration of i to a different concentration, while the environmental conditions remain identical to those in the reference state, i.e., if $\gamma_i = 1.0$ in both solutions. When the concentration of i changes, we expect no change in the internal bonding energy in molecules of i, so the first term contributing to $\Delta \overline{U}_i$ [the first term on the right in Equation (2.47a)] is zero. Since the environmental conditions in the initial and final systems are identical (reference state), the interactions between a molecule of i and its neighbors are also expected to be independent of c_i, so the second term is zero as well. Thus, $\Delta \overline{U}_i = 0$ when c_i changes, if γ_i remains unchanged.

The product $P \Delta \overline{V}_i$ in Equation (2.47b) can be interpreted by recalling that V_i is the partial volume of i, i.e., the portion of the system volume attributable to molecules of i, in the same way that the partial pressure of a gaseous species is the portion of the total pressure attributable to it. Correspondingly, $\overline{V}_i$ is the volume contributed by species i per mole of i in the system; i.e., it is V_i/n_i.

For an ideal gas at a fixed temperature and pressure, $\overline{V}_i$ is constant ($\overline{V}_i = V_i/n_i = RT/P$) regardless of the identity or the concentration of i. Therefore, for an ideal gas, $\Delta \overline{V}_i$ is zero for the change under consideration. For solutes, we have no expression comparable to the ideal gas law with which to calculate $\overline{V}_i$. However, since we have specified that the interactions of i with neighboring molecules are unchanged by the change in c_i (as is implied by the fact that the initial and final systems have reference state environmental conditions), it is reasonable to assume that a given increment in c_i will have the same (small) effect on system volume, regardless of the concentration of i. This situation is referred to as *ideal mixing*. A similar argument applies to pure liquids and solids, so we expect the value of $\overline{V}_i$ to be independent of c_i if γ_i does not change; i.e., we expect $P \Delta \overline{V}_i$, like $\Delta \overline{U}_i$, to be zero when c_i changes and γ_i does not.

The entropy contribution to $\Delta \overline{G}_i$ is split into changes in thermal and configurational entropy in Equation (2.47c). As described earlier in the chapter, $\overline{S}_{i,\text{therm}}$ is a measure of the number of ways that energy can be distributed within and among molecules of i and is independent of the concentration of i. Thus, $\overline{S}_{i,\text{therm}}$ is unaltered in the process under consideration, and $-T \Delta \overline{S}_{i,\text{therm}}$ is zero. On the other hand, $\overline{S}_{i,\text{config}}$ does depend on the concentration of i, as shown explicitly by Equation (2.16). Applying that equation to the change in system conditions under consideration, we find $-T \Delta \overline{S}_{i,\text{config}} = RT \ln(c_i/c_{\text{std.state}})$.

Based on the above discussion, we conclude that $\overline{G}_i$ in a system that has reference state environmental conditions but a non-reference state concentration differs from $\overline{G}_i^\circ$ only with respect to its configurational entropy. As a result, $\overline{G}_i$ in the non-standard state system can be expressed as

$$\overline{G}_i\big|_{\gamma_i=1.0} = \overline{G}_i^\circ + RT\ln\frac{c_i}{c_{\text{std.state}}} = \overline{G}_i^\circ + RT\ln c_i \qquad \textbf{(2.48)}$$

where the second equality applies if c_i is given in the units that are conventionally used to express $c_{\text{std.state}}$ (so that $c_{\text{std.state}} = 1.0$).

EFFECT OF ENVIRONMENTAL CONDITIONS ON $\overline{G}_i$ Next, consider how $\overline{G}_i$ changes when the environmental conditions are shifted away from those in the reference state. The only change of this type that we will consider is in the identities of the molecules surrounding a species i in solution.

Based on arguments very similar to those presented above, the only term in Equation (2.47) likely to be affected by a change in the molecules surrounding i is the second contribution to enthalpy, i.e., the molar energy of bonding between i and its neighbors.[5] As noted in Chapter 1, by far the most significant of these interactions are electrical in nature (ion-ion, ion-dipole, and dipole-dipole). Therefore, our objective can be restated as a desire to understand and model the effect of intermolecular electrical interactions on $\overline{G}_i$.

The analysis requires that we quantify the electrical interactions between molecules and translate that quantity into the same terms as are used to characterize chemical interactions. The derivation is a bit lengthy, but following it closely is worthwhile, as the end result provides significant insight into the connections between electrical and more traditional chemical ways of interpreting the behavior of aqueous solutes.

The electrical environment at a point is characterized quantitatively by the electrical potential Ψ, which was defined earlier in the chapter as the average electrical potential energy stored at a point per unit charge at that point (or, if no charge is present, the average energy per charge that would pertain if a differential amount of charge were placed there).

Every molecule in a solution is subjected to an electrical potential that reflects the combined effects of all the charged groups near it in the solution (e.g., ions and the charged regions of dipoles), as well as any contributions that originate from sources outside the solution phase. We will refer to the electrical potential originating within the solution, i.e., the electrical potential attributable to solvent and solute molecules, as Ψ_{chem} and to that originating externally as Ψ_{ext}. Typical sources of Ψ_{ext} include charged electrodes inserted into the solution to induce an oxidation/reduction reaction or to cause ions to migrate to a specific location (e.g., as part of a groundwater remediation process) or charged surfaces

---

[5]The partial volume of i might also change, but we will assume that any such change would have a much smaller effect on $\overline{G}_i$ than the change in intermolecular bonding.

with which the solution is in contact (e.g., soil particles, particles suspended in solution, or the charged surface of a membrane).

As noted earlier, absolute values of potential can be assigned only with respect to some arbitrary baseline. For Ψ_{chem}, this baseline is defined such that Ψ_{chem} experienced by a species is zero when the species is in its reference state.[6] According to this convention, water molecules are defined to experience $\Psi_{chem} = 0$ in pure water, in which case each molecule is surrounded by other, randomly oriented water molecules. By contrast, the Ψ_{chem} experienced by a Na^+ ion is defined to be zero when it is present in *its* reference state, i.e., at infinite dilution. In such a solution, the water molecules surrounding the ion are mostly oriented with the oxygen adjacent to the Na^+ ion, so the environment is slightly different from that where water molecules themselves are defined to experience zero electrical potential.[7]

The baseline for Ψ_{ext} is defined so that $\Psi_{ext} \equiv 0$ when no electrical potential is imposed from outside the solution phase. In most cases, our interest will be in systems where $\Psi_{ext} = 0$, and the following analysis focuses on such systems. However, systems in which Ψ_{ext} is non-zero are important and are considered in Chapters 9 and 10, so a brief analysis of those situations is also provided.

If a water molecule or an ion is *not* in its reference state, then the electrical potential it experiences might be different from zero. For instance, in a 0.1 M NaCl solution, most Na^+ ions will be surrounded by a combination of oriented water molecules and some Cl^- ions. Thus, from the perspective of a Na^+ ion, a shift from the reference state to a 0.1 M NaCl solution involves replacement of some H_2O dipoles near the Na^+ ion with Cl^- ions. This replacement of oriented dipoles with negatively charged ions causes the electrical potential experienced by the Na^+ ions to decrease. As a result, since the value of Ψ_{chem} for Na^+ ions in the reference state is zero by definition, its value in the 0.1 M NaCl solution will be less than zero. This situation is shown schematically in Figure 2.4.

Armed with at least a qualitative understanding of Ψ_{chem}, we can now consider how changes in Ψ_{chem} affect a solute's chemical potential ($\overline{G}_i$). That consideration turns out to be quite straightforward: as suggested by Equations (2.46) and (2.47a) the chemical potential of i includes a term accounting for the interaction of i with its neighbors, and Ψ_{chem} is a measure of the intensity of that interaction, so Ψ_{chem} contributes directly to $\overline{G}_i$. The only remaining task is to convert the units of Ψ_{chem} (typically, volts) into the conventional units used to express $\overline{G}_i$ (kilojoules per mole of i) so that the summations given in those two equations can be carried out.

To accomplish this conversion, we note that the electrical potential energy per molecule of i is given by the product $z_i e \Psi_{chem}$, where z_i is the charge number

---

[6]As was also noted earlier, the electrical potential is a property of a location, not a species. Nevertheless, because our interest here is strictly in the electrical potential experienced by a given chemical species when it is in different environments, it is acceptable (and convenient) to define a unique baseline for Ψ_{chem} for each species under consideration.

[7]It is important to recognize that Ψ_{chem}, as defined here, refers to the electrical potential in the microenvironment of the ion of interest. An electrode inserted into the solution integrates the potential across its entire sensing surface, effectively sampling a representative portion of the whole solution. Individual molecules in bulk solution experience an electrical potential that is integrated over a much smaller region (of molecular dimensions) and is therefore different from the value sensed by the electrode.

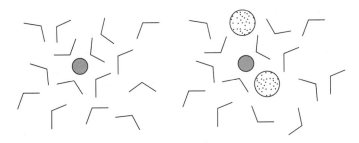

Figure 2.4   Schematic of molecular arrangements around a sodium ion (filled circle) in the infinite dilution reference state and in a solution containing dissolved Cl^- (lightly stippled circles). The value of Ψ_{chem} experienced by the Na^+ is zero by definition in the solution on the left and is less than zero in the one on the right.

of the ion (including sign), and e is the elementary charge (i.e., the charge on an electron, equal to 1.602×10^{-19} C). We can think of the units of z_i as being charges per molecule and those of e as being coulombs per unit charge, so the product $z_i e \Psi_{chem}$ has units of coulomb-volts per molecule of i.

By multiplying $z_i e \Psi_{chem}$ by Avogadro's number (N_A), we obtain the electrical potential energy per mole of i. Then, by applying the unit conversion 1 C-V = 10^{-3} kJ, we can express the quantity of interest in the desired units. Representing the electrical potential energy per mole of i as $\overline{\overline{PE}}_{i,elec}$,[8] we have

$$\overline{\overline{PE}}_{i,elec}\left(\frac{kJ}{mol\ of\ i}\right) = \left\{ z_i e N_A \left(\frac{C}{mol\ of\ i}\right)\right\} \left\{ 10^{-3}\ \frac{kJ}{C\text{-}V}\right\} \{\Psi_{chem}\ (V)\} \qquad \textbf{(2.49)}$$

The value product eN_A is called the **Faraday** or **Faraday constant** (F), which, when combined with the unit conversion shown in the middle set of braces, has a value of 96.48 $\dfrac{kJ/(mol\ of\ charge)}{V}$.[9] Thus, when all terms are expressed in their conventional units, Equation 2.49 can be written succinctly as:

$$\overline{\overline{PE}}_{i,elec} = z_i F \Psi_{chem} \qquad \textbf{(2.50)}$$

Assuming that the change in bond strength between a solute and its neighbors can be attributed primarily to electrical interactions, we can use Equation (2.50) to quantify the second term on the right in Equation (2.47a), i.e.,

$$\Delta \left(\begin{array}{c} \text{bonding energy between} \\ i \text{ and neighboring species,} \\ \text{per mole of } i \end{array} \right) = z F(\Delta \Psi_{chem}) \qquad \textbf{(2.51)}$$

---

[8]The double overbar is used here to distinguish $PE_{i,elec}$ normalized to 1 mol of i from $PE_{i,elec}$ normalized based on charge, which yields Ψ.

[9]Recall that one mole of charge is sometimes expressed as one equivalent. This terminology is commonly used to define the Faraday, i.e., the value of F is usually given as 96.48 $\dfrac{kJ}{equiv - V}$.

where $\Delta\Psi_{chem}$ is the difference between the value of Ψ_{chem} affecting solute i in the system of interest and in the reference state. However, since Ψ_{chem} in the reference state is zero by definition, the right-hand side of Equation (2.51) can be written simply as $z_i F\Psi_{chem}$.

Equation (2.48) was derived earlier to express $\overline{G}_i$ in a system where i is in its reference state ($\gamma_i = 1.0$). That equation can now be combined with Equation (2.51) so that it applies to a solute in *any* (reference or non-reference) state by adding the term $z_i F\Psi_{chem}$, i.e.:

$$\overline{G}_i = \overline{G}_i^o + RT \ln c_i + z_i F\Psi_{chem} \qquad (2.52)$$

In most solutions, z_i and Ψ_{chem} will have opposite signs; e.g., a cation ($z_i > 0$) will be preferentially surrounded by anions ($\Psi_{chem} < 0$), so $z_i F\Psi_{chem} < 0$. This result indicates that the cation has less potential energy, and hence is more stable, than in the reference solution. Thus, increasing the salt content of the solution decreases the electrical potential energy of i and decreases $\overline{G}_i$; i.e., it has the same effect on $\overline{G}_i$ as decreasing the concentration of i in the solution.

RELATING Ψ TO THE ACTIVITY COEFFICIENT γ_i   It is instructive to combine the last two terms in Equation (2.52) as follows:

$$\overline{G}_i = \overline{G}_i^o + RT \ln c_i + RT\frac{z_i F\Psi_{chem}}{RT} \qquad (2.53a)$$

$$\overline{G}_i = \overline{G}_i^o + RT \ln c_i + RT \ln\left(\exp\frac{z_i F\Psi_{chem}}{RT}\right) \qquad (2.53b)$$

$$\overline{G}_i = \overline{G}_i^o + RT \ln\left(c_i \exp\frac{z_i F\Psi_{chem}}{RT}\right) \qquad (2.53c)$$

Since $\overline{G}_i^o$ is a constant and we are considering only reactions that take place under conditions where T is constant, Equation (2.53c) indicates that $\overline{G}_i$ depends solely on the product $c_i \exp[z_i F\Psi_{chem}/(RT)]$. Thus, a species i has the same value of $\overline{G}_i$ in any solution with a given value of $c_i \exp[z_i F\Psi_{chem}/(RT)]$. Also, because $\overline{G}_i$ is a measure of chemical potential energy, we can associate equal values of $\overline{G}_i$ with equal reactivity of i.

In Chapter 1, we associated the reactivity of i directly with its chemical activity a_i, which can be represented as $c_i\gamma_i$. In other words, we concluded that i is equally reactive in any solution with a given value of $c_i\gamma_i$. Here, we conclude that i is equally reactive in any solution with a given value of $c_i \exp[z_i F\Psi_{chem}/(RT)]$. The equivalent implications of these two expressions suggest that we can equate γ_i with $\exp[z_i F\Psi_{chem}/(RT)]$ and rewrite Equation (2.53c) as follows:

$$\overline{G}_i = \overline{G}_i^o + RT \ln c_i\gamma_i \qquad (2.54a)$$

$$\overline{G}_i = \overline{G}_i^o + RT \ln a_i \qquad (2.54b)$$

The equivalence of γ_i and $\exp[z_i F\Psi_{chem}/(RT)]$ allows us to reinterpret γ_i as a term that accounts for the fact that electrical interactions with surrounding species can alter the molar Gibbs energy (the chemical potential) of a solute.

Applied to ions, this idea indicates that the electrical attraction between a central ion and surrounding ones decreases the electrical potential energy (and therefore the total potential energy) of the ion, compared to the situation that would apply if those surrounding ions were not there (infinite dilution). The decrease in total potential energy makes the ion less "active" (reactive) than it otherwise would be. The higher the concentration of ions in solution (the higher the ionic strength), the greater is this effect and the less active i becomes.

Although the idea presented in Chapter 1 that γ accounts for "shielding" of an ion by other ions might seem more intuitive than the idea that it accounts for a change in the electrical contribution to the chemical potential energy, explaining γ in the latter way is more fundamentally correct and more amenable to quantitative analysis. In fact, it was precisely by solving for Ψ_{chem} as a function of ionic strength that Debye and Huckel first developed the theoretical equation for γ that was presented in Chapter 1 (the Debye–Huckel limiting law).

Example 2.3

The value of $\overline{G}^o_{CO_3^{2-}}$ is -527.9 kJ/mol. Compute $\overline{G}_{CO_3^{2-}}$ in two solutions, each containing 10^{-5} mol/L CO_3^{2-}. In one solution, $\gamma_{CO_3^{2-}}$ is approximately 1.0, and in the other, the ionic strength is large enough (~ 0.04) to reduce $\gamma_{CO_3^{2-}}$ to 0.5.

Solution

The molar Gibbs energy of CO_3^{2-} in the two solutions of interest can be computed by using Equation (2.54):

$$\overline{G}_{CO_3^{2-}} = \overline{G}^o_{CO_3^{2-}} + RT \ln c_{CO_3^{2-}} + RT \ln \gamma_{CO_3^{2-}}$$

The results are $\overline{G}_{CO_3^{2-}} = -556.35$ kJ/mol and $\overline{G}_{CO_3^{2-}} = -558.06$ kJ/mol in the solutions with activity coefficients of 1.0 and 0.5, respectively. The chemical potential energy is lower in the system with the higher ionic strength, meaning the CO_3^{2-} is more stable (less reactive) in that system.

2.5.4 USING CHEMICAL POTENTIAL TO INTERPRET MOLECULAR DIFFUSION

Equation (2.54), combined with the idea that a gradient in potential can be interpreted as a force, is the basis for an important derivation that is used to assess the effective size of dissolved species. We can reproduce this derivation by considering a species i that is dispersed non-uniformly in the x direction of an ideal solution (i.e., $\gamma_i = 1.0$ and $a_i = c_i$). According to Equation (2.54), the chemical potential of i is greater in regions where c_i is higher. Therefore, we expect the chemical "force" on i ($F'_{i,x}$) to cause it to move from regions of high to low concentration, with the force equaling the negative gradient in chemical potential:

$$F'_{i,x} = -\frac{d\mu_i}{dx} = -\frac{d\overline{G}_i}{dx} \qquad \textbf{[2.55]}$$

The term $d\overline{G}_i/dx$ is the gradient in chemical potential energy per mole of i. In the current analysis, we are interested in the energy gradient applicable to a single

Like the balls shown in the previous photograph, molecules are subject to random forces that might push them in any direction. However, the net effect of these random moves is always to transport chemicals (i.e., cause them to diffuse) from regions of higher to lower chemical activity, as when tea diffuses throughout a pitcher of water.

| Michael Dalton/Fundamental Photographs.

molecule, so we need to divide by Avogadro's number N_A. Thus, the diffusive force on a molecule of i ($F_{i,x}$) can be written as

$$F_{i,x} = -\frac{1}{N_A}\frac{d\overline{G}_i}{dx} \tag{2.56}$$

Any object moving through a viscous fluid is resisted by a drag force that is proportional to the object's velocity. For a small sphere of radius r_i moving slowly through a uniform fluid of viscosity μ, the drag force is

$$F_{\text{drag}} = -6\pi\mu r_i v_x \tag{2.57}$$

where v_x is the velocity in the x direction and the minus sign indicates that the force is in the opposite direction from the velocity. Thus, if a force is applied to an object, causing it to accelerate, the counteracting drag force increases as the velocity increases. This process continues until the drag force equals the applied force, at which time the object neither accelerates nor decelerates, but proceeds at a steady, so-called terminal, velocity. Treating molecules of dissolved i as small spheres and assuming they have reached their terminal velocity, we can apply Equation (2.57) to them and write

$$F_{i,x} + F_{\text{drag},x} = 0 \tag{2.58}$$

$$-\frac{1}{N_A}\frac{d\overline{G}_i}{dx} - 6\pi\mu r_i v_x = 0 \tag{2.59}$$

$$v_x = -\frac{(1/N_A)(d\overline{G}_i/dx)}{6\pi\mu r_i} \tag{2.60}$$

Expressing $\overline{G}_i$ in terms of activity, $(1/N_A)(d\overline{G}_i/dx)$ can be written as

$$\frac{1}{N_A}\frac{d\overline{G}_i}{dx} = \frac{RT}{N_A}\frac{d\ln a_i}{dx} = \frac{RT}{N_A a_i}\frac{da_i}{dx} \tag{2.61}$$

As noted above, the ratio R/N_A is the Boltzmann constant k_B. Making that substitution, and substituting Equation (2.61) into Equation (2.60), we obtain

$$v_x = \frac{k_B T}{6\pi\mu r_i}\frac{1}{a_i}\frac{da_i}{dx} \tag{2.62}$$

If the assumption of an ideal solution is met, c_i can be substituted for a_i:

$$v_x = -\frac{k_B T}{6\pi\mu r_i}\frac{1}{c_i}\frac{dc_i}{dx} \tag{2.63}$$

Equation (2.63) is a quantitative prediction for the average, net velocity of molecules through water under the influence of a gradient in chemical potential that is generated by a concentration gradient. The minus sign indicates that movement is in the opposite direction from the concentration gradient; i.e., diffusion drives i from areas of high to low concentration.

By multiplying through by c_i, we can convert the left-hand side of Equation (2.63) to the flux of i, i.e., the rate at which i passes through a hypothetical plane perpendicular to the concentration gradient, per unit area of the plane. Designating the flux of i as J_i with dimensions of mass/area-time,

$$c_i v_x = J_i = -\frac{k_B T}{6\pi\mu r_i}\frac{dc_i}{dx} \tag{2.64}$$

According to Fick's law, which was derived from empirical studies of the rate of chemical diffusion under various conditions, the one-dimensional diffusive flux of any chemical is proportional to the chemical's concentration gradient, with the proportionality factor being defined as the diffusion coefficient of i, denoted D_i:

$$J_i = -D_i\frac{dc_i}{dx} \tag{2.65}$$

Comparison of Equations (2.64) and (2.65) indicates that the diffusion coefficient is related to the effective radius by

$$D_i = \frac{k_B T}{6\pi\mu}\frac{1}{r_i} \tag{2.66}$$

The derivation above (which follows that of Cussler[10]) is the basis for the Einstein–Stokes equation for estimating diffusion coefficients of molecules, or, equivalently, for estimating the effective radius of dissolved molecules from experimental measurements of their diffusion rates.

Note that the above result was obtained by treating the dissolved molecules *as though* they were subject to a chemical force equal to the negative gradient in

| [10]E. L. Cussler, *Diffusion: Mass Transfer in Fluid Systems*, Cambridge University Press, Cambridge, 1984.

chemical potential, causing them to diffuse toward regions of lower chemical activity. It would not be correct to think of this driving force "pulling" or "pushing" molecules in a given direction in the way that gravity or mechanical forces affect objects; nor is it valid to say that the molecules "sense" a concentration gradient. The molecules simply move about randomly, and this random motion tends to cause gradients to dissipate.

Example 2.4	At 25°C, the diffusion coefficient of Na^+ is 1.33×10^{-5} cm^2/s, and the viscosity of water is 10^{-2} g/(cm-s).

  a. Estimate the effective radius of a Na^+ ion.

  b. Two solutions containing $10^{-2.0}$ and $10^{-4.0}$ mol/L Na^+ are connected through a capillary tube that is 2.0 mm long. If a linear gradient of Na^+ concentration develops in the capillary, what is the Na^+ flux between the two solutions?

Solution

  a. The effective radius is found simply by inserting the diffusion coefficient into Equation (2.66). The result is that $r_{Na^+} = 0.16$ nm.

  b. The flux can be determined by substitution into Equation (2.65), with a gradient dc/dx of 9.9×10^{-3} $M/2$ mm. The computed flux is 6.58×10^{-10} mol/cm^2-s.

2.5.5   EFFECT OF EXTERNAL ELECTRIC FIELDS ON CHEMICAL REACTIVITY: ELECTROCHEMICAL POTENTIAL AND ELECTROCHEMICAL ACTIVITY

As noted above, electrical fields are sometimes imposed on solutions from sources outside the solution phase, either as part of an engineered process or as a consequence of the presence of charged particles in the water. Although we do not consider the details of systems where external potentials are present until later in the text, it is worthwhile to establish the framework for that analysis here, since it so closely parallels the analysis for potentials generated by molecules and ions in solution.

The effect of a given electrical potential on solute behavior is, of course, identical regardless of whether that potential is generated by dissolved ions, suspended colloids, or external circuitry—after all, an ion has no way of "knowing" the source of an electrical potential that it experiences. Therefore, we can use essentially the same approach to analyze the effects of electrical potential from any of these sources.

The sum of the chemical and electrical potential energy of a solute *i* is referred to as its **electrochemical potential energy,** which we will designate as $PE_{i,ec}$. As we did with the chemical potential energy $PE_{i,chem}$, we can normalize the electrochemical potential energy to one mole of *i* to generate an expression for the **electrochemical potential** of *i*, $\overline{\overline{PE}}_{i,ec}$. The only difference between the *electrochemical* potential $\overline{\overline{PE}}_{i,ec}$ and the *chemical* potential $\overline{G}_i$ is that $\overline{\overline{PE}}_{i,ec}$ includes a contribution from Ψ_{ext}. Therefore, we can express $\overline{\overline{PE}}_{i,ec}$ as the sum of $\overline{G}_i$ and a term containing Ψ_{ext}. Using Equation (2.50) as a template to convert Ψ_{ext} to the same units as $\overline{\overline{PE}}_{i,ec}$, we can compute $\overline{\overline{PE}}_{i,ec}$ as follows:

$$\overline{\overline{PE}}_{i,ec} = \overline{G}_i + z_i F \Psi_{ext} \tag{2.67}$$

Furthermore, by manipulating Equation (2.67) as we did Equation (2.52), we can express $z_i F \Psi_{ext}$ as an activity coefficient that accounts for electrical potentials arising outside of the solution, exactly as we did earlier for $z_i F \Psi_{chem}$ to account for electrical potentials arising within the solution. After doing so, we obtain an expression for the electrochemical potential that is almost identical to that for the chemical potential, except that it contains an additional activity coefficient:

$$\overline{\overline{PE}}_{i,ec} = \overline{G}_i^{\circ} + RT \ln{(c_i \gamma_{i,chem} \gamma_{i,ext})} \tag{2.68}$$

where $\gamma_{i,ext} \equiv \exp[z_i F \Psi_{ext}/(RT)]$. The term shown as $\gamma_{i,chem}$ in the above equation is the same as γ_i in previous equations; the additional subscript is simply for clarity.

Equation (2.68) can be written in a concise form that is analogous to Equation (2.54a) or (2.54b) as follows:

$$\overline{\overline{PE}}_{i,ec} = \overline{G}_i^{\circ} + RT \ln{(c_i \gamma_{i,ec})} = \overline{G}_i^{\circ} + RT \ln a_{i,ec} \tag{2.69}$$

where $\gamma_{i,ec} \equiv \gamma_{i,chem} \gamma_{i,ext}$ and $a_{i,ec} \equiv \gamma_{i,ec} c_i$. We can refer to $\gamma_{i,ec}$ and $a_{i,ec}$ as the **electrochemical activity coefficient** and the **electrochemical activity** of i, respectively.

The parameters $\gamma_{i,ec}$ and $a_{i,ec}$ are exactly analogous to $\gamma_{i,chem}$ and $a_{i,chem}$, respectively, except that the former pair of parameters is more general; i.e., they describe the overall activity coefficient and the chemical activity of i in solutions with or without electrical fields imposed from outside the solution phase, whereas the latter pair describes systems in which the only electrical fields affecting an ion are those generated by other species in the solution.

Although Ψ_{chem} affects the potential energy of a solute in the same way that Ψ_{ext} does, the two contributions are almost always treated separately, with Ψ_{chem} embedded in $\gamma_{i,chem}$, a_i, or $\overline{G}_i$. That is, the expression for the molar electrochemical potential of i is usually given in one of the following three forms:

$$\overline{\overline{PE}}_{i,ec} = \overline{G}_i^{\circ} + RT \ln{\left(c_i \gamma_{i,chem} \exp{\frac{z_i F \Psi_{ext}}{RT}} \right)} \tag{2.70a}$$

$$\overline{\overline{PE}}_{i,ec} = \overline{G}_i^{\circ} + RT \ln{\left(a_{i,chem} \exp{\frac{z_i F \Psi_{ext}}{RT}} \right)} \tag{2.70b}$$

$$\overline{\overline{PE}}_{i,ec} = \overline{G}_i + z_i F \Psi_{ext} \tag{2.70c}$$

In fact, Ψ_{ext} is usually just shown as Ψ, with the implicit understanding that it refers strictly to electrical potentials generated outside of the solution. Similarly, $a_{i,chem}$ and $\gamma_{i,chem}$ are understood to incorporate Ψ_{chem} and are typically shown simply as a_i and γ_i, respectively.

To summarize, the reactivity of a solute i is *always* determined by its electrochemical activity $a_{i,ec}$ or, equivalently, its electrochemical potential $\overline{\overline{PE}}_{i,ec}$.

These two parameters are related to each other via Equation (2.69). For convenience, each of these parameters is often divided into two components—one that is relevant in any solution, and another that is relevant only when the solute is subject to an electrical potential that originates outside the solution phase.

The parts of the electrochemical activity and electrochemical potential that are relevant in any solution are called the chemical activity a_i and chemical potential $\overline{G}_i$ (or μ_i), respectively. These two parameters are related via Equation (2.54b). Both a_i and $\overline{G}_i$ include contributions that account for electrical interactions among species in solution. These electrical interactions give rise to the chemical activity coefficient γ_i.

In cases where a solute is subject to an electrical potential Ψ_{ext} originating outside of the solution, $a_{i,ec}$ and $\overline{\overline{PE}}_{i,ec}$ can be computed via $a_{i,ec} \equiv a_i \exp\left(\dfrac{z_i F \Psi_{ext}}{RT}\right)$ and Equation (2.70), respectively. Inspection of these equations makes it clear that if $\underline{\underline{\Psi}}_{ext} = 0$ (i.e., if no external electrical potential is applied), then $a_{i,ec} = a_i$ and $\overline{PE}_{i,ec} = \overline{G}_i$.

Example 2.5 | Determine the electrical potential difference, in millivolts, that would have to be applied between the two solutions described in Example 2.4 to cause Na^+ to be driven from the low- (10^{-4} M) to the high- (10^{-2} M) concentration side of the system.

Solution

The Na^+ moves in the direction of decreasing electrochemical potential. Therefore, if the electrochemical potentials of the two solutions are equal, no Na^+ transport will occur. The difference in electrical potential that corresponds to this condition can be computed using Equation (2.70c). Designating the solutions containing 10^{-2} and 10^{-4} M Na^+ as solutions 1 and 2, application of that equation yields

$$\overline{G}_1 + z_{Na^+} F \Psi_{ext,1} = \overline{G}_2 + z_{Na^+} F \Psi_{ext,2}$$

$$\overline{G}^{\circ}_{Na^+} + RT \ln 10^{-2} + z_{Na^+} F \Psi_{ext,1} = \overline{G}^{\circ}_{Na^+} + RT \ln 10^{-4} + z_{Na^+} F \Psi_{ext,2}$$

$$z_{Na^+} F (\Psi_{ext,2} - \Psi_{ext,1}) = RT \ln \frac{10^{-2}}{10^{-4}}$$

Substituting values for z_{Na^+}, F, R, and T, we find $\Psi_{ext,2} - \Psi_{ext,1} = 0.118$ V $= 118$ mV. Therefore, if an electrical potential difference of 118 mV is imposed between the two solutions by external circuitry, with $\Psi_{ext,2} > \Psi_{ext,1}$, diffusion of Na^+ will be prevented. And if the imposed electrical potential difference is >118 mV, Na^+ will transfer from the low- to the high-concentration solution, i.e., in the opposite direction from that driven purely by chemical diffusion.

2.5.6 COMPARISON OF ALTERNATIVE EXPRESSIONS FOR $\overline{G}_i$

Although they look quite different from one another, Equations (2.42) and (2.54a) (repeated below) are closely related, both expressing the Gibbs energy of a species i as a function of other parameters that characterize i itself and its environment. Specifically, Equation (2.42) expresses $\overline{G}_i$ in terms of the classical thermodynamic parameters $\overline{H}_i$ and $\overline{S}_i$, while Equation (2.54a) expresses the same function in terms of its value under standard state conditions $(\overline{G}_i^o)$ and two other parameters $(c_i$ and $\gamma_i)$ that describe how different the actual conditions are from those in the standard state. The linkage between these two ways of expressing $\overline{G}_i$ is instructive and can be seen most easily by considering how each expression relates to the word expression we developed for $\overline{G}_i$, which is also repeated below.[11]

$$\Delta\overline{G}_i = \Delta\overline{H}_i - T\,\Delta\overline{S}_i \qquad\qquad (2.42)$$

$$\overline{G}_i = \overline{G}_i^o + RT\ln(c_i\gamma_i) \qquad\qquad (2.54a)$$

$$\Delta\overline{G}_i = \left[\Delta\left(\begin{array}{c}\text{bonding energy}\\\text{within } i\text{, per}\\\text{mole of } i\end{array}\right) + \Delta\left(\begin{array}{c}\text{bonding energy between}\\ i \text{ and neighboring species,}\\\text{per mole of } i\end{array}\right)\right.$$

$$\left. + \Delta\left(\begin{array}{c}\text{value of } PV_i\\\text{per mole}\\\text{of } i\end{array}\right)\right] - T\left[\Delta\left(\begin{array}{c}\text{thermal}\\\text{entropy per}\\\text{mole of } i\end{array}\right) + \Delta\left(\begin{array}{c}\text{configurational}\\\text{entropy per}\\\text{mole of } i\end{array}\right)\right]$$

$$(2.47)$$

The first three terms in Equation (2.47) represent enthalpic contributions to $\overline{G}_i$, while the last two are entropic. The discussion of those five terms suggested that only two of them (those relating to intermolecular bonding and configurational entropy) change significantly when the system composition changes, while the other three (those relating to the internal bonding energy, molar volume, and thermal entropy of the species) are, to a first approximation, independent of the system composition. Put another way, two of the five terms relate to the properties of the system, and the other three relate to properties inherent in i, regardless of what system it is in.

Presumably, the same five contributions to $\overline{G}_i$ are somehow embedded in Equation (2.54a). To see where each contribution resides, it is useful to split the final term in that equation into two, as follows:

$$\overline{G}_i = \overline{G}_i^o + RT\ln c_i + RT\ln\gamma_i \qquad\qquad (2.71)$$

Only the first term in Equation (2.71) is independent of the system composition; indeed, the last two terms are nothing *but* measures of the system composition. Thus, $\overline{G}_i^o$ tells us something about i regardless of what system it is in, while the logarithmic terms in Equation (2.71) tell us something about how i interacts with the rest of the particular system being analyzed. Accordingly, we conclude that the contributions from the three system-insensitive factors in Equation (2.47) (the internal bonding energy in molecules of i, the molar volume of i, and

---

[11]Equations (2.42) and (2.47) actually describes *changes* in $\overline{G}_i$, but they can be readily modified to yield the absolute value of $\overline{G}_i$ once the datum level for $\overline{G}_i$ is established.

the thermal entropy of i) are all embedded in $\overline{G}_i^\circ$ in Equation (2.71), and the other two factors contributing to $\overline{G}_i$ (configurational entropy and intermolecular bonding) are related to $RT \ln c_i$ and $RT \ln \gamma_i$.

The defining equation for configurational entropy [Equation (2.15)] indicates that that factor depends only on the concentration of i, and our analysis of activity coefficients associates γ_i with intermolecular bonding. Logically, then, the second and third terms on the right of Equation (2.71) must reflect the contributions of configurational entropy and intermolecular bonding to $\overline{G}_i$, respectively, and we can rearrange Equation (2.47) as follows to show explicitly how each factor contributing to $\overline{G}_i$ is incorporated into Equation (2.71):

$$\overline{G}_i^\circ \equiv \left(\begin{array}{c} \text{internal bonding} \\ \text{energy per} \\ \text{mole of } i \end{array}\right) + \left(\begin{array}{c} \text{value of } PV_i \\ \text{per mole of } i \end{array}\right)$$

$$- T\left(\begin{array}{c} \text{thermal entropy} \\ \text{per mole of } i \end{array}\right) + \text{constant}$$

$$RT \ln c_i \equiv -T\left(\begin{array}{c} \text{configurational entropy} \\ \text{per mole of } i \end{array}\right) + \text{constant} \qquad \textbf{(2.72)}$$

$$RT \ln \gamma_i \equiv \left(\begin{array}{c} \text{intermolecular} \\ \text{bonding energy} \\ \text{per mole of } i \end{array}\right) + \text{constant}$$

The constant on the right-hand side of each equation above is included to account for the fact that each expression on the left-hand side is defined relative to some arbitrary baseline condition that might be offset from the absolute value of the terms on the right.

Comparison of Equation (2.72) with those developed above demonstrates that the only difference between the expressions for $\overline{G}_i$ written in terms of classical thermodynamic parameters [Equation (2.42)] and descriptors of solution composition [Equation (2.54)] is the way that the factors contributing to $\overline{G}_i$ are grouped: the classical thermodynamic approach separates the contributions into enthalpic and entropic groups, while the approach expressing $\overline{G}_i$ in terms of solution parameters separates the same contributions into system-sensitive and system-insensitive groups. These simple connections between the two ways of calculating $\overline{G}_i$ are not widely recognized, but they prove to be very useful for understanding changes in G_{tot} of systems in which chemical reactions are occurring.

2.5.7   COMPARISON OF $\overline{G}_i$ OF DIFFERENT SPECIES

Equation (2.54) provides the critical link between a thermodynamic analysis of a system and its chemical composition. Among other things, it indicates that the Gibbs energy *per mole* of i increases as the activity of i increases. Put another way, the larger the activity of i in a system, the more energy that is required to put yet more i into it. Thus, it takes more energy to increase the concentration of a dissolved solute from 0.30 to 0.31 M than to increase it from 0.10 to 0.11 M. Applied to a gas phase, the equation indicates that more energy is required to increase the pressure by a given amount if the initial pressure is high than if it is low. The consequences of this result are obvious to anyone who has ever inflated a bicycle tire.

At 25°C, the product RT equals 2.48 kJ/mol. Substituting this value into Equation (2.54b) and replacing $\ln a_i$ by $2.303 \log a_i$, we can write the molar Gibbs energy of any species at 25°C as

$$\overline{G}_i = \overline{G}_i^{\circ} + (5.71 \text{ kJ/mol}) \log a_i \qquad \text{(2.73)}$$

A plot of Equation (2.73) applied to any element is shown in Figure 2.5. The relationship is the same for all elements, because all elements have the same value of $\overline{G}_i$ under standard conditions ($\overline{G}_i^{\circ} = 0$ for all elements) and they all obey the same relationship between $\overline{G}_i$ and activity for non-standard conditions [Equation (2.73)].

Figure 2.6 shows the relationship between the molar Gibbs energy of a few compounds and their activity at 25°C. Each compound i has a unique, non-zero value of $\overline{G}_i^{\circ}$, corresponding to the amount of chemical potential energy that is acquired (if $\overline{G}_i^{\circ} > 0$) or released (if $\overline{G}_i^{\circ} < 0$) when the compound is formed from the constituent elements, with all the chemicals in their standard state. However, the *change* in $\overline{G}_i$ for a given change in the logarithm of the activity (as indicated by the slope of the lines) is the same for all elements and compounds, because the molar Gibbs energy of all species depends on activity in the same way, as given by Equation (2.73).

Many of the ideas presented above are synthesized schematically in Figure 2.7 for a system containing elemental hydrogen, oxygen, and carbon, along with some

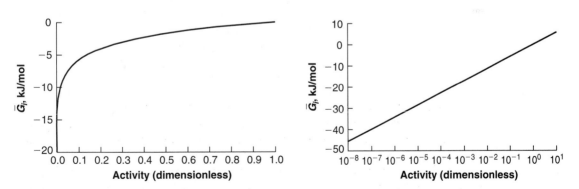

Figure 2.5   The molar Gibbs energy of elements at 25°C, when present in the same phase that they are in under standard conditions (e.g., gas for O_2, H_2, N_2, etc.; liquid for Hg; solid for Fe, Cu, Zn, etc.) The two graphs show the same relationship, the one on the left with an arithmetic scale for the abscissa and the one on the right with a logarithmic scale. In accord with Equation (2.54), $\overline{G}_i$ increases by 5.71 kJ/mol for each factor-of-10 increase in a_i (for $T = 25$°C).

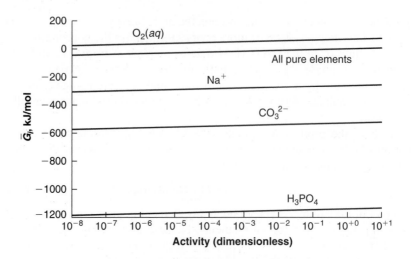

Figure 2.6   The molar Gibbs energy of various compounds at 25°C. The abscissa is the same as in Figure 2.5b, but the ordinate is compressed in order to show the data for compounds with widely varying $\bar{G}_i^\circ$ values (as indicated by the values of $\bar{G}_i$ corresponding to an activity of 1.0). The relationship for elements (from Figure 2.5) is shown for reference. For all species, $\bar{G}_i$ increases by 5.71 kJ/mol for each factor-of-10 increase in a_i.

of the compounds that can be formed from these elements. In the diagram, the standard molar Gibbs energy of $H_2(g)$, C(graphite), and $O_2(g)$ under standard conditions is shown as zero, in accord with convention. When these elements in their standard states combine to form $CO_2(g)$, $H_2O(l)$, or $H_2CO_3(aq)$ in *their* standard states, a substantial amount of Gibbs energy is released, so the standard molar Gibbs energy of these compounds is, in all cases, less than zero. These changes are designated by the arrows labeled R1, R2, and R3 on the figure, respectively.

Reducing the activity of $CO_2(g)$ lowers its molar Gibbs energy. In the diagram, the change in Gibbs energy when the $CO_2(g)$ partial pressure is reduced from its standard state value (1.0 bar) to its normal atmospheric value ($10^{-3.46}$ bar) is shown by arrow R4. The magnitude of the change is $RT \ln 10^{-3.46}$ or, at 25°C, -19.71 kJ/mol.

Next, consider the dissolution of $CO_2(g)$ at a partial pressure of 1.0 bar into water containing H_2CO_3 in its standard state (i.e., with an activity of 1.0). The total Gibbs energy of one mole of water plus one mole of $CO(g)$, each in its standard state, is $-237.18 + (-394.37)$, or -631.55 kJ. This summation is shown by the dashed lines in the figure. An input of 8.55 kJ of chemical potential energy is required to convert these compounds to one mole of H_2CO_3 in its standard state (arrow R5), so the standard molar Gibbs energy of formation of H_2CO_3 is $-631.55 + 8.55$, or -623.20 kJ/mol.

At H_2CO_3 activities other than 1.0, $\bar{G}_{H_2CO_3}$ can be computed as $\bar{G}_{H_2CO_3(aq)}^\circ + RT \ln \{H_2CO_3\}$. Values of $\bar{G}_{H_2CO_3(aq)}$ for H_2CO_3 activities of $10^{-3.0}$ and $10^{-4.94}$ are

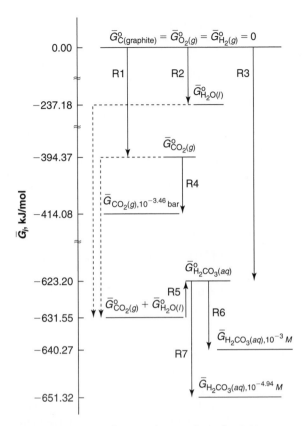

Figure 2.7 Schematic showing the molar Gibbs energy of formation of C(graphite), $O_2(g)$, $H_2(g)$, $H_2O(l)$, and $H_2CO_3(aq)$ under various standard and non-standard state conditions. Note that the Gibbs energy scale is discontinuous at three places.

shown on the diagram; the corresponding dilution process, starting with a solution containing H_2CO_3 in its standard state, is designated by arrows R6 and R7, respectively.

Example 2.6

The standard molar Gibbs energy of liquid water is -237.18 kJ/mol, and the partial pressures of $H_2(g)$ and $O_2(g)$ in the atmosphere are 6×10^{-7} and 0.21 bar, respectively. Compute the change in Gibbs energy when 2 moles of $H_2(g)$ combines with 1 mole of $O_2(g)$, each at its atmospheric pressure, to form 1 mole of pure liquid water.

Solution

The molar Gibbs energy of hydrogen and oxygen at their normal atmospheric pressures can be computed from Equation (2.73). Since the standard Gibbs energy of each of these gases is zero and the activity of each is its partial pressure,

$$\bar{G}_{H_2(g)} = \bar{G}_i^{\circ} + (5.71 \text{ kJ/mol}) \log (6 \times 10^{-7}) = -35.53 \text{ kJ/mol}$$

$$\bar{G}_{O_2(g)} = \bar{G}_i^{\circ} + (5.71 \text{ kJ/mol}) \log 0.21 = -3.87 \text{ kJ/mol}$$

The water that is formed is in its standard state, so its molar Gibbs energy in the system of interest equals its standard molar Gibbs energy. Therefore, when 2 moles of $H_2(g)$ combines with 1 mole of $O_2(g)$, the total Gibbs energy of the reactants is $2(-35.53) + 1(-3.87)$, or -74.93 kJ, and the Gibbs energy of the product is -237.18 kJ. The difference between these values, equal to 162.25 kJ, is the amount of Gibbs energy that is released when the reaction occurs.

2.6   INTERPRETATION OF CHEMICAL PROCESSES IN TERMS OF CHEMICAL POTENTIALS

2.6.1   THE DRIVING FORCE FOR CHEMICAL CHANGE AND THE GIBBS ENERGY OF REACTION

The discussion up to this point in the chapter has provided the underlying basis for understanding and quantifying chemical reactivity in terms of the Gibbs energy function. In particular, we have seen that, in the absence of other driving forces, chemical systems tend to undergo change in the direction of a downhill gradient in chemical potential energy, i.e., in G_{tot}. Correspondingly, we can consider the negative gradient in G_{tot} to be the driving force for chemical change in the system. For instance, as demonstrated in section 2.5.4, a spatial gradient in the chemical potential of a species can be viewed as a (virtual) force driving diffusion of that species from regions of higher to lower potential (i.e., higher to lower activity).

In systems where a chemical reaction can occur, we also expect that changes in the state of the system will be in the direction of decreasing G_{tot}. However, in these cases, the gradient of interest is the change in G_{tot} per unit movement along a coordinate axis describing the amount of reaction that has occurred. To develop such an axis, a variable called a **mole of stoichiometric reaction** can be defined such that, for the generic reaction, $aA + bB \leftrightarrow cC + dD$, 1 mole of stoichiometric reaction occurs when a moles of A plus b moles of B is converted to c moles of C plus d moles of D. For instance, for the following reaction, 1 mole of stoichiometric reaction corresponds to reaction of 2 moles of H_2 and 1 mole of O_2 to form 2 moles of H_2O:

$$2H_2(g) + O_2(g) \leftrightarrow 2H_2O(l)$$

Note that if the same reaction as above were written with the coefficients divided by 2, then 1 mol of stoichiometric reaction would correspond to reaction of 1 mol of H_2 and 0.5 mol of O_2 to form 1 mol of H_2O:

$$H_2(g) + \tfrac{1}{2} O_2(g) \leftrightarrow H_2O(l)$$

Therefore, when quantities are expressed in terms of moles of stoichiometric reaction, it is critical to specify the reaction completely, i.e., to specify not only the identities of the reactants and products, but also their stoichiometric coefficients.

The number of moles of stoichiometric reaction that occur in a given system is represented as $\Delta\eta$. Thus, for the water formation reaction written in the first way shown above, conversion of 0.1 mole of H_2 and 0.05 mole of O_2 to 0.1 mole of H_2O corresponds to 0.05 mol of stoichiometric reaction; i.e., $\Delta\eta = 0.05$ mole. When we extend this relationship to the generic reaction, we see that whenever a reaction proceeds by an amount $\Delta\eta$, the changes in the number of moles of A, B, C, and D in the system are $\Delta n_A = -a\,\Delta\eta$, $\Delta n_B = -b\,\Delta\eta$, $\Delta n_C = +c\,\Delta\eta$, and $\Delta n_D = +d\,\Delta\eta$, or in general, $\Delta n_i = \nu_i\,\Delta\eta$, where ν_i is the stoichiometric coefficient of i in the reaction of interest (positive for products, negative for reactants).

Now consider a system in which A, B, C, and D are each present and a differential amount of reactants is converted to products. Because the activity of each of the constituents remains virtually constant, each of the $\overline{G}_i$ values changes negligibly. The values of G_{tot} before and after the reaction (designated as states 1 and 2, respectively) and the differential change in Gibbs energy of the system (dG_{tot}) can therefore be written as follows:

$$G_{tot,1} = n_{A,1}\overline{G}_A + n_{B,1}\overline{G}_B + n_{C,1}\overline{G}_C + n_{D,1}\overline{G}_D$$
$$G_{tot,2} = n_{A,2}\overline{G}_A + n_{B,2}\overline{G}_B + n_{C,2}\overline{G}_C + n_{D,2}\overline{G}_D \qquad \textbf{(2.74)}$$
$$dG_{tot} = dn_A\,\overline{G}_A + dn_B\,\overline{G}_B + dn_C\,\overline{G}_C + dn_D\,\overline{G}_D$$

Substituting $\nu_i\,d\eta$ for each dn_i term in Equation (2.74) and then dividing through by $d\eta$ yields an expression for dG_{tot} normalized to the amount of stoichiometric reaction that has occurred:

$$dG_{tot} = (-a\,d\eta)\overline{G}_A + (-b\,d\eta)\overline{G}_B + (c\,d\eta)\overline{G}_C + (d\,d\eta)\overline{G}_D \qquad \textbf{(2.75)}$$

$$\Delta\overline{G}_r \equiv \frac{dG_{tot}}{d\eta} = -a\overline{G}_A - b\overline{G}_B + c\overline{G}_C + d\overline{G}_D \qquad \textbf{(2.76)}$$

or, more succinctly,

$$\Delta\overline{G}_r = \sum_i \nu_i\overline{G}_i \qquad \textbf{(2.77)}$$

The parameter $\Delta\overline{G}_r$ is called the **molar Gibbs energy of reaction.** As the name implies, $\Delta\overline{G}_r$ is the Gibbs energy change *per mole of (stoichiometric) reaction.* The definition of $\Delta\overline{G}_r$ as $dG_{tot}/d\eta$ indicates that it is the parameter we sought as an indicator of the chemical driving force for a reaction. That is, $\Delta\overline{G}_r$ is the gradient in the chemical potential energy of the system along a coordinate that measures the progress of the reaction.

If $\Delta\overline{G}_r$ is evaluated under conditions where all the reactants and products are in their standard states, it is called the **standard molar Gibbs energy of**

reaction, represented as $\Delta \bar{G}_r^\circ$:

$$\Delta \bar{G}_r^\circ = -a\bar{G}_A^\circ - b\bar{G}_B^\circ + c\bar{G}_C^\circ + d\bar{G}_D^\circ \qquad \textbf{(2.78a)}$$

$$\Delta \bar{G}_r^\circ = \sum_i \nu_i \bar{G}_i^\circ \qquad \textbf{(2.78b)}$$

Because $\bar{G}_i^\circ$ values for pure elements are all defined to be zero, and those for many compounds have been determined experimentally, the value of $\Delta \bar{G}_r^\circ$ for many reactions of interest can be computed from tabulated values.

Using Equation (2.54b) to expand the $\bar{G}_i$ term for each constituent in Equation (2.76) and substituting the definition of $\Delta \bar{G}_r^\circ$ into the result, we can rewrite Equation (2.76) as follows:

$$\Delta \bar{G}_r = -a(\bar{G}_A^\circ + RT\ln\{A\}) - b(\bar{G}_B^\circ + RT\ln\{B\})$$
$$+ c(\bar{G}_C^\circ + RT\ln\{C\}) + d(\bar{G}_D^\circ + RT\ln\{D\}) \qquad \textbf{(2.79a)}$$

$$\Delta \bar{G}_r = (-a\bar{G}_A^\circ - b\bar{G}_B^\circ + c\bar{G}_C^\circ + d\bar{G}_D^\circ) + RT(-a\ln\{A\}$$
$$- b\ln\{B\} + c\ln\{C\} + d\ln\{D\}) \qquad \textbf{(2.79b)}$$

$$\Delta \bar{G}_r = \Delta \bar{G}_r^\circ + RT\ln\frac{\{C\}^c\{D\}^d}{\{A\}^a\{B\}^b} \qquad \textbf{(2.79c)}$$

Finally, recalling that the ratio $\{C\}^c\{D\}^d/[\{A\}^a\{B\}^b]$ was defined in Chapter 1 as the reaction quotient Q, we have

$$\Delta \bar{G}_r = \Delta \bar{G}_r^\circ + RT\ln Q \qquad \textbf{(2.80)}$$

Equation (2.80) expresses the molar Gibbs energy of reaction in terms of a constant ($\Delta \bar{G}_r^\circ$) that is specific to the reaction but independent of the reaction conditions and a variable (Q) that describes those conditions. Most importantly, it allows us to calculate whether a differential amount of conversion of reactants to products will increase or decrease the total Gibbs energy of the system; i.e., it tells us whether the reaction is favorable under the given conditions. Equation (2.80) can be written using base-10 logarithms as

$$\Delta \bar{G}_r = \Delta \bar{G}_r^\circ + RT\ln Q = \Delta \bar{G}_r^\circ + 2.303RT\log Q \qquad \textbf{(2.81)}$$

or, at 25°C,

$$\Delta \bar{G}_r = \Delta \bar{G}_r^\circ + (5.71 \text{ kJ/mol})\log Q \qquad \textbf{(2.82)}$$

Example 2.7 | Compute the total Gibbs energy change in a system in which 15 μmol of atmospheric CO_2 combines with water to form 30 μmol of H^+ and 15 μmol of CO_3^{2-}, which are present at concentrations of 10^{-8} and 10^{-7} M, respectively, in a system at 25°C. The reaction and the standard Gibbs energies of formation of the constituents are given below. The partial pressure of $CO_2(g)$ in the atmosphere is 3×10^{-4} bar. Assume that the volume of solution is large enough that dissolution of this amount of CO_2 does not change the concentrations of dissolved H^+ and CO_3^{2-} significantly, and that all activity coefficients are approximately 1.0. Is the reaction thermodynamically favorable?

$$CO_2(g) \quad + \quad H_2O(l) \quad \rightarrow \quad 2H^+ \quad + \quad CO_3^{2-}$$

3×10^{-4} bar   nearly pure 10^{-8} mol/L 10^{-7} mol/L

Species	$\bar{G}_i^{\circ}$ kJ/mol
$CO_2(g)$	-394.37
$H_2O(l)$	-237.18
CO_3^{2-}	-527.9
H^+	0.00

Solution

The molar Gibbs energy of reaction can be computed using Equation (2.82). To use this equation, we first compute the standard Gibbs energy of reaction ($\Delta \bar{G}_r^{\circ}$) from Equation (2.78):

$$\Delta \bar{G}_r^{\circ} = 2(0 \text{ kJ/mol}) + 1(-527.9 \text{ kJ/mol}) - 1(-394.37 \text{ kJ/mol}) - 1(-237.18 \text{ kJ/mol})$$

$$= 103.65 \text{ kJ/mol}$$

Then, using Equation (2.82),

$$\Delta \bar{G}_r = \Delta \bar{G}_r^{\circ} + RT \ln Q$$

$$= 103.65 \frac{\text{kJ}}{\text{mol}} + \left(5.71 \frac{\text{kJ}}{\text{mol}} \right) \log \frac{(10^{-7})(10^{-8})^2}{(3 \times 10^{-4})(1.0)}$$

$$= -7.56 \text{ kJ/mol}$$

For the given stoichiometry, reaction of 15 μmol of $CO_2(g)$ corresponds to 15 μmol of stoichiometric reaction, i.e., $\Delta \eta = 15$ μmol. Therefore, the overall Gibbs energy change of interest is computed as

$$\Delta G_r = (\Delta \bar{G}_r)(\Delta \eta) = (-7.56 \text{ kJ/mol})(15 \, \mu\text{mol}) = -1.14 \times 10^{-4} \text{kJ}$$

Since the Gibbs energy change is negative, the reaction decreases the chemical potential energy of the system, so it is favorable.

2.6.2   THE RELATIONSHIP OF THE GIBBS ENERGY OF REACTION TO THE EQUILIBRIUM CONSTANT

In Chapter 1, we derived the equilibrium constant for a reaction as the ratio of the chemical activities of the products to those of the reactants in a system that has reached equilibrium. The derivation was based on reaction kinetics expressions and the recognition that equilibrium corresponds to the condition where the forward and reverse reactions proceed at equal rates. The principle that reactions always proceed in the direction that minimizes G_{tot} in the system provides a vehicle to derive the same equilibrium constant from thermodynamic considerations, as follows.

As shown in Equation (2.75), for the reaction $aA + bB \leftrightarrow cC + dD$, the change in Gibbs energy accompanying a differential amount of stoichiometric reaction can be expressed by

$$dG_{tot} = \sum_i \bar{G}_i \, dn_i = -a\bar{G}_A \, d\eta - b\bar{G}_B \, d\eta + c\bar{G}_C \, d\eta + d\bar{G}_D \, d\eta$$

We expect the reaction to proceed in the direction that decreases G_{tot} and to be at equilibrium when G_{tot} is at a minimum. The mathematical condition for G_{tot} to be minimized is that $dG_{tot}/d\eta = 0$.[12] Since $dG_{tot}/d\eta$ is, by definition, $\Delta\bar{G}_r$, at equilibrium

$$\left.\frac{dG_{tot}}{d\eta}\right|_{eq} = \Delta\bar{G}_r|_{eq} = 0 \qquad (2.83)$$

According to Equation (2.79b), the molar Gibbs energy of any reaction can be computed as

$$\Delta\bar{G}_r = \Delta\bar{G}_r^\circ + RT \ln Q \qquad (2.79b)$$

Inserting the information that $\Delta\bar{G}_r = 0$ for a system at equilibrium [Equation (2.83)] into Equation (2.79b), we obtain

$$0 = \Delta\bar{G}_r^\circ + RT \ln Q|_{eq} \qquad (2.84)$$

$$\Delta\bar{G}_r^\circ = -RT \ln Q|_{eq} \qquad (2.85)$$

The value of Q at equilibrium is, by definition, the equilibrium constant K_{eq}. Thus, the equilibrium constant can be related to thermodynamic data by

$$\Delta\bar{G}_r^\circ = -RT \ln K_{eq} \qquad (2.86a)$$

$$K_{eq} = \exp\left(-\frac{\Delta\bar{G}_r^\circ}{RT}\right) \qquad (2.86b)$$

By converting the natural logarithm to base 10, substituting the numerical value of R, and assuming $T = 25°C$, Equation (2.86a) becomes

$$\Delta\bar{G}_r^\circ = -(5.71 \text{ kJ/mol}) \log K_{eq} \qquad (2.87)$$

Equations (2.86) and (2.87) establish a link between the thermodynamics of a reaction and its equilibrium constant. Note that whereas the definition of K_{eq} developed in Chapter 1 requires experimental study of the specific reaction of interest, the corresponding thermodynamic equations allow K_{eq} to be evaluated for any hypothetical reaction for which all the $\bar{G}_i^\circ$ values are known (thereby allowing calculation of $\Delta\bar{G}_r^\circ$). This possibility is often of critical importance, since

---

[12]The condition of minimum G_{tot} could also be found by differentiation with respect to the number of moles of one of the species, say, $dG_{tot}/dn_A = 0$, or $dG_{tot}/dn_B = 0$, etc. Substituting $dn_i = \nu_i \, d\eta$ into any of these equations leads to Equation (2.83).

many reactions of interest are extremely difficult to study experimentally under equilibrium conditions, due to the slow approach to equilibrium and/or difficulties analyzing for all the reactants and products at the accuracy required.

One approach for removing dissolved nickel from solution is to precipitate it as $NiS(s)$. The standard molar Gibbs energies of Ni^{2+}, $NiS(s)$, and HS^- are -45.6, -86.2, and 12.0 kJ/mol, respectively. Use the given thermodynamic data to compute the equilibrium constant for the following reaction: **Example 2.8**

$$Ni^{2+} + HS^- \leftrightarrow NiS(s) + H^+$$

Solution

The standard molar Gibbs energy of H^+ is zero by definition, so the standard Gibbs energy of the given reaction is $[-86.2 + 0.0 - (-45.6) - 12.0]$ kJ/mol, or -52.6 kJ/mol. Using Equation (2.87), the equilibrium constant for the reaction is

$$\log K_{eq} = \frac{\Delta \overline{G}_r^o}{5.71 \text{ kJ/mol}} = -9.2 \qquad K_{eq} = 10^{-9.2}$$

Equation (2.87) also provides the basis for calculating $\overline{G}_i^o$ from the equilibrium constant for any reaction involving i if the molar Gibbs energies of formation of all the j other species in the reaction ($\overline{G}_{j \neq i}^o$) are known. For instance, if $\overline{G}_{H_2CO_3(aq)}^o$ were known from prior investigations, $\overline{G}_{HCO_3^-}^o$ could be found based on the experimentally determined equilibrium constant for the following reaction.

$$H_2CO_3 \leftrightarrow H^+ + HCO_3^-$$
$$\Delta \overline{G}_r^o = -RT \ln K_{eq} = \overline{G}_{HCO_3^-}^o + \overline{G}_{H^+}^o - \overline{G}_{H_2CO_3}^o$$

Since $\overline{G}_{H^+}^o \equiv 0$, we could compute $\overline{G}_{HCO_3^-}^o$ by

$$\overline{G}_{HCO_3^-}^o = \overline{G}_{H_2CO_3}^o - RT \ln K_{eq}$$

When air at 25°C, containing $O_2(g)$ at a partial pressure of 0.21 bar, equilibrates with pure water, the equilibrium concentration of dissolved oxygen ($O_2(aq)$) is 9.31 mg/L (2.91×10^{-4} M). Assuming that $\gamma_{O_2(aq)}$ is 1.0, compute $\overline{G}_{O_2(aq)}^o$. **Example 2.9**

Solution

Based on the given information, the equilibrium constant for the reaction $O_2(g) \leftrightarrow O_2(aq)$ is

$$K_{eq} = \frac{\{O_2(aq)\}}{\{O_2(g)\}} = \frac{2.91 \times 10^{-4}}{0.21} = 1.39 \times 10^{-3}$$

Using Equation (2.86), we can compute $\Delta \overline{G}_r^o$ for the reaction to be $+16.45$ kJ/mol. Since $\Delta \overline{G}_r^o = \nu_i \overline{G}_i^o = \overline{G}_{O_2(aq)}^o - \overline{G}_{O_2(g)}^o$, and $\overline{G}_{O_2(g)}^o$ is zero by definition, $\overline{G}_{O_2(aq)}^o$ can be equated with $\Delta \overline{G}_r^o$; i.e., $\overline{G}_{O_2(aq)}^o = +16.32$ kJ/mol.

2.6.3 THE GIBBS ENERGY OF REACTION UNDER NON-EQUILIBRIUM CONDITIONS

By substituting Equation (2.86a) into Equation (2.79), we can obtain the following expression for the molar Gibbs energy of any reaction as a function of the ratio of Q to K_{eq}, for the reaction, i.e., of the extent of disequilibrium characterizing the reaction:

$$\Delta \overline{G}_r = \Delta \overline{G}_r^\circ + RT \ln Q = -RT \ln K_{eq} + RT \ln Q = RT \ln \frac{Q}{K_{eq}} \qquad \textbf{(2.88)}$$

and, at 25°C,

$$\Delta \overline{G}_r = \left(5.71 \frac{kJ}{mol} \right) \log \frac{Q}{K_{eq}} \qquad \textbf{(2.89)}$$

Equation (2.89) indicates that if $Q > K$, then $\Delta \overline{G}_r$ is greater than zero. In such a case, converting reactants to products increases G_{tot} in the system, so the reaction would be thermodynamically unfavorable. In fact, the reverse reaction (products to reactants) would proceed spontaneously. On the other hand, if $Q < K$, then $\Delta \overline{G}_r$ is less than zero and the reaction proceeds to form products from reactants. If $Q = K$, then $\Delta \overline{G}_r$ equals zero and the reaction is at equilibrium. These relationships are summarized in Table 2.2.

It is worth noting that all the relationships shown in Table 2.2 have exact analogs in terms of electrochemical parameters and that the relationships *must* be evaluated in those terms if an electrical field is imposed on the system from outside the solution phase. In particular, if Ψ_{ext} is non-zero, the criterion for equilibrium is that $\Delta \overline{\overline{PE}}_r = 0$, not that $\Delta \overline{G}_r = 0$. Attempting to determine the

Table 2.2   Summary of relationships between Gibbs energy of reaction and system condition[1]

Relationship between standard Gibbs energy of reaction and the equilibrium constant:

$$\Delta \overline{G}_r^\circ = -RT \ln K_{eq}$$

Expressions for the Gibbs energy of reaction under equilibrium or non-equilibrium conditions:

$$\Delta \overline{G}_r = \Delta \overline{G}_r^\circ + RT \ln Q$$

$$\Delta \overline{G}_r = -RT \ln K_{eq} + RT \ln Q$$

$$\Delta \overline{G}_r = RT \ln \frac{Q}{K_{eq}}$$

Relationship between sign of Gibbs energy change and direction of reaction:

$\Delta G_r < 0$: Reaction proceeds to form products

$\Delta G_r > 0$: Reaction proceeds to form reactants

$\Delta G_r = 0$: Reaction is at equilibrium

---

[1]Assuming $\Psi_{ext} = 0$.

equilibrium condition using the $\Delta \bar{G}_r = 0$ criterion in such a case ignores one or more of the contributions to the total potential energy of the system and leads to an incorrect result. Such an approach is analogous to trying to determine the equilibrium location of the charged particle in Example 2.2 by considering gravity but ignoring the electrical field that affects the particle's behavior.

In Chapters 3 through 8, we will use $\Delta \bar{G}_r = 0$ as the criterion for equilibrium, implicitly assuming that the reactions are taking place in bulk solution with $\Psi_{ext} = 0$. However, it is useful to keep this assumption in mind while reading the chapters, so that when the assumption is abandoned in later chapters, that change can be placed in context.

Example 2.10

Historically, chromium (Cr) has been used in numerous industrial processes, including leather tanning and metal plating operations. In these applications, if the water is mildly acidic, the Cr is typically present as bichromate ($HCrO_4^-$) and/or dichromate ($Cr_2O_7^{2-}$) ions. These ions can be interconverted by the reaction

$$2HCrO_4^- \leftrightarrow Cr_2O_7^{2-} + H_2O$$

The standard Gibbs energies of formation of $HCrO_4^-$, $Cr_2O_7^{2-}$, and H_2O are -764.8, -1301, and -237.18 kJ/mol, respectively.

a. Compute the equilibrium constant for the conversion of $HCrO_4^-$ to $Cr_2O_4^{2-}$.

b. Compute G_{tot} as a function of the relative amounts of $HCrO_4^-$ and $Cr_2O_7^{2-}$ in a system containing 10^{-1} M total dissolved chromium ($c_{Cr,tot}$). Compare the condition of minimum chemical potential energy with the equilibrium condition as specified by the equilibrium constant. Assume all activity coefficients are 1.0.

c. Compute $\Delta \bar{G}_r$ as a function of the $HCrO_4^-$ concentration for the solution described in part (b).

Solution

a. The reaction of interest is $2HCrO_4^- \leftrightarrow Cr_2O_7^{2-} + H_2O$, for which the standard Gibbs energy of reaction is

$$\Delta \bar{G}_r^o = \bar{G}_{Cr_2O_7^{2-}}^o + \bar{G}_{H_2O}^o - 2\bar{G}_{HCrO_4^-}^o$$

$$\Delta \bar{G}_r = [-1301 + (-237.18) - 2(-764.8)] \text{ kJ/mol} = -8.58 \text{ kJ/mol}$$

Applying Equation (2.87), we obtain

$$\log K_{eq} = -\frac{-8.58 \text{ kJ/mol}}{5.71 \text{ kJ/mol}} = +1.51$$

$$K_{eq} = 32.5$$

b. The total Gibbs energy of the solution is the sum of the Gibbs energy contributions of the various constituents:

$$G_{tot} = n_{H_2O}(\bar{G}_{H_2O}^o + RT \ln a_{H_2O}) + n_{HCrO_4^-}(\bar{G}_{HCrO_4^-}^o + RT \ln a_{HCrO_4^-})$$
$$+ n_{Cr_2O_7^{2-}}(\bar{G}_{Cr_2O_7^{2-}}^o + RT \ln a_{Cr_2O_7^{2-}}) \qquad \textbf{(2.90)}$$

where n_i is the number of moles of i in the system. Assuming that the system volume (V) is independent of the distribution of Cr, Vc_i can be substituted for n_i for each species. Substituting molar concentrations for the activities of the chromium species, and setting the activity of water equal to its mole fraction ($a_{H_2O} = c_{H_2O}/\sum c_i$), we obtain:[13]

$$G_{tot} = Vc_{H_2O}\left(\bar{G}^o_{H_2O} + RT \ln \frac{c_{H_2O}}{\sum c_i}\right) + Vc_{HCrO_4^-}(\bar{G}^o_{HCrO_4^-} + RT \ln c_{HCrO_4^-})$$
$$+ Vc_{Cr_2O_7^{2-}}(\bar{G}^o_{Cr_2O_7^{2-}} + RT \ln c_{Cr_2O_7^{2-}}) \qquad (2.91)$$

$$\frac{G_{tot}}{V} = c_{H_2O}\left(\bar{G}^o_{H_2O} + RT \ln \frac{c_{H_2O}}{\sum c_i}\right) + c_{HCrO_4^-}(\bar{G}^o_{HCrO_4^-} + RT \ln c_{HCrO_4^-})$$
$$+ c_{Cr_2O_7^{2-}}(\bar{G}^o_{Cr_2O_7^{2-}} + RT \ln c_{Cr_2O_7^{2-}}) \qquad (2.92)$$

The concentrations of $HCrO_4^-$ and $Cr_2O_7^{2-}$ are related to each other by stoichiometry. Furthermore, since 1 L of solution contains 55.56 mol of water, the mole fraction of water can be computed as a function of the concentration of either of the other species:

$$c_{Cr_2O_7^{2-}} = 0.5(c_{Cr,tot} - c_{HCrO_4^-}) = 0.5(0.10 - c_{HCrO_4^-}) \qquad (2.93)$$

$$a_{H_2O} = \frac{55.56}{55.56 + c_{HCrO_4^-} + c_{CrO_7^{2-}}} = \frac{55.56}{55.56 + c_{HCrO_4^-} + 0.5(0.10 - c_{HCrO_4^-})}$$

$$= \frac{55.56}{55.56 + 0.05 + 0.5c_{HCrO_4^-}} \qquad (2.94)$$

Substitution of Equations (2.93) and (2.94) into (2.92) yields a long but not complex expression for G_{tot} per unit volume of solution in terms of a single variable, $c_{HCrO_4^-}$. The exact value of $c_{HCrO_4^-}$ where G_{tot}/V is minimized could be found by differentiation with respect to $c_{HCrO_4^-}$. A plot of G_{tot}/V versus $c_{HCrO_4^-}$ (Figure 2.8) indicates that the point of minimum Gibbs energy in the system (and hence the equilibrium condition) is at $c_{HCrO_4^-} = 0.0320$ mol/L, corresponding to $c_{Cr_2O_7^{2-}} = 0.034$ mol/L. The activity (mole fraction) of water under these conditions is close to 0.999, so the activity quotient for the reaction is

$$Q = \frac{\{Cr_2O_7^{2-}\}\{H_2O\}}{\{HCrO_4^-\}^2} = \frac{0.034(0.999)}{(0.032)^2} = 33.2 \qquad \ln Q = 3.50$$

The computed value of Q is essentially identical to the value of K_{eq} computed in part (a). That is, the activity ratio indicates that when the total chemical potential energy of the system is minimized, $Q = K$, i.e., the reaction is at equilibrium. (The small difference between the computed values of Q and K is due to round-off error.)

---

[13]Although the activity of water is very near 1.0 regardless of the distribution of Cr species, it does vary slightly when some of the Cr converts from one form to another, because the total number of moles in the system changes. While the variation in a_{H_2O} is very small, the term containing a_{H_2O} in Equation (2.90) is multiplied by a much bigger number (55.56) than are the other terms in the equation, and the small variation in a_{H_2O} turns out to have a significant effect on G_{tot} in the system.

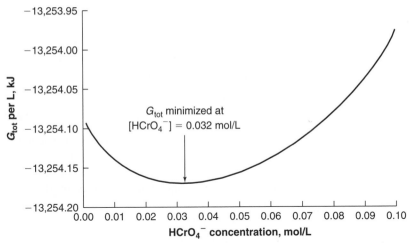

Figure 2.8   Total Gibbs energy per liter of a solution containing $HCrO_4^-$ and $Cr_2O_7^{2-}$ at a total Cr concentration of 10^{-1} mol/L, as a function of the $HCrO_4^-$ concentration.

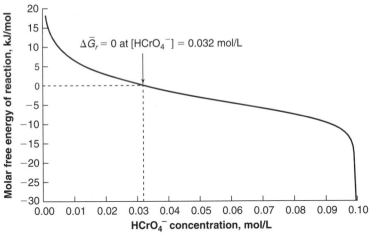

Figure 2.9   The molar Gibbs energy of reaction for conversion of $HCrO_4^-$ to $Cr_2O_7^{2-}$ in a system containing 0.1 M total Cr.

c. Figure 2.9 shows $\Delta \overline{G}_r$ as a function of $c_{HCrO_4^-}$. Comparison of Figures 2.8 and 2.9 makes it clear that the condition of minimum total Gibbs energy corresponds exactly with the condition $\Delta \overline{G}_r = 0$.

The relative amounts of reactants and products of a reaction in a specified system are sometimes represented in terms of a coordinate ε, called the **extent of reaction.** This coordinate has a value of 0.0 if the reaction has proceeded to the left (toward reactants) until one of the products is completely depleted, and it has a value of 1.0 if the reaction has proceeded to the right (toward products) until one of the reactants is completely depleted.[14] Since a change in ε represents some conversion of reactants to products (or vice versa), any change $\Delta\varepsilon$ can also be represented as a corresponding number of moles of stoichiometric reaction, i.e.,

$$\Delta\varepsilon = k\,\Delta\eta \qquad d\varepsilon = k\,d\eta \tag{2.95}$$

where the proportionality factor k depends on the total amounts of reactants and products in the system. The qualitative difference between η and ε is that η is a generic term that applies to a given reaction in any system, whereas ε is defined for a specific system containing known total concentrations of the various constituents.

According to Equation (2.83), equilibrium is the condition in which $dG_{tot}/d\eta = 0$. Since, according to Equation (2.95), any change $d\varepsilon$ is proportional to $d\eta$, the condition where $dG_{tot}/d\eta = 0$ must also be a condition where $dG_{tot}/d\varepsilon = 0$. The Gibbs energy changes in a system and the state of the system relative to its equilibrium state can therefore be expressed as a function of ε. For instance, in the example above describing the Cr distribution, Figures 2.8 and 2.9 could be redrawn as functions of ε simply by redefining the abscissa such that an $HCrO_4^-$ concentration of 0 mol/L was assigned a value of $\varepsilon = 0$, and an $HCrO_4^-$ concentration of 0.1 mol/L was assigned a value of $\varepsilon = 1$.

Example 2.11 | The initial concentrations of HOCl and OCl^- in a solution are $10^{-3}\,M$ and $10^{-4}\,M$, respectively, and the value of $\{H^+\}$ in the solution is fixed at 10^{-7}. Given the following thermodynamic data, plot $\Delta\bar{G}_r$ versus ε in the system for the dissociation of hypochlorous acid (HOCl) according to the following reaction. Assume that $\gamma = 1.0$ for all constituents. What are the initial and equilibrium values of ε?

$$HOCl \leftrightarrow H^+ + OCl^-$$

$$\bar{G}^\circ_{H^+} = 0.00\ \text{kJ/mol}$$

$$\bar{G}^\circ_{HOCl} = -79.9\ \text{kJ/mol}$$

$$\bar{G}^\circ_{OCl^-} = -36.8\ \text{kJ/mol}$$

---

[14]The conditions $\varepsilon = 0$ and $\varepsilon = 1.0$ correspond to the maximum extent that the reaction can proceed toward reactants and products, respectively. In some circumstances, certain constituents are assumed to be replenished as they are depleted and therefore are not considered as potentially limiting reactants. For instance, H^+ and OH^- are usually assumed to be available in virtually infinite quantities by dissociation of water ($H_2O \rightarrow H^+ + OH^-$), so the determination of the condition where $\varepsilon = 0$ or $\varepsilon = 1.0$ does not consider the possibility that all the H^+ or all the OH^- can be consumed. In other circumstances, one might assume that another constituent is replenished, depending on the specifics of the system of interest; e.g., oxygen would not be a potentially limiting reactant in a reaction if, in that system, oxygen could enter the solution from the atmosphere.

Solution

For the reaction of interest, $\Delta \overline{G}_r^{\,\circ}$ is

$$\Delta \overline{G}_r^{\,\circ} = \overline{G}_{H^+}^{\circ} + \overline{G}_{OCl^-}^{\circ} - \overline{G}_{HOCl}^{\circ}$$
$$= 0 + (-36.8) - (-79.9)$$
$$= +43.1 \text{ kJ/mol}$$

The value of $\Delta \overline{G}_r$ under the specified, non-standard conditions can then be computed from Equation (2.79b) as

$$\Delta \overline{G}_r = 43.1 \frac{kJ}{mol} + \left(5.71\frac{kJ}{mol}\right) \log \frac{\{H^+\}\{OCl^-\}}{\{HOCl\}}$$

For this system, the potential limiting conditions are for all the OCl (1.1×10^{-3}) to be in the form of HOCl ($\varepsilon = 0.0$) or for all of it to be in the form of OCl$^-$ ($\varepsilon = 1.0$). Thus, $\varepsilon = \{OCl^-\}/1.1 \times 10^{-3}$, and $\varepsilon_{\text{initial}}$ is 0.091. Substituting the given, fixed value of $\{H^+\}$ into the expression for $\Delta \overline{G}_r$ and writing $\{OCl^-\}$ and $\{HOCl\}$ in terms of ε, we find

$$\Delta \overline{G}_r = 43.1 \frac{kJ}{mol} + \left(5.71\frac{kJ}{mol}\right) \log \frac{10^{-7.0} (1.1 \times 10^{-3} \varepsilon)}{(1.1 \times 10^{-3})(1 - \varepsilon)}$$

$$\Delta \overline{G}_r = 43.1 \frac{kJ}{mol} + \left(5.71\frac{kJ}{mol}\right) \log \left(10^{-7.0} \frac{\varepsilon}{1 - \varepsilon}\right)$$

A plot of $\Delta \overline{G}_r$ against ε (Figure 2.10) indicates that the equilibrium condition ($\Delta \overline{G}_r = 0$) is at $\varepsilon = 0.22$. Thus, as the solution equilibrates, ε increases from 0.091 to 0.22; i.e., HOCl dissociates to form additional H$^+$ and OCl$^-$. The fact that $\{H^+\}$ remains constant in the solution means that the H$^+$ generated by the dissociation reaction must be consumed by some other reaction taking place in solution. Approaches for maintaining fixed $\{H^+\}$ in such solutions are discussed in Chapter 5.

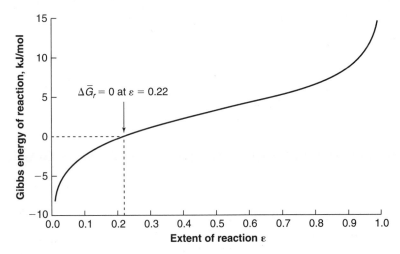

$\Delta \overline{G}_r = 0$ at $\varepsilon = 0.22$

Figure 2.10 The molar Gibbs energy of reaction for the conversion of HOCl to OCl$^-$ at a fixed $\{H^+\}$ of $10^{-7.0}$ as a function of the extent of conversion ε.

Note that in the above example $\Delta \bar{G}_r^\circ$ is a large positive number, indicating that under standard conditions, the driving force strongly favors formation of HOCl from H^+ and OCl^- (i.e., the *association* reaction, rather than the *dissociation*). However, under the conditions stated, the net driving force is in the opposite direction. Because the concentrations of most dissolved constituents of interest in environmental aquatic chemistry are orders of magnitude less than 1 mol/L, they are far from their standard state conditions. As a result, the standard molar Gibbs energy of reaction ($\Delta \bar{G}_r^\circ$) for a reaction involving these species is often very different from the molar Gibbs energy of reaction under the ambient conditions ($\Delta \bar{G}_r$), so the value of $\Delta \bar{G}_r^\circ$ is not a very good indicator of the direction in which the reaction will proceed.

One commonly encountered situation in environmental engineering where $\Delta \bar{G}_r^\circ$ is often an acceptable approximation of $\Delta \bar{G}_r$ is in the analysis of oxidation/reduction reactions, and in particular, reactions describing the aerobic biological degradation of organic substrates. The magnitude of $\Delta \bar{G}_r^\circ$ is often so huge for such reactions that adjustments to account for non-standard state conditions have only a small effect. In these cases, a fairly widespread practice has developed of approximating $\Delta \bar{G}_r$ under any conditions by its value in a system where all the reactants and products except H^+ are in their standard state, and the activity of H^+ is 10^{-7} (pH = 7.0). The $\Delta \bar{G}_r$ values computed in this way are often designated $\Delta \bar{G}_r^\circ$ (W), where the W stands for water (the idea being that the standard conditions have been adjusted to apply to pure water, which has a pH of 7.0).

Example 2.12

A water source contains 10 mg/L of dissolved organic carbon (DOC), of which approximately 20% is in compounds that are relatively easily biodegraded. The water also contains 8 mg/L dissolved oxygen ($O_2(aq)$) and 10^{-3} M HCO_3^-, and it is at pH 7.5. Estimate the molar Gibbs energy that would be available to microorganisms that consume the easily degraded compounds by the reaction shown below. Assume that the microorganisms are 100% efficient in their metabolism, and that the compounds can be treated as though they were glucose ($C_6H_{12}O_6$). Compare the result with $\Delta \bar{G}_r^\circ$(W). Standard molar Gibbs energies for glucose, $O_2(aq)$, and HCO_3^- are -908.01, 16.32, and -586.8 kJ/mol, respectively.

$$C_6H_{12}O_6 + 6O_2(aq) \rightarrow 6HCO_3^- + 6H^+$$

Solution

The standard molar Gibbs energy of H^+ is zero by definition, so the standard Gibbs energy of the reaction can be computed as

$$\Delta \bar{G}_r^\circ = [6(-586.8) + 6(0) - 1(-908.01) - 6(16.40)] \text{ kJ/mol}$$

$$= -2710.7 \text{ kJ/mol}$$

The molar concentration of oxygen in the solution is

$$[O_2(aq)] = \frac{8 \text{ mg/L}}{32,000 \text{ mg/mol}} = 2.5 \times 10^{-4} M$$

Treating the 2 mg/L of degradable organic carbon as if it were glucose (MW 180) the concentration of this material would be

$$[C_6H_{12}O_6] = \left(\frac{2 \text{ mg C/L}}{12,000 \text{ mg C/mol C}}\right)\left(\frac{1 \text{ mol } C_6H_{12}O_6}{6 \text{ mol C}}\right) = 2.78 \times 10^{-5}\,M$$

Assuming that $\gamma_i = 1.0$ for all the solutes, the molar Gibbs energy of reaction under the given conditions can then be computed as follows:

$$\Delta\bar{G}_r = \Delta\bar{G}_r^\circ + \left(5.71\frac{\text{kJ}}{\text{mol}}\right)\log\frac{(10^{-3})^6\,(10^{-7.5})^6}{(2.78 \times 10^{-5})(2.5 \times 10^{-4})^6}$$

$$= -2921 \text{ kJ/mol}$$

The calculation of $\Delta\bar{G}_r^\circ(W)$ is essentially identical to the calculation above, except that the pH is assumed to be 7.0, and the activities of all the other species participating in the reaction are assumed to be 1.0:

$$\Delta\bar{G}_r^\circ(W) = \Delta\bar{G}_r^\circ + \left(5.71\frac{\text{kJ}}{\text{mol}}\right)\log\frac{(1.0)^6\,(10^{-7.0})^6}{(1.0)(1.0)^6}$$

$$= -2951 \text{ kJ/mol}$$

The difference between $\Delta\bar{G}_r$ and $\Delta\bar{G}_r^\circ(W)$ is only ~1%, so the approximation that $\Delta\bar{G}_r = \Delta\bar{G}_r^\circ(W)$ is acceptable.

The key thermodynamic terms and relationships presented above are summarized in Table 2.3.

Table 2.3   Summary of some thermodynamic terms and relationships

Term	Description	Typical Units	Relationship
$\bar{G}_i$	Molar Gibbs energy of species i	$\dfrac{\text{kJ}}{\text{mol } i}$	$\bar{G}_i = \bar{G}_i^\circ + RT \ln a_i$
$\Delta\bar{G}_r^\circ$	Standard Gibbs energy of reaction	$\dfrac{\text{kJ}}{\text{mol stoichiometric reaction}}$	$\Delta\bar{G}_r^\circ = \sum \nu_i \bar{G}_i^\circ$
			($\nu_i > 0$ for products, $\nu_i < 0$ for reactants)
$\Delta\bar{G}_r$	Molar Gibbs energy of reaction	$\dfrac{\text{kJ}}{\text{mol stoichiometric reaction}}$	$\Delta\bar{G}_r = \sum \nu_i \bar{G}_i$
ΔG_r	Free energy change accompanying a given amount of reaction	kJ	$\Delta G_r = \Delta\bar{G}_r\,\Delta\eta$
			$= \sum_i \bar{G}_i \nu_i\,\Delta\eta = \sum_i \bar{G}_i\,\Delta n_i$
			($\Delta\eta$ = moles of stoichiometric reaction, Δn_i = moles of i reacted)
	Relationship among $\Delta\bar{G}$, $\Delta\bar{H}$, and $T\,\Delta\bar{S}$	$\dfrac{\text{kJ}}{\text{mol}}$	$\Delta\bar{G} = \Delta\bar{H} - T\,\Delta\bar{S}$

2.6.4 INTERPRETING CHANGES IN GIBBS ENERGY, ENTHALPY, AND ENTROPY IN CHEMICAL REACTIONS

The molar enthalpy and entropy of reaction can be computed in a manner analogous to the molar Gibbs energy of reaction, i.e., as the corresponding values of the products minus those of the reactants:

$$\Delta \bar{H}_r = \sum \nu_i \bar{H}_r \tag{2.96}$$

$$\Delta \bar{S}_r = \sum \nu_i \bar{S}_r \tag{2.97}$$

Furthermore, the molar Gibbs energy of reaction can be expressed in terms of the molar enthalpy and entropy of reaction in an equation analogous to Equation (2.42), i.e.,

$$\Delta \bar{G}_r = \Delta \bar{H}_r - T \Delta \bar{S}_r \tag{2.98}$$

or, under standard conditions,

$$\Delta \bar{G}_r^\circ = \Delta \bar{H}_r^\circ - T \Delta \bar{S}_r^\circ \tag{2.99}$$

It is instructive to consider the enthalpic and entropic contributions to the molar Gibbs energy of reaction, just as we did for the molar Gibbs energy of individual species. We can analyze those contributions using a slightly modified version of Equation (2.47), as follows:

$$\Delta \bar{G}_r = \left[\Delta \left(\begin{array}{c} \text{internal} \\ \text{bonding energy} \end{array} \right) + \Delta \left(\begin{array}{c} \text{energy of bonds with} \\ \text{neighboring species} \end{array} \right) + \Delta \left(\begin{array}{c} \text{value of} \\ \text{product of } PV_i \end{array} \right) \right]$$
$$- T \left[\Delta \left(\begin{array}{c} \text{thermal} \\ \text{entropy} \end{array} \right) + \Delta \left(\begin{array}{c} \text{configurational} \\ \text{entropy} \end{array} \right) \right] \tag{2.100}$$

where each Δ refers to a change per mole of stoichiometric reaction.

When a reaction proceeds, all five terms in Equation (2.100) are likely to change. However, some of those terms are expected to change by approximately the same amount regardless of the conditions under which the reaction occurs, while others are expected to be very sensitive to the system composition. Specifically, the products of the reaction are likely to have different internal bond energies, different molar volumes, and different numbers of accessible quantum states (i.e., different thermal entropy) from the reactants. These differences, corresponding to the first, third, and fourth terms on the right-hand side of Equation (2.100), all contribute to the Gibbs energy of reaction. However, the magnitude of each of these contributions is likely to be approximately the same regardless of the system composition, since each of them relates to the intrinsic properties of individual molecules. For instance, the change in the amount of energy stored in internal molecular bonds when a CO_2 molecule and an H_2O molecule are converted to an H_2CO_3 molecule would be expected to be the same no matter how much of each constituent happened to be in the system. We can infer, then, that the changes in internal bonding energy, molar volume, and thermal entropy that accompany a reaction contribute to the unchanging part of the Gibbs molar energy of reaction, i.e., they contribute to $\Delta \overline{G}_r^o$.

By contrast, the changes in intermolecular bond energy and configurational entropy that accompany a reaction [i.e., the second and fifth terms on the right-hand side of Equation (2.100)] are expected to depend strongly on the system composition. For instance, disappearance of a concentrated species and appearance of a more dilute species would increase the configurational entropy of the system. Similarly, disappearance of a neutral molecule and appearance of one positively and one negatively charged ion in its place would alter the intermolecular bond energy in the system. These changes contribute to the variable part of the Gibbs molar energy of reaction; i.e., they contribute to the $RT \ln Q$ term.

The overall Gibbs energy change accompanying a reaction is thus seen to be the sum of a group of terms that are insensitive to system composition and are embedded in $\Delta \overline{G}_r^o$ and another group of terms that depend on the composition and are embedded in the $RT \ln Q$ term. For any given reaction, any of the terms in either group might dominate the sum. Thus, a favorable reaction might be "driven" primarily by the formation of strong bonds in the product molecules ($\Delta \overline{H}_r^o$), an increase in the number of ways that energy can be distributed in product molecules compared to reactant molecules ($\Delta \overline{S}_{r,\text{therm}}^o$), or an increase in the number of ways that the product molecules can be distributed in space compared to the number of ways that the reactant molecules can be distributed ($\Delta \overline{S}_{r,\text{config}}$).

2.6.5 The Dependence of $\Delta \overline{G}_r^o$ and K_{eq} on Temperature and Pressure

Equation (2.86) ($\Delta \overline{G}_r^o = -RT \ln K_{eq}$) expresses the equilibrium constant for a reaction at standard temperature and pressure as a function of the standard Gibbs energy of the reaction. However, if we wish to determine the equilibrium constant at, say, 50°C in order to model the solution chemistry of a hot spring or some industrial waste stream, or at the elevated pressures extant near the ocean floor, then knowledge of the standard Gibbs energy of reaction in the conventional standard state (25°C and 1 bar) would not suffice. Rather, we would need to know the Gibbs energy of reaction for a system in which the reactants and products were all at unit activity, under the temperature and pressure conditions of interest. In other words, we would need to know how $\Delta \overline{G}_r^o$ varied with temperature and pressure.

The concept that $\Delta \overline{G}_r^o$ varies with temperature and pressure might seem a bit confusing at first; since $\Delta \overline{G}_r^o$ is defined for a specified set of environmental conditions, the idea of computing a value under different conditions and still referring to it as $\Delta \bar{G}_r^{\circ}$ seems paradoxical. The key to understanding this idea lies in recognizing that even though Equations (2.86a) and (2.86b) apply only under standard state conditions, it is not restricted to any particular choice of the standard state. That is, the equation is equally applicable at 50°C as at 25°C; however, to use it at 50°C, we would need to define the reference state to be at 50°C, and we would need data for $\Delta \bar{G}_r^{\circ}$ at that temperature. Since $\bar{G}_i^{\circ}$ values are normally available only at 25°C and 1.0 bar, the challenge is to estimate those values under different conditions.

Starting with Equation (2.86b), we can represent the change in $\ln K_{eq}$ as a function of temperature as follows:

$$\ln K_{eq} = -\frac{\Delta \bar{G}_r^{\circ}}{RT} \qquad \textbf{(2.86b)}$$

$$\frac{d}{dT} \ln K_{eq} = -\frac{d}{dT}\left(\frac{\Delta \bar{G}_r^{\circ}}{RT}\right) \qquad \textbf{(2.101)}$$

$\Delta \bar{G}_r^{\circ}$ might vary significantly with temperature. However, we have no fundamental or theoretical approach for predicting the details of that relationship, so evaluating the differential on the right-hand side of Equation (2.101) is nontrivial. This challenge has been dealt with by making certain assumptions about the major factors that affect $\Delta \bar{G}_r^{\circ}$, as follows.

According to Equation (2.99), for any change in system conditions, we can express $\Delta \bar{G}_r^{\circ}$ as

$$\Delta \bar{G}_r^{\circ} = \Delta \bar{H}_r^{\circ} - T\,\Delta \bar{S}_r^{\circ} \qquad \textbf{(2.99)}$$

As a first approximation, we might guess that while $\Delta \bar{H}_r^{\circ}$ and $\Delta \bar{S}_r^{\circ}$ could depend on temperature, the most dramatic effect of T in Equation (2.99) is as a multiplier of $\Delta \bar{S}_r^{\circ}$. In that case, we can treat $\Delta \bar{H}_r^{\circ}$ and $\Delta \bar{S}_r^{\circ}$ as remaining approximately constant when the reference state temperature changes, so we can equate them with the corresponding values at 25°C, i.e.,

$$\Delta \bar{H}_r^{\circ}\big|_{\text{any } T} \approx \Delta \bar{H}_r^{\circ}\big|_{25°C} \qquad \textbf{(2.102a)}$$

$$\Delta \bar{S}_r^{\circ}\big|_{\text{any } T} \approx \Delta \bar{S}_r^{\circ}\big|_{25°C} \qquad \textbf{(2.102b)}$$

Using Equation (2.99) to express $\Delta \bar{G}_r^{\circ}$ in terms of $\Delta \bar{H}_r^{\circ}$ and $\Delta \bar{S}_r^{\circ}$, and then substituting Equation (2.102) into the result, we obtain

$$\frac{d}{dT} \ln K_{eq} = -\frac{d}{dT}\left(\frac{\Delta \bar{H}_r^{\circ}}{RT} - \frac{\Delta \bar{S}_r^{\circ}}{R}\right) \qquad \textbf{(2.103)}$$

$$\frac{d}{dT} \ln K_{eq} \approx -\frac{d}{dT}\left(\frac{\Delta \bar{H}_{r,25°C}^{\circ}}{RT} - \frac{\Delta \bar{S}_{r,25°C}^{\circ}}{R}\right) \qquad \textbf{(2.104)}$$

$$\approx -\frac{\Delta \bar{H}_{r,25°C}^{\circ}}{R}\frac{d}{dT}\left(\frac{1}{T}\right) \qquad \textbf{(2.105)}$$

$$\approx \frac{\Delta \bar{H}_{r,25°C}^{\circ}}{R}\frac{1}{T^2} \qquad \textbf{(2.106)}$$

Reviewing what we have done, we note that Equation (2.103) involves no approximations and is valid at any temperature. However, to use it directly, we would need to insert the values of $\Delta \bar{H}_r^o$ and $\Delta \bar{S}_r^o$ at the temperature of interest, and chances are that these values would not be known at any temperature other than 25°C. By making the approximation that $\Delta \bar{H}_r^o$ and $\Delta \bar{S}_r^o$ are independent of temperature, Equation (2.103) can be used to derive Equation (2.106), which is easily integrated. Dropping the subscript "25°C" on the $\Delta \bar{H}_r^o$ term, we obtain

$$\int_{K_{eq,T_1}}^{K_{eq,T_2}} d \ln K_{eq} = \frac{\Delta \bar{H}_r^o}{R} \int_{T_1}^{T_2} \frac{dT}{T^2} \tag{2.107}$$

$$\ln K_{eq}\big|_{T_2} - \ln K_{eq}\big|_{T_1} = \frac{\Delta \bar{H}_r^o}{R}\left(\frac{1}{T_1} - \frac{1}{T_2}\right) \tag{2.108}$$

$$\ln \frac{K_{eq}\big|_{T_2}}{K_{eq}\big|_{T_1}} = \frac{\Delta \bar{H}_r^o}{R}\left(\frac{1}{T_1} - \frac{1}{T_2}\right) \tag{2.109}$$

We can use Equation (2.109), known as the van't Hoff equation, to evaluate $\Delta \bar{H}_r^o$ by plotting experimentally determined values of $\ln K_{eq}$ at various temperatures against $1/T$ (see Example 2.13 below). The same equation can then be used to compute K_{eq} at any other temperature. To repeat, the key assumption that allowed us to develop this equation is that, while $\Delta \bar{G}_r^o$ and K_{eq} vary significantly with temperature, $\Delta \bar{H}_r^o$ and $\Delta \bar{S}_r^o$ do not. Experimental studies have verified this assumption for many systems of interest. In cases where $\Delta \bar{H}_r^o$ and/or $\Delta \bar{S}_r^o$ varies too much over the range of temperatures of interest for the van't Hoff equation to apply, that variation can be characterized experimentally and incorporated into Equation (2.103), which can then be integrated numerically.[15]

| **Example 2.13** |

The equilibrium constants for the reaction in which bromoform ($CHBr_3$) transfers from the aqueous to the gas phase are 0.0071, 0.017, and 0.028 at 4, 20, and 30°C, respectively. Use the van't Hoff equation to estimate the standard enthalpy of the reaction and the equilibrium constant at 15°C.

Solution

When plotted as $\ln K_{eq}$ versus $1/T$ (Figure 2.11), the data fit the linear equation $\ln K_{eq} = -4390 \text{ K}/T + 10.9$. Equating the slope with $-\Delta \bar{H}_r^o/R$, we can compute $\Delta \bar{H}_r^o$ to be 36.5 kJ/mol. Plugging a value of $T = 15°C$ (288 K) into the equation of the line, we estimate K_{eq} at that temperature to be 0.013.

A derivation analogous to the one above yields a corresponding expression for the dependence of K_{eq} on pressure. For this derivation, the required assumption is that the molar volume of reaction (the change in system volume per mole

---

[15]See *Aquatic Chemistry* by Stumm and Morgan or *Aqueous Environmental Geochemistry* by Langmuir for a discussion of systems in which the variations in ΔH^o and ΔS^o as a function of temperature are significant, and for equations to model this dependence.

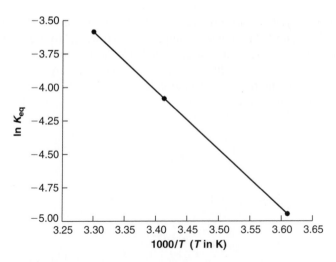

Figure 2.11   The van't Hoff relationship for transfer of bromoform from solution to the gas phase.

of stoichiometric reaction) be independent of pressure. That is, if we represent the molar volume of reaction as $\Delta\bar{V}_r$, the assumption is that $\Delta\bar{V}_r|_{any\ P} = \Delta\bar{V}_r|_{1.0\ bar} \equiv \Delta\bar{V}_r^{\circ}$. In such cases, the dependence of the equilibrium constant on pressure is as follows:

$$\ln\frac{K_{eq}|_{P_2}}{K_{eq}|_{P_1}} = -\frac{\Delta\bar{V}_r^{\circ}(P_2 - P_1)}{RT} \qquad \textbf{(2.110)}$$

Effectively, the restriction on $\Delta\bar{V}_r$ means that either the reaction involves only condensed phases (solids and liquids) or the number of moles of gaseous species is the same on the reactant and product sides of the reaction. For systems that meet the assumption of the derivation, $\Delta\bar{V}_r^{\circ}$ is exceedingly small unless the pressure change of interest is hundreds of bars or more. For instance, Stumm and Morgan show that the activity of water at a pressure of 1000 bar is larger than that at 1.0 bar by a factor of approximately 2.4, and that the equilibrium constants for several reactions at 1000 bar pressure differ from that at 1.0 bar by factors of approximately 1.4 to 7. Therefore, unless one is comparing a reaction at the earth's surface with one in the deep ocean or in a very high-pressure reactor, the effect of pressure on K_{eq} can generally be ignored.

2.6.6 THE ENTROPY, ENTHALPY, AND GIBBS ENERGY OF COMPOSITE REACTIONS

If a reaction can be represented as the sum of two or more other reactions, the Gibbs energy change for the composite reaction is simply the summation of the Gibbs energy changes of the component reactions. For instance, recall the composite reaction considered in Chapter 1, in which three H^+ ions transfer from phosphoric acid (H_3PO_4) to water molecules to form phosphate ion (PO_4^{3-}). The overall reaction can be derived by adding the following three reactions:

$$H_3PO_4 \; + H_2O \; \leftrightarrow H_2PO_4^- + H_3O^+ \qquad \Delta\overline{G}_{r,1}$$

$$H_2PO_4^- + H_2O \; \leftrightarrow HPO_4^{2-} + H_3O^+ \qquad \Delta\overline{G}_{r,2}$$

$$HPO_4^{2-} + H_2O \; \leftrightarrow PO_4^{3-} \; + H_3O^+ \qquad \Delta\overline{G}_{r,3}$$

$$\overline{\rule{3cm}{0.4pt}}$$

$$H_3PO_4 \; + 3H_2O \leftrightarrow PO_4^{3-} \; + 3H_3O^+ \qquad \Delta\overline{G}_{r,\text{overall}} = \; ?$$

where $\Delta\overline{G}_{r,i}$ is the molar Gibbs energy of the corresponding reaction under the extant experimental conditions.

Manipulation of the Gibbs energies of the various reactions leads to the following simple result:

$$\Delta\overline{G}_{r,1} = \overline{G}_{H_2PO_4^-} + \overline{G}_{H_3O^+} - \overline{G}_{H_3PO_4} - \overline{G}_{H_2O}$$

$$\Delta\overline{G}_{r,2} = \overline{G}_{HPO_4^{2-}} + \overline{G}_{H_3O^+} - \overline{G}_{H_2PO_4^-} - \overline{G}_{H_2O}$$

$$\Delta\overline{G}_{r,3} = \overline{G}_{PO_4^{3-}} + \overline{G}_{H_3O^+} - \overline{G}_{HPO_4^{2-}} - \overline{G}_{H_2O}$$

$$\Delta\overline{G}_{r,\text{overall}} = \overline{G}_{PO_4^{3-}} + 3\overline{G}_{H_3O^+} - \overline{G}_{H_3PO_4} - 3\overline{G}_{H_2O} = \Delta\overline{G}_1 + \Delta\overline{G}_2 + \Delta\overline{G}_3$$

Thus, the molar Gibbs energy of the overall reaction is the sum of the molar Gibbs energies of the reactions that are added to generate the overall reaction. Generalizing this result yields

$$\Delta G_{r,\text{overall}} = \sum_i \Delta G_{r,i} \qquad\qquad \textbf{(2.111)}$$

where the summation is over the reactions that are added to generate the overall reaction.

The same relationship applies, of course, to standard molar Gibbs energies of reaction. A simple extension of this argument leads to the conclusion that if the direction of a reaction is reversed (i.e., the reactants are replaced by the products, and vice versa), the magnitude of $\Delta\overline{G}_r$ is unchanged, but the sign is reversed. The $\Delta\overline{S}_r$ and $\Delta\overline{H}_r$ values for composite reactions can be computed in analogous ways.

The result for $\Delta\overline{G}_r^\circ$ of a composite reaction is consistent with the derivation in Chapter 1 showing that K_{eq} for an overall reaction equals $\prod_i K_{eq,i}$, where the $K_{eq,i}$ values are the equilibrium constants of the reactions that are combined to generate the overall reaction. For instance, for the example reactions involving phosphate species described above, we can write

$$\Delta\overline{G}_{r,\text{overall}}^\circ = -RT \ln K_{\text{overall}} \qquad\qquad \textbf{(2.112)}$$

$$\Delta\overline{G}_{r,\text{overall}}^\circ = -RT \ln (K_1 K_2 K_3) \qquad\qquad \textbf{(2.113)}$$

$$\Delta\overline{G}_{r,\text{overall}} = -RT \ln K_1 - RT \ln K_2 - RT \ln K_3 \qquad \textbf{(2.114)}$$

$$\Delta\overline{G}_{r,\text{overall}}^\circ = \Delta\overline{G}_1^\circ + \Delta\overline{G}_2^\circ + \Delta\overline{G}_3^\circ \qquad\qquad \textbf{(2.115)}$$

Note that the steps shown in Equations (2.112) through (2.115) represent algebraic manipulation of numbers; we could carry out those operations without having any idea that the terms in the equations or the final result had chemical significance.

Example 2.14 | The reaction for the dissociation of acetic acid (CH_3COOH) and the corresponding equilibrium constant are

$$CH_3COOH \leftrightarrow CH_3COO^- + H^+ \qquad K_{eq} = 1.72 \times 10^{-5}$$

The standard molar Gibbs energy of formation of acetate ion (CH_3COO^-) is -365.35 kJ/mol. Acetic acid and acetate ion are sometimes abbreviated as HAc and Ac^-, respectively.

a. Calculate $\bar{G}^\circ_{HAc}$.

b. Find $\Delta \bar{G}^\circ_r$ and K_{eq} for the above reaction "doubled," i.e.,

$$2CH_3COOH \leftrightarrow 2CH_3COO^- + 2H^+$$

c. Write out the form of the equilibrium constant for each of the above reactions [the one given in the problem statement and the one in part (b)].

Solution

a. Since $K_{eq} = 1.72 \times 10^{-5}$, $\log K_{eq} = -4.76$, and

$$\Delta \bar{G}^\circ_r = -(5.71 \text{ kJ/mol}) \log K = -(5.71 \text{ kJ/mol})(-4.76) = +27.07 \text{ kJ/mol}$$

$$\Delta \bar{G}^\circ_r = \bar{G}^\circ_{CH_3COO^-} + \bar{G}^\circ_{H^+} - \bar{G}^\circ_{CH_3COOH}$$

$$27.07 \text{ kJ/mol} = -369.4 \text{ kJ/mol} + 0 \text{ kJ/mol} - \bar{G}^\circ_{CH_3COOH}$$

$$\bar{G}^\circ_{CH_3COOH} = -396.5 \text{ kJ/mol}$$

b. Designate the reactions written with coefficients of 1 and 2 as reactions I and II, respectively. For reaction II,

$$\Delta \bar{G}^\circ_{r,II} = 2\bar{G}^\circ_{CH_3COO^-} + 2\bar{G}^\circ_{H^+} - 2\bar{G}^\circ_{CH_3COOH} = 2 \Delta \bar{G}^\circ_{r,I}$$

$$\Delta \bar{G}^\circ_{r,II} = 2(27.07 \text{ kJ/mol}) = 54.14 \text{ kJ/mol}$$

$$K_{II} = \frac{\{H^+\}^2 \{CH_3COO^-\}^2}{\{CH_3COOH\}^2}$$

$$K_{II} = K_I^2 = (1.74 \times 10^{-5})^2 = 3.03 \times 10^{-10}$$

c. The form of the equilibrium constant in the two cases is

$$K_I = 1.74 \times 10^{-5} = \frac{\{CH_3COO^-\}\{H^+\}}{\{CH_3COOH\}}$$

$$K_{II} = 3.03 \times 10^{-10} = \frac{\{CH_3COO^-\}^2 \{H^+\}^2}{\{CH_3COOH\}^2}$$

In the above example, although the numerical values of $\Delta \bar{G}_r^\circ$ and K_{eq} change depending on the stoichiometry, the derived relationship among the variables is the same. That is, if we take the square root of both sides of the equilibrium constant expression for reaction II, we obtain the result for reaction I. This outcome emphasizes the fact that we must refer values of $\Delta \bar{G}_r^\circ$ and K_{eq} to a specific reaction stoichiometry. The sentence "The standard Gibbs energy of reaction for dissociation of acetic acid is 27.07 kJ/mol" is, formally, ambiguous unless we specify that it refers to dissociation of 1 mol of CH_3COOH.

In reactions such as the dissociation of acetic acid, the chance of misinterpretation is minimal, since all the coefficients in the reaction are the same and would normally be assumed to have values of 1. However, in reactions in which the reactants and products have different stoichiometric coefficients, the reactions could reasonably be written with a number of stoichiometries, so the need for clarity is greater.

2.6.7 DISSOLUTION VERSUS REACTION WITH THE SOLVENT

Equation (2.78) indicates that the Gibbs energy of a reaction is the difference between the Gibbs energy of the products and the reactants. If water is a reactant or a product, its Gibbs energy must be considered in the overall change. However, if water is simply the solvent, then it is not viewed as participating in the reaction, so it is not included explicitly in the energy analysis (in such cases, any changes in the energy of the water are embedded in the Gibbs energy values of the solutes).

For instance, when gaseous oxygen dissolves in water, the water molecules rearrange to allow the oxygen molecules to fit in. However, no individual water molecules react with the oxygen, so the relevant reaction can be written as either reaction 1 or 2 below, but not as reaction 3. In fact, in a formal sense, reaction 3 is not balanced. On the other hand, when gaseous carbon dioxide dissolves, some of the $CO_2(aq)$ molecules react with water molecules to form a new, identifiable species (H_2CO_3, as shown in reaction 4). In this case, it is appropriate to include the Gibbs energy of water in evaluating the Gibbs energy of reaction.

$$O_2(g) + H_2O \leftrightarrow O_2(aq) + H_2O \qquad \text{reaction 1}$$

$$O_2(g) \leftrightarrow O_2(aq) \qquad \text{reaction 2}$$

$$O_2(g) + H_2O \leftrightarrow O_2(aq) \qquad \text{reaction 3}$$

$$CO_2(g) + H_2O \leftrightarrow H_2CO_3(aq) \qquad \text{reaction 4}$$

SUMMARY

The second law of thermodynamics postulates that systems always change spontaneously in the direction that increases universal entropy (S_{tot}). Entropy is a well-defined quantitative function characterizing how widely energy is dispersed among all the possible locations where it can be stored in a system. If many different distributions of energy at a microscopic scale cause the system to look the same macroscopically (i.e., to be in the same thermodynamic state), then that macroscopic appearance (state) will be observed frequently. On the other hand, macroscopic appearances that correspond to few energy distributions will be observed rarely. An observer of the macroscopic system would then characterize the commonly encountered appearance as stable and the infrequently encountered appearance as unstable. The former appearance is characterized as having greater entropy than the latter.

In chemical systems, entropy can be divided conceptually into two portions, one related to the distribution of energy within molecules (thermal entropy) and the other related to the distribution of molecules in space (configurational entropy). For any given chemical species, the thermal entropy per mole of that species depends on the system temperature but not the species' concentration, whereas the configurational entropy depends on the species' concentration but not the temperature.

Entropy is closely related to two other thermodynamic functions, the enthalpy (H) and Gibbs energy (G). Enthalpy is defined as the sum of the internal energy stored in the molecules in a system and the amount of energy (equal to PV) required to create a space for the system (i.e., to create a space of volume V against an external pressure P). Gibbs energy can be thought of as chemical potential energy. Entropy, enthalpy, and Gibbs energy are all state functions, meaning that their values are fully defined by the current state of a system, regardless of the system's history. The same statement applies to these functions when they are normalized to one mole of a particular species i, in which case they are represented as $\bar{S}_i$, $\bar{H}_i$, and $\bar{G}_i$, respectively.

The molar Gibbs energy $\bar{G}_i$ is also called the chemical potential of i and is often represented as μ_i. Changes in $\bar{G}_i$ and $\bar{H}_i$ are unambiguously measurable, but neither parameter can be quantified absolutely unless some datum level for energy is defined; by contrast, entropy is quantifiable without reference to any arbitrary datum. The datum levels for $\bar{G}_i$ and $\bar{H}_i$ that have been adopted are that $\bar{G}_i \equiv 0$ and $\bar{H}_i \equiv 0$ for pure elements in their standard states. Using these datum levels, $\bar{G}_i$ and $\bar{H}_i$ values for elements under non-standard conditions and compounds under any conditions can be computed.

The molar Gibbs energy of any species is related to the chemical activity of that species by $\bar{G}_i = \bar{G}_i^\circ + RT \ln a_i$. Also $\bar{G}_i$ is related to $\bar{H}_i$ and $\bar{S}_i$ by $\bar{G}_i = \bar{H}_i - T\bar{S}_i + T\bar{S}_i^\circ$. Both expressions for $\bar{G}_i$ include factors related to the intra- and intermolecular bonding of i, the molar volume of i, the energy distribution in molecules of i, and the distribution of molecules of i among the other molecules in the system. However, these factors are grouped differently in the two expressions.

The chemical activity coefficient (γ_{chem}) was introduced in Chapter 1 as a way of accounting for the shielding of an ion by others in its vicinity, due to the electrical interactions among the ions. In this chapter, we reinterpret γ_{chem} as a term that accounts for the fact that electrical interactions with surrounding ions decrease the electrochemical potential of an ion in solution. That is, because of the electrical attraction between the central ion and the surrounding ones, the central ion has less electrical potential energy (and therefore less total potential energy) than it would have if those surrounding ions were not there. The decrease in total potential energy makes the ion less "active" (reactive) than it otherwise would be.

Systems always change in the direction that leads to an increase in universal entropy S_{tot}, and in isolated systems at constant temperature and pressure, an increase in S_{tot} always corresponds to a decrease in total potential energy. In most systems of relevance for this text, the change of interest is the progression of a chemical reaction, and the only form of potential energy that changes significantly is the Gibbs energy. Therefore, analysis of the change in Gibbs energy when the reaction proceeds indicates whether the reaction is thermodynamically favorable.

The change in Gibbs energy accompanying a chemical reaction (the Gibbs energy of reaction) can be computed as the sum of the Gibbs energy changes associated with the changes in the concentrations of all the reactants and products in the system. This value is typically normalized to one mole of stoichiometric reaction, defined as the conversion of reactants to products in molar amounts that correspond to the stoichiometric coefficients of the reaction. If the molar Gibbs energy of a reaction under ambient conditions is negative, the reaction will tend to proceed by conversion of reactants to products; if it is positive, the reaction will proceed in the opposite direction. The reaction is at equilibrium if and only if the Gibbs energy of reaction is zero. The molar Gibbs energy of reaction is related quantitatively to the extent of disequilibrium by $\Delta \bar{G}_r = RT \ln Q/K_{eq}$.

The equilibrium constant for a reaction bears a simple relationship to the molar Gibbs energy change under standard conditions: $\Delta \bar{G}_r^{\circ} = -RT \ln K_{eq}$. Although K_{eq} and $\Delta \bar{G}_r^{\circ}$ change significantly with temperature, the corresponding changes in $\Delta \bar{H}_r^{\circ}$ and $\Delta \bar{S}_r^{\circ}$ are generally small. As a result, the effect of temperature on K_{eq} can be computed by

$$\ln K_{eq}|_{T_2} - \ln K_{eq}|_{T_1} = \frac{\Delta \bar{H}_r^{\circ}}{R} \left(\frac{1}{T_1} - \frac{1}{T_2} \right)$$

Problems

1. When a piece of metal is exposed to a strong acid, the metal corrodes and gas bubbles are sometimes formed. These gas bubbles are pure hydrogen (H_2), and they indicate that the partial pressure of $H_2(g)$ in the system is 1.0 bar. Consider the reaction when the metal is lead and the reaction taking place is as shown below:

$$Pb(s) + 2H^+ \leftrightarrow Pb^{2+} + H_2(g)$$

The Gibbs energy of reaction is determined to be -9.06 kJ/mol in a system that is at pH 4.0, that contains 1 mg/L, Pb^{2+}, and in which hydrogen bubbles are forming on the surface of the $Pb(s)$. What is the equilibrium constant for the reaction?

2. The following Gibbs energy data apply at 1 bar and 25°C:

Substance	$\bar{G}_i^o$, kJ/mol	Substance	$\bar{G}_i^o$, kJ/mol
H^+	0.00	Cl^-	-131.3
$AgBr(s)$	-96.90	$H_2PO_4^-$	-1130.4
$AgCl(s)$	-109.8	HPO_4^{2-}	-1089.3
Ag^+	77.12	PO_4^{3-}	-1018.8
Br^-	-104		

a. Determine the standard Gibbs energy change and the equilibrium constant for the following reactions, using the data given above:

 i. $AgBr(s) \leftrightarrow Ag^+ + Br^-$
 ii. $PO_4^{3-} + H^+ \leftrightarrow HPO_4^{2-}$
 iii. $HPO_4^{2-} + H^+ \leftrightarrow H_2PO_4^-$
 iv. $PO_4^{3-} + 2H^+ \leftrightarrow H_2PO_4^-$
 v. $AgBr(s) + Cl^- \leftrightarrow AgCl(s) + Br^-$
 vi. $AgCl(s) + Br^- \leftrightarrow AgBr(s) + Cl^-$

b. A solution contains $10^{-2}\, M\, Cl^-$, $10^{-5}\, M\, Ag^+$, and $10^{-6}\, M\, AgCl(s)$ initially. Is it an equilibrium solution? If not, what is the molar Gibbs energy of reaction for the formation of more $AgCl(s)$ under these conditions? What reaction will occur spontaneously? Find the concentrations of all the species, including the solid, when equilibrium is attained. Assume all activity coefficients are unity.

3. The odor of solutions containing ammonia is generated by volatilization, i.e., transfer into the gas phase, of NH_3 molecules; NH_4^+, like all ions, is essentially non-volatile. The equilibrium constant for reaction 1 below is $10^{-9.25}$, and the standard Gibbs energy of the volatilization reaction (reaction 2) is $+10.09$ kJ/mol.

$$NH_4^+ \leftrightarrow NH_3(aq) + H^+ \qquad \text{reaction 1}$$

$$NH_3(aq) \leftrightarrow NH_3(g) \qquad \text{reaction 2}$$

A solution containing $10^{-2}\, M$ total dissolved ammonia species ($TOT\,NH_3$) is at pH 7.25 and 25°C and is in contact with a gas phase containing a partial pressure of 10^{-6} bar NH_3.

a. Does the chemical driving force favor dissolution of ammonia, volatilization of ammonia, or is the system at equilibrium?

b. How much and in what direction (increase or decrease) would the total Gibbs energy of the system change if 10^{-6} mol of NH_3 was transferred from the gas to solution? You may assume that this transfer is carried out in such a way that the partial pressure of ammonia and the composition of the solution change negligibly.

4. The standard Gibbs energies of formation of $CaO(s)$, $CaCO_3$(calcite), and $CO_2(g)$ are -604.04, -1128.80, and -394.37 kJ/mol, respectively. Find the value of $\Delta \overline{G}_r^o$ and K_{eq} for the following reaction:

$$CaCO_3(s) \leftrightarrow CaO(s) + CO_2(g)$$

A dry mixture containing 1 g of each solid [$CaCO_3(s)$ and $CaO(s)$] is on a lab bench in contact with the atmosphere, which contains a partial pressure of $10^{-3.5}$ bar $CO_2(g)$. Does the equilibrium driving force favor conversion of one of the solids into the other, or are the solids equilibrated with one another?

5. A system with $\{H^+\} = 10^{-7.0}$ and in equilibrium with $NH_3(g)$ at a partial pressure of $10^{-6.0}$ bar is rapidly adjusted to a condition where $\{H^+\} = 10^{-6.0}$. Standard Gibbs energy values for some relevant species are $\overline{G}_{NH_3(aq)}^o = -26.57$ kJ/mol, $\overline{G}_{NH_4^+(aq)}^o = -79.37$ kJ/mol, and $\overline{G}_{H^+}^o = 0.00$ kJ/mol. The equilibrium constant for the reaction $NH_4^+ \leftrightarrow NH_3(aq) + H^+$ is $10^{-9.25}$.

   The Gibbs energy of the species $NH_3(aq)$ under the given initial conditions is -50.54 kJ/mol. Assuming that the reaction shown above equilibrates instantaneously, but that the exchange of NH_3 between the gas phase and dissolved phase is slow, compute the following values.

   a. The equilibrium constant for the reaction $NH_3(aq) \leftrightarrow NH_3(g)$. (This equilibrium constant is referred to as the Henry law constant for ammonia and is discussed extensively in Chapter 7.)

   b. The molar Gibbs energy of reaction for the dissolution of gaseous $NH_3(g)$ prior to the change in $\{H^+\}$.

   c. The molar Gibbs energy of reaction for the dissolution of gaseous $NH_3(g)$ after the change in $\{H^+\}$, but prior to any transfer of NH_3 into or out of solution.

6. The Boeing Co. has developed a process for removing copper from wastewaters generated during electroplating and printed-circuit board manufacturing. The central feature of the process is addition of scrap aluminum metal to an acidic solution containing dissolved copper ions. If the solution contains enough fluoride or chloride ions, a reaction proceeds in which the aluminum dissolves and the copper precipitates as metallic $Cu^o(s)$, which can then be removed from the suspension by settling and/or filtration. The relevant reaction can be written as follows:

$$3Cu^{2+} + 2Al^o(s) \leftrightarrow 3Cu^o(s) + 2Al^{3+}$$

   Standard Gibbs energies of formation of the ionic species in the above reaction are $\overline{G}_{Cu^{2+}}^o = 65.5$ kJ/mol and $\overline{G}_{Al^{3+}}^o = -489.4$ kJ/mol.

   If a wastewater initially containing 300 mg/L Cu^{2+} and no dissolved aluminum is dosed with 100 mg/L of Al scraps and the above reaction proceeds until equilibrium is reached, what will the final concentrations of Cu^{2+}, Al^{3+}, $Cu^o(s)$, and $Al^o(s)$ be?

7. Consider two containers (A and B) containing calcium ions dissolved in 1.0 and 2.0 L of water, respectively. The initial Ca^{2+} concentrations are 0.1 M in container A and 0.01 M in container B. The system is shown schematically in the figure below. Assume that both solutions are ideal, i.e., that $\gamma_{Ca^{2+}} = 1.0$ in both solutions. Compute $\bar{G}_{Ca^{2+}}$ in each container as a function of the mass of Ca^{2+} that diffuses from A to B, and use the results to determine the distribution of the Ca^{2+} at equilibrium. The value of $\bar{G}^{\circ}_{Ca^{2+}}$ is -553.54 kJ/mol. (*Note*: In theory, water would also diffuse from one side of the system to the other, since its chemical potential is slightly different on the two sides of the system. However, we can ignore this small transfer and assume that the volumes of water in the two containers are constant.)

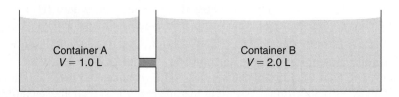

8. In Example 2.3, the Gibbs energy of CO_3^{2-} was evaluated in two hypothetical solutions, with the activity coefficient of CO_3^{2-} being 1.0 in one solution and 0.5 in the other. If the first of these solutions (the one with a low ionic strength) were at equilibrium, and then the ionic strength were increased, lowering the activity coefficients of CO_3^{2-} and H^+, would the following reaction proceed to the right or left, or would it still be at equilibrium?

$$CO_2(g) + H_2O \leftrightarrow 2H^+ + CO_3^{2-}$$

ACIDS AND BASES, PART 1

Acid/Base Speciation and Exact Solutions to Acid/Base Problems

---

CHAPTER OUTLINE

3.1   INTRODUCTION

This chapter introduces the chemistry of acids and bases, groups of species that play a critical role in water quality. Acids and bases influence water quality both directly by controlling solution pH and indirectly by, for example, controlling the dissolution and precipitation of solids, altering the solubility of gases, catalyzing many other reactions, and affecting the interactions of chemicals with organisms. For instance, as noted in the preceding chapters, solution pH controls the relative concentrations of hypochlorous acid (HOCl) and hypochlorite ion (OCl^-) in a solution; because HOCl is a much more effective disinfectant that OCl^-, the U.S. Environmental Protection Agency (EPA) has established different disinfectant dosing requirements for waters depending on their pH.

As another example, consider the following reaction for the abiotic oxidation of ferrous iron (iron with a +2 charge) to ferric iron (iron with a +3 charge) by dissolved oxygen.

$$Fe^{2+} + \tfrac{1}{4}O_2 + H^+ \leftrightarrow Fe^{3+} + \tfrac{1}{2}H_2O \qquad \textbf{(3.1)}$$

The rate of reaction (3.1) has been reported to increase by a factor of 100 for every unit increase in solution pH.[1] In addition to helping control the rate of the Fe^{2+}-to-Fe^{3+} conversion, the pH affects the ultimate solubility of the iron once the system equilibrates. For instance, at pH 7.0, the equilibrium solubility of ferrous iron is almost six orders of magnitude greater than that of ferric iron (Figure 3.1a).

Solution pH also plays a central role in determining the affinity of many dissolved species for oxide particles with which they come into contact. For instance, Figure 3.1b shows that the binding of dissolved organic matter to the surface of aluminum oxide particles increases dramatically with decreasing pH. A similarly strong dependence on pH characterizes the binding of many metals to oxide particles, except that in this case the trend is reversed: binding increases with increasing pH. In cases where it is desirable to remove these solutes from water (e.g., water and wastewater treatment plants), oxide particles are sometimes added to or generated in the raw water, and the solution pH is adjusted to the range where the target contaminant binds strongly to the particles' surfaces. The particles are then removed (along with the contaminants) by settling and/or filtration.

In addition to being important reactions in their own right, acid/base reactions provide an appropriate starting point for studying the equilibrium speciation of a much larger class of chemical reactions, for three reasons. First, almost all acid/base reactions are extremely fast, so an assumption that equilibrium is attained (and that the computed speciation is therefore the actual speciation in the system) is usually justified. Second, acid/base reactions are probably the simplest examples of a large class of reversible reactions in which two molecules collide and a portion of one of the molecules is transferred to the other one. In the case of acids and bases, the exchangeable unit is simply an H^+ ion. These reactions are prototypical of many other types of reactions in which the attachment/detachment is more complicated.

Finally, acids and bases are commonly encountered in everyday life. For instance, the widely used industrial chemical muriatic acid is the same material referred to in chemical contexts as hydrochloric acid (HCl); hypochlorite ion (OCl^-) is the major ingredient in bleach; acetylsalicylic acid (CH_3COO—C_6H_4—OH) is

---

[1]The oxidation is often mediated by microorganisms, in which case the rate of abiotic oxidation does not necessarily control the speciation of iron. However, the types and growth rates of microorganisms also respond to solution pH, so it has an indirect effect on iron speciation even when abiotic reactions do not dominate.

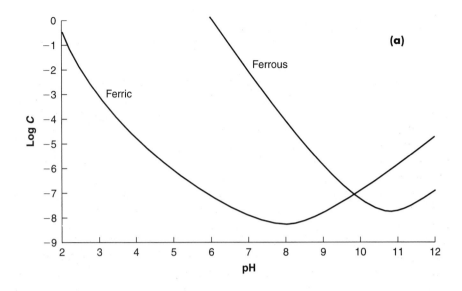

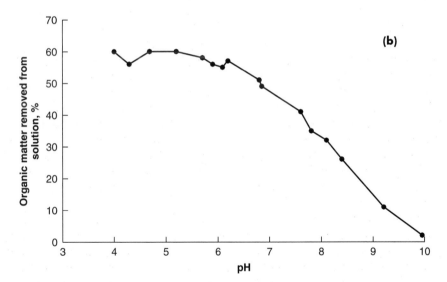

Figure 3.1 **(a)** The total concentration of dissolved iron in equilibrium with amorphous oxides of Fe^{2+} and Fe^{3+}, as a function of solution pH. **(b)** Effect of pH on binding of natural organic matter (4.7 mg DOC/L) onto γ-Al_2O_3.

| After Davis, J., *Geochim. Cosmochim. Acta* **48,** 679–691 (1982).

aspirin; ascorbic acid ($C_6H_8O_6$) is vitamin C; and carbonic acid (H_2CO_3) and phosphoric acid (H_3PO_4) are ingredients of many beverages.

The chapter begins with some definitions that identify the key characteristics of acids and bases, describes how the stability of those species is characterized, and concludes with several examples showing how acid/base speciation can be evaluated in some simple systems. The discussion and many of the examples are presented in the context of answering the question, If a known amount of such-and-such acid or base is added to pure water, what will the solution composition be at equilibrium? In practice, the situation is rarely that simple. Solutions of interest usually contain a multitude of chemicals, and when an acid or base is added, the changes in water quality go well beyond those that are addressed in this chapter. Nevertheless, the acid/base reactions explored here are often of central importance, and studying the behavior of very simple systems can provide insights that serve us well in trying to understand the more complex systems that we might encounter in natural or engineered environments.

3.2   THE DISSOCIATION OF WATER; K_w

In any aqueous solution, even the purest distilled water, some water molecules split apart in **hydrolysis** or **dissociation** reactions to form H^+ and OH^- ions. Like all equilibrium reactions, hydrolysis is reversible, so the reaction can be written as

$$H_2O \leftrightarrow H^+ + OH^- \tag{3.2}$$

The equilibration of water molecules with H^+ and OH^- ions is sufficiently rapid that the reaction is assumed to always be in equilibrium. The equilibrium constant for reaction (3.2) is given a special symbol, K_w. That is,

$$K_w \equiv \frac{\{H^+\}\{OH^-\}}{\{H_2O\}} \tag{3.3}$$

Since the activity of H_2O is very nearly 1.0 in almost all solutions of interest, the term $\{H_2O\}$ is often left out of the equilibrium constant expression, in which case the expression is written as $K_w = \{H^+\}\{OH^-\}$. The value of K_w is $10^{-14.0}$ at 25°C and increases with increasing temperature, varying by approximately a factor of 50 over the range of temperatures in environmental systems (see Table 3.1). Because $\{H^+\}$ and $\{OH^-\}$ are related through K_w, any increase in $\{H^+\}$ is accompanied by a decrease in $\{OH^-\}$, and vice versa. A solution in which $\{H^+\} > \{OH^-\}$ is called **acidic,** and a solution in which $\{H^+\} < \{OH^-\}$ is called **basic** or **alkaline.**

Although we write reaction (3.2) as though H^+ and OH^- existed as distinct entities in solution, H^+ ions are actually extremely unstable as free dissolved species, and almost every H^+ ion combines with a single water molecule or a cluster of water molecules to form ions such as H_3O^+, $H_5O_2^+$, and $H_7O_3^+$ (Figure 3.2). Thus, in reality, hydrolysis is not just the splitting of a water molecule;

Table 3.1   The value of K_w from 0 to 100°C

T, °C	0	10	20	25	30	40	50	100
$10^{14} \times K_w$	0.114	0.292	0.681	1.008	1.47	2.92	5.5	55

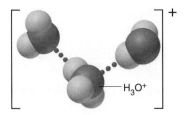

Figure 3.2   A hydrated H^+ ion, hydrogen-bonded to two other water molecules.

From M. S. Silberberg, *Chemistry: The Molecular Nature of Matter and Change.* Copyright © 2000 The Mc-Graw-Hill Companies, New York, NY. Reproduced by permission of The McGraw-Hill Companies.

it is the transfer of an H^+ ion from one water molecule to another, in a reaction such as the following:

$$H_2O + H_2O \rightarrow OH^- + H_3O^+ \qquad \textbf{(3.4)}$$

Under most circumstances, there is no need to distinguish among the various $(H_2O)_x$—H^+ species. Although we will continue to use the symbol H^+, it should be understood that this is simply a shorthand notation to describe all species of the form $(H_{2x+1}O_x)^+$, of which H_3O^+ is the most common. Collectively these ions are called *hydronium* ions. Somewhat incorrectly, hydronium ions are often referred to simply as protons, as though they were indeed isolated H^+ ions (an H^+ ion has one proton and no neutrons in its nucleus, and no electrons orbiting it, so it is indeed simply a proton).

3.3   THE STRUCTURE OF ACIDS AND BASES

In addition to the dissociation of H_2O molecules, H^+ and OH^- ions can be provided to solution by dissociation of chemicals that contain only one of these ions. For instance, addition of hydrochloric acid (HCl) or acetic acid (CH_3COOH) to a solution can increase the H_3O^+ concentration by the following **acid dissociation** reactions:

$$HCl + H_2O \leftrightarrow H_3O^+ + Cl^- \qquad K > 10^{-1} \qquad \textbf{(3.5)}$$

$$CH_3COOH + H_2O \leftrightarrow H_3O^+ + CH_3COO^- \qquad K = 1.74 \times 10^{-5} \quad \textbf{(3.6)}$$

Molecules that act as acids via these types of reactions (transfer of an H^+ ion) are sometimes referred to as **Bronsted acids.** Correspondingly, species that can combine with an H^+ ion, such as Cl^- and CH_3COO^- in the reverse of reactions (3.5) and (3.6), are referred to as **Bronsted bases.**

Some metal ions can also release H^+ (which can then combine with a water molecule to form H_3O^+) by reactions such as the following:

$$Cu^{2+} + H_2O \leftrightarrow CuOH^+ + H^+ \qquad K = 10^{-8.0} \qquad \textbf{(3.7)}$$

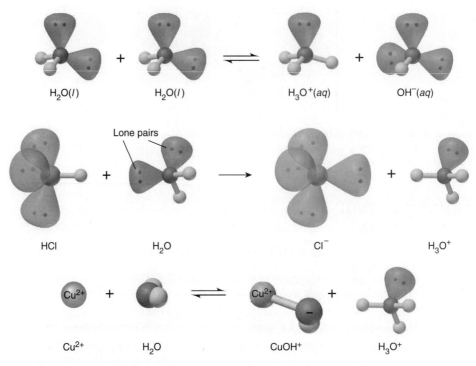

Figure 3.3   Schematic representation of the donation of H^+ from three acids—H_2O, HCl, and Al^{3+}—to water, forming a hydronium ion. The round lobes with two black dots represent molecular orbitals occupied by a pair of unshared electrons.

From M. S. Silberberg, *Chemistry: The Molecular Nature of Matter and Change.* Copyright © 2000 The McGraw-Hill Companies, New York, NY. Reproduced by permission of The McGraw-Hill Companies.

In reaction (3.7), the copper ion can be viewed as binding to a water molecule and then promoting its dissociation (hydrolysis). The net result is that copper causes the concentration (and activity) of H_3O^+ ions in solution to increase, even though the copper itself has no H^+ associated with it when it is initially added to solution. Acids that release H^+ by such a mechanism are sometimes called **Lewis acids,** and the corresponding product of the reaction [$CuOH^+$ in reaction (3.7)], which can combine with H^+ in the reverse reaction, is called a **Lewis base.**[2]

Based on the above discussion, acids (especially Bronsted acids) are sometimes referred to as proton donors (they "donate" protons, or H^+ ions, to water molecules to form H_3O^+) and bases as proton acceptors (they acquire protons from the donor molecules, i.e., the acids). Schematic representations of reactions (3.4), (3.5), and (3.7) are shown in Figure 3.3.

---

[2]The formal definition of a Lewis acid is a molecule that has the ability to accept a free electron pair. In the case of water binding to Cu^{2+}, the copper ion is an acid because it can accept an electron pair from the oxygen of the water molecule.

Extremely acidic solutions can be generated in nature in locations where sulfide-containing minerals are exposed to air and limited amounts of water. This photograph is of a solution from Richmond Mine in Iron Mountain, CA, where pH values as low as -3.6 were recorded.

| Photograph by Charles N. Alpers and D. Kirk Nordstrom, U.S. Geological Survey.

3.4 Strong and Weak Acids, K_a, and Conjugate Acid/Base Pairs

The tendency of different acids to release an H^+ ion to solution (or, equivalently, to donate an H^+ ion to a water molecule) covers a wide spectrum. Logically enough, the strength of an acid is defined by its tendency to participate in such a reaction. For instance, hydrochloric acid (H—Cl), acetic acid (CH_3COO—H) and hypochlorous acid (H—OCl) all have H^+ ions that can be released to solution when the compound dissolves (which is why all three are acids):

$$HCl \leftrightarrow H^+ + Cl^- \qquad\qquad \textbf{(3.5)}$$

$$CH_3COOH \leftrightarrow H^+ + CH_3COO^- \qquad\qquad \textbf{(3.6)}$$

$$HOCl \leftrightarrow H^+ + OCl^- \qquad\qquad \textbf{(3.8)}$$

However, when equal concentrations of these three acids are added to solution (either to separate solutions or all together), the relative activities (and concentrations) of the various species at equilibrium are as follows:

$$\{HCl\} < \{CH_3COOH\} < \{HOCl\} \quad \text{and} \quad \{Cl^-\} > \{CH_3COO^-\} > \{OCl^-\}$$

From this sequence, we can conclude that hydrochloric acid is stronger (has a greater tendency to release H^+) than acetic acid, which is in turn stronger than hypochlorous acid. By following this reasoning to its logical extreme, an infinitely strong acid would completely dissociate in solution, thereby donating all its available H^+ to other molecules.

Although the above comparison is useful, it is qualitative and not a particularly convenient way to discuss the relative strengths of acids. To address this issue, a shorthand has been developed by which the relative strengths of different acids can be assessed rapidly and quantitatively. Specifically, if HA is an acid, its **acid dissociation constant** or simply **acidity constant** is defined as the equilibrium constant for the reaction:

$$HA \leftrightarrow H^+ + A^- \tag{3.9}$$

$$K_a = \frac{\{H^+\}\{A^-\}}{\{HA\}} \tag{3.10}$$

The acidity constant is always designated K_a and is always written such that the acid is on the reactant side of the equation and one proton is released in the reaction. For instance, the equilibrium constants for reactions (3.5), (3.6), and (3.8) would all be called acidity constants, as would the reaction of Cu^{2+} and H_2O to form $CuOH^+$ and H^+ [reaction (3.7)]. It should be apparent that the larger K_a is for a given acid, the greater the acid's tendency to dissociate, i.e., the stronger the acid is. Acidity constants for several environmentally important acids are listed in Table 3.2.

Note that, according to reaction (3.4), water can be considered an acid with $A^- = OH^-$, as might be more apparent if reaction (3.2) were written as follows:

$$H\!-\!OH \leftrightarrow H^+ + OH^- \tag{3.11}$$

The acidity constant for this acid dissociation reaction is given by $K_a = \{H^+\}\{OH^-\}/\{H_2O\}$, which is, of course, K_w.

Some acids can donate more than one proton to solution (i.e., to water molecules). For these acids, acidity constants are subscripted $1, 2, \ldots$, starting with the most protonated species. Thus, for phosphoric acid (H_3PO_4):

$$H_3PO_4 \leftrightarrow H^+ + H_2PO_4^- \qquad K_{a1} = \frac{\{H^+\}\{H_2PO_4^-\}}{\{H_3PO_4\}} = 10^{-2.16}$$

$$H_2PO_4^- \leftrightarrow H^+ + HPO_4^{2-} \qquad K_{a2} = \frac{\{H^+\}\{HPO_4^{2-}\}}{\{H_2PO_4^-\}} = 10^{-7.20}$$

$$HPO_4^{2-} \leftrightarrow H^+ + PO_4^{3-} \qquad K_{a3} = \frac{\{H^+\}\{PO_4^{3-}\}}{\{HPO_4^{2-}\}} = 10^{-12.35}$$

In this system, the three species H_3PO_4, $H_2PO_4^-$, and HPO_4^{2-} can all act as acids, since each can release a proton. The K_a values indicate that H_3PO_4 is a stronger acid than $H_2PO_4^-$, which in turn is a stronger acid than HPO_4^{2-}.

Table 3.2   Chemical formulas and acidity constants of some important acids[3]

Name	Formula	pK_{a1}	pK_{a2}	pK_{a3}	pK_{a4}
Nitric acid	HNO_3	−1.30			
Trichloroacetic acid	CCl_3COOH	−0.5			
Hydrochloric acid	HCl	<0			
Sulfuric acid	H_2SO_4	<0	1.99		
Hydronium ion	H_3O^+	0.00	14.00		
Chromic acid	H_2CrO_4	0.86	6.51		
Oxalic acid	$(COOH)_2$	0.90	4.20		
Dichloroacetic acid	$CHCl_2COOH$	1.1			
Sulfurous acid	H_2SO_3	1.86	7.30		
Phosphoric acid	H_3PO_4	2.16	7.20	12.35	
Arsenic acid	H_3AsO_4	2.24	6.76		
Monochloroacetic acid	$CH_2ClCOOH$	2.86			
Salicylic acid	$C_6H_4OHCOOH$	2.97	13.70		
Citric acid	$C_3H_4OH(COOH)_3$	3.13	4.72	6.33	
Hydrofluoric acid	HF	3.17			
Benzoic acid	C_6H_5COOH	4.20			
Pentachlorophenol	C_6Cl_5OH	4.7			
Acetic acid	CH_3COOH	4.76			
Carbonic acid	H_2CO_3	6.35	10.33		
Hydrogen sulfide	H_2S	6.99	12.92		
Hypochlorous acid	$HOCl$	7.60			
Cupric ion	Cu^{2+}	8.00	5.68		
2-Chloro-phenol	C_6H_4ClOH	8.53			
Hypobromous acid	$HOBr$	8.63			
Zinc ion	Zn^{2+}	8.96	8.94		
Arsenous acid	H_3AsO_3	9.23	12.10		
Hydrocyanic acid	HCN	9.24			
Boric acid	H_4BO_4	9.24			
Ammonium ion	NH_4^+	9.25			
2,4-Dichloro-phenol	$C_6H_3Cl_2OH$	9.43			
Silicic acid	H_4SiO_4	9.84	13.20		
Phenol	C_6H_5OH	9.98			
Cadmium ion	Cd^{2+}	10.08	10.27	12.95	14.05
Calcium ion	Ca^{2+}	12.60			

[3]Throughout this text, the equilibrium constants given are consistent with the database in the chemical equilibrium computer program Mineql+. This choice has been made to facilitate correspondence of computations carried out manually with the output from that program. Equilibrium constants for reactions of interest that are not in the program's database have been selected from a variety of other sources. MINEQL+ is a registered trademark of Environmental Research Software.

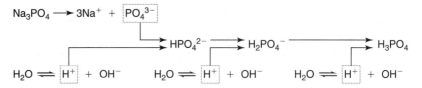

Figure 3.4   Schematic showing the stepwise protonation of phosphate ion, consuming dissolved H^+.

Since all chemical reactions are reversible,[4] the above reactions can also proceed from right to left, consuming H^+ instead of releasing it. Consider, for instance, a solution prepared by adding Na_3PO_4 (trisodium phosphate, commercially sold as "TSP") to water. When the Na_3PO_4 first dissolves, it dissociates into Na^+ and PO_4^{3-} ions. The Na^+ ions are immediately surrounded by water molecules and are unlikely to react further. The rest of the solution then contains PO_4^{3-} ions, along with H_2O, H^+, and OH^-, but no HPO_4^{2-} or $H_2PO_4^-$. Such a solution is out of equilibrium with respect to all three acid dissociation reactions shown above. To approach equilibrium, some PO_4^{3-} will combine with H^+ ions to form HPO_4^{2-}, some HPO_4^{2-} ions will combine with H^+ to form $H_2PO_4^-$, etc. The process is shown schematically in Figure 3.4.

The attachment of H^+ to PO_4^{3-}, HPO_4^{2-}, and $H_2PO_4^-$ reduces the activity of dissolved H^+, so the product $\{H^+\}\{OH^-\}$ decreases and becomes transiently less than K_w. This process, in turn, causes a disequilibrium in the H_2O dissociation reaction, which is alleviated by the splitting of additional water molecules to increase the concentrations of H^+ and OH^-. The latter process continues until the product $\{H^+\}\{OH^-\}$ is once again equal to K_w. The entire network of reactions proceeds until all the reactions are in equilibrium simultaneously.

In the scenario described, when the final equilibrium state is achieved, the concentrations of H^+ and OH^- will not be equal; $\{OH^-\}$ will be greater than $\{H^+\}$, because equal amounts of H^+ and OH^- were added to solution by the water dissociation reaction, some of the added H^+ was consumed by reactions with phosphate species, but no OH^- was consumed. We will shortly develop the equations necessary to characterize the final equilibrium condition quantitatively. For now, it is only important to note that the net effect of the addition of PO_4^{3-} is the removal of H^+ from solution to form various $H_xPO_4^{3-x}$ species. Because it removes H^+ from solution (or, put slightly differently, it accepts H^+ ions), we conclude that PO_4^{3-} acts as a base in this situation.

The species PO_4^{3-} and HPO_4^{2-} are interconverted simply by the exchange of a proton. When HPO_4^{2-} releases a proton to become a PO_4^{3-} ion, it is acting as an acid. The PO_4^{3-} ion, on the other hand, can combine with a proton to re-form HPO_4^{2-}, thereby acting as a base. In fact, the dissociation of any acid

---

[4]Reactions are often referred to as *irreversible* if the reverse reaction is so slow or so thermodynamically unfavorable that it does not proceed to a significant extent in the system of interest. Nevertheless, formally, all chemical reactions are reversible to some extent.

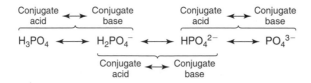

Figure 3.5 Conjugate acid/base relationships among phosphate species.

always generates a corresponding (potential) base, i.e., a species that, under different conditions, could recombine with an H^+ ion to regenerate the acid. Acid and base species that are related in this way are said to form a **conjugate acid/base pair.** Thus, HPO_4^{2-} is the conjugate base of the acid $H_2PO_4^-$, and $H_2PO_4^-$ is the conjugate acid of HPO_4^{2-}. Note that HPO_4^{2-} is also the conjugate acid of the base PO_4^{3-} (Figure 3.5).

By conservation of charge, an acid always has a charge that is greater by 1 than its conjugate base. This is true regardless of whether the acid actually contains the "detachable" H^+ (like HCl, CH_3COOH, and HOCl) or whether it releases an H^+ ion by hydrolyzing a water molecule and combining with the OH^- (like Cu^{2+}). The way in which the charge on the molecule becomes redistributed when the proton is released plays a major role in determining the relative stability of the acid and conjugate base, and hence the value of K_a, as follows.

The tendency of an atom to attract electrons is called its electronegativity. The relative electronegativity of most common elements has been characterized and quantified, so the tendency of electrons to associate primarily with one or another of the atoms in various compounds can be predicted. For instance, the partial negative charge that resides near the oxygen atom and partial positive charges that reside near the hydrogen atoms in a water molecule reflect the fact that oxygen is more electronegative than hydrogen. Flipping that argument around, we might say that highly electronegative elements are more "comfortable" carrying negative charge than are their less electronegative counterparts. Applying this concept to solutions is a bit risky because interactions with water might change the electronic distribution in molecules, but in general it provides a good perspective from which to explore the relative acidity of various molecules.

Some important acids have a very simple structure, consisting of one or two H^+ ions attached to a single electronegative atom (HCl, H_2S). The acidity of these species arises because when the H^+ ion(s) is (are) released, the element that is left behind is strongly electronegative and hence comfortable carrying the negative charge. However, the vast majority of environmentally significant acids are more complex molecules. In most cases, these acids contain one or more oxygen atoms, and it is to these oxygen atoms that the acidic H^+ ions are attached. Many inorganic acids (HNO_3, H_2SO_4, $HClO_4$, HClO, H_2CO_3) fall into this category, as well as the most environmentally significant group of organic acids (carboxylic acids). If the acid is neutral or anionic, the conjugate base is anionic, and the negative charge resides primarily on the oxygen atoms.

In addition to being useful in a myriad of biochemical and industrial applications, acids can be destructive, as when they attack and dissolve pipe materials. This photograph shows a sewer pipe that collapsed due to "crown corrosion," which occurs when hydrogen sulfide gas [$H_2S(g)$] is released from the sewage and dissolves into water droplets that have condensed on the top of the sewer pipe. Oxygen also dissolves into these droplets and reacts with the H_2S to form sulfuric acid (H_2SO_4), which attacks the pipe structure.

| HDR Engineering, Inc.

One molecule that is a good example of such a system and that has special significance of its own is carbonic acid, H_2CO_3. This diprotic acid (containing two releasable H^+ ions) can lose either or both of its H^+ ions to form bicarbonate or carbonate ion (HCO_3^- or CO_3^{2-}), respectively. All three of these species are planar; i.e., the carbon atom and the three oxygen atoms lie in a plane, with the oxygen atoms roughly at the corners of an equilateral triangle surrounding the carbon, as shown in Figure 3.6. Although simplistic stick-and-ball representations usually show the negative charge on the molecules residing entirely on one oxygen atom in HCO_3^- and on two oxygen atoms in CO_3^{2-}, in fact the negative charge is distributed among all three oxygen atoms. In HCO_3^-, most of the negative charge is split between the two oxygen atoms that are not bound to a

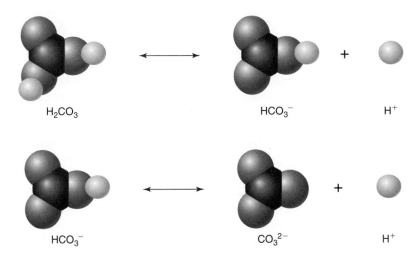

Figure 3.6 Schematic representation of the release of protons from carbonic acid to form bicarbonate and carbonate ions.

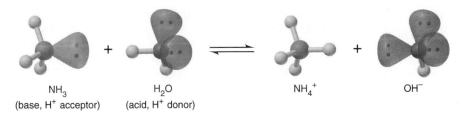

| NH₃ | H₂O | NH₄⁺ | OH⁻ |

NH$_3$ H$_2$O   NH$_4$$^+$ OH$^-$
(base, H$^+$ acceptor)   (acid, H$^+$ donor)

Figure 3.7 Structure of molecular ammonia, NH$_3$, and its protonation reaction.

From M. S. Silberberg, *Chemistry: The Molecular Nature of Matter and Change.* Copyright © 2000 The McGraw-Hill Companies, New York, NY. Reproduced by permission of The McGraw-Hill Companies.

hydrogen ion. However, some of the negative charge resides on the oxygen atom to which the H$^+$ is attached, increasing the strength of the O—H bond compared to the O—H bonds in H$_2$CO$_3$. The dilution of the charge over multiple atoms stabilizes the molecule, while the increased strength of the O—H bond causes HCO$_3$$^-$ to be a weaker acid than H$_2$CO$_3$.

Conceptually, much of the above discussion applies to bases as well, except that strong basicity is associated with electropositivity. The prototypical inorganic base is ammonia, with chemical formula NH$_3$. This molecule has a structure similar to that of water, except that a nitrogen atom is at the center of the tetrahedron instead of oxygen, and three corners of the tetrahedron are occupied by hydrogen atoms, instead of only two. The reaction in which ammonia acquires a fourth H$^+$ ion (thereby acting as a base, because it removes H$^+$ from solution) is very similar to that by which water is protonated to H$_3$O$^+$ (Figure 3.7). The NH$_4$$^+$ ion is referred to as *ammonium* ion (note the similarity with hydronium, H$_3$O$^+$).

3.5   A BRIEF DIVERSION: THE STRUCTURE OF SOME IMPORTANT ORGANIC ACIDS AND BASES

Aside from being a critical species in all environmental systems, carbonic acid can be considered a prototype of a major class of organic acids known as carboxylic acids. These acids have a structure similar to H_2CO_3, except that the bond to one of the OH groups in H_2CO_3 is replaced by a bond to another carbon atom, which might be part of a large organic structure. Conventionally, any such structure (large or small) is represented by the letter R. Also, the HCO_2- group is usually written as COOH, so the whole molecule is represented generically by the formula RCOOH. For instance, one of the simplest carboxylic acids can be thought of as being formed by replacing one OH group in H_2CO_3 by a methyl (CH_3) group, to form acetic acid (one of the acids mentioned above, with chemical formula CH_3COOH). A structural representation of acetic acid is shown in Figure 3.8.

The bulk of the organic matter in natural waters is derived from the microbial decay of plants. Many of the molecules generated by this decay process are used locally by microorganisms for growth and energy, but a small fraction are relatively resistant to further decay and can leach into rivers and lakes, from where they might ultimately be transported to the oceans. Other organic molecules found in the water bodies are generated in situ, e.g., by the decay of bacteria, algae, or other organisms. Regardless of its source, the organic matter that we are likely to detect in natural waters consists primarily of decay-resistant molecules, for the simple reason that the more degradable molecules have already been consumed.

Chemically, some of the earliest steps in the decay process break apart large organic molecules in the source material by an oxidation step that leaves a carboxyl group at the point where the new molecule had previously been joined to the original structure. As a result, many of the organic molecules found in natural waters (collectively referred to as *natural organic matter,* or NOM) contain one or more carboxyl groups. The molecules are also thought to contain phenolic groups (aromatic rings with OH groups attached) that are slightly acidic (the proton can dissociate from the OH group, converting the phenol functional group to a *phenolate* group). Most NOM molecules are therefore polyprotic acids, capable of carrying several anionic charges per molecule. Current research suggests that many of these molecules contain 20 to 30 carbon atoms, at least a few of which are part of carboxylic and phenolic groups (Figure 3.9). Although the acidity of these groups depends on the structure of the entire molecule, most

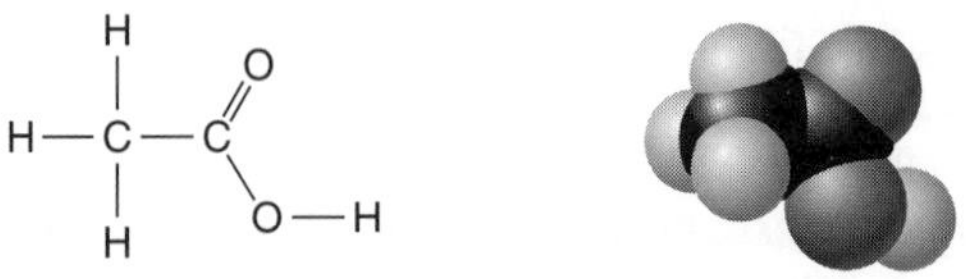

Figure 3.8   Structural representation of acetic acid.

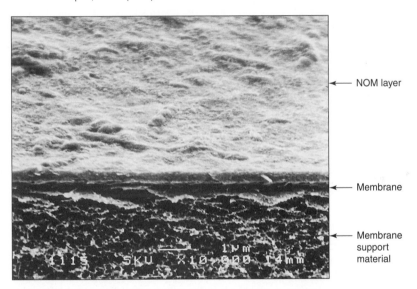

Figure 3.9   A conceptual model for a typical molecule of water-soluble natural organic matter.

Source: After Thurman, E. M. *Organic Geochemistry of Natural Waters,* Nijhoff, Boston (1985).

Natural organic matter can play either a beneficial or a harmful role in many water treatment processes. This scanning electron micrograph shows NOM accumulated on the raw water side of a membrane. The membrane is intended to reject particulate contaminants while letting clean water pass through. The accumulated NOM dramatically increases the resistance to permeation of water through the membrane.

Reprinted from *Journal AWWA,* Vol. 88, No. 12 (December 1996). By permission. Copyright © 1996, American Water Works Association.

carboxylic groups are thought to have K_a values of 10^{-3} to 10^{-6}, while the phenolic groups have K_a values of 10^{-9} to 10^{-13}.

As is the case for H_2CO_3 and organic acids, NH_3 is an inorganic analog for the most common type of organic bases. That is, most organic bases consist of an

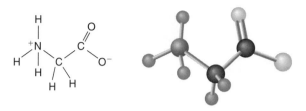

Figure 3.10   Structure of the amino acid glycine.

organic backbone attached at some point to a nitrogen atom, which is (in the simplest case) attached to two H^+ ions, forming a molecule that can be represented generically as $R—NH_2$. Attachment of a third H^+ ion to the nitrogen removes H^+ from solution (so the nitrogen is acting as a base) and forms RNH_3^+, an analog of NH_4^+. Organic nitrogen atoms sometimes are attached to other strands of the organic molecule, rather than just one strand. That is, molecules that might be represented as RNH_2, R_1R_2NH, and $R_1R_2R_3N$ all exist and are referred to as primary, secondary, and tertiary amines, respectively. All these types of molecules can act as bases and form species of the type RNH_3^+, $R_1R_2NH_2^+$, and $R_1R_2R_3NH^+$, respectively. The acidity of most primary amines is fairly close to that of ammonium ion ($K_a = 10^{-9.25}$), but the value does vary depending on the structure of the rest of the molecule; the acidity of secondary and tertiary amines is often in the same neighborhood, but can also vary considerably.

One important group of biomolecules is amino acids, which contain both an organic nitrogen base and a carboxylic acid group. The structure of glycine, a simple amino acid, is shown in Figure 3.10, with the amine group in its acidic form (with the acid H^+ attached) and the carboxyl group in its basic form. Additional information about these relatively simple molecules, which are the basic building blocks of proteins and enzymes, can be found in any organic chemistry or biochemistry text.

3.6   COMBINING ACIDITY REACTIONS AND THE DEFINITION OF BASICITY

When an acidity reaction is reversed, it describes a reaction in which an H^+ ion is consumed. By adding that reaction to the reaction for the dissociation of water, a reaction is generated which is the analog of an acidity reaction, but written to show the release of one OH^- ion rather than an H^+ ion:

Description	Reaction		Equilibrium Constant
Acidity reaction reversed:	$H^+ + A^- \leftrightarrow HA$	**(3.12)**	K_a^{-1}
Water dissociation:	$H_2O \leftrightarrow H^+ + OH^-$	**(3.13)**	K_w
(3.12) + (3.13):	$A^- + H_2O \leftrightarrow HA + OH^-$	**(3.14)**	$K_b = K_a^{-1}K_w$

Reaction (3.14) represents a base being converted to its conjugate acid by the release of one OH^- ion, just as reaction (3.9) represents an acid being converted to its conjugate base by the release of one H^+ ion. Reaction (3.14) is called a **basicity** reaction, and its equilibrium constant is designated K_b. Just as the magnitude of K_a indicates the strength of an acid, the magnitude of K_b indicates the strength of a base. As shown by the equilibrium constant for reaction (3.14), K_a and K_b for a conjugate acid/base pair are related by

$$K_a K_b = K_w \qquad \textbf{(3.15a)}$$

$$pK_a + pK_b = pK_w = 14.0 \qquad \textbf{(3.15b)}$$

According to Equations (3.15), the larger K_a is for an acid, the smaller K_b is for its conjugate base; i.e., strong acids always have weak conjugate bases, and vice versa. This conclusion is logical: since a strong acid is one that has a strong tendency to release H^+, it makes sense that the conjugate base of that acid would have only a slight tendency to acquire an H^+ or, equivalently, to release an OH^-.

The acidity constant for the second deprotonation reaction of phosphoric acid is $K_{a2} = 10^{-7.20}$. What is the corresponding basicity constant, and to what reaction does it apply? | **Example 3.1**

Solution

The value of the basicity constant is K_w divided by the acidity constant, so $K_b = 10^{-6.80}$. (Alternatively, we could note that pK_a is 7.20, so pK_b must be $14.0 - 7.20$, or 6.80.)

The acidity constant describes a reaction in which $H_2PO_4^-$ is the acid and HPO_4^{2-} is the conjugate base. The corresponding basicity reaction is for conversion of the base to the conjugate acid accompanied by release of one hydroxyl ion; i.e., it is for the reaction

$$HPO_4^{2-} + H_2O \leftrightarrow H_2PO_4^- + OH^-$$

When zinc perchlorate, $Zn(ClO_4)_2$, is added to water, the compound readily dissociates, and some of the Zn hydrolyzes to form species with the generic formula $Zn(OH)_x^{2-x}$. No dissolved species are known in which a Zn^{2+} is bound to one or more ClO_4^- ions, and ClO_4^- ions have negligible tendency to combine with H^+. Equilibrium constants for Zn-induced hydrolysis reactions are shown below. The symbols used to represent the reactions are conventional notation for various types of equilibrium constants and are defined in greater detail later in the text; for now, all that is important is that the values shown are the equilibrium constants for the corresponding reactions. | **Example 3.2**

$$Zn^{2+} + H_2O \leftrightarrow ZnOH^+ + H^+ \qquad {}^*K_1 = 10^{-8.96}$$

$$Zn^{2+} + 2OH^- \leftrightarrow Zn(OH)_2^0 \qquad \beta_2 = 10^{+11.10}$$

$$Zn(OH)_2^0 + OH^- \leftrightarrow Zn(OH)_3^- \qquad K_3 = 10^{+2.50}$$

$$Zn^{2+} + 4H_2O \leftrightarrow Zn(OH)_4^{2-} + 4H^+ \qquad {}^*\beta_4 = 10^{-41.20}$$

a. Using the above equilibria and K_w, determine the equilibrium constant for the following reaction (this equilibrium constant is commonly designated β_4):

$$Zn^{2+} + 4OH^- \leftrightarrow Zn(OH)_4^{2-} \qquad \beta_4 = ?$$

b. At what pH or in what pH range do the following relationships hold?
   (i) $\{Zn^{2+}\} = \{ZnOH^+\}$
   (ii) $\{Zn(OH)_2^\circ\} > \{Zn(OH)_4^{2-}\}$

Solution

a. The desired equilibrium constant can be derived by combining the fourth reaction shown above with 4 times the reverse of the water dissociation reaction, as follows:

$$Zn^{2+} + 4H_2O \leftrightarrow Zn(OH)_4^{2-} + 4H^+ \qquad {}^*\beta_4 = 10^{-41.20}$$

$$\underline{4H^+ + 4OH^- \leftrightarrow 4H_2O \qquad\qquad\qquad K_w^{-4} = 10^{+56.00}}$$

$$Zn^{2+} + 4OH^- \leftrightarrow Zn(OH)_4^{2-} \qquad\qquad \beta_4 = 10^{+14.80}$$

Note that since the dissociation of four water molecules is equivalent to combining four "normal" water dissociation reactions, the equilibrium constant for the reaction is K_w^4, and for the reverse reaction, $K = K_w^{-4}$.

b. (i) The relative concentrations of two species can be determined based on the equilibrium constant relating them to each other (in this case, the one designated ${}^*K_1$). Using the expression for ${}^*K_1$, we can derive the following relationship:

$$\frac{\{ZnOH^+\}}{\{Zn^{2+}\}} = \frac{K_1}{\{H^+\}} = \frac{10^{-8.96}}{\{H^+\}}$$

Based on this equation, $\{ZnOH^+\} = \{Zn^{2+}\}$ when $\{H^+\} = 10^{-8.96}$, i.e., at pH = 8.96.

(ii) We can follow a procedure analogous to that used in part (i), except that in this case we need to derive the equilibrium constant of interest. Combining the second reaction given in the problem statement with the result of part (a), we obtain

$$Zn^{2+} + 2OH^- \leftrightarrow Zn(OH)_2^\circ \qquad\qquad \beta_2 = 10^{+11.10}$$

$$\underline{Zn(OH)_4^{2-} \leftrightarrow Zn^{2+} + 4OH^- \qquad\qquad \beta_4^{-1} = 10^{-14.80}}$$

$$Zn(OH)_4^{2-} \leftrightarrow Zn(OH)_2^\circ + 2OH^- \qquad K = 10^{-3.70}$$

$$10^{-3.70} = \frac{\{Zn(OH)_2^\circ\}\{OH^-\}^2}{\{Zn(OH)_4^{2-}\}}$$

Then, manipulating the equilibrium constant expression to generate an expression for the relative activities of $Zn(OH)_2^\circ$ and $Zn(OH)_4^{2-}$ gives

$$\frac{\{Zn(OH)_2^\circ\}}{\{Zn(OH)_4^{2-}\}} = \frac{10^{-3.70}}{\{OH^-\}^2}$$

The resulting expression indicates that $\{Zn(OH)_2^\circ\} > \{Zn(OH)_4^{2-}\}$ when $\{OH^-\}^2 < 10^{-3.70}$, i.e., when $\{OH^-\} < 10^{-1.85}$. Based on K_w, a requirement that $\{OH^-\} < 10^{-1.85}$ is equivalent to $\{H^+\} > 10^{-12.15}$, so we conclude that $\{Zn(OH)_2^\circ\} > \{Zn(OH)_4^{2-}\}$ when pH < 12.15.

Carbonic acid (H_2CO_3) forms by hydration of dissolved carbon dioxide according to the following reaction:

$$CO_2(aq) + H_2O \leftrightarrow H_2CO_3(aq)$$

Example 3.3

Although H_2CO_3 and $CO_2(aq)$ are distinct chemical species, for many applications the distinction between them is unimportant, and they are often lumped together and treated as a single type of molecule. In such cases, the sum of the concentrations or activities of H_2CO_3 and $CO_2(aq)$ is often represented as $[H_2CO_3^*]$ or $\{H_2CO_3^*\}$, respectively, i.e.:

$$\{H_2CO_3^*\} \equiv \{CO_2(aq)\} + \{H_2CO_3\}$$

and

$$[H_2CO_3^*] \equiv [CO_2(aq)] + [H_2CO_3]$$

The first dissociation constant (K_{a1,H_2CO_3}) for carbonic acid, $H_2CO_3(aq)$, is approximately $10^{-3.50}$. The corresponding dissociation constant for $H_2CO_3^*$ is $10^{-6.35}$, i.e.,

$$H_2CO_3^* \leftrightarrow H^+ + HCO_3^- \qquad K_{a1,H_2CO_3^*} = 10^{-6.35}$$

From the given information, compute the equilibrium constant for hydration of $CO_2(aq)$ and comment on the relative concentrations of $CO_2(aq)$ and H_2CO_3 in an equilibrium solution. Assume activity coefficients are all 1.0.

Solution

The equilibrium constant that we seek is for the hydration of $CO_2(aq)$, i.e., it is for the following reaction:

$$CO_2(aq) + H_2O \leftrightarrow H_2CO_3(aq)$$

$$K = \frac{\{H_2CO_3(aq)\}}{\{CO_2(aq)\}\{H_2O\}} = \frac{\{H_2CO_3(aq)\}}{\{CO_2(aq)\}} = ?$$

The acidity constant of H_2CO_3 indicates that $\{H^+\}\{HCO_3^-\}/\{H_2CO_3(aq)\} = 10^{-3.50}$, and that for $H_2CO_3^*$ indicates that $\{H^+\}\{HCO_3^-\}/\{H_2CO_3^*(aq)\} = 10^{-6.35}$. Thus, at equilibrium, the ratio of $\{H_2CO_3^*\}$ to $\{H_2CO_3\}$ is

$$\frac{\{H_2CO_3^*(aq)\}}{\{H_2CO_3(aq)\}} = \frac{K_{a1,H_2CO_3}}{K_{a1,H_2CO_3^*}} = 10^{+2.85}$$

Expanding the numerator in the above expression and carrying out some simple algebra, we have

$$\frac{\{H_2CO_3(aq)\} + \{CO_2(aq)\}}{\{H_2CO_3(aq)\}} = 1 + \frac{\{CO_2(aq)\}}{\{H_2CO_3(aq)\}} = 10^{+2.85} = 708$$

$$\frac{\{CO_2(aq)\}}{\{H_2CO_3(aq)\}} = 707$$

The equilibrium constant that we are seeking is the inverse of this ratio, or 1.41×10^{-3}. The result indicates that the vast majority of the dissolved, undissociated carbon dioxide in an equilibrium solution is present as $CO_2(aq)$ and not as $H_2CO_3(aq)$.

In most texts and the technical literature, the term H_2CO_3 is used to represent not just true H_2CO_3, but rather $H_2CO_3^*$. This convention is followed in this text, unless explicitly stated otherwise.

The preceding examples reinforce the point made earlier that chemical activities and concentrations can be added, multiplied, or manipulated in any other mathematically valid manner, and the result will be equally valid. In previous examples, we multiplied equilibrium constants in order to obtain a new constant for an overall reaction, and in Example 3.3 we added the activities of two distinct species in order to characterize a pair of similar species as a single one. In each case, the validity of the result is ensured simply by the validity of the mathematical operations, without any reference to the chemical implications of the calculations.

3.7   ACID/BASE SPECIATION AS A FUNCTION OF pH; DIAGRAMS OF LOG C VERSUS pH

3.7.1   MORE CHEMICAL SHORTHAND: α NOTATION

The speciation of acids and bases is frequently described by a shorthand known as α (alpha) *notation*. To use the shorthand, we define the sum of the concentrations of all the species in solution that contain a chemical group A as $TOTA$.[5] For instance, if we are interested in the speciation of the phosphate group in a solution, A is identified as PO_4, and $TOTPO_4$ is the sum of the concentrations of H_3PO_4, $H_2PO_4^-$, HPO_4^{2-}, and PO_4^{3-}. Returning to the generic case, α_0 is defined as the fraction of $TOTA$ that is in the most protonated form, and α_i is defined as the fraction of the $TOTA$ that is in a form that has lost i protons. For phosphoric acid, the values of α for the species H_3PO_4, $H_2PO_4^-$, HPO_4^{2-}, and PO_4^{3-} would be defined as follows:

$$\alpha_0 = \frac{[H_3PO_4]}{TOTPO_4} \qquad \alpha_1 = \frac{[H_2PO_4^-]}{TOTPO_4} \qquad \alpha_2 = \frac{[HPO_4^{2-}]}{TOTPO_4} \qquad \alpha_3 = \frac{[PO_4^{3-}]}{TOTPO_4}$$

where $TOTPO_4 = [H_3PO_4] + [H_2PO_4^-] + [HPO_4^{2-}] + [PO_4^{3-}]$. By the definition of α, $\sum \alpha_i$ must always equal 1.0. As is shown next, the α values of acid/base species depend only on the solution pH (assuming all species behave ideally, so the numerical value of each species' concentration equals that of its activity). Thus, writing the concentration of a species A^- as $[A^-] = \alpha_1 (TOTA)$ represents it as the product of one term that depends on solution pH and is

---

[5]The total concentration of a substance is represented in different ways by different authors. The choice here follows that of Morel and Hering (1993), which is consistent with the way that the term is usually represented in software for calculating chemical equilibrium.

independent of $TOTA$ and another term that depends on $TOTA$ and is independent of solution pH.

3.7.2 ACID/BASE SPECIATION AS A FUNCTION OF pH

To explore the relationships between α_i and pH, we next prepare a chart showing, for a generalized monoprotic acid HA, the relative concentrations of HA and A^- as a function of pH and pK_a. By a slight manipulation of the expression defining K_a, we obtain the following relationship:

$$\frac{\{A^-\}}{\{HA\}} = \frac{K_a}{\{H^+\}} \tag{3.16}$$

Taking logarithms of both sides gives

$$\log \frac{\{A^-\}}{\{HA\}} = -\log \{H^+\} + \log K_a = pH - pK_a \tag{3.17}$$

Equations (3.16) and (3.17) are important results that we will use extensively. They indicate that every time the solution pH increases by 1, the value of $\log \{A^-\}/\{HA\}$ also increases by 1, so the ratio $\{A^-\}/\{HA\}$ increases by a factor of 10. This result applies to any two species related by a K_a expression, regardless of the value of K_a, the total concentration of A in the system, or the concentration of other species in the system.

Returning to Equation (3.16), adding a value of 1.0 to each side gives

$$\frac{\{A^-\}}{\{HA\}} + \frac{\{HA\}}{\{HA\}} = \frac{K_a}{\{H^+\}} + \frac{\{H^+\}}{\{H^+\}} \tag{3.18}$$

Assuming that $\gamma_{A^-} = \gamma_{HA} = 1.0$, the sum $\{A^-\} + \{HA\}$ can be equated to $TOTA$. Making this substitution and rearranging the terms, Equation (3.18) can be rewritten as follows:

$$\frac{TOTA}{\{HA\}} = \frac{K_a + \{H^+\}}{\{H^+\}} \tag{3.19}$$

Inverting both sides of Equation (3.19) yields an expression for α_0 as a function of $\{H^+\}$ and K_a. The derivation for α_1 is parallel to that for α_0. The resulting expressions are

$$\alpha_0 = \frac{\{H^+\}}{K_a + \{H^+\}} = \frac{1}{(K_a/\{H^+\}) + 1} \tag{3.20}$$

$$\alpha_1 = \frac{K_a}{K_a + \{H^+\}} = \frac{1}{1 + (\{H^+\}/K_a)} \tag{3.21}$$

Numerical calculations according to Equations (3.17), (3.20), and (3.21) are shown in Table 3.3, and the results are shown graphically in Figure 3.11. The key

Table 3.3   Values of α_0 and α_1 of a monoprotic acid as a function of the ratio of the acidity constant to the H^+ activity*

$\dfrac{K_a}{\{H^+\}}$	$\log \dfrac{K_a}{\{H^+\}}$ $= pH - pK_a$	$\left(1 + \dfrac{K_a}{\{H^+\}}\right)^{-1}$ $= \alpha_0$	$\left(1 + \dfrac{\{H^+\}}{K_a}\right)^{-1}$ $= \alpha_1$	$\dfrac{K_a}{\{H^+\}}$	$\log \dfrac{K_a}{\{H^+\}}$ $= pH - pK_a$	$\left(1 + \dfrac{K_a}{\{H^+\}}\right)^{-1}$ $= \alpha_0$	$\left(1 + \dfrac{\{H^+\}}{K_a}\right)^{-1}$ $= \alpha_1$
10^{-8}	-8	1.00	1.0×10^{-8}	0.03	-1.52	0.97	0.029
10^{-7}	-7	1.00	1.0×10^{-7}	0.06	-1.22	0.94	0.057
10^{-6}	-6	1.00	1.0×10^{-6}	0.08	-1.09	0.93	0.074
10^{-5}	-5	1.00	1.0×10^{-5}	0.10	-1.00	0.91	0.091
10^{-4}	-4	1.00	1.0×10^{-4}	0.30	-0.52	0.77	0.23
10^{-3}	-3	1.00	1.0×10^{-3}	0.50	-0.30	0.67	0.33
10^{-2}	-2	0.99	9.9×10^{-3}	0.70	-0.15	0.59	0.41
10^{-1}	-1	0.91	0.091	0.90	-0.046	0.53	0.47
1	0	0.500	0.500	1	0.00	0.50	0.50
10^{1}	1	0.091	0.91	3	0.48	0.25	0.75
10^{2}	2	9.9×10^{-3}	0.99	5	0.70	0.17	0.83
10^{3}	3	1.0×10^{-3}	1.00	6	0.78	0.14	0.86
10^{4}	4	1.0×10^{-4}	1.0	8	0.90	0.11	0.89
10^{5}	5	1.0×10^{-5}	1.0	10	1.00	0.091	0.91
10^{6}	6	1.0×10^{-6}	1.0	30	1.48	0.032	0.97
10^{7}	7	1.0×10^{-7}	1.0	50	1.70	0.020	0.98
10^{8}	8	1.0×10^{-8}	1.0	70	1.85	0.014	0.99

*NOTE: The final four columns are an expansion of the results in the region $0.01 < \dfrac{K_a}{\{H^+\}} < 100$.

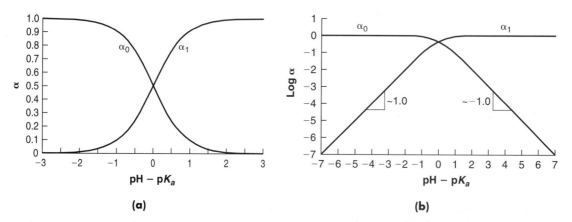

Figure 3.11   (a) Arithmetic and (b) logarithmic representations of the distribution of protonated (α_0) and deprotonated (α_1) forms of a monoprotic acid, as a function of the difference between solution pH and the negative logarithm of the acidity constant pK_a.

characteristics of the system are evident from the slopes and the point of intersection of the two lines. Under conditions where $\{H^+\}$ is larger than K_a by a factor of about 10 or more, corresponding to pH values 1 or more units less than
pK_a, Equations (3.20) and (3.21) indicate that α_0 is nearly 1.0 and α_1 is nearly 0.
Thus, almost all the A in the system is present as the species HA. Conversely, at
pH values significantly greater than pK_a (i.e., under conditions where $\{H^+\}$ is
much smaller than K_a), $\alpha_0 \approx 0.0$, $\alpha_1 \approx 1.0$, and almost all of the A in the system
is present as species A^-. Only in the region where $-1 < pH - pK_a < 1$ are both
species present in significant concentrations relative to $TOTA$. At pH $= pK_a$, α_0
and α_1 equal each other:

$$\text{At pH} = pK_a: \qquad \alpha_0 = \frac{\{H^+\}}{K_a + \{H^+\}} = \frac{\{H^+\}}{\{H^+\} + \{H^+\}} = 0.5 \qquad \textbf{(3.22a)}$$

$$\alpha_1 = \frac{K_a}{K_a + \{H^+\}} = \frac{\{H^+\}}{\{H^+\} + \{H^+\}} = 0.5 \qquad \textbf{(3.22b)}$$

Figure 3.11 can be converted to a plot of log α versus pH for a specific
acid/base conjugate pair by plugging in the appropriate value for pK_a and plotting
the corresponding values on the abscissa. For instance, log α versus pH for the
acetic acid/acetate system ($pK_a = 4.76$) is plotted in Figure 3.12. The great benefits of a graph such as that in Figure 3.12 are that (1) it summarizes a great deal of
speciation information (covering a range of 10^{14} in the H^+ activity) in a concise and
convenient format and (2) it is very easily converted to apply to any other

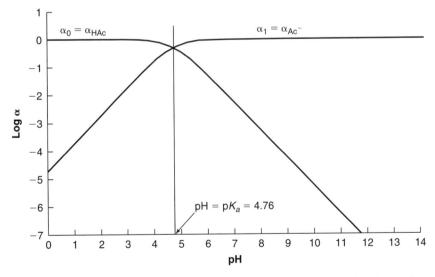

Figure 3.12   Log α versus pH for the acetic acid/acetate pair. Note that the graph is
essentially identical to that in Figure 3.11b, except that the x axis has been converted to a
pH scale, with the value of 4.76 plugged in for pK_a.

monoprotic acid/base group. For instance, to convert Figure 3.12 to a plot of log α versus pH for the hypochlorous acid/hypochlorite system ($pK_a = 7.60$), we simply change the K_a value in the spreadsheet calculations to $10^{-7.60}$, which, on the graph, has the effect of changing the pH at the intersection point of the two curves to 7.60.

3.7.3   LOG C–pH DIAGRAMS FOR MONOPROTIC ACIDS

Assuming that HA and A^- are ideal solutes (each having $\gamma = 1.0$), the activities of HA and A^- in a system can be found by multiplying the corresponding α values by $TOTA$; similarly, the logarithm of $\{HA\}$ and $\{A^-\}$ can be found by adding log $(TOTA)$ to log α.

$$\{HA\} = (TOTA)\alpha_0 = (TOTA)\frac{\{H^+\}}{K_a + \{H^+\}} \qquad \textbf{(3.23a)}$$

$$\log\{HA\} = \log(TOTA) + \log\alpha_0 \qquad \textbf{(3.23b)}$$

$$\{A^-\} = (TOTA)\alpha_1 = (TOTA)\frac{K_a}{K_a + \{H^+\}} \qquad \textbf{(3.24a)}$$

$$\log\{A^-\} = \log(TOTA) + \log\alpha_1 \qquad \textbf{(3.24b)}$$

According to Equations (3.23) and (3.24), modifying the calculations in Table 3.3 to compute $\{HA\}$ and $\{A^-\}$ requires no more than multiplying the appropriate α value from the spreadsheet by $TOTA$. Furthermore, Equations (3.23b) and (3.24b) indicate that a plot of log α versus pH (Figure 3.12) can be converted to one of species concentration versus pH simply by changing the values on the ordinate of the graph. A plot showing log $\{HA\}$ and log $\{A^-\}$ for a single, hypothetical acid with pK_a of 5.5, at three different values of $TOTA$, is shown in Figure 3.13.

It is important to recognize that the curves in Figures 3.11 and 3.13 are simply plots of Equations (3.23) and (3.24). The plots are useful because they allow us to see at a glance trends that occur over many orders of magnitude of change in H^+, but ultimately they contain no information that is different from what is in the equations. Because the concentrations of all the species of interest can be plotted as a function of H^+ (for a given acid at a given total concentration), H^+ is sometimes referred to as a **master variable** in such systems. Graphs such as Figure 3.13 are generically referred to as **log (concentration)–pH** or **log C–pH** graphs.

The point of intersection of the curves representing [HA] and [A^-] is the condition where these two species are present at equal concentrations. Based on Equation (3.16) or Equation (3.22), this equality is satisfied at $\{H^+\} = K_a$, or pH $= pK_a$. Since, under those conditions, each species accounts for one-half of $TOTA$:

At pH $= pK_a$: $[HA] = [A^-] = 0.5(TOTA)$

$$\log[HA] = \log[A^-] = \log 0.5 + \log(TOTA) \qquad \textbf{(3.25)}$$

$$= -0.3 + \log(TOTA)$$

That is, the point of intersection of the curves representing [HA] and [A^-] must be at pH $= pK_a$; and in a system in which these are the only two possible

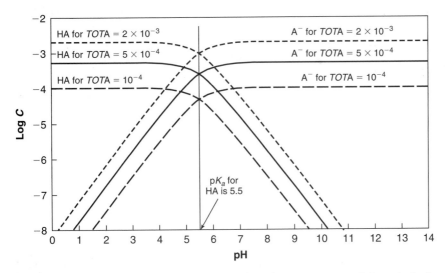

Figure 3.13 Speciation of an acid HA, with acidity constant pK_a = 5.5, in ideal solutions with three different concentrations of *TOTA*.

forms of A, it must be at a point that is 0.3 unit below *TOTA* when plotted on a (base-10) logarithmic scale.

Since the only equations used to prepare the above graphs are the acidity constant and the mass balance on A, the information in the graphs is very generic. That is, by choosing appropriate values for *TOTA* and K_a when carrying out the calculations, the same spreadsheet program can be used to determine the speciation of any monoprotic acid present at any concentration in a solution. Graphically, choosing appropriate values of *TOTA* and K_a corresponds to choosing appropriate scales for the ordinate and abscissa of the log *C*–pH diagram. Plots of the speciation of acetic acid, ammonia, and hypochlorous acid as a function of pH are all shown together in Figure 3.14. Consistent with the above discussion, the graphs for all three acids look essentially identical, but they are displaced from one another vertically, reflecting the different concentrations of the acids in system, and horizontally, reflecting the different strengths of the acids. In particular, the stronger the acid, the farther to the left the lines for HA and A$^-$ shift.

The graphical display of the dependence of speciation on pH provides additional insight into the definition of a strong acid. In particular, note that HOCl is almost completely dissociated at pH 10. Therefore, in a pH 10 solution, it behaves as a strong acid, releasing essentially all its protons. On the other hand, at pH 5, it is almost completely in the HOCl form; adding HOCl to a solution at pH 5 would hardly increase the dissolved H$^+$ concentration at all, because virtually none of the added HOCl would dissociate. Thus, HOCl might behave as a strong acid in one solution and as a weak acid in another.

Note that the slopes of the lines representing HA and A$^-$ are closely related: they change in tandem, so that when the slope of the {HA} line is 0, that of the A$^-$ line is 1, and when the former changes to -1, the latter changes to 0. This

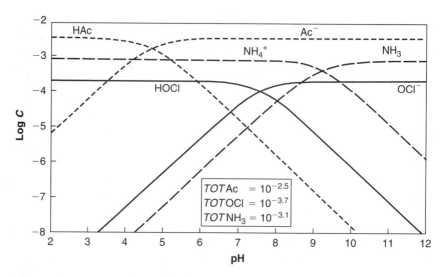

Figure 3.14   Distribution of acetate, hypochlorite, and ammonia species as a function of pH for a system containing $10^{-2.5}$ TOTAc, $10^{-3.7}$ TOTOCl, and $10^{-3.1}$ TOTNH$_3$. The pK_a values for these acids are 4.76, 7.60, and 9.25, respectively.

observation can be generalized to a broad range of systems by manipulating Equation (3.17) as follows:[6]

$$\log \frac{\{A^-\}}{\{HA\}} = -\log\{H^+\} + \log K_a = pH - pK_a \qquad \textbf{(3.17)}$$

$$\log\{A^-\} - \log\{HA\} = pH - pK_a \qquad \textbf{(3.26)}$$

$$\log\{A^-\} = \log\{HA\} - pK_a + pH \qquad \textbf{(3.27)}$$

Differentiating both sides of Equation (3.27) with respect to pH and noting that, since pK_a is a constant, $d\,(pK_a)/d\,(pH)$ is zero, we obtain

$$\frac{d\log\{A^-\}}{d\,pH} = \frac{d\log\{HA\}}{d\,pH} + 1 \qquad \textbf{(3.28)}$$

Equation (3.28) indicates that the slope of a curve of log $\{A^-\}$ versus pH is always greater by 1 than the slope of a curve of log $\{HA\}$ versus pH. Thus, when one of these slopes changes, the other must change as well. This result, as most of the others derived above, is simply a consequence of the mathematical form of the K_a expression and hence applies to all conjugate acid/base pairs under all conditions. We will see shortly that lines representing $\{HA\}$ and $\{A^-\}$ can have slopes larger or smaller than those shown thus far; however, in all cases, the *difference* in their slopes must be 1.0, in accord with Equation (3.28).

[6]Throughout the text, the assumption of ideal behavior is applied to all solutions unless explicitly stated otherwise. Therefore, the y axis on log C–pH diagrams can be interpreted as indicating either molar concentrations or activities. Both ways of interpreting those values are used in the discussion, depending on the particular point being made.

Based on Table 3.3 and Figure 3.10, acids dissociate almost completely if pH $>$ pK_a + 1. This fact, in conjunction with the patterns observed in the log C–pH diagram, make it clear why individuals working in different subfields of water chemistry might have quite different ideas about what constitutes a strong acid. Since systems of pH $<$ 3 are rare in natural systems, any acid with pK_a lower than about 2 is a relatively strong acid in that context; i.e., any acid with pK_a $<$ 2 will be essentially completely dissociated in any natural water system. On the other hand, solutions at pH near 0 are used fairly commonly in analytical aquatic chemistry, so an analytical chemist might consider an acid with pK_a of 2.0 to be very weak, because such an acid is almost fully protonated in many solutions with which that chemist works.

A similar comment applies to strong bases, which are defined as bases that release almost all the OH$^-$ that they contain (or combine completely with H$^+$). Bases that have K_b values of 10^{-1} or 10^{-2} might be considered very strong in some contexts (specifically, if they are present in systems at pOH $>$ 3, pH $<$ 11) and weak in others (e.g., at pOH $<$ 1, pH $>$ 13).

3.7.4 Log C–pH Diagrams for Multiprotic Acids

The relationships characterizing the concentrations and activities of the species comprising multiprotic acid/base systems can be derived by simple extensions of the discussion and derivations above, as shown in the following example.

	Example 3.4

Carbonic acid undergoes two dissociation reactions:

$$H_2CO_3 \leftrightarrow HCO_3^- + H^+ \qquad K_{a1} = 4.47 \times 10^{-7} = 10^{-6.35}$$

$$HCO_3^- \leftrightarrow CO_3^{2-} + H^+ \qquad K_{a2} = 4.65 \times 10^{-11} = 10^{-10.33}$$

Prepare a table showing log α_0, log α_1, and log α_2 as a function of pH for this acid, for pH values of 3.35, 4.35, 5.35, . . . , 13.35, and plot the log α_i values as a function of pH, assuming $\gamma_i = 1.0$ for all species.

Solution

From the acidity constants, the following relationships between the activities of the three carbonate species can be derived:

For the H$_2$CO$_3$/HCO$_3^-$ Pair		For the HCO$_3^-$/CO$_3^{2-}$ Pair	
$K_{a1} = \dfrac{\{H^+\}\{HCO_3^-\}}{\{H_2CO_3\}}$	**(3.29a)**	$K_{a2} = \dfrac{\{H^+\}\{CO_3^{2-}\}}{\{HCO_3^-\}}$	**(3.30a)**
$\log K_{a1} = \log \{H^+\} + \log \dfrac{\{HCO_3^-\}}{\{H_2CO_3\}}$	**(3.29b)**	$\log K_{a2} = \log \{H^+\} + \log \dfrac{\{CO_3^{2-}\}}{\{HCO_3^-\}}$	**(3.30b)**
$pK_{a1} = pH - \log \dfrac{\{HCO_3^-\}}{\{H_2CO_3\}}$	**(3.29c)**	$pK_{a2} = pH - \log \dfrac{\{CO_3^{2-}\}}{\{HCO_3^-\}}$	**(3.30c)**
$\log \dfrac{\{HCO_3^-\}}{\{H_2CO_3\}} = pH - pK_{a1}$	**(3.29d)**	$\log \dfrac{\{CO_3^{2-}\}}{\{HCO_3^-\}} = pH - pK_{a2}$	**(3.30d)**

The ratio of $\{CO_3^{2-}\}$ to $\{H_2CO_3\}$ can be obtained by combining the Equations (3.29d) and (3.30d), as follows:

$$\log \frac{\{CO_3^{2-}\}}{\{H_2CO_3\}} = \log \left(\frac{\{CO_3^{2-}\}}{\{HCO_3^-\}} \cdot \frac{\{HCO_3^-\}}{\{H_2CO_3\}} \right) \qquad \textbf{(3.31a)}$$

$$= \log \frac{\{CO_3^{2-}\}}{\{HCO_3^-\}} + \log \frac{\{HCO_3^-\}}{\{H_2CO_3\}} \qquad \textbf{(3.31b)}$$

$$= (pH - pK_{a2}) + (pH - pK_{a1}) \qquad \textbf{(3.31c)}$$

$$\log \frac{\{CO_3^{2-}\}}{\{H_2CO_3\}} = 2pH - pK_{a1} - pK_{a2} \qquad \textbf{(3.31d)}$$

To convert this information to α values, i.e., terms of the form $\{H_xCO_3\}/(TOTCO_3)$, we can manipulate the equations as follows:

$$\alpha_0 = \frac{\{H_2CO_3\}}{TOTCO_3} = \frac{\{H_2CO_3\}}{\{H_2CO_3\} + \{HCO_3^-\} + \{CO_3^{2-}\}}$$

$$= \frac{1}{\{H_2CO_3\}/\{H_2CO_3\} + \{HCO_3^-\}/\{H_2CO_3\} + \{CO_3^{2-}\}/\{H_2CO_3\}} \qquad \textbf{(3.32)}$$

$$\alpha_0 = \frac{1}{1 + \{HCO_3^-\}/\{H_2CO_3\} + \{CO_3^{2-}\}/\{H_2CO_3\}}$$

(Note that the assumption of ideal behavior is used when $TOTCO_3$ is equated to the sum of the activities of the carbonate species in the denominator of the first expression above.) Since the ratios in the denominator can be evaluated based on Equations (3.28a)

Table 3.4   Distribution of carbonate species as a function of pH

pH	$\log \frac{\{HCO_3^-\}}{\{H_2CO_3\}}$	$\log \frac{\{CO_3^{2-}\}}{\{H_2CO_3\}}$	$\log \frac{\{CO_3^{2-}\}}{\{HCO_3^-\}}$	$\log \alpha_0$	$\log \alpha_1$	$\log \alpha_2$
3.35	−3.00	−9.98	−6.98	−0.0004	−3.0004	−9.9804
4.35	−2.00	−7.98	−5.98	−0.0043	−2.0043	−7.9843
5.35	−1.00	−5.98	−4.98	−0.0414	−1.0414	−6.0214
6.35	0.00	−3.98	−3.98	−0.3010	−0.3010	−4.2810
7.35	1.00	−1.98	−2.98	−1.0418	−0.0418	−3.0218
8.35	2.00	0.02	−1.98	−2.0088	−0.0088	−1.9888
9.35	3.00	2.02	−0.98	−3.0436	−0.0436	−1.0236
10.35	4.00	4.02	0.02	−4.3112	−0.3112	−0.2912
11.35	5.00	6.02	1.02	−6.0596	−1.0596	−0.0396
12.35	6.00	8.02	2.02	−8.0241	−2.0241	−0.0041
13.35	7.00	10.02	3.02	−10.020	−3.0204	−0.0004

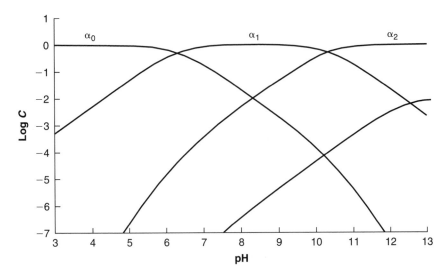

Figure 3.15   Distribution of carbonate species as a function of pH. Carbonic acid is a diprotic acid with $pK_{a1} = 6.33$ and $pK_{a2} = 10.35$. The distribution is represented as the fraction of $TOTCO_3$ present in the most protonated (H_2CO_3, α_0), partially deprotonated (HCO_3^-, α_1), and fully deprotonated (CO_3^{2-}, α_2) forms.

and (3.30a), the above expression for α_0 can be evaluated at any pH. Similarly,

$$\alpha_1 = \frac{1}{\{H_2CO_3\}/\{HCO_3^-\} + 1 + \{CO_3^{2-}\}/\{HCO_3^-\}} \tag{3.33}$$

$$\alpha_2 = \frac{1}{\{H_2CO_3\}/\{CO_3^{2-}\} + \{HCO_3^-\}/\{CO_3^{2-}\} + 1} \tag{3.34}$$

We can insert the above equations into a spreadsheet to prepare the desired table, which is shown as Table 3.4. The α values are plotted in Figure 3.15.

One important result that is apparent from the preceding example and that can be derived following the same logic that led to Equation (3.28) is that, for any multiprotic acid, the slopes of the lines for two species H_nA and H_mA differ by $m - n$ when plotted on a log C–pH diagram. For instance, based on Figure 3.14, the slope of the line representing $\{H_2CO_3\}$ is always 2 less than the slope of the line representing $\{CO_3^{2-}\}$: at pH 5.0, the slopes are 0 and $+2$, respectively; at pH 8.0, they are -1 and $+1$, respectively; and at pH 12.0, they are -2 and 0, respectively.

The relationships characterizing the phosphate system are shown below and graphed for a total PO_4 concentration of $10^{-2}\,M$ in Figure 3.16. All the calculations needed to derive these equations follow logically from the previous example. Given the K_a values for the individual deprotonation reactions (equilibrium reactions 1, 2, and 3 below), you should be able to derive all the other equilibrium

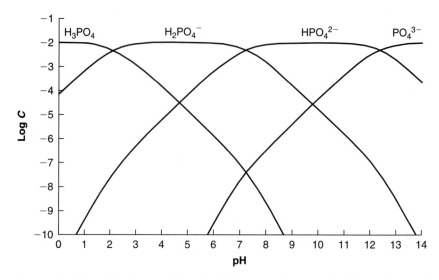

Figure 3.16   Distribution of phosphate species as a function of pH for a system containing 10^{-2} M $TOTPO_4$. Phosphoric acid is a triprotic acid with $pK_{a1} = 2.16$, $pK_{a2} = 7.20$, and $pK_{a3} = 12.35$.

constants and relationships shown. The equations for α derived above and those shown in the table can be generalized to show α_i for any acid that undergoes k acid dissociation reactions, as follows:

$$\alpha_i \equiv \frac{\{H^+\}^{-i} \prod_{j=0}^{i} K_{aj}}{\sum_{k=0}^{n} \left(\{H^+\}^{-k} \prod_{j=0}^{k} K_{aj} \right)} \tag{3.35}$$

where $K_{a0} \equiv 1.0$.

Equilibrium Reaction	Equilibrium Constant	Relationship among Species' Activities
1. $H_3PO_4 \leftrightarrow H_2PO_4^- + H^+$	K_{a1}	$\log \dfrac{\{H_2PO_4^-\}}{\{H_3PO_4\}} = \log K_1 + pH$
2. $H_2PO_4^- \leftrightarrow HPO_4^{2-} + H^+$	K_{a2}	$\log \dfrac{\{HPO_4^{2-}\}}{\{H_2PO_4^-\}} = \log K_2 + pH$
3. $HPO_4^{2-} \leftrightarrow PO_4^{3-} + H^+$	K_{a3}	$\log \dfrac{\{PO_4^{3-}\}}{\{HPO_4^{2-}\}} = \log K_3 + pH$
4. $H_3PO_4 \leftrightarrow PO_4^{3-} + 3H^+$	$K_{03} = K_{a1}K_{a2}K_{a3}$	$\log \dfrac{\{PO_4^{3-}\}}{\{H_3PO_4\}} = \log K_{03} + 3pH$
5. $H_2PO_4^- \leftrightarrow PO_4^{3-} + 2H^+$	$K_{13} = K_{a2}K_{a3}$	$\log \dfrac{\{PO_4^{3-}\}}{\{H_2PO_4^-\}} = \log K_{13} + 2pH$
6. $H_3PO_4 \leftrightarrow HPO_4^{2-} + 2H^+$	$K_{02} = K_{a1}K_{a2}$	$\log \dfrac{\{HPO_4^{2-}\}}{\{H_3PO_4\}} = \log K_{02} + 2pH$

7. $\alpha_0 = \dfrac{\{H_3PO_4\}}{\{H_3PO_4\} + \{H_2PO_4^-\} + \{HPO_4^{2-}\} + \{PO_4^{3-}\}}$

$= \dfrac{1}{\dfrac{\{H_3PO_4\}}{\{H_3PO_4\}} + \dfrac{\{H_2PO_4^-\}}{\{H_3PO_4\}} + \dfrac{\{HPO_4^{2-}\}}{\{H_3PO_4\}} + \dfrac{\{PO_4^{3-}\}}{\{H_3PO_4\}}}$

$= \dfrac{1}{1 + \dfrac{K_{a1}}{\{H^+\}} + \dfrac{K_{a1}K_{a2}}{\{H^+\}^2} + \dfrac{K_{a1}K_{a2}K_{a3}}{\{H^+\}^3}}$

8. $\alpha_1 = \dfrac{\{H_2PO_4^-\}}{\{H_3PO_4\} + \{H_2PO_4^-\} + \{HPO_4^{2-}\} + \{PO_4^{3-}\}}$

$= \dfrac{1}{\dfrac{\{H_3PO_4\}}{\{H_2PO_4^-\}} + 1 + \dfrac{\{HPO_4^{2-}\}}{\{H_2PO_4^-\}} + \dfrac{\{PO_4^{3-}\}}{\{H_2PO_4^-\}}}$

$= \dfrac{1}{\dfrac{\{H^+\}}{K_{a1}} + 1 + \dfrac{K_{a2}}{\{H^+\}} + \dfrac{K_{a2}K_{a3}}{\{H^+\}^2}}$

Alternatively, α_1 can be derived from α_0, as follows:

$\alpha_1 = \alpha_0 \dfrac{\{H_2PO_4^-\}}{\{H_3PO_4\}} = \alpha_0 \cdot \dfrac{K_{a1}}{\{H^+\}} = \dfrac{\alpha_0}{\{H^+\}/K_1}$

9. $\alpha_2 = \dfrac{\{HPO_4^{2-}\}}{\{H_3PO_4\} + \{H_2PO_4^-\} + \{HPO_4^{2-}\} + \{PO_4^{3-}\}}$

$= \dfrac{1}{\dfrac{\{H^+\}^2}{K_{a1}K_{a2}} + \dfrac{\{H^+\}}{K_{a2}} + 1 + \dfrac{K_{a3}}{\{H^+\}}}$

or, if one had already computed α_0, α_2 could be derived easily from it:

$\alpha_2 = \dfrac{\{HPO_4^{2-}\}}{TOTPO_4} = \alpha_0 \cdot \dfrac{\{HPO_4^{2-}\}}{\{H_3PO_4\}} = \alpha_0 \cdot \dfrac{K_{a1}K_{a2}}{\{H^+\}^2} = \dfrac{\alpha_0}{\{H^+\}^2/(K_{a1}K_{a2})}$

10. $\alpha_3 = \alpha_0 \cdot \dfrac{K_{a1}K_{a2}K_{a3}}{\{H^+\}^3} = \dfrac{1}{\dfrac{\{H^+\}^3}{K_{a1}K_{a2}K_{a3}} + \dfrac{\{H^+\}^2}{K_{a2}K_{a3}} + \dfrac{\{H^+\}}{K_{a3}} + 1}$

3.8   DETERMINING SPECIES AND RELEVANT EQUATIONS FOR SOLVING EQUILIBRIUM PROBLEMS

Many chemical equilibrium problems of interest can be posed in the form of questions such as: If known amounts of the following substances are added to an aqueous solution, what species will the system contain, and what will the concentrations of those species be once the system has reached equilibrium? Or conversely: If we wish to convert a known initial system to one containing a targeted amount of a specific species (e.g., if we wish to change solution pH from 5.0 to 7.0), what chemicals, and how much of those chemicals, should we add? These questions have both qualitative and quantitative components. The qualitative aspects can be dealt with based on a general understanding of the properties of the chemicals of interest, and this understanding is developed through study and experience. The immediate goal here, however, is to explore the quantitative aspects, a task that, it turns out, can be largely accomplished using basic algebra. Specifically, in most cases, the problem reduces to one of identifying the unique solution to a set of simultaneous algebraic equations. In this section, we begin to develop the skills for writing and solving those equations.

As a brief review, recall that a system of n independent equations in n unknowns can be solved uniquely if the equations are linear. If the equations are non-linear, there may be more than one solution. Most of the systems we will be dealing with are characterized by at least some non-linear equations; and therefore if we have n unknowns, n equations might not uniquely define the system. However, it will always be possible to choose among the mathematically correct answers by imposing the restriction that the answer involve physically meaningful quantities. For instance, the concentration of all constituents must be real and positive. With these restrictions, n equations will always be sufficient and necessary to define a system with n unknowns, regardless of whether the equations are linear or not.

Equilibrium constant expressions will be among the equations used to solve virtually any such problem of interest, so it is appropriate to review the conditions under which those equations are applicable. First, of course, equilibrium constants are applicable only if the reaction of interest is either known or assumed to reach equilibrium. While this is an important restriction, it is not one with which we concern ourselves here: we will always be interested in computing the expected speciation in a system that has reached equilibrium, whether we believe that state actually exists in the system or whether we are simply investigating a theoretical condition toward which the system moves.

Second, and equally obvious, we should consider only equilibrium reactions and expressions that are independent of one another. For example, if the two following reactions occur in a system

$$A + B \leftrightarrow C + D \qquad K_1 = \frac{\{C\}\{D\}}{\{A\}\{B\}}$$

$$B + C \leftrightarrow F \qquad K_2 = \frac{\{F\}}{\{B\}\{C\}}$$

then it is possible to write a completely valid third reaction by addition of the above two, as follows:

$$A + 2B \leftrightarrow D + F \qquad K_3 = \frac{\{D\}\{F\}}{\{A\}\{B\}^2} = K_1 K_2$$

An analysis of this system could consider the first two reactions, but the inclusion of the third reaction would be redundant, since it does not provide any new information about the system. Of course, the first and third equations, or the second and third, could be used just as well as the first and second. That is, any two of the above three reactions/equations completely describe the equilibrium relationships among A, B, C, D, and F; inclusion of the third equation is neither necessary nor helpful.

Finally, an equilibrium expression is applicable to a system if and only if all the constituents on each side of the reaction are present. In such a case, by the very nature of the equilibrium concept, some of those constituents will combine to form the species on the other side of the reaction, and the reaction will proceed until the equilibrium constant is satisfied. This process might involve conversion of a very small fraction of the existing constituents, almost all of them, or any amount in between; the actual amount is determined by the magnitude of the equilibrium constant. However, regardless of how *much* conversion occurs, we know that if the reaction can be characterized by an equilibrium constant, *some* conversion will take place. Indeed, one could take the preceding statement as a definition of an equilibrium reaction: equilibrium implies some type of balance between reactants and products, and if no conversion occurred, that would indicate that no such balance existed; i.e., the reactants and products do not really participate in an equilibrium reaction.

The preceding discussion suggests an approach for determining which equilibrium constants apply to a given system. Specifically, we can write all the constituents that we know have been added to a solution, consider what reactions occur when they dissolve, and then try to imagine all possible species that can form from the added constituents, either by combination of constituents with one another or by dissociation of one of the components. If we have reason to believe that a particular species can form by any of those processes, we must assume that it will do so.

What, then, is the criterion we should use to conclude that a species can form by some hypothetical reaction that we imagine? Somewhat paradoxically, at least for the purposes of this text, it is the availability of an equilibrium constant for the reaction. That is, the existence of an equilibrium constant *describing* the tendency for a reaction to occur is the only evidence that we can obtain that indicates the reaction *does* occur and that it reaches an equilibrium condition of interest. Of course, our inability to find the equilibrium constant for a reaction does not prove that the reaction cannot occur; such a situation might only indicate that the reaction has never been studied fully, or that the equilibrium lies so far toward the products that the reaction is conventionally treated as being irreversible (i.e., the equilibrium condition is approximated as complete conversion of reactants to products).

If the possibility that a particular reaction could occur were critical to the analysis and we had difficulty finding an equilibrium constant for it, we might first try deriving the equilibrium constant from tabulated thermodynamic values. (Recall that equilibrium constants can be calculated based on standard molar Gibbs energy values.) Alternatively, we could try to evaluate the equilibrium constant experimentally. However, with respect to reactions discussed in this text, we will assume that any equilibrium reaction of interest will be associated with a known or derivable equilibrium constant; i.e., we will take the absence of an equilibrium constant for a hypothetical reaction as an indication that the reaction is not known to occur, that it proceeds to such a slight extent that it can be ignored, or that it proceeds so extensively that it can be ignored (because a negligible amount of the reactants remains at equilibrium). This point is made explicit in the following examples.

Example 3.5

A solution is made by adding hydrogen cyanide, HCN, to water. The acidity constant for dissociation of this species is $K_a = 10^{-9.24}$. Which of the following reactions should be considered in analyzing the acid/base balance in the solution?

$$HCN \leftrightarrow H^+ + CN^-$$

$$H_2O \leftrightarrow H^+ + OH^-$$

$$CN^- \leftrightarrow C^{4+} + N^{3-}$$

$$OH^- \leftrightarrow H^+ + O^{2-}$$

Solution

The species in the first two reactions are known to be present in the solution, and equilibrium constants for those reactions (K_a and K_w) are available, so they should be considered. In the absence of a detailed understanding of the chemical nature and reactivity of CN^- and OH^-, the third and fourth reactions are entirely reasonable reactions to hypothesize. However, lacking any information about their tendency to occur (via values for the equilibrium constants), we will assume that they are not important reactions to consider in the analysis.

Example 3.6

A solution is made by adding sodium cyanide, NaCN, to water. The value of K_a for HCN is given above. The value of K_b for the base CN^- is $10^{-4.76}$. Which of the following reactions should be considered in analyzing the acid/base balance in the solution?

$$H_2O \leftrightarrow H^+ + OH^-$$

$$HCN \leftrightarrow H^+ + CN^-$$

$$NaCN \leftrightarrow Na^+ + CN^-$$

$$CN^- + H_2O \leftrightarrow OH^- + HCN$$

Solution

The salt NaCN dissociates in water to release Na^+ and CN^- ions. We assume this simply because we know that sodium is extremely stable in water and therefore has a tendency to "fall off" any compound to which it is attached when the compound dissolves in an aqueous solution. This assumption is made regarding Na^+, K^+, Cl^-, NO_3^-, and other salt ions throughout the text, unless otherwise indicated.

The same result could also be arrived at by noting that we expect at least some NaCN to dissociate to form Na^+ and CN^-. If the NaCN dissociated only partially, we would need an equilibrium constant to quantify its tendency to do so; in the absence of an equilibrium constant establishing this tendency, we assume that the dissociation is complete.[7]

Given that water is present and that K_w is well established, it is clear that the first reaction needs to be considered. Similarly, knowing that CN^- is generated by dissolution of

---

[7]Another possible assumption is that the NaCN does not dissociate at all. In the absence of any chemical knowledge, this assumption would appear to be just as likely as the assumption that dissolution is complete. Even with minimal chemistry background, however, the information in the preceding sections allows us to guess that the assumption of complete dissolution is a better one.

the NaCN and that H^+ is present from dissociation of water, we see that the constituents on the product side of the second reaction are all present. An equilibrium constant for this reaction is known (K_a), so some HCN is bound to form from the available H^+ and CN^-. As a result, the second reaction must be considered as well.

We know that the third reaction proceeds when NaCN enters the solution, but we have no equilibrium constant characterizing the reaction. We therefore assume that it proceeds essentially to completion, i.e., until no NaCN remains, and do not consider the reaction as a relevant one describing the final equilibrium solution.

The fourth reaction shown above is the basicity reaction for conversion of the base CN^- to its conjugate acid, HCN. This reaction certainly occurs in the system, and its equilibrium constant is known. However, the basicity reaction provides no independent information about the system. The reaction and the corresponding equilibrium constant can be derived by combining the acidity reaction and the dissociation reaction of water. As in the previous example, one could use any two of the three relevant reactions to completely define the system. Thus, it is acceptable to use K_b as one of the constants describing the system quantitatively, in conjunction with either K_w or K_a. Similarly, it is acceptable to use K_w and K_a together to describe the system quantitatively. However, it is not acceptable, or at least not useful, to try to consider all three equilibrium constants in the analysis of the system.

Say we wish to analyze the equilibrium speciation in a system made by adding known quantities of various chemicals to water. The equilibrium concentration of each species in the system represents one *unknown*. To solve for these unknowns, we need to write the same number of equations as there are unknowns; i.e., if there are five species at equilibrium, we need five equations. The following algorithm describes a process by which we can determine (1) how many species will be present at equilibrium and (2) how to write the requisite equations to solve for each concentration. Once these two things are done, the problem becomes simply an algebraic one.

I. Identify species expected to be present at equilibrium.
   A. The species H_2O, H^+, and OH^- are always present in any aqueous solution. However, in all solutions of interest, $\{H_2O\} \approx 1$. Thus, it is not an unknown. Call H^+ and OH^- *type a* species.
   B. For every acid (or base) to be considered, there is a conjugate base (or acid). If one of these is present in the solution, the other must be there also. If an acid or base can undergo multiple dissociation reactions, all the species resulting from these reactions will be present. In general, an acid that undergoes $n - 1$ acid dissociation reactions requires consideration of n species. For example, H_3PO_4 is a triprotic acid; therefore the PO_4 "group" comprises four species (n), which are related to one another via three ($n - 1$) reactions. Call these *type b* species.
   C. If one of the ionic species added to the system is an extremely weak acid or a weak base, it will be present at equilibrium as a salt ion, but the conjugate species will not be present in a significant concentration. For

instance, Cl^-, NO_3^-, Na^+, and K^+ do not combine measurably with H^+ or OH^-, so species such as HCl, HNO_3, NaOH, and KOH can generally be ignored in the equilibrated solution. Call salt ions that are extremely weak acids or bases *type c* species.

D. Having listed all the species resulting from dissociation reactions, and all species generated via acid/base reactions, we now look for species which might form from combinations of type *b* and type *c* species. If there is reason to suspect that a new species can be formed from those that are listed, it should be included. For the purposes of this discussion, if a species can be formed from some combination of previously listed species, an equilibrium constant for the reaction will be provided. Call species formed in this way type *d* species.

II. List equilibrium constants for reactions among the species. These include (referring to the various groups of species types)

Type *a*: One equation (K_w).

Type *b*: For each acid-base system, $n - 1$ acid/base equilibrium equations, where n is the number of species associated with that acid/base system, e.g.,

For HAc: One equation (species: HAc, Ac^-; $n = 2$)

For H_2CO_3: Two equations (species: H_2CO_3, HCO_3^-, CO_3^{2-}; $n = 3$)

For H_3PO_4: Three equations (species: H_3PO_4, $H_2PO_4^-$, HPO_4^{2-}, PO_4^{3-}; $n = 4$)

Type *c*: No equilibrium constants for these species.

Type *d*: One equilibrium constant for each species in this category.

III. List mass balances. Define an *ion type* as any ion or molecule in its most dissociated form. For instance, CO_3^{2-}, Na^+, Cu^{2+}, and NH_3° are ion types, but HCO_3^-, $CuCO_3^\circ$, and NH_4^+ are not. Although OH^- could be considered to be an ion type according to this definition, the $H^+/H_2O/OH^-$ group is present in all solutions at a total concentration that is so much larger than that of any other ion type that it is convenient to treat this group differently, so we will temporarily exclude it from the definition. The number of ion types in a system is equal to the number of weak acid/base systems (groups of type *b* species) plus the number of type *c* species.

A mass balance can be written for every ion type in a system. A mass balance is simply a mathematical expression of the principle of conservation of mass, which can be stated as follows: if something is present in a system initially or is subsequently added to the system, it must be present, in some form, at equilibrium (removing something from the system is included in this statement by considering the process to be a *negative addition*).

When one is dealing with solution chemistry, it is often convenient to define the aqueous phase as the system, in which case processes such as dissolution of a chemical from the gas phase or precipitation of a solid out of the solution would be considered an addition or a removal from the system, respectively. If the system is defined in this way, then for each ion type, a mass balance can be written with the following form:

$$\sum_i (c_i n_{iA})_{\substack{\text{initial} \\ \text{or added}}} = TOTA = \sum_i (c_i n_{iA})_{\substack{\text{at} \\ \text{equilibrium}}} \tag{3.36}$$

where A is an ion type, c_i is the concentration of a species i which contains ion type A, and n_{iA} is the number of A groups in each molecule of i. If we know how the system was prepared, the individual c_i and n_i terms in the first summation in Equation (3.36) are known, so $TOTA$ can be calculated. The mass balance consists of equating this value with the second summation in the equation, in which the individual c_i terms (the concentrations of the species at equilibrium) are the unknowns for which we wish to solve.

All species that are expected to be present in the equilibrium solution and that contain A must be included in the mass balance. For a type b species, these species include all forms of the weak acid or base and any type d species it forms. For a type c species, the mass balance includes the salt ion plus any type d species it forms.

IV. Write the charge balance. This equation, which simply says that all solutions are electrically neutral overall, can be written as $\sum_i c_i z_i = 0$, where c_i is the concentration of a species i in solution and z_i is the ionic charge of species i, including the sign. The summation must include all charged species in solution.[8] It is worth noting that since the units of c_i are moles of i per liter and those of z_i are moles of charge per mole of i, each $c_i z_i$ term in the charge balance has units of moles of charge per liter; i.e., the equation is a balance on the concentration of charge in the solution and not on the concentrations of individual chemical species. It is conventional to place all the cations on one side of the equation and all the anions on the other, in which case the charge balance can be expressed as

$$\sum_{\text{cations}} (c_i |z_i|) = \sum_{\text{anions}} (c_i |z_i|) \tag{3.37}$$

V. Write expressions for the ionic strength of the equilibrium solution and the activity coefficients of all species in it. These expressions are needed because the equilibrium constants that are written in step II express relationships among chemical activities, whereas the mass balances and charge balance in steps III and IV relate concentrations to one another. Although most of the examples presented in the text are simplified by assuming that all solutes behave ideally, that assumption is not usually valid in real systems of interest. An example showing the magnitude of the errors that can be introduced by the assumption of ideal behavior is presented later in the chapter.

---

[8]Alternatively, a mass balance on H or OH can be written and used instead of the charge balance in the problem solution. This approach is described in detail in Chapter 4.

Inclusion of activity coefficients in the calculations is routine in computer programs written specifically to determine the equilibrium speciation in these types of problems.

Example 3.7

List the equilibrium species, and write the relevant equations necessary to solve for the equilibrium speciation in solutions prepared with the following inputs. Assume ideal behavior of all solutes.

a. $10^{-4}\ M\ Na_2CO_3 + 10^{-3}\ M\ HCl + 10^{-3}\ M\ NaHCO_3 + H_2O$

Step I: List the species according to type:

Type a: H^+, OH^-

Type b: The compounds added comprise only one weak acid/base system (carbonate). Three species are associated with this system: CO_3^{2-}, HCO_3^-, and H_2CO_3.

Type c: Na^+, Cl^-

Type d: We need to look at various combinations of type b and type c species to see if they can combine to form a type d species. If a type d species can form, we must have an equilibrium constant to describe the reaction forming it. For this example, there are no type d species to consider.

Step II: Write the relevant equilibrium constants:

Type a species: K_w

Type b species: $H_2CO_3 \leftrightarrow HCO_3^- + H^+ \qquad K_{a1}$
$HCO_3^- \leftrightarrow CO_3^{2-} + H^+ \qquad K_{a2}$

Step III: Write a mass balance (MB) on each ion type other than H^+ or OH^-. These include MBs on CO_3, Na, and Cl. (Again, although MBs involve concentration terms, they are written here in terms of activities, taking advantage of the assumption of ideal behavior of the solutes.)

Ion Type	$\sum_i (c_i n_{iA})_{\text{initial or added}} = \sum_i (c_i n_{iA})_{\text{at equilibrium}}$
CO_3	$(10^{-4})(1) + (10^{-3})(1) = \{CO_3^{2-}\}(1) + \{HCO_3^-\}(1) + \{H_2CO_3\}(1)$
Cl	$(10^{-3})(1) = \{Cl^-\}(1)$
Na	$(10^{-4})(2) + (10^{-4})(1) + (10^{-3})(1) = \{Na^+\}(1)$

Step IV: Write the charge balance (the same comment as in step III applies regarding concentrations and activities):

$$\{Na^+\}(1) + \{H^+\}(1) = \{Cl^-\}(1) + \{OH^-\}(1) + \{HCO_3^-\}(1) + \{CO_3^{2-}\}(2)$$

As noted above, the coefficient 2 following the $\{CO_3^{2-}\}$ term reflects the fact that each mole of CO_3^{2-} accounts for two moles of negative charge.

Step V: Because we are assuming ideal behavior of all solutes, this step can be skipped.

Steps I through IV have identified nine species whose equilibrium concentrations are unknown and nine equations (four equilibrium equations, four mass balances, and one charge balance) that relate these concentrations to one another. Solving the equations simultaneously provides the concentration of all species at equilibrium.

b. $H_2O + 10^{-2} M$ $CdCl_2 + 3 \times 10^{-3} M$ $Cd(Ac)_2$ (cadmium acetate), considering the following reactions:

K_1: K_a for HAc (HAc $\rightarrow$ Ac$^-$ + H$^+$)

K_2: K for the reaction $Cd^{2+} + OH^- \leftrightarrow CdOH^+$

K_3: K for the reaction $CdOH^+ + OH^- \leftrightarrow Cd(OH)_2^{\,0}$

K_4: K for the reaction $Cd^{2+} + Cl^- \leftrightarrow CdCl^+$

Step I: List species:
      Type a: H$^+$, OH$^-$

      Type b: Ac$^-$, HAc
              Cd^{2+}, CdOH$^+$, $Cd(OH)_2^{\,0}$

      Type c: Cl$^-$

      Type d: CdCl$^+$

Step II: Write the equilibrium equations:
      Type a: K_w

      Type b: K_1, K_2, K_3 from the list at the beginning of the problem

      Type c: None

      Type d: K_4 from the list

Step III: Write mass balances on the following ion types: Ac, Cd, Cl.

Ion Type	$\sum_i (c_i n_{iA})_{\text{initial or added}}$	$= \sum_i (c_i n_{iA})_{\text{at equilibrium}}$
Ac	$(3 \times 10^{-3})(2)$	$= \{HAc\}(1) + \{Ac^-\}(1)$
Cd	$(10^{-2})(1) + (3 \times 10^{-3})(1)$	$= \{Cd^{2+}\}(1) + \{CdOH^+\}(1) + \{Cd(OH)_2^{\,0}\}(1) + \{CdCl^+\}(1)$
Cl	$(10^{-2})(2)$	$= \{Cl^-\}(1) + \{CdCl^+\}(1)$

Note that the type d species CdCl$^+$ had to be included in the mass balances for *both* Cd and Cl.

Step IV: Write the charge balance:

$$\{H^+\} + \{Cd^{2+}\}(2) + \{CdOH^+\} + \{CdCl^+\} = \{OH^-\} + \{Ac^-\} + \{Cl^-\}$$

Step V: Again, this step can be skipped.
      The problem consists of nine species (unknowns) and nine equations (five equilibrium equations, three mass balances, and the charge balance) and hence can be solved.

3.9 NUMERICAL APPROACHES FOR SOLVING ACID/BASE PROBLEMS

Having defined some of the nomenclature and stated a few basic principles of chemical equilibrium, we can now solve an impressively large array of important problems. In the following section, a few simple acid/base problems are solved analytically. In some cases, simplifying assumptions are introduced, and the reasoning behind them is discussed. In subsequent chapters, approaches are presented for solving the same types of problems graphically or numerically.

The solutions to the example problems involve the use of unsophisticated and, at times, fairly tedious algorithms for solving a set of simultaneous equations. Furthermore, once some fairly simple techniques are mastered, the solution approaches presented in subsequent chapters are simpler and faster than those presented here, especially for systems containing several acids and bases. In particular, we will see in Chapter 4 that diagrams of log α versus pH such as those shown previously represent concise and highly informative summaries of the information derived by carrying out much of the algebra; i.e., use of the diagrams eliminates the need to do the algebra at all.

Nevertheless, there is value in developing a firm understanding of the manual, analytical, solution algorithm, for two reasons. First, it leads to a deeper understanding of how and why the alternative, faster approaches work. Indeed, the reason that those techniques are more efficient is that they combine several steps of the manual method into one, or they use more sophisticated mathematics; however, these very factors also make them less transparent to the user. Second, some valuable insights can be gained from analysis of very simple systems, and those insights are most easily acquired by a step-by-step analysis, rather than by just considering the final result. The discussion accompanying each of the examples is intended to help provide some of those insights.

3.9.1 THE pH OF SOLUTIONS CONTAINING ONLY WATER AND STRONG ACIDS OR BASES

If a solution is prepared that contains only water and a strong acid or base, determining the pH and speciation at equilibrium is fairly trivial, as shown in the following examples.

Example 3.8	Determine the pH of pure water.

Solution

Species present: H^+, OH^- (type a)

Relevant reactions: $H_2O \leftrightarrow H^+ + OH^-$

Equilibrium relationships: $K_w = \{H^+\}\{OH^-\} = 10^{-14.0}$

Charge balance: $\{H^+\} = \{OH^-\}$

Substitution of the charge balance into the K_w expression yields $K_w = \{H^+\}^2$, which can be solved directly to give $\{H^+\} = 10^{-7.0}$, i.e., pH = 7.0; according to the charge balance, then, $\{OH^-\}$ is also $10^{-7.0}$, and the problem is solved.

Example 3.9	Determine the pH of a solution made by adding HCl or NaOH to water, each at a concentration of 10^{-2} mol/L.

Solution

Strong acids and bases completely dissociate in water, so that we need not consider the undissociated species as being present at equilibrium. Therefore, for the solution made by adding HCl, the system can be summarized as follows:

Species present: H^+, OH^- (type a); Cl^- (type c)

Relevant reactions: $H_2O \leftrightarrow H^+ + OH^-$

Unknowns: $\{H^+\}$, $\{OH^-\}$, $\{Cl^-\}$

Equilibrium relationships: $K_w = \{H^+\}\{OH^-\} = 10^{-14.0}$

Mass balances: $TOTCl = 10^{-2.0} = \{Cl^-\}$

Charge balance: $\{H^+\} = \{OH^-\} + \{Cl^-\}$

Since the solution has been made by adding an acid to water, we can assume tentatively that $\{H^+\} \gg \{OH^-\}$, so the charge balance simplifies to $\{H^+\} = \{Cl^-\}$. Since $\{Cl^-\}$ is known from the mass balance on Cl, the modified charge balance indicates that $\{H^+\} = 10^{-2.0}$. Substituting this value of $\{H^+\}$ into the K_w expression, we find $\{OH^-\} = 10^{-12.0}$. Comparing the computed concentrations of H^+ and OH^-, we see that indeed $10^{-2} \gg 10^{-12}$, so the assumption was correct. The solution pH is therefore $-\log 10^{-2.0}$, or 2.0.

The analysis of the system in which NaOH is added to water is almost the same, except that in this case we make the initial assumption that $\{OH^-\} \gg \{H^+\}$. The system summary is as follows:

Species present: H^+ and OH^- (type a); Na^+ (type c)

Relevant reactions: $H_2O \leftrightarrow H^+ + OH^-$

Unknowns: $\{H^+\}$, $\{OH^-\}$, $\{Na^+\}$

Equilibrium relationships: $K_w = \{H^+\}\{OH^-\} = 10^{-14.0}$

Mass balances: $TOTNa = 10^{-2.0} = \{Na^+\}$

Charge balance: $\{H^+\} + \{Na^+\} = \{OH^-\}$

The calculations are left as an exercise. The assumption that $\{OH^-\} \gg \{H^+\}$ is confirmed by the calculations, and the computed equilibrium concentrations are $\{H^+\} = 10^{-12.0}$, $\{OH^-\} = 10^{-2.0}$, $\{Na^+\} = 10^{-2.0}$. The pH of the solution is therefore 12.0.

Although it is not in the least surprising that addition of a strong acid (10^{-2} M HCl) lowers the solution pH, consideration of the details of the reactions leading to this result can be useful. Prior to the acid addition, the activities of H^+ and OH^- are both 10^{-7}. When the acid is added, almost all of it dissociates, increasing the H^+ concentration (and activity) to approximately $10^{-2} + 10^{-7}$. Assuming that process occurs almost instantaneously, the OH^- activity at that instant is still 10^{-7}, so the product $\{H^+\}\{OH^-\}$ is approximately 10^{-9}. Since this value is considerably larger than K_w, chemical reactions proceed that reduce the H^+ and OH^- concentrations, moving the $H^+/OH^-/H_2O$ system toward equilibrium. Specifically, H^+ and OH^- ions combine to form H_2O.

When water is formed, equal concentrations of H^+ and OH^- ions are consumed. Thus, if 10^{-7} mol/L of H^+ and OH^- combined to form 10^{-7} mol/L additional H_2O in this system, no OH^- would remain in solution, and the new value of the product $\{H^+\}\{OH^-\}$ would be zero, indicating that the reaction had "overshot" equilibrium. Thus, the amount of new H_2O that forms must be less than 10^{-7} mol/L. The corresponding loss of H^+ is therefore bound to be negligible compared to the amount present ($\sim 10^{-2}\,M$), so the final concentration of H^+ will remain close to $10^{-2}\,M$. To satisfy K_w, then, the OH^- concentration must be approximately $10^{-12}\,M$. Since the initial concentration of OH^- is 10^{-7} mol/L, the vast majority (99.999%) of the OH^- originally present combines with H^+ in this process. This conclusion is, of course, exactly the same one we came to above, using a purely mathematical approach to solve the problem.

3.9.2 The pH of Ideal Solutions Containing Weak Acids and Bases

If, instead of a strong acid, we add a weak acid to solution, a sequence of events similar to that described above ensues. However, since the acid dissociates incompletely in this case, the increase in the H^+ concentration and the corresponding decrease in the OH^- concentration are less than in the above example. Consider, for example, the speciation in a solution prepared by adding $10^{-3}\,M$ propionic acid to pure water.

Propionic acid is a three-carbon carboxylic acid ($H_3C—CH_2—COOH$) which we will abbreviate as HPr. It is an intermediate product in the biological

Propionic acid is a key chemical intermediate in the anaerobic degradation of organic matter. If propionic acid accumulates in the solution and is not neutralized or degraded, the treatment process fails. This photograph shows an egg-shaped anaerobic digester at a sewage treatment plant.

| Tom Pantages.

degradation of many waste compounds and is particularly important in anaerobic processes. When it dissolves, the carboxylic acid portion of the molecule can release its proton, in a reaction for which K_a is $10^{-4.87}$. A summary of the information needed to solve for the equilibrium speciation of the solution of interest is as follows.

Species present: H^+ and OH^- (type a)

                                  HPr and Pr^- (type b)

Relevant reactions: $H_2O \leftrightarrow H^+ + OH^-$

                                  $HPr \leftrightarrow H^+ + Pr^-$

Equilibrium relationships: $K_w = \{H^+\}\{OH^-\} = 10^{-14}$

$$K_a = \frac{\{H^+\}\{Pr^-\}}{\{HPr\}} = 10^{-4.87}$$

Mass balance: $TOT\,Pr = 10^{-3.0} = \{HPr\} + \{Pr^-\}$

Charge balance: $\{H^+\} = \{OH^-\} + \{Pr^-\}$

The mass balance is generated by knowing that $10^{-3}\,M$ HPr was added to the initial solution, and that all the Pr must be present as either HPr or Pr^- in the final, equilibrium solution. Note also that since Pr^- is a charged species, it must be included in the charge balance.

The equilibrium speciation can be determined most easily by substituting expressions that depend only on $\{H^+\}$ for the terms on the right-hand of the charge balance, specifically,

$$\{OH^-\} = \frac{K_w}{\{H^+\}}$$

$$\{Pr^-\} = \alpha_{1,HPr}(TOT\,Pr) = \frac{K_a}{\{H^+\} + K_a}(TOT\,Pr)$$

Thus, the charge balance can be rewritten as

$$\{H^+\} = \frac{K_w}{\{H^+\}} + \frac{K_a}{\{H^+\} + K_a}(TOT\,Pr)$$

Multiplying through by $(\{H^+\} + K_a)\{H^+\}$ gives

$$\{H^+\}^3 + K_a\{H^+\}^2 - [K_w + (TOT\,Pr)K_a]\{H^+\} - K_wK_a = 0 \qquad \textbf{(3.38)}$$

The above expression is a cubic equation in one unknown, $\{H^+\}$, and can be solved by trial and error or by software embedded in most spreadsheet applications. The correct solution is the only one that yields a positive, real value for $\{H^+\}$, specifically $\{H^+\} = 10^{-3.963}$; pH = 3.963.

Substituting this value of $\{H^+\}$ into the K_w expression, we can find $\{OH^-\} = 10^{-10.037}$. Then, from the charge balance, $\{Pr^-\} = 10^{-3.963}$. Finally, from the mass balance, $\{HPr\} = 10^{-3} - 10^{-3.963} = 10^{-3.05}$.

Recall that in the analysis of a solution prepared by adding only HCl to water, an assumption that $\{H^+\}$ was much greater than $\{OH^-\}$ simplified the algebra needed to solve the problem. The assumption was checked once the problem was solved, and its validity was confirmed. Making the same assumption in this example might have simplified the algebra similarly. To see if that is the case, we can solve the problem again making that assumption, and see how far off the result is from the exact result obtained above.

In this problem, the assumption $\{H^+\} \gg \{OH^-\}$ simplifies the charge balance to $\{H^+\} \cong \{Pr^-\}$. Substituting this approximate equality into the expression for K_a yields

$$K_a = \frac{\{H^+\}\{Pr^-\}}{\{HPr\}} = \frac{\{H^+\}\{Pr^-\}}{TOTPr - \{Pr\}} \cong \frac{\{H^+\}^2}{TOTPr - \{H^+\}} \qquad \textbf{(3.39)}$$

$$\{H^+\}^2 + K_a\{H^+\} - (TOTPr)K_a = 0$$

As before, the problem can be represented as a polynomial equation in one unknown, but now the polynomial is only second-order (quadratic), whereas before it was third-order (cubic). The solution to the quadratic is $\{H^+\} = 1.1 \times 10^{-4}$; pH = 3.960. That is, making the assumption caused almost no error in the final result. Since $\{OH^-\} = K_w/\{H^+\} = 10^{-10.04}$, it is clear that the assumption $\{H^+\} \gg \{OH^-\}$ was justified.

There is one further assumption we could have made. Knowing that HPr is a *weak* acid (not completely dissociated in typical solutions), we might try the assumption that $\{HPr\} \gg \{Pr^-\}$ in addition to $\{H^+\} \gg \{OH^-\}$. Then the mass balance on propionate would simplify to $\{HPr\} \cong TOTPr$, and the expression for K_a would be

$$K_a = \frac{\{H^+\}\{Pr^-\}}{\{HPr\}} = \frac{\{H^+\}^2}{TOTPr} \qquad \textbf{(3.40)}$$

$$\{H^+\} = \sqrt{10^{-3}K_a} = 10^{-3.93}$$

Then, using the K_w expression, we could find $\{OH^-\} = 10^{-10.07}$, and from the charge balance, $\{Pr^-\} = 10^{-3.93}$. The pH determined using these assumptions is in error by 0.06 unit. The general conclusion to be drawn from this exercise is that the more simplifying assumptions that are made, the easier it is to solve the problem and the less exact the result becomes. Depending on how important it is to determine the pH and solution composition exactly, a given assumption might or might not be acceptable.

An important observation from the above analysis is that both the exact result [Equation (3.38)] and the result obtained when the various simplifying assumptions are made [Equations (3.39) and (3.40)] can be generalized to solutions containing any weak acid at any concentration, simply by inserting the appropriate values of K_a and $TOTA$, as demonstrated in the following example.

Determine the pH of a solution of 10^{-3} M HOCl, using the exact equation developed above and the equations applicable when assumptions are made that $\{H^+\} \gg \{OH^-\}$ and that $\{HOCl\} \gg \{OCl^-\}$. Are the assumptions valid in this case? | **Example 3.10**

Solution

The various pH values can be computed by using Equations (3.38) to (3.40), and by substituting $10^{-7.60}$ for K_a and 10^{-3} for *TOTOCl*. The results are that pH is 5.30 in all cases, so both assumptions are good.

Next, consider the composition of a solution containing the same total amount of Pr (10^{-3} mol/L) as in the preceding analysis, but in which the Pr is added as the salt sodium propionate, NaPr. In this case, when the NaPr dissolves, the sodium is released and undergoes no further reaction. The Pr^- released can combine with H^+ to form HPr. This reaction removes H^+ from solution, so the solution becomes alkaline; i.e., the pH goes up. Following the same algorithm as previously, the system can be summarized as follows:

Species present: H^+ and OH^- (type *a*)

$\qquad\qquad\qquad$ HPr and Pr^- (type *b*)

$\qquad\qquad\qquad$ Na^+ (type *c*)

Without any knowledge of the tendency for Na^+ and Pr^- to interact, we cannot know whether type *d* species such as $NaPr^o$ or $NaPr_2^-$ might be present at equilibrium. If we suspected they might be important species, we would have to search for equilibrium constants describing their formation from the ions; in this case, having no such information, we assume that they are not significant.

Relevant reactions: $\qquad\qquad$ $H_2O \leftrightarrow H^+ + OH^-$

$\qquad\qquad\qquad\qquad\qquad$ $HPr \leftrightarrow H^+ + Pr^-$

Equilibrium relationships: $\quad K_w = \{H^+\}\{OH^-\} = 10^{-14.0}$

$\qquad\qquad\qquad\qquad\qquad K_a = \dfrac{\{H^+\}\{Pr^-\}}{\{HPr\}} = 10^{-4.87}$

Mass balances: $\qquad\qquad$ $TOTPr = 10^{-3.0} = \{HPr\} + \{Pr^-\}$

$\qquad\qquad\qquad\qquad\qquad$ $TOTNa = 10^{-3.0} = \{Na^+\}$

Charge balance: $\qquad\qquad$ $\{H^+\} + \{Na^+\} = \{OH^-\} + \{Pr^-\}$

Note that the reactions, equilibrium constants, and mass balance on Pr are all the same as in the case where HPr was the only constituent added to the solution. The system of equations is marginally more complex than in the previous example because $\{Na^+\}$ appears in the charge balance, but the mass balance on Na is so trivial that we can immediately substitute a known value for $\{Na^+\}$.

As before, we can substitute into the charge balance to derive an equation that contains $\{H^+\}$ as the only unknown, as follows:

$$\{H^+\} + \{Na^+\} = \frac{K_w}{\{H^+\}} + \alpha_1(TOTPr) \qquad \textbf{(3.41a)}$$

$$\{H^+\} + \{Na^+\} = \frac{K_w}{\{H^+\}} + \frac{K_a}{\{H^+\} + K_a}(TOTPr) \qquad \textbf{(3.41b)}$$

Multiplying Equation (3.41b) through by $\{H^+\}$ and then $\{H^+\} + K_a$, we obtain the following cubic equation, which can be solved by trial and error:

$$\{H^+\}^3 + (K_a + \{Na^+\})\{H^+\}^2 - [K_w + (TOTPr + \{Na^+\})K_a]\{H^+\} - K_w K_a = 0$$

$$\textbf{(3.42)}$$

The result for $TOTPr = \{Na^+\} = 10^{-3}$ is pH = 7.93. Substituting this value into the other expressions characterizing the system, we can find $\{OH^-\} = \{HPr\} = 8.7 \times 10^{-7}, \{Pr^-\} = 9.99 \times 10^{-4}$.

As above, simplifying assumptions could be used to reduce the order of the polynomial, if desired. For instance, since NaPr is a base, we expect the pH at equilibrium to be greater than 7.0, and hence we expect that $\{H^+\}$ will be much less than $\{Na^+\}$ (which we know is 10^{-3}). Assuming that $\{H^+\}$ is negligible in the charge balance leads to the following quadratic expression, for which the solution is also pH = 7.93:

$$\{Na^+\}\{H^+\}^2 - [K_w + (TOTPr + \{Na^+\})K_a]\{H^+\} - K_w K_a = 0 \qquad \textbf{(3.43)}$$

Note that the equations used to solve this problem, i.e., the mass balance on Pr and the K_a and K_w expressions, apply to any system containing water and propionate species, and that the charge balance applies to any system that contains H^+, OH^-, Pr^-, and Na^+ as the only ions. That is, this set of equations is not restricted to systems where $\{Na^+\} = TOTPr$, as is the case when NaPr is added, but rather applies to solutions made by any combination of HPr, NaPr, and NaOH. As long as the proper values are used for $\{Na^+\}$ and $TOTPr$, Equation (3.42) applies.

Example 3.11

What is the pH of a solution prepared by adding $6 \times 10^{-4}\ M$ HPr, $2 \times 10^{-4}\ M$ NaPr, and 3×10^{-4} NaOH to water? What is the speciation of the Pr at equilibrium?

Solution

The activity of Na^+ at equilibrium will include contributions from the NaPr and the NaOH, both of which are assumed to dissociate completely when they dissolve. Similarly, $TOTPr$ will include contributions from both the HPr and NaPr additions. The equilibrium pH can therefore be determined by plugging values of 5×10^{-4} and 8×10^{-4} for $\{Na^+\}$ and $TOTPr$, respectively, into Equation (3.43). The solution to the equation, found by trial and error, is pH = 5.11. The activities of the propionate species can be determined by computing α_0 and α_1 at this pH, and multiplying the respective values by $TOTPr$, yielding $\{HPr\} = 2.92 \times 10^{-4}$ and $\{Pr^-\} = 5.08 \times 10^{-4}$.

As shown in Example 3.10, the equilibrium pH of a solution of 10^{-3} M HOCl ($pK_a = 7.60$) is 5.30; carrying out the corresponding calculations for a solution of 10^{-3} M NaOCl indicates that such a solution would have pH 9.30 (this calculation is left as an exercise). The corresponding values for addition of HPr and NaPr, computed above, are 3.90 and 7.95, respectively.

Comparing the results for addition of HPr and HOCl, we note that both acids would be considered weak acids because they dissociate only partially in water. Since an acid can be defined as a substance that donates protons to solution, the greater the tendency to donate H^+, the stronger the acid. Thus propionic acid, which is 50% dissociated at pH 4.87 (its pK_a), is a stronger acid than hypochlorous acid, which is only ~0.2% dissociated at the same pH. This is also shown by the fact that 10^{-3} M HPr leads to a pH 3.90 solution and 10^{-3} M HOCl leads to a pH 5.30 solution. The greater acid strength of HPr is also indicated by the fact that its acidity constant is larger (pK_a is smaller) than that of HOCl.

Conversely, propionate and hypochlorite ions are bases, because they remove H^+ from solution. Adding 10^{-3} M NaPr or NaOCl to water raises the pH to 7.95 or 9.30, respectively. Thus, NaOCl must be a stronger base than NaPr. This result is consistent with the previous discussion which indicated the stronger the acid (HPr > HOCl), the weaker the conjugate base (NaPr < NaOCl).

3.9.3   A SIMPLE SPREADSHEET ANALYSIS FOR DETERMINING ACID/BASE SPECIATION

Although the approach presented above is straightforward and reasonably simple for systems in which only one monoprotic acid/base group is present, the order of the polynomial that must be derived increases with each additional acid/base group that is present, and also if acid/base groups that can exchange more than one H^+ ion (e.g., H_2CO_3 or H_3PO_4) must be considered. In such cases, an alternative numerical approach that avoids much of the algebra can be taken by solving the charge balance equation directly, after substituting expressions into it that characterize the ionic concentrations in terms of α values. This approach is particularly easy to use in conjunction with spreadsheet software.

For instance, as noted above, the charge balance for a generic solution containing H^+, OH^-, Pr^-, and Na^+ as the only ions can be written follows:

$$\{H^+\} + \{Na^+\} = \{OH^-\} + \{Pr^-\}$$

$$\{H^+\} + \{Na^+\} = \frac{K_w}{\{H^+\}} + (TOTPr)\, \alpha_{1,Pr} \qquad \textbf{(3.44)}$$

By preparing a spreadsheet with one column for pH (the master variable) and a separate column for each term in Equation (3.44), we can rapidly identify the pH where the equation is satisfied, along with the activities of each of the species. The relevant values can be computed based on the following equations, all developed earlier in the chapter:

$$\{H^+\} = 10^{-pH} \qquad \{OH^-\} = \frac{K_w}{\{H^+\}} \qquad \{Na^+\} = TOT\,Na$$

$$\{HPr\} = \alpha_0(TOT\,Pr) \qquad \{Pr^-\} = \alpha_1(TOT\,Pr)$$

where $\alpha_0 = \{H^+\}/(\{H^+\} + K_a)$ and $\alpha_1 = K_a/(\{H^+\} + K_a)$. Note that these calculations are essentially identical to those used earlier to develop the log C–pH diagram.

The spreadsheet is shown in Table 3.5 and includes a column for $\{HPr\}$, because the activity of this species is of interest, even though it is not needed to solve the charge balance equation. The final column in the spreadsheet shows the charge balance at each pH, i.e., the value of the expression $\{H^+\} + \{Na^+\} - \{OH^-\} - \{Pr^-\}$.

The values of $\{H^+\}$, $\{OH^-\}$, $\{Na^+\}$, $\{Pr^-\}$, and $\{HPr\}$ in any row of the spreadsheet satisfy the mass balances on Na and Pr and the K_a and K_w equations for the corresponding pH. However, the values in most rows are not consistent with the

Table 3.5   Speciation and net calculated charge as a function of pH in a solution of 10^{-3} M NaPr

pH	$\{H^+\}$	$\{OH^-\}$	$\{Na^+\}$	$\{Pr^-\}$	$\{HPr\}$	Net Charge
4.0	1.00E–04	1.00E–10	1.00E–03	1.19E–04	8.81E–04	9.81E–04
5.0	1.00E–05	1.00E–09	1.00E–03	5.74E–04	4.26E–04	4.36E–04
6.0	1.00E–06	1.00E–08	1.00E–03	9.31E–04	6.90E–05	7.00E–05
7.0	**1.00E–07**	**1.00E–07**	**1.00E–03**	**9.93E–04**	**7.36E–06**	**7.36E–06**
8.0	**1.00E–08**	**1.00E–06**	**1.00E–03**	**9.99E–04**	**7.41E–07**	**−2.49E–07**
9.0	1.00E–09	1.00E–05	1.00E–03	1.00E–03	7.41E–08	−9.92E–06
10.0	1.00E–10	1.00E–04	1.00E–03	1.00E–03	7.41E–09	−1.00E–04
11.0	1.00E–11	1.00E–03	1.00E–03	1.00E–03	7.41E–10	−1.00E–03
7.0	1.00E–07	1.00E–07	1.00E–03	9.93E–04	7.36E–06	7.36E–06
7.1	7.94E–08	1.26E–07	1.00E–03	9.94E–04	5.85E–06	5.81E–06
7.2	6.31E–08	1.58E–07	1.00E–03	9.95E–04	4.66E–06	4.56E–06
7.3	5.01E–08	2.00E–07	1.00E–03	9.96E–04	3.70E–06	3.55E–06
7.4	3.98E–08	2.51E–07	1.00E–03	9.97E–04	2.94E–06	2.73E–06
7.5	3.16E–08	3.16E–07	1.00E–03	9.98E–04	2.34E–06	2.05E–06
7.6	2.51E–08	3.98E–07	1.00E–03	9.98E–04	1.86E–06	1.49E–06
7.7	2.00E–08	5.01E–07	1.00E–03	9.99E–04	1.48E–06	9.96E–07
7.8	1.58E–08	6.31E–07	1.00E–03	9.99E–04	1.17E–06	5.58E–07
7.9	**1.26E–08**	**7.94E–07**	**1.00E–03**	**9.99E–04**	**9.32E–07**	**1.51E–07**
8.0	**1.00E–08**	**1.00E–06**	**1.00E–03**	**9.99E–04**	**7.41E–07**	**−2.49E–07**

charge balance equation. To find the equilibrium pH, i.e., the row in the spreadsheet that characterizes the system at equilibrium, we need to determine the pH at which the electroneutrality condition is satisfied. This is easily accomplished by scanning the final column to identify the pH where the charge balance expression equals zero.

The key rows in the spreadsheet are shown in boldface. The top portion of the spreadsheet has increments of 1.0 pH unit and indicates that the electroneutrality condition is satisfied between pH 3.0 and 4.0; the lower portion, with increments of 0.1 pH unit, narrows the range to between 3.9 and 4.0 (the interpolated value is 3.96), consistent with the result obtained above.

The spreadsheet approach offers the possibility of solving for the solution pH directly once the charge balance has been written. It has the substantial advantage over the algebraic approach presented above that, for each additional acid/base group in the system, we need only add more columns to the spreadsheet, without making any changes to those that have already been programmed. Furthermore, the similarity of the expressions for α for different acid/base groups means that once we have developed the expression for any given acid/base group, the expressions for other groups can be written by minor modifications of that template. A closely related approach that allows the equilibrium pH to be determined even more quickly is to program the charge balance expression into a cell in the spreadsheet and use embedded software (such as the Solver routine embedded in Excel[9]) to find the value of $\{H^+\}$ that causes the equation to be satisfied.

Next, we use this approach to address an important and slightly different type of acid/base problem: determining the amount of acid or base that must be added, or must have been added, to systems where the equilibrium pH is known. To explore this issue, we consider an example solution whose pH has been lowered to 4.4 by addition of HPr to pure water, and we ask ourselves: How much HPr was needed?

Although the variables whose values are known in this situation differ from those in the systems we investigated above (the final pH is known, but the total Pr concentration is not), the equations used to solve the problem (mass balance, charge balance, and equilibrium constants) are identical, i.e.,

Species present:	H^+ and OH^- (type a)
	HPr and Pr^- (type b)
Relevant reactions:	$H_2O \leftrightarrow H^+ + OH^-$
	$HPr \leftrightarrow H^+ + Pr^-$
Equilibrium relationships:	$K_w = \{H^+\}\{OH^-\} = 10^{-14.0}$
	$K_{a,HPr} = \dfrac{\{H^+\}\{Pr^-\}}{\{HPr\}} = 10^{-4.87}$
Charge balance:	$\{H^+\} = \{OH^-\} + \{Pr^-\}$

[9]Solver and Excel are registered trademarks of Microsoft Corp. (Redmond WA).

Since the equilibrium value of $\{H^+\}$ is known, $\{OH^-\}$ can be computed directly from K_w, allowing us to then compute $\{Pr^-\}$ directly from the charge balance:

$$\{OH^-\} = \frac{K_w}{\{H^+\}} = 10^{-9.6}$$

$$10^{-4.4} = 10^{-9.6} + \{Pr^-\}$$

$$\{Pr^-\} = 10^{-4.4} = 4.0 \times 10^{-5}$$

Finally, $\{HPr\}$ can be determined from K_a, and $TOTPr$ can be computed from the mass balance:

$$\text{From } K_a: \quad \{HPr\} = \frac{\{H^+\}\{Pr^-\}}{K_a} = \frac{(10^{-4.4})(10^{-4.4})}{10^{-4.87}} = 10^{-3.93} = 1.2 \times 10^{-4}$$

$$TOTPr = 4.0 \times 10^{-5} + 1.2 \times 10^{-4} = 1.6 \times 10^{-4} = 10^{-3.80}$$

The amount of HPr that must be added is $TOTPr$ $(1.6 \times 10^{-4}\ M)$, of which approximately 25% dissociates.

Example 3.12 | A solution is made by adding $10^{-3}\ M$ HOCl and $10^{-3}\ M$ NaPr to water. Then NaOH is added until the solution pH is 7.0. How much NaOH must be added?

Solution

Following the same procedure as in the preceding example, we write the system summary:

Species present:	H^+ and OH^- (type a)
	HPr and Pr^-; HOCl and OCl^- (type b)
	Na^+ (type c)
Relevant reactions:	$H_2O \leftrightarrow H^+ + OH^-$
	$HPr \leftrightarrow H^+ + Pr^-$
	$HOCl \leftrightarrow H^+ + OCl^-$
Equilibrium relationships:	$K_w = \{H^+\}\{OH^-\} = 10^{-14.0}$
	$K_{a,HPr} = \dfrac{\{H^+\}\{Pr^-\}}{\{HPr\}} = 10^{-4.87}$
	$K_{a,HOCl} = \dfrac{\{H^+\}\{OCl^-\}}{\{HOCl\}} = 10^{-7.6}$
Mass balances:	$TOTPr = 10^{-3.0} = \{HPr\} + \{Pr^-\}$
	$TOTOCl = 10^{-3.0} = \{HOCl\} + \{OCl^-\}$
	$TOTNa = ? = \{Na^+\}$
Charge balance:	$\{H^+\} + \{Na^+\} = \{OH^-\} + \{Pr^-\} + \{OCl^-\}$
Other known information:	pH = 7.0, i.e., $\{H^+\} = 10^{-7.0}$

In this problem, the total amount of Na in the system is unknown, since we do not know how much NaOH has to be added to adjust the pH to 7.0. Therefore, we have three equilibrium expressions, two mass balances, and a charge balance to work with (six independent equations). There are seven species in the system, but one of them is H^+, for which the final concentration is specified. Thus, the number of equations (6) match the number of unknowns, and the problem can be solved.

Substituting into the charge balance, we obtain the following equation, with $\{Na^+\}$ as the only unknown:

$$\{H^+\} + \{Na^+\} = \frac{K_w}{\{H^+\}} + \alpha_{1,HPr}(TOTPr) + \alpha_{1,HOCl}(TOTOCl)$$

The pH is known, so the α_1 value of each weak acid is also known; in this case, at pH 7.0, the values are

$$\alpha_{1,HOCl} = \frac{K_{a,HOCl}}{\{H^+\} + K_{a,HOCl}} = \frac{10^{-7.6}}{10^{-7.0} + 10^{-7.6}} = 0.201$$

$$\alpha_{1,HPr} = \frac{K_{a,HPr}}{\{H^+\} + K_{a,HPr}} = \frac{10^{-4.87}}{10^{-7.0} + 10^{-4.87}} = 0.993$$

Since $\alpha_{1,HOCl}$ is much less than $\alpha_{1,HPr}$, a smaller fraction of the HOCl ($\sim$20%) than the HPr ($\sim$99%) dissociates. This result is consistent with our expectations, since HOCl is a weaker acid than HPr.

Substituting the α values and the values for $TOTPr$, $TOTOCl$, K_w, and $\{H^+\}$ into the rewritten charge balance, we can solve for $\{Na^+\}$. The computed value is

$$\{Na^+\} = 1.194 \times 10^{-3} \, M$$

Since $1.0 \times 10^{-3} \, M \, Na^+$ was added with the NaPr, the increment that must be added with the NaOH is $1.94 \times 10^{-4} \, M$, and this is the amount of NaOH added.

3.9.4   THE pH AND SPECIATION OF SYSTEMS CONTAINING MULTIPROTIC WEAK ACIDS AND BASES

The analysis of solutions containing multiprotic acid/base groups is a simple and direct extension of the analysis for monoprotic systems, as is shown in the following example.

A solution is made by adding $2 \times 10^{-3} \, M \, NaHCO_3$ to water. What is the equilibrium pH of the solution? | **Example 3.13**

Solution

This problem is very similar to prior examples in which we found the pH of solutions to which HPr or NaPr had been added. The only difference is that in the current problem, we need to consider both acid dissociation reactions that relate the three carbonate species, and we need to consider all three carbonate species in the carbonate mass balance.

It should be apparent from the previous few examples that we could write the charge balance equation with $\{H^+\}$ as the only unknown and could solve for the equilibrium pH without even going through the preliminary steps of listing the species, reactions, equilibrium relationships, etc. Nevertheless, for the sake of completeness, we will carry out those steps:

Species present:

H$^+$ and OH$^-$ (type a)

H$_2$CO$_3$, HCO$_3$$^-$, and CO$_3$$^{2-}$ (type b)

Na$^+$ (type c)

Relevant reactions:

H$_2$O $\leftrightarrow$ H$^+$ + OH$^-$

H$_2$CO$_3$ $\leftrightarrow$ H$^+$ + HCO$_3$$^-$

HCO$_3$$^-$ $\leftrightarrow$ H$^+$ + CO$_3$$^{2-}$

Equilibrium relationships:

$K_w = \{H^+\}\{OH^-\} = 10^{-14.0}$

$$K_{a1} = \frac{\{H^+\}\{HCO_3^-\}}{\{H_2CO_3\}} = 10^{-6.35}$$

$$K_{a2} = \frac{\{H^+\}\{CO_3^{2-}\}}{\{HCO_3^-\}} = 10^{-10.33}$$

Mass balances:

$TOTCO_3 = 2 \times 10^{-3.0} = \{H_2CO_3\} + \{HCO_3^-\}$
$+ \{CO_3^{2-}\}$

$TOTNa = 2 \times 10^{-3.0} = \{Na^+\}$

Charge balance:

$\{H^+\} + \{Na^+\} = \{OH^-\} + \{HCO_3^-\} + 2\{CO_3^{2-}\}$

Substituting the known value of $\{Na^+\}$ and the expression for $\{OH^-\}$ into the charge balance, we get

$$\{H^+\} + 2 \times 10^{-3} = \frac{K_w}{\{H^+\}} + \alpha_1(TOTCO_3) + 2\alpha_2(TOTCO_3)$$

As in the preceding examples with monoprotic acid/base groups, all the terms in the charge balance can be written as expressions that contain only known values and $\{H^+\}$. These expressions can be programmed into a spreadsheet, and the pH at which the charge balance is satisfied can then be identified. In this case, the equation is satisfied at $\{H^+\} = 10^{-8.28}$, i.e., pH = 8.28. At that pH, almost all the carbonate (1.96×10^{-3}) is present as HCO$_3$$^-$, and the values of $\{H_2CO_3\}$ and $\{CO_3^{2-}\}$ are 2.06×10^{-5} and 1.87×10^{-5}, respectively.

3.9.5 ACID/BASE EQUILIBRIA IN NON-IDEAL SOLUTIONS

To this point in the chapter, we have considered only ideal solutes, i.e., solutes whose activity coefficients are 1.0, so that their activities can be equated with the numerical value of their molar concentrations. In most situations of interest, this simplification will not apply.

While the incorporation of activity coefficients in the calculations is mathematically trivial, the resulting equations become more complex-looking, potentially obscuring the key points being made. Nevertheless, it is important to acknowledge the role that non-ideal behavior can play in speciation, so that issue is addressed in a final example that closes the chapter.

Example 3.14

Hydrofluoric acid is a weak acid ($pK_a = 3.17$) that is used extensively in industry because F^- attaches strongly to Fe^{3+} and Al^{3+} ions. As a result, it is an excellent reagent for removing surface scales from iron and aluminum metal before painting or other surface finishing operations. Compare the pH and the concentrations of HF and F^- in a solution prepared by adding $10^{-3}\ M$ HF to pure water, with that of a solution prepared the same way, except using water containing $0.1\ M$ NaCl. Assume that the ionic strength is low enough in the NaCl-free solution that the solution behaves ideally, and use the Davies equation to estimate activity coefficients in the solution containing the NaCl. (*Note*: pH is defined as $-\log\{H^+\}$, *not* $-\log[H^+]$).

Solution

The mass balance (MB) on *TOT*F is identical in the two solutions:

MB: $$TOTF = 10^{-3} = [HF] + [F^-]$$

The charge balances are also effectively identical because, although Na^+ and Cl^- contribute to the total charge in the solution to which NaCl was added, these contributions exactly cancel each other. Therefore, in each case, the charge balance (CB) can be written as

CB: $$[H^+](1) + [\cancel{Na^+}](1) = [OH^-](1) + [F^-](1) + [\cancel{Cl^-}](1)$$

As a reminder, the mass balances and charge balance for a system are *always* based on concentrations (one Na^+ ion represents a charge of +1 regardless of the ionic strength of the solution). In previous examples, these equations were written in terms of species' activities, because the solutions were assumed to behave ideally. In this example, that assumption is not being made, so the equations must be written in terms of species' concentrations.

Equilibrium constants always relate species' activities. Thus, the relevant equilibrium constant expressions for this problem are

$$K_w = \{H^+\}\{OH^-\} = 10^{-14.0}$$

$$K_a = \frac{\{H^+\}\{F^-\}}{\{HF\}} = 10^{-3.17}$$

Because the charge balance and mass balance have already been written in terms of species' concentrations, it is convenient to write the equilibrium constants in those terms as well. This can be accomplished by writing the activity of each solute as the product of its concentration and its activity coefficient, i.e.,

$$K_w = \gamma_{H^+}[H^+]\gamma_{OH^-}[OH^-] = 10^{-14.0}$$

$$K_a = \frac{\gamma_{H^+}[H^+]\gamma_{F^-}[F^-]}{\gamma_{HF}[HF]} = 10^{-3.17}$$

Because all the activity coefficients are 1.0 in the solution that behaves ideally, it can be treated as having four unknowns (the concentrations of the four species). We can solve for these unknowns using the mass balance, charge balance, and two equilibrium expressions. The solution containing NaCl, on the other hand, has the same four unknowns, along with four unknown activity coefficients. Therefore, four additional equations are needed. These additional equations are, of course, the ones that relate the activity of each species to its concentration.

Because the ionic strength is known in the solution containing NaCl, the activity co-efficients of the three ions of interest (H^+, OH^-, and F^-) can be computed without knowing the ultimate speciation of *TOTF*. Using the Davies equation (which depends only on the charge of the ion), γ for all three ions is estimated to be 0.78. The value of γ_{HF} is assumed to be 1.0, since HF is uncharged. Thus, the four additional equations that characterize the solution with NaCl are

$$\{H^+\} = 0.78\,[H^+] \qquad \{OH^-\} = 0.78\,[OH^-] \qquad \{F^-\} = 0.78\,[F^-] \qquad \{HF\} = 1.0\,[HF]$$

Substituting these expressions into the K_w and K_a expressions yields *effective* or *conditional* equilibrium constants (^cK) that describe equilibrium relationships among species concentrations, rather than activities:

$$^cK_w \equiv [H^+][OH^-] = \frac{10^{-14.0}}{(0.78)^2} = 10^{-13.78}$$

$$^cK_a \equiv \frac{[H^+][F^-]}{[HF]} = \frac{10^{-3.17}}{(0.78)^2} = 10^{-2.95}$$

The constants are conditional in the sense that they apply only under the conditions specified (0.1 M ionic strength), whereas the activity-based equilibrium constants apply at any ionic strength.

The considerations described above show that once activity coefficients are taken into account, the system with NaCl reduces to four simultaneous equations, just like the system with no NaCl. The mass balance and charge balance equations are completely identical in the two solutions, and the equilibrium constant expressions have the same form in the two systems, but the numerical values of the constants are different. Each set of equations can be solved using the approach presented above. The results are summarized in the following table.

Species	[NaCl] = 0		[NaCl] = 0.1 M	
	Concentration	Activity	Concentration	Activity
H^+	5.49×10^{-4}	5.49×10^{-4}	6.39×10^{-4}	4.98×10^{-4}
OH^-	1.82×10^{-11}	1.82×10^{-11}	2.62×10^{-11}	2.04×10^{-11}
HF	4.51×10^{-4}	4.51×10^{-4}	3.61×10^{-4}	3.61×10^{-4}
F^-	5.49×10^{-4}	5.49×10^{-4}	6.39×10^{-4}	4.98×10^{-4}
pH	3.26		3.19	

The results for the two cases are quite similar. In the water with the NaCl, the concentrations of H^+ and OH^- are both larger than in the less salty solution, reflecting the fact that the product of the activities of these two species must be 10^{-14} in both solutions, but that the activity coefficients are lower in the salty water. This fact alone, however, has no effect on the solution pH; i.e., in water containing 0.1 M NaCl and no other solutes, the pH would be 7.0, just as it is in salt-free water.

The water with the NaCl has a slightly lower pH and higher F^- concentration than the other solution because the salt decreases the activity coefficient of F^-. Since the activity coefficient of HF is 1.0 in both solutions, the decrease in γ_{F^-} forces more HF to dissociate to satisfy the K_a expression, and this extra dissociation releases more H^+ to solution. It is this effect that causes the pH to be lower in the solution with the NaCl. The effect of the salt would be larger if more highly charged species were involved (for example, CO_3^{2-} or PO_4^{3-}).

SUMMARY

Acids and bases are, respectively, substances that increase and decrease the activity of H^+ in solution. Many acids are species that have one or more "detachable" H^+ ions, and these ions are the source of the compound's acidity. Other acids cause H^+ to enter solution by facilitating the splitting of water (hydrolysis) and then combining with the OH^- ion that is released.

The strength of an acid, i.e., its tendency to cause the H^+ activity of a solution to increase, is quantified by its acid dissociation constant K_a. Multiprotic acids have several K_a values, one for each acidity reaction they undergo. Each acid has a conjugate base; the stronger the acid, the weaker the conjugate base. The strength of the base is quantified by a basicity constant K_b, and the product $K_a K_b$ is the dissociation constant for water K_w.

The fractional distribution of a compound among its acid and base forms depends on solution pH and is independent of the total amount of the compound in the system. If the solution pH happens to equal the pK_a of an acid/base conjugate pair, then the acidic (protonated) and basic (deprotonated) forms of the compound are present at equal activities in an equilibrium solution. At pH more acidic than pK_a (i.e., at pH $<$ pK_a), the protonated form has a greater activity than the deprotonated form, and vice versa.

The determination of the pH in an equilibrium solution that has been prepared with known inputs, or the determination of solution speciation in a solution of known pH, involves writing and solving a fairly straightforward set of simultaneous equations. These equations include mass balances on various species that have been added to solution, equilibrium constants for reactions among those species, and a charge balance for the whole solution. Chemical information is embedded in the values of the various equilibrium constants, but the solution of the equations is a purely mathematical exercise, requiring no specialized chemical knowledge.

Problems

1. Identify the strongest acid and the strongest base among the four species in the equilibrium equation below, using only the K value given.

$$H_2S + OCl^- \leftrightarrow HOCl + HS^- \quad K = 10^{+0.6}$$

2. The transfer of a proton from butyric acid (stomach acid) to acetate can be described by the following reaction:

$$CH_3CH_2CH_2COOH + CH_3COO^- \leftrightarrow CH_3CH_2CH_2COO^- + CH_3COOH$$

$$\text{Butyric acid} \quad + \text{ Acetate} \quad \leftrightarrow \quad \text{Butyrate} \quad + \text{ Acetic acid}$$

   a. The equilibrium constant for the above reaction is $K = 0.87$, and K_a for acetic acid is 1.74×10^{-5}. Using those values and K_w, compute the acidity constant for butyric acid.

   b. Write the reaction and determine the value of K_b (the basicity constant) for butyrate.

   c. A solution is made by adding some acetic acid and some sodium butyrate to water. List all the species in the equilibrium solution that can act as acids, and rank them from most acidic to least acidic. Do the same for all species that can act as bases.

3. Citric acid, $HOOC—CH_2—C(OH)(COOH)—CH_2—COOH$, which we can abbreviate as H_3Cit, is a triprotic carboxylic acid with acidity constants $pK_{a1} = 3.13$, $pK_{a2} = 4.72$, and $pK_{a3} = 6.33$.

   a. Consider a solution made by adding lemon juice to water until the pH of the solution is 2.2. Assuming all the acidity is from dissociation of citric acid, find the total concentration of citrate species, i.e., *TOT*Cit, and the concentration of $HCit^{2-}$.

   b. The solution in part (*a*) is diluted 1:10 and partly neutralized by addition of sodium bicarbonate ($NaHCO_3$). Designating the total concentration of carbonate species added as *TOT*CO$_3$, write out all the equations necessary to compute the new pH. You need not solve the equations.

   c. If the final pH of the solution in part *b* is 6.0, what is the ratio of $\{HCO_3^-\}$ to $\{H_2CO_3\}$ in the solution? What is the ratio of $\{CO_3^{2-}\}$ to $\{H_2CO_3\}$? Would these ratios change if the pH were still 6.0, but *TOT*Cit were doubled?

4. Dissolved copper ion (Cu^{2+}) can bond to several different molecules in aqueous solution, including some weak bases. Following are some of the equilibrium reactions in which Cu^{2+} participates.

$$Cu^{2+} + Cl^- \leftrightarrow CuCl^+ \qquad\qquad \log K = 0.43$$
$$Cu^{2+} + 2Cl^- \leftrightarrow CuCl_2{}^\circ \qquad\qquad \log K = 0.16$$
$$Cu^{2+} + CO_3{}^{2-} \leftrightarrow CuCO_3{}^\circ \qquad\quad \log K = 6.73$$
$$Cu^{2+} + OH^- \leftrightarrow CuOH^+ \qquad\qquad \log K = 6.00$$
$$Cu^{2+} + NH_3 \leftrightarrow CuNH_3{}^{2+} \qquad\quad \log K = 5.8$$
$$Cu^{2+} + 2NH_3 \leftrightarrow Cu(NH_3)_2{}^{2+} \qquad \log K = 10.7$$

   a. What is the equilibrium constant for the following reaction?

$$Cu(NH_3)^{2+} + NH_4{}^+ \leftrightarrow Cu(NH_3)_2{}^{2+} + H^+$$

   b. List the species that would be present at equilibrium if 10^{-3} *M* $Cu(HCO_3)_2$ + $10^{-3.3}$ *M* NaCl were added to water, and write out the equations that would be used to solve for the concentrations of all these species at equilibrium. Include numerical values in the equations, if they are known.

   c. Solve for the equilibrium speciation of the mixture in part (*b*). Use the Davies equation to estimate activity coefficients.

   d. What is K_b for the base $CuOH^+$?

5. Several simple acids and bases are common constituents of household items. Among these are acetic acid (vinegar), bicarbonate of soda ($NaHCO_3$), ammonia, ascorbic acid (vitamin C, $C_5H_9O_5$—$COOH$), hypochlorous acid (bleach), and trisodium phosphate (in many cleansers).

   a. A 6-year-old playing chemist finds some of these items in a cupboard and decides to mix some vinegar, vitamin C, bleach, and cleanser, hoping to make something exciting happen (like an explosion, or at the very least some serious fizzing). Alas, nothing dramatic occurs, and she decides to devote her energy to theory rather than experimentation. Hence, she decides to list:

      i. All the chemical species that she thinks were in the mixture.
      ii. The species that could act as acids and those that could act as bases, listing each from weakest acid to strongest.
      iii. The equations necessary to solve for the solution pH.

      Being rather precocious, she does this perfectly. Reproduce the lists that she prepared.

   b. Having opened the bottle of ammonia, she finds the odor offensive and decides to eliminate it. The odor, she knows, is from evaporation of NH_3°, because ammonium ions (NH_4^+) cannot enter the gas phase in significant concentrations. Having already used most of the other "reagents" available to her, she has only bicarbonate of soda, vinegar, and some cleanser containing trisodium phosphate remaining. Which should she add? Explain briefly.

6. A solution containing $4 \times 10^{-3}\,M$ of an acid HX has a pH of 2.5. What is the pH of a solution containing $4 \times 10^{-3}\,M$ of the Na salt of its conjugate base, i.e., NaX?

7. How much HOCl must be added to pure water to make a solution of pH 4.3? pH 6.5? Do not assume $\{H^+\} \gg \{OH^-\}$ at pH 6.5.

8. Cyanide ion, CN^-, is important in metal plating industries because it can keep metals dissolved under conditions where they would otherwise form solids (precipitate) and settle out of solution. The details of how this occurs are presented in Chapter 8. It is important to maintain pH > 10.5 in these solutions to avoid forming toxic hydrogen cyanide gas. If a solution is prepared by dissolving $10^{-2}\,M$ NaCN in water, will the pH be in the region where the solution is safe? What is $\{HCN\}$ in the solution?

9. The equilibrium constants for dissociation of water and of H_2CO_3 were derived using major ion seawater as the reference state in Example 1.6. Compare the carbonate speciation in seawater containing $[H^+] = 10^{-8.1}$ and $2 \times 10^{-3}\,M\;TOTCO_3$ with that in river water with the same concentrations of H^+ and $TOTCO_3$, if the infinite dilution assumption applies to the river.

4

ACIDS AND BASES, PART 2

Use of Log C–pH Diagrams

4.1   GRAPHICAL SOLUTIONS FOR SETS OF SIMULTANEOUS EQUATIONS

In the previous chapter, the central equations that characterize acid/base equilibria were presented, along with an algorithm for solving those equations. As noted, the algorithm could be programmed into a spreadsheet that allows us to identify the approximate equilibrium condition simply by scanning a column that characterizes how closely the charge balance is satisfied at various pH values. Log *C*–pH diagrams were also introduced as convenient tools for visualizing acid/base speciation over a wide range of conditions, and the point was made that these diagrams are developed using essentially the same information as is in the spreadsheet. In this chapter, we combine those two ideas to demonstrate how the equilibrium pH (and indeed, the speciation of the whole solution) can be determined using log *C*–pH diagrams in conjunction with the charge balance equation. Before computers were so widely accessible, this approach was used almost universally to solve complex acid/base problems, using hand-drawn log *C*–pH diagrams.

For many types of problems, the graphical approach can be carried out very rapidly and can yield very good approximations for the equilibrium speciation, even using rough back-of-the-envelope sketches. However, in systems where both acid/base and other types of reactions are important, this approach can become impractical, and software packages written specifically for solving chemical equilibrium problems offer the only practical means of solving the problems. This type of software is introduced in Chapter 6, and both approaches are used to analyze problems later in the text.

Of course, simple problems as well as complex ones can be solved using the software that is introduced in Chapter 6, and for this reason the graphical approach to problem solving has been abandoned in some water chemistry courses and textbooks. However, because many students find it easier to develop an intuitive feel for the key chemical relationships that control solution behavior in acid/base systems using the graphical approach, it is presented in detail in this chapter.

4.2 USING LOG *C*–pH DIAGRAMS TO SOLVE WEAK ACID/BASE PROBLEMS

In Chapter 3, we determined algebraically that the equilibrium pH of a solution of 10^{-3} *M* HPr was 3.963. Consider now how we might solve the same problem graphically. The graphical solution must, of course, satisfy the same set of equations as a numerical solution. In the current case, these equations are the relevant equilibrium constant expressions, the mass balances on the various species, and the charge balance.[1]

We start by showing the speciation of HPr and Pr^- on a log *C*–pH diagram, using the techniques developed in Chapter 3. The dependence of log $\{H^+\}$ and log $\{OH^-\}$ on pH is trivial, and we plot those values on the same log *C*–pH graph. The result is a single plot, shown in Figure 4.1, that tells us what the equilibrium values of $\{H^+\}$, $\{OH^-\}$, $\{HPr\}$, and $\{Pr^-\}$ would be at any pH in a system with the given *TOT*Pr.

To expand on a point that was made in Chapter 3, the log *C*–pH diagram displays exactly the same information, no more and no less, as is contained in the K_a and K_w expressions and in the mass balance on Pr. That is, just as a straight line on a graph is completely equivalent in terms of information content to an equation of the form $y = mx + b$, the group of lines representing $\{HPr\}$, $\{Pr^-\}$, $\{H^+\}$, and $\{OH^-\}$ on a log *C*–pH diagram has equivalent information content to the set

---

[1]As has been noted in previous chapters, the equilibrium constant expression relates activities, while the mass and charge balances relate concentrations. Nevertheless, throughout this chapter and the remainder of the text, terms in the mass and charge balances representing the concentrations of solutes will be written as activities, based on the assumption that these solutes behave ideally (i.e., that they have activity coefficients of 1.0) in the solutions being analyzed.

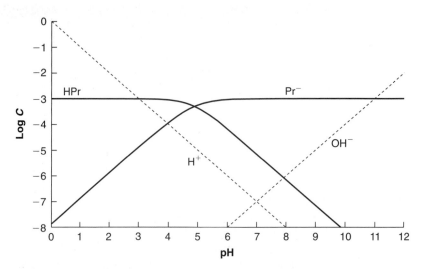

Figure 4.1   A log C–pH diagram for a solution containing 10^{-3} M TOTPr.

of K_a, K_w, and mass balance equations. (The same can be said for the first five columns of the spreadsheet in Table 3.5.)

The equilibrium pH can now be determined by solving the charge balance equation. Whereas in Chapter 3 we accomplished that task by adding a column to the spreadsheet, in the graphical analysis we plot the terms in the charge balance on the log C–pH diagram, as follows.

The charge balance equation is

$$\{H^+\} = \{Pr^-\} + \{OH^-\} \tag{4.1}$$

If we draw curves on the log C–pH diagram representing the left-hand side (LHS = $\{H^+\}$) and right-hand side (RHS = $\{OH^-\}$ + $\{Pr^-\}$) of the charge balance, the point where those curves intersect will be the pH where the charge balance is satisfied; i.e., they will intersect at an x value corresponding to the equilibrium pH.

The curve for the LHS is trivial to draw: it simply overlays the $\{H^+\}$ line. The curve for the RHS is a bit trickier, since we want to represent a linear summation on a logarithmic scale; i.e., we want to plot log (RHS), which is equivalent to log ($\{OH^-\}$ + $\{Pr^-\}$). (Recall from basic math that log ($\{OH^-\}$ + $\{Pr^-\}$) does **not** equal log $\{OH^-\}$ + log $\{Pr^-\}$!) It turns out, however, that this function can be plotted to a very close approximation without carrying out any additional calculations. To understand why, consider the data in Table 4.1, showing the results of the relevant calculations at a few pH values.

The values in the final column of the table correspond to log (RHS); i.e., they are the values we want to plot. Comparison of these values with log $\{OH^-\}$ and log $\{Pr^-\}$ shows that, over most of the pH scale, log (RHS) is approximately equal to either log $\{OH^-\}$ or log $\{Pr^-\}$, whichever is larger (the values shown in boldface type in the table). As a result, over most of the graph, a curve representing log (RHS) simply overlays the higher of the log $\{OH^-\}$ and log $\{Pr^-\}$ curves.

Table 4.1 Values of log $(\{OH^-\} + \{Pr^-\})$ at various pH values for a system containing $10^{-3.0}$ *M TOTPr*

pH	log $\{OH^-\}$	log $\{Pr^-\}$	log $(\{OH^-\} + \{Pr^-\})$
6.0	−9.00	**−3.24**	−3.24
8.0	−6.00	**−3.00**	−3.00
10.0	−4.00	**−3.00**	−2.96
11.0	−3.00	−3.00	−2.70
12.0	**−2.00**	−3.00	−1.96
14.0	**0.00**	−3.00	0.0

The only region where this approximation does not hold is where $\{OH^-\} \approx \{Pr^-\}$, in this example near pH 11. Exactly at the pH where $\{OH^-\} = \{Pr^-\}$, we can compute that log $(\{OH^-\} + \{Pr^-\})$ equals log $(2\{OH^-\})$, which is the same as $0.3 + $ log $\{OH^-\}$. Thus, at that pH, the curve representing log (RHS) is exactly 0.3 log unit above the intersection of the other two curves. Given this information, we can plot RHS in regions away from the point where $\{OH^-\} = \{Pr^-\}$ and also at the point where $\{OH^-\}$ and $\{Pr^-\}$ exactly equal each other; we can then sketch in the curve nearby the intersection point, to obtain a good approximation of RHS across the whole pH range. Both LHS and RHS are shown in bold in Figure 4.2a, and their intersection is circled. Figure 4.2b shows an expanded view of the region of the plot where the RHS curve shifts from being dominated by Pr^- to being dominated by OH^-.

The intersection of the RHS and LHS lines is approximately at pH = 3.95. At this pH, the concentrations of all the species in solution can be read off the graph as

$$\{H^+\} = 10^{-3.95} \qquad \{HPr\} \cong 10^{-3.03}$$

$$\{OH^-\} = 10^{-10.05} \qquad \{Pr^-\} \cong 10^{-3.95}$$

The accuracy of the result can be checked by evaluating the charge balance (CB) and the K_a expression using these estimates. As shown below, both equations are indeed satisfied to a close approximation:

Check CB: $\qquad 10^{-3.95} \overset{?}{=} 10^{-3.95} + 10^{-10.05}$

Check K_a: $\qquad 10^{-4.87} \overset{?}{=} \dfrac{(10^{-3.95})(10^{-3.95})}{10^{-3.03}} = 10^{-4.87}$

If we needed a more accurate answer than the approximate one determined using the log *C*–pH diagram (which is unlikely), we could at least use the graph to note that at the equilibrium condition $\{OH^-\} \ll \{H^+\}$ and then solve the problem numerically as shown in Chapter 3. In such a case, the graph would provide a quick way to determine which assumptions are reasonable; more often, however, an answer within ± 0.05 of the exact pH is obtainable and entirely satisfactory, so the graphical method can be used to solve the problem without carrying out any calculations at all.

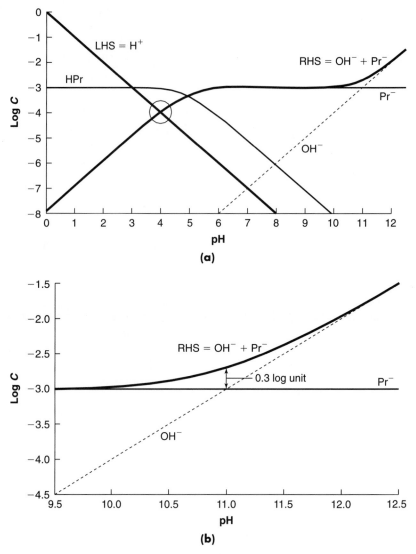

Figure 4.2   **(a)** A reproduction of Figure 4.1, with additional lines (in bold) indicating the values of the left-hand side (LHS) and right-hand side (RHS) of the charge balance equation for a solution of 10^{-3} *M* HPr. **(b)** An expanded view of the pH region where {OH$^-$} is close to {Pr$^-$}.

The preceding discussion demonstrates (not surprisingly) that the spreadsheet, graphical, and analytical solutions all yield essentially the same information. However, the graphical approach makes it much easier to visualize the behavior of the various species as a function of pH than does either of the other two approaches, because in the analytical and numerical solutions, all the information about the system is embedded in equations or lists of numbers. This advantage of the graphical approach is marginal in the very simple example system analyzed above, but it becomes dramatic when one is analyzing more complex systems, as shown in the remainder of the chapter.

The convenience of the graphical representation of the system stems from the fundamental similarity of the equations characterizing speciation in different systems. That is, the equations for α_0 and α_1 are essentially identical for any monoprotic acid/base system, differing only in the value of K_a, and of course the equations describing $\{H^+\}$ and $\{OH^-\}$ as a function of pH are the same in all systems. As a consequence, the graphs "look" very similar for any monoprotic acid or base system. That is, the lines for $\{H^+\}$ and $\{OH^-\}$ always have slope -1 and $+1$, respectively, and always intersect at $(7.0, 7.0)$, and the curves for $\{HA\}$ and $\{A^-\}$ have consistent, characteristic shapes, regardless of which acid/base pair is under consideration. Increasing the total concentration of the acid or base present moves both the HA and the A^- curves "up" the graph, and increasing pK_a moves both to the right (higher pH). Regardless of the total concentration and pK_a value, however, the shapes of these curves and their relationship to one another never change: they always intersect at pH $=$ pK_a, and at a value on the ordinate that is 0.3 log unit below log $(TOTA)$.

Furthermore, it is very common for each side of the charge balance equation to be dominated by a single term at any given pH, as was the case in the above example. In such cases, the LHS or RHS curve is just a linked series of sections, each overlapping the species that makes the largest contribution to the LHS or RHS summation in that pH region. Occasionally, two or more terms contribute significantly to the charge balance over a fairly wide pH range. A simple algorithm for drawing the LHS and RHS lines in such situations is described in Appendix 4A.

An additional advantage of the graphical approach is demonstrated in the following example.

| Use a graphical analysis to determine the equilibrium concentrations of all species in a solution of 10^{-3} M NaPr. | **Example 4.1** |

Solution

As in the analysis of the 10^{-3} M HPr solution, we start by drawing the log C–pH diagram that characterizes the system. When we identify the equations that we need to consider to draw the H^+, OH^-, HPr, and Pr^- curves on the graph, however, we reach an extremely useful conclusion: the K_a, K_w, and mass balance on Pr are identical to those in the previous problem, and those are the *only* equations that are used to draw the four curves. As a result, all the lines in Figure 4.1 are applicable to this problem as well, and we can use the same graph to solve the problem. Since we want to represent the concentrations of all dissolved species on the graph, we need to add a line for $\{Na^+\}$, but this is trivial, because $\{Na^+\} = 10^{-3}$, independent of pH.

We also need to include Na^+ in the charge balance, which becomes

$$\{Na^+\} + \{H^+\} = \{OH^-\} + \{Pr^-\}$$

As before, we can draw lines on the graph to represent the LHS and RHS of the charge balance, and again, in each case, the LHS or RHS line simply overlays the higher of the two lines in the corresponding summation. Unfortunately, though, as shown in Figure 4.3a, the lines do not identify a single pH value where the charge

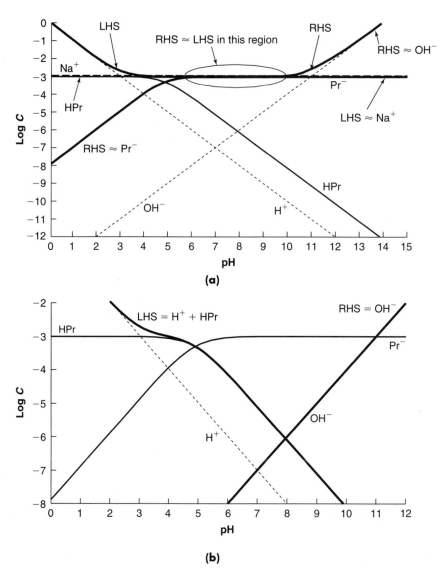

Figure 4.3   **(a)** A log C–pH diagram for a solution of 10^{-3} M NaPr, with the RHS and LHS of the charge balance shown in bold. **(b)** The same diagram as in part (a), but with RHS and LHS for an equation developed by combining the charge balance with the mass balance on Pr.

balance is satisfied. Rather, they indicate that the equation is approximately satisfied over the entire pH range $6 \leq pH \leq 10$. Thus, the graph is not sufficiently precise to allow us to solve the problem using the CB in its present form. Note that the graph is still a perfectly valid representation of the system, and that the RHS and LHS lines actually do intersect at a single point which characterizes the equilibrium condition. The problem we are encountering is strictly a practical one, related to the fact that we cannot draw lines thin enough to pick out the exact point of intersection.

   The difficulty in identifying the pH at which RHS = LHS can be circumvented in at least two ways. One approach would be to return to using a spreadsheet to solve the problem, taking advantage of the fact that the numerical processing in the spreadsheet is much more precise than can be displayed graphically. Although that approach is effective, it would still be nice to take advantage of the features in the graphical approach for solving the problem. This can be accomplished quite easily by combining the charge balance equation with two of the mass balances. Specifically, if we equate the mass balances for Na and Pr (since $TOT\text{Na} = TOT\text{Pr} = 10^{-3}$) and then substitute the resulting expression for $\{\text{Na}^+\}$ in the charge balance, we obtain the following expression:

CB:	$\{\text{Na}^+\} + \{\text{H}^+\} = \{\text{OH}^-\} + \{\text{Pr}^-\}$
Equating the two MBs:	$\{\text{Na}^+\} = \{\text{Pr}^-\} + \{\text{HPr}\}$
Substituting the above equality into the CB:	$\cancel{\{\text{Pr}^-\}} + \{\text{HPr}\} + \{\text{H}^+\} = \{\text{OH}^-\} + \cancel{\{\text{Pr}^-\}}$
Modified CB:	$\{\text{HPr}\} + \{\text{OH}^-\} = \{\text{OH}^-\}$

By plotting the left- and right-hand sides of this modified charge balance equation on the graph (Figure 4.3b), it is easy to identify their point of intersection. Based on this intersection point, the equilibrium pH is 7.95, consistent with the spreadsheet analysis, and the concentrations of the various species at equilibrium are as follows.

$$\{\text{Pr}^-\} = 10^{-3} \qquad \{\text{H}^+\} = 10^{-7.95} \qquad \{\text{Na}^+\} = 10^{-3.0}$$
$$\{\text{HPr}\} = 10^{-6.05} \qquad \{\text{OH}^-\} = 10^{-6.05}$$

   Checking the value of K_a as a way of confirming the result, we see that the computed concentrations do indeed meet the criterion for equilibrium:

$$K_a = 10^{-4.90} \overset{?}{=} \frac{(10^{-7.95})(10^{-3})}{10^{-6.05}} = 10^{-4.90}$$

   The preceding example points out that although the graphical analysis might provide the best big-picture view of the relationships of interest, it suffers from the drawback that, at least in some cases, we need to combine the charge balance with one or more mass balances to solve the problem. Two obvious questions arise from this outcome: (1) How can we know which equation(s) to combine with the charge balance to obtain the best equation to use with the graph?; and (2) Can we write the desired equation directly, without going through the algebra of combining multiple equations?

   The answer to the first question is fairly straightforward. For the NaPr system, each side of the charge balance consists of the sum of two numbers:

$$\{\text{Na}^+\} + \{\text{H}^+\} = \{\text{OH}^-\} + \{\text{Pr}^-\}$$

The problem we encounter when trying to solve this equation directly is that, in a fairly wide region near where it is satisfied, one of the numbers on each side ($\{\text{Na}^+\}$ on the left, $\{\text{Pr}^-\}$ on the right) is much larger than the other. As a result, even though the equation is satisfied exactly at only one pH, it is satisfied approximately over a wide range where $\{Na^+\} \approx \{Pr^-\}$. To "home in" on the pH where the equation is satisfied exactly, we need to eliminate the large number from each side of the equation. The desire to eliminate $\{Na^+\}$ from the left-hand side and $\{Pr^-\}$ from the right-hand side was the basis for using the mass balances on Na and Pr to convert the charge balance into a more useful equation for the graphical analysis. Generalizing this result, whenever a charge balance appears to be satisfied over a wide pH range, we need to combine the charge balance with mass balances that allow us to eliminate the largest number on each side of the charge balance. Additional examples employing this approach and extending the analysis to systems containing mixtures of acids and bases or multiprotic acids are provided below; we defer answering the second question above until after those examples.

Example 4.2	Find the equilibrium concentrations of H^+, HOCl, OH^-, and OCl^- when the following chemicals are added to pure water.

    a. $10^{-3} M$ HOCl
    b. $10^{-3} M$ KOCl
    c. $10^{-4} M$ KOCl $+ 9 \times 10^{-4} M$ HOCl

Solution

    a. The charge balance for part (a) is

        CB: $\{H^+\} = \{OCl^-\} + \{OH^-\}$

The left- and right-hand sides of the charge balance are plotted in Figure 4.4, along with the lines for the various species' concentrations. The charge balance is satisfied

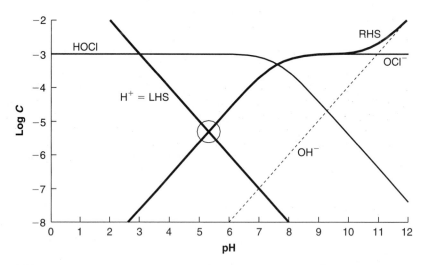

Figure 4.4   Log C–pH diagram for a solution containing $10^{-3} M$ TOTOCl. The LHS and RHS of the charge balance are shown for a solution made by adding all the OCl as HOCl.

at pH = 5.30, corresponding to

$$\log\{OH^-\} = -8.70 \qquad \log\{OCl^-\} = -5.30 \qquad \log\{HOCl\} = -3.00$$

$$K_a = 10^{-7.60} \overset{?}{=} \frac{(10^{-5.30})(10^{-5.30})}{10^{-3.00}} = 10^{-7.60} \qquad (K_a \text{ equation is satisfied})$$

b. The CB for a solution of 10^{-3} M KOCl is:

$$\{K^+\} + \{H^+\} = \{OCl^-\} + \{OH^-\}$$

If we plot the two sides of the charge balance directly, we run into the same problem as in the NaPr system above: the unique intersection point of the RHS and LHS cannot be discerned from the graph. Noting that $\{K^+\}$ and $\{OCl^-\}$ are the largest terms on the left- and right-hand sides of the charge balance, respectively, we can equate the mass balances on K and OCl and subtract the resulting equation from the charge balance to obtain the desired equation:

CB: $\qquad\qquad\qquad\qquad \{K^+\} + \{H^+\} = \{OCl^-\} + \{OH^-\}$

$-$ MB: $\qquad\qquad\qquad\quad - \quad\; \{K^+\} = \{OCl^-\} + \{HOCl\}$

$$\{H^+\} + \{HOCl\} = \{OH^-\}$$

Plotting the left-hand side (LHS) and right-hand side (RHS) of the modified charge balance equation (Figure 4.5), we find their intersection at

$$pH = 9.30 \qquad pOH = 4.70 \qquad p\{HOCl\} = 4.70 \qquad p\{OCl^-\} = 3.00$$

$$K_a \overset{?}{=} \frac{(10^{-9.30})(10^{-3.00})}{10^{-4.70}} = 10^{-7.60}$$

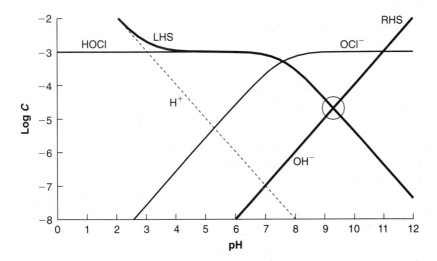

Figure 4.5   Log C–pH diagram for a solution containing 10^{-3} M TOTOCl. The LHS and RHS of a modified charge balance are shown for a solution made by adding all the OCl as KOCl.

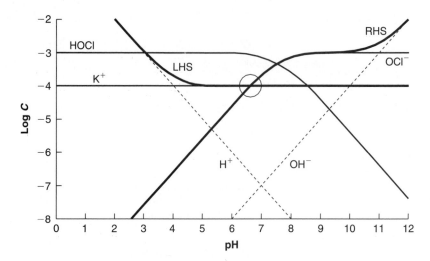

Figure 4.6   Log C–pH diagram for a system containing 10^{-3} M TOTOCl. The RHS and LHS of the charge balance are shown in bold for a system made by adding 10^{-4} M KOCl and 9×10^{-4} M HOCl to pure water.

c. The charge balance for a solution of 10^{-4} M KOCl $+ 9 \times 10^{-4}$ M HOCl is shown below, and the two sides of the equation are plotted in Figure 4.6.

CB: $\{K^+\} + \{H^+\} = \{OCl^-\} + \{OH^-\}$

Although this charge balance has the same form as the one in part (b), it differs in that the K^+ concentration is only 10^{-4}. As a result, the intersection point of the LHS and RHS is easily determined on the graph without any modification of the equation. The equilibrium condition is such that $\{K^+\} \approx \{OCl^-\}$, at which point the composition of the solution is as follows:

$$\log \{K^+\} = -4.0 \qquad \log \{OCl^-\} = -4.0$$
$$pH = 6.60 \qquad \log \{HOCl\} = -3.04 \qquad \log \{OH^-\} = -7.40$$

Example 4.3 | Prepare log C–pH graphs for the following solutions, and determine the speciation in each system. The pK_a values for NH_4^+ and HOCl are 9.25 and 7.60, respectively.

    a. 10^{-3} M HOCl $+ 5 \times 10^{-4}$ M NH_4Cl
    b. 10^{-3} M NaOCl $+ 5 \times 10^{-4}$ M NH_3
    c. 10^{-3} M NaOCl $+ 2.5 \times 10^{-4}$ M $NH_3 + 2.5 \times 10^{-4}$ M NH_4Cl

Solution

Because all three solutions of interest contain 10^{-3} M TOTOCl and 5×10^{-3} M TOTNH$_3$, a single graph can be used to determine the equilibrium composition of all of them (Figure 4.7). However, in all three cases, the charge balances have to be modified by combining them with various mass balances before the equilibrium pH can be determined. The various manipulations of the equations and the final results are sum-

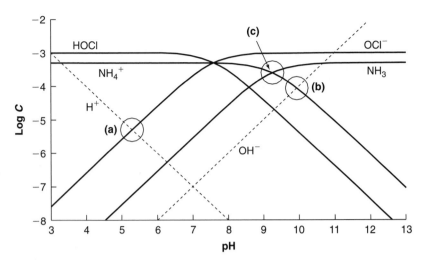

Figure 4.7 Log *C*–pH diagram for a system containing 10^{-3} *M TOTOCl* and 5×10^{-4} *M TOTNH_3*.

marized below.

a.

$$\{H^+\} + \{NH_4^+\} = \{OH^-\} + \{OCl^-\} + \{Cl^-\}$$

$$+ \quad \{Cl^-\} = \{NH_4^+\} + \{NH_3\}$$

$$\overline{\{H^+\} = \{OH^-\} + \{OCl^-\} + \{NH_3\}}$$

pH $\approx$ 5.35 (where $\{H^+\} + \{OCl^-\}$)

b.

$$\{H^+\} + \{NH_4^+\} + \{Na^+\} = \{OH^-\} + \{OCl^-\}$$

$$+ \quad \{OCl^-\} + \{HOCl\} = \{Na^+\}$$

$$\overline{\{H^+\} + \{NH_4^+\} + \{HOCl\} = \{OH^-\}}$$

pH $\approx$ 9.95 (where $\{OH^-\} \approx \{NH_4^+\}$)

c. $\{NH_4^+\} + \{Na^+\} + \{H^+\} = \{OCl^-\} + \{OH^-\} + \{Cl^-\}$

$$+ \quad \{OCl^-\} + \{HOCl\} = \{Na^+\}$$

$$\overline{\{NH_4^+\} + \{H^+\} + \{HOCl\} = \{OH^-\} + 2.5 \times 10^{-4}} \quad \text{(since } \{Na^+\} = 2.5 \times 10^{-4})$$

pH $\approx$ 9.25 $\quad$ (= pK_a, where $\{NH_3\} = \{NH_4^+\} = 2.5 \times 10^{-4}$)

Note that, in Example 4.3c, the $\{HOCl\}$ term contributes negligibly to the LHS of the modified charge balance at the point where the equation is satisfied, and $\{OCl^-\}$ does not appear in the equation at all. As a result, that term has essentially no effect on the result; i.e., the pH is the same in this solution as it would be if we had added the ammonia species but no OCl at all. The log C–pH diagram makes it clear why this is so. At the equilibrium pH, almost all the OCl in the system is present as hypochlorite ion, OCl^-. Since that is also the form in which all the OCl was added to the solution, it apparently entered solution and underwent no further reaction, i.e., a negligible fraction of the OCl^- ions that entered solution combined with H^+ to form HOCl. In essence, then, in this system, NaOCl acts not like a base but rather like an inert salt. This result applies because the NH_3 increases the pH to a value where the OCl^- has no significant tendency to become protonated.

Example 4.4

    a. What are the concentrations of all species in solutions of (*i*) 10^{-2} *M* Na_2CO_3; (*ii*) 10^{-2} *M* $NaHCO_3$; (*iii*) 10^{-2} *M* H_2CO_3?

    b. A solution has 3 times as much H_2CO_3 as CO_3^{2-}. What is its pH? What is the concentration of HCO_3^- relative to H_2CO_3?

Solution

    a. As in the previous examples, the chemical relationships in this system can be characterized by a combination of mass balances, equilibrium constants, and a charge balance. The mass balance on *TOTCO*$_3$ and the equilibrium equations are the same for all three solutions, so they can be characterized by a single log C–pH graph, which is shown in Figure 4.8.

      The charge balances for the three systems can be analyzed as shown below:

    (*i*)   CB: $\{Na^+\} + \{H^+\} = \{OH^-\} + \{HCO_3^-\} + 2\{CO_3^{2-}\}$

This charge balance differs from the ones analyzed in the previous examples because, under conditions where the third term on the right dominates that side of the equation, the relevant approximation is RHS $\approx 2\{CO_3^{2-}\}$. Thus, under these conditions, the curve representing RHS will not overlap the $\{CO_3^{2-}\}$ curve, but rather will coincide with a curve representing $2\{CO_3^{2-}\}$. To determine where

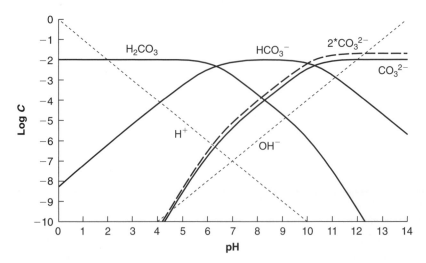

Figure 4.8   Log C–pH diagram for a system containing 10^{-2} *M* *TOTCO*$_3$.

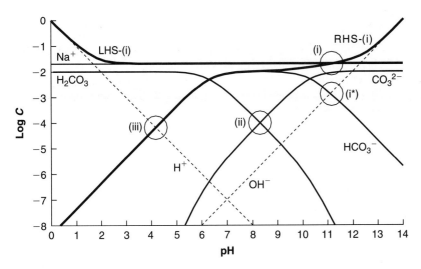

Figure 4.9   Log *C*–pH diagram for Example 4.4. The bold lines show the two sides of the unmodified charge balance equation for part (i). The solutions for part (i) using the modified charge balance and for parts (ii) and (iii) are also shown.

such a curve lies on the graph, we can rewrite the approximation as follows:

$$\log \{RHS\} \approx \log (2\{CO_3^{2-}\}) = \log 2 + \log \{CO_3^{2-}\} = 0.3 + \log \{CO_3^{2-}\}$$

The result indicates that, in the region of the graph where the RHS is dominated by the term $2\{CO_3^{2-}\}$, the curve representing the RHS is 0.3 log unit above the curve representing $\{CO_3^{2-}\}$. This curve is shown in Figure 4.8 for reference, and the two sides of this charge balance are shown in Figure 4.9, labeled as RHS-(i) and LHS-(i), respectively.

It is not too difficult to estimate the equilibrium pH by using the charge balance in this case, but the pH could be identified more accurately if the largest terms in the charge balance were eliminated. Based on the intersection of the RHS-(i) and LHS-(i) curves, the largest terms in the equation at the equilibrium pH are $\{Na^+\}$ and $2\{CO_3^{2-}\}$, respectively. We can follow the same approach as above for eliminating these terms from the equation. This is easily done, since $TOTNa = 2(TOTCO_3)$. Subtracting that equation from the CB, we obtain

$$\{Na^+\} + \{H^+\} = \{OH^-\} + \{HCO_3^-\} + 2\{CO_3^{2-}\}$$

$$-\quad \{Na^+\} = 2\{H_2CO_3\} + 2\{HCO_3^-\} + 2\{CO_3^{2-}\}$$

$$\overline{\{HCO_3^-\} + 2\{H_2CO_3\} + \{H^+\} = \{OH^-\}}$$

The intersection of the two sides of the modified equation is at pH 11.15 [indicated by the circle labeled (i*) in the figure]. At this pH,

$$\{H^+\} = 10^{-11.15} \qquad \{OH^-\} = 10^{-2.85}$$
$$\{H_2CO_3\} \approx 10^{-7.7} \qquad \{HCO_3^-\} = 10^{-2.85}$$
$$\{CO_3^{2-}\} \approx TOTCO_3 - \{HCO_3^-\} = 10^{-2.07}$$

(*ii*) CB: $\{Na^+\} + \{H^+\} = \{OH^-\} + \{HCO_3^-\} + 2\{CO_3^{2-}\}$

Once again, identifying the exact pH where the RHS and LHS of the charge balance equal each other is difficult, and we can overcome the problem by substituting to eliminate the largest terms in the CB. In this case, the Na and CO_3 mass balances are related by $TOT\text{Na} = TOT\text{CO}_3$. Combining this expression with the CB yields

$$\{Na^+\} + \{H^+\} = \{OH^-\} + \{HCO_3^-\} + 2\{CO_3^{2-}\}$$
$$-\qquad \{Na^+\} = \{H_2CO_3\} + \{HCO_3^-\} + \{CO_3^{2-}\}$$
$$\overline{\{H_2CO_3\} + \{H^+\} = \{OH^-\} + \{CO_3^{2-}\}}$$

The equation is satisfied at $\{H_2CO_3\} \approx \{CO_3^{2-}\}$, pH 8.3. This point is circled and labeled (ii) in Figure 4.9.

$$\{H^+\} = 10^{-8.3} \qquad \{OH^-\} = 10^{-5.7}$$
$$\{HCO_3^-\} \approx 10^{-2.0} \qquad \{H_2CO_3\} = 10^{-4.0}$$
$$\{CO_3^{2-}\} = 10^{-4.0}$$

(*iii*) CB: $\{H^+\} = \{OH^-\} + \{HCO_3^-\} + 2\{CO_3^{2-}\}$

The point where this CB is satisfied can be identified directly on the graph. The result is that the equilibrium pH is 4.35, where $\{HCO_3^-\} \approx \{H^+\}$.

$$\{H^+\} = 10^{-4.35} \qquad \{OH^-\} = 10^{-9.65} \qquad \{H_2CO_3\} = 10^{-2.0}$$
$$\{HCO_3^-\} = 10^{-4.35} \qquad \{CO_3^{2-}\} = 10^{-9.15}$$

b. In this problem, we know the relative concentrations of two species and want to determine the solution pH. One approach for doing this is to generate a reaction relating the two species of interest and H^+, by adding the reactions for dissociation of H_2CO_3 and HCO_3^-. Recalling that when we add reactions, we must multiply equilibrium constants, we find:

$$H_2CO_3 \leftrightarrow CO_3^{2-} + 2H^+ \qquad K_{a02} = K_{a1}K_{a2} = 10^{-16.68}$$

Thus, $\{CO_3^{2-}\}\{H^+\}^2/\{H_2CO_3\} = 10^{-16.68}$. Substituting the given information that $\{H_2CO_3\}/\{CO_3^{2-}\} = 3$, we can compute the solution pH.

$$\{H^+\}^2 = 3 \times 10^{-16.68} = 10^{-16.20}$$
$$\{H^+\} = 10^{-8.10} \qquad \text{pH} = 8.10$$

The relative concentrations of HCO_3^- and H_2CO_3 can be determined from the known pH and K_{a1}. From the K_{a1} value,

$$\frac{\{HCO_3^-\}}{\{H_2CO_3^\circ\}} = \frac{10^{-6.33}}{\{H^+\}}$$

At pH 8.10, $\dfrac{\{HCO_3^-\}}{\{H_2CO_3\}} = \dfrac{10^{-6.33}}{10^{-8.10}} = 10^{-1.77} = 0.017$

4.3   THE MASS BALANCE ON H (THE PROTON CONDITION OR *TOT*H EQUATION)

The preceding examples demonstrate how a log C–pH diagram can be used to determine the pH of a system with known inputs. The graph contains information about the relevant acidity constants (indicated by the intersection points of conjugate acid/base pairs), the dissociation of water (indicated by the relative positions of the H^+ and OH^- lines), and the total amount of each weak acid/base group in the system (indicated by the largest value that log C approaches for the individual acid and base species). Once the graph is drawn, all we need to do to solve for the equilibrium speciation of the system is to write the charge balance equation and find the unique pH where that equation is satisfied.

In some cases, we have been able to determine the equilibrium pH from the graph directly from the charge balance; however, we have also encountered several cases in which it was necessary to derive a new equation by combining the charge balance with the mass balances before the graphical analysis could be used. While knowing how to convert the charge balance into a more useful form is helpful, it would be even better if we could write the final form of the equation directly, without going through the charge balance at all. To accomplish that goal, we reconsider the mass balance on H, which we chose not to evaluate when we first analyzed the system. The reason for this decision is most easily understood by contrasting the mass balance on H with that for other substances in the system.

Consider a hypothetical system containing 10^{-6} mol/L total arsenate (75 μg AsO_4/L). The arsenate acid/base group is triprotic, forming arsenic acid (H_3AsO_4) when it is fully protonated, and its general chemistry is very similar to that of the phosphate group. However, whereas phosphate is a major component of nucleic acids and adenosine triphosphate (ATP, the key chemical used to store energy in most cells), arsenate is a poison at relatively low levels. It has been used as an herbicide in the past, but its use in such applications and in industrial processing is now severely limited.[2]

The acidity constants of arsenic acid are close to those of phosphoric acid, so that the forms most likely to be found in natural waters are the mono- and divalent ions $H_2AsO_4^-$ and $HAsO_4^{2-}$, with arsenic acid and arsenate ion (AsO_4^{3-}) present at much lower concentrations. Nevertheless, in a formal mass balance, we should consider the contributions of all four species to *TOT*AsO_4. Thus, in the

---

[2]Most current environment concerns about arsenic in the developed world are related to its presence in drinking water and to the cleanup of a few severely contaminated but localized sites of soil pollution. Because the health effects of long-term, low-level exposure to arsenic are unclear, there is a good deal of controversy regarding the appropriate regulatory limit for arsenic in drinking water. As of late 2000, the maximum contaminant level (MCL) for arsenic in the United States was 50 μg/L, but the U.S. Environmental Protection Agency (EPA) has proposed reducing the MCL by an order of magnitude. In a few locations in the world, water that has been used as a potable source has arsenic concentrations that far exceed safe levels, leading to disastrous epidemiological consequences. In those cases, the source of the contamination has usually been natural, arsenic-bearing rocks in the aquifer.

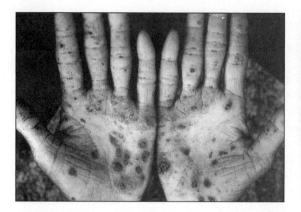

The long-term effects of arsenic in drinking water are causing the EPA to consider lowering the arsenic maximum contaminant level (MCL). In Bangladesh, natural concentrations of arsenic in groundwater have had a devastating effect on some communities. The photograph on the left shows one of the symptoms of arsenic poisoning. Arsenic can also enter the water supply by careless disposal of industrial wastes. The photograph on the right shows sheep being dipped in an arsenic-based solution to kill external parasites.

| Left: Richard Wilson/Harvard University; Right: G. R. Roberts/G. R. "Dick" Roberts Photo Library.

example solution, the mass balance on arsenic would be

$$TOTAsO_4 = 10^{-6} = \{H_3AsO_4\} + \{H_2AsO_4^-\} + \{HAsO_4^{2-}\} + \{AsO_4^{3-}\}$$

The baseline arsenic concentration that we use when we do this accounting is zero. That is, the total amount of arsenic present is quantified by comparison with a solution containing zero arsenic. This is so logical and obvious that it is normally taken for granted. However, there are times when it is more useful to compare the total amount of some substance in solution to a baseline other than zero. The easiest way to understand this concept is to consider a non-chemical example, such as banking.

Banks provide monthly statements that tell people how much money they have in their accounts, and these statements are based on a balance of zero as the baseline. However, let's say you have an account that provides free checking if you maintain a balance of at least $500. If you want to make sure that you always maintain the minimum balance in your checkbook, you might choose to keep track of your *available* funds, meaning the balance above $500. In this case, a balance of $650 on the bank's statement would show up as a balance of only $150 in your checkbook. Furthermore, if your *real* balance did dip below the minimum (say, to $350), your records would show that your balance had become negative (−$150). This scenario is perfectly rational (at least I think so; I used to use it!), and it is as accurate and reliable as a more conventional accounting practice. The only critical requirement is that you need to know what the accounting rules are in order to interpret the numerical data correctly.

In the absence of a good reason to define a non-zero baseline for a chemical mass balance, we generally do not do so, and as noted above, we generally do not even take note of the fact that we are choosing "zero substance" as the baseline condition. When writing a mass balance on H, however, we do have a good reason for using a non-zero baseline. Specifically, the fact that the number of H atoms associated with H_2O molecules is enormous ($2 \times 55.6 \, M$) compared to the number present in other forms in solution makes it very inconvenient to try to keep track of all the H in the system. Rather than count the total number of H ions in solution, it is much easier to just keep track of the relatively small *changes* in the concentrations of H-bearing species. This goal can be accomplished by defining a relatively large absolute concentration of H (approximately equal to the H concentration present as part of H_2O molecules, but also including a few other terms) as the baseline concentration in the system.

The mass balance on H using a non-zero baseline is sometimes referred to as a **proton condition** (PC). In the terminology that goes along with the proton condition, species or solutions that contain more H than the baseline level are said to have an *excess* of H, and those that contain less than the baseline level are said to have an H *deficiency*. The net concentration of H contained in all the species in solution (i.e., the sum of all excess H minus all H deficiencies) is designated as *TOT*H, and the proton condition is sometimes called the *TOT*H balance or *TOT*H equation.

To write the proton condition, we carry out all the steps that we would carry out to write a mass balance on any ion group in the system, but we write the concentrations of all species relative to a non-zero baseline concentration. The steps involved in doing this can be formalized as follows.

Defining the Baseline Concentration of H   Recall that, in Chapter 3, the species present in solution at equilibrium were categorized into a group consisting of H^+, H_2O, and OH^- (which were designated type *a* species); weak acid/base groups such as HAc and Ac^-, or H_2CO_3, HCO_3^-, and CO_3^- (type *b* species); salts, such as Na^+ or Cl^- (type *c* species); and species formed by combinations of types *b* and *c* species (type *d* species). In writing the proton condition, we select one species from each type *a* or *b* group to be the *reference species* for that group. In addition, we select all salts to be reference species.

The decision as to which particular species from a given acid/base group is chosen as the reference level species is arbitrary, but for reasons that will become apparent, it is best to choose the species that is expected to be present at the highest concentration in the equilibrium solution. The reason is that, as shown subsequently, this choice eliminates the problem of trying to find the intersection of the LHS and RHS lines in a pH range where they are almost overlapping. Because different species are dominant in different systems, the best choices for reference species vary from one system to another. However, H_2O is always dominant over H^+ and OH^-, so it is always chosen as the reference species for the $H^+/H_2O/OH^-$ group. Collectively, the reference species define the baseline for determining *TOT*H. That is, a solution made up entirely of baseline species is defined as having *TOT*H $= 0$.

Defining the H⁺ Excess or Deficiency of Every Species in the Equilibrium Solution, and Writing an Expression for *TOT*H in the Solution   For each species *i* present at equilibrium, we can write a reaction of the following form:

$$\text{Various reference species} + n_i H^+ \rightarrow 1 \text{ Species } i \qquad \textbf{(4.2)}$$

That is, we can represent each of the species present in solution at equilibrium as a combination of reference species and H^+ ions. Since H^+ must be conserved, and since the reference species have been defined as having zero $TOTH$, each molecule of i contains n_i excess protons (if n_i is negative, i has a proton deficiency).

For instance, if a solution is made by adding $10^{-2}\ M$ HAc and $3 \times 10^{-3}\ M$ Na$_2$CO$_3$ to water, we need to choose four reference species: H$_2$O; one species from the HAc/Ac$^-$ group; one from the H$_2$CO$_3$/HCO$_3^-$/CO$_3^{2-}$ group; and Na$^+$. Choosing HAc and HCO$_3^-$ as the reference species for the corresponding groups, we could write the desired reactions for forming the nine chemical species present at equilibrium as follows:

$$1\ H_2O\ +\ 0\ H^+ \rightarrow 1\ H_2O$$

$$0\ H_2O\ +\ 1\ H^+ \rightarrow 1\ H^+$$

$$1\ H_2O\ +\ -1\ H^+ \rightarrow 1\ OH^-$$

$$1\ HAc\ +\ 0\ H^+ \rightarrow 1\ HAc$$

$$1\ HAc\ +\ -1\ H^+ \rightarrow 1\ Ac^-$$

$$1\ HCO_3^-\ +\ -1\ H^+ \rightarrow 1\ CO_3^{2-}$$

$$1\ HCO_3^-\ +\ 0\ H^+ \rightarrow 1\ HCO_3^-$$

$$1\ HCO_3^-\ +\ 1\ H^+ \rightarrow 1\ H_2CO_3$$

$$1\ Na^+\ +\ 0\ H^+ \rightarrow 1\ Na^+$$

The value of $TOTH$ at equilibrium, denoted by $TOTH_{eq}$, can then be calculated as $\sum c_{i,eq} n_i$, where $c_{i,eq}$ is the concentration of species i in the equilibrium solution, and the summation is over all species present at equilibrium. That is, in the example system, $TOTH_{eq}$ is given by

$$TOTH_{eq} = (c_{H_2O,eq} n_{H_2O} + c_{H^+,eq} n_{H^+} + c_{OH^-,eq} n_{OH^-}) + (c_{HAc,eq} n_{HAc} + c_{Ac^-,eq} n_{Ac^-})$$
$$+ (c_{H_2CO_3,eq} n_{H_2CO_3} + c_{HCO_3^-,eq} n_{HCO_3^-} + c_{CO_3^{2-},eq} n_{CO_3^{2-}}) + (c_{Na^+,eq} n_{Na^+})$$

For formality, the above expression for $TOTH_{eq}$ has been written to include all species present in the equilibrium solution. However, several of those species are reference species and hence have $n_i = 0$, so they do not contribute to $TOTH_{eq}$. If we rewrite the expression without these species, substitute the values of the other n_i's, and write $c_{i,eq}$ as $\{i\}_{eq}$, the equation becomes much simpler:

$$TOTH_{eq} = \{H^+\}_{eq}(1) + \{OH^-\}_{eq}(-1) + \{Ac^-\}_{eq}(-1) + \{H_2CO_3\}_{eq}(1)$$
$$+ \{CO_3^{2-}\}_{eq}(-1)$$
$$= \{H^+\}_{eq} - \{OH^-\}_{eq} - \{Ac^-\}_{eq} + \{H_2CO_3\}_{eq} - \{CO_3^{2-}\}_{eq}$$

COMPUTING THE AMOUNT OF H$^+$ THAT WAS INPUT TO THE SOLUTION RELATIVE TO THE BASELINE   To compute the value of $TOTH$ of the chemicals that were used to prepare the solution ($TOTH_{in}$), we need to determine the proton excess or deficiency associated with every chemical that is in the initial solution or is added subsequently. The calculations are exactly parallel to those shown above for the equilibrated solution. That is, we determine the number of excess or deficient protons for each chemical that is present initially or is added later by writing a reaction that has only reference species and H^+ as reactants and that has the chemical of interest as the only product. $TOT H_{in}$ is then determined as $\sum c_{i,in} n_i$, where the summation is over all chemicals present initially and all input chemicals.

In the example system described above, the system was prepared by combining water, acetic acid, and sodium carbonate. The reactions forming these input species, again using H_2O, HAc, Na^+, and HCO_3^- as reference species, are as follows:

Reference species	+	n_i H^+ $\rightarrow$ Input or added chemical
1 H_2O	+	0 H^+ $\rightarrow$ 1 H_2O
1 HAc	+	0 H^+ $\rightarrow$ 1 HAc
2 Na^+ + 1 HCO_3^-	+ −1	H^+ $\rightarrow$ 1 Na_2CO_3

The first two reactions shown above are trivial; they are included simply to reinforce the fact that H_2O and HAc are reference species and therefore have no excess or deficiency of H^+. The third reaction specifies that one molecule of the input chemical Na_2CO_3 can be made by removal of one H^+ from a combination of two reference species (Na^+ and HCO_3^-); therefore Na_2CO_3 is a one-proton deficient species.

It should be obvious that the proton "status" of a chemical depends on the choice of reference species. For instance, if H_2CO_3 were chosen as a reference species instead of HCO_3^-, Na_2CO_3 would have a two-proton rather than a one-proton deficiency. However, once a set of reference species is chosen, the proton excess or deficiency of every input chemical and every species present at equilibrium is fully determined.

For the example system, $TOT H_{in}$ is

$$TOT H_{in} = c_{H_2O\ added} n_{H_2O} + c_{HAc\ added} n_{HAc} + c_{Na_2CO_3\ added} n_{Na_2CO_3}$$
$$= (55.6)(0) + (10^{-2})(0) + (3 \times 10^{-3})(-1) = -3 \times 10^{-3} \quad \textbf{(4.3)}$$

Thus, the inputs to the given solution have -3×10^{-3} M excess protons (or, alternatively, they represent a 3×10^{-3} M proton deficiency) relative to a solution containing only reference species.

Note that, when carrying out this step, we must include as inputs all chemicals in the solution prior to the time when it reaches equilibrium. In many cases (as in the above example), the original solution is simply pure water, and all the subsequent chemical inputs are known. In others, we might know the complete composition of the solution at some initial point without knowing exactly what chemicals were added to prepare that solution, and also know the amounts of any chemicals added subsequently. In such cases, we can consider the ultimate solution to be a combination of the initial solution and the additional chemicals, and we can calculate $TOT H_{in}$ based on that conceptualization. An example showing such a calculation is provided below.

EQUATING $TOTH_{IN}$ WITH $TOTH_{EQ}$ The various reactions used to determine the proton excesses and deficiencies of the chemicals present in the initial solution, any added chemicals, and the equilibrium species can be presented concisely in a table, as shown below. In the table, all the chemicals having a given n_i appear in a column, and all those in a given acid/base group appear in a row. The easiest way to prepare the table is to first list the reference species in the $n_i = 0$ column and then to distribute the other species as appropriate. The table also includes a column showing the concentration of the chemicals used to prepare the solution.

| | **H^+ Excess** | | | |
	−1	**0**	**+1**	**Concentration**
Species present at equilibrium	OH^- Ac^- CO_3^{2-}	H_2O HAc HCO_3^- Na^+	H^+ H_2CO_3	
Initial and/or input species	Na_2CO_3	H_2O HAc		55.6 10^{-2} 3×10^{-3}

As the solution changes from the pre-equilibrium composition to the equilibrium condition, H^+ ions may transfer from one species to another, but the total concentration of protons must remain constant. Therefore, $TOTH_{eq}$ must equal $TOTH_{in}$. The proton condition is the mathematical expression of this concept. Equating the expressions developed above for $TOTH_{eq}$ and $TOTH_{in}$, and dropping the subscript "eq," we obtain the proton condition or $TOTH$ equation. For the example system, the equation is

$$TOTH_{in} = TOTH_{eq} \qquad \textbf{(4.4)}$$

$$-3 \times 10^{-3} = \{H^+\} - \{OH^-\} - \{Ac^-\} + \{H_2CO_3\} - \{CO_3^{2-}\} \qquad \textbf{(4.5)}$$

Note that the coefficients on the right-hand side of the above equation are simply the values of n_i corresponding to each species, and they are also the coefficients heading the various columns in the PC table. Thus, the upper part of the table summarizes the contributions to $TOTH_{eq}$, and the lower part summarizes the contributions to $TOTH_{in}$.

COMPARISON OF THE PROTON CONDITION AND THE $TOTH$ EQUATION   Whether the non-zero-based mass balance on H is referred to as the proton condition or the $TOTH$ equation for the system is mostly a semantic issue. However, to avoid any potential confusion about these terms, it is worthwhile to clarify their relationship.

The idea of the proton condition was developed in the 1960s for use in conjunction with log C–pH diagrams. The proton condition table was developed as a convenient way to formulate the equation, and the terms *proton excess, proton deficiency,* and *reference species* were coined to describe key parameters in that approach. Also, the resulting equation is easiest to apply if all the terms are positive, so the proton condition is always written as a summation of positive terms on each side of the equation. That is, for the example system described above, the proton condition is written as

$$\{H^+\} + \{H_2CO_3\} + 3 \times 10^{-3} = \{OH^-\} + \{Ac^-\} + \{CO_3^{2-}\} \quad \textbf{(4.6)}$$

With the development of software to carry out chemical equilibrium computations, the use of graphical analyses to obtain highly accurate information about system composition diminished. Currently, the graphical approach is used primarily as a way to carry out quick and approximate analyses when a computer is not handy. As shown in Chapter 6, the input to these computer programs is most easily compiled using the algorithm described earlier, in which each species present in the equilibrium system is defined by a reaction involving the reference level species and H^+. In the terminology used with the software, these species (the reference species and H^+) are referred to as *components,* and if the dominant species are selected as reference species, they are referred to as the *principal components* of the solution. Finally, in the software, the H mass balance is most conveniently expressed as in Equation (4.5), with $TOTH_{in}$ on the left-hand side of the equation and the species present in the equilibrium solution on the right.

A comparison between the proton condition and the *TOTH* equation is provided in Table 4.2. The table makes clear that the two ways of writing the H^+ mass balance contain identical information and are, essentially, the same equation.

With a little practice, preparing the proton excess/deficiency table and writing the reactions defining *TOTH* of each species become very easy, and the conversion of the data into either a proton condition or a *TOTH* equation is straightforward. This process is demonstrated in the following examples.

Table 4.2 Comparison of terminology and other conventions associated with the proton condition and the *TOTH* equation

	Proton Condition	**TOTH Equation**
Name of baseline (proton-neutral) species	Reference species	Components
Species identified as best choices for baseline species	Dominant species	Principal components
Format for preparing information	Table with H^+ excess and deficient species; n_i is the heading for columns in the table	Reactions forming species from components; n_i is the coefficient for H^+ in the reaction
Species with $n_i > 0$	Proton-excess species	Not named explicitly
Species with $n_i < 0$	Proton-deficient species	Not named explicitly
Format of equation	Summation on both sides of equation, arranged so that all terms are positive	Summation of $TOTH_{in}$ on left, $TOTH_{eq}$ on right; some terms negative
Most direct application	In conjunction with log C–pH graph	In conjunction with chemical equilibrium software

| **Example 4.5** | Write the H mass balance for the following solutions, in the form of either a proton condition or the *TOT*H equation. |

   a. Pure H_2O

This solution contains only type *a* species, so H_2O is the only reference species. The proton condition table is therefore trivial to prepare:

	−1	0	+1	Concentration
Equilibrium species	OH^-	H_2O	H^+	
Initial and/or input species		H_2O		55.6

Since no proton excess or deficient species were added to prepare the solution, $\sum c_{i,in} n_i$ is zero. The equilibrium solution must therefore also have no net proton excess or deficiency, i.e., $\sum c_{i,eq} n_i = 0$. The formal computations show that the input includes 55.6 *M* of a chemical (H_2O) with zero excess protons.[3]

$$TOTH_{in} = \sum_i c_{i,in} n_i = (55.6)(0) = 0$$

$$TOTH_{eq} = \sum_i c_{i,eq} n_i = \{H^+\}(1) + \{OH^-\}(-1)$$

The PC equates $\sum c_{i,eq} n_i$ with $\sum c_{i,in} n_i$, yielding

PC: $\{H^+\} = \{OH^-\}$

The only difference between the PC equation and the *TOT*H equation is that, by convention, the latter is written with the numerical value of $TOTH_{in}$ on the left-hand side and the concentrations of all species that contribute to $TOTH$ in the equilibrium solution on the right.

*TOT*H equation: $0 = \{H^+\} - \{OH^-\}$

   b. Water $+ \ 10^{-3} \ M$ HOCl

At equilibrium, this system will contain the three type *a* species and two type *b* species (HOCl and OCl^-). Water is always chosen as the reference species for the type *a* group, and we can choose either HOCl or OCl^- as the reference species for the type *b* group. Choosing (arbitrarily) HOCl for this role, we can calculate the proton excess or deficiency of each species in the system based on the following reactions:

Reference species $+$	$n_i H^+$	$\to$ Species of interest
1 H_2O	$+$ 0 H^+	$\to$ 1 H_2O
None	$+$ 1 H^+	$\to$ 1 H^+
1 H_2O	$+$ $-1 H^+$	$\to$ 1 OH^-
1 HOCl	$+$ 0 H^+	$\to$ 1 HOCl
1 HOCl	$+$ $-1 H^+$	$\to$ 1 OCl^-

---

[3]Note that the concentration and activity of H_2O differ from one another even in an ideal solution, and it is the concentration that must be used to compute the contribution of H_2O to the proton condition.

The proton condition table and equation are as follows:

	−1	0	+1	Concentration
Equilibrium species	OH^- OCl^-	H_2O $HOCl$	H^+	
Initial and/or input species		H_2O $HOCl$		55.6 10^{-3}

$$\sum_i c_{i,in} n_i = (55.6)(0) + (10^{-3})(0) = 0$$

$$\sum_i c_{i,eq} n_i = \{H^+\}(1) + \{OH^-\}(-1) + \{OCl^-\}(-1)$$

PC:　　　　$\{H^+\} = \{OH^-\} + \{OCl^-\}$

It should be clear that if H_2O and HOCl are chosen as components, the *TOT*H equation is simply the PC rewritten with a value of zero on the left and the $\{H^+\}$ term on the right.

*TOT*H equation:　　　　　$0 = \{H^+\} - \{OH^-\} - \{OCl^-\}$

c. Water + 10^{-3} M HOCl + 10^{-4} M NaOCl

If we again choose HOCl as the reference species for the $HOCl/OCl^-$ group, the proton balance table is very similar to that in part (b). The only differences are that we add Na^+ as a reference species, and we list NaOCl as a one-proton-deficient input species, based on the following reaction.

$$1 Na^+ + 1 HOCl + -1 H^+ \rightarrow 1 NaOCl$$

Note that, although OCl is added to the solution in two different forms, only one OCl species can be selected as the reference species for the OCl group.

The proton balance table and PC equation are:

	−1	0	+1	Concentration
Equilibrium species	OH^- OCl^-	H_2O $HOCl$ Na^+	H^+	
Initial and/or input species		H_2O $HOCl$		55.6 10^{-3}
	NaOCl			10^{-4}

$$\sum_i c_{i,in} n_i = (55.6)(0) + (10^{-3})(0) + (10^{-4})(-1)$$

$$\sum_i c_{i,eq} n_i = \{H^+\}(1) + \{OH^-\}(-1) + \{OCl^-\}(-1)$$

PC:　　$\{H^+\} + 10^{-4} = \{OH^-\} + \{OCl^-\}$

If we choose HOCl as a component, the calculation of $TOTH_{in}$ and the $TOTH$ equation are as follows:

$$TOTH_{in} = (10^{-3})(0) + (10^{-4})(-1) = -10^{-4}$$

$TOTH$ equation: $-10^{-4} = \{H^+\} - \{OH^-\} - \{OCl^-\}$

d. Water $+ 10^{-4} M$ HOCl $+ 2 \times 10^{-3} M$ NaOCl $+ 10^{-3} M$ NH$_3$ $+ 5 \times 10^{-4} M$ NH$_4$Cl $+ 10^{-3} M$ NaOH

Choosing OCl$^-$ and NH$_3$ as reference species for the two type b groups in this solution, and including the two type d species (Na$^+$ and Cl$^-$) as reference species as well, we derive the following table and proton condition. The determination of the proton excess or deficiency of each species should be becoming intuitive by now, so we skip the step of writing out the corresponding reactions explicitly.

	−1	0	+1	Concentration
Equilibrium species	OH$^-$	H$_2$O	H$^+$	
		OCl$^-$	HOCl	
		NH$_3$	NH$_4$$^+$	
		Na$^+$		
		Cl$^-$		
Initial and/or		H$_2$O		55.6
input species			HOCl	10^{-4}
		NaOCl		2×10^{-3}
		NH$_3$		10^{-3}
			NH$_4$Cl	5×10^{-4}
	NaOH			10^{-3}

$$\sum_i c_{i,in} n_i = (55.6)(0) + (10^{-4})(1) + (2 \times 10^{-3})(0) + (10^{-3})(0)$$
$$+ (5 \times 10^{-4})(1) + (10^{-3})(-1)$$
$$= -4.0 \times 10^{-4}$$

$$\sum_i c_{i,eq} n_i = \{H^+\}(1) + \{NH_4^+\}(1) + \{OH^-\}(-1) + \{OCl^-\}(-1)$$

PC: $\{H^+\} + \{NH_4^+\} + 4.0 \times 10^{-4} = \{OH^-\} + \{OCl^-\}$

Again, if the same species are chosen as components for the $TOTH$ equation as were chosen to be reference species for the PC, the $TOTH$ equation is simply the PC equation rearranged, i.e.,

$TOTH$ equation: $-4.0 \times 10^{-4} = \{H^+\} + \{NH_4^+\} - \{OH^-\} - \{OCl^-\}$

e. H$_2$O $+ 10^{-1} M$ Na$_2$CO$_3$ $+ 10^{-2} M$ H$_2$CO$_3$

The carbonate acid/base system is diprotic. That is, it can exchange two protons with solution. Nevertheless, it represents only one acid/base group, and we choose a single reference species for it. If we choose CO$_3$$^{2-}$ as the reference species, then each

H_2CO_3 molecule has two excess protons, and the concentration of excess protons associated with H_2CO_3 molecules equals $2\{H_2CO_3\}$. That is, 100 H_2CO_3 molecules represent 200 excess protons.

The proton condition table is therefore as follows:

	−1	0	+1	+2	Concentration
Equilibrium species	OH^-	H_2O CO_3^{2-} Na^+	H^+ HCO_3^-	H_2CO_3	
Initial and/or input species		Na_2CO_3			10^{-1}
			H_2CO_3		10^{-2}

We can skip the formality of including H_2O in the calculation of $\sum c_{i,in}n_i$, since its contribution is always zero. The PC and *TOT*H equations that we derive are:

$$\sum_i c_{i,in}n_i = (10^{-1})(0) + (10^{-2})(2) = 2 \times 10^{-2}$$

$$\sum_i c_{i,eq}n_i = \{H^+\}(1) + \{HCO_3^-\}(1) + \{H_2CO_3\}(2) + \{OH^-\}(-1)$$

PC: $\{H^+\} + \{HCO_3^-\} + 2\{H_2CO_3\} = 2 \times 10^{-2} + \{OH^-\}$

*TOT*H equation: $2 \times 10^{-2} = \{H^+\} + \{HCO_3^-\} + 2\{H_2CO_3\} - \{OH^-\}$

f. The same system as in part (e), choosing HCO_3^- as the reference species for the carbonate system in the PC.

	−1	0	+1	Concentration
Equilibrium species	OH^- CO_3^{2-}	H_2O HCO_3^- Na^+	H^+ H_2CO_3	
Initial and/or input species		Na_2CO_3		10^{-1}
			H_2CO_3	10^{-2}

$$\sum_i c_{i,in}n_i = (10^{-1})(-1) + (10^{-2})(1) = -9 \times 10^{-2}$$

$$\sum_i c_{i,eq}n_i = \{H^+\}(1) + \{OH^-\}(-1) + \{H_2CO_3\}(1) + \{CO_3^{2-}\}(-1)$$

PC: $\{H^+\} + \{H_2CO_3\} + 9 \times 10^{-2} = \{OH^-\} + \{CO_3^{2-}\}$

*TOT*H equation: $-9 \times 10^{-2} = \{H^+\} + \{H_2CO_3\} - \{OH^-\} - \{CO_3^{2-}\}$

In parts (e) and (f) of the preceding example, the PCs derived using different species as the reference level species look different from one another: the PC in part (f) is identical to the charge balance, whereas that in part (e) contains different coefficients and variables. However, the two forms are equally valid representations of the proton mass balance in the same solution; i.e., they are different representations of the same basic equation. One proof of this statement is that either equation can be converted to the other by combining it with the mass balance on $TOTCO_3$. In fact, in all the above examples, the proton condition that is shown can be derived by combining the charge balance with one or more mass balances.

In the above examples, the initial solution was pure water, and all the subsequent chemical inputs were specified. As noted earlier, though, we often encounter situations in which the initial solution already contains species that have a proton excess or deficiency. The analysis of such systems is only slightly more complicated, as demonstrated by the following example.

Example 4.6

A wastewater at pH 7.0 contains 125 mg/L total organic carbon (TOC) and $1.5 \times 10^{-3} M$ total inorganic carbon (TIC). The TOC is distributed among a variety of molecular forms. However, since the waste is thought to be mostly carbohydrates and is highly biodegradable, you have decided to treat all the TOC as if it were the simple sugar glucose $(C_6H_{12}O_6)$. Write the proton condition and the $TOTH$ equation for the solution after microorganisms respire essentially all the TOC, converting it to carbonic acid by the reaction $C_6H_{12}O_6 + 6O_2 \rightarrow 6H_2CO_3$. Use HCO_3^- as the reference species for the proton condition. Assume that the TOC in the original solution does not have any significant interactions with H^+ ions.

Solution

At pH 7.0, the alpha values for the carbonate system are $\alpha_0 = 0.18$, $\alpha_1 = 0.82$, and $\alpha_2 = 10^{-3.33}$. Therefore, the concentrations of H_2CO_3, HCO_3^-, and CO_3^{2-} in the initial solution are $0.27 \times 10^{-3} M$, $1.23 \times 10^{-3} M$, and $7.0 \times 10^{-7} M$, respectively. In the remainder of the example, we will treat the initial CO_3^{2-} concentration as negligible.

The statement that the solution contains 125 mg/L TOC means that 125 mg C/L is present in the solution as part of organic molecules. If all this TOC is respired, 125 mg C/L is added to the solution as part of H_2CO_3 molecules, corresponding to the addition of $1.04 \times 10^{-2} M H_2CO_3$:

$$\left(125 \frac{\text{mg C}}{\text{L}}\right)\left(\frac{1 \text{ mol C}}{12{,}000 \text{ mg C}}\right) = 1.04 \times 10^{-2} \frac{\text{mol C}}{\text{L}}$$

Therefore, after the biological treatment step, the solution can be thought of as a combination of the initial solution containing $0.27 \times 10^{-3} M H_2CO_3$ and $1.23 \times 10^{-3} M HCO_3^-$, and a subsequent addition of $1.04 \times 10^{-2} M H_2CO_3$.

The proton condition table for the system, with H_2O and HCO_3^- selected as reference species, is shown below. Note that the H_2CO_3 and HCO_3^- in the initial solution are included in the lower part of the table along with the H_2CO_3 that was added by respiration, because we are interested in the net proton excess in the solution when it equilibrates. The fact that some of this excess was there before we encountered the solution and that we do not know exactly what chemicals were used to generate that solution is irrelevant, as long as we have some way of determining the net proton excess at that time.

	−1	0	+1	Concentration
Equilibrium species	OH^- CO_3^{2-}	H_2O HCO_3^-	H^+ H_2CO_3	
Initial and/or input species		HCO_3^-	$H_2CO_3^{\dagger}$ $H_2CO_3^{\ddagger}$	0.27×10^{-3} 1.23×10^{-3} 1.04×10^{-2}

†From TIC present initially.

‡From respiration of TOC.

$$\sum_i c_{i,in} n_i = (0.27 \times 10^{-3})(1) + (1.23 \times 10^{-3})(0) + (1.04 \times 10^{-2})(1) = 1.07 \times 10^{-2}$$

$$\sum_i c_{i,eq} n_i = \{H^+\}(1) + \{OH^-\}(-1) + \{H_2CO_3\}(1) + \{CO_3^{2-}\}(-1)$$

PC: $\{H^+\} + \{H_2CO_3\} = \{OH^-\} + \{CO_3^{2-}\} + 1.07 \times 10^{-2}$

*TOT*H equation: $1.07 \times 10^{-2} = \{H^+\} + \{H_2CO_3\} - \{OH^-\} - \{CO_3^-\}$

Note that in the preceding example, we could not have written a detailed charge balance on the initial solution, because we had no information about cations or anions in that solution other than H^+, OH^-, and the species of the carbonate system. That is, while the problem statement specifies the pH and allows us to calculate the concentrations and activities of the carbonate-containing species, it does not imply that those species are the only ones in solution. Indeed, for the specified constituents to be present, there must be at least one additional cation present, and there might be many additional cations and anions.

We could write a generic charge balance that acknowledges our limited information about the ionic composition of the solution as follows:

$$\{H^+\}_{init} + \underset{\substack{\text{all other} \\ \text{cations } i}}{\sum} c_i z_i = \{OH^-\}_{init} + \{HCO_3^-\}_{init} + 2\{CO_3^{2-}\}_{init} + \underset{\substack{\text{all other} \\ \text{anions } j}}{\sum} c_j z_j$$

(4.7)

where c_i and c_j are the concentrations of the cations and anions, respectively, that are in the solution but are not represented explicitly in the rest of the equation, and z_i and z_j represent the charges on those ions. The summations $\underset{\substack{\text{all other} \\ \text{cations } i}}{\sum} c_i z_i$ and $\underset{\substack{\text{all other} \\ \text{anions } j}}{\sum} c_j z_j$ are commonly represented as C_B and C_A, respectively.[5]

[5]These designations are based on the assumption that all unspecified cations are salt ions that entered the solution associated with a strong base (B), and all unspecified anions are salt ions that entered the solution associated with a strong acid (A). The way that Equation (4.7) represents the contributions of unspecified cations and anions to the charge balance is more general, and it also makes the point explicitly that the summation focuses on charge, not mass.

Although the values of the individual $c_i z_i$ and $c_j z_j$ terms are not known, the net positive charge of all unspecified ions in the solution (given by the difference $\sum_{\substack{\text{all other} \\ \text{cations } i}} c_i z_i - \sum_{\substack{\text{all other} \\ \text{anions } j}} c_j z_j$) must balance the net negative charge on the ions that *are* specified, so

$$\sum_{\substack{\text{all other} \\ \text{cations } i}} c_i z_i - \sum_{\substack{\text{all other} \\ \text{anions } j}} c_j z_j = \{OH^-\}_{\text{init}} + \{HCO_3^-\}_{\text{init}} + 2\{CO_3^{2-}\}_{\text{init}} - \{H^+\}_{\text{init}}$$

(4.8a)

$$= 10^{-7.0} + 1.23 \times 10^{-3} + 2(7.0 \times 10^{-7}) - 10^{-7.0}$$

(4.8b)

$$= 1.23 \times 10^{-3}$$

The values shown in Equation (4.8b) are those in the initial solution (i.e., prior to the respiration of the TOC). However, since the respiration of the organics does not add any unspecified cations or anions to the solution, the value of $\sum_{\substack{\text{all other} \\ \text{cations } i}} c_i z_i - \sum_{\substack{\text{all other} \\ \text{anions } j}} c_j z_j$ computed in Equation (4.8b) also applies after the biological reaction. Substituting this value into the charge balance for the solution at that time yields

$$1.23 \times 10^{-3} = \{OH^-\}_{\text{final}} + \{HCO_3^-\}_{\text{final}} + 2\{CO_3^{2-}\}_{\text{final}} - \{H^+\}_{\text{final}} \quad \textbf{(4.9)}$$

The result shown in Equation (4.9) is a correct formulation of the charge balance in the solution. As such, it could be used in conjunction with a spreadsheet analysis or a log C–pH diagram to determine the speciation after the respiration and equilibration steps. However, the equation is somewhat more tedious to derive than the PC or *TOT*H equation developed in the example, and it turns out that an equation identical to Equation (4.9) can be developed as the PC, without explicitly considering the unspecified cations and anions at all, if H_2CO_3 is chosen as the reference species. Thus, while not being a charge balance itself, the PC is a way to characterize the same information as is in the CB even if not all the charged species in solution are known.

The proton condition or *TOT*H equation can also be used to evaluate the dose of a chemical that is required to adjust a solution to a specified endpoint, as illustrated in the following example.

Example 4.7 | Phenol is a weak organic acid with chemical formula C_6H_6O. Its conjugate base is phenolate ($C_6H_5O^-$), with $pK_b = 4.0$. A solution at pH 10.1 contains 10^{-4} total phenol (*TOT*Ph). You wish to add enough phosphoric acid to lower the pH to 7.0. How much H_3PO_4 must be added per liter of solution?

Solution

Since pK_b for phenolate (which we will represent as PhO$^-$) is 4.0, pK_a for phenol (PhOH) is $14.0 - 4.0$, or 10.0. The α values of PhOH and PhO$^-$ at pH 10.1 are

$$\alpha_{\text{PhOH}} = \frac{\{H^+\}}{\{H^+\} + K_a} = \frac{10^{-10.1}}{10^{-10.1} + 10^{-10.0}} = 0.44 \qquad \alpha_{\text{PhO}^-} = 1 - \alpha_{\text{PhOH}} = 0.56$$

After the solution pH has been adjusted to 7.0, the speciation of the phenol and phosphate groups will be as follows. Since pK_a for phenol is 10.0, essentially all the phenol will be protonated ($\alpha_{\text{PhO}^-} \approx 0.001$). Because pK_{a1} for H_3PO_4 is 2.16, H_3PO_4 represents a negligible fraction of *TOT*PO$_4$ at pH 7.0. Similarly, PO$_4^{3-}$ is negligible, since pK_{a3} is 12.35. Thus, at pH 7.0, the *TOT*PO$_4$ will be distributed primarily between $H_2PO_4^-$ and HPO$_4^{2-}$, in the following proportions:

$$\alpha_{H_2PO_4^-} = \frac{1}{\{H^+\}/K_{a1} + 1 + K_{a2}/\{H^+\} + K_{a2}K_{a3}/\{H^+\}^2}$$

$$= \frac{1}{10^{-4.84} + 1 + 10^{-0.20} + 10^{-5.55}} = 0.61$$

$$\alpha_{HPO_4^{2-}} \approx 1.0 - \alpha_{H_2PO_4^-} = 0.39$$

The proton condition table for the solution is shown below, with the dominant species at pH 7.0 (PhOH and $H_2PO_4^-$) chosen as the reference species, and the unknown input of H_3PO_4 designated as x.

The concentrations of H^+, OH$^-$, PhOH, and PhO$^-$ in the initial solution are included in the lower portion of the table, along with the unknown input concentration of H_3PO_4. The concentrations of H^+, OH$^-$, PhOH, and PhO$^-$ in the final (pH 7.0) solution are also known, so they can be inserted into the expression for the net proton excess at equilibrium; the contribution of phosphate species to this value can be represented as the product of the respective α values and *TOT*PO$_4$, which in this case is simply x. The value of x can then be computed by equating the net proton excess of the initial solution and input chemicals with that of the equilibrated solution. As shown, the required dose of H_3PO_4 is 1.12×10^{-4}.

	-2	-1	0	$+1$	Concentration
Equilibrium species		OH$^-$	H_2O	H^+	
	PO$_4^{3-}$	HPO$_4^{2-}$	$H_2PO_4^-$	H_3PO_4	
		PhO$^-$	PhOH		
Initial and/or input species				H^+	1.0×10^{-10}
		OH$^-$			1.0×10^{-4}
			PhOH		0.56×10^{-4}
		PhO$^-$			0.44×10^{-4}
				H_3PO_4	x

$$\sum_i c_{i,\text{in}} n_i = (10^{-10})(1) + (10^{-4.0})(-1) + (0.44 \times 10^{-4})(0) + (0.56 \times 10^{-4})(-1)$$

$$+ (x)(1)$$

$$= -1.56 \times 10^{-4} + x$$

$$\sum_i c_{i,eq} n_i = \{H^+\}(1) + \{OH^-\}(-1) + \{H_3PO_4\}(1) + \{HPO_4^{2-}\}(-1) + \{PO_4^{3-}\}(-2)$$
$$+ \{PhO^-\}(-1)$$

$$= 10^{-7.0} - 10^{-7.0} + (\alpha_{0,H_3PO_4})_{pH\,7.0}\,x - (\alpha_{2,H_3PO_4})_{pH\,7.0}\,x - 2(\alpha_{3,H_3PO_4})_{pH\,7.0}\,x$$
$$- (0.001)(10^{-4})$$

$$\approx -0.39\,x$$

$$-1.56 \times 10^{-4} + x = -0.39x$$

$$x = 1.12 \times 10^{-4}$$

Thus, $1.12 \times 10^{-4}\,M\;H_3PO_4$ must be added to lower the pH to 7.0.

4.4   IDENTIFYING DOMINANT SPECIES

Recall that the primary reason for developing the proton condition equation was that we wanted an equation that could be used in conjunction with a log C–pH diagram to identify the equilibrium pH of a system, without going through the effort of writing and then modifying the charge balance. (Again, the *TOT*H equation tells us the same information as the PC, but it is written in a form that is less convenient to use directly with a log C–pH diagram.) Recall also that, to be most useful, the final equation must not contain terms for species that are dominant at equilibrium. Since the species that we choose to be proton-neutral (i.e., reference species) do not appear in the final proton condition, it appears that we can achieve our objective by writing a proton condition for the system choosing the dominant species as reference species.

As noted earlier, H_2O is always dominant over H^+ and OH^-, and it is always chosen as the reference species for the $OH^-/H_2O/H^+$ group. Similarly, salt ions (type c species) are almost always the dominant form of the corresponding group (e.g., Na^+ is almost always dominant over type d species containing Na), so they are also chosen as reference species. However, for groups of weak acids and bases like the carbonate, hypochlorite, or acetate groups, it is not necessarily apparent initially which species will be dominant at equilibrium. In the next section, a procedure is described for making good guesses about which species will be dominant at equilibrium, so that we can make appropriate choices for reference level species when writing the PC, i.e., choices that will allow us to solve acid/base problems graphically using the PC directly. In addition to assisting in the choice of a good set of reference species for the PC, the procedure provides insight into the behavior of acid/base systems when those systems are titrated with other acids or bases. Those reactions, which are central to many water treatment processes and are frequently used in chemical analysis protocols, are discussed in detail in Chapter 5.

The basis for the procedure is the idea that equilibrium solutions have no "memory." That is, once a solution has reached equilibrium, it is impossible to tell which chemicals were added, or the order in which they were added, to prepare the solution; all that one can say for sure is that all the species in the equilibrium system were added in some form. To be more explicit, if a solution contains 3×10^{-3} M HOCl, 1×10^{-3} M OCl$^-$, and 1×10^{-3} M Na$^+$ at equilibrium, one can be certain that 4×10^{-3} M *TOT*OCl was added to the solution, but not whether that *TOT*OCl was added entirely as HOCl (perhaps followed by some NaOH), as a mixture of HOCl and NaOCl (followed by a smaller amount of NaOH), or as a mixture of HOCl and NaOCl in amounts such that no NaOH was needed to establish the equilibrium distribution.

A corollary of the idea that equilibrium solutions have no memory is that we are free to imagine that the solution was concocted in any way we like and, as long as we account for all the species that are present in the actual solution, our imaginary concoction will behave identically to the real system. In fact, in the procedure about to be described, we will imagine that solutions can be prepared in a totally unrealistic way: by adding everything but the H$^+$ ions first, then adding the H$^+$ later. In reality, it is impossible to add individual ions to solution; anytime we add an anion, we must add an equal and opposite amount of cationic charge. Nevertheless, this imaginary procedure proves useful for predicting the dominant species in the system, and because solutions have no memory, we can analyze systems as though they were prepared in this way without any risk of drawing incorrect conclusions.

For the purposes of this procedure, the key attribute of acid/base systems is that, except at pH values quite near pK_a, each conjugate acid/base pair is present almost entirely as the acid or almost entirely as the base. For instance, in a solution at pH 6 containing some OCl species and some Ac species, almost all the OCl is protonated (i.e., *TOT*OCl $\cong$ {HOCl}), whereas almost all the Ac is dissociated (i.e., *TOT*Ac $\cong$ {Ac$^-$}). Put another way, OCl$^-$ ions have a stronger affinity for protons than Ac$^-$ ions do, so the OCl$^-$ ions outcompete Ac$^-$ ions for any available protons; only after virtually every OCl$^-$ ion has acquired an H$^+$ can Ac$^-$ ions begin to bind them. This, of course, defines OCl$^-$ as a stronger base than Ac$^-$ (recall that bases are sometimes called proton acceptors; the greater the tendency to accept a proton, the greater the strength of the base).

Generalizing from the above discussion, we can make the approximation that the H$^+$ ions added to a system combine sequentially with the various bases in the system, starting with the strongest base and working toward the weakest. Returning to our hypothetical procedure of adding all the bases to solution and then adding the H$^+$, we imagine that the strongest bases will combine with the first doses of H$^+$ that we add. The protonation of bases continues, with successively weaker bases becoming protonated, until we have added all the H$^+$ available. At that point, we can review the speciation of each acid/base conjugate pair and determine which of the two species is dominant.

Consider, for example, a system made by adding 10^{-3} M HAc, 2×10^{-3} M NaAc, 3×10^{-4} M NaOCl, and 5×10^{-4} M NaOH to water. We could imagine that all these species were added in the following sequence: inert salt ions, then bases, and then H$^+$ ions. Once the salts and bases had been added, the solution would contain 2.8×10^{-3} M Na$^+$, 3.0×10^{-3} M Ac$^-$, 3.0×10^{-4} M OCl$^-$, and 5×10^{-4} M OH$^-$, and we would have 10^{-3} M H$^+$ still "waiting" to be added. Listing the bases in the system from strongest to weakest (higher to lower pK_a), we develop the following table:

Acid/Base	pK_a	TOTA
H_2O/OH^-	14.00	5×10^{-4}
$HOCl/OCl^-$	7.60	3×10^{-4}
HAc/Ac^-	4.76	3×10^{-3}

The 10^{-3} mol/L of available H^+ is expected to combine first with the OH^-, converting the OH^- to its conjugate acid, H_2O. This conversion requires 5×10^{-4} mol/L H^+, so once it is complete, we still have 5×10^{-4} mol/L H^+ to add. Conversion of the OCl^- to $HOCl$ consumes 3×10^{-4} mol/L of these H^+ ions, leaving 2×10^{-4} mol/L. These remaining protons are expected to combine with Ac^-, but since $TOTAc$ is 3×10^{-3} M, the available H^+ can convert only about 7% of the Ac^- to HAc. We therefore conclude that, at equilibrium, the dominant species of the weak acid/base groups are likely to be $HOCl$ and Ac^-.

| **Example 4.8** | Predict the acid/base species that will be dominant at equilibrium in a system prepared by combining 10^{-3} M HCl, 10^{-2} M NH_3, 10^{-2} M $NaHCO_3$, and 10^{-3} M butyric acid (HBut). Butyric acid is a four-carbon carboxylic acid, in the same chemical family as but larger than formic acid (one carbon), acetic acid (two carbons), and propionic acid (three carbons). It is the highly distinctive, foul-smelling constituent of stomach acid that can induce (or be an unpleasant consequence of) nausea. It also sometimes accumulates in anaerobic waste treatment systems that are not operating properly, to the dismay of all around. The pK_a for HBut is 4.73, very close to that of its smaller analogs. |

Solution

A table listing the acid/base conjugate pairs in the system in order of decreasing basicity and the corresponding $TOTA$ values is shown below. Note that the two acid/base pairs of the carbonate system are listed independently, and that water is listed as a base that could potentially be protonated to form H_3O^+.

As above, we imagine that the solution initially contains only inert salt ions (10^{-3} M Cl^- and 10^{-2} M Na^+) and the bases that were added to the solution (10^{-2} M NH_3, 10^{-2} M CO_3^{2-}, and 10^{-2} M But^-). The concentration of H^+ available to add to this hypothetical solution is 1.2×10^{-2} M, consisting of the protons added with the $NaHCO_3$, HBut, and HCl. A column has been added to the table indicating the concentration of H^+ available in our hypothetical reservoir before any of the base in that row has been protonated.

Base/Acid	pK_a	TOTA	H^+ Available
CO_3^{2-}/HCO_3^-	10.33	10^{-2}	1.2×10^{-2}
NH_3/NH_4^+	9.25	10^{-2}	0.2×10^{-2}
HCO_3^-/H_2CO_3	6.35	10^{-2}	0
$But^-/HBut$	4.73	10^{-3}	—
H_2O/H_3O^+	0.00	55.6	—

The strongest base in the solution is CO_3^{2-}, with a total concentration of $10^{-2} M$. We assume that $10^{-2} M$ protons combine with the CO_3^{2-} to convert it all to HCO_3^-, while all the other bases remain completely deprotonated. This leaves $0.2 \times 10^{-2} M$ exchangeable protons to combine with the other bases. The next strongest base is NH_3. It would take $10^{-2} M$ protons to convert all the NH_3 to NH_4^+, but there are only $0.2 \times 10^{-2} M$ protons available. We therefore assume that the available protons convert 20% of the NH_3 to NH_4^+. Approximately 80% of the NH_3 and 100% of the HCO_3^- and But^- remain in the basic form. Thus, we would guess that the dominant species in this system at equilibrium are HCO_3^-, NH_3, and But^-.

Which species would you expect to be dominant in solutions prepared by adding the following chemicals to water?   **Example 4.9**

a. $5 \times 10^{-3} M$ $KH_2PO_4 + 10^{-2} M$ $NaOH + 10^{-2} M$ Na_2HPO_4

As before, we list bases, K_a values, and concentrations.

Base/Acid	pK_a	TOTA	H$^+$ Available
OH^-/H_2O	14.00	10^{-2}	2×10^{-2}
PO_4^{3-}/HPO_4^{2-}	12.35	1.5×10^{-2}	1×10^{-2}
$HPO_4^{2-}/H_2PO_4^-$	7.20	1.5×10^{-2}	0
$H_2PO_4^-/H_3PO_4$	2.16	1.5×10^{-2}	—

The available proton concentration is

$$[H^+]_{avail} = 2(5 \times 10^{-3}) + 1(1 \times 10^{-2}) = 2 \times 10^{-2}$$

The protons will bind first to the OH^- ($1 \times 10^{-2} M$) and will then convert $\sim\frac{2}{3}$ of the PO_4^{3-} to HPO_4^{2-}. We therefore surmise that HPO_4^{2-} will be the dominant form of phosphate at equilibrium.

b. $10^{-4} M$ $H_2CO_3 + 10^{-4} M$ $NH_3 + 10^{-2} M$ HCl

Base/Acid	pK_a	TOTA	H$^+$ Available
CO_3^{2-}/HCO_3^-	10.33	10^{-4}	1.02×10^{-2}
NH_3/NH_4^+	9.25	10^{-4}	1.01×10^{-2}
HCO_3^-/H_2CO_3	6.35	10^{-4}	1.0×10^{-2}
H_2O/H_3O^+	0.00	55.6	0

$$[H^+]_{avail} = (2)(1 \times 10^{-4}) + 10^{-2} = 1.02 \times 10^{-2}$$

The protons will bind to the weak bases in the system in the following sequence and quantities:

$1 \times 10^{-4} M$ to convert CO_3^{2-} to HCO_3^-
$1 \times 10^{-4} M$ to convert NH_3 to NH_4^+
$1 \times 10^{-4} M$ to convert HCO_3^- to H_2CO_3

After all these weak bases are converted to their fully protonated forms, the dominant species are NH_4^+ and H_2CO_3, and there are still 0.99×10^{-2} M protons available. These protons would combine with the next-weakest base in the system, H_2O, forming H_3O^+. A row showing the possible protonation of H_2O has been included in the table, because H_2O is available in any solution to acquire H^+ that is "left over" after all the other weak bases in the system have been protonated. The concentration of H_2O is 55.6 M, so clearly its capacity to acquire H^+ will never be exhausted. Therefore, not only can we predict that the dominant species will be NH_4^+ and H_2CO_3, but we can also estimate that the final solution will contain 0.99×10^{-2} M H_3O^+, so the pH will be approximately 2.0.

The term identified as $[H^+]_{avail}$ in the above examples is closely related to the term we defined as $TOTH$ earlier in the chapter. Specifically, $[H^+]_{avail}$ is the value of $TOTH_{in}$ (or, equivalently, the net proton excess) that we would compute if we chose the water and fully deprotonated bases as reference species. Thus, the determination of $[H^+]_{avail}$ is really just the same process as the determination of $TOTH_{in}$ for a specific choice of reference species. Recognition of this fact is helpful in the analysis of acid/base titrations in Chapter 5.

The method described above will almost always correctly identify the dominant species in a system. The only time it may fail is when there are several acid/base pairs in the system with nearly identical K_a values. In that case, it will be evident when we try to evaluate the proton condition graphically which species should have been chosen as reference species. We can then rewrite the PC and solve the problem.

4.5 PUTTING IT ALL TOGETHER: USE OF LOG C–pH DIAGRAMS AND THE PC OR TOTH EQUATION TO SOLVE FOR SOLUTION SPECIATION

Having developed the basic procedures for drawing log C–pH diagrams, identifying the species expected to be dominant at equilibrium, and writing the PC equation or TOTH balance, we are in a position to determine the equilibrium composition of virtually any mixture of acids and bases. A couple examples of this process are provided below. In reading through the examples, it is worthwhile to mentally compare this approach with the numerical approaches presented in Chapter 3. In particular, consider that you could sketch the log C–pH diagram freehand to get a sense of how speciation varies with pH, quickly compare the strength and concentration of the bases that were added with the concentration of available H^+ to make good guesses about the dominant species, write the PC equation, and then draw curves for LHS and RHS on the graph to determine the equilibrium solution composition. With a little practice, these steps can be carried out very rapidly. By contrast, the numerical methods are mathematically equivalent, but the result is rarely intuitive, it is difficult to obtain without carrying out lengthy algebra or using a computer, and the solution procedures provide no overview of how speciation depends on pH.

Write proton conditions using the dominant species as the reference level species for each of the following solutions. Then construct a log C–pH diagram for a system containing 0.10 M $TOTPO_4$, and use it along with the PCs to determine the equilibrium pH and the concentration of all species in the solutions.

| **Example 4.10** |

a. 0.10 M NaH_2PO_4
b. 0.10 M Na_2HPO_4
c. 0.10 M Na_3PO_4

Solution

Using the approach described above, we surmise that $H_2PO_4^-$, HPO_4^{2-}, and PO_4^{3-} are likely to be the dominant PO_4-containing species in parts (a), (b), and (c), respectively. (Carry out the calculation to convince yourself of this result.) Proton conditions for the three solutions of interest using the species anticipated to be dominant as the reference species are derived below.

a. 0.10 M NaH_2PO_4

	−2	−1	0	+1	Concentration
Equilibrium species	PO_4^{3-}	OH^- HPO_4^{2-}	H_2O $H_2PO_4^-$ Na^+	H^+ H_3PO_4	
Initial and/or input species			NaH_2PO_4		0.10

$$TOTH_{in} = \sum_i c_{i,in} n_i = (0.10)(0) = 0$$

$$\sum_i c_{i,eq} n_i = \{H^+\}(1) + \{OH^-\}(-1) + \{H_3PO_4\}(1) + \{HPO_4^{2-}\}(-1) + \{PO_4^{3-}\}(-2)$$

PC: $\quad \{H^+\} + \{H_3PO_4\} = \{OH^-\} + \{HPO_4^{2-}\} + 2\{PO_4^{3-}\}$

b. 0.10 M Na_2HPO_4

	−1	0	+1	+2	Concentration
Equilibrium species	OH^- PO_4^{3-}	H_2O HPO_4^{2-} Na^+	H^+ $H_2PO_4^-$	H_3PO_4	
Initial and/or input species		Na_2HPO_4			0.10

$$TOT\,H_{in} = \sum_i c_{i,in} n_i = (0.10)(0) = 0$$

$$\sum_i c_{i,eq} n_i = \{H^+\}(1) + \{OH^-\}(-1) + \{H_3PO_4\}(2) + \{H_2PO_4^-\}(1) + \{PO_4^{3-}\}(-1)$$

PC: $\{H^+\} + 2\{H_3PO_4\} + \{H_2PO_4^-\} = \{OH^-\} + \{PO_4^{3-}\}$

c. $0.10\ M\ Na_3PO_4$

	−1	0	+1	+2	+3	Concentration
Equilibrium species	OH^-	H_2O PO_4^{3-} Na^+	H^+ HPO_4^{2-}	$H_2PO_4^-$	H_3PO_4	
Initial and/or input species		Na_3PO_4				0.10

$$TOT\,H_{in} = \sum_i c_{i,in} n_i = (0.10)(0) = 0$$

$$\sum_i c_{i,eq} n_i = \{H^+\}(1) + \{OH^-\}(-1) + \{H_3PO_4\}(3) + \{H_2PO_4^-\}(2) + \{HPO_4^{2-}\}(1)$$

PC: $\{H^+\} + 3\{H_3PO_4\} + 2\{H_2PO_4^-\} + \{HPO_4^{2-}\} = \{OH^-\}$

A single log C–pH diagram describes all three systems (Figure 4.10). The only difference among the systems is in the proton conditions. Curves representing the right- and left-hand sides of each PC are shown in bold in the figure, and the equilibrium pH and concentration of all species are summarized in the following table:

Part	pH	Log $\{H_3PO_4\}$	Log $\{H_2PO_4^-\}$	Log $\{HPO_4^{2-}\}$	Log $\{PO_4^{3-}\}$
(a)	4.7	−3.5	−1.0	−3.5	−11.2
(b)	9.7	−11.0	−3.5	−1.0	−3.5
(c)	12.5	−17.1	−6.8	−1.5	−1.2

Note that in parts (b) and (c), the lines representing $\{H_3PO_4\}$ and $\{H_2PO_4^-\}$ are drawn in the "usual" way when preparing the log C–pH graphs, but the LHS of the PC contains terms that are 2 or 3 times those values; i.e., the PC contains terms like $3\{H_3PO_4\}$ and $2\{H_2PO_4^-\}$. As before, we incorporate these coefficients in the graphical analysis by noting that

$$\log (3\{H_3PO_4\}) = \log 3 + \log \{H_3PO_4\} = 0.5 + \log \{H_3PO_4\}$$

$$\log (2\{H_2PO_4^-\}) = \log 2 + \log \{H_2PO_4^-\} = 0.3 + \log \{H_2PO_4^-\}$$

Thus, curves representing $2\{H_3PO_4\}$ and $3\{H_3PO_4\}$ are, respectively, 0.3 and 0.5 log unit above the $\{H_3PO_4\}$ curve. The effect of the coefficient on the $\{H_3PO_4\}$ term is somewhat difficult to see in the graphs, but the effect on the $\{H_2PO_4^-\}$ term in part (c) is quite apparent in the range $3 < pH < 7$.

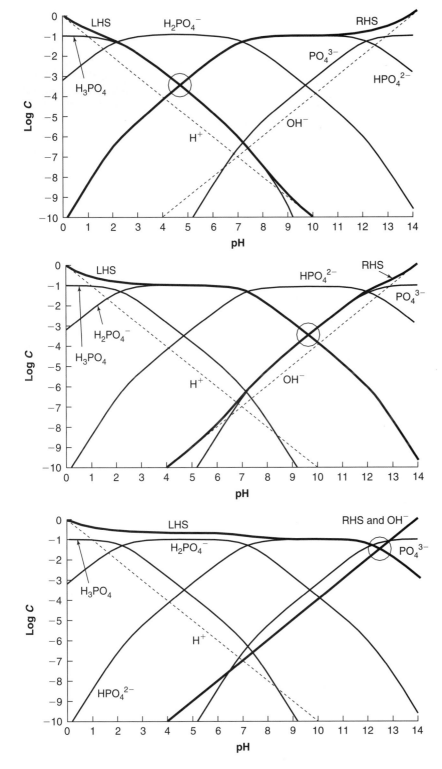

Figure 4.10 The log C–pH diagram for a system containing 0.1 M $TOTPO_4$, with the RHS and LHS of the proton condition from parts (a), (b), and (c) of the example drawn in bold.

Example 4.11 | A model anaerobic waste solution is prepared as a mixture of $10^{-2}\ M$ NaHCO$_3$, $10^{-4}\ M$ Na$_2$S, and $4 \times 10^{-3}\ M$ HAc. Determine the solution composition at equilibrium. Upon dissolution, sodium sulfide (Na$_2$S), releases sulfide ion (S^{2-}), a relatively strong base that can protonate to form bisulfide ion (HS$^-$) or hydrogen sulfide (H$_2$S). Hydrogen sulfide can enter the gas phase to generate a rotten egg smell that can be detected even at extremely low concentrations. Organic sulfides (R–SH) also tend to have strong, unpleasant odors and are also generated in anaerobic environments, as well as being found in some industrial wastes (e.g., from pulping). The values of pK_{a1} and pK_{a2} for H$_2$S are 6.99 and 12.92, respectively.

Solution

The species expected to be dominant at equilibrium can be determined by imagining the solution to be prepared by adding the salts and deprotonated bases (Na$^+$, CO$_3^{2-}$, S^{2-}, and Ac$^-$) and then adding the available H$^+$, in this case equal to $10^{-2}\ M$ from the NaHCO$_3$ plus $4 \times 10^{-3}\ M$ from HAc. Based on that value and the following summary table, the dominant species are likely to be H$_2$S, HCO$_3^-$, and Ac$^-$.

Base/Acid	pK_a	TOTA	H$^+$ Available
S^{2-}/HS$^-$	12.92	10^{-4}	1.40×10^{-2}
CO$_3^{2-}$/HCO$_3^-$	10.33	10^{-2}	1.39×10^{-2}
HS$^-$/H$_2$S	6.99	10^{-4}	0.39×10^{-2}
HCO$_3^-$/H$_2$CO$_3$	6.35	10^{-2}	0.38×10^{-2}
Ac$^-$/HAc	4.73	4×10^{-3}	0

The proton condition table and equation using the expected dominant species as reference species are as follows:

	−2	−1	0	+1	Concentration
Equilibrium species		OH$^-$ CO$_3^{2-}$	H$_2$O HCO$_3^-$ Ac$^-$	H$^+$ H$_2$CO$_3$ HAc	
	S^{2-}	HS$^-$	H$_2$S Na$^+$		
Initial and/or input species			NaHCO$_3$		10^{-2}
				HAc	4×10^{-3}
	Na$_2$S				10^{-4}

$$\sum_i c_{i,\text{in}} n_i = (10^{-2})(0) + (4 \times 10^{-3})(1) + (10^{-4})(-2) = 3.8 \times 10^{-3}$$

$$\sum_i c_{i,\text{eq}} n_i = \{H^+\}(1) + \{OH^-\}(-1) + \{H_2CO_3\}(1) + \{CO_3^{2-}\}(-1)$$
$$+ \{HAc\}(1) + \{HS^-\}(-1) + \{S^{2-}\}(-2)$$

$$3.8 \times 10^{-3} = \{H^+\} - \{OH^-\} + \{H_2CO_3\} - \{CO_3^{2-}\} + \{HAc\} - \{HS^-\} - 2\{S^{2-}\}$$

$$\{H^+\} + \{H_2CO_3\} + \{HAc\} = 3.8 \times 10^{-3} + \{OH^-\} + \{CO_3^{2-}\} + \{HS^-\} + 2\{S^{2-}\}$$

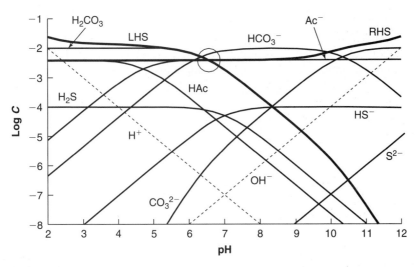

Figure 4.11　Log C–pH diagram for a system consisting of 10^{-2} M NaHCO$_3$, 4×10^{-3} M HAc, and 10^{-4} M Na$_2$S.

The log C–pH diagram is shown in Figure 4.11; the equilibrium pH is 6.57, and the composition of the solution is as follows:

$$\{H^+\} = 10^{-6.57} \qquad \{OH^-\} = 10^{-7.43}$$
$$\{H_2CO_3\} = 10^{-2.42} \qquad \{HCO_3^-\} = 10^{-2.20} \qquad \{CO_3^{2-}\} = 10^{-5.96}$$
$$\{HAc\} = 10^{-4.24} \qquad \{Ac^-\} = 10^{-2.40}$$
$$\{H_2S\} = 10^{-4.14} \qquad \{HS^-\} = 10^{-4.56} \qquad \{S^{2-}\} = 10^{-10.91}$$

4.6　SOME POINTS ABOUT NON-OBVIOUS PROTON CONDITIONS

The PC equation and *TOT*H balance are simply mathematical statements that protons, like other species, must be conserved when chemicals combine or break apart. For instance, if NH$_3$ is defined as a reference level species and H$^+$ is a one-proton-excess species, then NH$_4^+$, being the combination of something with no excess protons and something else with one excess proton, must itself have one excess proton. Of course, that is also obvious simply from a comparison of the number of protons in the NH$_3$ and NH$_4^+$ species. However, there are cases where the proton excess is not so obvious from inspection. In such cases, use of the table defining proton-excess and proton-deficient species, or the algorithm of writing reactions to form each species from a combination of reference species and protons, will always allow the correct PC equation and/or *TOT*H balance to be derived.

For instance, consider a system made by adding 10^{-3} M copper chloride ($CuCl_2$) and 2.5×10^{-3} M NH_3 to water. The copper chloride completely dissociates, but the copper ions then combine with some of the hydroxide and some of the ammonia molecules to form the following species: $CuOH^+$, $Cu(OH)_2^\circ$, $Cu(OH)_3^-$, $Cu(NH_3)^{2+}$, $Cu(NH_3)_2^{2+}$, $Cu(NH_3)_3^{2+}$, and $Cu(NH_3)_4^{2+}$. Of course, Cu^{2+}, NH_3, NH_4^+, H^+, and OH^- will all also be present in the equilibrium solution. What is the proton excess or deficiency associated with each of these species?

As always, we first need to choose reference level species. Let's choose H_2O, Cu^{2+}, and NH_3 as reference level (case I). We know immediately that H^+ and NH_4^+ are one-proton-excess species, and OH^- is one-proton-deficient. Since protons must be conserved when species combine, it is also clear that $CuOH^+$, which is formed by combining a zero-level species and a one-proton-deficient species, must be one-proton-deficient. Similarly, $Cu(OH)_2^\circ$ and $Cu(OH)_3^-$ are two and three protons deficient, respectively. The species containing one Cu^{2+} ion and various numbers of NH_3 molecules are all proton-neutral (reference level), because they are formed by combinations of two, three, four, or five proton-neutral components. Although the $Cu(NH_3)_x^{2+}$ species do not appear in the final PC or $TOTH$ equation in this case, they are not being ignored. Rather, they are being considered explicitly, and the consideration leads to the conclusion that they must be reference level species. The proton condition table and $TOTH$ equation for this system are shown below.

Case I

	−3	−2	−1	0	+1	Concentration
Equilibrium species	$Cu(OH)_3^-$	$Cu(OH)_2^\circ$	OH^- $CuOH^+$	H_2O Cu^{2+} NH_3 $Cu(NH_3)_x^{2+}$ Cl^-	H^+ NH_4^+	
Initial and/or input species				$CuCl_2$ NH_3		10^{-3} 2.5×10^{-3}

$$0 = \{H^+\} + \{NH_4^+\} - \{OH^-\} - \{CuOH^+\} - 2\{Cu(OH)_2^\circ\} - 3\{Cu(OH)_3^-\}$$

Next, we develop the $TOTH$ equation for the same system, but choose $Cu(OH)_2^\circ$ and NH_4^+ as reference species (case II). In this case, NH_3 would be a one-proton-deficient species, as would $Cu(OH)_3^-$; $CuOH^+$ and Cu^{2+} would be counted as having one and two excess protons, respectively. In this scenario, the $Cu(NH_3)_x^{2+}$ species are not reference-level species, and, in fact, each one of these species has a different proton excess or deficiency. For instance, $Cu(NH_3)^{2+}$ has one excess proton, because it is generated by combining a two-proton-excess species (Cu^{2+}) with one one-proton-deficient species (NH_3). Similarly, $Cu(NH_3)_2^{2+}$ is a reference level species, $Cu(NH_3)_3^{2+}$ is one-proton-deficient, and

$Cu(NH_3)_4^{2+}$ is two-protons-deficient. The PC table and equation for this choice of reference species are shown below.

Case II

	−2	−1	0	+1	+2	Concentration
Equilibrium species		OH^- $Cu(OH)_3^-$ NH_3	H_2O $Cu(OH)_2^\circ$ NH_4^+	H^+ $CuOH^+$	Cu^{2+}	
	$Cu(NH_3)_x^{2+}$	$Cu(NH_3)_3^{2+}$	$Cu(NH_3)_2^{2+}$ Cl^-	$Cu(NH_3)_3^{2+}$		
Initial and/or input species		NH_3			$CuCl_2$	10^{-3} 2.5×10^{-3}

$$(10^{-3})2 - 2.5 \times 10^{-3} = \{H^+\} - \{OH^-\} + 2\{Cu^{2+}\} + \{CuOH^+\} - \{Cu(OH)_3^-\}$$
$$- \{NH_3\} + \{Cu(NH_3)^{2+}\} - \{Cu(NH_3)_3^{2+}\} - 2\{Cu(NH_3)_4^{2+}\}$$

Thus, species that must be considered in the proton condition include not only those that can combine with or release protons directly, but also any species that can combine with *others* that react with protons. Furthermore, the assignment of proton levels is arbitrary for a limited number of species (in the present system, this number is three). Once proton reference levels for the H_2O, NH_3, and Cu groups are assigned, the proton level of every species in the system is fixed.

We can use the same approach to determine the proton levels of species that enter or exit the aqueous phase by exchange with a gas or suspended solid phase. For instance, say that ammonia is dissolving into a solution from an overlying gas phase. If $NH_3(aq)$ has been defined as a one-proton-deficient species, then each mole of NH_3 gas dissolving in the water is identical to a mole of NH_3 that was added initially. Thus, the amount of NH_3 dissolving (in moles per liter) must be treated as part of the proton deficiency input to the system. Similarly, if $Cu(OH)_2(s)$ precipitates out of solution during the equilibration process, and if Cu^{2+} is a reference level species and OH^- is one-proton-deficient, then each mole of $Cu(OH)_2(s)$ that forms accounts for a deficit of two moles of protons. From the perspective of a balance on protons, the molecules that form the solid are identical to any other two-proton-deficient species present at equilibrium, and they must be included in the PC if it is to yield a correct result.

SUMMARY

This chapter demonstrates the development and use of graphical approaches for analyzing acid/base systems. The graphs that are used in these analyses highlight certain universal features of such systems. In particular, they emphasize the fact that the acid form of an acid/base conjugate pair is present at a larger activity than the basic form at any solution pH less than pK_a, and the reverse is true at pH values greater than pK_a. At a pH equal to the pK_a, the acid and base are present at equal activities. If the assumption of ideal behavior of the solutes applies, then these same statements apply to the species' concentrations.

The representation of the system characteristics on a log C–pH graph is such that, at any pH, the mass balances on the acid/base groups that are present, the K_a expressions, and the K_w expression are all satisfied. Thus, the only additional equation that must be satisfied to determine the equilibrium composition of the system is the charge balance. This situation allows a large number of solutions of interest to be analyzed quite easily.

The pH at which the charge balance is satisfied is sometimes difficult to identify directly on the log C–pH graph. In such cases, the difficulty can usually be overcome by writing a new equation that can be derived by combining the charge balance with one or more mass balances. The resulting equation is, in effect, a mass balance on protons using a non-zero baseline concentration. This equation is known as a proton condition or a TOTH balance. The equation can be written directly by choosing a few species to be reference species, in essence defining a solution containing only those species as the baseline at which the net proton excess in the solution, or TOTH, is zero. The proton excess or proton deficiency of all other species present in the equilibrium solution or added to prepare the solution can then be determined. The proton condition is a statement that the net proton excess in the equilibrium solution is the same as that in the initial solution plus that of any subsequent chemical additions, i.e., that $TOT\text{H}_{in} = TOT\text{H}_{eq}$.

A simple algorithm can be used to predict which acid/base species will be dominant once equilibrium is attained. This algorithm treats the components of the system as though they could be added sequentially to solution, even if such a procedure would be impossible to carry out in a real system. In the imaginary process, all the bases are added to the initial solution, and then protons are added subsequently. An approximation is made that the available protons bind sequentially, first to the strongest base in the system and then to progressively weaker ones, until all the protons that are present in the real system have been taken into account. The basis of this algorithm is that solutions have no memory, so that any approach that generates the same total inputs of species yields the same predicted equilibrium condition, regardless of how or in what order the chemicals were added.

If the species that are expected to be dominant at equilibrium are chosen as the reference species in the proton condition, the PC equation will not contain terms for the dominant species and will therefore be easy to use in conjunction with the log C–pH diagram to determine the equilibrium solution composition.

Problems

1. Determine the solution pH in each of the following cases, using a log C–pH diagram. Check your result by substituting the computed concentrations of all the species into the charge balance equation.

   a. $1.0 \times 10^{-3}\ M$ NaAc.

   b. $1.0 \times 10^{-4}\ M$ NH$_4$Cl.

   c. $5.0 \times 10^{-3}\ M$ NaCN.

    d. $5.0 \times 10^{-3}\ M$ HOCl.

    e. $1.0 \times 10^{-3}\ M$ HAc.

2. The acidity constants for citric acid are $pK_{a1} = 3.13$, $pK_{a2} = 4.72$, and $pK_{a3} = 6.33$.

    a. Write expressions for α_0, α_1, α_2, and α_3 as a function of the K_a values and $\{H^+\}$.

    b. Identify the region where each species is dominant.

    c. Determine which terms in the denominators for the α values are significant at pH 7.5 (terms contributing less than 5% to the summation can be considered negligible). For a solution with $10^{-1}\ M$ total citrate, write out an approximate equation of the form $\log C_i = \text{constant} + n(\text{pH})$ for each citrate species at pH 7.5, approximating the dominant species' concentration as *TOT*Cit.

    d. Draw a log *C*–pH diagram for a solution with $10^{-1}\ M$ *TOT*Cit. Include the pH range $1.0 < \text{pH} < 8.0$.

    e. Find the pH of the following solutions:

       i. $10^{-1}\ M$ H$_3$Cit + water.

       ii. $10^{-1}\ M$ Na$_2$HCit + water.

       iii. $0.05\ M$ Na$_3$Cit + $0.05\ M$ NaH$_2$Cit + water.

3. What are the concentrations of all species in a solution of $10^{-2}\ M$ Na$_2$CO$_3$? $10^{-2}\ M$ NaHCO$_3$? $10^{-2}\ M$ H$_2$CO$_3$?

4. Below is a log *C*–pH diagram for $10^{-3}\ M$ hydrogen sulfide (H$_2$S), a diprotic acid, with $pK_{a1} = 6.99$ and $pK_{a2} = 12.92$.

    a. Label the lines with the species they represent, and assign correct values to the axes.

    b. Draw lines for H$^+$ and OH$^-$ on the diagram.

    c. What is the pH of a solution made by mixing $10^{-4}\ M$ NaHS and $9 \times 10^{-4}\ M$ H$_2$S?

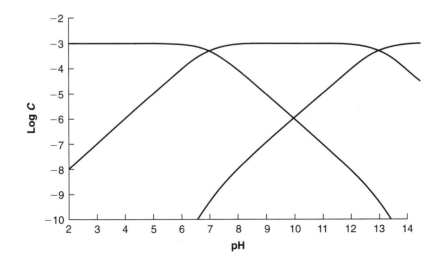

   d. Would a solution made by adding $0.5 \times 10^{-4}\ M\ Na_2S$ and $9.5 \times 10^{-4}\ M\ H_2S$ be more acidic, more alkaline, or the same as the solution in part (c)? Explain your answer in one or two sentences or equations.

5. Consider an acid H_2A, with $pK_{a1} = 5.5$ and $pK_{a2} = 9.5$, in a solution of $10^{-2}\ M\ TOTA$.

   a. At what pH or in what pH range (if any) will the following conditions be met?

      i. The activity of A^{2-} increases by approximately a factor of 100 for every increase of one pH unit.

      ii. The activity of HA^- decreases by approximately a factor of 10 for every increase of one pH unit.

      iii. The activity of H_2A increases by approximately a factor of 10 for every increase of one pH unit.

      iv. The ratio $\{H_2A\}/\{A^{2-}\}$ decreases by approximately a factor of 100 for every increase of one pH unit.

      v. The value of $p\{HA^-\}$ is approximately 2.3.

      vi. The value of $p\{H_2A\}$ is approximately 4.0.

   b. What is the pH of a solution of $0.005\ M\ H_2A + 0.005\ M\ NaHA$?

   c. Write the proton condition and determine the pH of a solution of $5 \times 10^{-3}\ M$ $Na_2A + 5 \times 10^{-3}\ M\ H_2A$.

   d. Will a solution of $10^{-2}\ M\ NaHA$ be acidic, neutral, or alkaline?

   e. Write the equilibrium equation and the form and the value of K_b for the base HA^-.

6. A wastewater is simulated as a solution containing $10^{-2.7}\ M\ NaHCO_3$ and $10^{-3}\ M\ NH_4Cl$.

   a. List the dissolved species that you expect to be present in solution.

   b. Prepare a log C–pH diagram for the system, and calculate the pH of the solution.

   c. In order to remove nitrogen by volatilizing NH_3 gas, 148 mg/L $Ca(OH)_2$ (lime) will be added to raise the pH and convert NH_4^+ to the $NH_3(aq)$ form. Calculate the pH after lime addition. (Assume that no $CaCO_3$ precipitation or ammonia loss takes place during lime addition.)

   d. If $CaCO_3$ precipitation did occur during lime addition, would it raise, lower, or not change the pH of the solution? (*Hint:* Consider what species are removed from solution by this reaction, and what the response would be of the species that remain.)

   e. Would removal of $NH_3(aq)$ by volatilization raise, lower, or not change the pH of the solution? Explain your reasoning briefly.

7. The figure at the top of p. 233 represents a log C–pH diagram for oxalic acid.

   a. What is $TOTOx$? What are pK_{a1} and pK_{a2}?

   b. Over what range of pH values is HOx^- the predominant oxalate species?

   c. Is H_2Ox a relatively strong or weak acid?

   d. If $10^{-2}\ M\ NaHOx$ were added to water, what would the solution pH be?

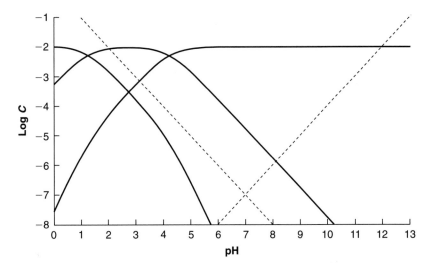

8. A log C–pH diagram for an acid/base system containing $10^{-2.7}$ M $TOTA$ is shown below. The s values indicate the slope of the curve in the given region, and the pH values indicate the intersection points of the various curves. What is the equilibrium constant for the reaction $H_3A \leftrightarrow HA + 2H^+$? What is the value of K_b for a reaction in which HA is the acid?

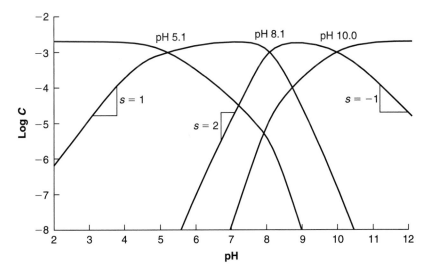

9. Write a proton condition for each of the following systems, using the dominant species as reference level species.

   a. 0.02 M HCl
   b. 10^{-3} M NaCN
   c. 10^{-3} M HOCl

  d. $10^{-2} M$ NaCl
  e. $10^{-4} M$ NaHCO$_3$
  f. $10^{-4} M$ NH$_4$Cl
  g. $10^{-3} M$ NaCN + $10^{-3} M$ HOCl

10. Write the proton condition for the following systems.

  a. $10^{-3} M$ Na$_3$PO$_4$ + $10^{-4} M$ Na$_2$HPO$_4$, using PO$_4^{3-}$ as the basis species.
  b. Same system as in (a), using HPO$_4^{2-}$ as the basis species.
  c. $10^{-3} M$ NH$_4$Cl + $5 \times 10^{-3} M$ HNO$_3$ + $2 \times 10^{-3} M$ NaOH, using dominant species as the reference species. (HNO$_3$ is nitric acid, a strong acid.)

11. Nickel forms three hydroxo complexes with the equilibrium constants shown below.

$$\text{Ni}^{2+} + \text{OH}^- \leftrightarrow \text{NiOH}^+ \qquad \log K = 4.14$$
$$\text{Ni}^{2+} + 2\text{OH}^- \leftrightarrow \text{Ni(OH)}_2^{\,0} \qquad \log K = 9.00$$
$$\text{Ni}^{2+} + 3\text{OH}^- \leftrightarrow \text{Ni(OH)}_3^{\,-} \qquad \log K = 12.00$$

  a. Combine these reactions with K_w so that they have the form of K_a's. Determine which species is dominant at any given pH in the range from 0 to 14.
  b. Draw a log C–pH diagram for a solution containing $10^{-2} M$ nickel in the pH range 5 to 14. (*Hint:* The diagram might be difficult to draw at first, because the acidity of various nickel species is unusual in a way that will be apparent from the K_a values. Despite the unusual features, the diagram must reflect the basic equations that characterize any acid/base system: the mass balances and equilibrium constants. If you have difficulty drawing the graph, use a spreadsheet to compute the α values of the different species over a range of pH values, and plot the results. Once you have drawn the graph, you will be able to see how and, hopefully, why the appearance of this graph is slightly different from that of log C–pH diagrams you have seen previously.)
  c. What is the pH of a solution of $10^{-2} M$ Ni(NO$_3$)$_2$?

12. Chromic acid (H$_2$CrO$_4$) is a diprotic acid which has important uses in industrial processes and in the preparation and analysis of environmental samples. Under certain conditions it can "dimerize" according to the following reaction:

$$2\text{HCrO}_4^- \leftrightarrow \text{Cr}_2\text{O}_7^{2-} + \text{H}_2\text{O} \qquad \log K = 1.54$$

The product shown is called *dichromate* and is the deprotonated form of a quite strong diprotic acid. The acidity constants for chromate are p$K_{a1} = -0.86$ and p$K_{a2} = 6.51$, and those for dichromate are p$K_{a1} < 0$; p$K_{a2} = -0.07$. Find the pH and composition of the following two solutions:

  a. $0.2 M$ Na$_2$CrO$_4$ + H$_2$O
  b. $0.2 M$ NaHCrO$_4$ + H$_2$O

Hint: Because of the dimerization reaction, the total *molar* concentration of dichromate species in a system where most of the Cr is dimerized is different from the total *molar* concentration of chromate in the system when most of the Cr is present as the monomer. Write out the mass balance and equilibrium equations as the first step in your analysis. Then either solve those equations or develop a log C–pH diagram by substituting into the mass balance on Cr until it contains H^+ and one Cr-containing species as the only variables. You can then solve that equation at various pH values and use the results to determine the concentrations of the other Cr-containing species, so that the complete diagram can be drawn. Also, in writing the proton condition to solve for the equilibrium pH, note that only one Cr-containing species can be chosen arbitrarily as a reference species.

13. In the late 1970s, acid drainage from Holden Mine near Holden Village, WA, made Railroad Creek uninhabitable by fish. A consultant recommended dissolving limestone, $CaCO_3$, in Railroad Creek to neutralize the acid. The pH of the drainage water is 3.2 and results entirely from sulfuric acid formed during the oxidation of CuS. The acidity constants for H_2SO_4 are $pK_{a1} = -3$ and $pK_{a2} = 1.99$, and that for Ca^{2+} (to form its conjugate base, $CaOH^+$) is 12.60.

   a. What is the concentration of $TOTSO_4$ in the drainage water?

   b. Considering that Ca^{2+} is a weak acid and CO_3^{2-} is a weak base, write the proton condition for the system after enough $CaCO_3$ has been added to raise the stream's pH to 7.0. Choose the dominant species as zero-level species. Assume all the limestone added to the stream dissolves and designate its concentration by x.

   c. What is the concentration of Ca^{2+} in the stream after neutralization?

   d. The analysis thus far ignores the fact that $CaCO_3^\circ$ may form in the water. Rewrite the PC from part (b) to include this consideration. If significant amounts of $CaCO_3^\circ$ form in the water, would you expect the pH to be greater than, less than, or equal to 7.0 after the limestone is added?

APPENDIX 4A: ARITHMETIC OPERATIONS AND LOG-LOG DIAGRAMS

One of the keys to using either a charge balance or a PC equation in conjunction with a log C–pH graph is recognizing that each side of the equation is often dominated by a single term, so that the summation can be approximated as that term alone. Although that situation is common, there are times when two or more terms must be considered over a fairly wide pH range. Most often, this complication arises when two or more lines on a log C–pH graph are parallel and fairly close to one another. In such a case, it might not be an acceptable approximation to ignore the contribution of the lower line to the summation.

The task of drawing a new line to represent the sum of the other two is complicated by the fact that the values on the graph are logarithmic, while the summation is arithmetic. To compute the sum correctly, we need to take the antilogarithm of the log C values for the two lines, add these values, and then take the logarithm of this sum and plot it on the graph. The equation of the line representing the summation can be derived as follows.

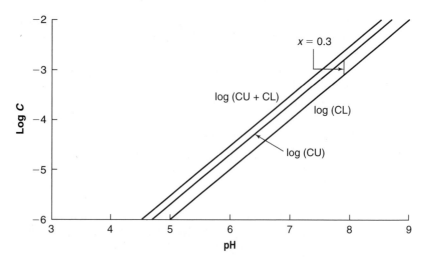

Figure 4.12   A logarithmic plot of the values of two individual terms (CL and CU) and the value of their sum. In this plot, CU is always twice CL.

Let CU be the concentration represented by the upper line and CL be the concentration represented by the lower line. Define x as the vertical distance between them on a logarithmic scale; $\log(\text{CL}) = \log(\text{CU}) - x$ (Figure 4.12).

Taking the antilogarithm of each side of the above equation gives

$$10^{\log(\text{CL})} = 10^{\log(\text{CU})-x}$$
$$= 10^{\log(\text{CU})}(10^{-x})$$
$$\text{CL} = (\text{CU})10^{-x}$$

Therefore,

$$\text{CU} + \text{CL} = (\text{CU})(1 + 10^{-x})$$
$$\log(\text{CU} + \text{CL}) = \log[\text{CU}(1 + 10^{-x})]$$
$$= \log(\text{CU}) + \log(1 + 10^{-x})$$

Thus, a line representing the logarithm of the sum of the two concentrations is displaced vertically from the upper line by an amount $\log(1 + 10^{-x})$. If $x = 0$, then the two lines are "on top of each other," and the vertical displacement is $\log(1 + 1)$, or 0.3 unit. This is the maximum displacement possible. If the lower line is one log unit below the upper one, the shift is log 1.1, or 0.04 log unit. This is about the smallest displacement we would logically be concerned about. Thus, if two lines are separated by less than one log unit, we probably want to consider that their sum is somewhat larger than the value of the upper line alone. On the other hand, if the lower line is more than one log unit below the upper one, we can probably ignore it without affecting our conclusions substantially.

CHAPTER OUTLINE

5.1   INTRODUCTION

In Chapters 3 and 4, our focus was on the determination of the pH and speciation in a solution with known input composition. This information is useful because, at times, we do have a pretty good idea of the major acid/base components used to prepare a solution, and also because those types of systems provide a convenient vehicle for learning some general principles about acid/base chemistry. However, both natural waters and process waters typically contain mixtures of strong and weak acids and bases that are too complex to analyze completely; i.e., it is not possible to identify and quantify every acid and base present. In this chapter, our focus changes just slightly, as we explore ways to estimate the concentration and strength of acidic and basic components of waters with unknown composition and to answer the question: For a solution with a given initial composition, how much acid or base must be added to adjust the solution to a desired final pH? The key data needed to address these issues are collected by conducting acid/base titrations.

In an acid/base titration, known quantities of an acid or base are added to a solution, and the pH response of the solution is characterized. Based on this response, we can make reasonable inferences about the (unknown) acids and bases present in the original solution. Titrations can also help us predict the pH change that will result from the mixing of two waters, e.g., a waste stream and a river, or from the addition of a process chemical to a water undergoing treatment, e.g., addition of HOCl or NaOCl to disinfect the water, $Al_2(SO_4)_3$ addition to coagulate particles and remove some dissolved contaminants (phosphate, natural organic matter, metals), or air injection to strip out ammonia. All these processes involve the mixing of substances with differing composition, and an understanding of titrations will help us predict the characteristics of those mixtures. Also, many natural geochemical cycles are controlled by processes that can be viewed as titrations of basic rocks by volatile acidic components released into the atmosphere by volcanic activity, life processes, or pollution. Since life developed under conditions controlled by these balances, and since many human activities alter the balances (e.g., generation of acid precipitation), it serves us well to understand them, so that we can evaluate how large an effect our activities have.

The types of questions we will address include these:

What makes lakes sensitive or resistant to acid inputs in precipitation?

Why does pH stay steady and then change precipitously in some titrations?

Why can two solutions with the same initial pH respond very differently to acid or base addition?

What is alkalinity?

5.2   REACTIONS OCCURRING DURING A TITRATION: QUALITATIVE CONSIDERATIONS

Consider what reactions occur when a solution containing a mixture of weak acids and bases is titrated with, say, a strong acid. For this exercise, it is useful to recall the approach we used to predict which acid/base species would be dominant in a solution with known inputs. In that case, we imagined that the solution could be treated as though it had been prepared by first adding all the bases (i.e., the unprotonated species) to water and then adding H^+ ions incrementally. The simplifying assumption we made was that the H^+ ions combined with the bases sequentially, converting all of one base to its conjugate acid before beginning to combine with a different base.

In such a process, H^+ ions attach to stronger bases first, then to progressively weaker bases, until all the H^+ in the system is accounted for. For instance, if we add strong acid to a pH 10 solution containing 10^{-4} M TOTAc, a certain amount of acid will be required to lower the pH to, say, 6.76 ($pK_a + 2$). However, the amount of acid required to do this will not be affected significantly by the acetate in the system, because essentially all the acetate in the solution is present as Ac^- at pH 10 and is still in this form at pH 6.76, as illustrated in Figure 5.1. Because

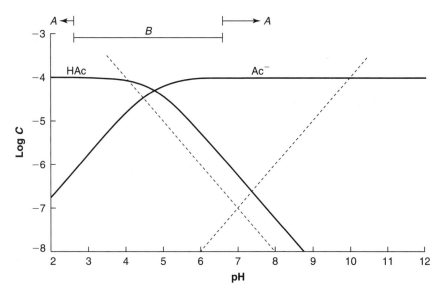

Figure 5.1 Log C–pH diagram for a solution containing 10^{-4} M TOTAc. The HAc/Ac$^-$ speciation changes little in the pH regions far from pK_a (labeled A), but it changes substantially in the region near pK_a (labeled B).

the acetate does not react significantly with added protons over this pH range, its presence has no more effect on the solution than would the presence of an inert anion such as chloride. On the other hand, if we continue adding acid until the pH decreases to, say, 4.76 ($= pK_a$), one-half of the acetate present will be converted to HAc. During this process, the protonation of dissolved Ac$^-$ ions consumes many of the H$^+$ ions added. As a result, many more protons must be added to lower the pH than would be required if the acetate were not present, or if chloride were present instead of acetate.

If the acid addition continued, almost all the acetate ions would be converted to acetic acid molecules as the pH was lowered another two pH units, to 2.76. Once again, the number of H$^+$ ions required to accomplish this pH change would be substantially greater than if the solution contained chloride instead of acetate. On the other hand, once the pH reached 2.76, the acetate would be distributed as 99% HAc and 1% Ac$^-$. Acidifying the solution further would not be significantly different in the example system than in one containing chloride instead of acetate, because both the acetate and chloride would be inert, i.e., would not change speciation, as H$^+$ is added. Comparing the two systems from start to finish, we see that the total amount of H$^+$ needed to titrate the system with acetate from pH 10.0 to 2.76 would be approximately equal to the total amount needed to titrate the system without acetate, plus the amount of H$^+$ needed to convert all the Ac$^-$ to HAc, i.e., TOTAc. Furthermore, the effect of the acetate on the titration is manifested almost entirely in the pH range near pK_a.

If a different weak base had been present, say, OCl^-, it should be clear that exactly the same argument would apply, except that the "extra" acid needed would be required to move from approximately pH 9.60 ($pK_a + 2$) to pH 5.60 ($pK_a - 2$); once the system reached pH 5.60, further additions of H^+ would cause the pH to change about as much whether OCl was present or not. It should also be clear that these systems are completely reversible. That is, the amount of base needed to bring the pH back to 10.0 from 2.76 would be the same as the amount of acid needed to lower the pH initially, and the region where "extra" acid was needed to lower the pH in the systems with weak acids (near the pK_a) would also be a region where "extra" base would be needed to raise the pH.

The response of solution pH to additions of strong acid or base is analogous in many ways to the response of water temperature when energy in the form of heat is added or removed. Specifically, consider the change in temperature when heat is extracted from pure water initially at 25°C. At first, the removal of heat energy causes the temperature to decline steadily, at a rate determined by the heat capacity of liquid water. However, when the temperature reaches 0°C, further extraction of heat causes the water to freeze, while the temperature stays constant. Only after all the water has frozen does the temperature start dropping again, this time at a rate determined by the heat capacity of ice.

In the physical system, heat energy can be added or removed, and temperature is a measure of the instantaneous state of the system. Similarly, in a chemical system, protons can be added or removed, and the H^+ activity (or the pH) is a measure of the instantaneous state of the system. Over most ranges of temperature (all temperatures other than 0°C, in the example), all the heat added or removed affects the temperature; however, in a limited range, the heat exchange mediates a transition in the form of the water, and none of it is available to alter the bulk temperature of the system. Similarly, in the chemical system, almost all the H^+ that is added to or removed from the system under most conditions leads to a change in the concentration of free H^+ and therefore in the pH. However, in some ranges (near the pK_a values of dissolved acid/base groups), added H^+ is used mostly to mediate the conversion of A^- to HA, and little remains as free H^+, so the pH change is minimal (if H^+ is being removed, the conversion is, of course, from HA to A^-).

In the physical system, a second transition exists at 100°C, and in the chemical system, transitions exist at the pK_a value of every dissolved acid. Also, in both systems, the amount of the "titrant" (energy or H^+) needed to pass through the transition is proportional to the amount of material being converted (the mass of H_2O in the physical system, $TOTA$ in the chemical system), and this amount of titrant might be much greater than the amount needed to move between transitions. The main difference between the physical and chemical processes is that water freezes and/or melts at a single, fixed temperature, whereas HA and A^- are interconverted more gradually, over a small range of pH values, rather than entirely at a single pH.

The imaginary process of holding H^+ ions out of the solution and then adding them after all the bases have been added is very similar to the real process of carrying out a titration with a strong acid. The only differences are that the initial solution might already have a significant concentration of protonated species in it, and that the H^+ ions are added in conjunction with an inert salt anion (e.g., Cl^-, NO_3^-, or SO_4^{2-}). The addition of the salts has no effect on pH (assuming that they do not dramatically change the activity coefficients of species already in solution), and the fact that some bases might be protonated before any acid is added simply means that the real titration starts in the middle of the imaginary one. These are minor differences, and it should be apparent that the key results of the imaginary process analysis are applicable to the real one.

The two major weak acid/base couples in a waste at pH 10.5 are pentachlorophenol (a wood preservative sometimes referred to as PCP or simply *penta*, with formula C_6Cl_5OH and $pK_a = 4.7$), and *p*-chlorophenol (C_6H_4ClOH, $pK_a = 9.43$). These two constituents are present at total concentrations of 4×10^{-3} *M* and 2×10^{-3} *M*, respectively. It is necessary to neutralize the water to pH $\leq$ 8 to prepare the waste for further treatment. Which of the two acid/base groups exerts a greater demand for acid during the neutralization process?

Example 5.1

Solution

Both acid/base couples are substantially deprotonated (>90%) in the initial solution. However, *p*-chlorophenol is a much weaker acid than PCP, so the deprotonated *p*-chlorophenol (*p*-chlorophenate) is a much stronger base than deprotonated PCP. Whereas *p*-chlorophenate will become mostly protonated as the pH is lowered to 8, consuming approximately 2×10^{-3} *M* H^+, the speciation of the PCP will barely change. Therefore, the *p*-chlorophenol species will exert almost all the demand for acid in the neutralization process. Note that we can draw this conclusion without having any information about how the solution was originally prepared, i.e., whether the species of interest were added as acids or bases, and whether other strong acids or bases might also have been added. All we need to know is the initial and final pH values.

From the above discussion and our prior knowledge, we can draw several important qualitative conclusions related to acid/base titrations. First, of course, it is possible to change the pH of any system by addition of strong acid or base. Second, when a weak acid or base is present in the system, it affects the amount of strong acid or base needed to cause a given pH change only to the extent that the weak acid/base couple reacts with the H^+ or OH^- that is being added, by the following reactions:

Titration with acid: $A^- + H^+ \rightarrow HA$ **(5.1)**

Titration with base: $HA + OH^- \rightarrow A^- + H_2O$ **(5.2)**

Because the speciation of a weak acid/base pair (HA/A^-) shifts most dramatically at pH values near the pK_a, the presence of some A in solution has the greatest effect on a titration when the titration moves through the pH region near $pK_{a,A}$. Finally, if the titration goes from an initial pH at which the A is almost fully deprotonated to one at which it is almost fully protonated, the extra acid needed compared to a system with no A present is approximately $TOTA$, and the same amount of extra base would be needed to reverse the process. Armed with this general understanding of what happens during a titration, we next consider the quantitative details of titration curves.

5.3 QUANTITATIVE INTERPRETATION OF TITRATION DATA

5.3.1 TITRATION OF A STRONG ACID WITH A STRONG BASE

Consider the conditions during the titration of a 10^{-3} M HCl solution with strong base (NaOH). In the most general sense, we can identify $\{H^+\}$, $\{OH^-\}$, and $\{Cl^-\}$ as unknowns in the initial solution. We can solve for these unknowns using the mass balance on Cl, the charge balance (CB) or some form of the mass balance on H (the PC or $TOTH$ equation), and K_w. Since HCl is a strong acid, all the added HCl dissociates, and the result is that the solution pH is 3.0.

We could evaluate the pH at any point during a titration of the 10^{-3} M HCl solution with NaOH by considering $\{Na^+\}$ as an additional unknown and identifying the mass balance on Na as an additional independent equation characterizing the system. If we use the charge balance as one of the equations to describe the system, the formal description of the system after some NaOH has been added is:

Unknowns:	$\{Na^+\}$, $\{Cl^-\}$, $\{H^+\}$, $\{OH^-\}$
Equations that can be solved to determine values of the unknowns	
Equilibrium:	$K_w = \{H^+\}\{OH^-\}$
MB on Na:	$TOTNa = \{Na^+\}$
MB on Cl:	$TOTCl = \{Cl^-\}$
Charge balance:	$\{Na^+\} + \{H^+\} = \{Cl^-\} + \{OH^-\}$

Rewriting the CB as $\{H^+\} = \{Cl^-\} - \{Na^+\} + \{OH^-\}$ and then substituting values for $\{Cl^-\} - \{Na^+\}$, we can solve for the pH after various additions of base. The results are summarized in Table 5.1 and are shown in Figure 5.2.

A significant amount of NaOH must be added to increase the pH from 3 to 4, but only a small amount of additional NaOH is then required to increase the pH to 10. This observation can be explained as follows. To titrate the solution from pH 3.0 to pH 4.0 ($\{H^+\} = 1.0 \times 10^{-3}$ to 0.1×10^{-3}), 90% of the free H^+ ions

Table 5.1   Charge balance and solution pH for various additions of NaOH to a solution containing 10^{-3} M HCl

OH$^-$ Added	Na$^+$	Cl$^-$	Charge Balance (with Values Substituted for Na$^+$ and Cl$^-$)	pH
0	0	10^{-3}	$\{H^+\} = 1.0 \times 10^{-3} + \{OH^-\}$	3.00
1.0×10^{-4}	1.0×10^{-4}	10^{-3}	$\{H^+\} = 0.9 \times 10^{-3} + \{OH^-\}$	3.05
3.0×10^{-4}	3.0×10^{-4}	10^{-3}	$\{H^+\} = 0.7 \times 10^{-3} + \{OH^-\}$	3.15
5.0×10^{-4}	5.0×10^{-4}	10^{-3}	$\{H^+\} = 0.5 \times 10^{-3} + \{OH^-\}$	3.30
7.0×10^{-4}	7.0×10^{-4}	10^{-3}	$\{H^+\} = 0.3 \times 10^{-3} + \{OH^-\}$	3.52
9.0×10^{-4}	9.0×10^{-4}	10^{-3}	$\{H^+\} = 0.1 \times 10^{-3} + \{OH^-\}$	4.00
1.0×10^{-3}	1.0×10^{-3}	10^{-3}	$\{H^+\} = \{OH^-\}$	7.00
1.1×10^{-3}	1.1×10^{-3}	10^{-3}	$0.1 \times 10^{-3} + \{H^+\} = \{OH^-\}$	10.00
1.3×10^{-3}	1.3×10^{-3}	10^{-3}	$0.3 \times 10^{-3} + \{H^+\} = \{OH^-\}$	10.48
1.5×10^{-3}	1.5×10^{-3}	10^{-3}	$0.5 \times 10^{-3} + \{H^+\} = \{OH^-\}$	10.70
1.7×10^{-3}	1.7×10^{-3}	10^{-3}	$0.7 \times 10^{-3} + \{H^+\} = \{OH^-\}$	10.85
2.0×10^{-3}	2.0×10^{-3}	10^{-3}	$1.0 \times 10^{-3} + \{H^+\} = \{OH^-\}$	11.00

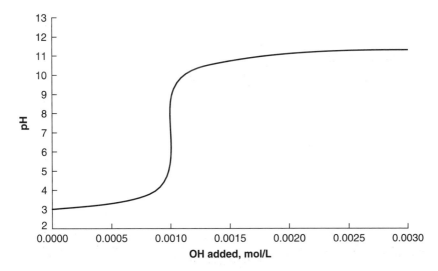

Figure 5.2   Titration curve for a solution containing 10^{-3} M HCl, titrated with strong base.

must be neutralized, consuming 0.9×10^{-3} M OH$^-$. To then increase the pH to 7.0 requires only that most of the remaining 10% of the H$^+$ added originally be neutralized, which can be accomplished with an additional 0.1×10^{-3} M OH$^-$. Then increasing the pH to 10 requires an additional 0.1×10^{-3} M OH$^-$ to build up the free OH$^-$ concentration from 10^{-7} to 10^{-4} M (the simultaneous reduction of the H^+ concentration from 10^{-7} to 10^{-10} M consumes negligible OH^-). Thus, the pH of the system jumps from pH 4 to 10 with the addition of approximately 22% as much base as is required to increase it from 3 to 4 (0.2×10^{-2} M compared to 0.9×10^{-3} M).

5.3.2   TITRATION OF A WEAK ACID WITH A STRONG BASE

Now consider what would happen if we carried out the same titration process as described above, but with a solution that initially contained 10^{-3} M HCl plus 10^{-3} M HOCl. In this initial solution, essentially all the HCl is dissociated, while almost all the HOCl is protonated. As a result, the initial pH is once again around 3.0. (This result can be confirmed by drawing the log C–pH diagram and solving the CB, PC, or TOTH equation.) Given that {HOCl} and {H^+} are both approximately 10^{-3} in the initial solution, {OCl^-} can be calculated from the value of K_a. The result is {OCl^-} = $10^{-7.6}$ M.

As OH^- is added to this solution, it neutralizes both H_3O^+ and HOCl in such a way that the two equilibrium constants (K_w and K_a) are satisfied. That is, each time an increment of base is added, some OH^- ions participate in each of the following reactions:

$$HOCl + OH^- \rightarrow OCl^- + H_2O \tag{5.3}$$

$$H_3O^+ + OH^- \rightarrow 2H_2O \tag{5.4}$$

In the system discussed earlier containing only HCl, all the OH^- added either neutralized H^+ or remained in the solution as OH^- ions. Because, in the system with HOCl, some OH^- is consumed in the neutralization of HOCl, a given addition of NaOH increases the pH less in the solution containing HOCl than in the OCl-free solution. Equivalently, for a given increase in pH, we need to add more NaOH to the solution containing OCl than to the OCl-free solution. To quantify this effect, we represent the system conditions when it is titrated with a strong base by the following information:

Unknowns	{Na^+}, {Cl^-}, {H^+}, {OH^-}, {HOCl}, {OCl^-}
Equations	
Equilibrium:	$K_w = \{H^+\}\{OH^-\}$
Equilibrium:	$K_a = \dfrac{\{H^+\}\{OCl^-\}}{\{HOCl\}}$
MB on Na:	$TOT\text{Na} = \{Na^+\}$
MB on Cl:	$TOT\text{Cl} = \{Cl^-\}$
MB on OCl:	$TOT\text{OCl} = \{HOCl\} + \{OCl^-\}$
Charge balance:	$\{Na^+\} + \{H^+\} = \{Cl^-\} + \{OH^-\} + \{OCl^-\}$

Table 5.2   Charge balance and solution pH for various additions of NaOH to a solution containing 10^{-3} M HCl and 10^{-3} M HOCl.

OH⁻ Added	Na⁺	Cl⁻	Charge Balance (with Values Substituted for Na⁺ and Cl⁻)	pH
0	0	10^{-3}	$\{H^+\} = 1.0 \times 10^{-3} + \{OH^-\} + \{OCl^-\}$	3.00
1.0×10^{-4}	1.0×10^{-4}	10^{-3}	$\{H^+\} = 0.9 \times 10^{-3} + \{OH^-\} + \{OCl^-\}$	3.05
3.0×10^{-4}	3.0×10^{-4}	10^{-3}	$\{H^+\} = 0.7 \times 10^{-3} + \{OH^-\} + \{OCl^-\}$	3.15
5.0×10^{-4}	5.0×10^{-4}	10^{-3}	$\{H^+\} = 0.5 \times 10^{-3} + \{OH^-\} + \{OCl^-\}$	3.30
7.0×10^{-4}	7.0×10^{-4}	10^{-3}	$\{H^+\} = 0.3 \times 10^{-3} + \{OH^-\} + \{OCl^-\}$	3.52
9.0×10^{-4}	9.0×10^{-4}	10^{-3}	$\{H^+\} = 0.1 \times 10^{-3} + \{OH^-\} + \{OCl^-\}$	4.00
1.0×10^{-3}	1.0×10^{-3}	10^{-3}	$\{H^+\} = \{OH^-\} + \{OCl^-\}$	5.30
1.1×10^{-3}	1.1×10^{-3}	10^{-3}	$0.1 \times 10^{-3} + \{H^+\} = \{OH^-\} + \{OCl^-\}$	6.65
1.3×10^{-3}	1.3×10^{-3}	10^{-3}	$0.3 \times 10^{-3} + \{H^+\} = \{OH^-\} + \{OCl^-\}$	7.23
1.5×10^{-3}	1.5×10^{-3}	10^{-3}	$0.5 \times 10^{-3} + \{H^+\} = \{OH^-\} + \{OCl^-\}$	7.60
1.7×10^{-3}	1.7×10^{-3}	10^{-3}	$0.7 \times 10^{-3} + \{H^+\} = \{OH^-\} + \{OCl^-\}$	7.97
2.0×10^{-3}	2.0×10^{-3}	10^{-3}	$1.0 \times 10^{-3} + \{H^+\} = \{OH^-\} + \{OCl^-\}$	9.30
2.2×10^{-3}	2.2×10^{-3}	10^{-3}	$1.1 \times 10^{-3} + \{H^+\} = \{OH^-\} + \{OCl^-\}$	10.30
2.4×10^{-3}	2.4×10^{-3}	10^{-3}	$1.3 \times 10^{-3} + \{H^+\} = \{OH^-\} + \{OCl^-\}$	10.60
2.6×10^{-3}	2.6×10^{-3}	10^{-3}	$1.5 \times 10^{-3} + \{H^+\} = \{OH^-\} + \{OCl^-\}$	10.80
2.8×10^{-3}	2.8×10^{-3}	10^{-3}	$1.7 \times 10^{-3} + \{H^+\} = \{OH^-\} + \{OCl^-\}$	10.90
3.0×10^{-3}	3.0×10^{-3}	10^{-3}	$2.0 \times 10^{-3} + \{H^+\} = \{OH^-\} + \{OCl^-\}$	11.00

The pH after various additions of NaOH can be computed by solving the six equations simultaneously, which is accomplished most easily by drawing a log C–pH diagram for the system and identifying the pH at which the CB is satisfied for each increment of base added. The results of such an analysis are summarized in Table 5.2 and plotted in Figure 5.3.

As can be seen by comparison of the solid curves in Figure 5.3, the titration curve is very similar to that in the OCl-free system up to the point where a little more than 0.95×10^{-3} M OH⁻ has been added, corresponding to pH near 6: the pH in both systems increases gradually up to pH 4 and then starts increasing more steeply. However, whereas the steep increase in solution pH continues to around pH 10 in the system with no OCl, it ceases at a pH slightly below the pK_a in the system with OCl. At that point, the titration curve becomes quite flat again; i.e., a large amount of OH⁻ must be added to generate even a slight increase in pH, because much of the OH⁻ added to the system is consumed as it converts HOCl to OCl⁻. This flat portion of the curve extends until enough OH⁻ has been added to raise the pH to around 8.5, at which point the steep increase in pH begins again. The amount of OH⁻ required to adjust the pH to 11.0 is 2.0×10^{-3} in the system without OCl and 3.0×10^{-3} in the system with it. The difference,

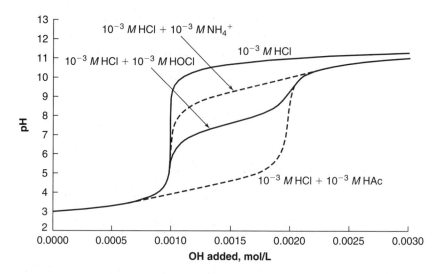

Figure 5.3   Titration curves for four solutions containing either 10^{-3} M HCl alone or in addition to 10^{-3} M of a weaker acid.

1.0×10^{-3}, is the amount of OH^- that is consumed in the conversion of essentially 100% of the hypochlorite from HOCl to OCl^-; i.e., it is *TOT*OCl.

The two broken lines in Figure 5.3 are for titrations of systems initially containing either 10^{-3} *M TOT*Ac or 10^{-3} *M TOT*NH_3 in addition to 10^{-3} *M* HCl. The similarities among all the titration curves and the effect of a change in the pK_a of the weak acid present are obvious.

Figure 5.4 shows the speciation of the weak acids as a function of the amount of base added for the three systems in Figure 5.3 that contain weak acid/base groups. Despite the fact that the titration curves are all quite distinct, after a given addition of base, the speciation of the weak acids is very similar in the three systems. For instance, all the weak acids are split equally between their acid and base forms when 1.5×10^{-3} *M* OH^- has been added, even though the pH in the three systems at this point varies from 4.7 to 9.2.

The reason for this similarity becomes clear when we think about the fate of the OH^- added during the titration. Initially, all the systems are at pH near 3.0, i.e., $\{H^+\} = 10^{-3}$. Free H^+ is the strongest acid in the system, and according to the approximation discussed previously, added OH^- is likely to neutralize the majority of this strong acid before reacting significantly with any weaker acid. Therefore, the conversion of any of the weak acids to their conjugate bases is negligible until about 10^{-3} *M* OH^- has been added. Then, as the next 10^{-3} *M* OH^- is added, it reacts almost entirely with the weak acid (NH_4^+, HOCl, or HAc), converting each to its conjugate base. The pH range over which most of this deprotonation occurs is approximately $pK_a - 1$ to $pK_a + 1$; i.e., it is different in the three systems. In each case, after most of the acid has been

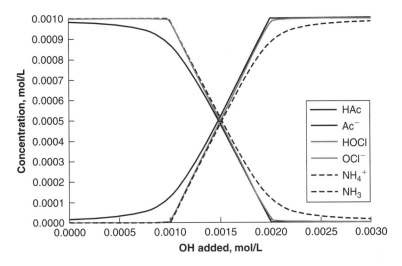

Figure 5.4 Speciation of acetate, hypochlorite, and ammonia during titration of samples characterized in Figure 5.3.

deprotonated, the pH increase accompanying further additions of base is more or less as in the blank. Thus, the speciation of $TOTA$ between the forms HA and A$^-$ is very similar in the three systems as a function of the amount of base added, even though the pH at which most of the conversion from HA to A$^-$ occurs is quite different.

Figure 5.3 indicates that titration curves are relatively flat (i.e., the solution pH changes very gradually) when the pH is near the pK_a of a conjugate acid/base pair that is in solution. Extending the analysis, it should be apparent that if more than one weak acid is present in a solution, the titration curve will have flat portions near each pK_a value. These flat portions may be quite easily distinguishable from one another, or they may be indistinct if the pK_a's are close enough to one another that the flat portions overlap.

Figure 5.5a shows the titration curve for a system made by mixing $1.0 \times 10^{-3} M$ HCl, $0.4 \times 10^{-3} M$ HAc, $0.25 \times 10^{-3} M$ HOCl, and $0.35 \times 10^{-3} M$ NH$_4$Cl. The graph is relatively flat near the pK_a's, and the width of each flat region is proportional to $TOTA$ for the corresponding acid, although the flat portions are not easily distinguished. Figure 5.5b shows the speciation of the three weak acids during this titration and makes the point very dramatically that each acid is virtually completely converted to its conjugate base before the next weaker acid begins to deprotonate. That is, almost none of the acetate becomes deprotonated until most of the H$^+$ has been neutralized (OH$^-$ added $= 10^{-3} M$), almost none of the hypochlorous acid deprotonates until almost all the acetic acid has been converted to acetate (OH$^-$ added $= 1.4 \times 10^{-3} M$), and almost none of the ammonium deprotonates until almost all the hypochlorous acid has been converted to hypochlorite (OH$^-$ added $= 1.65 \times 10^{-3} M$).

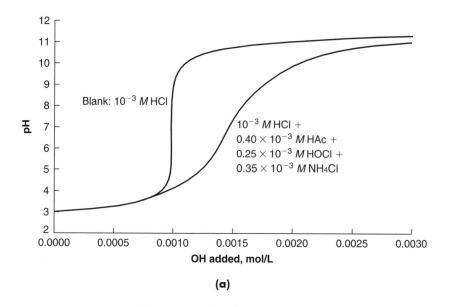

(a)

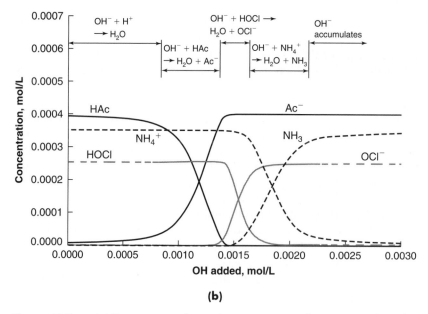

(b)

Figure 5.5 **(a)** Titration curves for a solution containing only a strong acid and for one containing a mixture of weak acids in addition to the strong acid. **(b)** Speciation in the system with several weak acids characterized in (a). The dominant reaction of the added OH^- in each pH range is shown at the top of the figure.

Describe qualitatively and semi-quantitatively the features of the titration curve if strong acid is added to a solution containing 4×10^{-3} M PCP ($pK_a = 4.7$) and 2×10^{-3} M p-chlorophenate ($pK_a = 9.43$) to decrease the pH from 10.5 to 4.0.

Example 5.2

Solution

At pH 10.5 (1.07 pH units above pK_a of p-chlorophenol), approximately 8% of this acid/base group is protonated ($\alpha_0 = 0.08$). As acid is added, the solution pH will change gradually due to both the relatively large concentration of free OH^- in solution (3.1×10^{-4} M) and the protonation of the p-chlorophenate. The gradual change continues until almost all the p-chlorophenate has been converted to its conjugate acid, C_6H_4ClOH, at pH ~8.4 (one pH unit below pK_a, at which the group is approximately 91% protonated). To reach this pH requires addition of approximately 3.00×10^{-4} M H^+ to neutralize 99% of the OH^- in the initial solution and 1.66×10^{-3} M H^+ to protonate ~83% of the p-chlorophenate, for a total of 1.96×10^{-3} M H^+.

Thereafter, the pH will decrease rapidly until the PCP begins to protonate significantly, at pH ~5.7. To lower the pH from 8.2 to 4.0 requires an amount of acid approximately equal to the remaining 9% of the p-chlorophenate plus about two-thirds of TOTPCP (2.67×10^{-3} M), since α_0 of PCP is 0.67 at pH 4.0. In addition, 10^{-4} M H^+ is needed to supply the free H^+ at pH 4.0. Therefore, the titration from pH 10.5 to 4.0 requires the addition of approximately 4.92×10^{-3} M H^+. The fate of the added H^+ can be summarized as follows:

H^+ to neutralize OH^-	0.31×10^{-3}
H^+ to protonate p-chlorophenate	1.84×10^{-3}
H^+ to protonate PCP	2.67×10^{-3}
Free H^+	0.10×10^{-3}
	4.92×10^{-3}

5.3.3   TITRATIONS OF MULTIPROTIC ACIDS AND BASES

Titration curves for systems containing a multiprotic acid are similar in many respects to those for systems with two or more independent monoprotic acid/base pairs. As each pK_a is passed, an additional amount of base equal to TOTA is required to convert H_3A to H_2A, or H_2A to HA, or HA to A. As is the case when chemically distinct acids are present, if the pK_a's of the multiprotic acid are widely separated, each will lead to a distinct flat region of the titration curve, whereas if the pK_a's are close to one another, the flat regions may merge into a single, somewhat broad and indistinct zone. A titration curve for a system with 10^{-3} M HCl plus 10^{-3} M TOTCO$_3$ is shown in Figure 5.6.

5.3.4   TITRATION OF SOLUTIONS CONTAINING UNKNOWN ACIDS AND BASES

The preceding discussion indicates that titration curves give information about the pK_a values of acids dissolved in a system and the quantities of those acids present. For instance, consider the titration curve of a hypothetical sample shown

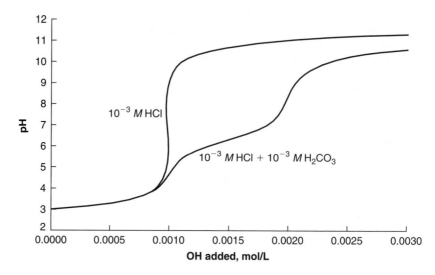

Figure 5.6   Titration curves for solutions containing 10^{-3} M HCl or 10^{-3} M HCl plus 10^{-3} M H_2CO_3 with strong base.

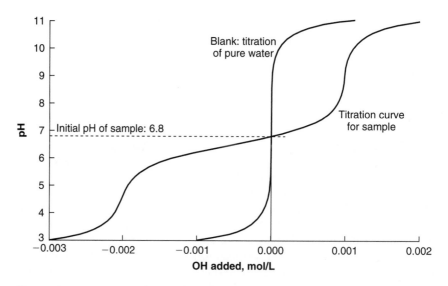

Figure 5.7   Titration of a sample containing an unknown acid. Additions of strong acid are represented as negative additions of OH^-.

in Figure 5.7. Imagine that the solution, with an initial pH of 6.8, is split and titrated with acid and base to pH 3.0 and 11.0, respectively. As indicated in the figure, the requirements of acid and base to reach these pH values are 0.003 and 0.002 M, respectively, so 0.005 M acid would be required to titrate the sample from pH 11 to 3. The titration curve for a blank (pure water) is also shown, indicating that additions of 0.001 M HCl and 0.001 M NaOH are required to reach pH 3.0 and 11.0, respectively. Thus, adjusting the sample between pH 3 and 11 requires 0.003 M more acid or base than a corresponding adjustment of the blank.

Based on these results, we could infer that the sample contains a total of 0.003 M of unidentified weak acids or bases that become protonated between pH 11 and 3. Furthermore, since the titration curve is flattest at pH near 6.5, we could infer that the dominant weak acid in the system has a pK_a value near 6.5. (A more precise approach to estimating K_a values from a titration curve is discussed below.) This type of information can be obtained without having any idea, in advance, of the identities of the acids in the solution. It is this aspect of titrations that makes them valuable analytical techniques.

A titration curve for a solution is shown below. The analyst hypothesizes that the solution contains a diprotic acid with pK_a values of 5.5 and 8.5. Do you agree with this interpretation? Why or why not? | **Example 5.3**

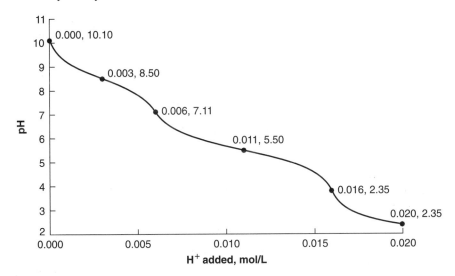

Solution

Based on the preceding discussion and as shown in Figure 5.7, the concentration of H^+ needed to titrate a blank solution (one containing no weak acid/base groups) between pH 10 and 4 is on the order of only 2×10^{-4} M, and that needed to titrate a blank from pH 4 to 3 is on the order of 10^{-3} M. Titration of the sample from pH 10 to 4 required the addition of far more H^+ than a blank (0.016 M), but the further titration of the sample from pH 4 to 3 required about as much H^+ as a blank.

The "extra" H^+ required to titrate the sample is attributable to the presence of weak acid/base groups that become protonated during the titration. Apparently, these groups are almost fully protonated once the pH has been lowered to 4, since no extra H^+ is required to continue the titration to pH 3. Visually, the curve suggests that the weak bases in the sample become protonated in two fairly distinct steps: one protonation step is centered near pH 8.5 and consumes 0.006 M H^+, and another is centered near pH 5.5 and consumes 0.010 M H^+. Since the protonation reaction occurs in a range approximately one to two pH units on either side of pK_a, it is reasonable to infer that the solution contains weak bases with pK_a values near 8.5 and 5.5, as the analyst suggests.

The concentration of each weak acid can be correlated with the concentration of extra H^+ that must be added to titrate the sample compared to the blank. If the only weak acid/base group in the sample were diprotic (e.g., H_2A), then the extra H^+ needed to convert A^{2-} to HA^- (near pH 8.5) would be identical to the amount needed to convert HA^- to H_2A (near pH 5.5). Since the concentration of extra H^+ required is different in the regions near the two apparent pK_a values, the analyst is apparently incorrect in attributing the shape of the titration curve to the presence of a single diprotic acid. Rather, it appears that two different weak monoprotic acids might be present, one with pK_a near 5.5 and at a total concentration of 0.010 M, and the other with pK_a near 8.5, present at a concentration of 0.006 M. Alternatively, the solution might contain 0.006 M of a diprotic acid with pK_a's near 5.5 and 8.5, but it would then also have to contain a monoprotic acid with pK_a near 8.5, at a concentration of 0.004 M. In either case, the titration curve is not completely consistent with the analyst's interpretation.

5.3.5 UNITS FOR QUANTIFYING TITRATION DATA: EQUIVALENTS AND mg/L AS CaCO₃

Additions of titrant are often expressed in terms of *acid equivalents* or simply *equivalents*. This term is used in a number of ways in chemistry, but in all cases the implication is that an equivalent of substance A reacts with an equivalent of substance B; i.e., equivalents express information about reaction stoichiometry. For acids and bases, equivalents are defined with reference to H^+. That is, one *acid equivalent* is defined as 1 mole of H^+, and one equivalent (1 equiv) of any other substance is the amount of that substance that can react with 1 mole of H^+. Similarly, the *equivalent weight* (EW) of a substance is the weight (or, more often, mass) of the substance that can react with 1 mole of H^+. For instance, 1 mole of a triprotic acid can release 3 equiv of H^+, so 1/3 mole of the acid is "equivalent" to 1 mole of H^+. The equivalent weight of the acid, i.e., the mass of the material that provides one equivalent of reactivity, is therefore one-third of its molecular weight. Since 1 mole of OH^- can react with 1 mole of H^+, 1 mole of OH^- is 1 equiv. Thus, in all the example titrations described above, the number of moles of H^+ or OH^- added per liter could also be represented as the number of equivalents added per liter.

One must be careful in interpreting equivalency, since the number of equivalents that actually react in a given situation is not always the number that could potentially react. Consider, for instance, the definition of equivalency for phosphoric acid, a triprotic acid with $pK_{a1} = 2.16$, $pK_{a2} = 7.20$, and $pK_{a3} = 12.35$. If phosphate were present in a solution initially at pH 9.0, almost all of it would be present as HPO_4^{2-}. If that solution were then titrated to a final pH of 4.5, almost all the HPO_4^{2-} would be converted to $H_2PO_4^-$.[1] Thus, each mole of phosphate

---

[1] Titrations to pH in the range of 4.3 to 4.7 are commonly used in environmental engineering to assess the alkalinity of solutions. Alkalinity is defined and discussed in detail later in the chapter.

(a)

(b)

A simple, low-technology approach to neutralization of acidic runoff is to dose the water with lime, CaO(s). Slow dissolution of the solid combines with hydrolysis to release OH⁻, which consumes some of the acidity in the water. These photographs show a river in Sweden being dosed with lime and lime being sprayed into an acidified lake.

| (a) Martin Bond/Science Photo, Library/Photo Researchers, Inc. (b) R. Wright/Visuals Unlimited.

would consume one equivalent of H^+ during the titration, despite the fact that each mole of HPO_4^{2-} (i.e., each mole of the substance initially present) could potentially react with two moles of H^+.

It is therefore ambiguous to state that a solution contains one equivalent per liter of HPO_4^{2-}, because it is not clear whether the statement is referring to the amount of H^+ with which the HPO_4^{2-} combines in a given situation, or the amount with which it could potentially combine if the titration were carried out until the species was maximally protonated. The potential confusion engendered by this situation emphasizes the importance of specifying the reaction of interest when expressing masses as equivalents, or concentrations as equivalents per liter. If a reaction is not specified, the usual assumption is that equivalents are being computed on the basis of ionic charge; i.e., an ion with charge $+n$ or $-n$ is defined as having n equivalents per mole and therefore having an equivalent weight equal to its molecular weight divided by n.

A second common way to express titration data (especially when values are reported for alkalinity, which is discussed below) is as milligrams per liter *as CaCO₃*. This type of representation was discussed in Chapter 1, and it would be worthwhile to review that section if you have forgotten the details. Briefly, expressing a concentration of acid or base in this way indicates that the acid or base is present in the given solution at the equivalent concentration (i.e., the same number of equivalents per liter) as the indicated concentration of $CaCO_3$. The concentration of $CaCO_3$ is used as the basis for this type of reporting both for historical reasons and because of the convenient coincidence that its molecular weight is 100. Since each molecule of $CaCO_3$ can potentially bind two protons (when the molecule dissociates, the Ca^{2+} does not react further, and the CO_3^{2-} can potentially be converted to H_2CO_3), the equivalent weight of $CaCO_3$ is 50. Thus a concentration of base reported as 50 mg/L *as CaCO₃* corresponds to 1.0 meq/L of the base.

5.3.6   TITRATIONS WITH WEAK ACIDS OR BASES

Titrations with strong acids or bases are common analytical procedures, and addition of strong acids or bases to change the pH of process waters is also common. Permits for many industries require that the wastewaters that they discharge be within one to three pH units of neutrality, and addition of a strong acid or base is often the most convenient and inexpensive way to meet that requirement. However, if the wastewater contains only small amounts of weak acids or bases in addition to some strong acids or bases, it might have a titration curve like that of the solution containing only HCl in Figure 5.2, in which case adjusting the pH to the desired range could be quite tricky. The difficulty is exacerbated if the composition of the water changes irregularly, as is the case in many industries.

The difficulty of neutralizing a solution whose pH is highly sensitive to the addition of strong acid or base can be largely overcome by titrating with a weak acid or base instead. For instance, say we wished to neutralize a solution containing 5×10^{-3} HCl to a pH between 6.0 and 9.0. If we did so with NaOH, we would need to add at least 4.999×10^{-3} M NaOH, but no more than $5.010 \times 10^{-3} M$; the difficulty of maintaining control of such a system is obvious.

On the other hand, if we used Na_2CO_3 to neutralize the acidity instead of NaOH, the behavior of the system would be quite different. The first doses of Na_2CO_3 would consume two H^+ ions per molecule of base added, as the carbonate was converted to carbonic acid:

$$Na_2CO_3 + 2H^+ \rightarrow H_2CO_3 + 2Na^+ \qquad \text{(pH less than about 5.5)}$$

However, as more base was added and the pH increased, the carbonic acid that formed previously would begin to dissociate, neutralizing some of the CO_3^{2-} added. In effect, the titration curve would become flatter due to the presence of a weak acid in the system, the weak acid being the previously added titrant. With each addition of Na_2CO_3, the acidity of the solution is partially neutralized, but the concentration of weak acids in solution (H_2CO_3 and HCO_3^-) also increases.

Curves characterizing the titration of the hypothetical solution with NaOH and with Na_2CO_3 are shown together in Figure 5.8, from which the potential advantages of titrating with carbonate are apparent. The range of Na_2CO_3 additions yielding a final pH between 6 and 9 is 1.2×10^{-3} to 2.1×10^{-3} mol/L (2.4×10^{-3} to 4.2×10^{-3} eq/L). This range of acceptable reagent additions is more than 200 times as large as that when NaOH is used to neutralize the acid. Clearly, at least in this example, the pH can be controlled much more easily and can be maintained at a more stable value by using carbonate rather than hydroxide as the titrant.

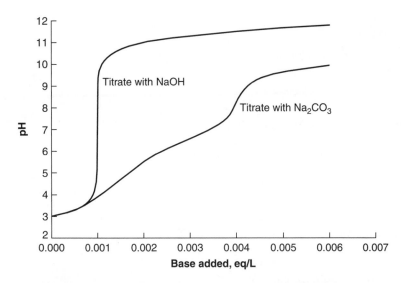

Figure 5.8   Titration of a solution containing 10^{-3} M HCl with a strong base (NaOH) and with Na_2CO_3. The values on the abscissa are in equivalents of base per liter. The amount of NaOH added in moles per liter is the same as the value in equivalents per liter; the number of moles per liter of Na_2CO_3 added can be computed by dividing the value in equivalents per liter by 2.

Compute the pH and speciation in a solution after 2×10^{-3} eq/L of base is added to 10^{-3} M HCl if the base is added as NaOH. Repeat the calculation for a system in which the same number of equivalents per liter of base is added as Na_2CO_3.

Example 5.4

Solution

Each mole of NaOH corresponds to one equivalent, so an addition of 2×10^{-3} eq/L of NaOH corresponds to addition of 2×10^{-3} mol/L. The first 1×10^{-3} mol/L of NaOH would exactly neutralize the acidity of the initial solution, and the second 1×10^{-3} mol/L would dissociate to Na^+ and OH^-, increasing the solution pH to 11.0. On the other hand, each mole of Na_2CO_3 corresponds to two equivalents of base, so addition of 2×10^{-3} eq/L of Na_2CO_3 corresponds to addition of 1×10^{-3} mol/L. The pH of the solution after addition of this amount of Na_2CO_3 can be determined by the techniques developed in previous chapters. The fact that the Na_2CO_3 is added in a titration *after* the HCl, of course, is irrelevant to the calculations (because the solution has no memory); the problem is identical to one in which 1×10^{-3} M Na_2CO_3 and 10^{-3} M HCl are mixed initially. The final pH is computed to be 8.3.

The pH is lower when Na_2CO_3 is added than when an "equivalent" amount of NaOH is added. This is not surprising, since we know that CO_3^{2-} is a weaker base than OH^-. But in truth, the difference between the two systems is not that CO_3^{2-} is a weaker base than OH^-, but rather that HCO_3^- is a weaker base than OH^-. That is, essentially every CO_3^{2-} ion that is added to the system combines with an H^+, just as every OH^- that is added does. However, once the CO_3^{2-} is converted to HCO_3^-, the HCO_3^- acts as a virtually inert ion, rather than a base: almost none of the HCO_3^- combines with an H^+ to form H_2CO_2. Thus,

This photograph shows a bank of reverse osmosis (RO) membrane modules for treating an industrial wastewater. The water being treated contains hydrofluoric acid, which until recently was neutralized with sodium hydroxide upstream of the RO process. However, it proved very difficult to control the pH of the water, leading to problems with the RO treatment step. When the company switched to sodium carbonate as the neutralizing agent, the process control problems were overcome.

| Courtesy of the Boeing Company.

it is the weakness of HCO_3^- as a base that causes the two systems to behave differently. Put another way, even though each CO_3^{2-} ion has the *potential* to react with two H^+ ions (and is equivalent to two OH^- ions in that respect), in fact the CO_3^{2-} ions react with only one H^+ under the specified conditions; i.e., the CO_3^{2-} is acting as a monoprotic base in this system.

5.3.7   THE EFFECT OF THE ACIDITY OF WATER ON TITRATION CURVES

All the titration curves shown thus far are relatively flat at pH < 4 and pH > 10, suggesting that some aspect of acid/base chemistry overrides the effect of weak acids and bases and causes all systems to behave similarly in these pH ranges. The key to understanding this feature of all titrations curves lies in remembering that pH is a logarithmic function. As a result, when the pH changes from 3.0 to 2.0, the change in H^+ activity is 1000 times as great as when pH changes from 6.0 to 5.0.

Most of the example systems considered above contain weak acids or bases at total concentrations on the order of 10^{-2} to 10^{-4} M. In the region extending from one pH unit above to one pH unit below pK_a, the acid/base group consumes an amount of H^+ or OH^- equal to 0.82 ($TOTA$) (because α_0 changes from 0.09 to 0.91). As a result, the consumption of H^+ or OH^- by the group is approximately 0.82×10^{-4} to 0.82×10^{-2} M over a two-unit range of pH.

When the pH of any solution is lowered from 4.0 to 2.0, the amount of H^+ needed just to increase the concentration of H_3O^+ is 0.99×10^{-2} M for a two-unit change in pH; i.e., it is comparable to or greater than the H^+ requirement to titrate a weak acid/base group present at a concentration $\leq 10^{-2}$ M, even if the pK_a of that group is fairly low. As a result, regardless of whether weak acid/base groups are present, and regardless of their pK_a values, the shape of the titration curve at pH less than approximately 4.0 is usually dominated by the effect of free H^+ ions, and all titration curves are therefore very similar in this pH range.

At relatively high pH ($\sim$10 or greater), one can make a similar argument about the relatively large dose of free OH^- required to generate a pH change of one or two units. These ideas are elaborated and developed mathematically later in the chapter, in the section on pH buffering.

5.4   PRACTICAL ASPECTS OF TITRATIONS: EQUIVALENCE POINTS AND ALKALINITY

5.4.1   EQUIVALENCE POINTS

Consider, once again, solutions prepared by adding 10^{-3} M HAc, HOCl, or NH_4^+ (perhaps as NH_4Cl) to water. By solving the proton condition, $TOTH$ equation, or charge balance equation in conjunction with the equilibrium constants and mass balances for these solutions, we can determine that the pH values of the solutions would be 3.88, 5.30 and 6.15, respectively. We can also compute the expected change in pH as each solution is titrated with a strong base, yielding the results shown in Figure 5.9.

When 10^{-3} M NaOH has been added to these solutions, the titration is said to be at the *equivalence point,* because the amount of base added (expressed in equivalents) is the same as the amount of acid initially present. At that point, the most convenient proton condition for determining the pH would use A^- as the reference species, yielding the following proton balance table and PC equation.

	−1	0	+1	Concentration
Equilibrium species	OH^-	H_2O A^- Na^+	H^+ HA	
Initial and/or input species	NaOH		HA	10^{-3} 10^{-3}

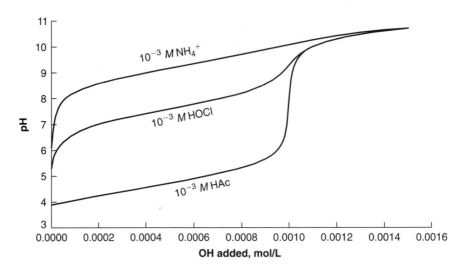

Figure 5.9   Titration curves for addition of strong base to solutions containing $10^{-3} M$ HAc, HOCl, or NH_4^+.

$$TOTH_{in} = \sum_i c_{i,in} n_i = (10^{-3})(1) + (10^{-3})(-1) = 0$$

$$\sum_i c_{i,eq} n_i = \{H^+\}(1) + \{OH^-\}(-1) + \{HA\}(1)$$

$$PC: \qquad \{H^+\} + \{HA\} = \{OH^-\}$$

Inspection of the proton condition at the equivalence point indicates that it is identical to the one we would derive if we had simply added 10^{-3} M NaA to pure water. (Review what the proton balance table would look like for a solution of 10^{-3} M NaA to convince yourself of this fact.) The mass balance on A is, of course, also identical in a system prepared with 10^{-3} M NaA and in one prepared with 10^{-3} M HA and 10^{-3} M NaOH. Because the proton condition and the mass balance on A are identical in these two systems, the solution pH and speciation must also be the same. Generalizing this result to any concentration and any acid/base group, we conclude that a solution made by adding an acid HA to water and then titrating the solution with strong base to the equivalence point is identical, in terms of its acid/base properties, to one to which the same total concentration of A is added as the conjugate base instead of as the acid.

Upon reflection, this result should not be surprising. Written as individual ions, the chemical additions to the system that has been titrated to its equivalence point include 10^{-3} M each of H^+, A^-, Na^+, and OH^-. Because a solution has no way of "knowing" the form in which constituents were added, we can imagine these constituents to have been added in any order and in any combination that we like. In particular, we can treat the solution as though, instead of adding HA and NaOH, the ions were added as 10^{-3} M NaA plus 10^{-3} M HOH (i.e., H_2O). Obviously, the addition of $10^{-3} M H_2O$ has no effect on the solution pH, so the solution is identical to one prepared by the addition of $10^{-3} M$ NaA to pure water.

Multiprotic acids have one equivalence point for each proton that they can release. That is, if a solution of $10^{-3} M H_3A$ is prepared and titrated with base, the conditions when 10^{-3}, 2×10^{-3}, and $3 \times 10^{-3} M$ NaOH have been added are referred to as the first, second, and third equivalence points, respectively. By writing out the mass balance and PC equations, you should be able to convince yourself that adding an amount of base equal to $TOTA$ to a multiprotic acid H_3A yields a solution identical to one made by adding NaH_2A to pure water; adding an amount of base equal to $2(TOTA)$ yields a solution identical to one prepared by adding Na_2HA to pure water; and adding an amount of base equal to $3(TOTA)$ yields a solution identical to one prepared by adding Na_3A to pure water.

The amount of acid or base added during a titration is sometimes quantified by comparison with $TOTA$. Specifically, the *fraction titrated* f is assigned a value of zero in a solution made by adding $TOTA$ in its most protonated form to water. Then, at various points during a titration with base, f is defined by

$$f \equiv \frac{c_{OH^- \text{ added}}}{TOTA} \tag{5.5}$$

Thus, $f = 1.0$ at the (only) equivalence point for a monoprotic acid, or at the first equivalence point for a multiprotic acid; $f = 2.0$ at the second equivalence point for a multiprotic acid; etc.

When $f = 0.5$, the amount of OH^- added is equal to one-half of $TOTA$. Following the same reasoning as above, a system that is titrated to $f = 0.5$ is identical to one in which, instead of adding all the $TOTA$ as HA and then adding a concentration of NaOH equal to $0.5(TOTA)$, the A was added one-half as HA and one-half as NaA [along with an amount of HOH equal to $0.5(TOTA)$]. Adding equal amounts of an acid and its conjugate base to pure water yields a solution with pH close to pK_a, so we expect the pH to be very near pK_a when $f = 0.5$. Review of Figure 5.9 shows that this expectation is correct for the three acids shown.

In the abstract, characterizing the amount of acid or base that would have to be added to a solution to reach various equivalence points ($f = 0, 1, 2$) would seem to be quite informative. Unfortunately, such a characterization is impossible to carry out for any real solution of interest, because such solutions always contain a variety of acid/base groups. Since the pH of the equivalence point is different for each acid/base group (depending on its K_a values and total concentration), a solution containing a mixture of acid/base groups does not have a unique, well-defined set of equivalence points. Under the circumstances, by convention, the characterization is based on the equivalence points computed for a system containing only the carbonate acid/base group, at a concentration that is typical for environmental solutions. The formalities of carrying out this characterization are described next.

5.4.2   EQUIVALENCE POINTS IN THE CARBONATE SYSTEM

In many environmental aquatic systems, the dominant weak acid/base group is the carbonate group, and $TOTCO_3$ and pH are in the range of 10^{-4} to 10^{-2} M and 6.0 to 9.0, respectively. Temporarily ignoring the effects of other acid/base systems, we can determine the range of equivalence points for these systems using a log C–pH diagram and an appropriate proton condition. For instance, for 10^{-2} M $TOTCO_3$, a system in which $f = 2.0$ could be prepared by adding 10^{-2} M H_2CO_3 and 2×10^{-2} M NaOH to water. Representing these chemicals as bases plus available H^+, we could develop the following table to identify the species that are expected to be dominant in the solution:

Base/Acid	pK_a	$TOTA$	H^+ Available
OH^-/H_2O	14.00	2×10^{-2}	2×10^{-2}
CO_3^{2-}/HCO_3^-	10.33	10^{-2}	0
HCO_3^-/H_2CO_3	6.35	10^{-2}	0

The total available H^+ from the H_2CO_3 addition is 2×10^{-2} M, all of which would be consumed by the OH^-. Therefore, we expect the dominant carbonate-containing species at equilibrium to be CO_3^{2-}. The PC table and equation using CO_3^{2-} as a reference species are shown below, and the equilibrium condition (pH 11.13) is identified by an open circle in Figure 5.10. The graph also shows the solution to the PC equation for systems with 10^{-3} and 10^{-4} M $TOTCO_3$ and for f values of 0.0, 1.0, and 2.0. The equilibrium pH values for these and a few

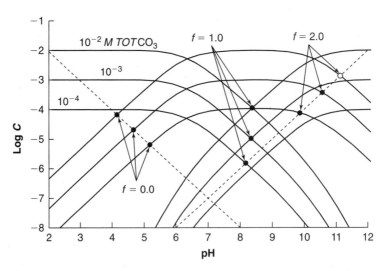

Figure 5.10   Log C–pH diagram for solutions containing 10^{-4}, 10^{-3}, or 10^{-2} M $TOTCO_3$. The solid points indicate the solutions to the proton condition for addition of all the carbonate as Na_2CO_3, $NaHCO_3$, or H_2CO_3, corresponding to the equivalence points for $f = 2.0$, $f = 1.0$, and $f = 0.0$.

Table 5.3 pH of solutions containing various amounts of H_2CO_3, during titrations with strong base

	f				
$TOTCO_3$	0.0	0.5	1.0	1.5	2.0
$10^{-4.0}$	5.19	6.36	8.09	9.55	9.87
$10^{-3.0}$	4.68	6.35	8.30	10.10	10.56
$10^{-2.0}$	4.18	6.35	8.34	10.30	11.13

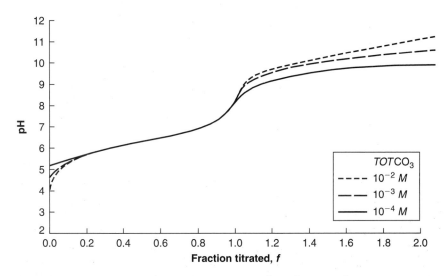

Figure 5.11 Titration curves for solutions of 10^{-4}, 10^{-3}, and 10^{-2} M H_2CO_3, using strong base as titrant.

other conditions are summarized in Table 5.3, and the complete titration curves for a few systems are shown in Figure 5.11.

	−1	0	+1	+2	Concentration
Equilibrium species	OH^-	H_2O CO_3^{2-} Na^+	H^+ HCO_3^-	H_2CO_3	
Initial and/or input species	$NaOH$			H_2CO_3	10^{-3} 2×10^{-3}

$$(2)\{H_2CO_3\}_{in} + (-1)\{NaOH\}_{in} = (1)\{H^+\} + (2)\{H_2CO_3\}$$
$$+ (1)\{HCO_3^-\} - (1)\{OH^-\}$$
$$(2)(10^{-3}) - 2 \times 10^{-3} = \{H^+\} + 2\{H_2CO_3\} + \{HCO_3^-\} - \{OH^-\}$$
$$\{H^+\} + 2\{H_2CO_3\} + \{HCO_3^-\} = \{OH^-\}$$

5.4.3   THE CONCEPTUAL BASIS AND DEFINITION OF ALKALINITY AND ACIDITY

Because many anthropogenic inputs to water bodies are acidic, the response of the receiving water to acidification is typically of greater interest than its response to bases. Table 5.3 shows that, for typical environmental solutions, acid addition to the point where $f = 0$ for the carbonate group yields pH's in the range of 4.7 ± 0.5. Also, as a general rule, waters with pH's lower than about 4.5 have significantly different and less diverse biological populations than those at higher pH values. For these reasons, titrations to pH endpoints of 4.5 to 4.7 have become very common analytical tools as a measure of the capacity of solutions to absorb acid without major ecological consequences.

The amount of strong acid needed to titrate a solution to a preselected pH near 4.7 is called the **alkalinity** (ALK). Alternatively, alkalinity is sometimes defined specifically in terms of the carbonate group; i.e., it is defined as the amount of strong acid needed to titrate the solution to the pH where $f = 0$ for the carbonate system, for the value of $TOTCO_3$ in that particular sample.[2] The former definition is used most often in practice, both because it is more convenient to carry out the titration to the same pH endpoint in all cases than to compute a solution-specific endpoint for each sample, and because the resulting value is more meaningful for solutions containing several acid/base groups. The latter definition, on the other hand, has some theoretical utility, as described below. As shown in Table 5.3, the pH endpoint using the latter definition of alkalinity can be as much as 0.5 pH unit away from 4.7 for typical environmental samples. A comparison of the numerical values of alkalinity computed using the two approaches is provided in an example later in the chapter. Unless stated otherwise, the former definition of ALK (titration to a predetermined, consistent pH endpoint) is used in this chapter.

The fact that the critical pH for significant ecological effects is usually very near the pH corresponding to $f = 0$ for carbonate-dominated systems is, of course, not purely coincidental. For many millions of years, biological evolution has proceeded in aquatic systems whose pH has been substantially controlled by the carbonate acid/base group. As shown later in this chapter, the presence of this group helps maintain (buffer) the pH at values where $f > 0$. Therefore, most aquatic organisms have not developed mechanisms to protect themselves from pH values lower than those corresponding to $f = 0$, i.e., lower than approximately 4.7. For this reason, the alkalinity is sometimes referred to as the *acid-neutralizing capacity* (ANC) of the water, based on the assumption that once this capacity is used up, severe ecological effects become much more likely.

The alkalinity of a solution is sometimes divided into a few sub-categories defined by titrations to pH values corresponding roughly to f values of 2.0, 1.0,

---

[2]When this definition is used, the pH at $f = 0$ typically must be determined *after* the titration has been carried out, since it depends on $TOTCO_3$, which is usually not known in advance. A numerical strategy known as the *Gran titration* technique can be used to determine the pH endpoint based on the titration curve. This technique is described in *Aquatic Chemistry Concepts*, by Pankow (Lewis Publishers, Chelsea, MI, 1991).

and 0.0 for the carbonate system. (The approximate pH values for such solutions are given in Table 5.3.) Alkalinities based on these endpoints are called the **caustic alkalinity,** the **carbonate** or **phenolphthalein alkalinity,** and the **total alkalinity,** respectively. That is, if the initial pH is high, the amount of H^+ required to lower pH to ~10.5 (near $f = 2.0$ for the carbonate system) is called *caustic alkalinity.* The term *phenolphthalein alkalinity* is sometimes used to describe the H^+ requirement to titrate the solution to pH near 8.3 (i.e., near $f = 1.0$), because that is the pH at which phenolphthalein indicator changes color. Before pH meters and electrodes were as inexpensive as they are now, this indicator was widely used to signal the condition when $f = 1.0$. If the term *alkalinity* is used without a qualifier, it usually refers to the total alkalinity (defined by a titration endpoint of ~4.7).

A concept that is exactly analogous to alkalinity but that measures the base-neutralizing capacity of a solution is the **acidity.** This quantity represents the concentration of OH^- ions that must be added to titrate the solution to pH near 10.5 (near $f = 2.0$ for a solution dominated by the carbonate group). If the initial pH is less than the endpoint for the alkalinity titration, the OH^- addition required to raise the pH to the $f = 0$ endpoint is identified as the **mineral acidity.**

5.4.4   COMPUTING ALKALINITY FROM TITRATION DATA OR BASED ON SOLUTION COMPOSITION

To demonstrate the calculation of alkalinity from experimental data, consider again the titration curve shown in Figure 5.7, which is reproduced in Figure 5.12. Based on the amount of acid required to titrate the sample to pH 4.7, the alkalinity

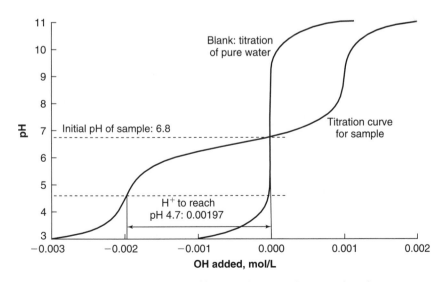

Figure 5.12   The concentration of base (eq/L) required to titrate the solution to pH 4.7 is defined as the alkalinity.

of the water is 1.97×10^{-3} eq/L. (The alkalinity is sometimes defined as the *extra* acid needed to titrate the solution to pH 4.7, compared to the amount needed to titrate a blank—one containing no weak acids or bases—over the same pH range. However, the correction for the blank is small if the initial pH is less than 10, so it is often ignored.)

The preceding paragraph demonstrates the calculation of alkalinity from experimental data for titration of a sample of unknown composition. It is also instructive to consider the alkalinity of a solution whose composition is known, as shown in the following example.

Example 5.5	Determine the pH and alkalinity in milliequivalents per liter of a solution prepared by adding 1.5×10^{-3} M Na_2CO_3 and 5×10^{-4} M NaOCl to water. How would this alkalinity be expressed in mg/L *as $CaCO_3$*? Interpret the alkalinity qualitatively, in terms of how protons are consumed during the titration from the initial pH to pH 4.7. You may assume that the titration is carried out by addition of HCl to the sample.

Solution

Using the techniques developed in previous chapters, we determine the pH of the original solution to be 10.66. The proton balance table for the system both before and after the alkalinity titration, using Na^+, CO_3^{2-}, OCl^-, and Cl^- as proton-neutral species, is shown below. The data shown in normal text are part of the proton balance both before and after the titration; the data shown in *italic* apply only after the titration.

	−1	0	+1	+2	Concentration
Equilibrium species	OH^-	H_2O CO_3^{2-} OCl^- Na^+ Cl^-	H^+ HCO_3^- $HOCl$	H_2CO_3	
Initial and/or input species		Na_2CO_3 $NaOCl$			1.5×10^{-3} 5.0×10^{-4}
			HCl		*ALK*

The value of *TOT*H before the titration, i.e., *TOT*H$_{init}$, is zero, and its value after the titration is equal to the amount of HCl added. However, the concentration of HCl added during the titration is also, by definition, ALK. Therefore, *TOT*H$_{final}$ = ALK.

The value of *TOT*H$_{final}$ can be calculated from the known final concentrations of HOCl, HCO_3^-, and H_2CO_3, so we can determine the alkalinity of the original solution as follows:

$$TOTH_{final} = \{H^+\}_{pH\,4.7} + 2\{H_2CO_3\}_{pH\,4.7} + \{HCO_3^-\}_{pH\,4.7} - \{OH^-\}_{pH\,4.7} + \{HOCl\}_{pH\,4.7}$$

$$= 3.48 \text{ meq/L} = \text{ALK}$$

One meq ALK corresponds to 50 mg $CaCO_3$. Therefore, the alkalinity of the solution could be reported as 174 mg/L *as $CaCO_3$*.

Considering this result qualitatively, we note that the initial solution was prepared with Na_2CO_3 and $NaOCl$. Since the Na^+ ions break away from the rest of the molecule in both salts and are unreactive thereafter, the input species that can contribute to the alkalinity are CO_3^{2-} and OCl^-. The alkalinity titration brings the pH down to 4.7, under which conditions the CO_3^{2-} and OCl^- are converted almost completely to H_2CO_3 and $HOCl$, respectively. Thus, we expect each mole of carbonate ion that was added to the solution to consume two moles of H^+ and thereby contribute two equivalents of alkalinity, and each mole of hypochlorite ion to contribute one equivalent of alkalinity. The total alkalinity of the solution can therefore be estimated based on the original chemical additions to be $2\{Na_2CO_3\}_{added} + \{NaOCl\}_{added}$. This sum equals 3.50 meq/L, which is indeed very close to the computed result; the difference is simply due to the fact that the ions are not truly 100% protonated at the final pH.

Next, imagine a solution at an initial pH of 8.0, containing an unknown mixture of weak acid/base groups. These groups might include carbonate species, phosphate species, ammonia species, and some other, unknown species which we will represent as a diprotic acid group $H_2A^+/HA/A^-$. Even though we don't know anything about the original inputs to the solution, we can write an equation for $TOTH$ in this solution. In this case, it is convenient to choose H_2CO_3 as the reference species for the $CO_3^{2-}/HCO_3^-/H_2CO_3$ group. In addition, we choose PO_4^{3-}, NH_4^+, HA, and, of course, H_2O as proton-neutral species. The proton balance table for such a system is shown below.

	−2	−1	0	+1	+2	+3	Concentration
Equilibrium species	CO_3^{2-}	OH^- HCO_3^- NH_3 A^-	H_2O H_2CO_3 PO_4^{3-} NH_4^+ HA	H^+ HPO_4^{2-} H_2A^+	$H_2PO_4^-$	H_3PO_4	
Initial and/or input species	?			?	?	?	? ? ?

The bottom (input) portion of the table is devoid of information, because we have no idea what inputs were used to prepare the solution; in fact, we do not even know the identities of some of the species in the solution. However, that information is not critical at the moment. Using the subscript *init* to designate values before the alkalinity titration, we can represent $TOTH_{init}$ by the following summation:

$$TOTH_{init} = \{H^+\}_{init} - \{OH^-\}_{init} - \{HCO_3^-\}_{init} - 2\{CO_3^{2-}\}_{init} + 3\{H_3PO_4\}_{init}$$
$$+ 2\{H_2PO_4^-\}_{init} + \{HPO_4^{2-}\}_{init} - \{NH_3\}_{init} + \{H_2A^+\}_{init} - \{A^-\}_{init}$$

When the solution is titrated to the alkalinity endpoint, the amount of strong acid added is, by definition, ALK. The added H^+ contributes to $TOTH$, and since that is the only new material added to the solution, $TOTH$ increases by an amount exactly equal to ALK. That is,

$$TOTH_{final} = TOTH_{init} + ALK \qquad (5.6)$$

$TOTH_{final}$ can also be written as a summation analogous to that for $TOTH_{init}$, i.e.,

$$
\begin{aligned}
TOTH_{final} = \{H^+\}_{final} &- \{OH^-\}_{final} - \{HCO_3{}^-\}_{final} - 2\{CO_3{}^{2-}\}_{final} \\
&+ 3\{H_3PO_4\}_{final} + 2\{H_2PO_4{}^-\}_{final} + \{HPO_4{}^{2-}\}_{final} - \{NH_3\}_{final} \\
&+ \{H_2A^+\}_{final} - \{A^-\}_{final}
\end{aligned}
$$

During the titration, the concentration of every species in the summation for $TOTH$ changes, some increasing and others decreasing. Based on Equation (5.6), the alkalinity can be defined as the difference between $TOTH_{final}$ and $TOTH_{init}$, so it can be computed as follows:

$$
\begin{aligned}
ALK = \{H^+\}_{init}^{final} &- \{OH^-\}_{init}^{final} - \{HCO_3{}^-\}_{init}^{final} - 2\{CO_3{}^{2-}\}_{init}^{final} \\
&+ 3\{H_3PO_4\}_{init}^{final} + 2\{H_2PO_4{}^-\}_{init}^{final} + \{HPO_4{}^{2-}\}_{init}^{final} \\
&- \{NH_3\}_{init}^{final} + \{H_2A^+\}_{init}^{final} - \{A^-\}_{init}^{final} \qquad (5.7)
\end{aligned}
$$

We can also express the concentration of each species that is part of a weak acid/base group as a product of the form $\alpha(TOTA)$. Carrying out this conversion, and putting in values for the pH at the beginning and end of the titration, we can write Equation (5.7) as follows:

$$
\begin{aligned}
ALK = \{H^+\}_{pH\,8.0}^{pH\,4.7} &- \{OH^-\}_{pH\,8.0}^{pH\,4.7} - (TOTCO_3)(\alpha_{1,CO_3} + 2\alpha_{2,CO_3})_{pH\,8.0}^{pH\,4.7} \\
&+ (TOTPO_4)(3\alpha_{0,PO_4} + 2\alpha_{1,PO_4} + \alpha_{2,PO_4})_{pH\,8.0}^{pH\,4.7} - (TOTNH_3)(\alpha_{1,NH_3})_{pH\,8.0}^{pH\,4.7} \\
&+ (TOTA)(\alpha_{0,H_2A} - \alpha_{2,H_2A})_{pH\,8.0}^{pH\,4.7} \qquad (5.8)
\end{aligned}
$$

Although Equations (5.7) and (5.8) look cumbersome, we will simplify them shortly into more usable forms. They are written out in detail above because, in that form, they highlight several important points about the alkalinity concept. Most importantly, they make it clear that every weak acid or base in solution affects the alkalinity; when the pH is lowered to the alkalinity endpoint, each acid/base group becomes increasingly protonated, and all the protons needed to carry out this protonation process contribute to the alkalinity. Further, the contributions of the different acid/base groups are strictly additive. That is, if a solution contains 2×10^{-4} M $TOTNH_3$ and is at pH 8.0, the NH_3 will contribute $(TOTNH_3)(\alpha_{1,NH_3})_{pH\,8.0}^{pH\,4.7}$ to the alkalinity of that solution, regardless of how much or what types of other acid/base groups are in solution.

Now, consider the relative values of some of the terms in Equations (5.7) and (5.8) and how those values change in response to changes in the total concentration of an acid/base group and the pH of the initial solution. The general rule is that the contribution of each group is proportional to the total concentration of that group and to the extent to which the group's speciation changes when the solution is titrated to the alkalinity endpoint. That is, if $TOTA$ is very small, the contribution of A-containing species to ALK will be small, no matter what its speciation is at the beginning and end of the titration. This point should be intuitive; a very small amount of A^- can consume only a small amount of H^+ during the titration, regardless of the pK_a value of HA.

Note also, however, that the contribution of A-containing species to ALK will also be small if the speciation of $TOTA$ changes only slightly during the alkalinity titration (i.e., if the change in its α_i values during the titration is small), *regardless of how large $TOTA$ is*. This point is obvious from the defining equations for alkalinity [such as Equations (5.7) and (5.8) for the example system], but it is easily missed if one thinks of alkalinity as a summation of species initially present *in* solution, rather than as a reflection of how speciation *changes* when the solution is titrated.

To see how the above observations apply to the example system, imagine that the solution is a domestic wastewater with the following total concentrations of weak acid/base groups: 5×10^{-3} M $TOTCO_3$, 1×10^{-3} M $TOTNH_3$, 1×10^{-4} M $TOTPO_4$, and 4×10^{-4} M $TOTA$. Assume also that the H_2A^+/HA/ A^- group has pK_{a1} and pK_{a2} values of 9.0 and 4.7, respectively.

The contribution of the carbonate acid/base group to ALK is significant because it is present at a substantial total concentration and because its speciation changes from almost all HCO_3^- to almost all H_2CO_3 when the solution is titrated from pH 8.0 to 4.7. Thus, essentially every CO_3-containing molecule in the original solution consumes one H^+ ion during the titration.

The total concentration of the NH_3 group is less than that of the CO_3 group by a factor of five in the original solution. However, the contribution of the NH_3 group to ALK is *much* less than can be accounted for simply by that factor of five. The reason is that almost all the $TOTNH_3$ is present as NH_4^+ in the initial solution and also at the end of the titration (~95% initially, >99% at pH 4.7). Because so small a portion of $TOTNH_3$ is present as NH_3 initially, the NH_4^+/NH_3 group has a very small capacity to combine with H^+ ions during the titration, so this group consumes very little H^+; i.e., it contributes very little to ALK.

The situation is reversed for $TOTPO_4$, but the result is the same. That is, most of the PO_4 in the original solution is present as HPO_4^{2-} in the original solution, and most of it is converted to $H_2PO_4^-$ during the titration, so one H^+ ion is consumed for almost every PO_4 molecule initially present (similar to the case for CO_3-containing molecules). However, since $TOTPO_4$ is only 2% as large as $TOTCO_3$, the contribution of the carbonate species to ALK dominates over that of the phosphate species, just as it does over the contribution of the ammonia species.

The result for A-containing species (which might represent the weak acid/base groups in organic molecules in the solution) is largely the same as that for phosphate and ammonia groups; they also contribute negligibly to ALK compared to the carbonate group, both because of their relatively low concentration and because their α values change less than the α values of the carbonate species do during the titration.

5.4.5 ALKALINITY AS A CONSERVATIVE QUANTITY

Because alkalinity is conservative, if a given solution contains x meq of alkalinity and a substance is added that has an alkalinity of y meq, the alkalinity of the new solution is exactly $x + y$ meq. Similarly, if two solutions with different alkalinities are mixed, the alkalinity of the mixture is the weighted average of the alkalinities of the original solutions (weighted according to their mass or volume). This property of alkalinity is demonstrated mathematically by Equations (5.7) and (5.8) and is also apparent from computation of the alkalinity of solutions containing various combinations of weak acids and bases. For instance, Table 5.4 shows the computed pH and ALK of a few relatively simple solutions. The calculation of pH in each case has been carried out using a proton condition in conjunction with a log C–pH diagram, and ALK was computed using an approach analogous to that used in Example 5.5 [the solution analyzed in that example is shown as solution (g) in the table].

Table 5.4   pH and alkalinity of various ideal solutions[†]

Solution	pH	ALK, meq/L
(a) No added chemicals (pure water)	7.00	0.02
(b) $1.5 \times 10^{-3} M \, Na_2CO_3$	10.66	2.98
(c) $5 \times 10^{-4} M \, NaOCl$	8.94	0.52
(d) $5 \times 10^{-4} M \, NaAc$	7.73	0.28
(e) $2 \times 10^{-3} M \, NaAc$	8.03	1.08
(f) $10^{-3} M \, HOCl$	5.30	0.002
(g) $1.5 \times 10^{-3} M \, Na_2CO_3 + 5 \times 10^{-4} M \, NaOCl$	10.66	3.48
(h) $1.5 \times 10^{-3} M \, Na_2CO_3 + 10^{-3} M \, HOCl$	9.90	2.97
(i) $1.5 \times 10^{-3} M \, Na_2CO_3 + 5 \times 10^{-4} M \, NaAc$	10.66	3.26

[†]Alkalinity defined by titration to pH 4.7.

The additivity of the alkalinities is apparent from the results in the table as follows. Subtraction of the alkalinity of solution (a) from that of solution (b) indicates that the addition of 1.5×10^{-3} mol of Na_2CO_3 to 1.0 L of water increases the alkalinity of the solution by 2.96 meq/L. A similar calculation for solution (c) indicates that addition of 5×10^{-4} mol of NaOCl to 1.0 L of water contributes 0.50 meq/L of alkalinity. Because ALK is conservative, the alkalinity of solution (g) is therefore 0.02 + 2.96 + 0.50, or 3.48 meq/L.

In contrast to the alkalinity, it is clear from all the examples in this and preceding chapters that pH is a highly non-linear, non-conservative parameter. That is, one cannot predict the pH of a solution generated by mixing two other solutions based simply on the pH values of the original solutions; the complete acid/base composition of those samples must be known in order to predict the pH of the mixture.

Students often find alkalinity to be a very challenging concept, though it need not be. When confusion arises, the contrast between the conservative nature of alkalinity and the non-conservative nature of pH is very often at its core. This contrast, for instance, leads to the counterintuitive result that the addition of some chemicals to water can affect the alkalinity without changing the pH, change the pH without affecting the alkalinity, or sometimes even cause the alkalinity and pH to change in opposite directions. These possibilities emphasize the key conceptual difference between pH and alkalinity: pH describes the current acid/base state of a solution, while alkalinity describes how difficult it is to change that state to a fixed, different state.

How does addition of the following substances affect the alkalinity and pH of the solution, if ALK is defined by titration to pH 4.7? Assume that the initial solution contains only water and strong acid or base. | **Example 5.6**

a. 10^{-4} M NH_3 if the initial pH is 10.0.
b. 10^{-4} M NH_3 if the initial pH is 6.0.
c. 10^{-4} M NH_4Cl if the initial pH is 8.0.

Solution

Because alkalinity is a conservative quantity, the increment in alkalinity associated with addition of a given chemical does not depend on the initial pH; it depends only on the identity of the species being added and the pH at the endpoint of the alkalinity titration. The effect on solution pH, on the other hand, depends on how the chemical interacts with protons when it is added to the solution.

a. At pH 10, most of the TOTNH$_3$ in solution is present as the neutral species $NH_3(aq)$. Any additional NH_3 put in the solution does not protonate very significantly, so it does not affect the solution pH very much. However, NH_3 does become almost fully protonated when the solution is titrated to pH 4.7. Therefore, each NH_3 molecule added to solution combines with one H^+ by the time the alkalinity titration has been completed, regardless of the initial pH. As as result, the addition of 10^{-4} M NH_3 contributes 10^{-4} eq/L of alkalinity.

b. Because the effect of a chemical addition on alkalinity depends on the identity of the chemical but not on the initial pH, addition of 10^{-4} M NH_3 contributes 10^{-4} eq/L of alkalinity in this solution, just as in part (a). At pH 6.0, however, almost all the NH_3 added to a solution becomes protonated, so the pH of the solution is likely to increase much more in this solution than in the solution in part (a).

c. Addition of 10^{-4} M NH_4Cl adds essentially no alkalinity to a solution, because the ions that are added (NH_4^+, Cl^-) are in the same form at the end of the alkalinity titration as the form in which they were added. If the initial pH of the solution is 8.0, addition of NH_4Cl will also not induce a significant pH change, since the added ions do not combine with or release protons to a significant extent at that pH.

The fact that alkalinity is conservative can be extremely useful for solving certain types of problems, especially those related to systems in which carbon dioxide is exchanged between a solution and a gas phase. Gas-liquid equilibrium is described in some detail in Chapter 7, but it is worthwhile to explore the effect of CO_2 exchange on solution pH and alkalinity at this point.

When CO_2 enters solution and becomes hydrated, it forms carbonic acid (H_2CO_3), which can then deprotonate. As a result, dissolution of CO_2 tends to lower solution pH. Similarly, evolution of CO_2 from solution tends to cause the solution pH to increase. Now consider what happens if some $CO_2(g)$ dissolves and the solution is then titrated to pH 4.7. At the end of the titration, virtually 100% of the carbonate in the system is present as H_2CO_3; i.e., essentially all the carbon that entered solution from the gas phase is in a form that has neither gained nor released H^+ compared with the form in which it originally entered the solution. As a result, its presence has no effect on the amount of H^+ required to carry out the alkalinity titration.[3]

The effect of CO_2 or H_2CO_3 addition on alkalinity is analogous to that for addition of NH_4Cl in the preceding example. However, unlike NH_4^+, H_2CO_3 dissociates substantially at near-neutral pH, so it causes the solution pH to drop, even though it does not alter the alkalinity. Because this result seems to be difficult for many students to grasp, it might be helpful to consider how an analogous argument might be applied to a non-chemical system.

Say, for instance, that we were interested in the amount of thermal energy that must be removed from a solution to freeze it and lower its temperature to $-5°C$. We'll call this quantity the *freezalinity* of the system.

If we had a 10-L batch of water at 10°C whose freezalinity we wished to characterize, we could carry out the necessary calculations based on the heat capacities of water and ice and the latent heat of melting. If we added warm water to the initial solution, the temperature of the system would increase, and the freezalinity would increase as well, because more heat would have to be extracted to get to the freezalinity endpoint. In fact, even if we added ice at 0°C and the temperature of the system *decreased,* the freezalinity would increase, because heat would have to be extracted from the added material during the freezalinity "titration."

On the other hand, consider what would happen if we added ice at $-5°C$ to the system. At least some of the added ice would melt, and the temperature of the system would decrease. However, when the system was titrated to $-5°C$, the ice that had been added would revert to exactly its original condition. As a result, its presence in the system would neither increase nor decrease the amount of heat needed to drive the system to the freezalinity endpoint.

Adding ice at $-5°C$ to the physical system described above is, of course, analogous to adding H_2CO_3 to a solution whose alkalinity is of interest. The H_2CO_3

[3]The effect on ALK of addition or removal of H_2CO_3 is exactly zero if ALK is determined by titration to $f = 0$ for the carbonate system. In that case, the endpoint of the titration is slightly different for the solution before and after the H_2CO_3 exchange. If ALK is defined by titration to a fixed pH endpoint, the effect of H_2CO_3 addition or removal is almost always negligible, but it is not exactly zero. Specifically, addition of n mol of H_2CO_3 to a solution *decreases* ALK by $\alpha_1 n$ equivalents, where α_1 is $\{HCO_3^-\}/TOTCO_3$, evaluated at the endpoint of the titration. For instance, if the titration endpoint is 4.7, each mole of H_2CO_3 added to the solution decreases ALK by 0.023 eq.

partially dissociates when it is first added, causing the pH to drop; this process is analogous to what happens when some of the ice melts and the solution temperature decreases. However, the H_2CO_3 re-forms during the alkalinity titration, so that in the end it has no effect on the quantity of H^+ needed to reach the endpoint.

A second source of the difficulty that many students encounter when trying to understand alkalinity can result from the use of an approximation that is numerically valid, but conceptually confusing. To derive the approximate equation, we can return to consideration of the example solution discussed earlier, containing a mixture of carbonate species and other weak acids and bases. The discussion of that solution indicated that the contribution of carbonate species to its alkalinity is likely to be much larger than the contributions of other weak acids and bases. As a result, the terms for the non-carbonate weak acid/base groups in the alkalinity equation [given by Equation (5.7)] can be omitted, yielding the following approximation:

$$ALK = \{H^+\}_{init}^{final} - \{OH^-\}_{init}^{final} - \{HCO_3^-\}_{init}^{final} - 2\{CO_3^{2-}\}_{init}^{final} \qquad \textbf{(5.9)}$$

$$ALK = (\{H^+\}_{final} - \{H^+\}_{init}) - (\{OH^+\}_{final} - \{OH\}_{init})$$
$$- (\{HCO_3^-\}_{final} - \{HCO_3^-\}_{init}) - 2(\{CO_3^{2-}\}_{final} - \{CO_3^{2-}\}_{init}) \qquad \textbf{(5.10)}$$

Furthermore, the final concentrations of OH^-, HCO_3^-, and CO_3^{2-} (at pH 4.7) are usually much less than their concentrations in the initial solution, and the final concentration of $H^+(10^{-4.7})$ is almost always negligible compared to the other terms, so Equation (5.10) can be simplified even more, as follows:[4]

$$ALK = -\{H^+\}_{init} + \{OH\}_{init} + \{HCO_3^-\}_{init} + 2\{CO_3^{2-}\}_{init} \qquad \textbf{(5.11)}$$

Finally, for the range of alkalinities of natural waters, the $\{H^+\}$ and $\{OH^-\}$ contributions to the alkalinity can usually be ignored in Equation (5.11), and if the solution pH is less than ~9.0, the CO_3^{2-} term can be ignored as well. In such cases, all the complicated equations shown above reduce to the trivial approximate equality $ALK \approx \{HCO_3^-\}$.

It is easy to see why the approximation $ALK \approx \{HCO_3^-\}$ leads to confusion about the effect of H_2CO_3 addition on alkalinity. After all, H_2CO_3 addition usually leads to an increase in $\{HCO_3^-\}$, and the approximation suggests that such an increase should increase ALK. However, if we apply the slightly less extreme assumptions associated with Equation (5.11) and carry out the relevant calculations to determine the resulting changes in solution speciation, we see that the release of HCO_3^- and CO_3^{2-} accompanying H_2CO_3 dissociation and the simultaneous release of H^+ and consumption of OH^- exactly counteract one another, so the reactions do not affect ALK.

---

[4]The choices about which terms to ignore are not entirely consistent, since we are retaining some terms that are usually smaller than those that we are discarding. This choice is made to yield a convenient final expression. Even though we could write that final expression with even fewer terms, the form in which it is written is acceptable, since the terms we are excluding are indeed negligible in most cases.

Equation (5.11) is widely cited as the defining equation for calculating alkalinity. It does, in fact, yield the exact value of ALK if all the alkalinity of the solution is attributable to the carbonate group, and if ALK is defined based on a titration endpoint corresponding to $f = 0.0$, i.e., if the pH endpoint is based on $TOTCO_3$ and is not arbitrarily set at 4.7 for all solutions. A comparison of the alkalinity computed using the two conventions and an illustration of how alkalinity data might be misinterpreted if acid/base groups other than the carbonate group are present in significant quantities are provided in the following examples.

Example 5.7

How different is the alkalinity determined by a titration to pH 4.7 from the alkalinity determined by Equation (5.11) for a solution containing 2×10^{-3} M $TOTCO_3$ that is initially at pH 8.8? Assume that no weak acid/base groups other than the carbonate group contribute significantly to the alkalinity.

Solution

The α values for the carbonate system at pH 8.8 are $\alpha_0 = 0.003$, $\alpha_1 = 0.968$, and $\alpha_2 = 0.029$. The corresponding values at pH 4.7 are $\alpha_0 = 0.978$, $\alpha_1 = 0.022$, $\alpha_3 \ll 0.001$. The amount of acid needed to titrate the system to pH 4.7 can be computed by using these values in conjunction with Equation (5.10). The computation indicates that 2.04×10^{-3} M H^+ must be added to titrate the solution to pH 4.7. The alkalinity computed according to Equation (5.11) is

$$\text{ALK} = -\{H^+\} + \{OH^-\} + (TOTCO_3)(\alpha_1 + 2\alpha_2) = 2.06 \times 10^{-3} \text{ eq/L}$$

The difference between the two estimates of alkalinity is only about 1%.

Example 5.8

A leachate solution from a landfill is at pH 6.6 and contains a group of organic acids whose behavior under acidic conditions can be approximately represented as being equivalent to 2×10^{-3} M $TOTAc$. The solution also contains 1.5×10^{-3} M $TOTCO_3$.

a. What is the alkalinity of the solution, using pH 4.5 as the endpoint for the alkalinity titration?

b. If one conducted the titration to pH 4.5 and assumed that Equation (5.11) applied, what estimate would be obtained for $TOTCO_3$ in the system?

c. Knowing that the solution contains a substantial amount of acetate species, an analyst uses Equation (5.11) as a template to develop the following equation and uses the equation to interpret the titration results. What is wrong with this reasoning?

$$\text{ALK} = -\{H^+\}_{init} + \{OH\}_{init} + \{HCO_3^-\}_{init} + 2\{CO_3^{2-}\}_{init} + \{Ac^-\}_{init}$$

Solution

a. As in the previous example, the speciation of the two acid/base systems in the initial solution can be determined, since the initial pH is known. The amount of base needed to titrate this solution to pH 4.5 can then be computed using equations analogous to Equation (5.8), modified to apply to the solution of interest. The top portion of the proton condition table for the system is shown below, with HAc as the reference species for the acetate group. An equation analogous to Equation (5.8) that can be used to compute ALK is shown below the table.

	−2	−1	0	+1
Equilibrium species	CO_3^{2-}	OH^- HCO_3^- Ac^-	H_2O H_2CO_3 HAc	H^+

$$ALK = \{H^+\}_{pH\,6.6}^{pH\,4.5} - \{OH^-\}_{pH\,6.6}^{pH\,4.5} - (TOTCO_3)(\alpha_{1,CO_3} + 2\alpha_{2,CO_3})_{pH\,6.6}^{pH\,4.5}$$
$$- (TOTAc)(\alpha_{1,HAc})_{pH\,6.6}^{pH\,4.5}$$

The result is that ALK is 2.20×10^{-3} eq/L.

b. The initial pH is 6.6, at which $\{HCO_3^-\} \gg \{CO_3^{2-}\}$. If one assumed that the alkalinity was entirely attributable to carbonate species according to Equation (5.11), the approximation $\{HCO_3^-\} \approx ALK$ would apply; i.e., we would estimate $\{HCO_3^-\}$ to be 2.20×10^{-3}. Given the pH and this estimate of $\{HCO_3^-\}$, the H_2CO_3 concentration could be computed based on K_{a1} as 1.10×10^{-3} M. The estimate of $TOTCO_3$ would then be $\{HCO_3^-\} + \{H_2CO_3\}$, or 3.30×10^{-4} M. This value is more than twice the actual value of $TOTCO_3$. Thus, in this example, Equation (5.11) would not apply because a substantial fraction of the alkalinity is attributable to acetate.

c. When the solution is titrated to pH 4.5, it is reasonable to make the approximation that all the carbonate species are converted to H_2CO_3. As a result, each HCO_3^- ion in the original solution consumes one H^+, and each CO_3^{2-} ion consumes two. However, at pH 4.5, only ~61% of the total acetate is protonated to HAc (i.e., $\alpha_0 = 0.61$). Therefore, each mole of Ac^- in the original solution consumes only 0.61 mole of H^+ during the alkalinity titration; i.e., it contributes only 0.61 eq of alkalinity. The correct equation is therefore as follows [note that this equation is actually just a simplification of the equation shown in part (a), based on the assumption that several of the α values are negligible]:

$$ALK = -\{H^+\}_{init} + \{OH\}_{init} + \{HCO_3^-\}_{init} + 2\{CO_3^{2-}\}_{init} + 0.61\{Ac^-\}_{init}$$

5.4.6 pH AND ALK IN CARBONATE-DOMINATED SYSTEMS

Since the CO_2 system is of such overriding importance in most natural and many engineered water systems, a number of approaches have been developed to show the relationship among pH, $TOTCO_3$, and ALK in systems where the carbonate system provides essentially all the alkalinity. It should be clear that in such a system, only two independent variables are needed to completely define the system. For instance, knowing pH and $TOTCO_3$, we could compute the concentration of all species and ALK. Knowing $\{CO_3^{2-}\}$ and $\{ALK\}$, or $\{HCO_3^-\}$ and $\{H_2CO_3\}$, we could do the same.

Log C–pH graphs are very useful for quickly assessing the effect of a change in pH on speciation in the system, but they are not particularly convenient for evaluating the effect of changes in other system variables. For instance, to show the effect on solution composition of adding Na_2CO_3, we would have to prepare a new log C–pH diagram for each Na_2CO_3 addition, because each system would have a different $TOTCO_3$.

Using the same set of equations as are used to prepare log C–pH graphs, Deffeyes and others have prepared graphs which are easy to use to determine changes in solution composition in response to the addition or removal of a known quantity of H^+, OH^-, CO_3^{2-}, $TOTCO_3$, etc. These graphs, which are simply a different way of representing the mass balance, charge balance, and equilibrium equations, are easier to use than log C–pH diagrams for some applications, but they have the disadvantage that they are not applicable if any other weak acid/base group affects the solution acid/base behavior. An example of such a graph is shown in Figure 5.13. A detailed description of the chemistry of systems dominated by the carbonate group has been prepared by Butler.[5]

5.4.7 ALKALINITY OF NATURAL WATERS

Aside from the carbonate system, the most important group of weak acids in most natural waters is the diverse group of molecules known as *natural organic matter* (NOM), which was described in Chapter 3. Acidity titrations are used extensively to characterize the acid/base properties of NOM. A large fraction of the molecules that comprise NOM are polymeric products that contain a variety of acidic functional groups, the most numerous of which are carboxyls. These carboxyl groups can be attached to either aliphatic or aromatic portions of the molecule. Because each carboxyl (or other acidic) group is in a slightly different chemical environment, the complete complement of acid groups in the molecules can have K_a values that cover a wide range. Therefore, titration curves of NOM are usually relatively featureless, with few sections that are dramatically steeper or flatter than others. A set of curves showing the change in anionic charge of NOM molecules due to protonation/deprotonation reactions during a titration is presented in Figure 5.14.

Alkaline titration curves of natural waters (i.e., titrations of the sample with a strong base, such as NaOH) are sometimes categorized in terms of *strong acidity,* defined as the amount of base needed to titrate the sample from its initial value to around pH 8, and *weak acidity,* the amount of base needed to titrate the sample from pH 8 to around pH 11 (the cutoff is defined differently by different investigators). In the past, efforts have been made to correlate the strong acidity with carboxyl groups and the weak acidity with phenolic groups, but these efforts have not met with great success, and referring to the groups simply as strong and weak acids is preferable.

5.4.8 SUMMARY OF THE ALKALINITY CONCEPT

The alkalinity is a simple, widely used parameter to characterize the capacity of a solution to accept acid inputs without becoming "too acidic." The definition of "too acidic" typically refers to a cutoff in the pH range of 4.5 to 4.7 and is based

---

[5]James N. Butler, *Carbon Dioxide Equilibria and Their Applications*, Lewis Publishers, Chelsea MI, 1991.

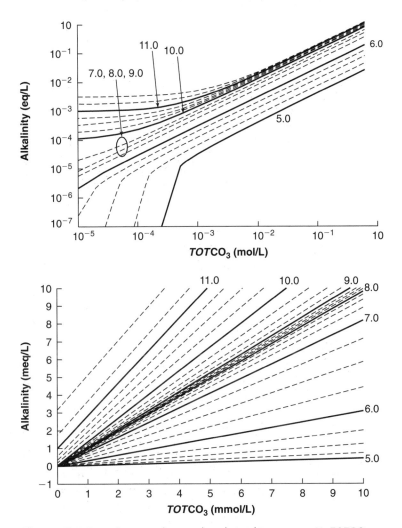

Figure 5.13   Diagram showing the relationships among pH, $TOTCO_3$, and ALK for a solution whose acid/base properties are dominated by the carbonate system. Unlike log C–pH diagrams, this diagram is easy to apply to systems with changing $TOTCO_3$.

| Adapted from K. S. Deffeyes, *Limnology and Oceanography*, Vol. 10: 412–426, 1965.

on both the chemistry of the carbonate acid/base group and empirical observations about the ecology of natural systems. In many natural waters and process waters, the alkalinity can be approximated as the summation of the bicarbonate concentration and twice the carbonate concentration. Although this approximation is often valid, in any real system, weak bases other than those that comprise the carbonate system will contribute to the alkalinity. In such cases, after an alkalinity titration has been carried out, some weak acid/base groups will be almost fully protonated, and others will be protonated to a much lesser extent.

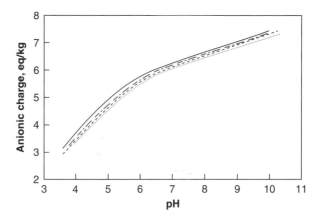

Figure 5.14   Titration of a sample of NOM isolated from a "pristine" groundwater source in Denmark shows the anionic charge on the molecules gradually increasing as the pH increases and the molecules deprotonate. The different curves are for titrations in solutions of different ionic strengths.

Adapted from J. B. Christensen, E. Tipping, D. G. Kinniburgh, C. Grøn, and T. H. Christensen, *Environmental Science and Technology*, Vol. 32, 3346–3355. Copyright 1998 American Chemical Society.

The pH and alkalinity both provide information about the acid/base properties of a solution, and although the two parameters are related, they are independent. Specifically, since the pH characterizes the activity of H^+ in solution, it is an indicator of the current state of the system. By contrast, the alkalinity characterizes the stability of the system, i.e., its ability to stay at or near the original pH when other substances are added. Unlike pH, alkalinity is a conservative parameter, so it is quite easy to compute the alkalinity of a solution prepared by mixing other solutions or by adding known chemicals to an existing solution. One consequence of this property is that addition of H_2CO_3 or CO_2 to solution, or its removal from solution, does not affect the alkalinity, because H_2CO_3 has no alkalinity itself; i.e., it cannot combine with H^+, and at pH 4.7 it does not release H^+ to a significant extent.

5.5   BUFFERS

5.5.1   DEFINITION OF THE BUFFER INTENSITY

Because the presence of a weak acid/base system makes it more difficult to change the pH when the pH is near the pK_a, the weak acid/base is said to *buffer* the pH near the pK_a. Thus, ammonia is a good pH buffer near pH 9 but not at pH 5, and acetate is a good buffer near pH 5 but not at pH 9. If one wished to maintain a solution at pH 5.0, addition of acetate species would be an effective approach for doing so: if protons subsequently entered the solution, some of the Ac^- ions would react with them, and if OH^- ions entered solution, some HAc molecules would dissociate and provide protons to neutralize a portion of the OH^-.

Obviously, the larger the concentration of weak acid or base that is present, the more proton-donating and proton-accepting species would be available to respond to addition of hydroxide or protons, and the greater would be the resistance of the system to pH changes.

The ability of a system to resist pH changes is quantified by a term called the **buffer intensity** β, defined as

$$\beta = \frac{dc_{OH^-\,added}}{d\text{pH}} = -\frac{dc_{H^+\,added}}{d\text{pH}} \qquad \textbf{(5.12)}$$

where $dc_{OH^-\,added}$ and $dc_{H^+\,added}$ represent differential additions of strong base and strong acid to the sample, respectively.[6] In words, the buffer intensity is the incremental amount of base that must be added to a solution per unit increase in pH. A large buffer intensity means that a large amount of base must be added (or, equivalently, a large amount of protons must be consumed) in order to increase solution pH even slightly. Typical units of buffer intensity are equivalents per liter per pH unit.

If a titration curve is plotted with pH on the ordinate and the amount of strong acid or base added on the abscissa, Equation (5.12) indicates that the buffer intensity is the inverse slope of the titration curve (the negative inverse slope, if the titration is with acid). Thus, the buffer intensity is large when the titration curve is flat (small slope) and small when the titration curve is steep (large slope).

5.5.2 THE BUFFER INTENSITY OF WATER AND WEAK ACID/BASE GROUPS

To compute and/or interpret the buffer intensity of a solution, it is useful to consider the three possible fates of OH^- ions added to a solution containing a weak acid: reaction with H^+, reaction with the weak acid, or no reaction (increasing the OH^- concentration in solution). Although these three changes are all linked chemically, they can be evaluated independently.

Consider, for example, titration with a strong base of a system that contains only water and a single weak acid/base pair. Assume that the pK_a of the weak acid is 6.5, the total concentration of the acid/base pair in the solution is $10^{-4}\,M$, and the initial pH is 6.0. We can compute the amount of OH^- required to titrate the solution from pH 6.0 to 6.1 by considering the amount of OH^- needed to carry out each of the three processes described above.

The speciation of the system at the two pH values can be determined analytically, using the techniques developed previously, to give the results shown in Table 5.5.

---

[6]The expression $\beta \equiv d(TOTH)/d(pH)$ is sometimes presented as a definition of buffer intensity. However, that expression has different values depending on which acid is used to provide the increment in $TOTH$, with the buffer intensity being the value of the expression only when a *strong* acid or base is added. In most other contexts, $TOTH$ is defined to include species other than strong acids and bases. To avoid any confusion over that point, Equation (5.12) is used as the definition of buffer intensity here.

Table 5.5   Changes in speciation when a system with 10^{-4} M TOTA is adjusted from pH 6.0 to 6.1 ($pK_{a,HA} = 6.5$)

	Species Concentration			
	{H⁺}	{OH⁻}	{HA}	{A⁻}
pH 6.0	$10^{-6.0}$	$10^{-8.0}$	0.760×10^{-4}	0.240×10^{-4}
pH 6.1	$10^{-6.1}$	$10^{-7.9}$	0.715×10^{-4}	0.285×10^{-4}
Change	-2.06×10^{-7}	$+2.59 \times 10^{-9}$	-0.045×10^{-4}	$+0.045 \times 10^{-4}$

Note that the changes in the H^+ and OH^- concentrations as the solution pH is changed from 6.0 to 6.1 would be the same regardless of the presence or absence of HA in the system: whenever the pH of any solution is changed from 6.0 to 6.1, 2.06×10^{-7} M H^+ is consumed (by reacting with OH^- to form H_2O) and the concentration of OH^- in solution increases by 2.59×10^{-9} M. The amount of OH^- that must be added to meet these requirements is the sum of these two values, or 2.09×10^{-7} M. Because these calculations depend only on K_w and are independent of any other acids or bases in the system, this sum is sometimes referred to as the amount of OH^- needed to titrate the water. Dividing this value (2.09×10^{-7} M) by the pH change (0.1 pH unit), we obtain an estimate of the contribution of these two processes (neutralization of H^+ and accumulation of OH^-) to the buffer intensity of the solution:

$$\frac{OH^- \text{ added to titrate } H^+ \text{ and } OH^-}{\Delta pH} = \frac{2.09 \times 10^{-7} \text{ equiv/L}}{0.1 \text{ pH unit}}$$

$$= 2.09 \times 10^{-6} \frac{\text{equiv/L}}{\text{pH unit}}$$

The above calculations can be extended over the entire pH range of interest to estimate the buffer intensity of water (designated β_w) at any pH, yielding the result shown by the thick line in Figure 5.15. The curve is symmetric around pH 7.0, where it passes through a minimum. The steep rise in this curve at low and high pH corresponds to the flat portions of titration curves that appear at pH greater than approximately 10 and lower than approximately 4 regardless of which (if any) weak acid/base pairs are present in solution.

Table 5.5 also shows that 0.045×10^{-4} M OH^- is needed to convert a portion of the HA to A^- when the pH is changed from pH 6.0 to 6.1.[7] Whereas the computation of β_w depends only on K_w, this calculation depends on the pK_a of the

[7]This calculation is listed twice in Table 5.5. Addition of 0.045×10^{-4} M base causes the HA concentration to decrease by 0.045×10^{-4} and the A^- to increase by the same amount. Since both changes are part of the same process, the amount of base consumed is 0.045×10^{-4}, not twice this value. The calculations for the neutralization of H^+ and the increase in OH^- are listed separately because those reactions are for two different conjugate acid/base pairs (H_3O^+/H_2O and H_2O/OH^-).

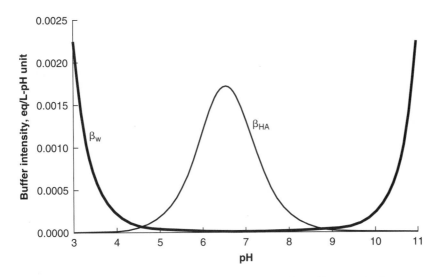

Figure 5.15 Buffer intensity contributed by water only and by HA only, for an acid HA with $pK_a = 6.5$ present at $TOTA = 3 \times 10^{-3}$ M.

HA/A$^-$ conjugate pair and on $TOTA$. The calculation for the amount of base needed for this conversion could be written generically as follows:

$$\text{OH}^- \text{ required to convert HA to A}^- \text{ when pH increases from 6.0 to 6.1} = (TOTA)(\alpha_{1,\text{pH }6.1} - \alpha_{1,\text{pH }6.0})$$

or, extrapolating to a pH change between any two pH values,

$$\text{OH}^- \text{ required to convert HA to A}^- \text{ when pH increases from pH}_{\text{init}} \text{ to pH}_{\text{final}} = (TOTA)(\alpha_{1,\text{pH}_{\text{final}}} - \alpha_{1,\text{pH}_{\text{init}}})$$

As we did above, we can divide this expression by ΔpH (i.e., pH$_{\text{final}}$ − pH$_{\text{init}}$) to estimate the buffer intensity contributed by the weak acid

$$\beta_{\text{HA}} = (TOTA)\frac{\alpha_{1,\text{pH}_{\text{final}}} - \alpha_{1,\text{pH}_{\text{init}}}}{\Delta\text{pH}} \qquad \textbf{(5.13)}$$

(Note the similarity between the presentation here and the previous discussion of the alkalinity of a mixture of weak acids and bases. The arguments are essentially the same, except that to determine the buffer intensity, we consider small increments of acid or base to the system instead of the total acid requirement to adjust the sample to the pH endpoint, and we divide these increments by the incremental change in pH.)

Carrying out this calculation for small values of ΔpH across a wide pH range yields the thin curve labeled β_{HA} in Figure 5.15. The curve is symmetric around pH = $pK_{a,\text{HA}}$ and passes through a maximum at that point, indicating, as expected, that the greatest tendency for a weak acid to resist pH changes is at the pH equal to its pK_a value. Equation (5.13) indicates that increasing $TOTA$

increases β_{HA} proportionately. Furthermore, since plots of α_1 versus pH for monoprotic acids are all identical except for being offset from one another along the pH axis, a curve showing the buffer intensity contributed by a different acid, HB, would have an identical shape as the one for HA, but it would be shifted so that its peak was at $pK_{a,HB}$.

The above discussion can be generalized to yield the following expression for the amount of OH^- needed to titrate a solution containing a single, monoprotic acid/base couple between any two pH values.

OH^- required to titrate a solution containing $TOTA$ from pH_{init} to pH_{final}

$$= -(\{H^+\}_{final} - \{H^+\}_{init}) + (\{OH^-\}_{final} - \{OH^-\}_{init}) + (TOTA)(\alpha_{1,final} - \alpha_{1,init})$$

$$\textbf{(5.14a)}$$

$$= -(\{H^+\}_{final} - \{H^+\}_{init}) + (\{OH^-\}_{final} - \{OH^-\}_{init}) - (TOTA)(\alpha_{0,init} - \alpha_{0,final})$$

$$\textbf{(5.14b)}$$

The terms in the first two sets of parentheses in either form of Equation (5.14) account for the titration of water, and the term in the final set of parentheses accounts for the titration of the HA/A^- couple. If both sides of Equation (5.14a) are divided by $pH_{final} - pH_{init}$ and then $pH_{final} - pH_{init}$ is made differentially small, we obtain the buffer intensity of the solution as follows:

$$\frac{dc_{OH^- added}}{dpH} \equiv \beta = \left\{-\frac{d\{H^+\}}{dpH} + \frac{d\{OH^-\}}{dpH}\right\} + (TOTA)\frac{d\alpha_1}{dpH} \quad \textbf{(5.15a)}$$

$$\beta_{tot} = \beta_w + \beta_{HA} \quad \textbf{(5.15b)}$$

Because $\alpha_0 = 1 - \alpha_1$, we can substitute $-d\alpha_0$ for $d\alpha_1$, and Equation (5.15a) can be written in terms of α_0 as follows:

$$\beta = \left\{-\frac{d\{H^+\}}{dpH} + \frac{d\{OH^-\}}{dpH}\right\} - (TOTA)\frac{d\alpha_0}{dpH} \quad \textbf{(5.15c)}$$

Based on the identities that $\log x = \ln x/\ln 10 \approx (\ln x)/2.303$ and $d(\ln x) = dx/x$, we can rewrite dpH as follows:

$$dpH = d\left(-\frac{1}{2.303}\ln\{H^+\}\right) = -\frac{1}{2.303}\frac{d\{H^+\}}{\{H^+\}} \quad \textbf{(5.16)}$$

Then, by writing $\{OH^-\}$ and α_1 in terms of $\{H^+\}$ and substituting for dpH according to Equation (5.16), the three differentials on the right-hand side of Equation (5.15a) can be expressed as follows:

$$-\frac{d\{H^+\}}{dpH} = -\frac{d\{H^+\}}{-(1/2.3)(d\{H^+\}/\{H^+\})} = 2.3\{H^+\} \quad \textbf{(5.17)}$$

$$\frac{d\{OH^-\}}{dpH} = \frac{d(K_w/\{H^+\})}{-(1/2.3)(d\{H^+\}/\{H^+\})}$$

$$= \frac{-(K_w/\{H^+\}^2)\,d\{H^+\}}{-(1/2.3)(d\{H^+\}/\{H^+\})}$$

$$= 2.3\frac{K_w}{\{H^+\}} = 2.3\{OH^-\} \qquad \textbf{(5.18)}$$

$$\frac{d\alpha_1}{dpH} = \frac{d\left\{\dfrac{K_{a,HA}}{(K_{a,HA} + \{H^+\})}\right\}}{-(1/2.3)(d\{H^+\}/\{H^+\})}$$

$$= \frac{-K_{a,HA}/(K_{a,HA} + \{H^+\})^2\,d\{H^+\}}{-(1/2.3)(d\{H^+\}/\{H^+\})}$$

$$= 2.3\frac{\{H^+\}K_{a,HA}}{(K_{a,HA} + \{H^+\})^2} = 2.3\alpha_0\alpha_1 \qquad \textbf{(5.19)}$$

Finally, substituting Equations (5.17) through (5.19) into Equation (5.16), we obtain a very concise expression for the buffer intensity:

$$\beta_w = 2.3(\{H^+\} + \{OH^-\}) \qquad \textbf{(5.20a)}$$

$$\beta_{HA} = 2.3(TOTA)\alpha_0\alpha_1 \qquad \textbf{(5.20b)}$$

$$\beta = 2.3(\{H^+\} + \{OH^-\} + (TOTA)\alpha_0\alpha_1) \qquad \textbf{(5.20c)}$$

The first term in Equation (5.20) (including both $\{H^+\}$ and $\{OH^-\}$) is β_w, and the final term is β_{HA}. This equation establishes formally what we have already noted from visual inspection of buffer intensity curves: the buffer intensity becomes large when $\{H^+\}$ or $\{OH^-\}$ becomes large, regardless of whether or how much A is present, and the contribution of A to the buffer intensity is directly proportional to its total concentration.

Equation (5.20) can be extended to systems containing two or more weak acids simply by adding terms for β_{HB}, β_{HC}, etc. Each such term would have the same form as that for β_{HA}, i.e.,

$$\beta_{HB} = (TOTB)\frac{d\alpha_{1,HB}}{dpH} = (TOTB)\alpha_{0,HB}\alpha_{1,HB}$$

Thus, the buffer intensity in a system containing several monoprotic acid/base systems can be represented as follows:

$$\beta = 2.3\left\{\{H^+\} + \{OH^-\} + \sum_{\substack{\text{all HX/X}^- \\ \text{pairs}}} (TOTX)\alpha_{0,HX}\alpha_{1,HX}\right\} \qquad \textbf{(5.21a)}$$

$$= \beta_w + \sum_{X = A,B,C,\dots} \beta_{HX} \qquad \textbf{(5.21b)}$$

Equation (5.21b) indicates that the buffer intensity is a conservative quantity; i.e., the total buffer intensity of a solution containing several acid/base groups is simply the summation of the buffer intensities attributable to the individual groups.

Example 5.9

Estimate the buffer intensity of a solution of 5×10^{-3} M NH_4Cl and 2×10^{-3} M sodium benzoate (a common food preservative) at pH 9.25 (the pK_a of NH_4^+). Benzoate is the conjugate base of benzoic acid, a monoprotic acid with $pK_a = 4.20$.

Solution

We can use Equation (5.21a) to estimate the buffer intensity. At the given pH of 9.25, the benzoic acid/benzoate pair (which we will represent as HBz/Bz^-) is almost completely deprotonated, while the $NH_4^+/NH_3(aq)$ pair is split evenly between its two forms. Since the pH is 5.05 units above pK_a for HBz/Bz^-, we can estimate the concentrations of the four weak acid/base species as

$$\{HBz\} = 10^{-5.05} \, TOTBz = 10^{-7.75} \qquad \{Bz^-\} \approx TOTBz = 10^{-2.70}$$

$$\{NH_4^+\} = \{NH_3(aq)\} = 2.5 \times 10^{-3}$$

Inserting these values into Equation (5.21a), we obtain β:

$$\beta = 2.3(10^{-9.25} + 10^{-4.75}) + 2.3(10^{-2.7})(10^{-5.05})(1) + 2.3(10^{-2.3})(0.5)(0.5)$$

$$= 2.88 \times 10^{-3} \, \frac{eq/L}{pH \; unit}$$

Evaluation of the terms in the buffer intensity equation indicates that more than 99% of the overall buffer intensity is contributed by the $NH_4^+/NH_3(aq)$ couple.

A plot of β versus pH for the solution described in Example 5.3 is shown in Figure 5.16. Note that the precise pH and the magnitude of the peaks in the buffer intensity are a good deal easier to evaluate than are the corresponding flat regions of the titration curve. For this reason, analysis of buffer intensity sometimes can provide information about the types and concentrations of weak acids that is almost impossible to ascertain from visual inspection of titration curves. In general, acids present at total concentrations somewhat larger than 10^{-4} M can be detected as peaks in plots of buffer intensity, with the ease of detection increasing as the pK_a value of the acid gets closer to 7.

5.5.3   THE BUFFER INTENSITY OF MULTIPROTIC ACIDS

The buffer intensity in a system with multiprotic acids and bases can be analyzed by using essentially the same approach as for monoprotic groups. The derivation is presented in Appendix 5A.

The results for mono-, di-, and triprotic systems are shown in Table 5.6, along with a generic expression applicable to any acid/base group, regardless of how many protons it can release. The contributions of multiple multiprotic acids to the buffer intensity are additive, just as the contributions of monoprotic acids are.

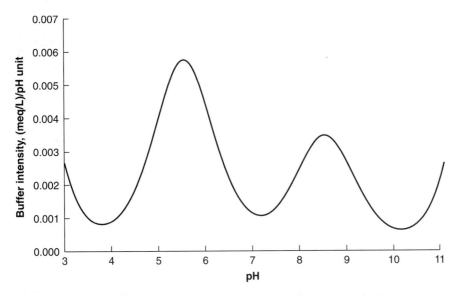

Figure 5.16   Buffer intensity curve for the sample described in Example 5.3.

Table 5.6   Expressions for the contribution of weak acids to the buffer intensity

	Buffer Intensity Contributed by the H_nA Acid/Base Group (β_A)	
Monoprotic, HA	$2.3(TOTA)\alpha_0\alpha_1$	**(5.22a)**
Diprotic, H_2A	$2.3(TOTA)(\alpha_0\alpha_1 + 4\alpha_0\alpha_2 + \alpha_1\alpha_2)$	**(5.22b)**
Triprotic, H_3A	$2.3(TOTA)(\alpha_0\alpha_1 + 4\alpha_0\alpha_2 + 9\alpha_0\alpha_3 + \alpha_1\alpha_2 + 4\alpha_1\alpha_3 + \alpha_2\alpha_3)$	**[5.22c]**
Any multiprotic, H_nA	$2.3(TOTA)\sum\limits_{j>i}^{n}\sum\limits_{i=0}^{n}(j-i)^2\alpha_i\alpha_j$	**(5.23)**

The equations shown in Table 5.6 indicate that the buffer intensity contributed by an acid/base group is proportional to its total concentration and to a summation that includes all possible pairwise products of the α values. In almost all cases, only one or at most two of these pairwise products (the ones corresponding to the two most dominant species and to the most dominant and third most dominant species) are significant. As a result, although the complete summation might include many terms, it can be approximated almost exactly by considering only one or two terms.

The buffer intensity of a solution containing 0.002 M $TOTCO_3$ is shown in Figure 5.17. For this system β_{tot} is similar to that for a system containing two independent monoprotic acids, both present at a total concentration $TOTA$. In the case of carbonate species, the pK_a's for the sequential deprotonation reactions are sufficiently separated that the two peaks in the buffer intensity curve are easily distinguished. If the pK_a's were close to one another, the two peaks would overlap noticeably and would be additive.

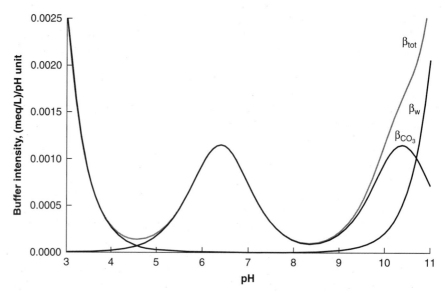

Figure 5.17   Buffer intensity of a solution containing 2×10^{-2} M $TOTCO_3$, showing the individual contributions of water and carbonate species as well as the overall buffer intensity.

Example 5.10

a. Based on qualitative reasoning, assess whether a solution containing 10^{-2} M ammonium phosphate, $(NH_4)_3PO_4$, would be more strongly buffered at pH 7.20 or pH 9.25, or whether the buffer intensity would be about the same at these two values.

b. Estimate β at the two pH values specified to check your assessment in part (a).

Solution

a. The system contains acid/base couples that have pK_a values at both pH's of interest. However, the acid/base pair providing the buffering at pH 9.25 (ammonia) is three times as concentrated as the one providing the buffering at pH 7.20 (phosphate). Therefore, the solution is more highly buffered at pH 9.25 than at pH 7.20.

b. The buffer intensity can be estimated by using the equations collected in Table 5.5. The $TOTA$ and α values of the various species at pH 7.20 are summarized in the table below. The table also shows the contributions of water, $TOTNH_3$, and $TOTPO_4$ to the buffer intensity at each pH, based on the following equations:

$$\beta_w = 2.3(\{H^+\} + \{OH^-\})$$

$$\beta_{NH_3} = 2.3(3 \times 10^{-2})(\alpha_0 \alpha_1)$$

$$\beta_{PO_4} = 2.3(10^{-3.0})(\alpha_0 \alpha_1 + \alpha_1 \alpha_2 + \alpha_2 \alpha_3 + 4\alpha_0 \alpha_2 + 4\alpha_1 \alpha_3 + 9\alpha_0 \alpha_3)$$

The total buffer intensities at pH 7.20 and 9.25 are 6.43×10^{-3} and 1.75×10^{-2} eq/L, respectively, confirming our qualitative expectation. The phosphate group accounts for approximately 89% of β_{tot} at pH 7.20, and the ammonia group accounts for approximately 98% at pH 9.25.

	H$_2$O at pH:		PO$_4$ at pH:		NH$_3$ at pH:	
	7.20	**9.25**	**7.20**	**9.25**	**7.20**	**9.25**
TOTA			10^{-2}	10^{-2}	3×10^{-2}	3×10^{-2}
α_0			$10^{-5.3}$	$10^{-9.1}$	0.01	0.5
α_1			0.5	0.01	0.99	0.5
α_2			0.5	0.99		
α_3			$10^{-5.4}$	$10^{-3.1}$		
β_i	5.10×10^{-7}	4.09×10^{-5}	5.75×10^{-3}	2.47×10^{-4}	6.83×10^{-4}	1.73×10^{-3}

Example 5.11

An anaerobic solution from a stagnant portion of sewer pipe can be represented as a mixture of $5 \times 10^{-3} M$ Na(NH$_4$)$_2$PO$_4$ and $5 \times 10^{-4} M$ NaHS.

a. Sulfuric acid has formed on the upper surface of the pipe and is dripping into the solution. Prepare a plot of pH versus H$_2$SO$_4$ added for additions of 0 to $5 \times 10^{-3} M$ H$_2$SO$_4$. Assume that, over the time frame of interest, sulfate and sulfide species are not interconvertible.

b. Prepare a plot of buffer intensity versus pH for the same range considered in part (a). For completeness, also show the data for addition of base to pH 11. At what pH is the solution most highly buffered during the titration?

c. How much acid can be added before the H$_2$S concentration reaches $1 \times 10^{-4} M$?

Solution

a. The titration can be modeled either by programming the equations into a spreadsheet or by preparing a log C–pH diagram and solving the proton condition for various input H$^+$ concentrations. The resulting profile of pH versus acid added is shown below. The plot reflects the fact that each mole of H$_2$SO$_4$ adds two moles of H$^+$.

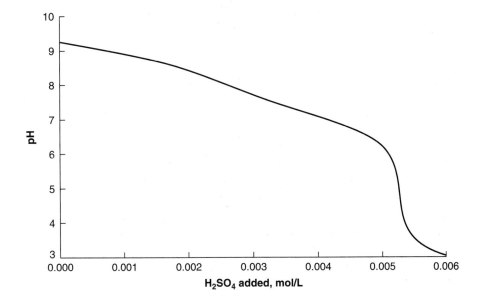

b. The result in part (a) indicates the pH response to different additions of H^+. The buffer intensity can be evaluated either by the equations in Table 5.5 or numerically, by approximating the differentials in the fraction $-dc_{H^+added}/d$pH based on the small discrete steps used to obtain the titration curve in part (a). The result is shown in the following graph. As expected, β has local maxima at the pH values corresponding to the pK_a's of the weak acids in the system. The largest buffer intensity is found at pH 9.25, right at the beginning of the titration and near the pK_a of the ammonia acid/base pair. Note again that the pK_a's of the acids in solution are identified much more easily from the buffer intensity plot than from the titration curve.

c. Since *TOTS* (i.e., the sum of the concentrations of H_2S, HS^-, and S^{2-}) is 5×10^{-4}, $\{H_2S\}$ is 10^{-4} when $\alpha_0 = 0.2$. For H_2S, pK_{a1} and pK_{a2} are 6.99 and 12.92, respectively. Using the defining equation for α_0 from Chapter 3, we compute that $\alpha_0 = 0.2$ at pH 7.7. The titration data indicate that the pH is reached when 3.05×10^{-3} M H_2SO_4 (6.10×10^{-3} M H^+) has entered solution.

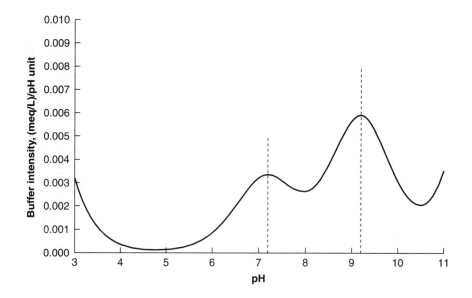

SUMMARY

This chapter explores the qualitative and quantitative characteristics of acid/base titration curves. In a titration, known amounts of known acids or bases are added to solutions of unknown composition, and the pH response of the solutions is used to infer the types and quantities of acids and bases in them.

When OH^- ions are added to a system containing a strong acid, some of those ions react with the H^+ that was already in solution, forming H_2O, and others remain in solution as OH^-, increasing the OH^- concentration so that K_w is satisfied (since some H^+ is consumed, the OH^- concentration must increase). In such cases, the pH changes slowly until the amount of base added is close to the total amount of acid initially present. It then increases extremely rapidly, passing through neutrality to an alkaline value.

If a weak acid HA is present in the initial solution, some of the added OH^- has another fate: it reacts with H^+ ions that are released into solution as a result of the partial dissociation of HA, a reaction that must proceed in order to maintain the HA/A^- equilibrium. In this case, the rapid increase in pH is mitigated, especially in the pH region near the pK_a of the weak acid. The pH region where the rate of pH change is mitigated and the extent of the mitigation can be used to infer information about the strength of the dissolved acid (i.e., its pK_a value) and its concentration.

Titrating a solution with a weak acid or base might cause the pH to change less drastically than it would if a strong acid or base were added. However, the *effective* strength of an acid or base depends on the solution to which it is added. For instance, sodium carbonate acts as a strong base (with each carbonate ion acquiring two protons) at pH 5.0, but it acts as a weaker base at pH 8 (each carbonate acquiring one proton) and an even weaker one at pH 10.

The amount of acid that must be added to a water sample to reach a pH near 4.7 is called the alkalinity. Alkalinity is conservative, and in systems dominated by the carbonate group, it can be approximated as the summation of the bicarbonate and twice the carbonate concentrations. Despite the validity of this approximation, it is important to recognize that the definition of alkalinity is based on the amount of acid needed to titrate the solution to a given endpoint, and not on the amounts of specific acid/base groups present in the initial solution.

Solutions are said to be buffered in regions where relatively large amounts of strong acids or bases are required to induce significant pH changes. Weak acid/base pairs buffer solutions near their pK_a values. Therefore, we can buffer a solution (i.e., maintain the pH in a fairly narrow band) by adding large amounts of an acid/base pair whose pK_a is in the pH region of interest. The carbonate acid/base system provides significant buffering to natural aquatic systems, reducing the likelihood that their pH will decrease below approximately 6.0 in response to the addition of moderate amounts of acid. If enough acid is added to a natural water to overcome this buffering and lower the pH substantially below 6.0, ecological consequences tend to be significant.

The buffer intensity, defined as the incremental addition of base required per unit increase in pH, is a quantitative measure of the pH stability of a solution. Buffer intensity is always large at low and high pH due to the large concentrations of H^+ and OH^- present under these conditions. This contribution to the buffer intensity is referred to as the buffer intensity of water β_w. The buffer intensity contributed by weak acid/base couples in the system passes through a maximum at the pH value corresponding to their pK_a's. The presence of a weak acid in solution is often easier to detect based on curves of buffer intensity than by inspection of unmodified titration curves (pH versus base added).

The analysis of titration data represents a synthesis of most of the key concepts developed in Chapters 1, 3, and 4: the concept of chemical equilibrium, the relative strengths of acids and bases, the relationship of acid/base speciation to solution pH, and the idea that solutions have no memory. The analysis also reinforces the usefulness of log C–pH diagrams as tools for understanding speciation, and the fact that even if we do not have much intuition about the qualitative chemistry occurring in a solution, many of the questions we might ask can be answered by writing and solving a set of algebraic equations. Having mastered these relatively simple concepts, we can explain behavior that, at first glance, is very complex (e.g., the shape of titration curves in systems containing multiple acid/base groups). In subsequent chapters, these same principles are applied to several other groups of chemical reactions.

Problems

1. The following solutions have been prepared:

   Solution 1: H_2O plus sufficient HCl and/or NaOH to attain pH 8.0.

   Solution 2: H_2O plus 10^{-3} M TOTOCl plus sufficient HCl and/or NaOH to attain pH 8.0.

   Solution 3: H_2O plus 10^{-2} M TOTOCl plus sufficient HCl and/or NaOH to attain pH 8.0.

   a. In Solutions 2 and 3, one-half of the TOTOCl was added as HOCl and one-half as NaOCl. Find the amount of HCl or NaOH needed to adjust the pH of all three of these solutions to 8.0.

   b. Compute and compare the pH when 10^{-3} mole of HCl is added to 1 L of each of the solutions.

2. *a.* How much does the alkalinity of a water sample change upon addition of 10^{-4} mol/L of the following chemicals? Assume alkalinity is determined by titration to pH 4.7 and that the water is initially at pH 7.5 due to the presence of an unknown mixture of weak acids and bases. Briefly explain your reasoning.

      i. HCl
      ii. NaOH
      iii. Na_2CO_3
      iv. $NaHCO_3$
      v. CO_2
      vi. NaH_2Cit (Cit = citrate, pK_{a1} = 2.94; pK_{a2} = 4.14; pK_{a3} = 5.82)
      vii. Na_2SO_4

   b. Would your answers to parts (iii) and (vi) above be different, and if so how, if the water was initially at pH 9.0? What if the water was initially at pH 4.7?

3. Plot a log C–pH diagram for a system containing 10^{-2} M TOTPO$_4$. For a solution made by adding 10^{-2} M Na$_3$PO$_4$ to water, prepare a table showing the pH and the concentration of all phosphate species for HCl additions of 0, 0.5, 1.0, 1.5, 2.0, 2.5 and 3.0 × 10^{-2} M.

4. Attempt to arrange the following solutions in order of increasing buffer intensity without carrying out any calculations, and then compute the buffer intensities numerically.

10^{-3} M TOTNH$_3$ at pH 7.05.
10^{-3} M TOTNH$_3$ at pH 9.25.
10^{-3} M TOTCO$_3$ at pH 8.25.
10^{-3} M TOTCO$_3$ at pH 6.35.

5. A solution is made by adding 3×10^{-4} M Ca(OCl)$_2$, 4×10^{-4} M CaAc$_2$, 7×10^{-4} M NH$_4$Ac, and 2×10^{-4} M H$_2$SO$_4$ to water. In the pH range of 4 to 10, at what pH do you expect the buffer intensity to be largest? Explain briefly.

6. Develop a titration curve, i.e., a plot of pH versus lime added, for the system described in Problem 8 of Chapter 4. Consider addition of 0 to 200 mg/L lime. Compare the result with what the titration curve would be for the same system without any NH$_4$Cl. In particular, compare the amounts of OH$^-$ needed to titrate the two solutions from the initial pH to pH 11.

7. An industrial wastewater from a pickling process contains 0.4 M TOTSO$_4$ and 0.1 M TOTFe at pH 0.8. (Pickling involves exposing metal parts to a strong acid solution in order to remove surface deposits that may have formed, thereby preparing the surface for painting or other modification.) There may be other ions in solution, but they are not acids or bases. Compute the amount of lime, Ca(OH)$_2$, in moles per liter, needed to bring the solution to pH 5.5 for discharge. Compute the pH after 25%, 50%, and 75% of the required lime dose (to reach pH 5.5) has been added. Plot the buffer intensity as a function of pH up to pH 10, and indicate the major reactions providing the buffer intensity at various pH values. Relevant equilibria are

H$_2$SO$_4$/HSO$_4$$^-$/SO$_4$$^{2-}$	pK_{a1} = -3.0	pK_{a2} = 1.99
Fe^{3+} + H$_2$O $\leftrightarrow$ FeOH^{2+} + H$^+$	p*K_1 = 2.19	
FeOH^{2+} + H$_2$O $\leftrightarrow$ Fe(OH)$_2$$^+$ + H$^+$	p*K_2 = 3.48	
Fe(OH)$_2$$^+$ + H$_2$O $\leftrightarrow$ Fe(OH)$_3$$^\circ$ + H$^+$	p*K_3 = 7.93	
Fe(OH)$_3$$^\circ$ + H$_2$O $\leftrightarrow$ Fe(OH)$_4$$^-$ + H$^+$	p*K_4 = 8.00	

8. Colorimetric indicators are convenient for evaluating the pH of a solution within a few tenths of a unit. They are also very useful for indicating the endpoint of a titration. Consider a model indicator "In", which can exist as H$_2$In$^+$, HIn$^\circ$, or In$^-$, with pK_{a1} = 6.0 and pK_{a2} = 9.0. H$_2$In$^+$ is red, HIn$^\circ$ is colorless, and In$^-$ is blue, with the red and blue colors becoming noticeable when the solution contains at least 10^{-7} M of the respective species.

a. What are the pH and color of a solution made by adding 5×10^{-7} M H$_2$InCl to water?

b. A solution contains 10^{-3} M ammonium carbonate, (NH$_4$)$_2$CO$_3$, and 5×10^{-4} M lime, Ca(OH)$_2$. Then 5×10^{-7} M H$_2$InCl is added. What are the color and pH of the solution?

    c. Nitric acid (HNO_3) is added to the solution in part (b) until the solution turns red. What is the pH? How much nitric acid was added? (*Hint:* Write and solve the proton condition for the final condition, using dominant species as reference species and identifying the amount of HNO_3 added as *x*.)

    d. If the solution in part (b) were adjusted to pH 12.0 with strong base and were then titrated with HNO_3 to pH 4.0, would the blue-to-colorless or colorless-to-red change be sharper? That is, which would occur over a smaller range of acid added? Why? Assume that both color changes occur over a pH range of about 0.1 pH unit.

9. The acid sulfite pulping process uses sulfurous acid, H_2SO_3, to attack the lignin that holds the wood fibers together. The fibers are then released and processed into paper products, and a hot solution containing acetic acid, sulfurous acid, and some larger organic molecules is generated as a waste. This waste solution is partially evaporated and then condensed, with the smaller molecules being transferred into the condensate and the larger ones remaining behind in an organic-rich solution that can be burned to recover its energy content. The condensate can be treated in an anaerobic biological treatment process, but only if the pH is maintained near neutrality (pH 7). When the process is operating successfully, the microorganisms mediate the following reactions:

$$CH_3COOH \leftrightarrow CH_4 + CO_2$$

$$4H_2SO_3 + 3CH_3COOH \leftrightarrow 4H_2S + 6H_2CO_3$$

    a. Find the pH of a solution of 5000 mg/L HAc and 300 mg/L H_2SO_3 *as S*.

    b. What would the solution pH be if all the HAc and all the H_2SO_3 were destroyed according to the above reactions?

    c. How much NaOH, in moles per liter, would be needed to bring the initial solution pH into the range of good treatability, say, pH 7.0? How much Na_2CO_3 would be required to accomplish the same result? Comment on the relative requirements of these two bases, considering their relative strengths and the number of protons each can accept.

    d. If the solution pH were increased to 7.0 with Na_2CO_3 and the biological reaction then proceeded to destroy one-half of the sulfite and all the acetate according to the reactions shown, what would the final solution pH be?

10. Compare the titration curve that you would obtain if you titrated a solution of 10^{-2} *M* $NiCl_2$ from pH 1.5 to 13.0 with that of a blank (water plus strong acid) that is titrated over the same pH range. In making the comparison, recall that the acidity of nickel ion (Ni^{2+}) is unusual in that $pK_{a1} > pK_{a2}$ (see Problem 4.11). You need not do any calculations for exact pH values, but you should describe the pH region(s) where the curve for the system with Ni is flatter or steeper than the blank and indicate the width (ΔOH^- added) of the regions. Describe your reasoning briefly.

11. A solution is initially at pH 8.3 and has an alkalinity (as determined by titration to pH 4.7) of 120 mg/L *as CaCO₃*, due primarily to species of the carbonate group. Estimate the alkalinity of this solution if 4×10^{-4} NaH_2PO_4 is added to it and enough acid or base is also added to keep the solution pH at 8.3.

12. A log C–pH diagram for a solution containing $7 \times 10^{-3}\,M$ of a triprotic acid H_3A is shown below. Note that HA^{2-} is never a significant species compared to the other forms of the acid.

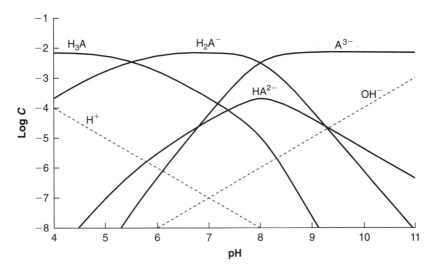

a. What are the pK_a values of the acid?

b. A solution at pH 6.8 containing an unknown amount of $TOTA$ is titrated with strong base to pH 12.0; the titration curve is provided below. What is $TOTA$ in the system being titrated?

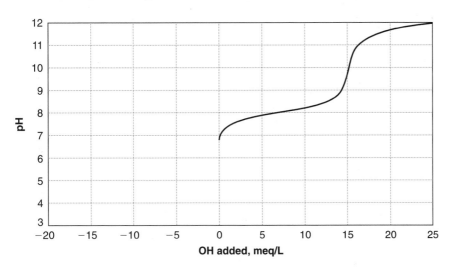

c. Hydrochloric acid (HCl) is gradually added to the solution that has been titrated to pH 12.0 until the pH decreases to 4.5. Sketch a plot of pH versus HCl added on the graph, starting at the point where the previous titration ended, i.e., at (0.025, 12.0). The amount of acid added can be represented as negative additions of base.

13. An acid present in a solution at a concentration of 10^{-2} M is titrated with NaOH. The pH at the endpoint of the titration, i.e., after 10^{-2} M NaOH has been added, is 8.5.

   a. Estimate the pK_a of the acid.
   b. What was the solution pH before the titration?

14. Natural organic matter can enter water supplies as a result of biological reactions in the water (e.g., release of algal and other microbial metabolic products) or by leaching of organic molecules released by degradation of terrestrial plants in the watershed. Many of these NOM molecules are acidic, and water that contains NOM and low concentrations of dissolved inorganic species can be quite corrosive to drinking water distribution systems. One approach to overcoming this problem is to contact the water with limestone ($CaCO_3$). This process causes some $CaCO_3$ to dissolve, neutralizing some of the acidity of the water and increasing its buffer intensity.

   Assume that the NOM molecules in a particular water supply can be represented as diprotic acids with $pK_{a1} = 4.8$ and $pK_{a2} = 7.7$, and that they have an average molecular weight of 250. The water, which contains negligible Ca^{2+} initially, is treated in a limestone contactor. The water exiting the contactor contains 4.5 mg/L Ca^{2+} and is at pH 7.9. Estimate (a) the NOM concentration and (b) the alkalinity (defined by a titration to pH 4.5) of the treated water, expressing the former as milligrams NOM per liter and the latter as milligrams per liter *as $CaCO_3$*.

15. You wish to use an anaerobic biological treatment process to degrade the acetate in a solution containing 4000 mg/L HAc. Anaerobic bacteria can convert acetic acid to carbon dioxide and methane by the following reactions, but the bacteria will not function unless the pH of the solution is at least 5.2. (This question is continued in Chapter 7, Problem 7.)

$$CH_3COOH + H_2O \rightarrow CH_4 + H_2CO_3$$
$$CH_3COO^- + H_2O \rightarrow CH_4 + HCO_3^-$$

   a. Draw the log C–pH diagram for the original solution. What is the initial pH?
   b. What concentration of NaOH must be added to adjust the solution to pH 5.2? What concentration of Na_2CO_3 could be used as an alternative?
   c. Say you neutralized the original solution with 6×10^{-2} M NaOH. The reaction then proceeds to convert 80% of the *TOT*Ac to end products. Draw the log C–pH diagram for the resultant solution. Find the pH before and after the biological conversion.

APPENDIX 5A: DERIVATION OF THE EXPRESSION FOR BUFFER INTENSITY OF MULTIPROTIC ACIDS

Consider a solution containing a multiprotic acid H_nA at a total concentration of *TOT*A. With H_nA as the reference species, the proton balance table and the *TOT*H equation for the solution are as shown below. We do not know how the initial solution was prepared, so we leave out the bottom (input) portion of the table.

	$-n$	$-i$	-3	-2	-1	0	$+1$	Concentration
Equilibrium species					OH^-	H_2O	H^+	
	A^{n-}	$H_{n-i}A^{i-}$	$H_{n-3}A^{3-}$	$H_{n-2}A^{2-}$	$H_{n-1}A^-$	H_nA		

$$TOTH = \{H^+\} - \{OH^-\} - \{H_{n-1}A^-\} - 2\{H_{n-2}A^{2-}\} - \cdots$$
$$- i\{H_{n-i}A^{i-}\} - \cdots - n\{A^{n-}\}$$
$$= \{H^+\} - \{OH^-\} - \sum_{i=1}^{n} i\{H_{n-i}A^{i-}\} \qquad \textbf{(5.24)}$$

If we titrate the solution from pH_{init} to pH_{final} with strong acid, the difference in $TOTH$ at the two conditions is simply the amount of acid added, i.e.,

$$\Delta c_{H^+ \text{added}} = TOTH_{final} - TOTH_{init}$$

Because addition of H^+ to the solution is equivalent to removal of OH^-, the OH^- requirement for the same pH change can be expressed as the opposite of the H^+ requirement:

$$\Delta c_{OH^- \text{added}} = -(TOTH_{final} - TOTH_{init}) \qquad \textbf{(5.25)}$$

Substituting the expression for $TOTH$ from Equation (5.24) into Equation (5.25), we obtain

$$\Delta c_{OH^- \text{added}} = -\{H^+\}_{init}^{final} + \{OH^-\}_{init}^{final} + \sum_{i=1}^{n} i\{H_{n-i}A^{i-}\}_{init}^{final} \qquad \textbf{(5.26)}$$

Dividing both sides of Equation (5.26) by ΔpH (i.e., $pH_{final} - pH_{init}$) and then shrinking ΔpH to a differential, we obtain an expression for β:

$$\frac{\Delta c_{OH^- \text{added}}}{\Delta pH} = \frac{-\{H^+\}_{init}^{final} + \{OH^-\}_{init}^{final} + \sum_{i=1}^{n} i\{H_{n-i}A^{i-}\}_{init}^{final}}{\Delta pH} \qquad \textbf{(5.27a)}$$

$$\frac{dc_{OH^- \text{added}}}{dpH} \equiv \beta = -\frac{d\{H^+\}}{dpH} + \frac{d\{OH^-\}}{dpH} + \sum_{i=1}^{n} i \frac{d\{H_{n-i}A^{i-}\}}{dpH} \qquad \textbf{(5.27b)}$$

Equation (5.27b) can be manipulated in essentially the same way as in the analysis of a monoprotic system as shown in the main chapter, using the identities $\log x = \ln x / \ln 10$ and $d \ln x = dx/x$. The algebra gets a bit long, but it is straightforward, leading to the results shown in Table 5.6.

6

SOFTWARE FOR SOLVING CHEMICAL EQUILIBRIUM PROBLEMS

CHAPTER OUTLINE

6.1 INTRODUCTION

In the preceding chapters, numerical and graphical approaches for determining the equilibrium composition of solutions containing acid/base groups have been presented. The types of problems that have been analyzed can be divided into two broad categories. In one, the concentrations of all chemicals used to prepare the system are known, and the final equilibrium concentrations are sought. In the other, at least one of the final conditions is known, (e.g., pH), and we wish to infer something about the chemical inputs needed to get to that point (e.g., how much acid or base must be added to adjust the solution to a particular pH).

The approach to solving either type of problem involves identifying the unknowns in the system, writing a set of independent equations that relate the unknowns to one another, and solving the equations simultaneously. In all cases, the relevant equations include mass balances, chemical equilibrium expressions, and a charge balance, proton condition, or *TOT*H equation.

The extremely consistent way in which the equations characterizing these systems are formulated makes them ideally suited for a computerized solution algorithm, and such algorithms have been available since at least the 1970s. Although the applications of the software have expanded greatly over the years and the interfaces available to run it have become more user-friendly, the underlying solution approach has hardly changed at all. Basically, the user inputs information about the types of chemicals in the system and the total mass input of any chemicals for which that information is known. Information about the relevant equilibrium expressions can be input manually or retrieved from a database that the program can access. The program then initiates a numerical routine that is carried out iteratively until all the equations are solved to within an acceptable tolerance.

A number of programs with various user interfaces are in common use today. These programs use terminology that is fairly consistent from one to the next, but not entirely so. Terminology that is used by these programs to characterize problems involving acid/base equilibria and titrations is described in this chapter; other terminology is introduced in the discussion of specific types of chemical reactions in later chapters. Nevertheless, because it is useful to collect all the terminology in a single location, a summary of all the terms is provided in Appendix 6A.

In addition to using slightly different jargon, different programs have drawn on different databases to obtain equilibrium constant data, and they have different built-in preferences for the chemical species that are used to describe the makeup of a system. In this chapter and throughout the text, the conventions used by the MINEQL family are adopted. These programs include MINEQL, MINTEQ, and their derivatives (e.g., MINEQL$+^1$ and Visual MINTEQ). However, the basic information in the chapter is common to the majority of programs used for evaluating equilibrium chemical speciation. The presentation here is meant to supplement, and cannot replace, the detailed user instructions that accompany specific software packages.

The central organizational element of most of the software in common use is a concise matrix, widely referred to as a *tableau,* that contains information about the chemical inputs to the system and the reactions they undergo. The first part of this chapter focuses on the development of such tableaus. The framework for the tableaus has already been introduced via the proton condition tables that

---

[1]MINEQL+ is a registered trademark of Environmental Research Software.

were developed in Chapter 4. The full tableau for a system is, in essence, an extension of this framework to all the system components.

While the chapter does not delve into the details of how the software manipulates the tableaus to solve for the equilibrium solution composition, it does provide an outline of how the programs work, so that you, the user, can better understand both the potential applications and the limitations of the software. As is often the case, developing an understanding of the computer algorithm provides insights into some of the steps that we carry out implicitly when solving the problems manually.

6.2   DEFINING THE COMPOSITION OF THE SYSTEM: COMPONENTS AND SPECIES

As is the case when we solve a chemical equilibrium problem manually, an essential first step when we use a chemical equilibrium software package is to identify all the chemicals expected to be present in the equilibrium system. In many software packages, all these chemicals are referred to as *species*. A smaller set of chemicals characterizing the system is defined as *components*. The term "components" was introduced in Chapter 4 in the development of the *TOT*H equation to refer to the set of reference or baseline chemicals that were chosen to be proton-neutral. That definition is adequate if we are interested only in the proton balance, but here we need to define it more generally and more rigorously. Specifically, the *components* in a system are defined as a set of chemicals that meets the following two criteria: (1) combinations of the *components* can be used to generate all the *species* in the system, and (2) none of the *components* can be generated solely by a combination of other *components*. Those familiar with the terminology of linear algebra will recognize these restrictions as requiring that the *components* span the space of *species*.[2] Formation of a group of four *species* from three *components* is shown schematically in Figure 6.1.

As indicated in the figure, it is allowable to either add or subtract *components* from one another to form *species*. For instance, *Spec*3 is formed by combining 1 unit of *Comp*2 with −1 unit of *Comp*3, i.e., by removing one *Comp*3 from a *Comp*2. An example of such a reaction would be the formation of the *species* OH^- by removing one H^+ ion (*Comp*3) from a water molecule (*Comp*2).

The figure also indicates that *Spec*1 consists of 1 unit of *Comp*1 and nothing else. Thus, *Spec*1 and *Comp*1 are chemically identical. However, the program views *components* and *species* as different types of variables, applying

---

[2]To distinguish between times when the words components and species are being used with the restrictive meanings given here and times when they have their more general meanings, they will be italicized throughout the text when they are meant to be interpreted in the context of chemical equilibrium software.

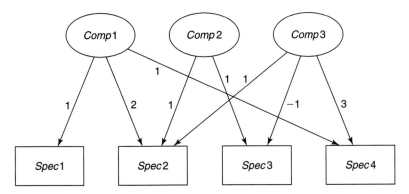

Figure 6.1 Schematic showing the formation of four *species* from various combinations of three *components*. The numbers next to the arrows indicate the stoichiometric coefficients for forming each *species*.

different mathematical operations to the two groups, and, of course, it has no knowledge of the underlying chemistry that we are trying to model. Therefore, even though we might know that *Spec*1 and *Comp*1 are the same chemical entity, it is necessary to maintain a distinction between them in the software. *Species* like *Spec*1 that have the same chemical composition as a *component* are often referred to as Type 1 *species,* and those that have compositions that are different from any *component* (for example, *Spec*2, *Spec*3, and *Spec*4) are called Type 2 *species.*

To make the process of defining *species* and *components* more explicit, consider a system made by adding 2×10^{-3} M HAc to water. The chemicals expected to be present at equilibrium (i.e., the *species* that must be considered) include H_2O, H^+, Ac^-, OH^-, and HAc. Say we chose H_2O, H^+, and Ac^- as *components*. Three of the *species* (H_2O, H^+, and Ac^-) are chemically identical to *components* (i.e., they are Type 1 *species*), so obviously those *species* could be formed from the *components* directly. The remaining two *species* can be generated by appropriate combinations of *components*. Specifically, OH^- can be made by "subtracting" an H^+ from an H_2O molecule, and HAc can be made by adding an H^+ to an Ac^-. When written as chemical reactions, these ways of generating OH^- and HAc bear a close and obvious relationship to chemical reactions for the dissociation of water (the K_w reaction) and the acetic acid dissociation reaction, respectively:

Reaction forming *species* as a combination of *components*	Reaction written conventionally	
$H_2O - H^+ \rightarrow OH^-$	$H_2O \leftrightarrow H^+ + OH^-$	**(6.1)**
$H^+ + Ac^- \rightarrow HAc$	$HAc \leftrightarrow H^+ + Ac^-$	**(6.2)**

The (conceptual) ability to form all the *species* by various combinations of the *components* satisfies the first criterion for using H_2O, H^+, and Ac^- as a set of *components* for the system. Since it is not possible to generate any of the *components* by a combination of the other *components* (for instance, it would not be possible to form H^+ by any combination of H_2O and Ac^-, without simultaneously generating or removing other species), the second criterion is satisfied as well, and the proposed set of *components* is acceptable. We will refer to this example system and this choice of *components* frequently throughout the remainder of the chapter, identifying it as the *baseline* system.

Once the *species* have been identified and the *components* chosen, the stoichiometry for forming each *species* from the *components* must be formalized. The simplest way to accomplish this task is to fill in the coefficients for each *species i* ($Spec_i$), in the following generic reaction:

$$\underline{\quad} \ Comp1 \ + \ \underline{\quad} \ Comp2 \ + \ \underline{\quad} \ Comp3 + \cdots + \leftrightarrow \underline{1} \ \ Spec_i$$

It should be clear that the appropriate coefficients for the baseline system are as follows:

$$\underline{\text{Combination of } components} \ \leftrightarrow \ \underline{1 \ Species \ i}$$

$$\underline{1} \ H_2O + \ \underline{0} \ H^+ + \underline{0} \ Ac^- \ \leftrightarrow \ \underline{1} \ H_2O \qquad \textbf{(6.3)}$$

$$\underline{0} \ H_2O + \ \underline{1} \ H^+ + \underline{0} \ Ac^- \ \leftrightarrow \ \underline{1} \ H^+ \qquad \textbf{(6.4)}$$

$$\underline{0} \ H_2O + \ \underline{0} \ H^+ + \underline{1} \ Ac^- \ \leftrightarrow \ \underline{1} \ Ac^- \qquad \textbf{(6.5)}$$

$$\underline{1} \ H_2O + \underline{-1} \ H^+ + \underline{0} \ Ac^- \ \leftrightarrow \ \underline{1} \ OH^- \qquad \textbf{(6.6)}$$

$$\underline{0} \ H_2O + \ \underline{1} \ H^+ + \underline{1} \ Ac^- \ \leftrightarrow \ \underline{1} \ HAc \qquad \textbf{(6.7)}$$

Although the choice of H_2O, H^+, and Ac^- as *components* is acceptable for the baseline system, this set is not unique; several other choices are equally acceptable. For instance, the group comprised of H_2O, H^+, and HAc meets the criteria for an acceptable *component* set, as does the group H_2O, HAc, and Ac^-. On the other hand, it would not be allowable to choose H^+, HAc, and Ac^- as a *component* set, both because it is possible to generate Ac^- by a reaction involving only the other *components* (by HAc dissociation) and because it is impossible to generate H_2O or OH^- solely by a combination of these three *components*. These examples make it clear that no individual chemical entity is inherently acceptable or unacceptable as a *component,* and that *component* sets are not unique. *Components* can only be acceptable or unacceptable as complete sets, and any system can be represented as a combination of many different such sets.

It is even possible to choose a group of chemicals that are not present in the system or that are not realistic molecules as an acceptable *component* set. For instance, if we thought it would be useful for some reason, we could specify ways to make all the *species* in the baseline system using H_2O, Ac^- and H_4Ac^{3+} as *components,* even though H_4Ac^{3+} does not exist in the equilibrium solution. The stoichiometry for forming the five *species* from this set of *components* would be

$$\underline{1}\, H_2O + \underline{0} \quad Ac^- + \underline{0} \quad H_4Ac^{3+} \leftrightarrow \underline{1}\, H_2O$$

$$\underline{0}\, H_2O + \underline{1} \quad Ac^- + \underline{0} \quad H_4Ac^{3+} \leftrightarrow \underline{1}\, Ac^-$$

$$\underline{1}\, H_2O + \underline{0.25}\, Ac^- + \underline{-0.25}\, H_4Ac^{3+} \leftrightarrow \underline{1}\, OH^-$$

$$\underline{0}\, H_2O + \underline{-0.25}\, Ac^- + \underline{0.25}\, H_4Ac^{3+} \leftrightarrow \underline{1}\, H^+$$

$$\underline{0}\, H_2O + \underline{0.75}\, Ac^- + \underline{0.25}\, H_4Ac^{3+} \leftrightarrow \underline{1}\, HAc$$

Thus, if we include the possibility of using unrealistic *components,* an infinite number of acceptable *component* sets can be used to describe any system. However, once a *component* set is chosen, there is only one way to write the reaction forming each *species* from those *components;* i.e., the stoichiometric coefficient for each *species* is uniquely defined once the *components* have been chosen.

Each reaction forming a *species* from the *components* can be assigned a unique equilibrium constant. In the baseline system, the equilibrium constants for forming H_2O, H^+, and Ac^-, i.e., for reactions (6.3) through (6.5), are trivial: they must be 1.0, as is obvious if the reactions are rewritten leaving out the *components* that have coefficients of zero. These reactions are repeated below, with the *components* shown in bold to distinguish them from the corresponding *species.*

$$\underline{1}\, \mathbf{H_2O} \leftrightarrow \underline{1}\, H_2O \qquad K_{H_2O} = 1.0 \qquad \textbf{(6.3)}$$

$$\underline{1}\, \mathbf{H^+} \leftrightarrow \underline{1}\, H^+ \qquad K_{H^+} = 1.0 \qquad \textbf{(6.4)}$$

$$\underline{1}\, \mathbf{Ac^-} \leftrightarrow \underline{1}\, Ac^- \qquad K_{Ac^-} = 1.0 \qquad \textbf{(6.5)}$$

Keep in mind that, in terms of our formal algorithm (which is the way the computer "understands" the problem), these reactions are not just saying that one water molecule forms one water molecule [reaction (6.3)] or that one acetate ion forms one acetate ion [reaction (6.5)]. Rather, reaction (6.5) indicates that the *species* Ac^- is formed by combining 1 unit of the *component* $\mathbf{Ac^-}$ plus 0 units of the *component* $\mathbf{H_2O}$ plus 0 units of the *component* $\mathbf{H^+}$, and then applying an equilibrium constant of 1.0 to that reaction.

The equilibrium constant for reaction (6.6) can be derived formally as follows:

$$H_2O \leftrightarrow H^+ + OH^- \qquad K_{(6.8)} = K_w \qquad \textbf{(6.8)}$$

$$- \qquad\qquad H^+ \leftrightarrow H^+ \qquad\qquad K_{(6.9)} = 1.00 \qquad \textbf{(6.9)}$$

$$\overline{\underline{1}\, \mathbf{H_2O} + \underline{-1}\, \mathbf{H^+} \leftrightarrow \underline{1}\, OH^-} \qquad K_{(6.6)} = K_{(6.8)}/K_{(6.9)} = 10^{-14.0} \quad \textbf{(6.6)}$$

Similarly, the equilibrium constant for reaction (6.7) is

$$\underline{1}\, \mathbf{Ac^-} + \underline{1}\, \mathbf{H^+} \leftrightarrow \underline{1}\, HAc \qquad K_{(6.7)} = K_a^{-1} = 10^{+4.76} \qquad \textbf{(6.7)}$$

The five equilibrium constant equations relating the *species* to the *components* in the baseline system are summarized in both algebraic and logarithmic form in Table 6.1. The same information is also presented in a more concise

Table 6.1a   Formal algebraic expressions for relating *species* activities to *component* activities

Reaction	Algebraic Equation to Compute *Species* Activity at Equilibrium
(6.3)	$\{H_2O\} = K_{(6.3)}\,\{H_2O\} = K_{(6.3)}\,\{H_2O\}^{1.0}\,\{H^+\}^{0.0}\,\{Ac^-\}^{0.0}$
(6.4)	$\{H^+\} = K_{(6.4)}\,\{H^+\} = K_{(6.4)}\,\{H_2O\}^{0.0}\,\{H^+\}^{1.0}\,\{Ac^-\}^{0.0}$
(6.5)	$\{Ac^-\} = K_{(6.5)}\,\{Ac^-\} = K_{(6.5)}\,\{H_2O\}^{0.0}\,\{H^+\}^{0.0}\,\{Ac^-\}^{1.0}$
(6.6)	$\{OH^-\} = K_{(6.6)}\,\dfrac{\{H_2O\}}{\{H^+\}} = K_{(6.6)}\,\{H_2O\}^{1.0}\,\{H^+\}^{-1.0}\,\{Ac^-\}^{0.0}$
(6.7)	$\{HAc\} = K_{(6.7)}\,\{H^+\}\{Ac^-\} = K_{(6.7)}\,\{H_2O\}^{0.0}\,\{H^+\}^{1.0}\,\{Ac^-\}^{1.0}$

Table 6.1b   Logarithmic forms of the equations in Table 6.1a

Reaction	Logarithmic Equation
(6.3)	$\log\{H_2O\} = \log K_{(6.3)} + 1.0\log\{H_2O\} + 0.0\log\{H^+\} + 0.0\log\{Ac^-\}$
(6.4)	$\log\{H^+\} = \log K_{(6.4)} + 0.0\log\{H_2O\} + 1.0\log\{H^+\} + 0.0\log\{Ac^-\}$
(6.5)	$\log\{Ac^-\} = \log K_{(6.5)} + 0.0\log\{H_2O\} + 0.0\log\{H^+\} + 1.0\log\{Ac^-\}$
(6.6)	$\log\{OH^-\} = \log K_{(6.6)} + 1.0\log\{H_2O\} - 1.0\log\{H^+\} + 0.0\log\{Ac^-\}$
(6.7)	$\log\{HAc\} = \log K_{(6.7)} + 0.0\log\{H_2O\} + 1.0\log\{H^+\} + 1.0\log\{Ac^-\}$

Table 6.2   Tableau showing the key features of the equilibrium expressions in Table 6.1

Species	Stoichiometric Coefficient*			Log K
	H_2O	H^+	Ac^-	
H_2O	1	0	0	0.00
H^+	0	1	0	0.00
Ac^-	0	0	1	0.00
OH^-	1	-1	0	-14.00
HAc	0	1	1	4.76

|*For forming the *species* in a given row from the *components*.

matrix format in Table 6.2, with each row in the matrix containing the stoichiometric coefficients and log K value for forming a particular *species* from the *components*. This matrix is the core of the tableau mentioned earlier in the chapter.

Example 6.1

Prepare a tableau like that shown in Table 6.2 for the same system (addition of HAc to water), but using H_2O, OH^-, and HAc as *components*.

Solution

The reactions and equilibrium constants for forming the five *species* are shown below, and the information is summarized in the following matrix. You should convince yourself that the equilibrium constants are correct for the reactions shown.

$$1\,H_2O + \underline{0}\,OH^- + \underline{0}\,HAc \leftrightarrow 1\,H_2O \qquad K = 1.00$$

$$\underline{0}\,H_2O + \underline{1}\,OH^- + \underline{0}\,HAc \leftrightarrow 1\,OH^- \qquad K = 1.00$$

$$\underline{0}\,H_2O + \underline{0}\,OH^- + \underline{1}\,HAc \leftrightarrow 1\,HAc \qquad K = 1.00$$

$$\underline{-1}\,H_2O + \underline{1}\,OH^- + \underline{1}\,HAc \leftrightarrow 1\,Ac^- \qquad K = 10^{9.24}$$

$$\underline{1}\,H_2O + \underline{-1}\,OH^- + \underline{0}\,HAc \leftrightarrow 1\,H^+ \qquad K = 10^{-14.00}$$

| Species | Stoichiometric Coefficient | | | |
	H_2O	OH^-	HAc	Log K
H_2O	1	0	0	0.00
OH^-	0	1	0	0.00
HAc	0	0	1	0.00
Ac^-	−1	1	1	9.24
H^+	1	−1	0	−14.00

Example 6.2

Prepare the tableau for a solution of $10^{-3}\,M$ NaAc, using H_2O, Na^+, H^+, and Ac^- as *components*. (Note that in this case, some sodium-containing chemical must be added as a *component*, in order to be able to "make" the *species* Na^+. Sodium ion is the simplest choice for such a *component*.)

Solution

The matrix is shown below, dropping the labels *species* and *stoichiometric coefficients*. The confirmation of the values of the equilibrium constants for the reactions shown is left as an exercise.

	H_2O	Na^+	H^+	Ac^-	Log K
H_2O	1	0	0	0	0.00
Na^+	0	1	0	0	0.00
H^+	0	0	1	0	0.00
Ac^-	0	0	0	1	0.00
OH^-	1	0	−1	0	−14.00
HAc	0	0	1	1	4.76

6.3 COMPLETING THE INPUT MATRIX: ESTIMATING *COMPONENT* CONCENTRATIONS AND COMPUTING *SPECIES* CONCENTRATIONS

Once the stoichiometry and equilibrium constants for forming the *species* from the *components* are established, specifying the activity of all the *components* allows one to compute the corresponding equilibrium activities of all the *species*. For instance, say we wish to determine the speciation of a solution of $2 \times 10^{-3}\,M$ HAc.

For convenience, we assume that the solution is ideal, so the activities of the solutes equal their concentrations, and the activity of H_2O is 1.0. If we made guesses for the equilibrium activities of the three *components,* the corresponding equilibrium activities of the *species* could be calculated from either the algebraic or the logarithmic form of the equations in Table 6.1. For instance, if our guesses were $\{H_2O\}$ = 1.0, $\{H^+\} = 10^{-8.0}$, and $\{Ac^-\} = 10^{-3.0}$, the computed activities of the five *species* would be as follows:

$$\{H_2O\} = 1.0 \qquad \{H^+\} = 10^{-8.0} \qquad \{Ac^-\} = 10^{-3.0}$$
$$\{OH^-\} = 10^{-6.0} \qquad \{HAc\} = 10^{-6.24}$$

The computed activities of the various *species* do not represent the final equilibrium composition of the solution, because they are inconsistent with the mass balance on acetate (the sum of the concentrations of HAc and Ac^- does not equal the total acetate input). Nevertheless, the computed *species'* activities are consistent with all the equilibrium constants. This situation is similar to that for a given pH on a log *C*–pH diagram: at any pH we select, the concentration values indicated on the graph are the correct values that apply at that pH. The equilibrium pH of a given system might not be the one we choose, but the concentration values we can read from the graph are consistent with the mass balance and equilibrium constants at that pH.

Since similar calculations could be carried out for any arbitrary guesses for the activities of the *components,* we conclude that, for any given values of the *components'* activities, the information in Table 6.1 or Table 6.2 could be used to compute activities of the *species* that satisfy all the equilibrium constants characterizing the system.

Given that we can calculate values for the *species'* activities that are consistent with any set of guessed *component* activities, how can we decide which values for the *components'* activities are the right ones? The answer, of course, is that only one set of *component* activities satisfies all the mass balances that characterize the system (just as, when a log *C*–pH graph is used, only one pH is consistent with the proton condition). So, the better question is: How can we test the computed *species* activities against the mass balances? We address that question next.

6.4   EVALUATING MASS BALANCES
ON THE *COMPONENTS*

To develop the proton condition or *TOT*H equation, we expressed every species that was present at equilibrium as a combination of reference species and protons. We then did the same for all the input species, and we equated the *TOT*H of the inputs with that of the species in the equilibrium solution. Chemical equilibrium software carries out essentially the same process, but it does so on all the *components* at once. Therefore, we need to provide to the program information that is essentially the same for every *component* as the information about H^+ that is embedded in the *TOT*H equation. We do this by describing the formation of every input chemical and every *species* in the equilibrium solution in terms of the *components*. The program can then iterate on the *component* activities until a set of equations of the form $\sum c_{i,\text{in}} n_i = \sum c_{i,\text{eq}} n_i$ is satisfied not just for H, but for every *component i*.

The steps involved in this process can be summarized as follows:

1. By representing the input chemicals as an equivalent mixture of *components,* determine the total input of each *component* to the system; i.e., for each *component* X, determine $TOTX_{\text{in}}$.

2. Make an initial guess for the equilibrium activities of all the *components*.

3. Using the guessed *component* activities, compute the corresponding activities of the *species*.

4. Represent the computed concentrations of the *species* as an equivalent mixture of *components*.

5. Determine whether the total amount of each *component* in the input ($TOTX_{\text{in}}$, per step 1) equals that in the computed concentrations of the *species* ($TOTX_{\text{eq}}$, per step 4). If so, the computed solution composition is the equilibrium composition, and the problem has been solved.

6. If the mass balance equations are not satisfied, use numerical techniques to improve upon the guesses for the *component* activities, and repeat the whole process until the guesses cause all the mass balance equations to be satisfied.

The solution procedure is characterized schematically in Figure 6.2 for a system in which the activity coefficients of all *components* and *species* are 1.0; adjustments that are needed to account for non-ideality are discussed below.

Note that, when we developed the *TOT*H equation, we carried out steps 1 and 4 above for H^+, but not for the other components. Steps 2, 3, 5, and 6 are the computerized analogs of searching for the pH where the proton balance equation is satisfied. That is, when we use the log C–pH diagram, we guess a pH value, compare $TOTH_{\text{in}}$ and $TOTH_{\text{eq}}$, and if they do not equal one another, we use some rules to make a better guess. We might not be aware of those steps, but that is really what we are doing when we look at the LHS and RHS curves and scan to find the location where they intersect. In the computer, this scanning is done in n-dimensional space (to find the conditions where all the mass balances are satisfied simultaneously) using matrix algebra, but at its core, the process is the same as the one we use when looking at the graph.

Now, consider how we can determine and easily input a value of $\sum c_{i,\text{in}} n_i$ for each *component*. To make the exercise specific, we return to the baseline system of 2×10^{-3} M HAc. An expanded tableau for this system is shown in Table 6.3, this time with information added at the bottom of the table expressing the inputs (HAc and H_2O) in terms of *components*. The input concentration of each chemical used to prepare the solution is also shown in the bottom portion of the table (in the final column). Note that the values in the $\mathbf{H^+}$ column are just the coefficients n_i characterizing the *TOT*H equation for a system in which the *components* are chosen as the reference (proton-neutral) chemicals. That is, H_2O and Ac^- have values of 0 in the $\mathbf{H^+}$ column, reflecting the fact that they are proton-neutral;

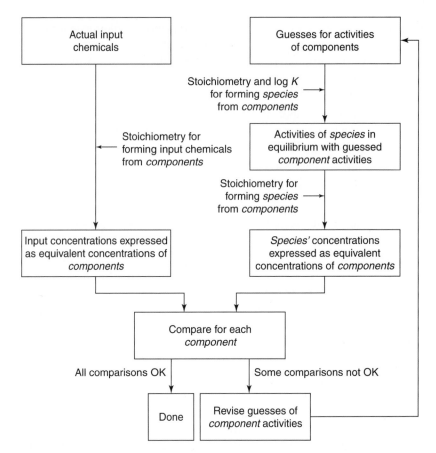

Figure 6.2   Flowchart of calculations used by many computer software packages to compute equilibrium speciation in a system containing only soluble *species*, and assuming ideal behavior (all activity coefficients equal 1.0).

H^+ and HAc each have one excess proton and have values of 1 in the H^+ column; and OH^- is a one-proton-deficient species with a coefficient of -1. However, whereas the proton balance table showed this type of information only for H^+, the tableau shows the comparable information for all the *components*.

To understand how the solution procedure is implemented, let's step through the first iteration that the computer might carry out for this system. We will assume that all the information in Table 6.3 has been input into the program correctly. Based on some internal rules, the program chooses an initial set of guesses for the *component* activities. For this example, we'll assume that the guesses are the ones we made earlier (1.0 for H_2O, 1×10^{-8} for H^+, and 1×10^{-3} for Ac^-).

Using the information in the upper part of Table 6.3 and the guessed values of the *component* activities, the program computes the activity of each *species*. If the *component* activities are expressed as logarithms, each calculation is a simple

Table 6.3 Expanded tableau showing the concentration and formulation of inputs used to prepare the system

	H_2O	H^+	Ac^-	Log K
H_2O	1	0	0	0.00
H^+	0	1	0	0.00
Ac^-	0	0	1	0.00
OH^-	1	-1	0	-14.00
HAc	0	1	1	4.76
Inputs					**Input Concentration**
H_2O	1	0	0		55.56
HAc	0	1	1		2.00×10^{-3}

summation. For instance, the activity of HAc is computed based on the values in the HAc row in the tableau as follows:

$$\log \{HAc\} = \underline{0} \log \{H_2O\} + \underline{1} \log \{H^+\} + \underline{1} \log \{Ac^-\} + 4.76$$
$$= (0)(1.00) + (1)(-8.00) + (1)(-3.00) + 4.76 = -6.24$$

The activities of the other *species* can be computed in the same way. In the revised tableau below, a column with these activities has been added; the guesses for the activities of the *components* are also shown, for reference.

	H_2O	H^+	Ac^-	Log K	**Computed Concentration**
	0.0*	**-8.00***	**-3.00***		
H_2O	1	0	0	0.0	$55.6^\dagger$
H^+	0	1	0	0.0	$10^{-8.00}$
Ac^-	0	0	1	0.0	$10^{-3.00}$
OH^-	1	-1	0	-14.00	$10^{-6.00}$
HAc	0	1	1	4.76	$10^{-6.24}$
Inputs					**Input Concentration**
H_2O	1	0	0		55.56
HAc	0	1	1		2.0×10^{-3}

*Guess of Log a_j.

†Water is treated specially by the programs, so that an activity of 1.0 is always associated with a concentration of 55.6 M.

The computed *species* concentrations are then converted to equivalent concentrations of *components,* and the total concentration of each *component* is computed, using an equation like the following one for each *component j*:

$$TOTComp_{j,est'd} = \sum_{\text{All species } i} n_{i,j} c_i \tag{6.10}$$

In Equation (6.10), $n_{i,j}$ is the stoichiometric coefficient of $Comp_j$ in the reaction forming $Spec_i$ and c_i is the computed concentration of $Spec_j$. For instance, for the current iteration, the total concentration of $\mathbf{Ac}^-$ in the computed solution is

$$
\begin{aligned}
TOTAc_{est'd} &= \sum_{\text{All species } i} n_{Ac,i}c_i \\
&= (0)[H_2O]_{est'd} + (1)[Ac^-]_{est'd} + (0)[OH^-]_{est'd} \\
&\quad + (0)[H^+]_{est'd} + (1)[HAc]_{est'd} \\
&= (0)(55.56) + (1)(10^{-3}) + (0)(10^{-6.0}) + (0)(10^{-8.0}) + (1)(10^{-6.24}) \\
&= 1.0006 \times 10^{-3}
\end{aligned}
\tag{6.11}
$$

The subscript *est'd* has been added to all the concentration terms in the above equation to emphasize that the concentrations are those that have been computed based on the current estimates of the *components'* activities.

As in previous chapters, the assumption of ideal behavior is made, so that the concentrations of H^+ and Ac^- are equated with their activities. However, the concentration of H_2O in an ideal solution is 55.6 *M*, and this is the value used in the calculation of $TOTComp_{j,est'd}$ for each *component*.

The known input concentration of *component* $\mathbf{Ac}^-$ can be represented as $TOTAc_{in}$ and is computed similarly to $TOTAc_{est'd}$, using the values in the bottom two rows of the matrix:

$$
\begin{aligned}
TOTAc_{in} &= \sum_{\substack{\text{All } k \text{ input} \\ \text{chemicals}}} n_{Ac,k}c_k \\
&= (0)[H_2O]_{in} + (1)[HAc]_{in} \\
&= (0)(55.5) + (1)(2 \times 10^{-3.0}) \\
&= 2.00 \times 10^{-3}
\end{aligned}
$$

Comparing $TOTAc_{est'd}$ with $TOTAc_{in}$, we see that the mass balance is not satisfied: the estimate based on the guesses of the *component* concentrations [from Equation (6.11)] is less than the amount of total acetate that was added [from Equation (6.12)]. A similar calculation for $\mathbf{H}^+$ indicates that the guesses also fail to satisfy the mass balance on that *component,* with $TOTH_{est'd}$ exceeding $TOTH_{in}$ in this case.

Based on the magnitude of the errors in the mass balances and information about how changes in the activities of $\mathbf{Ac}^-$ and $\mathbf{H}^+$ affect the mass balances (information that is also derived from the matrix), the program is able to make better guesses for the activities of the *components*. We do not want the program to revise the estimates for the activity and concentration of water (which we want to specify as 1.0 and 55.56 *M*, respectively), so we invoke an option that tells the program to keep using these values for all iterations. Typically, this option is included by default for H_2O; as is shown later, the same option turns out to be useful for other *components* under certain conditions, and in those cases we can invoke the option manually.

Once the program attempts to improve the estimates of the activities of $\mathbf{Ac}^-$ and $\mathbf{H}^+$, it repeats the above process, continuing to iterate until the errors in the mass balance equations are tolerably small. The actual calculation and iteration processes are carried out using matrix algebra and are quite efficient. Details of these calculations are provided in the source manuals.

Reviewing the above process, we see that the computer solves six equations in six unknowns in the example problem. The six unknowns are the activities of the *components* **Ac⁻** and **H⁺** and of the *species* Ac⁻, OH⁻, H⁺, and HAc. The six equations that are used are the equilibrium expressions for forming the four *species* and the mass balances on the two *components*. As noted, the activity of **H₂O** and the concentration H₂O have known values that are fixed throughout the calculations.

Set up the system tableau for a solution comprised of $1.0 \times 10^{-3}\ M\ Na_2CO_3$, $5.0 \times 10^{-4}\ M$ NaAc, and $7.0 \times 10^{-4}\ M$ HCl, using **H₂O, Na⁺, Ac⁻, Cl⁻, H⁺**, and **CO₃²⁻** as *components*. (Verify for yourself that these chemicals satisfy the criteria for an acceptable *component* set.) Write out the *TOT*H equation that is implied by the table.

Example 6.3

Solution

The *species* expected to be present at equilibrium include chemicals that are identical to all six *components* plus HCO_3^-, H_2CO_3, OH⁻, and HAc. Equilibrium reactions for forming all the *species* from the *components* can be written, and the corresponding equilibrium constants can be determined. The tableau for the system is shown below. Note that the reaction for forming H_2CO_3 from the *components* is the sum of the reversed K_{a1} and K_{a2} reactions ($2H^+ + CO_3^{2-} \leftrightarrow H_2CO_3$), so its equilibrium constant is $(K_{a1}K_{a2})^{-1}$, and the log of the corresponding equilibrium constant is +16.68.

	H₂O	Na⁺	Ac⁻	Cl⁻	H⁺	CO₃²⁻	Log K
H₂O	1	0	0	0	0	0	0.00
Na⁺	0	1	0	0	0	0	0.00
Ac⁻	0	0	1	0	0	0	0.00
Cl⁻	0	0	0	1	0	0	0.00
H⁺	0	0	0	0	1	0	0.00
CO₃²⁻	0	0	0	0	0	1	0.00
HCO₃⁻	0	0	0	0	1	1	10.33
H₂CO₃	0	0	0	0	2	1	16.68
OH⁻	1	0	0	0	-1	0	-14.00
HAc	0	0	1	0	1	0	4.76

Inputs							Input Concentration
H₂O	1	0	0	0	0	0	55.56
Na₂CO₃	0	2	0	0	0	1	1×10^{-3}
NaAc	0	1	1	0	0	0	5×10^{-4}
HCl	0	0	0	1	1	0	7×10^{-4}
*TOT*Comp$_{in}$†	55.6	2.5×10^{-3}	5×10^{-4}	7×10^{-4}	7×10^{-4}	1×10^{-3}	

†***TOT*Comp$_{in}$** is the total input concentration of each component, based on the known amount of each chemical added.

The mass balance on H^+ is carried out by comparing the H^+ concentration in the input chemicals with that in the *species*. These two quantities are computed using equations analogous to (6.11) and (6.12), as follows:

$$TOTH_{in} = 0[H_2O]_{in} + 0[Na_2CO_3]_{in} + 0[NaAc]_{in} + 1[HCl]_{in}$$
$$= 7.0 \times 10^{-4}$$

$$TOTH_{eq} = 0[H_2O]_{eq} + 0[Na^+]_{eq} + 0[Ac^-]_{eq} + 0[Cl^-]_{eq} + 1[H^+]_{eq} + 0[CO_3^{2-}]_{eq}$$
$$+ 1[HCO_3^-]_{eq} + 2[H_2CO_3]_{eq} - 1[OH^-]_{eq} + 1[HAc]_{eq}$$

When we equate the expressions for the *TOT*H in the input chemicals and the equilibrium solution, the mass balance becomes

$$7.0 \times 10^{-4} \overset{?}{=} [H^+]_{eq} + [HCO_3^-]_{eq} + 2[H_2CO_3]_{eq} - [OH^-]_{eq} + [HAc]_{eq}$$

Because water is always chosen as a *component* and we never concern ourselves with its mass balance, it is sometimes left out of the system tableau altogether and is simply assumed to be available to form other *species*. For example, the stoichiometry for forming OH^- is sometimes described as being simply $-1\ H^+$, with the necessity to have one H_2O molecule involved in the reaction assumed implicitly. Similarly, salt ions that do not participate in any reaction other than dissolution are often left out of the tableau. Thus, Na^+ and Cl^- might be left out of the tableau for the system described in the preceding example, since the only information the tableau provides about these *species* is that the activities of Na^+ and Cl^- ions at equilibrium equal the activities of all Na^+ and Cl^- ions that were added initially. The only drawback to this practice is that the computed equilibrium solution might not appear to satisfy a charge balance; i.e., to write a valid charge balance, one must consider ions that are not included in the output table.

6.5   INCORPORATING ION ACTIVITY COEFFICIENTS IN THE CALCULATIONS

All modern chemical equilibrium models are capable of incorporating activity coefficients into the calculations, often giving the user a choice of which model to use for the calculations (e.g., the extended Debye–Huckel equation, the Davies equation, etc.). Internally, the calculations are usually carried out by computing the ionic strength at the end of an iteration, computing the corresponding activity coefficients for all the *components,* and then converting the equilibrium constant for formation of each *species* to a concentration ratio (K^{conc}) by incorporating the various γ_i values into the value of K_{eq}. This is essentially the same process we used in Example 3.14 to define the conditional equilibrium constant for the HF/F^- group in a solution of known ionic strength.

For instance, for the baseline system, after an iteration is complete, the program can compute the ionic strength of the solution and use that value to compute the activity coefficients γ_{H^+}, γ_{OH^-}, and γ_{Ac^-}. Then, for the next iteration, the water dissociation and HAc dissociation reactions are represented as follows:

$$K_w = 10^{-14.00} = \{H^+\}\{OH^-\} = \gamma_{H^+}\gamma_{OH^-}[H^+][OH^-]$$

$$K_w{}^{conc} \equiv [H^+][OH^-] = \frac{K_w}{\gamma_{H^+}\gamma_{OH^-}}$$

$$K_{a,HAc} = 10^{-4.75} = \frac{\{H^+\}\{Ac^-\}}{\{HAc\}} = \frac{\gamma_{H^+}\gamma_{Ac^-}[H^+][Ac^-]}{\gamma_{HAc}[HAc]}$$

$$K_{a,HAc}{}^{conc} \equiv \frac{[H^+][Ac^-]}{[HAc]} = \frac{\gamma_{HAc}}{\gamma_{H^+}\gamma_{Ac^-}}K_{a,HAc}$$

The calculations described previously are then carried out, using the concentration-based (conditional) equilibrium constants. Thus, the program continually updates the values of the conditional equilibrium constants so that they always represent concentration ratios for the current estimate of the solution ionic strength. Since these adjustments are all done in the background, the user is generally not even aware of them.

6.6 DETERMINING THE EQUILIBRIUM SPECIATION OF A SYSTEM UNDER SEVERAL SIMILAR CONDITIONS: SIMULATING TITRATIONS WITH CHEMICAL EQUILIBRIUM SOFTWARE

One of the great advantages of using computer software to solve for the equilibrium speciation of a system is that changes in the system composition can be easily tracked as one or more parameters are gradually changed. For instance, if we wished to simulate a titration with NaOH of a solution containing 2×10^{-3} M HAc, we could first solve for the solution speciation in the initial solution and then solve essentially the same problem repeatedly, each time changing the specified input concentration of $\mathbf{H^+}$ slightly (decreasing its value, since an input of OH^- is equivalent to removal of H^+).

That is, if $\mathbf{H^+}$, $\mathbf{Ac^-}$ and $\mathbf{Na^+}$ were selected as *components,* the inputs to the program would specify that $TOTH_{in}$ and $TOTAc_{in}$ were both 2×10^{-3} for the first problem to be solved and that $TOTNa_{in}$ was 0. However, rather than stop after solving for the speciation in that system, the computer could be instructed to solve the same problem ten more times, each time decreasing $TOTH_{in}$ by 2×10^{-4} and increasing $TOTNa_{in}$ by the same amount. In such a case, the final calculation would be for a system prepared with $TOTAc_{in} = 2 \times 10^{-3}$, $TOTH_{in} = 0$ and $TOTNa = 2 \times 10^{-3}$; that is, it would simulate a system that contained 2×10^{-3} M HAc initially and was then titrated to $f = 1.0$ with NaOH.

Most modern chemical equilibrium software programs have the capability to carry out simulations like that described above. By using this approach, the programs can simulate not only conventional titrations with a strong or weak acid, but also mixing of two waters in different ratios, which are often much more difficult to simulate using the techniques described in previous chapters.

Example 6.4

Wastewaters in which the pH is controlled primarily by the HF/F$^-$ acid/base group ($pK_a = 3.17$) are generated in certain metal finishing processes. An industry is considering treating such a waste by using reverse osmosis (RO), a pressure-driven, membrane-based process that allows water and HF but not salt ions such as Na$^+$ and F$^-$, to pass through the membrane. Consider a solution that initially contains 2×10^{-3} M HF and 1×10^{-3} M NaF and that is to be treated by RO. Assuming that the wastewater is on one side of the membrane and pure water is initially on the other, compute the pH of each solution as HF is driven across the membrane, until 90% of the TOTF has transferred. Although water would pass through the membrane as well as HF, assume for now that equal volumes of water are present on each side of the membrane throughout the process.

Solution

This problem would be tedious to solve using a log C–pH diagram in conjunction with a charge balance, proton condition, or TOTH equation because the values of TOTF on each side of the membrane are constantly changing (and so a different log C–pH diagram would be required for each step). By using available software, it can be solved directly. The easiest way to analyze the problem would be to choose HF as a *component* and consider how the speciation in the two solutions changes as the total concentration of that *component* changed. However, many software packages have a set of built-in choices for the *components,* and in most cases these choices include F$^-$ and not HF. Therefore, the problem will be solved here choosing **Na$^+$**, **H$^+$**, and **F$^-$** as *components*.

　　The initial inputs to the program for the wastewater solution are shown in the following tableau. The process of removing HF from that solution is carried out by instructing the program to solve for the speciation several times, each time decreasing the values of both TOTH$_{in}$ and TOTF$_{in}$ by 3×10^{-4} M. The composition of the solution on the product side of the RO membrane is computed in the same way, except that the initial values of TOTH$_{in}$ and TOTF$_{in}$ are both zero (since the initial solution on that side is pure water), and the concentrations of those two *components* are increased by 3×10^{-4} M in each successive run. The tenth problem that the computer solves is for the wastewater solution after 90% of the TOTF has been removed.

	H$_2$O	Na$^+$	F$^-$	H$^+$	Log K
H$_2$O	1	0	0	0	0.00
Na$^+$	0	1	0	0	0.00
F$^-$	0	0	1	0	0.00
H$^+$	0	0	0	1	0.00
HF	0	0	1	1	4.13
OH$^-$	1	0	0	-1	-14.00

Inputs					Input Concentration
H$_2$O	1	0	0	0	55.56
HF	0	0	1	1	2×10^{-3}
NaF	0	1	1	0	1×10^{-3}
TOTComp$_{in}$	55.6	1×10^{-3}	3×10^{-3}	2×10^{-3}	

The values $TOTH_{in}$ and $TOTF_{in}$ used in each calculation are shown in the table below, and the pH of each solution is shown as a function of the amount of HF transferred in the graphs. Note that the value of $TOTH_{in}$ becomes negative after two-thirds of the $TOTF$ has transferred. The transfer does not stop at this point, even though the amount of HF that has passed through the membrane equals the amount that was initially present, because the HF/F$^-$ speciation continually adjusts as the transfer proceeds. Therefore, there is some HF present no matter how much has already transferred. However, the pH on the wastewater side of the membrane increases dramatically after two-thirds of the $TOTF$ has transferred, so the concentration of HF on the wastewater side of the membrane

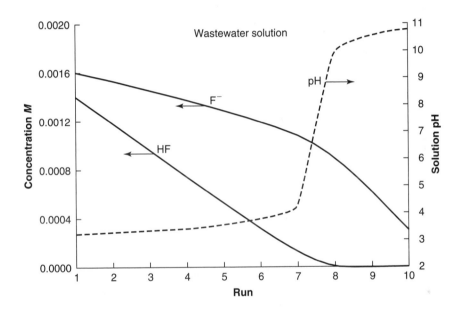

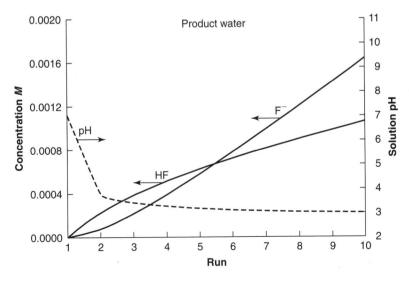

and the rate of transfer of HF to the clean water side would both decline substantially at that point.

Run	TOTF	TOTH
1	0.0030	0.0020
2	0.0027	0.0017
3	0.0024	0.0014
4	0.0021	0.0011
5	0.0018	0.0008
6	0.0015	0.0005
7	0.0012	0.0002
8	0.0009	−0.0001
9	0.0006	−0.0004
10	0.0003	−0.0007

6.7 SYSTEMS IN WHICH THE EQUILIBRIUM ACTIVITIES OF ONE OR MORE *SPECIES* ARE KNOWN

The preceding discussion describes the information needed to solve a problem in which all the inputs are known and the final solution composition is unknown. If the final concentration of one or more *species* is known in advance, we can follow almost the same procedure, but we need to indicate somehow that the program should force the activities of these *species* to the known values.

For every constraint that is added to the mathematical problem (e.g., specifying the final pH), another one must be removed to maintain the balance between the number of unknowns and the number of equations. In a case where the pH is fixed, the constraint that cannot be used in the solution algorithm is the mass balance on H^+, i.e., the statement that $TOTH_{in} = TOTH_{eq}$. Of course, eliminating consideration of this constraint does not mean that the mass balance is not satisfied—that balance is a statement of the conservation of matter, and it cannot be violated. Rather, what we are saying is that the equation is of no use to us because, if we fix the solution pH, we cannot know in advance how much H^+ must be added or removed to maintain that pH. That is, the equation $TOTH_{in} = TOTH_{eq}$ applies, but we don't know either $TOTH_{in}$ or $TOTH_{eq}$, so the equation does not help us solve for the other unknowns.

The mathematical argument presented above is, in a sense, a realistic reflection of what is known and not known about an actual solution that has been forced to a pre-determined pH. For instance, say we prepared a solution of 1.6×10^{-3} M HAc plus 0.4×10^{-3} M NaAc and then adjusted the solution to pH 7.5 by adding strong acid or base. If we wanted to characterize the system mathematically, we could fully describe the input information about Ac and Na by mass balances. However, we could not write a comparable mass balance on H, because we don't know exactly how much H^+ or OH^- was added. On the other hand, we can say for sure that the activity of H^+ at the end of the process is $10^{-7.5}$, and by combining these three equations with a charge balance, we could solve for the complete speciation of the solution. At that point we could write the mass balance on H^+ and infer how much acid or base had been added. Thus, the mass balance on H^+ applies to the system and is useful to us, but only *after* we have solved for the solution composition using other equations.

In theory, the steps needed to fix the activity of a *species* could be carried out in the background of the program, so that the user could simply specify which *species* were present at known, fixed activities, and the program would assure that that constraint was met. As noted above, virtually all available programs do include such a feature to specify that the activity of H_2O is always 1.0. Many programs also include a comparable feature allowing the user to specify the equilibrium solution pH in cases where that value is known in advance. However, there are times when one might want to specify a value for the equilibrium activity of some other *species*. The approach for instructing the programs to meet such a constraint is not particularly intuitive, but it is straightforward in a mechanical sense. Specifically, to indicate that the activity of a *species* is fixed, one defines a new, dummy *species* that has the same stoichiometry as the "real" *species* whose equilibrium activity is known. The new (dummy) *species* is labeled a *fixed entity* or *Type 3 species*, or, in this text, a *Type 3a species*.

The information that we need to specify for Type 3a *species* includes the stoichiometry for forming the *species* from the *components* and an equilibrium constant for the corresponding reaction; i.e., we need to specify the same information as for other *species*. However, while the rules for defining the stoichiometry of Type 3a *species* are the same as those for other *species*, the equilibrium constants are computed differently. Specifically, **the log K value that must be input for a Type 3a *species* equals the logarithm of the equilibrium constant for forming the corresponding real *species*, minus the logarithm of the known, fixed activity of the *species*.** The basis for this rule is described in the source manuals provided with the programs; for the purposes of using the software, knowing the rule suffices. The calculation of the appropriate value of log K for a Type 3a *species* is easier to demonstrate than it is to state in a sentence, as is apparent from the following example.

Specify the input data for computer analysis of a solution of 1.6×10^{-3} M HAc plus 0.4×10^{-3} M NaAc that has been adjusted to pH 7.5. Use $\mathbf{Na^+}$, $\mathbf{Ac^-}$, and $\mathbf{H^+}$ as *components*. | **Example 6.5**

Solution

The stoichiometry for forming the *species* H^+, OH^-, HAc, Ac^-, and Na^+ and the corresponding log K values are identical in this case to those in a system in which the pH is not known in advance. Similarly, the input concentrations of the *components* $\mathbf{Na^+}$ and $\mathbf{Ac^-}$ can be computed as in previous problems, yielding values of 0.4×10^{-3} M and 2.0×10^{-3} M, respectively. It is also useful to compute the concentration of $\mathbf{H^+}$ that is added as part of the known inputs and to list this value as $TOTH_{in}$, even though this calculation does not yield the true value of $TOTH_{in}$ (the program adjusts $TOTH_{in}$ as it carries out the iterations, in order to force pH to remain at 7.5).[3]

Thus, all parts of the tableau prepared for a system with unknown pH also apply to this problem. The only difference between the inputs for the two types of problems is that, to fix the pH at a specified value, we need to define a new, Type 3a *species*. This *species* is "made" in the same way as dissolved H^+, that is, by the following combination of components:[4]

$$\underline{1}\,H^+ + \underline{0}\,Na^+ + \underline{0}\,HAc^- \rightarrow \underline{1}\,\overline{H^+}$$

The value of $\log K$ that we must input for forming $\overline{H^+}$ equals the $\log K$ value for the real species H^+ minus the logarithm of the fixed activity of H^+. Thus, the correct input value for $\log K$ of $\overline{H^+}$ is $0 - (-7.5) = +7.5$.

The tableau for the system is shown below. A column has been added to specify the type of each species listed.

Species	Type	Component H⁺	Ac⁻	Na⁺	Log K
H^+	1	1	0	0	0.00
Ac^-	1	0	1	0	0.00
Na^+	1	0	0	1	0.00
OH^-	2	-1	0	0	-14.00
HAc	2	1	1	0	4.75
$\overline{H^+}$	3a	1	0	0	7.50

Inputs					Input Concentration
HAc		1	1	0	1.6×10^{-3}
NaAc		0	1	1	0.4×10^{-3}
$TOT\,Comp_{in}$		1.6×10^{-3}	2.0×10^{-3}	0.4×10^{-3}	

The program treats Type 3a *species* very much like additional input chemicals. However, whereas the amounts of other inputs are specified by the user and are not altered by the program, the program is free to adjust the input concentration of Type 3a *species* as necessary in order to reach the known equilibrium activity. This procedure can be illustrated by modifying the tableau for the example system to show $\overline{H^+}$ explicitly as an input chemical in the lower portion of the table, added at an unknown concentration of x mol/L.

---

[3]In essence, the input value of $TOTH_{in}$ is overridden by the specification of a Type 3a *species* containing H^+. However, the value of $TOTH_{in}$ associated with known inputs is useful for other reasons, as shown subsequently.

[4]In this text, dummy *species* are written with an overbar to distinguish them from other *species* that have the same stoichiometry and are really present in solution.

Species	Type	Component			Log K
		H^+	Ac^-	Na^+	
H^+	1	1	0	0	0.00
Ac^-	1	0	1	0	0.00
Na^+	1	0	0	1	0.00
OH^-	2	-1	0	0	-14.00
HAc	2	1	1	0	4.75
$\overline{\overline{H^+}}$	3a	1	0	0	7.50

Inputs					Input Concentration
HAc		1	1	0	1.6×10^{-3}
NaAc		0	1	1	0.4×10^{-3}
$\overline{\overline{H^+}}$		1	0	0	x
$TOT\,\text{Comp}_{in}$		$1.6 \times 10^{-3}\,(+x)$	2.0×10^{-3}	0.4×10^{-3}	

The input of $x\,M\,\overline{\overline{H^+}}$ is listed in parentheses in the $TOT\text{Comp}_{in}$ row, because we do not consider x when we actually set up the input to the program. That is, when we input a value for $TOTH_{in}$, the value we use is 1.6×10^{-3}. However, we (and the program) know that the true value of $TOTH_{in}$ equals the sum of 1.6×10^{-3} and an as yet undetermined amount of H^+ that will be added in the course of the calculations. We know this because we understand the chemistry of the system, and the program "knows" it because it has been told that a Type 3a *species* containing H^+ is to be considered. After the program has solved for the equilibrium speciation, the total concentration of each *component* in the equilibrated solution ($TOT\text{Comp}_{eq}$) can be computed. For instance, carrying out the analysis of the example system, we find that the equilibrium composition of the solution is as shown below, in yet one more version of the tableau.

Species	Type	Component			Log K	Equil. Concentration
		H^+	Ac^-	Na^+		
H^+	1	1	0	0	0.00	$10^{-7.50}$
Ac^-	1	0	1	0	0.00	2.00×10^{-3}
Na^+	1	0	0	1	0.00	0.40×10^{-3}
OH^-	2	-1	0	0	-14.00	$10^{-6.5}$
HAc	2	1	1	0	4.75	3.63×10^{-6}
$\overline{\overline{H^+}}$	3a	1	0	0	7.50	N/A
$TOT\,\text{Comp}_{eq}$		3.35×10^{-6}	2.0×10^{-3}	0.4×10^{-3}		

Inputs						Input Concentration
HAc		1	1	0		1.6×10^{-3}
NaAc		0	1	1		0.4×10^{-3}
$\overline{\overline{H^+}}$		1	0	0		x
$TOT\text{Comp}_{in}$		$1.6 \times 10^{-3}\,(+x)$	2.0×10^{-3}	0.4×10^{-3}		

A row has been added to the table to show the total concentrations of the *components* computed based on the equilibrium speciation. The value shown for the total concentration of **H⁺** at equilibrium is computed as

$$TOTH_{eq} = (1)[H^+] + (-1)[OH^-] + (1)[HAc]$$
$$= 10^{-7.5} - 10^{-6.5} + 3.63 \times 10^{-6} = 3.35 \times 10^{-6}$$

As is required, the mass balances are satisfied for the two *components* whose input concentrations were well known (**Na⁺** and **Ac⁻**). The mass balance on **H⁺** was not utilized by the algorithm when solving for the equilibrium speciation of the system; but, as noted earlier, this does not mean that it was violated. Rather, it was not used because we simply did not have enough information for it to be helpful. Now that we know how much **H⁺** is actually present in the *species* once equilibrium has been attained, we can use that information to determine how much **H⁺** <u>must</u> have been added when the system was prepared. That is, equating the input and equilibrium total concentrations of **H⁺**, we find

$$TOTH_{in} = TOTH_{eq}$$
$$1.600 \times 10^{-3} + x = 3.35 \times 10^{-6}$$
$$x = -1.597 \times 10^{-3}$$

Thus, the mass balance on **H⁺** does indeed turn out to be extremely useful, but it is useful after the calculation of speciation rather than during the calculation. In particular, the mass balance tells us the amount of **H⁺** (strong acid) that would have to be added to the specified inputs to bring the solution to pH 7.5. In this case, the amount added is negative, meaning that acid would have to be removed from solution or, equivalently, that 1.597×10^{-3} *M* OH⁻ would have to be added.

Example 6.6 | When H_2CO_3 enters a solution [which can be accomplished by bubbling $CO_2(g)$ through the water], some of the H_2CO_3 dissociates, generating HCO_3^- and CO_3^{2-} and releasing H⁺. In a particular system of interest, H_2CO_3 is added to a solution containing 10^{-4} *M* NaOH until the H_2CO_3 activity is $10^{-5.0}$. How much H_2CO_3 was added, and what is the pH of the solution?

Solution

We can analyze the system by using **H⁺**, **Na⁺**, and **CO₃²⁻** as *components*. The activity of H_2CO_3 in the equilibrium solution is known, so a Type 3a *species* with stoichiometry identical to H_2CO_3 must be input. The reaction for forming this *species* is as follows:

$$\underline{2}\,H^+ + \underline{1}\,CO_3{}^{2-} \rightarrow \underline{1}\,\overline{H_2CO_3}$$

Log K for forming $\overline{H_2CO_3}$ equals the value for the corresponding real *species* minus log K of the fixed activity of H_2CO_3. Therefore, log K for $\overline{H_2CO_3}$ is $16.68 - (-5.0)$, or 21.68. By using this information and the information about other inputs, the problem can be solved. A table showing both the input information and the final solution composition is shown on the next page.

| Species | Type | Component | | | Log K | Equil. Concentration |
		H⁺	Na⁺	CO₃²⁻		
H^+	1	1	0	0	0.00	4.48×10^{-8}
Na^+	1	0	1	0	0.00	1.00×10^{-3}
CO_3^{2-}	1	0	0	1	0.00	1.04×10^{-7}
OH^-	2	−1	0	0	−14.00	2.24×10^{-7}
H_2CO_3	2	2	0	1	16.68	1.00×10^{-5}
HCO_3^-	2	1	0	1	10.33	9.96×10^{-5}
$\overline{H_2CO_3}$	3a	2	0	1	21.68	N/A
$TOT\text{Comp}_{eq}$		1.196×10^{-4}	1.00×10^{-4}	1.096×10^{-4}		

Inputs						Input Concentration
NaOH		−1	1	0		1.0×10^{-4}
$\overline{H_2CO_3}$		2	0	1		x
$TOT\text{Comp}_{in}$		$-1.0 \times 10^{-4}\,(+2x)$	1.0×10^{-4}	$0\,(+x)$		

Note that the $TOT\text{Comp}_{in}$ values include inputs from NaOH ($10^{-4}\,Na^+$ and $-10^{-4}\,H^+$) and from addition of an unknown amount x of H_2CO_3 (contributing $2x\,H^+$ and $x\,CO_3^{2-}$). When the equilibrium speciation is determined, the program reports that the pH is 7.35 and that the total amount of the *component* CO_3^{2-} is 1.096×10^{-4}, so we conclude that this is the amount of H_2CO_3 that has been added (since the known inputs did not include any CO_3). Even though $TOT\text{CO}_3$ is 1.096×10^{-4}, the equilibrium concentration of the species H_2CO_3 is only 1.0×10^{-5}, as specified in the input. Furthermore, the total concentration of the component H^+ at equilibrium equals the amount input with the NaOH (-1.0×10^{-4}) plus the amount that entered with the H_2CO_3 ($2 \times 1.096 \times 10^{-4}$), so the program correctly accounts for the fact that the carbonate entered solution as carbonic acid.

6.8   USING CHEMICAL EQUILIBRIUM SOFTWARE TO DEVELOP LOG *C*–pH DIAGRAMS

A procedure similar to that described above for modeling a titration (Example 6.4) can be used to prepare a log *C*–pH diagram using the software packages. However, in this case, rather than change the total input concentration of H^+, we define a Type 3a *species* ($\overline{H^+}$) and solve for the system speciation repeatedly, each time changing the equilibrium constant for formation of that *species*. For instance, if we wish to fix the solution pH at 2.0, we can define a Type 3a dummy *species* ($\overline{H^+}$) and, following the steps in Example 6.5, assign it a log *K* value of 2.0. Running the program with that input, we would obtain output that indicates the concentration of all dissolved species at pH 2.0.

By following the same procedure, but changing log *K* for formation of $\overline{H^+}$ from 2 to 12 in forty, 0.25-unit increments, we could determine the solution speciation at pH 2.00, 2.25, 2.50, . . . , 12.00. That is, the program would

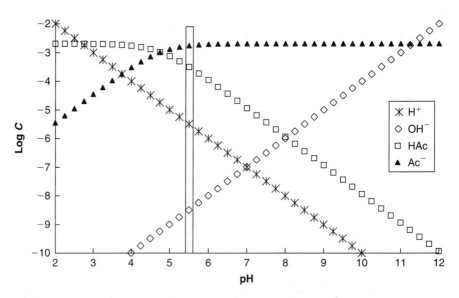

Figure 6.3 Log C–pH diagram for a system containing 2×10^{-3} M $TOTAc$. The diagram was generated by computer software, solving the appropriate mass balance and equilibrium constant equations at 41 different, fixed pH values. The result of one set of analyses, at pH 5.5, is shown in the box.

generate 41 sets of results, each of which applies at a specified (fixed) pH value. The results generated by such a process for a solution containing 2×10^{-3} M $TOTAc$, for all the pH values investigated, are shown in Figure 6.3. By simply connecting the dots representing the concentrations of a particular *species* across the pH range, we can generate a curve for the concentration of that *species* as a function of pH. Because the calculations are highly repetitive, the computer carries them out very quickly, and even very complex diagrams can be prepared easily. All modern software for determining chemical speciation is designed to make it easy to carry out these types of calculations for gradually changing input conditions.

SUMMARY

Software specifically designed to solve for the equilibrium speciation of chemical systems has been available for a few decades and is used widely to simulate chemical systems. The inputs to the software are essentially the same as those needed to solve for speciation manually; however, much of the data regarding plausible reactions and the associated equilibrium constants are available in the programs' databases, so that the effort required to run the programs (and the likelihood of errors in data input) is reduced.

Many of the available programs organize the data in a tableau, which is filled in partially by the user and partially from the program's database. To set up the tableau, we need

- A "recipe" describing the chemical inputs to the system.
- A list of *species* expected to be present in the system once equilibrium is attained.
- A list of the independent reactions that can occur in the system, along with their equilibrium constants.

Once we have the above information, we can choose a set of *components*. We then rewrite the equilibrium reactions in such a way that, in each reaction, one *species* appears on the right-hand side of the reaction with a coefficient of 1.0, and all other terms in the reaction are *components*. We can write similar reactions for each compound used to prepare the system. The system tableau is then prepared. The mass (mole) balances and equilibrium expressions (mass action laws) for this system are implied by the data in the table.

Software written to solve these types of problems includes algorithms by which guesses for the equilibrium activities of all the *components* are made and then used to compute the concentrations of all the *species*. These computations are carried out in such a way that the equilibrium constants describing the system are automatically satisfied. The software then compares the total amount of each *component* added to prepare the system with that in the computed solution. If the comparisons are satisfactory, the mass balances characterizing the system are satisfied and the problem is solved. If the mass balances are not satisfied, the software makes better guesses for the activities of the *components,* and the process is repeated. In most cases, the software converges to a correct solution very quickly.

The software is also capable of solving for speciation in a system where the activity of one or more *species* is known in advance. In these situations, the input is modified by defining a Type 3a *species,* using a slightly different rule to compute log K than is used for computing log K values of Type 1 and 2 *species*. Type 3a *species* are treated as input chemicals whose input concentration is to be determined by the program. Once the composition of the equilibrium solution is determined, we can extract information from the result to infer the amount of each Type 3a *species* that was added (or more properly, that would have to be added to get to the specified condition).

By carrying out repetitive calculations for systems with gradually varying inputs, the software can simulate titrations or develop log C–pH diagrams quickly and efficiently, even for solutions containing a complex mix of input chemicals.

Problem

1. Re-solve several of the problems you solved in previous chapters, using chemical equilibrium software. Develop log C–pH diagrams, carry out simulated titrations, and determine the speciation in systems with one or more species present at known, fixed activities. Compare the program output with the results you obtained manually.

APPENDIX 6A: TERMINOLOGY USED IN CONJUNCTION WITH CHEMICAL EQUILIBRIUM SOFTWARE

In any equilibrium chemical system, we can view each *species* that is present at equilibrium as a combination of chemicals taken from a smaller subset of molecules or ions. This latter group of chemicals is referred to almost universally in chemical equilibrium programs as *components*. The equilibrium constants of certain reactions (most notably oxidation/reduction and adsorption reactions, described in Chapters 9 and 10, respectively) contain mathematical functions that are manipulated by the programs in exactly the same way that chemical *components* are, even though those functions do not represent chemicals at all. It is expedient to lump these functions together with the chemical components in the numerical code, so the complete set of *components* includes both real chemicals and these mathematical pseudo-chemicals.

All chemicals that are present in the system once it reaches equilibrium are called *species*. Historically, the models have segregated these species into six groups, each of which is identified by both a number (Type 1, Type 2, etc.) and a name. While the latter approach was intended to be more descriptive, the names that were chosen were not as intuitive as one would hope, and in some cases they can actually be misleading. Each *species* type is described briefly below.

Type 1 and Type 2 *species* are chemicals that are present in solution at equilibrium; the only distinction between these two groups is that Type 1 *species* have the same chemical formula as one of the *components,* whereas Type 2 *species* have a chemical formula that is different from that of the *components*. All mathematical manipulations that are applied to dissolved species are applied in identical ways to Type 1 and Type 2 *species*.

Unfortunately, the names that have been assigned to Type 1 and Type 2 *species* sometimes obscure their basic similarity. Specifically, Type 1 *species* are referred to as components in many programs, and Type 2 *species* are labeled *aqueous species* or *aqueous complexes*. While it is true that Type 1 *species* have the same chemical composition as *components,* the program manipulates *components* and Type 1 *species* differently. Also, as noted above, the *component* set can include pseudo-chemicals that are not Type 1 *species,* so the labeling of Type 1 *species* as components can be confusing. Similarly, referring to Type 2 *species* as aqueous species seems to imply that other types of *species* are <u>not</u> dissolved, which is not the case (Type 1 *species* are dissolved as well). In Visual MINTEQ, no distinction is made between Type 1 and Type 2 *species;* both groups are simply identified as *aqueous species*.

Unlike Type 1 and 2 *species,* Type 3 *species* are chemicals that are <u>not</u> in the equilibrium solution. Rather, Type 3 *species* are dummy *species* that are used to impose some external constraint on the solution's equilibrium composition. For instance, we can use Type 3 *species* to force the computed solution composition to be in equilibrium with another phase (either a gas or a solid) or to have a particular dissolved chemical (a Type 1 or Type 2 *species*) present at a preestablished activity (e.g., if we wish to fix the equilibrium pH at a certain value). Type 3 *species* are usually named either fixed entities or fixed solids, with the former name being more inclusive and more accurate.

In this text, Type 3 *species* are separated into three groups: Type 3a *species* are those used to fix the activity of a dissolved *species* (either Type 1 or Type 2); Type 3b *species* are used to describe a separate phase with which the solution is in equilibrium; and Type 3c *species* are purely imaginary entities that do not represent any real chemical at all but are useful for inputting information about oxidation/reduction (redox) and adsorption reactions. Since the input data are computed slightly differently for these three groups of Type 3 *species,* distinguishing among them can avoid a potential source of confusion and error. However, once the input calculations are made, all Type 3 *species* are processed identically. In Visual MINTEQ, a toggle switch allows the user to specify that any *component* is present at a known, fixed activity. Also, separate menus are provided for inputting information about solids and gases that are in equilibrium with the solution, and about redox and adsorption reactions that the user wants the program to consider. These options, in effect, distinguish among Type 3a, 3b, and 3c *species*.

Type 4 *species* are solids that, based on a current, interim calculation, are expected to be present once the system reaches equilibrium. This group is closely related to Type 5 *species,* which are solids that might be present in the equilibrated system but whose presence is considered unlikely based on the current calculations. As the program iterates, it constantly re-evaluates the likelihood that each potential solid will be present at equilibrium, shuttling each solid between the Type 4 and Type 5 groups as appropriate. If, once the iterations are complete and all the relevant equations have been satisfied, a solid ends up as a Type 4 *species,* that solid is present in the equilibrium system; if it ends up as a Type 5 *species,* it is not present in the system. Type 4 and Type 5 *species* are sometimes named precipitated solids and dissolved solids, respectively.

Type 4 *species* differ from solids that are input as Type 3b *species* in that Type 4 *species* represent solids that might or might not be present at equilibrium; if they are present, the solution is in equilibrium with them, but if not, the reaction relating the solid to solution is ignored (and the solid is switched to a Type 5 *species,* as described above). By contrast, if a solid is listed as a Type 3b *species,* the model calculations force equilibrium between the solid and the solution under all circumstances.

The Type 6 group is simply a placeholder for chemicals (or, in the case of adsorption reactions, the mathematical functions that are manipulated as pseudo-chemicals) whose final concentration is not relevant to the calculation of the equilibrium composition in the current problem. The program ignores Type 6 *species* when carrying out mass balances, but it nevertheless calculates the concentration of each of these *species*. That concentration is sometimes of interest to us, even though it is often not the concentration of any real species in the equilibrium solution. Type 6 *species* are usually referred to as species not considered.

GAS/LIQUID EQUILIBRIUM

7.1 OVERVIEW

Equilibrium between gases and aqueous solutions drives numerous important environmental processes, including the dissolution of gases in cloud droplets, the stripping of sulfur dioxide from flue gas, the dissolution of oxygen into water during biological treatment processes, the release of methane from solution in anaerobic treatment processes, and the carbonation/decarbonation of waters undergoing treatment, to name just a few. Systems in which a solution can exchange material with a gas phase are referred to as *open systems,* and those where such exchange is precluded are called *closed systems.*[1] If the gases of interest can undergo other reactions once they dissolve, they can affect solution pH, the solubility and behavior of metals, and biological productivity. In this chapter, we explore gas/liquid equilibrium, with special emphasis on how gases affect the pH and buffering intensity of solutions.

---

[1]In the chemical engineering literature, the term *open system* is often used to describe any system in which mass crosses the system boundaries. In that case, if the system is defined as including only the liquid phase, then any exchange of material with a gas or solid, or any entry or departure of water across the system boundaries, makes the system open. The definition provided here is widely used in the environmental science and engineering literature. Although truly closed environmental systems are rare, the approximation that a system is closed is often useful.

7.2 BASIC CONCEPTS AND TERMINOLOGY FOR GAS TRANSFER REACTIONS

7.2.1 GAS-PHASE CONCENTRATIONS AND THE IDEAL GAS LAW

Concentrations in solution are usually expressed in units of mass per volume (mg/L, μg/L, or mol/L). The same units are often used for gas-phase concentrations, but units of partial pressures, mole fractions, or mass/volume expressed as micrograms per cubic meter are also common. In applications where the gas-phase concentrations are low and the primary focus is on the liquid phase, it is conventional to use units for gas-phase composition that are similar to those used for liquid-phase concentrations, e.g., μg/L or μg/m^3. Regulations setting limits for concentrations of gaseous pollutants in ambient air are commonly specified in micrograms per cubic meter.

Under normal environmental conditions, all gases share certain properties that relate their molar concentration and the pressure that they exert on their surroundings to the ambient temperature. These relationships are expressed concisely by the ideal gas law:

$$c_{G,i} = \frac{n_i}{V} = \frac{P_i}{RT} \tag{7.1}$$

where $c_{G,i}$ is the gas-phase concentration of i in moles per unit volume, n_i is the number of moles of i in the gas phase, V_G is the gas volume, P_i is the pressure exerted by i in the gas phase (the *partial pressure* of i), T is temperature, and R is the universal gas constant. At 25°C, RT is 24.15 bar-L/mol.[2]

Because of the broad applicability of the ideal gas law, gas-phase concentrations are frequently expressed in terms of partial pressures, rather than directly in mass/volume units. According to Equation (7.1), the partial pressure exerted by any ideal gas i, regardless of its chemical identity, is given by

$$P_i = RTc_{G,i} \tag{7.2}$$

The total pressure exerted by all the gases in a system is the sum of the partial pressures exerted by the individual gases. Therefore, for a gas phase containing N different species

$$P_{\text{tot}} = \sum_{j=1}^{N} P_j = RT \sum_{j=1}^{N} c_{G,j} \tag{7.3}$$

$$\frac{P_i}{P_{\text{tot}}} = \frac{RTc_{G,i}}{RT \sum_{j=1}^{N} c_{G,j}} = \frac{c_{G,i}}{\sum_{j=1}^{N} c_{G,j}} \equiv y_i \tag{7.4}$$

[2] The SI unit for expressing pressure is the *bar*, where 1 bar = 100 kPa. These units replace *atmospheres*, which had long been the conventional unit for expressing gas-phase pressures (1 bar = 0.987 atm). Gas-phase pressures are still reported in atmospheres in many engineering publications.

y_i is the dimensionless mole fraction of i in the gas phase, defined as follows:

$$y_i \equiv \frac{\text{moles of } i \text{ in gas phase}}{\text{total moles of all species in gas phase}} \qquad \textbf{(7.5)}$$

The mass concentration of i in the gas phase, which we represent as $c'_{G,i}$ (mass/volume), is the product of the molar concentration and the molecular weight (MW). Thus, the relationships among partial pressure P_i, mole fraction y_i, molar concentration $c_{G,i}$, and mass concentration $c'_{G,i}$ of a species i in the gas phase can be summarized as follows:

$$P_i = y_i P_{\text{tot}} = RTc_{G,i} = \frac{RTc'_{G,i}}{\text{MW}_i} \qquad \textbf{(7.6)}$$

Note that the mole fraction is not a direct measure of the volumetric concentration of a constituent in the gas phase, but rather must be multiplied by P_{tot} to obtain the concentration. Nevertheless, the use of mole fractions to represent gas-phase concentrations is fairly common, especially in the chemical engineering literature. This usage is almost always associated with the implicit assumption that $P_{\text{tot}} = 1$ bar, an assumption that is often justifiable but might be inappropriate in some cases. For instance, in relatively deep tanks or natural water bodies, the hydrostatic pressure might increase the total pressure significantly above atmospheric levels.

7.2.2   The Chemical Activity of Gas-Phase Constituents

By analogy with solutes, the chemical activity of a gaseous species i can be computed as the product of an activity coefficient and the concentration of i normalized to the standard state concentration:

$$a_i = \gamma_i \frac{c_i}{c_{i,\text{std. state}}} \qquad \textbf{(7.7)}$$

However, because of the nature of the gas phase, the activity coefficients of gas-phase constituents are virtually always assumed to be 1.0. The justification for this assumption hearkens back to the explanation of the activity coefficient given in Chapter 1. Specifically, the role of the activity coefficient is to account for any difference in the reactivity of a molecule of i between the actual environment of interest and the reference state environment.

For gases, the reference state is typically defined as an ideal gas at standard temperature and pressure (STP, i.e., 25°C and 1.0 bar). Molecules of an ideal gas are modeled as hard spheres that fly about unencumbered by one another except for fantastically brief periods when two molecules collide, change direction, and separate again. That is, for systems at normal temperatures and pressures, it would be fair to say that at any instant the overwhelming majority of gas molecules are unaware of one another's presence. As a result, the reactivity of a gas molecule is independent of the gas-phase composition and the total pressure over the range of interest in environmental systems, and gas-phase activity coefficients are always very close to 1.0.[3] Applying this assumption, Equation (7.7) can be simplified to

$$a_i = 1.0 \frac{c_i}{c_{i,\text{std. state}}} = \frac{c_i}{c_{i,\text{std. state}}} \tag{7.8}$$

Furthermore, the standard state concentration can be defined to have a magnitude of 1.0 and units that correspond to whatever units we wish to use to quantify the gas-phase concentration. If the standard state concentration is defined in this way, Equation (7.7) can be simplified even further, yielding

$$a_i = \frac{c_i}{c_{i,\text{std. state}}} = \frac{c_i\,(\text{any reasonable units})}{1.0\,(\text{same units as numerator})} = c_i\,(\text{magnitude, no units}) \tag{7.9}$$

According to Equation (7.9), the activity of a gas-phase species i has the same numerical value as the gas-phase concentration of i; a_i and c_i differ only in that c_i has units, while a_i is dimensionless, consistent with thermodynamic conventions. However, to interpret a reported value of a_i correctly, it is necessary to know the units being used to define the standard state concentration. (That is, does an activity of 10^{-2} refer to a concentration of 10^{-2} bar or $10^{-2}\ \mu g/m^3$?) Because gas-phase concentrations are commonly reported using a variety of different units, it is necessary to state the units being used to define the standard state concentration whenever a gas-phase activity is reported. This situation contrasts with that of solutes, for which a dimensionless value of activity is always understood to be based on a concentration of $1.0\ M$ in the standard state.

The need to specify the units of the standard state concentration for gases is often met by reporting the activity and the concentration units together; i.e., the activity of a gaseous species is often reported as x bar, or $y\ \mu g/L$. The correct way to interpret these values is that the activity of the species is x or y, respectively, and is dimensionless; the units should be interpreted as a separate "addendum" that indicates the units of c_i in the standard state. In other words, even though the activity of a gaseous species is commonly written in a way that looks exactly like a concentration, it is formally not a concentration, but rather a dimensionless value of activity plus an addendum. Not surprisingly, this subtlety has not always been appreciated by writers or readers. In truth, misinterpreting a gas-phase concentration as an activity or vice versa does not usually lead to any error in the numerical analysis of a system. Still, there is value in

[3] Reactivity does depend on temperature, but that effect is modeled directly, rather than as a change in activity coefficient.

adhering rigorously to the thermodynamic principle that activities are dimensionless, and for that reason the correct interpretation of reported gas-phase activities is important.[4]

7.2.3   QUANTIFYING GAS/LIQUID EQUILIBRIUM: HENRY'S LAW

The equilibrium reaction for a molecule transferring between a solution and a gas phase can be written with the dissolved molecule as either the reactant or the product:

$$A \cdot nH_2O \leftrightarrow A(g) + nH_2O \qquad \textbf{(7.10)}$$

$$A(g) + nH_2O \leftrightarrow A \cdot nH_2O \qquad \textbf{(7.11)}$$

In the above reactions, $A \cdot nH_2O$ represents a molecule of A dissolved in the aqueous solution, i.e., surrounded by water molecules, and $A(g)$ represents a molecule of A in the gas phase. Usually, $A \cdot nH_2O$ is abbreviated simply as $A(aq)$. The transfer of a molecule from solution to a gas phase is called *volatilization* or *stripping,* and the transfer from a gas phase to solution is called *absorption*. Qualitatively, the tendency of a molecule to reside in the gas phase is referred to as its *volatility.*

In some cases, dissolved A molecules form strong enough bonds with water that a water molecule is actually considered to be part of the dissolved species. For instance, $CO_2(aq)$ and $SO_2(aq)$ are often written as H_2CO_3 and H_2SO_3, respectively. Although there are times when it is important to know whether a water molecule is chemically bonded or just adjacent to a dissolved gas molecule, the distinction between these two situations is usually unimportant for our purposes, and we can consider them identical. As shown in Example 3.3, at equilibrium and 25°C, the concentration of $CO_2(aq)$ is a few hundred times as large as that of H_2CO_3; but these two species are almost always treated as being identical, with the sum of their concentrations often written as $H_2CO_3^*$.

The equilibrium constant characterizing gas/liquid equilibrium is called the Henry's law constant. It is unfortunate, but a fact of life, that two opposing conventions are in widespread use regarding the definition of this constant. In the environmental engineering community, Henry's law is usually defined by reference to reaction (7.10), using the letter H to represent the equilibrium constant. This terminology is used almost universally by those studying the stripping of trace organic contaminants in water and wastewater treatment operations. On the other hand, researchers studying natural aquatic systems tend to define Henry's law by reference to reaction (7.11), especially when describing the equilibrium of important atmospheric gases such as O_2 and CO_2. In the literature of that field, Henry's constant is most often represented by the symbol K_H. When these conventions

---

[4]This situation is identical to that discussed in Chapter 1, where it was pointed out that the equation $a_i = \gamma_i c_i$ is numerically correct, but dimensionally incorrect; the correct relationship is $a_i = \gamma_i c_i / c_{\text{std. state}}$, but because the numerical value of $c_{\text{std. state}}$ is 1.0, the shorter form of the equation is commonly presented.

are used, $H = 1/K_H$. However, even within each body of literature, the conventions have not been universally adopted, so it is important to always be aware of which convention a particular author is following. In this text, Henry's constant is defined based on reaction (7.10), so that the higher Henry's constant, the more favorable it is for the species to reside in the gas phase, and the less favorable it is for the species to be dissolved.

Since the activity of water is very close to 1.0, the equilibrium constant for reaction (7.10) can be written as follows:

$$H = \left. \frac{\{A(g)\}}{\{A(aq)\}} \right|_{eq} \tag{7.12}$$

where the subscript eq has been added to emphasize that the ratio equals H only if the system is at equilibrium. Because only uncharged species can exist in the gas phase in significant concentrations, the activity coefficient of a gaseous species when it dissolves (i.e., $\gamma_{A(aq)}$) is typically very close to 1.0. Therefore, the relationship shown in Equation (7.12) is often presented as a ratio of chemical concentrations, rather than activities, as in Equation (7.13):

$$H = \frac{\gamma_{A(g)}\{[A(g)]_{eq}/[A(g)]_{\text{std. state}}\}}{\gamma_{A(aq)}\{[A(aq)]_{eq}/[A(aq)]_{\text{std. state}}\}} = \left. \frac{[A(g)]}{[A(aq)]} \right|_{eq} \tag{7.13}$$

In the middle expression in Equation (7.13), the values of the activity coefficients in both phases and the standard state concentrations in both phases are crossed out because they typically have numerical values of 1.0 and therefore have no effect on the value of H. However, the concentration terms do have the effect of canceling the units of $[A(g)]$ and $[A(aq)]$.

As is done with gas-phase activities, when values of Henry's constants are reported, the units being used to define the standard state concentrations are commonly shown as an addendum to the constant. For example, Henry's constant for CO_2 is commonly reported as $10^{1.48}$ bar/(mol/L). Again, the correct interpretation is that Henry's constant for CO_2 has a value of $10^{1.48}$ (dimensionless) when the standard state concentration for the gas is 1.0 bar and that for dissolved CO_2 (i.e., H_2CO_3) is 1.0 mol/L. As is the case for gas-phase activities, the distinction between the two parts of H is often overlooked, and much of the literature represents Henry's constants as actually having dimensions. Some typical sets of standard state units used when reporting H values are shown in Table 7.1.

The ways of expressing Henry's law shown in the first three rows of Table 7.1 are widely used for applications in environmental engineering and science; the expressions in the first row are most frequently used to describe equilibria of volatile organic compounds (VOCs), while those in the second and third rows are more common for characterizing gas/liquid equilibria of species that are present in natural, unpolluted systems. The expressions in the fourth and fifth rows are commonly used in the chemical engineering literature. However, as noted above, no convention is universally applied in any particular field.

Table 7.1   Units commonly used for standard state concentrations when reporting Henry's law constants

Dimensions of Aqueous-Phase Standard State Concentration	Dimensions of Gas-Phase Standard State Concentration	Implied Dimensions of $H^\dagger$
Mass concentration (e.g., micrograms of i per liter of solution)	Gas-phase mass concentration (e.g., micrograms of i per liter of gas)	$\dfrac{\text{L liquid}}{\text{L gas}}\left(\dfrac{L_L}{L_G}\right)$
Mass concentration (milligrams of i per liter of solution)	Pressure (e.g., bar)	$\dfrac{\text{bar-}L_L}{\text{mg } i}$
Molar concentration (e.g., moles of i per liter of solution)	Pressure (e.g., bar)	$\dfrac{\text{bar-}L_L}{\text{mol } i}$
Molar concentration (e.g., moles of i per liter of solution)	Gas-phase mole fraction (moles of i per total moles of gas)	$\dfrac{L_L}{\text{mol of gas}}$
Aqueous-phase mole fraction (moles of i per mole of solution)	Gas-phase mole fraction (moles of i per total moles of gas)	$\dfrac{\text{mol of solution}}{\text{mol of gas}}$

†The symbols L_L and L_G refer to volumes (liters) in the liquid and gas phases, respectively.

Bubbling a gas through water can be an effective way to transfer some chemicals into solution. In this photograph, ozone in being bubbled through a column of potable water. The ozone that dissolves acts as a powerful disinfectant.

| Tom Pantages.

Example 7.1

Henry's constant for oxygen at 20°C is 0.73 bar O_2/(mol O_2/m$_L^3$). This constant would normally be written simply as 0.73 bar/(mol/m³). The designation O_2 has been added to emphasize that the numerator refers to the partial pressure of $O_2(g)$ (as opposed to the total pressure), and the denominator refers to the concentration of O_2 in solution (as opposed to the total number of moles per unit volume in the solution). The need for this distinction is made clear below.

     Express Henry's constant, using the following dimensions for the standard state concentrations in the gas and liquid, respectively.

a. bar O_2 and mol O_2/L_L
b. bar O_2 and mg O_2/L_L
c. mg O_2/L_L and mg O_2/L_G
d. Mole fraction in liquid and mole fraction in gas (for this part, assume that the molar density of solution is 55.6 mol/L and that the total pressure is 1.0 bar.)

Solution

The conversions among various ways of expressing gas-phase concentrations are described in Equation (7.6), and the conversion between mass/volume and mole/volume dimensions in solution simply involves division by the molecular weight. Therefore, the value of Henry's constant in the different ways requested can be computed as follows:

a. $\left(0.73\dfrac{\text{bar } O_2}{\text{mol } O_2/\text{m}_L^3}\right)\left(1000\dfrac{L_L}{\text{m}_L^3}\right) = 730\dfrac{\text{bar } O_2}{\text{mol } O_2/L_L}$

b. $\left(0.73\dfrac{\text{bar } O_2}{\text{mol } O_2/\text{m}_L^3}\right)\left(\dfrac{\text{mol } O_2}{32{,}000 \text{ mg } O_2}\right)\left(\dfrac{1000 \text{ } L_L}{\text{m}_L^3}\right) = 0.023\dfrac{\text{bar } O_2}{\text{mg } O_2/L_L}$

c. The value of H for implied units of mg O_2/L_L for the liquid-phase concentration and bars for the gas-phase concentration was computed in part (b). To determine H for implied units of mg O_2/L in both phases, we can start with the result in part (b) and just convert the units for the gas-phase concentration. A partial pressure in bars can be converted to milligrams per liter by dividing by RT and multiplying by the molecular weight. Therefore, the corresponding value of H is given by

$$\dfrac{\left(0.023\dfrac{\text{bar } O_2}{\text{mg } O_2/L_L}\right)\left(\dfrac{32{,}000 \text{ mg } O_2}{\text{mol } O_2}\right)}{\left(0.82\dfrac{\text{bar } O_2\text{-}L_G}{\text{mol } O_2\text{-K}}\right)(293 \text{ K})} = 30\dfrac{\text{mg } O_2/L_G}{\text{mg } O_2/L_L} = 30\dfrac{L_L}{L_G}$$

d. The result in part (a) indicates that an $O_2(g)$ partial pressure of 1 bar is in equilibrium with $\dfrac{1}{730}$ mol O_2/L in solution. If the total pressure is 1 bar, then such a gas has a mole fraction of $O_2(g)$ equal to 1.0. The mole fraction of O_2 in the solution is

$$\dfrac{\dfrac{1}{730}\text{ mol } O_2}{55.6 \text{ total mol in solution}} = 2.46 \times 10^{-5}\dfrac{\text{mol } O_2}{\text{mol solution}}$$

Henry's constant with implied concentrations given in mole fractions in both phases is therefore

$$\dfrac{1 \text{ mol } O_2/\text{mol gas}}{2.46 \times 10^{-5} \text{ mol } O_2/\text{mol solution}} = 4.01 \times 10^4 \dfrac{\text{mol solution}}{\text{mol gas}}$$

When the dimensions shown in the first row of Table 7.1 are used to describe gas/liquid equilibrium, the same units are used for mass and volume in both phases. Many authors refer to the associated Henry constant as dimensionless. In doing so, they imply that the units being used to define the standard state concentrations are the same for the gaseous and dissolved species. This implied identity is, unfortunately, incorrect; as shown in the table, the volumes of interest are V_L in the numerator and V_G in the denominator, and the distinction between these two terms turns out to be important.

The best way to demonstrate this point is by comparison of the apparent units in the first and last rows of Table 7.1, corresponding to parts (c) and (d) of Example 7.1. Like those in the first row, the units in the last row might be referred to as dimensionless, because the concentration in each phase is given in moles of i per mole of material in the bulk phase. However, as is apparent in parts (c) and (d) of the example, the value of Henry's constant based on mass/volume units in both phases is different from the value based on mole fractions. As a result, ambiguity arises when either of these forms of Henry's constant is reported as dimensionless, because it is not clear which "dimensionless" constant is being referred to. In the environmental engineering literature, references to dimensionless Henry constants are very common, and they almost always imply that both the gas- and liquid-phase activities are based on standard state concentrations in mass/volume units (mg/L, μg/L, or mol/L). However, in the chemical engineering literature, that generalization does not necessarily apply. To avoid this ambiguity, units of L_L/L_G or moles of solution per mole of gas should always be used when Henry's constant is reported in these ways.

The numerical value of H corresponding to implied units of L_L/L_G is particularly convenient for getting a sense of the physical significance of Henry's constant, based on the following analysis. Consider a closed vial containing species A equilibrated between the aqueous and gas phases. The ratio $[A(g)]/[A(aq)]$ in such a system has a numerical value equal to H and units corresponding to those used to define the standard state concentrations:

$$H\left(\frac{L_L}{L_G}\right) = \frac{[A(g)]}{[A(aq)]} = \frac{m_G/V_G}{m_L/V_L} \tag{7.14}$$

If the same mass of A is in each phase, then $m_G = m_L$, so $H = V_L/V_G$. That is, when the standard state concentrations used to express Henry's constant are mass/volume in both phases, H equals the liquid-to-gas volume ratio that causes the same mass of the constituent to be in each phase at equilibrium. For example, trichloroethylene (TCE), an organic contaminant found in many groundwater systems, has a Henry constant of 0.41 L_L/L_G. One interpretation of this value is that, at equilibrium, each liter of gas contains the same mass of TCE as is present in 0.41 L of solution. On the other hand, Henry's constant for benzene is only 0.22 L_L/L_G, so (again, at equilibrium) each liter of gas contains as much benzene as 0.22 L of solution. Thus, TCE is more volatile than benzene, and a higher Henry constant corresponds to greater volatility.

Although Henry's law fully characterizes the relative activities of a species in a gas and a solution at equilibrium, equilibration of gases with solutions typically is a much slower process than equilibration of acid/base reactions or of many other reactions that involve only dissolved species. As a result, it is not always appropriate to assume that equilibrium conditions apply to a system. This difference can be attributed in part to the fact that the equilibration rate in gas/liquid systems depends not only on the extent of chemical disequilibrium, but also on physical factors such as the amount of interfacial area available for gas transfer and the rate at which the gas and liquid near the interface are mixed with the bulk fluids. We will not concern ourselves here with the rate at which gas/liquid systems approach equilibrium; nevertheless, it is important to appreciate the fact that such systems are often not equilibrated in solutions of interest.

Recall that the activity quotient Q for a reaction is defined as the ratio of terms corresponding to the equilibrium constant, but for a system that is not necessarily at equilibrium. Thus, for reaction (7.10), Q is given by the following ratio:

$$Q = \frac{\{A(g)\}}{\{A(aq)\}}\bigg|_{\substack{\text{in system, not}\\ \text{necessarily at}\\ \text{equilibrium}}} \qquad \textbf{(7.15)}$$

If $Q < H$, the aqueous solution is said to be *supersaturated* with the gas, and molecules of A will spontaneously transfer from solution to the gas phase. If $Q > H$, the solution is said to be *undersaturated,* and molecules of A will dissolve from the gas phase into solution. These conditions are characterized schematically in Figure 7.1.

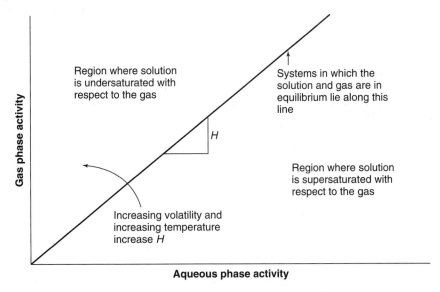

Figure 7.1 Schematic illustration of the relationship described by Henry's law, showing regions where the solution is undersaturated or supersaturated, and the effect of temperature on Henry's law constant.

7.2.4 PHYSICAL AND CHEMICAL FACTORS AFFECTING THE HENRY LAW CONSTANT

Recall the general discussion in Chapter 1 regarding factors that affect dissolution of molecules in water and pointing out that the tendency to dissolve is favored by the formation of strong A-to-H_2O bonds, and opposed by formation of strong A-to-A bonds in the undissolved phase. Because molecules in the gas phase are assumed to interact negligibly with one another (A-to-A bonds can be ignored), the relative tendency of different gases to dissolve can be related primarily to the strength of A-to-H_2O bonds.

For most molecules of interest to us, A-to-water bonds are primarily electrostatic. These bonds form as a result of the attraction of charged portions of the solute molecules to the oppositely charged portions of the water dipole. Bonds between solutes that carry a net charge (ions) and water are extremely strong, making all ions very hydrophilic and effectively precluding their existence in the gas phase. As a result, Henry's constants of ions are all essentially zero. Henry's constants of neutral molecules are higher than those of ions, but are still small if the molecules are polar, because such molecules can form relatively strong bonds to water. The less polar a molecule is, the weaker its bonds with water molecules and the larger its Henry constant tends to be. Also, although molecules that are non-polar have weak interactions with water molecules, those interactions are nevertheless favorable. Larger molecules have more favorable interactions with water than small molecules do, so (again, other factors being equal) large molecules tend to be less volatile than small ones.

As shown in Table 7.2, values of H for relatively small, non-polar molecules such as O_2, H_2, N_2, CO, and CH_4 are on the order of 1000 bar-L/mol. More polar

Table 7.2   Henry's constants of some environmentally important gases

Compound	$H^\dagger$	Compound	$H^\dagger$
Nitrogen	1560	Hydrogen sulfide	9.8
Hydrogen	1260	Chloroform	4.0
Carbon monoxide	1050	Sulfur dioxide	0.81
Oxygen	790	Bromoform	0.70
Methane	776	Benzene	0.22
Ozone	107	Hydrogen cyanide	0.040
Carbon dioxide	28.8	Ammonia	0.017
Tetrachloroethylene	20	Acetic acid	0.0013
Trichloroethylene	11		

| †Based on standard state concentrations given in units of bars and moles per liter at 20°C.[5]

[5]Values from Stumm and Morgan (*Aquatic Chemistry*, Wiley, 1996), Langmuir (*Aqueous Environmental Geochemistry*, Prentice-Hall, 1997), Staudinger et al. [*Journal of the American Water Works Association 82*, 1, 73–79 (1990)], and references therein.

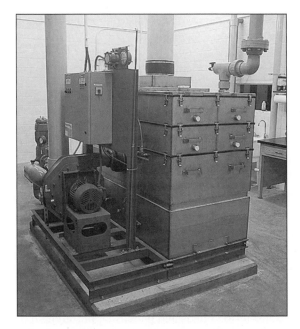

If the concentration of a dissolved species exceeds that in equilibrium with a gas phase that is in contact with the solution, the constituent will be driven into the gas phase. In this apparatus, radioactive, gaseous radon is stripped from a groundwater supply to prevent the radon from escaping into the air inside the building.

| Tom Pantages.

molecules such as CO_2 and H_2S have H values that are one to two orders of magnitude smaller, and ammonia, which is quite polar, has a Henry constant of only 0.017 bar-L/mol.

As we might expect from everyday experience with evaporation processes, transfer of molecules from solution to a gas phase is sensitive to temperature, with volatility increasing dramatically as temperature increases. This trend can be modeled using the van't Hoff equation (developed in Chapter 2) and indicates that the standard enthalpy of volatilization ($\Delta \overline{H}^{\circ}_{aq \to gas}$) of all gases is positive, i.e., volatilization is endothermic.

Example 7.2

Henry's law constant for oxygen is 588 bar-L/mol at 10°C. What is the concentration of oxygen dissolved in the epilimnion (surface layer) of a lake at that temperature if the water has equilibrated with the atmosphere? Estimate the standard enthalpy of volatilization of oxygen based on a comparison of this result with the value for $H_{O_2(g)}$ at 20°C given in Example 7.1.

Solution

The partial pressure of $O_2(g)$ in the atmosphere is 0.21 bar. The equilibrium dissolved oxygen concentration can be computed from Henry's law as follows:

$$[O_2(aq)] = \frac{P_{O_2(g)}}{H_{O_2}} = \frac{0.21 \text{ bar}}{588 \text{ bar/(mol/L)}} = 3.57 \times 10^{-4} \frac{\text{mol}}{\text{L}} = 11.4 \frac{\text{mg}}{\text{L}}$$

The enthalpy of volatilization can be estimated from the van't Hoff equation, which was presented in Chapter 2 and is repeated below:

$$\ln \frac{K_{eq}|_{T_2}}{K_{eq}|_{T_1}} = \left(\frac{\Delta \overline{H}_r^{\circ}}{R} \frac{1}{T_1} - \frac{1}{T_2}\right)$$

Rearranging the equation, we obtain

$$\Delta \overline{H}_r^{\circ} = R \frac{T_1 T_2}{T_2 - T_1} \ln \frac{K_{eq}|_{T_2}}{K_{eq}|_{T_1}}$$

Plugging in the given values of K_{eq} and T (in kelvins), we compute the standard enthalpy of the volatilization reaction to be $+14.9$ kJ/mol.

The volatility of a molecule can also be affected by other dissolved species. For instance, Henry's constant tends to increase with increasing salt concentration (i.e., with increasing ionic strength). In environmental engineering, this effect has been investigated most thoroughly with respect to equilibrium between gaseous and dissolved oxygen, for which the trend is apparent in Figure 7.2.

The basis for this effect is that the salt ions help orient the water molecules, forming strong water-to-ion bonds and increasing the strength of water-to-water

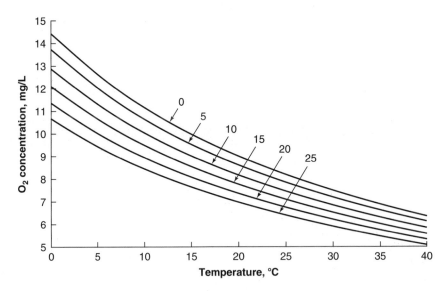

Figure 7.2   The solubility of oxygen in water as function of temperature and salt content. The numbers on the graph refer to the chloride concentration in grams per kilogram of solution. For reference, the Cl^- concentration in seawater is 19.3 g/kg. The data are for solutions containing that concentration of Cl^- and other salts in the same ratio to Cl^- as their ratio in seawater.

After *Standard Methods for Examination of Water and Wastewater*, 20th ed., American Public Health Association, 1998.

bonds. Dissolved gas molecules disrupt this structure. The higher the salt content, the greater the orienting force is, and the greater the tendency to exclude the "disruptive" gas molecules from the environment.

Other solutes can affect the solubility of gases in ways that are difficult to predict. Conventionally, in the waste treatment literature, the ratio of the oxygen saturation concentration in a particular water to the corresponding concentration in clean water at the same temperature is given the symbol beta (β):

$$c_{O_2,eq,actual} = \beta c_{O_2,eq,clean} \qquad \textbf{(7.16)}$$

7.2.5   HENRY'S LAW AND AQUEOUS-PHASE SPECIATION

As noted above, only neutral molecules can exist in the gas phase, so Henry's law describes equilibrium between those molecules and their dissolved neutral counterparts. For example, Henry's law describes the equilibria between $CO_2(aq)$ and $CO_2(g)$, $H_2S(aq)$ and $H_2S(g)$, and $NH_3(aq)$ and $NH_3(g)$. Since, for our purposes, $CO_2(aq)$ is identical to $H_2CO_3^*$, it is equally valid to say that Henry's law relates the concentrations of $H_2CO_3^*$ in solution and $CO_2(g)$ in a gas phase. On the other hand, Henry's law does not describe the relationship between gaseous CO_2 and dissolved species which result when dissolved CO_2 undergoes a reaction. Thus, it does not directly relate $CO_2(g)$ to HCO_3^- or total dissolved carbonate species $TOTCO_3$. One can, of course, derive such relationships if one wishes. For example, the equilibrium relationship between $CO_2(g)$ and HCO_3^- is given by

$$CO_2(g) + H_2O \leftrightarrow H_2CO_3(aq) \qquad K = 1/H \qquad \textbf{(7.17)}$$

$$H_2CO_3(aq) \leftrightarrow HCO_3^- + H^+ \qquad K = K_{a1} \qquad \textbf{(7.18)}$$

$$\overline{CO_2(g) + H_2O \leftrightarrow HCO_3^- + H^+ \qquad K = K_{a1}/H} \qquad \textbf{(7.19)}$$

Reaction (7.19) and the corresponding equilibrium constant are valid and represent the overall gas transfer reaction that dominates when solution pH is in the range $6.3 < pH < 10.3$. However, that equilibrium constant would not properly be referred to as Henry's law constant.

7.2.6   EFFECT OF GAS/LIQUID EQUILIBRATION
ON GAS-PHASE COMPOSITION

In many systems, the vastness of the gas phase means that transfer of gas molecules between it and solution has virtually no effect on the composition of the gas phase. This approximation is particularly common when the gas phase of interest is the bulk atmosphere. That is, the absorption of oxygen, carbon dioxide, nitrogen, etc., by a solution, or the release of these gases from solution, is assumed to have no effect on their partial pressures if the gas phase of interest is the bulk atmosphere. This approximation is sometimes described as an assumption that the gas phase is infinite (and that therefore its composition is unchangeable in response to finite additions or subtractions of a specific gas).

There are situations, of course, in which the assumption of an invariant gas-phase composition is not valid. For instance, gases present at extremely low concentrations in the atmosphere, such as SO_2 in areas where human influence is minimal, can be substantially removed from the atmosphere by contacting slightly alkaline salts onto which cloud water condenses.[6] Similarly, the concentrations of oxygen, carbon dioxide, and trace gases in a bubble rising through the water column of a biological treatment process might change significantly as the bubble ascends. In such a case, one can take the change in gas-phase composition into account by writing a mass balance on the gas phase and linking that to a mass balance on the solution as part of the analysis, as shown in the following example.

Example 7.3

One hundred mL of air is contacted with 1 L of a solution at 10°C in a closed vessel. The water is initially devoid of oxygen. What are the concentration of dissolved oxygen and the partial pressure of oxygen in the air once the system comes to equilibrium?

Solution

Because the air cannot be replenished, it is not clear that the equilibrium partial pressure of oxygen will be close to the initial value. We can write a mass balance on oxygen in the entire system (water plus air) and combine it with the equilibrium expression to determine the final system composition.

The unchanging number of moles of oxygen in the system is

$$n_{tot,O_2} = n_{gas,O_2}|_{init} = \frac{P_{O_2(g)}V}{RT}\bigg|_{init} = \frac{(0.21 \text{ bar})(0.1 \text{ L})}{[0.082 \text{ bar-L/(mol-°C)}](283°C)} = 9.04 \times 10^{-4} \text{ mol}$$

At equilibrium, this mass of oxygen is distributed between the gas and aqueous phases in accordance with Henry's law:

$$n_{O_2,tot} = n_{O_2,gas} + n_{O_2,solution} = \frac{P_{O_2(g)}V_{gas}}{RT} + \frac{P_{O_2(g)}}{H_{O_2}}V_{solution}$$

$$9.04 \times 10^{-4} \text{ mol} = \frac{P_{O_2(g)}(0.1 \text{ L})}{[0.082 \text{ bar-L/(mol-K)}](283 \text{ K})} + \frac{P_{O_2(g)}}{588 \text{ bar-L/mol}}(1.0 \text{ L})$$

$$P_{O_2(g)} = 0.15 \text{ bar} \qquad \{O_2(aq)\} = \frac{P_{O_2}}{H_{O_2}} = 8.2 \text{ mg/L}$$

Thus, in this case, the oxygen partial pressure declines by about 29% when the two phases equilibrate.

---

[6]This phenomenon, and the effect of gas/liquid equilibrium on the pH of precipitation, was addressed in a classic paper by Charlson and Rodhe, "Factors Controlling the Acidity of Natural Rainwater," *Nature* 295: 683–685, 1982.

7.3   Effect of Ionization of Dissolved Gases on Gas/Liquid Equilibrium

Next, consider the implications of Henry's law for a gaseous species such as CO_2, which can undergo an acid/base reaction in solution. The partial pressure of CO_2 in the atmosphere is approximately 3.5×10^{-4} bar ($10^{-3.46}$ bar), and its Henry constant is $10^{1.48}$ bar-L/mol. As a result, any solution in equilibrium with the normal atmosphere has a dissolved CO_2 concentration of $P_{CO_2(g)}/H_{CO_2}$, or 1.15×10^{-5} mol/L (i.e., $10^{-4.94}$ mol/L). Solutions that contain more than 1.15×10^{-5} mol/L H_2CO_3, such as carbonated beverages, are supersaturated with respect to atmospheric CO_2, causing them to release carbon dioxide when they are exposed to the atmosphere. Similarly, solutions containing less than 1.15×10^{-5} mol/L H_2CO_3 are undersaturated and absorb CO_2 if they are in contact with the atmosphere. We next explore how the acid/base reactions of dissolved carbonate species affect the overall equilibrium by developing the log C–pH diagram for the system.

Based on our knowledge that a solution in equilibrium with the atmosphere contains 1.15×10^{-5} mol/L H_2CO_3, a curve representing the H_2CO_3 concentration in the system is trivial to plot on the log C–pH diagram: $\log\{H_2CO_3\}$ must equal -4.94 at all pH values. Any deviation from a value of -4.94 would not satisfy Henry's law and would therefore not represent an equilibrium system. However, this line does not fully characterize the carbonate speciation in the system, because in any solution containing H_2CO_3, the deprotonated species HCO_3^- and CO_3^{2-} must also be present, in concentrations that satisfy the K_a expressions. When we combine those expression with the fixed activity of H_2CO_3, we obtain the following equations for the activities of the deprotonated species:

$$\{HCO_3^-\} = \frac{K_{a1}\{H_2CO_3\}}{\{H^+\}} = \frac{(10^{-6.35})(10^{-4.94})}{\{H^+\}} = \frac{10^{-11.29}}{\{H^+\}}$$

$$\log\{HCO_3^-\} = -11.29 - \log\{H^+\} = -11.29 + pH \qquad \textbf{(7.20)}$$

$$\{CO_3^{2-}\} = \frac{K_{a1}K_{a2}\{H_2CO_3\}}{\{H^+\}^2} = \frac{(10^{-6.35}10^{-10.33})(10^{-4.94})}{\{H^+\}^2}$$

$$\log\{CO_3^{2-}\} = -16.68 - 4.94 - 2\log\{H^+\} = -21.62 + 2\,pH \quad \textbf{(7.21)}$$

Equations (7.20) and (7.21) indicate that the curves representing bicarbonate and carbonate ion on the log C–pH diagram are straight lines with slopes of $+1$ and $+2$, respectively. The diagram, including lines for H^+ and OH^-, is plotted in Figure 7.3. Note that the lines for H_2CO_3 and HCO_3^- intersect at pH $= pK_{a1} = 6.35$. We should have anticipated this result: the equality of the concentrations of these two species at pH $= pK_{a1}$ is a consequence of the equilibrium between them and must apply, regardless of whether the solution is in equilibrium with a gas phase (as in this diagram) or not (as in diagrams that we considered previously).

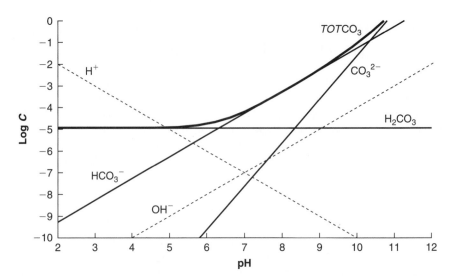

Figure 7.3   Log C–pH diagram for a solution in equilibrium with atmospheric $CO_2(g)$.

Similarly, the HCO_3^- and CO_3^{2-} lines intersect at pH = pK_{a2} (= 10.33). Note also that, as in systems in which gas/liquid equilibrium is not considered, carbonic acid is the dominant carbonate-containing species at pH < 6.35, bicarbonate is dominant at 6.35 < pH < 10.33, and carbonate is dominant at pH > 10.33; and the slopes of the lines are always related by

$$\frac{d\{CO_3^{2-}\}}{d(pH)} = \frac{d\{HCO_3^-\}}{d(pH)} + 1 = \frac{d\{H_2CO_3\}}{d(pH)} + 2$$

Now consider the total concentration of dissolved carbonate-containing species in the solution as a function of pH. This total concentration is the summation of the concentrations of the three carbon-containing species and is approximately equal to the concentration of the dominant species at each pH. Unlike previous systems we have considered, the total dissolved carbon varies tremendously as a function of pH. This result is the consequence of the requirement that all the equilibria be satisfied simultaneously. Since the H_2CO_3 concentration does not depend on pH, the only way for the constraints of these equilibria to be met (i.e., for H_2CO_3 to be the dominant carbonate species at low pH and non-dominant at higher pH) is for both of the other carbonate-containing species to be present at concentrations less than 1.15×10^{-5} mol/L at pH < 6.35, and for one or both to be present at concentrations greater than 1.15×10^{-5} mol/L at pH > 6.35.

Put another way, in the systems considered in previous chapters, the total concentration of all carbonate species was fixed, so an increase in the concentration of any of the species required a decrease in the concentration of at least one of the others. By contrast, in the system equilibrated with a gas phase, the concentration of an individual species (H_2CO_3) is fixed, so an increase in the concentrations of the other species requires that the total concentration of dissolved carbon increase. In a real system that includes a solution that is physically in contact with the atmosphere, this increase is made possible by the dissolution of more and more CO_2 as the solution pH is increased.

As noted above, Henry constants always relate the concentration of dissolved neutral species to their activity in the gas. In the case of the carbonate system, the neutral dissolved species is an acid which can dissociate to form its conjugate base and a second, more basic species. The concentrations of these latter species therefore increase as pH increases. If the neutral, volatile species is a base, then it can be converted to more acidic species upon dissolution, and the concentrations of these acidic species decrease with increasing pH. For instance, a solution in equilibrium with ammonia gas ($H = 0.017$ bar-L/mol) at a partial pressure of 10^{-7} bar would contain $NH_3(aq)$ and NH_4^+ at concentrations given by the following equations:

$$\{NH_3(aq)\} = \frac{P_{NH_3}}{H_{NH_3}} = \frac{10^{-7}\,\text{bar}}{0.017\,\text{bar}/(\text{mol/L})} = 5.48 \times 10^{-6}\,\frac{\text{mol}}{\text{L}}$$

$$\log\{NH_3(aq)\} = -5.23$$

$$\{NH_4^+\} = K_a\{NH_3\}\{H^+\} = (10^{+9.25})(10^{-5.23})\{H^+\} = 10^{+4.02}\{H^+\}$$

$$\log\{NH_4^+\} = 4.02 + \log\{H^+\} = 4.02 - \text{pH}$$

A log C–pH diagram for this system is shown in Figure 7.4. It has some of the same features as the diagram characterizing equilibrium with carbon dioxide, but the total concentration of dissolved ammonia increases with decreasing pH instead of with increasing pH. As noted above, this result is a consequence of the fact that NH_3 is a base, whereas H_2CO_3 is an acid.

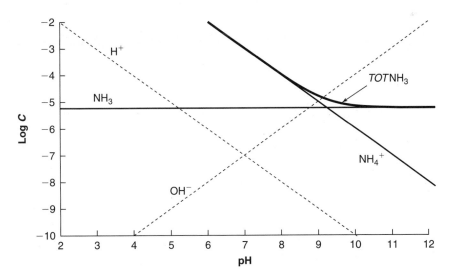

Figure 7.4   Log C–pH diagram for a solution in equilibrium with $NH_3(g)$ at a partial pressure of 10^{-7} bar.

7.3.1   THE pH OF SOLUTIONS IN EQUILIBRIUM WITH ACIDIC OR BASIC GASES

The necessary and sufficient condition for determining the equilibrium speciation of a solution is that the equations expressing the mass balances, the equilibrium relationships, and the charge balance, proton condition, or *TOT*H equation be satisfied. In problems that we have solved previously, we have typically been able to write a mass balance for each weak acid/base system in solution. If a solution is in equilibrium with a gas phase containing an acidic or basic gas, we have the same number of unknowns as if the gas were not there (one unknown corresponding to the concentration of each species in the system). However, it is not possible to write a mass balance specifying the total concentration of the species of interest, because we do not know how much gas will dissolve into or evolve out of the solution as the system equilibrates. This situation means we have one fewer mass balance than in the past. However, we also have one additional equilibrium expression that must be satisfied: the equilibrium specified by Henry's law. As a result, we still have the same number of equations as unknowns, and the equations can be solved to determine the solution composition at equilibrium, as shown in the following examples.

Example 7.4 | Compute the solution pH and speciation for a system containing pure water equilibrated with atmospheric CO_2.

Solution

The charge balance for the system is as follows:

$$\{H^+\} = \{OH^-\} + \{HCO_3^-\} + 2\{CO_3^{2-}\}$$

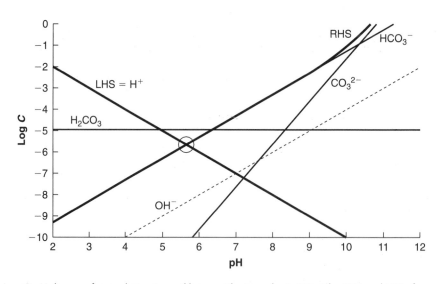

Log C–pH diagram for a solution in equilibrium with atmospheric CO_2. The RHS and LHS of the charge balance are shown in bold for a solution that has equilibrated with the gas phase and that contains no other acids or bases.

The log C–pH diagram for the system was shown in Figure 7.3 and is reproduced above, with the right- and left-hand sides of the charge balance highlighted. The equilibrium pH is 5.64, and the equilibrium concentrations of the carbonate species are $10^{-4.94}$ M, $10^{-5.64}$ M, and $10^{-10.30}$ M for H_2CO_3, and HCO_3^-, and CO_3^{2-}, respectively. Most of the dissolved carbonate is present as H_2CO_3, and $TOTCO_3$ is 1.38×10^{-5} M.

Compute the equilibrium composition of the system in Example 7.4, if 10^{-3} M NaAc is added to the solution. Assume that the solution remains in contact with the atmosphere and re-equilibrates with atmospheric CO_2 after the NaAc is added. Ignore volatilization of HAc. | **Example 7.5**

Solution

The charge balance for this system must be modified from that in the previous example to account for the presence of Na^+ and Ac^-:

$$\{Na^+\} + \{H^+\} = \{OH^-\} + \{Ac^-\} + \{HCO_3^-\} + 2\{CO_3^{2-}\}$$

A combined mass balance on Na and Ac yields the equality $\{Na^+\} = \{HAc\} + \{Ac^-\}$. When we substitute this equality into the charge balance, the equation becomes

$$2\{HAc\} + \cancel{\{Ac^-\}} + \{H^+\} = \{OH^-\} + \cancel{\{Ac^-\}} + \{HCO_3^-\} + 2\{CO_3^{2-}\}$$

$$\{HAc\} + \{H^+\} = \{OH^-\} + \{HCO_3^-\} + 2\{CO_3^{2-}\} \qquad \textbf{(7.22)}$$

A log C–pH diagram for the system is shown below. The lines for the carbonate species are the same as in the preceding example, and lines for the acetate species have been added. The RHS and LHS of Equation (7.22) are highlighted, indicating that the

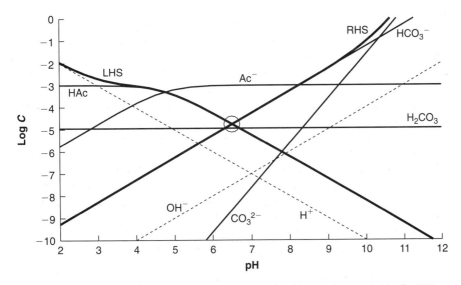

Log C–pH diagram for 10^{-3} M TOTAc that is equilibrated with atmospheric $CO_2(g)$. The RHS and LHS refer to Equation (7.22), which is the charge balance for a system in which the acetate is added as NaAc.

equilibrium pH is 6.50. The solution speciation is as follows:

$$\{H^+\} = 10^{-6.50} \qquad \{OH^-\} = 10^{-7.50}$$

$$\{H_2CO_3\} = 10^{-4.94} \qquad \{HCO_3^-\} = 10^{-4.76} \qquad \{CO_3^{2-}\} = 10^{-8.57}$$

$$\{HAc\} = 10^{-4.77} \qquad \{Ac^-\} = 10^{-3.01}$$

Note that the total dissolved carbonate in this solution (2.87×10^{-5} M) is more than twice as much as before the acetate was added (Example 7.4). The reason for this is that sodium acetate is a base, so when it is added, the pH of the solution increases. This pH increase, in turn, causes more CO_2 to dissolve. The subsequent dissociation of some of this CO_2 partially counteracts the effect of the acetate, so the pH does not increase as much as it would if the opportunity for re-equilibration with the atmosphere were not present.

Example 7.6

Compute the equilibrium composition of the solution in the previous example if it contacts air containing not only CO_2 at its normal partial pressure, but also NH_3 at a partial pressure of 10^{-7} bar.

Solution

The activity of dissolved $NH_3(aq)$ at equilibrium in this system is given by $P_{NH_3(g)}/H_{NH_3}$, or $10^{-5.23}$ ($= 5.88 \times 10^{-6}$) at all pH values. A horizontal line at this value can be added to the log C–pH diagram to represent $NH_3(aq)$. Then a straight line that has a slope of -1 and that intersects the $NH_3(aq)$ line at pH $= pK_a = 9.25$ can be drawn to represent NH_4^+. The diagram is shown below.

Rather than write a mass balance, we will solve for the speciation of this solution by writing the *TOT*H equation. If we choose Ac^-, H_2CO_3, and NH_3 as proton-neutral

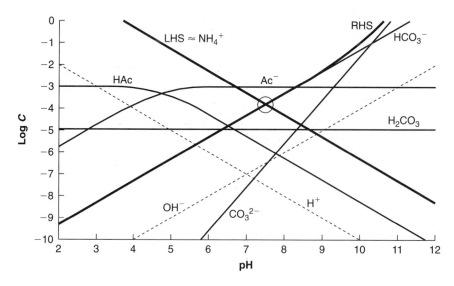

Log C–pH diagram for a solution containing 10^{-3} M NaAc and in equilibrium with a gas containing CO_2 at its normal atmospheric concentration and NH_3 at a partial pressure of 10^{-7} bar.

species, along with H_2O, the proton balance table is as follows:

	−2	−1	0	+1	Concentration
Equilibrium species		OH^-	H_2O Ac^-	H^+ HAc	
	CO_3^{2-}	HCO_3^-	H_2CO_3 NH_3 Na^+	NH_4^+	
Input species			H_2CO_3 NH_3 $NaAc$		? ? 10^{-3}

Since only proton-neutral species were used to make up the solution (dissolution of CO_2 is equivalent to adding H_2CO_3), all the terms contributing to the initial proton excess and deficiency (i.e., all terms contributing to $TOTH_{in}$) are zero, so the $TOTH$ equation is as follows:

$$TOTH_{in} = TOTH_{eq}$$

$$0 = \{H^+\} + \{HAc\} + \{NH_4^+\} - \{OH^-\} - \{HCO_3^-\} - 2\{CO_3^{2-}\}$$

$$\{H^+\} + \{HAc\} + \{NH_4^+\} = \{OH^-\} + \{HCO_3^-\} + 2\{CO_3^{2-}\}$$

The LHS and RHS of this equation are shown on the log C–pH diagram on p. 342. As would be expected considering that NH_3 is a base, the equilibrium pH is higher than in the previous example, equaling 7.62. Concentrations of all the dissolved species are as follows:

$$\{H^+\} = 10^{-7.62} \qquad \{OH^-\} = 10^{-6.38}$$

$$\{H_2CO_3\} = 10^{-4.94} \qquad \{HCO_3^-\} = 10^{-3.67} \qquad \{CO_3^{2-}\} = 10^{-6.38}$$

$$\{HAc\} = 10^{-5.86} \qquad \{Ac^-\} = 10^{-3.00}$$

$$\{NH_3\} = 10^{-5.23} \qquad \{NH_4^+\} = 10^{-3.67}$$

In this case, $TOTCO_3$ is $2.26 \times 10^{-4}\ M$, more than an order of magnitude greater than was dissolved before the bases (NaAc and NH_3) entered solution (as in Example 7.4).

Example 7.7

$10^{-3}\ M\ Na_2CO_3$ is added to the solution described in the preceding example. The solution remains in contact with the gas phase. What is the composition of the equilibrated solution? Does CO_2 enter the solution, leave it, or not exchange with the atmosphere in response to this perturbation?

Solution

To solve for the equilibrium speciation in the current system, we can use the same log C–pH diagram as in Example 7.6. The lines representing the species' concentrations in that figure are not affected by the addition of Na_2CO_3, because this addition does not alter any of the equilibria or mass balances used to develop the diagram. (Recall that no mass balance on carbonate species was used in preparing the diagram.) However, the addition of Na_2CO_3 does alter the $TOTH$ equation, because CO_3^{2-} is a two-proton-deficient species. Specifically,

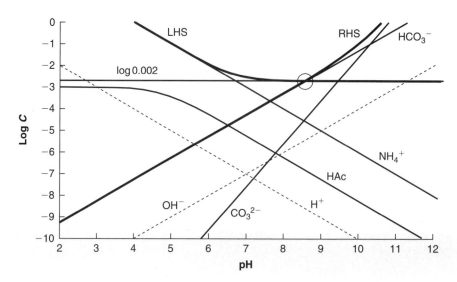

Log C–pH diagram for the system shown in Figure 7.6 after 10^{-3} M Na_2CO_3 is added to the solution, and the solution re-equilibrates with the gas phase.

$TOTH_{in}$ (and therefore $TOTH_{eq}$) becomes -2×10^{-3}, leading to the following $TOTH$ balance and proton condition.

$$-2 \times 10^{-3} = \{H^+\} + \{HAc\} + \{NH_4^+\} - \{OH^-\} - \{HCO_3^-\} - 2\{CO_3^{2-}\}$$

$$\{H^+\} + \{HAc\} + \{NH_4^+\} + 2 \times 10^{-3} = \{OH^-\} + \{HCO_3^-\} + 2\{CO_3^{2-}\}$$

The log C–pH diagram is shown again above with the revised LHS. The equilibrium solution composition in this case is

$$\{H^+\} = 10^{-8.58} \qquad \{OH^-\} = 10^{-5.42}$$

$$\{H_2CO_3\} = 10^{-4.94} \qquad \{HCO_3^-\} = 10^{-2.71} \qquad \{CO_3^{2-}\} = 10^{-4.46}$$

$$\{HAc\} = 10^{-6.82} \qquad \{Ac^-\} = 10^{-3.00}$$

$$\{NH_3\} = 10^{-5.23} \qquad \{NH_4^+\} = 10^{-4.63}$$

Now $TOTCO_3$ is 2.00×10^{-3} M, representing an increase of 1.87×10^{-3} M compared to the previous example. Of this increase, 1×10^{-3} M is attributable to the Na_2CO_3 that was added, and the remainder entered by dissolution of atmospheric CO_2. The increase in pH also causes NH_3 to exit solution, as evidenced by the fact that $TOTNH_3$ is less in this system than in the system before the Na_2CO_3 was added. Thus, in response to the addition of a base (Na_2CO_3), an acidic gas (CO_2) was absorbed into solution and a basic gas (NH_3) exited. These gas exchanges partially counteract the effect of the sodium carbonate on solution pH; in the absence of these gaseous constituents, the pH change due to addition of Na_2CO_3 would have been much more substantial.

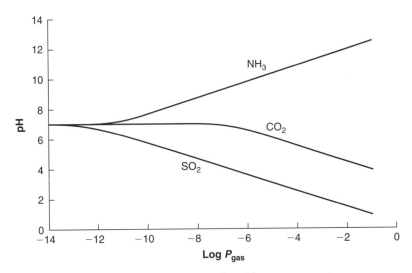

Figure 7.5 The effect of dissolution and acid/base reactions of various gases on the pH of pure water.

The equilibrium pH of solutions in equilibrium with various pressures of $CO_2(g)$, $SO_2(g)$, and $NH_3(g)$ are shown in Figure 7.5. SO_2 is much more acidic and more soluble than CO_2, so it has a much greater effect on pH, even when it is present at low pressures. The basicity constant of ammonia ($K_b = 10^{-4.75}$) is intermediate between the acidity constants of CO_2 and SO_2, as is its solubility. At a given pressure, it therefore affects the pH more than CO_2 and less than SO_2 (and, of course, its effect is in the opposite direction).

Sulfur dioxide emitted from the burning of fossil fuels can react with atmospheric oxygen and water to form sulfuric acid, which can later fall to the earth as acid precipitation. Such precipitation can dramatically increase the corrosion of statues, buildings, and other structures, in addition to damaging forests and disrupting poorly buffered aquatic ecosystems.

| Monika Andersson/Stock Boston.

7.3.2   THE PROTON CONDITION OR *TOT*H EQUATION FOR A SOLUTION IN EQUILIBRIUM WITH A GAS

In the preceding examples, the proton condition and *TOT*H equation were written with the neutral form of the volatile species as the reference species. The reason for this choice becomes obvious if we try to write these equations using a different reference species. For instance, say that we thought Ac^- and HCO_3^- would be the dominant species in the solution containing 10^{-3} M NaAc and in contact with the atmosphere (Example 7.5). We might then write the PC using those species as reference species, with the results shown below:

	−1	0	+1	Concentration
Equilibrium species	OH^- CO_3^{2-}	H_2O HCO_3^- Ac^-	H^+ H_2CO_3 HAc	
Input species		NaAc		10^{-3}
			H_2CO_3	?

$$TOTH_{in} = 0(10^{-3}) + (1)(?)$$

$$TOTH_{eq} = 1\{H^+\} + (1)\{H_2CO_3\} + (1)\{HAc\} - (1)\{OH^-\} - (1)\{CO_3^{2-}\}$$

Equating $TOTH_{in}$ with $TOTH_{eq}$ generates an equation that is of no use to us:

$$? = \{H^+\}_{eq} + \{H_2CO_3\}_{eq} + \{HAc\}_{eq} - \{OH^-\}_{eq} - \{CO_3^{2-}\}_{eq}$$

The question mark on the LHS arises because H_2CO_3 is a one-proton-excess species, so that each molecule of CO_2 that enters solution from the gas phase contributes one proton to *TOT*H, but we do not know how much CO_2 enters the solution as it equilibrates. Thus, it seems that we have to know how much CO_2 dissolves before we can write a meaningful *TOT*H equation to solve the problem; yet we have to solve the problem to know how much CO_2 dissolves! This dilemma can be resolved by choosing dissolved CO_2 (i.e., H_2CO_3) as a reference species. In that way, we ensure that the CO_2 that enters the solution by dissolution of gas or that leaves solution by volatilization of gas will have no excess or deficiency of protons, and the amount of gas transferred will not appear as a variable in the final equation. As was shown in Example 7.5, use of H_2CO_3 as a reference species allows the *TOT*H equation to be solved and the equilibrium speciation to be determined in this system.

Example 7.8 | In drinking water treatment operations, "hard" water containing unacceptably large concentrations of calcium is often "softened" by adding enough base and carbonate to precipitate much of the calcium as the mineral calcite, $CaCO_3(s)$. The operation is carried out at high pH, after which the solid is separated from solution. Often, the solution pH is then readjusted to near neutral by *recarbonation,* i.e., passing air through the solution to dissolve $CO_2(g)$.

Consider a softened solution at pH 10.5 that contains 7.40×10^{-5} M each of $TOTCO_3$ and TOTCa. If the solution is then equilibrated with air in a recarbonation step, how much $CO_2(g)$ dissolves (moles CO_2 per liter of solution), and what is the final pH? *Note:* pK_a for Ca^{2+} (to form its conjugate base $CaOH^+$) is 12.60, so we can assume that Ca^{2+} does not participate significantly in acid/base reactions during recarbonation.

Solution

At pH 10.5, the dissolved carbonate is comprised almost entirely of HCO_3^- and CO_3^{2-}; i.e., $\{H_2CO_3(aq)\}$ is negligible. At this pH, α_1 and α_2 for H_2CO_3 are 0.40 and 0.60, respectively, so $\{HCO_3^-\}$ and $\{CO_3^{2-}\}$ are 40% and 60% of $TOTCO_3$, or 4.44×10^{-5} and 2.96×10^{-5}, respectively.

The pH after equilibration with the atmosphere can be determined by drawing a log C–pH diagram for an open system in contact with atmospheric $CO_2(g)$ and identifying the pH at which the $TOTH$ equation is satisfied. To write a useful $TOTH$ equation, we must choose H_2CO_3 as the reference species for the carbonate system. As a result, the initial concentrations of $\{CO_3^{2-}\}$ and $\{HCO_3^-\}$ represent an initial net deficiency of protons. The proton condition table and corresponding equation are shown below.

	-2	-1	0	$+1$	**Concentration**
Equilibrium species	CO_3^{2-}	OH^- HCO_3^-	H_2O H_2CO_3 Ca^{2+}	H^+	
Initial and/or input species			H_2CO_3		?
		HCO_3^-			2.97×10^{-5}
	CO_3^{2-}				4.46×10^{-5}
			Ca^{2+}		7.43×10^{-5}

$$TOTH_{in} = (2.96 \times 10^{-5})(-1) + (4.44 \times 10^{-5})(-2)$$
$$+ (7.40 \times 10^{-5})(0) = -1.18 \times 10^{-4}$$

$$TOTH_{eq} = \{H^+\} - \{OH^-\} - \{HCO_3^-\} - 2\{CO_3^{2-}\}$$

$$\{H^+\} + 1.18 \times 10^{-4} = \{OH^-\} + \{HCO_3^-\} - 2\{CO_3^{2-}\}$$

The right- and left-hand sides of the equation are shown on the diagram on the next page, and their intersection indicates that the equilibrium pH is 7.34. At this pH, $TOTCO_3$ is 1.29×10^{-4}. The amount of $CO_2(g)$ that dissolves is $TOTCO_{3,final} - TOTCO_{3,init}$, or 5.5×10^{-5} moles per liter of solution.

7.3.3   BUFFERING IN SOLUTIONS EQUILIBRATED WITH WEAK ACID AND WEAK BASE GASES

Recall that pH buffers work by releasing H^+ (deprotonating) in response to an increase in pH and by acquiring H^+ (protonating) in response to a decrease in pH. The examples shown above indicate that acidification of a solution that is in contact with a gas phase causes acidic gases to exit solution and basic gases to enter solution. These processes, in effect, transport H^+ out of solution and drive species that consume H^+ into the solution, respectively. The response to addition of a base is exactly the opposite. As a result, gases that act as acids or bases when they dissolve can be very effective pH buffers.

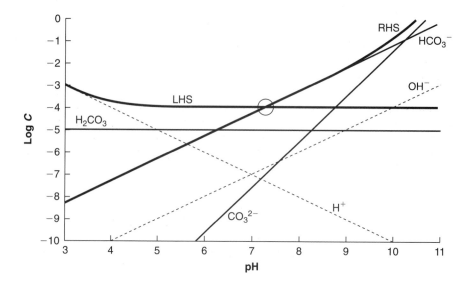

In fact, gases can provide buffering over a much wider pH range than do species that are always dissolved. This is because dissolved species buffer the solution pH only by protonating and deprotonating, so they are effective buffers if and only if the solution pH is near the pK_a. On the other hand, gases can buffer a solution by entering or leaving the dissolved phase, a process that can be significant under any condition where the equilibrium concentration of total dissolved gas is a strong function of pH.

For instance, in a system equilibrated with atmospheric CO_2, $TOTCO_3$ is sizable and depends strongly on pH at all pH values greater than about 7.5. Therefore, this gas serves as a good pH buffer at any pH > 7.5. Similarly, $NH_3(g)$ can serve as a good buffer at any pH < 9.0 if P_{NH_3} is large enough to cause $TOTNH_3$ to be significant. Keep in mind, though, that transfer of gas into or out of solution tends to be a much slower process than protonation/deprotonation of dissolved species, so the efficacy of buffering a system by a gas might not be as great as is implied by equilibrium considerations.

The buffer intensity of solutions equilibrated with a gas phase can be computed using the same concepts as described in Chapter 5, but the details of the buffer intensity equation must be modified to reflect the fact that the total dissolved concentration of the weak acid/base group changes when solution pH changes. Recall from Chapter 5 that if a solution contains a multiprotic weak acid/base group A, the concentration of OH^- that must be added to change the solution pH from pH_{init} to pH_{fii} can be represented as follows:

$$\Delta(c_{OH^-})_{added} = -(\{H^+\}_{fii} - \{H^+\}_{init}) + (\{OH^-\}_{fii} - \{OH^-\}_{init})$$

$$+ \sum_{i=1}^{n} i(\{H_{n-i}A^{i-}\}_{fin} - \{H_{n-i}A^{i-}\}_{init}) \tag{7.23}$$

In Chapter 5, we simplified the last term in Equation (7.23) by representing the concentration of each species i as $\alpha_i(TOTA)$ and then taking the $TOTA$ term out of the summation:

$$\Delta(c_{OH^-})_{added} = -(\{H^+\}_{fin} - \{H^+\}_{init}) + (\{OH^-\}_{fin} - \{OH^-\}_{init})$$

$$+ TOTA \sum_{i=1}^{n} i(\alpha_{i,fin} - \alpha_{i,init}) \tag{7.24}$$

The buffer intensity was then defined as $\Delta(c_{OH^-})_{added}/\Delta pH$, evaluated in the limit where ΔpH becomes differentially small.

If $TOTA$ changes as a function of pH, as it does when the solution is equilibrated with an A-containing gas, Equation (7.23) still applies, but the simplification in Equation (7.24) does not. Rather, we need to keep the $TOTA$ term in the argument of the summation and to account explicitly for the fact that $\{TOTA\}_{init} \neq \{TOTA\}_{fin}$. As in closed systems, β can then be calculated as the limit of $\Delta(c_{OH^-})_{added}/\Delta pH$ at very small ΔpH.

A solution is at pH 7.2 and has equilibrated with atmospheric CO_2. It is then dosed with 10^{-3} mol/L of NaOH. Compare the pH change if the system is closed with that if it is open. What are the buffer intensities of the closed and open systems at pH 7.2?   **Example 7.9**

Solution
The initial conditions in the solution can be read from the log C–pH diagram for a system in equilibrium with the atmosphere or computed using Henry's constant for CO_2 and the two K_a values for the system. The result is that $\{H_2CO_3\} = 1.15 \times 10^{-5}$, $\{HCO_3^-\} = 8.14 \times 10^{-5}$, and $\{CO_3^{2-}\} = 6.03 \times 10^{-8}$. Choosing H_2CO_3 as the reference species, we can write the proton balance table and the $TOTH$ equation for the closed system as follows:

	−2	−1	0	+1	Concentration
Equilibrium species	CO_3^{2-}	OH^- HCO_3^-	H_2O H_2CO_3 Na^+	H^+	
Initial and/or input species		NaOH			10^{-3}
			H_2CO_3		1.15×10^{-5}
		HCO_3^-			8.14×10^{-5}
	CO_3^{2-}				6.03×10^{-8}

$$TOTH_{in} = \{H^+\}_{init} - \{OH^-\}_{init} - \{HCO_3^-\}_{init} - 2\{CO_3^{2-}\}_{init} - [NaOH]_{added}$$

$$= 10^{-7.2} - 10^{-6.8} - 8.14 \times 10^{-5} - 2(6.03 \times 10^{-8}) - 10^{-3} = -1.08 \times 10^{-3}$$

$$TOTH_{eq} = \{H^+\} - \{OH^-\} - \{HCO_3^-\} - 2\{CO_3^{2-}\}$$

Note that the calculation of $TOTH_{in}$ involves adding the proton deficiency associated with the NaOH addition to the $TOTH$ value of the original solution. As discussed in Chapter 4, we can carry out this calculation without knowing how the original solution was prepared, because all we are interested in is the $TOTH$ value of the final mixture and because $TOTH$ is conservative. As a result, we can compute its value based on the composition at any instant and the inputs added subsequently. Equating $TOTH_{in}$ with $TOTH_{eq}$

and rearranging, we obtain the following proton condition for the closed system:

Closed-system PC: $\{H^+\} + 1.08 \times 10^{-3} = \{OH^-\} + \{HCO_3^-\} + 2\{CO_3^{2-}\}$

Now consider how we would write the corresponding equation for an open system. Because we expect $CO_2(g)$ to dissolve in response to the base addition, $TOTCO_3$ will be different in this system from that in the closed system. However, addition of $CO_2(g)$ does not change the proton condition in this problem, because dissolved CO_2 (i.e., H_2CO_3) is a reference level species. Therefore the proton condition for the open system is the same as that shown above for the closed system.

Open-system PC: $\{H^+\} + 1.08 \times 10^{-3} = \{OH^-\} + \{HCO_3^-\} + 2\{CO_3^{2-}\}$

Thus the same proton condition can be used to determine the pH in each system; however, it is used in conjunction with different diagrams. Specifically, it is used in conjunction with a log C–pH diagram for a system with fixed $TOTCO_3 = 9.30 \times 10^{-5}$ for the closed system, and with a log C–pH diagram for a system in equilibrium with atmospheric CO_2 for the open system. The resulting pH's are 11.0 in the former case but only 8.28 in the latter, pointing out once again the effectiveness of gas exchange as a way of buffering the pH. In the open system, the final $TOTCO_3$ is 9.91×10^{-4} M, i.e., about an order of magnitude larger than in the closed system.

The buffer intensity of the open and closed solutions can be computed using Equations (7.23) and (7.24), respectively, for a small pH change, say, 0.1 pH unit (for a pH change from 7.2 to 7.3). The computations yield values of 2.08×10^{-4} and 2.29×10^{-5} (mol/L)/pH unit, respectively. Thus, the possibility of gas exchange with the atmosphere increases the buffer intensity by a factor of 9.1 in this example.

7.4 CO_2 DISSOLUTION, ALKALINITY, AND ACIDITY

Recall that the alkalinity of a solution is defined as the concentration of H^+ that is required to titrate the solution to some arbitrarily defined endpoint near pH 4.5, and that carbonate species are often the dominant ones contributing to the alkalinity of a solution. However, as shown in Chapter 5, despite the fact that absorption or release of $CO_2(g)$ by a solution always affects solution pH, *exchange of $CO_2(g)$ between a solution and a gas phase has a negligible effect on the solution's alkalinity.*[7]

The effect on alkalinity when other gases enter or leave solution depends on whether, at the endpoint of the alkalinity titration, the solute is predominantly present in a form that has the same number of protons as the gaseous species. If it does, then the addition or removal of the gas does not change the number of H^+ ions needed to carry out the alkalinity titration, so the gas exchange does not affect the alkalinity.

Both of the above points are illustrated in the following examples.

---

[7] As a reminder, if the carbonate group were the only acid/base group in a sample, and if one always titrated to the endpoint corresponding to $f = 0$ (changing the endpoint in response to changes in $TOTCO_3$), then dissolution or volatilization of $CO_2(g)$ would have absolutely no effect on the alkalinity. As a practical matter, neither of these conditions ever applies, so the experimentally measured alkalinity does change slightly in response to addition or removal of $CO_2(g)$. However, the change is extremely small.

On a summer morning, a solution at 20°C whose acid/base behavior is dominated by the carbonate group is at pH 8.10 and is in equilibrium with atmospheric $CO_2(g)$.

Example 7.10

a. Determine $TOTCO_3$ in the water and compute the pH at which $f = 0$.
b. During the course of the day, algal photosynthesis removes 75% of the $TOTCO_3$ from the solution by converting H_2CO_3 and H_2O to algal biomass and O_2. Assuming that a negligible amount of CO_2 dissolves from the atmosphere, determine the new pH and the pH at which $f = 0$ at the end of the day.

Solution

a. The H_2CO_3 activity in the solution will be the value in equilibrium with the atmosphere; we have computed this value in prior examples as $10^{-4.94}$. The HCO_3^- and CO_3^{2-} activities can be computed from that value and the given pH, yielding $\{HCO_3^-\} = 10^{-3.19}$ and $\{CO_3^{2-}\} = 10^{-5.42}$. $TOTCO_3$ is the sum of the concentrations of the three species; assuming the solution is ideal, this sum is 6.61×10^{-4} M. The pH at which $f = 0$ is the pH of a solution in which all the $TOTCO_3$ is added as H_2CO_3. Solving for this pH either numerically or graphically, we find pH = 4.77.
b. Three-fourths of the $TOTCO_3$ is removed from the solution, so the $TOTCO_3$ at the end of the day is 1.65×10^{-4} M ($10^{-3.78}$ M). Removal of CO_2 (or, equivalently, H_2CO_3) from solution does not change the alkalinity. Therefore, the alkalinity of the solution in the evening will be the same as in the morning. Since the acid/base behavior of the solution is controlled by the carbonate group, the alkalinity can be approximated using Equation (5.11) as

$$ALK = \{HCO_3^-\} + 2\{CO_3^{2-}\} + \{H^+\} - \{OH^-\} = 6.55 \times 10^{-4}$$

The pH of algal-dominated waters can change dramatically according to a diurnal cycle, especially during warm, sunny days. During the day, photosynthesis removes H_2CO_3 from solution, and the pH increases; at night, respiration reverses the process.

| Photo Researchers, Inc.

In the evening, the alkalinity will have the same components, but $TOTCO_3$ will be less. We can express ALK in the evening as

$$\text{ALK} = TOTCO_3\,(\alpha_1 + 2\alpha_2) + \{H^+\} - \{OH^-\}$$

$$6.65 \times 10^{-4} = (1.65 \times 10^{-4})(\alpha_1 + 2\alpha_2) + \{H^+\} - \{OH^-\}$$

The variables in the above equation all depend only on pH. The equation can be solved by trial and error to find that the pH in the evening is 10.58. Note that OH^- contributes significantly to the alkalinity at this time.

Example 7.11

What is the effect of dissolution of $10^{-4}\,M$ NH_3 and $2 \times 10^{-5}\,M$ H_2S on the alkalinity of a solution that is initially at pH 8.0, if alkalinity is defined by a titration to pH 4.5?

Solution

Since the initial pH is 8.0, the first NH_3 molecules that dissolve will almost all become protonated to NH_4^+, consuming one H^+ per molecule. This reaction removes H^+ from solution, so the pH will increase, but we do not have enough information to know how large the increase will be. If the solution is well buffered, its pH will increase only slightly, in which case almost all the NH_3 entering solution will become protonated. If, on the other hand, the solution is poorly buffered, then the first additions of NH_3 might cause the pH to increase enough that only a fraction of the NH_3 molecules entering subsequently become protonated.

However, regardless of the extent to which the molecules become protonated initially, by the time the alkalinity titration is completed, essentially every NH_3 molecule that entered solution will be protonated. As a result, at that time, an amount of H^+ will have been added to solution that equals the amount of H^+ that would have been required in the absence of NH_3 dissolution, plus one additional H^+ for each molecule of NH_3 that dissolved. Dissolution of $10^{-4}\,M$ NH_3 therefore increases the alkalinity of the solution by 10^{-4} eq/L.

In contrast to NH_3, H_2S is almost fully protonated at pH 4.5. When H_2S molecules first enter the solution at pH 8.0, most of them will release one H^+, thereby mitigating the effect of the NH_3 on pH. However, when the solution is titrated to pH 4.5, every HS^- that was generated earlier by release of an H^+ will recombine with an H^+, so that the net effect of the H_2S on alkalinity will be nil.

Thus, the dissolution of the NH_3 will increase the alkalinity, but the dissolution of the H_2S will not, and the net effect will be that the alkalinity increases by 10^{-4} eq/L.

7.5 INCLUDING GAS/LIQUID EQUILIBRIUM IN COMPUTER EQUILIBRIUM PROGRAMS

Software for computing the equilibrium speciation of aqueous solutions can incorporate gas/liquid equilibrium in a variety of ways. Currently, most programs can compute the speciation in systems that are equilibrated with a gas at a fixed partial pressure, but they do not allow for the possibility that equilibration of the two phases will substantially change the gas-phase composition; modifications to allow that possibility are relatively straightforward and will undoubtedly be incorporated in the future.

In systems where $P_{A(g)}$ is fixed, the equilibrium activity of A(aq) is also fixed and equals $P_{A(g)}/H_A$. Therefore, one approach for modeling such a solution is to calculate {A(aq)} and proceed in the same way as we did when modeling a system with fixed pH in Chapter 6. That is, since A(aq) is present in the equilibrium solution, it must be a Type 1 or Type 2 *species*. As described in Chapter 6, we can cause the software to force the activity of this *species* to have the desired value by defining a dummy Type 3a *species* $\overline{A(aq)}$ and assigning a log K value to the dummy *species* equal to log K for the "real" *species* A(aq) minus the logarithm of the fixed activity of A(aq).

Alternatively, there is an equivalent approach that we can use that recognizes the existence of the gas phase more explicitly. In this approach, rather than compute the fixed activity of A(aq) directly, we treat the gas-phase species A(g) as the dummy *species*. Of course, A(g) is a real chemical, but it is a dummy *species* in terms of the software, because the calculations carried out by the software treat all the *species* as being dissolved. In this text, *species* that are not dissolved but are in equilibrium with the solution are identified as Type 3b *species*. The rules for defining the stoichiometry of Type 3b *species* are identical to those for forming Types 1, 2, and 3a *species,* but the rule for computing the log K value is slightly different. Specifically, **the input log K value for a constituent in a solid or gas phase that is in equilibrium with a solution equals the log K for forming that constituent from the *components* minus the logarithm of the activity of the constituent in the non-aqueous phase.** If the constituent of interest is in the gas phase, its activity is simply its partial pressure.

As explained in Chapter 6, in addition to calculating the equilibrium composition of the solution, the software can keep track of how much of each Type 3 *species* must be added in order for all the equilibrium constant equations to be satisfied. In the case of gases, this calculation indicates the amount of gas that dissolves into or exits from solution during the equilibration process.

Prepare the data input tableau for the system described in Example 7.7, i.e., a solution containing 10^{-3} M NaAc and 10^{-3} M Na$_2$CO$_3$ that is in equilibrium with 10^{-7} bar NH$_3$(g) and atmospheric ($10^{-3.46}$ bar) CO$_2$. Determine the total amounts of CO$_2$(g) and NH$_3$(g) that have dissolved at equilibrium (compared to a starting point of pure water).

Example 7.12

Solution

We can make arbitrary choices for the *components* that we will use to represent the system; for this example, we choose H$^+$, Na$^+$, CO$_3$$^{2-}$, NH$_4$$^+$, and Ac$^-$, because these four *components* are built-in choices in many available programs.

To specify that both CO$_2$(g) and NH$_3$(g) are in equilibrium with the solution, we need to input two Type 3b *species,* which we will designate $\overline{CO_2(g)}$ and $\overline{NH_3(g)}$, respectively. The appropriate stoichiometric coefficients and log K values for these *species* can be determined by writing reactions forming the gases from the *components,* as follows:

$$2H^+ + 1CO_3^{2-} \leftrightarrow H_2CO_3(aq) \qquad \log\{K\} = 16.68$$

$$\underline{H_2CO_3\,(aq) \leftrightarrow CO_2(g) + H_2O} \qquad \log\{H_{CO_2}\} = 1.48$$

$$\underline{2}\,H^+ + \underline{1}\,CO_3^{2-} - \underline{1}\,H_2O \leftrightarrow \underline{1}\,CO_2(g) \qquad \log\{K\} = 18.16$$

$$1NH_4^+ \leftrightarrow NH_3(aq) + 1H^+ \qquad \log\{K_a\} = -9.25$$

$$\underline{NH_3\,(aq) \leftrightarrow NH_3(g)} \qquad \log\{H_{NH_3}\} = -1.77$$

$$\underline{1}\,NH_4^+ - \underline{1}\,H^+ \leftrightarrow \underline{1}\,NH_3(g) \qquad \log\{K\} = -11.02$$

The log K values computed above indicate the true equilibrium constants for formation of the two gases of interest. The log K values that we need to input to the program for the dummy *species* $\overline{CO_2(g)}$ and $\overline{NH_3(g)}$ equal these values minus the fixed activities (i.e., the partial pressures) of those gases. Therefore, the log K values for the Type 3b *species* are as follows:

$$\log K_{\overline{CO_2(g)}} = +18.16 - (-3.46) = +21.62$$

$$\log K_{\overline{NH_3(g)}} = -11.02 - (-7.0) = -4.02$$

In addition to the log K values, we need to input the concentration of each *component.* The correct input concentrations are 0 for **H$^+$**, 3×10^{-3} for **Na$^+$**, 1×10^{-3} for **CO$_3^{2-}$**, 0 for **NH$_4^+$**, and 10^{-3} for **Ac$^-$**, which correspond to the inputs from the known amounts of NaAc and Na$_2$CO$_3$ that were added. These inputs "tell" the software about the addition of all species other than the gases; inputting the Type 3b *species* tells the program that enough of each gas will dissolve or be removed from solution for gas/liquid equilibrium to be established, but that at this point we cannot specify the amount of each gas that enters or leaves solution.

The input data are summarized in the following table, with the inputs of gaseous species designated as x_1 and x_2 for CO$_2$(g) and NH$_3$(g), respectively. The program output for solution speciation is shown in the final column, and that for the total concentration of each component is shown in the row labeled ***TOTComp***$_{eq}$.

Species	Type	Component					Log K	Equil. Concentration
		H$^+$	**Na$^+$**	**CO$_3^{2-}$**	**NH$_4^+$**	**Ac$^-$**		
H$^+$	1	1	0	0	0	0	0.00	2.62×10^{-9}
Na$^+$	1	0	1	0	0	0	0.00	3.00×10^{-3}
CO$_3^{2-}$	1	0	0	1	0	0	0.00	3.48×10^{-5}
NH$_4^+$	1	0	0	0	1	0	0.00	2.75×10^{-5}
Ac$^-$	1	0	0	0	0	1	0.00	1.00×10^{-3}
OH$^-$	2	-1	0	0	0	0	-14.00	3.83×10^{-6}
H$_2$CO$_3$	2	2	0	1	0	0	16.68	1.15×10^{-5}
HCO$_3^-$	2	1	0	1	0	0	10.33	1.95×10^{-2}
NH$_3$(aq)	2	-1	0	0	1	0	-9.25	5.86×10^{-6}
HAc	2	1	0	0	0	1	4.76	1.51×10^{-7}

(continued)

| Species | Type | Component | | | | | Log K | Equil. Concentration |
		H^+	Na^+	CO_3^{2-}	NH_4^+	Ac^-		
$\overline{CO_2(g)}$	3b	2	0	1	0	0	21.62	N/A
$\overline{NH_3(g)}$	3b	−1	0	0	1	0	−4.02	N/A
$TOT\text{Comp}_{eq}$		1.97×10^{-2}	3.0×10^{-3}	2.00×10^{-3}	3.33×10^{-5}	1.0×10^{-3}		

Inputs								Input Concentration
NaAc		0	1	0	0	1		1.0×10^{-3}
Na_2CO_3		0	2	1	0	0		1.0×10^{-3}
$\overline{CO_2(g)}$		2	0	1	0	0		x_1
$\overline{NH_3(g)}$		−1	0	0	1	0		x_2
$TOT\text{Comp}_{in}$		$0 + (2x_1 - x_2)$	3.0×10^{-3}	$1.0 \times 10^{-3}(+x_1)$	$0 (+x_2)$	1.0×10^{-3}		

The total inputs of $CO_2(g)$ and $NH_3(g)$ can be computed by equating the expressions for the inputs of these *components* (including the variables x_1 and x_2) with the program output for the total concentrations of the *components* ($TOTCO_3$ and $TOTNH_3$, respectively) at equilibrium. That is, for the mass balances on $TOTCO_3$ and $TOTNH_3$, we can write

$$TOTCO_{3,in} = TOTCO_{3,eq}$$

$$1.0 \times 10^{-3} + x_1 = 2.0 \times 10^{-3}$$

$$TOTNH_{3,in} = TOTNH_{3,eq}$$

$$0 + x_2 = 3.33 \times 10^{-5}$$

Solving for x_1 and x_2, we conclude that 1.0×10^{-3} mole of $CO_2(g)$ and 3.33×10^{-5} mole of $NH_3(g)$ dissolve per liter of solution. Note that these values also satisfy the mass balance on H^+, i.e., $TOTH_{in} = 2x_1 - x_2$, and $TOTH_{eq} = 1.97 \times 10^{-2}$.

By the nature of gas/liquid equilibria, every gas-phase constituent that we are interested in is equilibrated with a dissolved species with the same chemical formula. As a result, the reaction for forming the gas-phase constituent can be written as a combination of the reaction for forming the corresponding dissolved species and the reaction for Henry's law. Thus, in the preceding example, the reactions for forming $CO_2(g)$ and $NH_3(g)$ were generated by adding the reactions forming the corresponding dissolved *species* [H_2CO_3 and $NH_3(aq)$] and the reactions corresponding to Henry's law for the two gases. Generalizing this result, we can restate the rule for determining the log K value for forming gaseous Type 3b *species* as follows: **Log K for a Type 3b gas *species* equals log K for the corresponding Type 1 or 2 dissolved species, plus the logarithm of Henry's constant for the gas, minus the logarithm of the partial pressure of the gas.**

In symbols,

$$\log K_{\substack{\text{Type 3b} \\ \text{gas } i}} = \log K_{\substack{\text{corresponding} \\ \text{Type 1 or 2 } species}} + \log H_i - \log P_i \qquad \textbf{(7.25)}$$

Log K for the dissolved *species* with the same stoichiometry as the gas [the first term in Equation (7.25)] must be determined as one of the program inputs and is available directly from the database embedded in many of the available chemical equilibrium software packages, and Henry's constant and the partial pressure of interest (the latter two terms in the equation) are starting points for solving the problem. As a result, all the terms in Equation (7.25) are typically readily available, so it is often more convenient to use that equation for computing log K values for gases than the more general rule for Type 3 *species* given earlier.

Example 7.13 | Derive the log K values for the Type 3b *species* in Example 7.12, using Equation (7.25).

Solution

The log K values for the Type 2 *species* H_2CO_3 and $NH_3(aq)$ are 16.68 and -9.25, respectively. These values are embedded in the databases of most programs and so do not need to be calculated. The corresponding log H values are 1.48 and -1.77, respectively, and the P_i values of interest are -3.46 and -7.0. The correct input values for log K of the Type 3b *species* are therefore as follows:

$$\log K_{\overline{CO_2(g)}} = \log K_{H_2CO_3(aq)} + \log H_{CO_2} - \log P_{CO_2}$$
$$= \log 10^{16.68} + \log 10^{1.48} - \log 10^{-3.46} = 21.62$$

$$\log K_{\overline{NH_3(g)}} = \log K_{NH_3(aq)} + \log H_{NH_3} - \log P_{NH_3}$$
$$= \log 10^{-9.25} + \log 10^{-1.77} - \log 10^{-7.0} = -4.02$$

The results are, of course, identical to those obtained above, but the calculations leading to the result are easier (no reactions had to be added to generate the values).

SUMMARY

Several important chemical species that can be present in solution can also exist in a gas phase. The exchange of such species between the two phases can be represented as a chemical reaction, and their partitioning at equilibrium can be characterized by an equilibrium constant known as Henry's constant. All chemical species tend to become more volatile (less soluble, corresponding to higher Henry's constant) with increasing temperature. Henry's constant also tends to increase with decreasing molecular weight and decreasing ionic or polar character of the molecule. Although Henry's constant, like all equilibrium constants, is a ratio of activities, the constant is commonly reported as a partition coefficient, i.e., as a ratio of concentrations.

Acidic and basic volatile species affect the acid/base balance of a solution that is in contact with a gas. When a solution is equilibrated with an acidic or basic gas, the soluble concentration of the neutral form of the acid/base group (e.g., NH_3 or H_2CO_3) is independent of pH. In such cases, the dissolved concentrations of species that have more protons than the neutral species (e.g., NH_4^+) decrease with increasing pH. Conversely, the concentrations of deprotonated forms of the neutral species (e.g., HCO_3^- and CO_3^{2-}) increase with increasing pH. The net result is that the total dissolved concentration of a volatile acidic or basic gas can be very sensitive to solution pH.

Acidic gases tend to enter solution and basic gases to leave solution when some other reaction raises the pH, thereby mitigating the effect of the other reaction. Because of this, gases can serve as excellent pH buffers. Also, the nature of gas/liquid equilibrium allows acidic and basic gases to serve as buffers at any pH where the neutral species is not the dominant dissolved species of the acid/base group. The farther the pH is from the region where the neutral species is dominant, the better the gas can buffer the solution.

Although acidic and basic gases can alter solution pH, they might or might not alter the alkalinity of the solution when they dissolve into or evolve from solution. In particular, CO_2 has a negligible effect on the alkalinity when it enters or exits solution.

Systems equilibrated with gases can be modeled using chemical equilibrium software, via the definition of a Type 3b *species* that represents the gaseous species. The input value of log K for formation of the Type 3b *species* depends on both the equilibrium constant for forming the gas from the *components* and the partial pressure of the gas. Once the equilibrium composition of the solution has been determined, the amount of the gaseous *species* that had to enter or leave solution to reach equilibrium can be computed.

Problems

1. Compute the dissolved concentration of carbon monoxide, CO, that would be in equilibrium with a gas phase containing 0.1 ppm CO by volume. Give your answer in mg CO/L.

2. Calculate the pH of water in a cloud that has equilibrated with 1.0 ppm by volume SO_2.

3. A solution contains 2 mg/L CN^- and is at pH 8.5. The gas above the solution has a partial pressure of 10^{-6} bar HCN.

   a. What is the concentration of HCN(*aq*) in solution?

   b. Is HCN transferring from the gas phase into solution, from the solution into the gas, or is the system at equilibrium?

4. A solution at pH 7.0 is in equilibrium with a gas containing ammonia at a partial pressure of 7×10^{-8} bar. Enough base is added quickly to raise the pH to 11.0. Assume that acid/base reactions in solution are rapid, but that gas/liquid exchange is slow. What is the molar free energy of reaction for transfer of NH_3 from the solution to the gas phase shortly after the base is

added, i.e., at a time when the NH_3/NH_4^+ pair has come to equilibrium, but no gas exchange has occurred yet? The $\overline{G}^o$ values for H^+, $NH_3(aq)$, and NH_4^+ are given in Table 2.1 and Appendix A.

5. A groundwater supply is initially at pH 8.0 and contains 2×10^{-3} eq/L alkalinity.

   a. If the alkalinity is all attributable to the carbonate system, determine whether the water is undersaturated or supersaturated with respect to atmospheric CO_2.

   b. How much CO_2 must be dissolved into or volatilized out of solution for the system to reach equilibrium with the atmosphere?

   c. What will the pH and alkalinity be once equilibrium is attained?

6. A popular soft drink has a pH around 3 due partially to its high H_2CO_3 content.

   a. If all the acidity in the beverage were generated by bubbling it with pressurized CO_2, what would P_{CO_2} above the beverage be?

   b. What would the pH of the solution be if it were in equilibrium with a P_{CO_2} of 2 bar?

   c. The pH of the solution is actually kept low by a combination of dissolved CO_2 plus phosphoric acid. Assuming P_{CO_2} is 2 bar in the bottle, what concentration of H_3PO_4 must be added to the solution to reach pH 3.0?

7. If you have not done so already, solve Problem 15 in Chapter 5. If the solution described in Part C of that problem is exposed to air and equilibrates with the CO_2, what is the final pH of the solution? What is the alkalinity of the solution after equilibration with the atmosphere if the endpoint of the alkalinity titration is pH 4.76 (pK_a for HAc)?

8. Sulfur dioxide, SO_2, is a major air pollutant emitted during the burning of sulfur-bearing fossil fuels. Consider a stack gas containing 15% CO_2 and 2000 ppm SO_2 (2000×10^{-6} L SO_2 per liter of gas, or 2×10^{-3} bar). Treatment must lower the SO_2 concentration to 250 ppm SO_2. It is proposed to absorb the SO_2 into an alkaline solution made by adding 10^{-1} mol/L $Ca(OH)_2$ to water. The operation is countercurrent, as shown in the diagram on p. 359, so the liquid effluent is in contact with the untreated influent gas. Assume the gas and liquid phases come to equilibrium rapidly, so at every point in the stack, equilibrium calculations apply. Answer the following questions to explore the feasibility and some of the potential problems associated with this process.

   a. What will be the pH of the influent water? Do not consider $Ca(OH)_2$ to be a strong base for this part of the problem (pK_{a1} for Ca^{2+} is 12.60).

   b. What will be the pH and composition of the water as it exits the reactor, i.e., in equilibrium with the influent gas? You may ignore possible precipitation of solids.

   c. Do a mass balance on sulfur (flow rate in $\times$ concentration in $=$ flow rate out $\times$ concentration out) to compute the required ratio of gas to liquid flow rates (Q_G/Q_L) in liters of gas per liter of liquid. Assume STP.

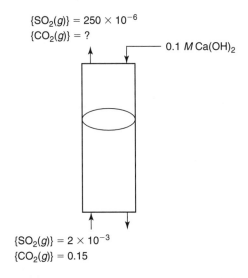

$\{SO_2(g)\} = 250 \times 10^{-6}$
$\{CO_2(g)\} = ?$

— 0.1 $M\,Ca(OH)_2$

$\{SO_2(g)\} = 2 \times 10^{-3}$
$\{CO_2(g)\} = 0.15$

   d. Assuming $TOTCO_3$ in the influent water is negligible, do another mass balance to determine what fraction of the CO_2 will be absorbed from the gas phase.

   e. The point of the process is to remove SO_2 from the gas. If most of the base added is used to neutralize CO_2 rather than SO_2, the process may be considered inefficient, at least in terms of reagent utilization. Of the total hydroxide consumed in the reactor, what fraction neutralizes SO_2, and what fraction neutralizes CO_2?

9. Hydrogen sulfide gas at a partial pressure of $10^{-4.82}$ bar is in contact with a solution at pH 8.5. The concentration of dissolved sulfide ion (S^{2-}) in the solution is $10^{-6.67}\,M$.

   a. Prepare a log C–pH diagram for the $H_2S/HS^-/S^{2-}$ system in equilibrium with $10^{-4.82}$ bar $H_2S(g)$.

   b. Assuming that the S-containing species are the only contributors to alkalinity, what is the alkalinity of the solution? Assume that the endpoint of the alkalinity titration is pH 4.5.

   c. Assuming that the system in part (*a*) remains in continuous equilibrium with the gas phase, write out the proton condition and find the pH after $5 \times 10^{-3}\,M$ H_3PO_4 has been added to the solution. What is the change in total dissolved sulfide concentration when the H_3PO_4 is added?

10. A solution containing $10^{-3}\,M\,TOTPO_4$ as the only weak acid/base system is initially at pH 7.20, at which pH essentially all the PO_4 in the system is present as either $H_2PO_4^-$ or HPO_4^{2-}.

     An additional $10^{-3}\,M\,Na_2HPO_4$ is added to the solution, which is then bubbled with air containing $CO_2(g)$ at a partial pressure of $10^{-3.5}$ bar. Prepare a table showing the inputs you would use to instruct a chemical equilibrium computer program to determine the final pH.

11. The microorganisms in a waste treatment pond are exposed to 35 mg/L of easily degradable organic compounds with a typical composition $C_5H_{10}O_5$. The standard molar Gibbs energy of formation of these compounds is -2797 kJ/mol. The pond is at pH 7.3, its temperature is 25°C, and its alkalinity is 2×10^{-3} eq/L. The overall reaction for the organic matter is

$$C_5H_{10}O_5 + 5O_2(aq) \leftrightarrow 5H_2CO_3(aq)$$

   a. Due to oxygen consumption by the above reaction, the dissolved oxygen concentration in the water is only 2 mg/L. What is the maximum amount of energy that an organism could obtain by oxidizing 1 μmol of the organic matter?

   b. The operator of the treatment facility has suggested that the treatment efficiency could be improved by bubbling air through the water to encourage biological activity. Would the organisms find the environment in the untreated water more favorable, less favorable, or identical (from an energy standpoint) after equilibration with air? Explain briefly.

   c. Determine the pH of the system after it has been bubbled with air.

12. Phosphoric acid (H_3PO_4) is an additive in many soft drinks. Imagine a soft drink containing no acidic or basic constituents, to which H_3PO_4 is then added to adjust the solution pH to 2.2. The solution is next equilibrated with compressed air at a total pressure of 2 bar and placed in cans. A social deviant has plans to wreck the company's business by dumping enough $Mg(OH)_2$ into the beverage solution so that it becomes flat, i.e., so that it does not bubble when the can is opened. $Mg(OH)_2$ can be considered a strong base that dissociates completely to Mg^{2+} and OH^- ions.

   a. How much phosphoric acid is added to the solution in the first step?

   b. The villain has plans to sneak into the bottling facility, carrying only the $Mg(OH)_2$ and a pH meter, and to add enough $Mg(OH)_2$ to cause the solution to be in equilibrium with atmospheric CO_2. What is the pH to which the solution must be adjusted to accomplish the dastardly goal?

   c. What concentration of $Mg(OH)_2$ must be added?

13. Two waters are in equilibrium with air, one at pH 9.5 and one at pH 7.3. These waters are mixed in 1:1 proportions. Assuming that carbonate species dominate the acid/base behavior of both solutions and that no gas transfer occurs when they are mixed, determine the alkalinity and pH of the mixture. Will the mixture be undersaturated, supersaturated, or in equilibrium with atmospheric CO_2?

14. The raw water supply for a community is at pH 7.5 and has ALK = 40 mg/L *as* $CaCO_3$. Essentially all the alkalinity is attributable to the carbonate acid/base system. The water is treated by addition of 60 mg alum ($Al_2(SO_4)_3 \cdot 14H_2O$) per liter of water, causing $Al(OH)_3(s)$ to precipitate and aiding in the removal of colloidal matter from suspension. The precipitation of the alum generates acid by the reaction shown below:

$$Al_2(SO_4)_3 \cdot 14H_2O \rightarrow 2Al(OH)_3(s) + 3SO_4^{2-} + 6H^+ + 12H_2O$$

    a. It is desired to maintain a pH of at least 6.5 in the treated water (after coagulation and removal of the precipitates by settling and filtration). Will chemical (base) addition be necessary when the alum is added? If so, how much? Assume that gas exchange during the various treatment steps is slow enough that the solution can be treated as a closed system.

    b. If no base is added to the solution, will it be possible to reach the target pH by bubbling the treated water with air? How much CO_2 must be stripped out of solution to reach the target pH? (*Hint:* Recall what happens to alkalinity when CO_2 is added to or removed from a solution.)

15. Compare the buffer intensity of an open system with that of a closed system containing a fixed $TOTCO_3$ of 10^{-3} over the pH range of 3 to 11.

16. A solution containing 1.5 meq/L alkalinity at pH 8.3 is dosed with 4×10^{-4} M NaH_2PO_4 and enough base to return the pH to 8.3 (this solution was discussed in Example 5.11). The resulting solution is then dosed with 10^{-3} mol/L $(NH_4)_2CO_3$ and subsequently equilibrated with air containing $NH_3(g)$ at a partial pressure of 10^{-8} bar and CO_2 at its normal atmospheric concentration.

    a. Compute the final pH, and determine whether the equilibration with the gas causes NH_3 to enter or leave solution.

    b. How much NH_3 is transferred between the two phases as the water and gas equilibrate? Give your answer in moles transferred per liter of solution.

    c. Would the final pH be higher, lower, or the same if the $(NH_4)_2CO_3$ had not been added? Explain this result.

17. You have received a sealed vial containing a water sample, which you are to analyze for a volatile organic compound present at trace levels. The sample volume is 10 mL. You notice a 0.25-mL air bubble in the vial. If Henry's constant of the compound is 350 bar/(mol/L), and if the sample originally contained 10^{-7} M of the analyte, what concentration of the analyte will you measure in the solution, assuming that it has equilibrated with the air bubble? The vial and its contents are at 20°C.

8

CHEMISTRY OF METALS IN AQUEOUS SYSTEMS

CHAPTER OUTLINE

8.1   INTRODUCTION

The behavior of metals in environmental systems is a topic of ongoing interest
and concern. Historically, the use of metals has been closely linked with indus-
trialization and development. As a result, the natural biogeochemical cycling of
many metals has been greatly accelerated, exposing some organisms or entire
ecosystems to unhealthy, and at times lethal, levels of these chemicals.

There is no universally accepted definition distinguishing metals from non-
metals. Any definition of metals includes those elements that are refined, used in
commerce, and referred to colloquially as such (e.g., Cu, Fe, Ag), but virtually all
elements in the third and lower rows of the periodic table (sodium and all elements with atomic numbers larger than that of sodium) other than the halogens
and noble gases share some properties that could be ascribed to metals. In this
chapter, we consider what *metallic behavior* is in terms of the reactions that some
elements undergo in aquatic environments.

One key characteristic of metals in aquatic systems is their ability to exist
as either dissolved or solid species and to combine with numerous other species.
That is, whereas virtually all the acetate in an aquatic system is likely to be present as either acetate ion or acetic acid, copper might be present as a constituent
in dozens of different dissolved species. The speciation of metals has a profound
effect on their physical and chemical behavior, their bioavailability, and their tox-
icity. The primary goal of this chapter is to provide a context for predicting and
interpreting that speciation.

8.2   SPECIATION OF DISSOLVED METALS

8.2.1   COMPLEXATION WITH H_2O OR OH^-

As noted in previous chapters, even though water molecules are electrically
neutral, their charges are not distributed symmetrically. Rather, they are dipoles,
with a small amount of positive charge residing near the hydrogen atoms and a balancing negative charge (associated with unshared pairs of electrons) on the other
side of the oxygen atom from the hydrogen atoms. Water molecules surrounding a
(positively charged) metal ion orient themselves with the unshared electrons adjacent to the metal. Some of the negative charge in these orbitals is shared with the
metal ion, forming a bond that might range from very weak to very strong.

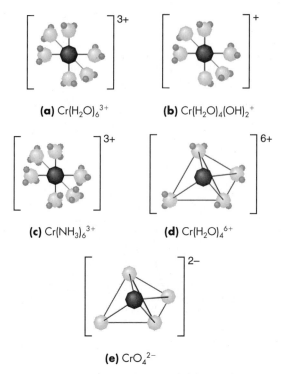

Figure 8.1   Chromium can exist in both a trivalent (Cr^{3+}) and hexavalent (Cr^{6+}) oxidation state. **(a)** A fully hydrated Cr^{3+} ion, i.e., a $Cr(H_2O)_6^{3+}$ complex. **(b)** A $Cr(H_2O)_4(OH)_2^+$ complex, i.e., a species similar to that in (a), except that one H^+ ion has been lost by each of two of the waters of hydration. **(c)** A Cr^{3+} ion in which all six waters of hydration have been replaced by ammonia molecules, forming the complex $Cr(NH_3)_6^{3+}$. **(d)** A fully hydrated $Cr(H_2O)_4^{6+}$ ion in a tetrahedral complex. This ion is theoretically plausible, but it is so acidic that it is never found in aqueous solutions. **(e)** A fully deprotonated chromate ion (CrO_4^{2-}), which is the dominant form of Cr(VI) in neutral or alkaline solutions.

Depending on the size of the metal ion, four to eight oriented water molecules may surround it. Since these water molecules are weakly (or not so weakly) bonded to the metal ion, the water plus the metal can be thought of as a single entity, and a reasonable representation of the metal in solution would be $Me(OH_2)_x^{n+}$, where $4 \leq x \leq 8$ (Figure 8.1a). The group of water molecules immediately surrounding the metal is called the **inner hydration sphere** or **inner coordination sphere.** As is described subsequently in the chapter, a variety of other molecules can replace the water molecules in the hydration spheres. All such molecules are called **ligands,** and the entire cluster is called a **metal-ligand complex** or simply a **complex.** A few inorganic ligands of importance and some metals to which they bind are listed in Table 8.1.

The exchange of water molecules in the hydration sphere with other molecules (either water molecules from bulk solution or different ligands) is generally quite fast, so that the formation of complexes proceeds rapidly and can be considered to reach equilibrium within seconds in most cases. However, in a few

Table 8.1   Some important inorganic ligands and the metals to which they bind

Ligand	Metals	Environment
H_2O	All	Any aquatic system
OH^-	Virtually all	Any aquatic system
F^-	Fe^{3+}, Al^{3+}	Some natural systems; industrial systems where HF is used to treat metal surfaces
Cl^-	Cu^{2+}, Cu^+, Pb^{2+}, Cd^{2+}	Estuaries, seawater; corrosion of metals
CN^-	Fe^{3+}, Fe^{2+}, Cu^+, Cu^{2+}, Ni^{2+}, Ag^+	Metal plating
NH_3	Cu^+, Cu^{2+}, Cd^{2+}, Ni^{2+}	Metal finishing
$S_2O_3^{2-}$	Ag^+	Photofinishing
$P_2O_7^{4-}$, $P_3O_{10}^{5-}$	Ca^{2+}, Mn^{2+}, Fe^{3+}	Detergents, corrosion inhibitors

cases (especially for more highly charged metals), the waters of hydration exchange slowly, and equilibration with ligands in solution might take much longer. For instance, the characteristic residence time of water in the hydration sphere of Cr^{3+} ranges from several seconds to several days, depending on the solution pH.

The number of water molecules or other ligand groups to which the metal binds is referred to as the metal's **coordination number.** Depending on the strength and extent of the electric field generated by the metal ion, a second sphere of water molecules (an **outer hydration sphere**) may also be bonded to the complex. In general, the smaller and more highly charged the central metal ion is, the more strongly it is able to bond to and orient water molecules around it, and the larger the complex becomes. Hydrated metal ions are often referred to as **free metal ions** or **uncomplexed ions** even though, in a formal sense, they are complexed with water molecules.

The ionic charge on a metal is called its **oxidation number** or **oxidation state.** Many metal ions can exist in stable forms in two or more oxidation states (e.g., Fe^{2+} or Fe^{3+}, Cr^{3+} or Cr^{6+}). In such cases, it is common to refer to all the species in which the metal has a given charge by that charge (e.g., divalent or trivalent iron; trivalent or hexavalent chromium). The same concept is often represented by writing the charge on the metal ion (not the complex) in Roman numerals after the metal's name [e.g., Fe(II) or Fe(III); Cr(III) or Cr(VI)]. Cr species in the trivalent and hexavalent oxidation states are shown in Figure 8.1. The significance of the different charges is discussed in Chapter 9, which deals with oxidation-reduction reactions.

The attraction of the unshared electron pairs on water molecules in the inner hydration sphere to the metal ion causes electrons to be withdrawn from the water's O—H orbitals, weakening those bonds. As a result, the H^+ ions that are part of the water molecules in the hydration sphere surrounding a metal ion are more easily dissociated than those on water molecules in bulk solution, causing metal ions to act as acids. Although the actual reaction taking place is best represented as:

$$Me(OH_2)_x^{n+} \leftrightarrow Me(OH_2)_{x-1}(OH)^{n-1} + H^+ \tag{8.1}$$

the reaction is most often abbreviated in one of the two following forms:

$$Me^{n+} + H_2O \leftrightarrow MeOH^{n-1} + H^+ \qquad K_{(8.2)} \tag{8.2}$$

$$Me^{n+} + OH^- \leftrightarrow MeOH^{n-1} \qquad\qquad K_{(8.3)} \tag{8.3}$$

Regardless of how the reaction is written, the net effect is replacement of a water molecule by a hydroxide ion as one of the ligands in the inner hydration sphere (Figure 8.1b). The equilibrium constant written for a reaction in which a metal-ligand complex forms is called a **stability constant** or **formation constant.** In the above reactions, the stability constant $K_{(8.2)}$ is also the acidity constant K_a for a reaction in which the free metal ion Me^{n+} is the acid and $Me(OH)^{n-1}$ is the conjugate base; similarly, $K_{(8.3)}$ is the inverse of a basicity constant.

Many metal ions can act as multiprotic acids, releasing protons from water molecules in the inner hydration sphere until they are surrounded by OH^- ions instead of H_2O molecules. In some cases, the remaining protons are bonded very tightly to the oxygen atoms and are essentially non-releasable; in others, the protons that are part of the OH^- ligands in the inner hydration sphere can be released, leaving the metal ion bonded to one or more oxide (O^{2-}) ions. For instance, $Fe(OH)_6^{3-}$ is a stable species that can act as a base [forming its conjugate acid, $Fe(OH)_5^{2-}$, in the process] but not as an acid. On the other hand, As^{3+}, Se^{4+}, and Cr^{6+} can cause surrounding water molecules to deprotonate completely, to form AsO_3^{3-}, SeO_3^{2-}, and CrO_4^{2-} ions, respectively (Figure 8.1d).

For most metals, no more than four acid/base species are of practical significance. Thus, although species of the form $Cr(OH_2)_3(OH)^{5+}$ are theoretically plausible, they are so acidic as to be unimportant in all systems of interest; rather, when one is considering hydrolyzed forms of hexavalent Cr, only $CrO_2(OH)_2^0$, $CrO_3(OH)^-$, and CrO_4^{2-} are considered (Figure 8.1e). [The first two of these three Cr(VI)-containing species are chromic acid and bichromate ion, commonly written as H_2CrO_4 and $HCrO_4^-$, respectively.] In the case of Cr(VI) and most other metals that bind so strongly to water that the water molecule releases both its hydrogen ions, the $Me-O^{2-}$ bonds are too strong to be broken under normal conditions. In such cases, rather than Me^{n+} and O^{2-} being considered to form a complex, they are considered to form a stable molecule, sometimes referred to as a *metal oxyanion*.

Just as the equilibrium constants for deprotonation of an acid (K_a) and for release of a gas from solution (H) are given special designations for convenience, conventions have been adopted for representing the stability constants of metal-ligand complexes. The designations can be summarized as follows:

K_i: Equilibrium constant for a reaction between two dissolved species, adding *one* ligand to a complex with $i - 1$ ligands, to form a complex with i ligands.

β_i: Equilibrium constant for a reaction between an uncomplexed ion (i.e., an aquo-complexed ion) and i ligands to form a complex of the type MeL_i.

β_{ij}: Equilibrium constant for a reaction between MeL_i and $j - i$ ligands to form a complex of the type MeL_j.

$*K_i$ or $*\beta_i$: Same as K_i or β_i, except that the reaction is written with the ligand in a protonated form, usually requiring inclusion of H^+ on the product side.

Reactions for the formation of several hydroxo complexes of cadmium are shown below, along with the conventional way of representing their stability constants.

$$Cd^{2+} + OH^- \leftrightarrow CdOH^+ \qquad\qquad \log K_1 = 3.92 \qquad \textbf{(8.4)}$$

$$CdOH^+ + OH^- \leftrightarrow Cd(OH)_2^\circ \qquad \log K_2 = 3.73 \qquad \textbf{(8.5)}$$

$$Cd(OH)_2^\circ + OH^- \leftrightarrow Cd(OH)_3^- \qquad \log K_3 = 1.05 \qquad \textbf{(8.6)}$$

$$Cd(OH)_3^- + OH^- \leftrightarrow Cd(OH)_4^{2-} \qquad \log K_4 = -0.05 \qquad \textbf{(8.7)}$$

$$Cd^{2+} + 2OH^- \leftrightarrow Cd(OH)_2^\circ \qquad\quad \log \beta_2 = 7.65 \qquad \textbf{(8.8)}$$

$$Cd^{2+} + 4OH^- \leftrightarrow Cd(OH)_4^{2-} \qquad \log \beta_4 = 8.65 \qquad \textbf{(8.9)}$$

$$CdOH^+ + H_2O \leftrightarrow Cd(OH)_2^\circ + H^+ \qquad \log *K_2 = -10.27 \quad \textbf{(8.10)}$$

$$Cd^{2+} + 3H_2O \leftrightarrow Cd(OH)_3^- + 3H^+ \qquad \log *\beta_3 = -33.3 \qquad \textbf{(8.11)}$$

If there is any chance of confusion about which ligand is being considered, its identity can be added to the subscript; e.g., K_1 might be written as $K_{OH,1}$.

Not all the stability constants listed above are independent. For instance, β_i is related to the various K_i's by

$$\beta_n = K_1 K_2 K_3 \cdots K_n = \prod_{i=1}^{n} K_i \qquad \textbf{(8.12)}$$

and β_4 and $*\beta_4$ are related through the following reactions:

$$Cd^{2+} + 4OH^- \leftrightarrow Cd(OH)_4^{2-} \qquad\qquad \beta_4 \qquad \textbf{(8.9)}$$

$$\underline{\qquad\qquad 4H_2O \leftrightarrow 4H^+ + 4OH^- \qquad\qquad K_w^4 \qquad \textbf{(8.13)}}$$

$$Cd^{2+} + 4H_2O \leftrightarrow Cd(OH)_4^{2-} + 4H^+ \qquad *\beta_4 \qquad \textbf{(8.14)}$$

Thus, $*\beta_4 = \beta_4 K_w^4$ or, in general, for formation of hydroxide complexes

$$*\beta_i = \beta_i K_w^i \qquad \textbf{(8.15)}$$

and

$$*K_i = K_i K_w \qquad \textbf{(8.16)}$$

A list of stability constants written in various ways for complexation of several metals by OH^- is provided in Table 8.2.

Table 8.2   Stability constants for complexation of metals by OH^-

	i	Log K_i	Log $*K_i$	Log β_i	Log $*\beta_i$
Ag^+	1	2.00	−12.00	2.00	−12.00
	2	2.00	−12.00	4.00	−24.00
Al^{3+}	1	9.01	−4.99	9.01	−4.99
	2	8.89	−5.11	17.90	−10.10
	3	8.10	−5.90	26.00	−16.00
	4	7.00	−7.00	33.00	−23.00
Ca^{2+}	1	1.40	−12.60	1.40	−12.60
Cd^{2+}	1	3.92	−10.08	3.92	−10.08
	2	3.73	−10.27	7.65	−20.35
	3	1.05	−12.95	8.70	−33.30
	4	−0.05	−14.05	8.65	−47.35
Co^{2+}	1	4.80	−9.20	4.80	−9.20
	2	4.90	−9.10	9.70	−18.30
	3	1.10	−12.90	10.80	−31.20
Cr^{3+}	1	10.00	−4.00	10.00	−4.00
	2	8.38	−5.62	18.38	−9.62
	3	6.87	−7.13	25.25	−16.75
	4	2.98	−11.02	28.23	−27.77
Cu^{2+}	1	6.00	−8.00	6.00	−8.00
	2	8.32	−5.68	14.32	−13.68
	3	0.78	−13.22	15.10	−26.90
	4	1.30	−12.70	16.40	−39.60
Fe^{2+}	1	4.50	−9.50	4.50	−9.50
	2	2.93	−11.07	7.43	−20.57
	3	3.57	−10.43	11.00	−31.00
Fe^{3+}	1	11.81	−2.19	11.81	−2.19
	2	10.52	−3.48	22.33	−5.67
	3	6.07	−7.93	28.40	−13.60
	4	6.00	−8.00	34.40	−21.60
Hg^{2+}	1	10.60	−3.40	10.60	−3.40
	2	11.30	−2.70	21.90	−6.10
	3	10.26	−3.74	20.86	−21.14
Mg^{2+}	1	2.21	−11.79	2.21	−11.79
Ni^{2+}	1	4.14	−9.86	4.14	−9.86
	2	4.86	−9.14	9.00	−19.00
	3	3.00	−11.00	12.00	−30.00
Pb^{2+}	1	6.29	−7.71	6.29	−7.71
	2	4.59	−9.41	10.88	−17.12
	3	3.06	−10.94	13.94	−28.06
	4	2.36	−11.64	16.30	−39.70
Zn^{2+}	1	5.04	−8.96	5.04	−8.96
	2	6.06	−7.94	11.10	−16.90
	3	2.50	−11.50	13.60	−28.40
	4	1.20	−12.80	14.80	−41.20

8.2.2 LOG C–pH DIAGRAMS FOR DISSOLVED METALS

Recognizing metals as polyprotic acids, we can develop log C–pH diagrams for metals by following the same procedures we used previously for other acids. For instance, consider a system containing $10^{-4} M$ TOTCd. Up to four protons can be released from the waters of hydration in the free aquo Cd^{2+} complex; i.e., Cd^{2+} is a tetra-protic acid, with K_a values as follows.

$$Cd^{2+} + H_2O \leftrightarrow Cd(OH)^+ + H^+ \qquad \log K_{a1} = \log {}^*K_1 = -10.08$$

$$CdOH^+ + H_2O \leftrightarrow Cd(OH)_2^{\,0} + H^+ \qquad \log K_{a2} = \log {}^*K_2 = -10.27$$

$$Cd(OH)_2^{\,0} + H_2O \leftrightarrow Cd(OH)_3^{\,-} + H^+ \qquad \log K_{a3} = \log {}^*K_3 = -12.95$$

$$Cd(OH)_3^{\,-} + H_2O \leftrightarrow Cd(OH)_4^{\,2-} + H^+ \qquad \log K_{a4} = \log {}^*K_4 = -14.05$$

The constants are designated as ${}^*K_i$ values; other than that change of symbols, everything else is as in any other system with a tetra-protic acid. The corresponding log C–pH diagram is shown in Figure 8.2.

The fraction of the total dissolved metal that is present as a given species is designated by an α, as in other acid/base systems. Thus, in the Cd system, $\alpha_{Cd^{2+}}$ is

$$\alpha_{Cd^{2+}} = \frac{[Cd^{2+}]}{TOTCd} \qquad \textbf{(8.17)}$$

Assuming that the solutes behave ideally, the total dissolved Cd concentration can be written in terms of the concentration of free Cd^{2+}, the stability constants

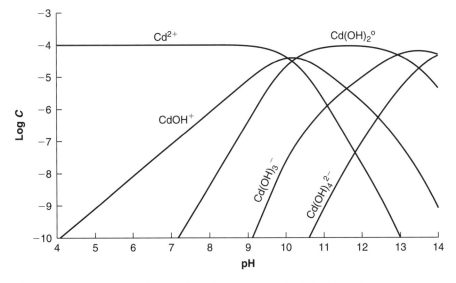

Figure 8.2 Log C–pH diagram for Cd in a system with $10^{-4} M$ TOTCd.

for formation of the complexes, and the hydroxide ion activity, as follows:

$$TOTCd = \{Cd^{2+}\} + \{CdOH^+\} + \{Cd(OH)_2^{\,0}\} + \{Cd(OH)_3^{\,-}\} + \{Cd(OH)_4^{\,2-}\}$$

$$\text{(8.18a)}$$

$$= \{Cd^{2+}\}\left[1 + \frac{\{CdOH^+\}}{\{Cd^{2+}\}} + \frac{\{Cd(OH)_2^{\,0}\}}{\{Cd^{2+}\}}\right.$$

$$\left. + \frac{\{Cd(OH)_3^{\,-}\}}{\{Cd^{2+}\}} + \frac{\{Cd(OH)_4^{\,2-}\}}{\{Cd^{2+}\}}\right] \qquad \text{(8.18b)}$$

$$= \{Cd^{2+}\}(1 + K_1\{OH^-\}^1 + \beta_2\{OH^-\}^2 + \beta_3\{OH^-\}^3 + \beta_4\{OH^-\}^4)$$

$$\text{(8.18c)}$$

$$= \{Cd^{2+}\}\left(1 + \sum_{i=1}^{4}\beta_i\{OH^-\}^i\right) = \{Cd^{2+}\}\left(1 + \sum_{i=1}^{4}\frac{{}^*\beta_i}{\{H^+\}^i}\right) \qquad \text{(8.18d)}$$

8.2.3 COMPLEXES WITH OTHER LIGANDS

In addition to water and hydroxide, other dissolved species can act as ligands and form soluble complexes with metal ions. In all such cases, at least a portion of the ligand molecule must have a local excess of electronic (negative) charge which it donates or shares with the metal ion, thereby forming a chemical bond. For instance, ammonia can form strong complexes with some metal ions because the nitrogen atom in an ammonia molecule has an orbital containing an unshared pair of electrons (see Figure 3.7). These electrons, which cause the nitrogen atom to carry a slight negative charge in free NH_3 molecules (much as the oxygen atom does in water molecules), can be shared with metal ions to form the $Me-NH_3$ bonds (Figure 8.1c). The most common *donor atoms* (i.e., atoms that can donate electronic charge to the metal) of environmental significance are O, N, and S.

If the bonds formed between a ligand and a metal ion are strong enough that the ligand replaces water in the inner hydration sphere, the reaction is referred to as **ligand exchange,** and the ligand is said to form an **inner sphere complex,** e.g.,

$$Me(H_2O)_x^{\,n+} + NH_3 \leftrightarrow Me(H_2O)_{x-1}(NH_3)^{n+} + H_2O \qquad \text{(8.19)}$$

The above reaction is more commonly written as

$$Me^{n+} + NH_3 \leftrightarrow Me(NH_3)^{n+} \qquad \text{(8.20)}$$

Yet other ligands bond to the metal ion but remain separated from it by a water molecule. These **outer sphere complexes** are relatively weak and are called **ion pairs.**

The rules for naming and using stability constants for ligands other than hydroxide are essentially identical to those listed above. For instance, for ammonia complexes of Zn^{2+}, the reactions and labels assigned to the various stability constants are as follows:

$$Zn^{2+} + NH_3 \leftrightarrow Zn(NH_3)^{2+} \qquad\qquad \log K_1 = 2.2 \quad \textbf{(8.21)}$$

$$Zn(NH_3)^{2+} + NH_3 \leftrightarrow Zn(NH_3)_2^{2+} \qquad \log K_2 = 2.3 \quad \textbf{(8.22)}$$

$$Zn(NH_3)_2^{2+} + NH_3 \leftrightarrow Zn(NH_3)_3^{2+} \qquad \log K_3 = 2.5 \quad \textbf{(8.23)}$$

$$Zn(NH_3)_3^{2+} + NH_3 \leftrightarrow Zn(NH_3)_4^{2+} \qquad \log K_4 = 2.0 \quad \textbf{(8.24)}$$

$$Zn^{2+} + 2NH_3 \leftrightarrow Zn(NH_3)_2^{2+} \qquad\qquad \log \beta_2 = 4.5 \quad \textbf{(8.25)}$$

$$Zn^{2+} + 3NH_3 \leftrightarrow Zn(NH_3)_3^{2+} \qquad\qquad \log \beta_3 = 7.0 \quad \textbf{(8.26)}$$

$$Zn^{2+} + NH_4^+ \leftrightarrow Zn(NH_3)^{2+} + H^+ \qquad \log {}^*K_1 = -7.0 \quad \textbf{(8.27)}$$

Note that, in this case, the ligand is neutral, so all the complexes are divalent (as is the case when water is the ligand).

8.2.4 MIXED LIGAND COMPLEXES AND CHELATING AGENTS

Complexes that contain more than one type of ligand (other than water) can also form. Such complexes are called **mixed-ligand complexes.** For instance, a reaction leading to formation of a mixed, ammonia-chloro complex is as follows:

$$Me(H_2O)_x^{n+} + Cl^- + NH_3 \leftrightarrow Me(H_2O)_{x-2}(NH_3)Cl^{n-1} + 2H_2O \quad \textbf{(8.28)}$$

Also, some ligands have more than one "donor" atom, in which case one ligand molecule might expel more than one water molecule from the inner hydration sphere. These ligands are called **multidentate** ligands and tend to form strong complexes. If the complexes are very strong, they are called **chelates,** and the ligands are called **chelators** or **chelating agents.** Commercially, the most important chelating agent is the tetraprotic acid EDTA (ethylene diamine tetra-acetic acid), which, in its most dissociated form, has the structure shown in Figure 8.3.

EDTA can bind to a metal to form a hexadentate chelate by replacing six water molecules in the hydration sphere and forming bonds with the metal ion by electron sharing via the two nitrogen atoms and the four carboxylate groups. Thus, a single EDTA molecule can act like six individual monodentate ligands. The binding of EDTA to many metals is so strong that, under many circumstances, EDTA—Me complexes form until either the free EDTA or free metal is essentially 100% complexed. Thus, adding excess EDTA to a solution is an effective way to maintain the activity of free (i.e., uncomplexed) metal ions at extremely low levels.

In addition to applications in industry, EDTA is used extensively as a food additive, at least in part to "tie up" metal ions that might come in contact with the food and prevent them from participating in other reactions with the food. It is also used medically in the treatment of acute metal poisoning, as a way of collecting metal ions from organs and holding them in solution so that they can be excreted.

The fact that metal—ligand complexes form in a system indicates that the potential energy of the system is lower (i.e., the system is more stable) with those complexes present than if the metal and ligand remained separate. Therefore, it is

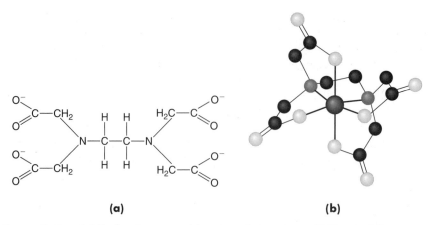

(a)   **(b)**

Figure 8.3   **(a)** The bond structure of the strong chelating agent EDTA, in its fully protonated form (H_4EDTA). **(b)** The structure of a $(Pb\text{-}EDTA)^{2-}$ chelate. In part (b), the stippled circle at the center is the Pb^{2+} ion, and the white, gray, and black circles are oxygen, nitrogen, and carbon atoms, respectively. Hydrogen atoms have been omitted to minimize clutter. The six bonds between the EDTA and Pb are shown in bold.

not surprising that the formation of complexes reduces the likelihood that the metal will undergo other reactions (at least from a thermodynamic perspective; the kinetics might be a different story). Mathematically, this effect is obvious from the fact that when complexes form, the concentration, activity, and molar free energy of the free metal ion decrease. The stronger the complex, the greater the reduction in free metal ion activity, so chelates can reduce metal reactivity very substantially.

One consequence of the lower reactivity of complexed metals compared to the same concentration of free metal is that complexation or chelation tends to stabilize metals in solution at the expense of metals associated with solids. For instance, complexation reduces the tendency for metals to precipitate out of solution as solids or to bind (adsorb) to the surface of solids in contact with the solution. The former fact is exploited extensively in industry, where chelating agents are used to prepare solutions with very high concentrations of metals, under conditions where most of the metal would precipitate if the chelating agent were not present. Similarly, adsorption of metals tends to diminish when they become complexed (Figure 8.4), although there are circumstances (discussed in Chapter 10) under which complexation enhances metal sorption. The same effect (lower chemical activity, and therefore lower reactivity) generally causes uptake of chelated metals by organisms to be much less efficient than uptake of free metal ions, and hence the chelated metals are less toxic, at a given total metal concentration. Stability constants for a variety of metals and ligands are listed in Table 8.3.

8.2.5   METAL SPECIATION IN SYSTEMS CONTAINING COMPLEXING LIGANDS

The equilibrium speciation of metals in systems containing ligands bears a close connection to the speciation of simple acids and bases and can be

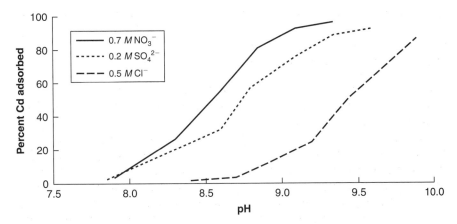

Figure 8.4 Adsorption of Cd^{2+} to $SiO_2(s)$ is diminished when Cl^- or SO_4^{2-} is added to the solution and free Cd^{2+} ions are converted to Cd-Cl or Cd-SO_4 complexes. All solutions contained 30 g/L $SiO_2(s)$ and 5×10^{-7} M TOTCd and had an ionic strength of 0.7 M.

From M. M. Benjamin and J. O. Leckie, *Environmental Science and Technology 16*, 162. Copyright 1982, American Chemical Society. Reprinted with permission.

represented on $\log C$–$\log$ {ligand} diagrams that bear many similarities to $\log C$–pH graphs. Consider, for instance, the speciation of Cd over a range of Cl^- activities in a system containing 10^{-4} M TOTCd. In the preliminary analysis, we will assume that the pH is low enough that OH^- complexes of Cd^{2+} can be ignored. The stability constants for formation of Cd—Cl complexes are given below.

$$Cd^{2+} + Cl^- \leftrightarrow CdCl^+ \qquad \log K_1 = 1.98 \qquad \textbf{(8.29)}$$

$$CdCl^+ + Cl^- \leftrightarrow CdCl_2°(aq) \qquad \log K_2 = 0.62 \qquad \textbf{(8.30)}$$

$$CdCl_2°(aq) + Cl^- \leftrightarrow CdCl_3^- \qquad \log K_3 = -0.20 \qquad \textbf{(8.31)}$$

By writing the equilibrium expression for formation of $CdCl^+$ as ${CdCl^+}/{Cd^{2+}} = K_1{Cl^-}$, the relationships shown in Table 8.4 are easily derived. The table includes analogous expressions for an acid/base pair for comparison.

The only significant difference between the equations for the acid/base and metal/complex systems is that the crossover point is at pH = pK_a in the acid/base system and at pCl = $\log K_1 = -pK_1$ in the metal/complex system. This difference in sign arises because K_a is defined as the equilibrium constant for a *dissociation* reaction, whereas K_1 is defined for an *association* reaction.

Using a spreadsheet, we can determine the concentrations of all Cd species at any given value of {Cl^-}, much as we did in Chapter 3 when developing the $\log C$–pH diagram for a multiprotic acid. The values in the spreadsheet, in essence, represent the simultaneous solutions to the three equilibrium-constant expressions [Equations (8.29) to (8.31)] and the mass balance on Cd. Plotting the

Table 8.3 Stability constants for some metal–ligand complexes. Values shown correspond to $\log \beta$.

	CO_3^{2-}	SO_4^{2-}	Cl^-	F^-	NH_3	PO_4^{3-}	EDTA	CN^-	HS^-
Ag^+		AgL 1.29	AgL 3.27 AgL$_2$ 5.27 AgL$_3$ 5.29 AgL$_4$ 5.51	AgL 0.36			AgL 7.36 AgHL	AgL$_2$ 20.38 AgL$_3$ 21.4	AgL 14.05 AgL$_2$ 18.45
Al^{3+}		AlL 4.92		AlL 7.01 AlL$_2$ 12.75 AlL$_3$ 17.02 AlL$_4$ 19.72 AlL$_5$ 20.8 AlL$_6$ 20.5			AlL 19.8 AlHL 22.5		
Ca^{2+}		CaL 2.91		CaL 0.94		CaHL 15.08			
Cd^{2+}	CdL 5.4 CdL$_3$ 6.22 CdHL 12.4	CdL 2.46 CdL$_2$ 3.5	CdL 1.98 CdL$_2$ 2.6 CdL$_3$ 2.4	CdL 1.1 CdL$_2$ 1.5		CdL 3.9	CdL 16.28 CdHL 2.9	CdL 5.32 CdL$_2$ 10.37 CdL$_3$ 14.83 CdL$_4$ 18.29	CdL 10.17 CdL$_2$ 16.53 CdL$_3$ 18.71 CdL$_4$ 20.9
Co^{2+}		CoL 2.5	CoL 0.5				CoL 18.6 CoHL 21.6		
Cr^{3+}		CrL 1.34	CrL −0.25 CrL$_2$ −0.96	CrL 4.92		CrH$_2$L 22.29			
Cu^{2+}	CuL 6.73 CuL$_2$ 9.83 CuHL 13.6	CuL 2.31	CuL 0.43 CuL$_2$ 0.16 CuL$_3$ −2.29 CuL$_4$ −4.59	CuL 1.26	CuL 5.8 CuL$_2$ 10.7 CuL$_3$ 14.7 CuL$_4$ 17.6	CuHL 16.6	CuL 18.78 CuHL 11.2		CuL$_3$ 25.9
Fe^{2+}		FeL 2.25	FeL 0.90 FeL$_2$ 0.04			FeH$_2$L 22.25	FeL 16.7 FeHL 20.1	FeL$_6$ 52.44 FeHL$_6$ 50 FeH$_2$L$_6$ 45.61	FeL$_2$ 8.95 FeL$_3$ 10.99

Fe³⁺		FeL 3.92; FeL_2 5.42	FeL 1.48; FeL_2 2.13; FeL_3 1.13	FeL 6.2; FeL_2 10.8; FeL_3 14		$FeHL$ 17.78	FeL 27.8; $FeHL$ 29.4	FeL_6 52.63	
Hg²⁺		HgL 1.39	HgL 6.75; HgL_2 13.12; HgL_3 14.02; HgL_4 14.43	HgL 1.98	HgL 8.76; HgL_2 17.43; HgL_3 18.4; HgL_4 19.17			HgL 18.07; HgL_2 34.55; HgL_3 38.3; HgL_4 41.31	HgL_2 37.72
Mg²⁺									
Ni²⁺	NiL 6.87; NiL_2 10.11; $NiHL$ 12.47	NiL 2.29; NiL_2 1.02	NiL 0.4; NiL_2 0.96	NiL 1.3			NiL 20.33; $NiHL$ 11.56	NiL_2 14.59; NiL_3 22.64; NiH_3L_3 43.95	
Pb²⁺	PbL 7.24; PbL_2 10.64; $PbHL$ 13.2	PbL 2.75; PbL_2 3.47	PbL 1.6; PbL_2 1.8; PbL_3 1.7; PbL_4 1.38	PbL 1.25; PbL_2 2.56; PbL_3 3.42; PbL_4 3.1			PbL 17.86; $PbHL$ 9.68; PbH_2L 6.22	PbL_4 10.6	PbL 15.27; PbL_2 16.57
Zn²⁺	ZnL 5.3; ZnL_2 9.6; $ZnHL$ 12.4	ZnL 2.37; ZnL_2 3.28	ZnL 0.43; ZnL_2 0.45; ZnL_3 0.5; ZnL_4 0.2	ZnL 1.15	ZnL_4 44.54	$ZnHL$ 15.7	ZnL 16.44; $ZnHL$ 9	ZnL_2 11.07; ZnL_3 16.05; ZnL_4 16.72	ZnL 14.94; ZnL_2 16.1

Table 8.4   Comparison of relationships between an acid/base and a
metal/complex system

Acid/Base Pair (HA/A⁻)			Metal/Complex Pair (Me/MeL)		
$\{HAc\} = \{Ac^-\}$	at	$pH = pK_a$	$\{CdCl^+\} = \{Cd^{2+}\}$	at	$pCl^- = \log K_1$
$\{HAc\} > \{Ac^-\}$	at	$pH < pK_a$	$\{CdCl^+\} > \{Cd^{2+}\}$	at	$pCl^- < \log K_1$
$\{HAc\} < \{Ac^-\}$	at	$pH > pK_a$	$\{CdCl^+\} < \{Cd^{2+}\}$	at	$pCl^- > \log K_1$

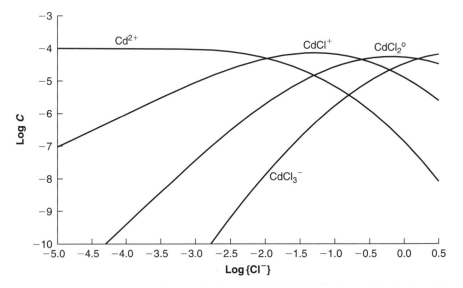

Figure 8.5   Log C–log $\{Cl^-\}$ diagram for a system containing 10^{-4} M *TOT*Cd, at a pH
low enough that formation of Cd-OH complexes is negligible.

corresponding values against $\log\{Cl^-\}$ yields the $\log C$–$\log\{Cl^-\}$ diagram shown
in Figure 8.5.

The sum of the dissolved cadmium species in this system can be represented
as a function of only $\{Cd^{2+}\}$, $\{Cl^-\}$, and the stability constants as follows:

$$TOTCd = \{Cd^{2+}\} + \sum_{i=1}^{3}\{CdCl_i\}$$

$$TOTCd = \{Cd^{2+}\}(1 + \beta_{Cl,1}\{Cl^-\}^1 + \beta_2\{Cl^-\}^2 + \beta_3\{Cl^-\}^3) \qquad \textbf{(8.32)}$$

$$= \{Cd^{2+}\}\left(1 + \sum_{i=1}^{n}\beta_{Cl,i}\{Cl^-\}^i\right) \qquad \textbf{(8.33)}$$

The fraction of *TOT*Cd that is in a given form can then be easily computed. For instance, the fraction of *TOT*Cd that is present as free Cd^{2+} ion is

$$\alpha_{Cd^{2+}} = \frac{[Cd^{2+}]}{TOTCd} = \left(1 + \sum_{i=1}^{3} \beta_{Cl,i}\{Cl^-\}^i\right)^{-1} \qquad \textbf{(8.34)}$$

The activity of Cl^- in seawater is approximately 0.5. Plugging this value and the appropriate values for $\beta_{Cl,i}$ into Equation (8.34), we find $\alpha_{Cd^{2+}} = 5.57 \times 10^{-3}$. In other words, only about 0.5% of the dissolved Cd in seawater is present as Cd^{2+} ions. (In truth, $\alpha_{Cd^{2+}}$ in seawater is even less than this value due to formation of complexes with other ligands in the solution, as described below.) The same approach can be used to compute α values of the other Cd species in the system.

Next, consider a system in which complexation by chloride is significant, but the assumption that the hydroxo complexes are negligible does not necessarily apply. Specifically, say we wished to analyze the Cd speciation in a solution containing 10^{-4} *M TOT*Cd in which $\{Cl^-\}$ is fixed at 0.5 *M*, but the pH is variable. Such a system can be represented by a log *C*–pH diagram for Cd in the system. However, we anticipate that the diagram will be different from Figure 8.2 because of the presence of Cl^-.

We first evaluate the relative concentrations of Cd^{2+} and the various Cd—Cl complexes. Because the reactions among these species do not involve exchange of H^+ or OH^-, the relative concentrations of these species must be independent of pH. This conceptual result can be demonstrated formally by manipulation of the relevant stability constant expressions:

$$Cd^{2+} + iCl^- \leftrightarrow CdCl_i^{2-i} \qquad \textbf{(8.35)}$$

$$\beta_{Cl,i} = \frac{\{CdCl_i^{2-i}\}}{\{Cd^{2+}\}\{Cl^-\}^i} \qquad \textbf{(8.36)}$$

$$\{CdCl_i^{2-i}\} = \beta_{Cl,i}\{Cd^{2+}\}\{Cl^-\}^i$$

$$\log\{CdCl_i^{2-i}\} = \log\beta_{Cl,i} + \log\{Cd^{2+}\} + i\log\{Cl^-\} \qquad \textbf{(8.37)}$$

According to Equation (8.37), if $\{Cl^-\}$ is fixed, as it is in the system of interest, the expression for the activity of each $CdCl_i^{2-i}$ complex can be written in the following form:

$$\log\{CdCl_i^{2-i}\} = \log\{Cd^{2+}\} + \text{constant} \qquad \textbf{(8.38)}$$

The absence of a term containing $\{H^+\}$ in Equation (8.38) indicates that the curve representing any $CdCl_i^{2-i}$ species on a log *C*–pH diagram will be identical to the curve for Cd^{2+}, except that it will be offset vertically from the Cd^{2+} curve by a constant amount. Hence, all the $CdCl_i^{2-i}$ lines are parallel to the Cd^{2+} line.

Figure 8.3 indicates that, at $\{Cl^-\} = 0.5$ (i.e., $\log\{Cl^-\} = -0.3$), the concentrations of all three chloro— complexes are greater than that of Cd^{2+}. As a result, the free, uncomplexed Cd^{2+} ion is never the dominant Cd species in this system. The same reasoning leads us to conclude that $CdCl^+$ and $CdCl_3^-$ will never be dominant, because the activity of $CdCl_2^{\circ}$ is always greater than that of these two species.

Now consider the relative concentrations of Cd^{2+} and the Cd—OH complexes. Because the relationships among these species are independent of $\{Cl^-\}$, the relative concentrations of those species at a given pH are the same as when no Cl^- is present. That is, $\{Cd^{2+}\} = \{CdOH^+\}$ at $pH = p^*K_{OH,1} = 10.08$, $\{CdOH^+\} = \{CdOH_2^{\circ}\}$ at $pH = p^*K_{OH,2} = 10.27$, etc.

Thus, we know the relationships among the activities of Cd^{2+} and the Cd—OH complexes, and those among the activities of Cd^{2+} and the Cd—Cl complexes; the only thing we have not yet determined is how the system speciation arranges itself such that all these equilibrium relationships are satisfied simultaneously. We can explore that issue by calculating the speciation at a particular pH value and then extrapolating the result to the entire range of pH values of interest.

Consider the conditions at pH 9.0. At this pH, the activity of each Cd—OH complex can be expressed in terms of the activity of Cd^{2+} as

$$\{Cd(OH)_i^{2-i}\} = \beta_{OH,i}\{OH^-\}^i = \beta_{OH,i}(10^{-5.0})^i$$

Similarly, since $\{Cl^-\}$ is fixed at 0.5, the activity of each Cd—Cl complex can be expressed in terms of $\{Cd^{2+}\}$ as

$$\{CdCl_i^{2-i}\} = \beta_{Cl,i}\{Cl^-\}^i = \beta_{Cl,i}(0.5)^i$$

Carrying out the various calculations associated with the above equations, we can summarize the relative activities of all the Cd species in solution as follows:

$$\{CdOH^+\} = 8.32 \times 10^{-2}\{Cd^{2+}\}$$
$$\{Cd(OH)_2^{\circ}\} = 4.47 \times 10^{-3}\{Cd^{2+}\}$$
$$\{Cd(OH)_3^-\} = 5.01 \times 10^{-7}\{Cd^{2+}\}$$
$$\{Cd(OH)_4^{2-}\} = 4.47 \times 10^{-12}\{Cd^{2+}\}$$
$$\{CdCl^+\} = 47.7\{Cd^{2+}\}$$
$$\{CdCl_2^{\circ}\} = 99.5\{Cd^{2+}\}$$
$$\{CdCl_3^-\} = 31.4\{Cd^{2+}\}$$

A mass balance on total Cd in the system can therefore be written as follows:

$$TOTCd = \{Cd^{2+}\} + \{CdOH^+\} + \{Cd(OH)_2^{\circ}\} + \{Cd(OH)_3^-\}$$
$$+ \{Cd(OH)_4^{2-}\} + \{CdCl^+\} + \{CdCl_2^{\circ}\} + \{CdCl_3^-\}$$
$$= 179.8\{Cd^{2+}\}$$
$$10^{-4} = 179.8\{Cd^{2+}\}$$
$$\{Cd^{2+}\} = 5.56 \times 10^{-7}$$

Having determined $\{Cd^{2+}\}$, we can plug into the relationships derived above to compute the activity of each of the other Cd-containing species:

$$\{CdOH^+\} = 4.63 \times 10^{-8} \qquad \{CdCl^+\} = 2.66 \times 10^{-5}$$

$$\{Cd(OH)_2^{\circ}\} = 2.48 \times 10^{-9} \qquad \{CdCl_2^{\circ}\} = 5.54 \times 10^{-5}$$

$$\{Cd(OH)_3^-\} = 2.79 \times 10^{-13} \qquad \{CdCl_3^-\} = 1.75 \times 10^{-5}$$

$$\{Cd(OH)_4^{2-}\} = 2.48 \times 10^{-18}$$

If we carried out analogous calculations at other pH values, we could determine the speciation of the system and plot the log C–pH diagram, which is shown in Figure 8.6.

As expected based on the discussion above, the lines for all $CdCl_i^{2-i}$ species are parallel to the Cd^{2+} line. Also, as must be the case, the various acid/base conjugate pairs of the Cd—OH system have equal activities at the corresponding $p*K_{OH,i}$ values, e.g., $\{Cd^{2+}\} = \{CdOH^+\}$ at pH $= p*K_1 = 10.08$. However, at these intersection points, each of the two species does not necessarily account for 50% of $TOTCd$. For instance, at pH 10.08, $\{Cd^{2+}\} = \{CdOH^+\}$, but each of these species represents only about 0.31% of $TOTCd$. The reason for this result, of course, is that the vast majority of the Cd in the system is present as Cd—Cl complexes. The fact that pH $= p*K_{OH}$ does not ensure that each of the acid/base species related by the $*K_{OH}$ expression will account for 50% of the total concentration in the system, only that the acid and conjugate base will be present at equal activities. Thus, the result that $\{Cd^{2+}\} = \{CdOH^+\} = 0.0031 TOTCd$ is entirely consistent with the constraint imposed by the equilibrium constant.

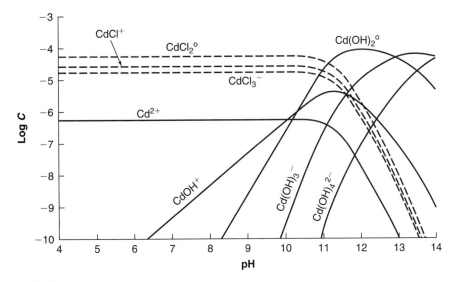

Figure 8.6 Log C–pH diagram for a system containing 10^{-4} M $TOTCd$ and $\{Cl^-\} = 0.5$.

The acidic waters that drain from mines can contain large concentrations of metals. When these waters mix with others or are neutralized by some other process, large amounts of metals can precipitate in the stream.

| Inga Spencer/Visuals Unlimited.

To summarize the key points of the above discussion, the presence of complexing ligands in a solution containing a weak acid metal group forces us to consider more equilibrium expressions and more terms in the metal mass balance than in their absence. Correspondingly, the $\log C$–pH diagram contains curves for more species. However, the relationships among those species are the same as in simpler systems. The mass balance for the example system and for a generic system with m different ligands (including OH^-) can be written as follows:

Cl^- and OH^- as ligands:

$$TOTCd = \{Cd^{2+}\} + \sum_{i=1}^{4} \{Cd(OH)_i\} + \sum_{i=1}^{3} \{CdCl_i\}$$

$$= \{Cd^{2+}\}\left(1 + \sum_{i=1}^{4} \beta_{OH,i}\{OH^-\}^i + \sum_{i=1}^{3} \beta_{Cl,i}\{Cl^-\}^i\right) \quad \textbf{(8.39)}$$

m ligands:

$$TOTCd = \{Cd^{2+}\}\left(1 + \sum_{j=1}^{m}\sum_{i=1}^{n} \beta_{L_j,i}\{L_j\}^i\right) \qquad (8.40)$$

Because Equation (8.40) is simply a concise way of writing the mass balance on Cd in a system with multiple ligands, it applies to any system containing Cd and ligands, regardless of what the ligands are or what other reactions affect the ligands' concentrations.

Comparing Figures 8.1 and 8.4, we see that the addition of 0.5 M Cl^- to the solution eliminates the dominance region of $CdOH^+$ and shifts that of $Cd(OH)_2^{\,o}$ to a narrower pH region. We might explain this observation as resulting from the fact that the addition of Cl^- and the corresponding formation of relatively strong Cd—Cl complexes (relative to Cd—OH complexes, that is) diminishes the importance of all Cd—OH complexes in the system, and that in some cases this effect prevents a Cd—OH complex from becoming dominant under conditions where it otherwise would be.

8.3 Use of Chemical Equilibrium Software to Model Systems Containing Metals, Ligands, and Complexes

The analysis of metal speciation using chemical equilibrium software is a trivial extension from the analysis of acid/base speciation. Obviously, the software does not "know" whether two *components* that are combining are a base and an H^+ ion or a metal ion and a ligand. Furthermore, the types of equations that need to be solved to determine the speciation are the same in the two groups of systems, comprising a group of equilibrium constant equations and some mass balances. Therefore, all the procedures for providing the software with information about the total concentrations of *components,* the stoichiometry and equilibrium constants for forming *species,* and the chemical inputs used to prepare the solution are identical in the two types of systems. (The situation does get slightly more complicated in cases where a metal can precipitate out of solution, and we will consider how the software deals with those situations later in the chapter.)

Prepare the input tableau for use of chemical equilibrium software to determine the speciation in the Cd-Cl-OH example system described above (pH 9.0, fixed Cl^- activity of 0.5, $TOTCd = 10^{-4}\,M$).

| **Example 8.1**

Solution

The tableau is prepared exactly as it would be for a system that contained only acid/base species. We can choose Cd^{2+}, Cl^-, and H^+ as *components* (in addition to H_2O, which is included in the *component* list implicitly). All three of the *components* will be present in the equilibrium solution as corresponding Type 1 *species.* In addition, all the Cd—OH and Cd—Cl complexes will be present as Type 2 *species.*

Because we wish to fix the activity of H^+ and Cl^-, we also need to include dummy *species,* which we designate as $\overline{H^+}$ and $\overline{Cl^-}$, as Type 3a *species.* The equilibrium constants for these *species* are computed as the equilibrium constant for the corresponding real *species* H^+ and Cl^- minus the logarithm of the fixed activity that we wish to specify. Because Type 1 *species* always have log K values of 0.0, the log K values of the Type 3a *species* are

$$\log K_{\overline{H^+}} = 0.0 - (-9.00) = +9.00$$

$$\log K_{\overline{Cl^-}} = 0.0 - \log 0.5 = +0.30$$

The reactions forming the Cd—Cl complexes are input as combinations of Cd^{2+} and $i\,Cl^-$ ions, so the appropriate equilibrium constants correspond to $\beta_{Cl,i}$ values. The hydroxo complexes must also be input using the *components* as reactants. However, since OH^- is not a *component,* the input log K values for these complexes do not correspond to log β_{OH^-} values. Rather, for each complex $Cd(OH)_i^{2-i}$, the input log K is the one applicable to the following reaction:

$$-i\,\mathbf{H^+} + 1\,\mathbf{Cd^{2+}} + 0\,\mathbf{Cl^-} + i\,\mathbf{H_2O} \leftrightarrow 1Cd(OH)_i^{2-i}$$

The equilibrium constants for these reactions are the ones designated $*\beta_{OH,i}$, so those are the ones that must be specified in the input data. The complete input data tableau is shown below.

| Species | Type | Component | | | Log K |
		H^+	Cd^{2+}	Cl^-	
H^+	1	1	0	0	0.00
Cd^{2+}	1	0	1	0	0.00
Cl^-	1	0	0	1	0.00
OH^-	2	−1	0	0	−14.00
$CdOH^+$	2	−1	1	0	−10.08
$Cd(OH)_2^\circ$	2	−2	1	0	−20.35
$Cd(OH)_3^-$	2	−3	1	0	−33.30
$Cd(OH)_4^{2-}$	2	−4	1	0	−47.35
$CdCl^+$	2	0	1	1	1.98
$CdCl_2^\circ$	2	0	1	2	2.60
$CdCl_3^-$	2	0	1	3	2.40
$\overline{H^+}$	3a	1	0	0	9.00
$\overline{Cl^-}$	3a	0	0	1	0.30
TOT Comp$_{eq}$		x_1	10^{-4}	x_2	

Inputs				Input Concentration
H^+	1	0	0	x_1
Cd^{2+}	0	1	0	1.0×10^{-4}
Cl^-	0	0	1	x_2

Although the stoichiometric coefficients and the log K values for all the Type 1 and 2 *species* of interest have been computed for inclusion in the tableau most available software packages obtain this information from a database automatically. As a result, all we need to specify is the identities of the components, and the program fills in all the rows for the Type 1 and 2 *species* automatically. In some software packages, we have to input the reactions and log K values for the Type 3a *species,* to specify the fixed pH and Cl^- activity of the solution.

The input concentrations for H^+ or Cl^- have been listed in the tableau as x_1 and x_2, respectively. No information is provided about how much of these *components* was added to prepare the solution. As shown in previous chapters, that information could be obtained once the equilibrium speciation of the solution was known, by equating $TOTH_{in}$ and $TOTCl_{in}$ with $TOTH_{eq}$ and $TOTH_{eq}$, respectively.

Most software packages require that some value be input for the total concentration of each *component.* If we were dealing with such a program, we could input a value of zero for $TOTH$ and $TOTCl$, indicating that the original solution was prepared without addition of H^+ or Cl^-. The program would then "add" as much H^+ or Cl^- as necessary to reach the specified, fixed activities of these *species*. In the current example, the program would determine that Cl^- must be added, but that the required addition of H^+ is negative; i.e., H^+ must be removed from solution (OH^- must be added) to reach pH 9.1.

8.4   Metal Ion Buffers

As shown in Chapter 5, an understanding of acid/base chemistry can help us choose appropriate chemicals to add to a solution if we wish to buffer the pH at certain values. Given the similarities between acid/base and metal/complex chemistry, it is not surprising that the activity of metal ions in solution can be buffered just as the activity of H^+ can be. Buffering of the activity of free metal ions is especially important in biological systems, because this activity can control many metabolic reactions and because free metal ions are thought to be more toxic than most complexes. As a result, organisms have developed the capability to maintain the activity of Me^{n+} within narrow bounds, even if the concentration of total Me in the system changes quite dramatically.

The critical feature of a good acid/base buffering system is the presence of a species that can readily combine with or release H^+. Furthermore, that species must be present at a concentration that is much greater than the desired activity of free H^+. The conditions for optimal buffering are that one-half of the buffering species is bound to an H^+ ion, while the other one-half is not. In this way, there is a reservoir of H^+ that can be released from the buffer if the H^+ activity in solution declines, and there is a reservoir of sites where H^+ can attach if more H^+ enters solution.

Similarly, a chemical that can serve as a good buffer for Me^{n+} must be a ligand that is present at a much higher concentration than the desired concentration of free Me^{n+} and that can combine with or release Me^{n+} readily if the total Me concentration in solution changes. We expect that a given ligand will serve as an effective buffer only for certain limited ranges of $\{Me^{n+}\}$, just as a given acid group is an effective buffer only over one or more limited pH ranges (near the pK_a values).

Specifically, we expect a ligand to buffer the Me^{n+} activity most effectively under conditions where the ligand is approximately equally split between species with different Me—L stoichiometries. For instance, for a metal and ligand that form only a single complex MeL, the buffering would be expected to be maximized when one-half of the ligand was free L and the other half was complexed with the metal. This situation applies when $\{Me^{n+}\} = 1/K_1$ (i.e., $\log \{Me^{n+}\} = pK_1$).

Example 8.2

Ethylene diamine (EN) is a relatively strong complexing agent for many metals. EN contains the structural core of EDTA, but lacks the four carboxyl groups that make EDTA such a strong chelator. The amine (ammonia-like) groups in EN can protonate, so EN is a weak base. The corresponding diprotic acid, H_2EN, has pK_{a1} and pK_{a2} values of 6.89 and 9.96, respectively. The deprotonated molecule can form complexes with Cu^{2+} with $\log K_{EN,1} = 10.49$ and $\log \beta_{EN,2} = 19.62$.

A $\log C$–$\log \{Cu^{2+}\}$ diagram for a solution at pH 7.0 that contains $10^{-3} M$ TOTEN is shown below. Based on the diagram, under what conditions would EN be a good buffer for free Cu^{2+}?

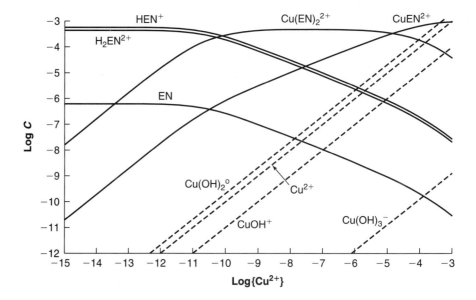

Solution

The Cu^{2+} activity is well buffered under conditions where addition of Cu causes some complexing group in solution to change speciation and substantially increase the amount of Cu^{2+} that it binds. According to the $\log C$–$\log \{Cu^{2+}\}$ diagram, this condition is met near pCu^{2+} values of ~10.1 and ~4.9. The plot on the following page showing the response of $\log \{Cu^{2+}\}$ to various additions of Cu (increases in TOTCu) confirms this expectation. Let's consider why the titration curve has the shape it does.

At $\log \{Cu^{2+}\}$ significantly less than -10 (very low Cu^{2+} activities), most of the EN is protonated but not complexed with Cu^{2+}. A large fraction of TOTCu is complexed ($\alpha_{Cu^{2+}} = 10^{-7.2}$), so the activity of Cu^{2+} is very small. More importantly, there is a large

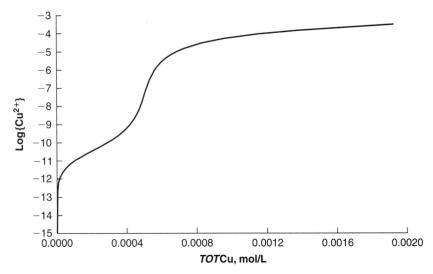

pool of uncomplexed EN in the system, so a great deal of Cu-complexing capacity is unused. When log $\{Cu^{2+}\}$ increases to approximately -11, this pool begins to be used, and by the time log $\{Cu^{2+}\}$ has increased to -9, most of the EN is present in $Cu(EN)_2^{2+}$ complexes. This transition consumes one Cu ion for every two EN molecules in the system, so the titration curve has a "flat" portion with a width of approximately 0.5 *TOT*EN, or $5 \times 10^{-4} M$.

Once essentially all the EN is complexed, further additions of Cu again have a larger effect on the Cu^{2+} activity, and the titration curve steepens, until log $\{Cu^{2+}\}$ gets large enough to induce a second dramatic change in EN speciation. This change converts most of the EN in solution from $Cu(EN)_2^{2+}$ to $Cu(EN)^{2+}$, a process that consumes an additional $5 \times 10^{-4} M$ Cu. The titration curve therefore has a second flat region, near the transition from $Cu(EN)_2^{2+}$ to $Cu(EN)^{2+}$, near log $\{Cu^{2+}\} = -4.9$.

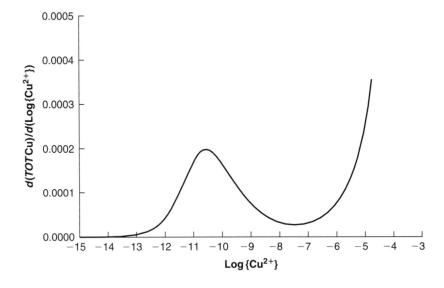

At slightly higher log $\{Cu^{2+}\}$ values (≈ -4), the capacity of EN molecules to complex Cu^{2+} ions has been fully utilized, and Cu species that are not complexed to EN $[Cu(OH)_2{}^o$ and $Cu^{2+}]$ become the dominant species. Under those conditions, the presence of the EN becomes irrelevant. The titration curve in this region would be flat regardless of the presence or absence of EN for the same reason that acid titration curves tend to be flat at very high H^+ activity (low pH).

The buffer intensity plot shown on the preceding page highlights the buffering of the EN near log $\{Cu^{2+}\}$ of -10.1 and also reinforces the fact that any buffering provided by the EN near log $\{Cu^{2+}\}$ of -4.9 blends with the non-EN buffering at higher Cu^{2+} activities.

Thus, the Cu activity is well buffered under conditions where the amount of Cu bound to EN species is undergoing a significant transition, but the EN provides minimal buffering under conditions where its speciation is relatively stable. The similarities to acid/base buffering are clear.

As the preceding example illustrates, strong complexing agents can buffer the activity of free metal ions at extremely low values—the stronger the complexing agent, the lower the Me^{2+} activity that it buffers—making the ligands powerful metabolic tools. That is, by producing an appropriate complexing agent, organisms can retain a pool of an essential metal ion, and yet maintain the activity of the free metal ion at a very low and stable level. If, for some reason, the organism acquires additional metal, the complexant binds it and prevents the free metal ion activity from increasing to potentially toxic levels. By the same token, if the environment becomes depleted in metals, the complexes dissociate to release free metal ions and reduce the danger that the organism will suffer from a metal ion deficiency. As is the case for acid/base buffers, the buffering intensity provided by a complexing agent is directly proportional to its concentration.

8.5   PREDOMINANCE AREA DIAGRAMS

Throughout this and previous chapters, we have displayed information about the speciation of a system as a function of pH on log C–pH diagrams. However, any such diagram can only reflect a single set of conditions for all variables other than pH. For instance, the log C–pH diagrams for Cd speciation shown in Figures 8.1 and 8.4 are for Cl^- activities of 0 and 0.5, respectively. Similarly, Figure 8.3 displays the speciation of Cd as a function of Cl^- activity, but it is applicable only for pH values at which the activity of free Cd^{2+} is much greater than that of any Cd—OH complex.

If we wished to display speciation graphically as a function of two independent variables, we could do so by using a three-dimensional plot with, for instance, pH and log $\{Cl^-\}$ varying along the x and y axes, respectively, and log C plotted on the z (vertical) axis. Any vertical plane slicing through such a plot at a fixed value of $\{Cl^-\}$ would be a conventional log C–pH diagram for that Cl^- activity, and any vertical plane slicing through the plot at a fixed pH would be a conventional log C–log $\{Cl^-\}$ diagram for that pH (Figure 8.7).

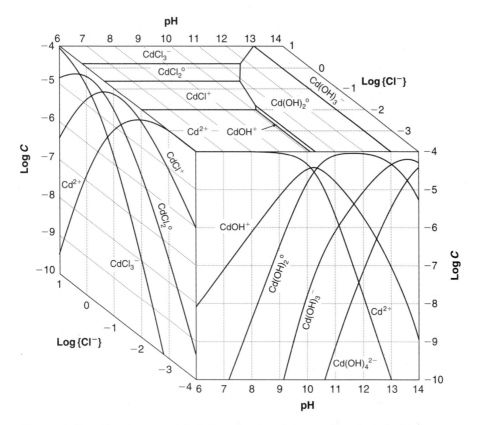

Figure 8.7 Two-dimensional slices through a three-dimensional log C–pH–log{L} diagram.

While such three-dimensional diagrams are conceptually very informative, as a practical matter they are difficult to prepare, visualize, and use. On the other hand, drawing a separate diagram for each value of pH and/or Cl^- concentration of interest also has obvious drawbacks. As an alternative, consider the information that can be displayed by a horizontal slice through the three-dimensional diagram described above, at a fixed value of log C equal to log (*TOT*Cd). Such a slice would have axes representing values of log $\{Cl^-\}$ and pH. In contrast to the diagrams we have used previously, the information in the body of the plot would not indicate the concentrations of various species in the system. Rather, at each point on the plot (i.e., at each point in log $\{Cl^-\}$–pH space), the diagram would identify only the species present at a concentration approximately equal to *TOT*Cd; i.e., it would identify the dominant Cd-containing species in the system. Such diagrams are called **predominance area diagrams.**

Although predominance area diagrams lack detailed information about the concentrations of non-dominant species in the system, they can display information over a large range of activities of two different ligands, identifying

Table 8.5 Predominance relationships among carbonate species

Species being compared	H_2CO_3, HCO_3^-	HCO_3^-, CO_3^{2-}	H_2CO_3, CO_3^{2-}
Equilibrium relationship	$K_{a1} = \dfrac{\{HCO_3^-\}\{H^+\}}{\{H_2CO_3\}}$	$K_{a2} = \dfrac{\{CO_3^{2-}\}\{H^+\}}{\{HCO_3^-\}}$	$K_{a1}K_{a2} = \dfrac{\{CO_3^{2-}\}\{H^+\}^2}{\{H_2CO_3\}}$
Conditions when species have equal activities	$pH = pK_{a1} = 6.35$	$pH = pK_{a2} = 10.33$	$pH = \dfrac{pK_{a1} + pK_{a2}}{2} = 8.34$
Comparison of species' activities at various pH values	At pH < 6.35, $\{H_2CO_3\} > \{HCO_3^-\}$; at pH > 6.35, $\{H_2CO_3\} < \{HCO_3^-\}$	At pH < 10.33, $\{HCO_3^-\} > \{CO_3^{2-}\}$; at pH > 10.33, $\{HCO_3^-\} < \{CO_3^{2-}\}$	At pH < 8.34, $\{H_2CO_3\} > \{CO_3^{2-}\}$; at pH > 8.34, $\{H_2CO_3\} < \{CO_3^{2-}\}$

the species accounting for most of the mass of the metal (or other constituent of interest) anywhere in that two-dimensional space. In this section, we develop a predominance area diagram for a model metal-ligand system. As an introduction, however, it is useful to consider the analogous, but much simpler, one-dimensional predominance diagram that characterizes speciation of an acid/base group as a function of pH.

For instance, to characterize dominance in the carbonate acid/base group, we need consider only three species. The three pairwise comparisons among these species are summarized in the three right-hand columns of Table 8.5. Each comparison divides the pH scale into two portions (summarized in the bottom two rows of the table) that differ with respect to which of the species being compared is dominant over the other. The two regions are separated by a single point at which the activities of the two species are equal. Note that the transition point at which dominance switches between H_2CO_3 and CO_3^{2-} (pH 8.34) can be determined by writing the equilibrium constant for the reaction between these species ($K_{02} = K_{a1}K_{a2}$), even though it turns out that neither species is dominant at that pH when the whole carbonate group is considered.

Consideration of all three pairwise comparisons indicates the pH region where each species is dominant. For instance, the activity of HCO_3^- is greater than that of H_2CO_3 at all pH values greater than 6.35, and it is greater than that of CO_3^{2-} at all pH values less than 10.33, so it is dominant over all (i.e., both) other possible species if and only if the pH is between these two values. Similar considerations for the other two species allow us to prepare what might be called a predominance line diagram for the system, shown below.

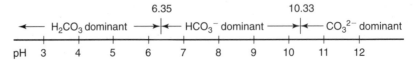

To extend the above analysis to systems in which two variables are considered (e.g., pH and chloride activity), we note that the conditions under which predominance switches between any two species can be determined by evaluating the equilibrium expression relating that pair of species. The region where a particular species is dominant over all the other species can then be identified by considering, simultaneously, the comparisons between that species and all others.

Consider, for example, a system containing Fe(II) at some fixed but unspecified total concentration and two complexing ligands (Cl^- and OH^-) whose activities might vary widely. The reactions characterizing such a system are as follows:

$$Fe^{2+} + H_2O \leftrightarrow FeOH^+ + H^+ \qquad *K_{OH,1} = 10^{-9.50}$$

$$Fe^{2+} + 2H_2O \leftrightarrow Fe(OH)_2{}^\circ + 2H^+ \qquad *\beta_{OH,2} = 10^{-20.57}$$

$$Fe^{2+} + 3H_2O \leftrightarrow Fe(OH)_3{}^- + 3H^+ \qquad *\beta_{OH,3} = 10^{-31.0}$$

$$Fe^{2+} + Cl^- \leftrightarrow FeCl^+ \qquad K_{Cl,1} = 10^{0.90}$$

$$FeCl^+ + Cl^- \leftrightarrow FeCl_2{}^\circ \qquad K_{Cl,2} = 10^{0.04}$$

By combining the top three equations in the above list, we can compute $p*K_{OH,1}$, $p*K_{OH,2}$, and $p*K_{OH,3}$ for Fe^{2+} as 9.50, 11.07, and 10.43, respectively. Based on these values, pairwise comparisons among the $Fe(OH)_x{}^{2-x}$ species can be made, yielding the results shown schematically below. Note that $Fe(OH)_2{}^\circ$ is never simultaneously dominant over its conjugate acid and conjugate base, so it is not the dominant Fe species under any conditions. Rather, the dominant Fe—OH complex switches from $FeOH^+$ directly to $Fe(OH)_3{}^-$ at pH 10.75.

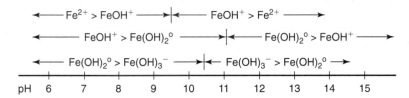

When Fe—Cl complexes are included, the list of possible dominant species includes Fe^{2+}, $FeOH^+$, $Fe(OH)_3{}^-$, $FeCl^+$, and $FeCl_2{}^\circ$. Pairwise comparisons of all ten possible combinations of these species are summarized in Table 8.6. The table shows the conditions under which the two species being compared have equal activities as well as the equation of a line in log $\{Cl^-\}$–pH space along which those conditions are met.

We can choose any of these species to start the analysis, so we begin with the simplest species, Fe^{2+}. The four pairwise comparison between the activity of Fe^{2+} and the activities of other possible dominant species correspond to lines 1, 2, 4, and 5 in the table. These lines are plotted in Figure 8.8. Lines 1 and 2 are vertical (i.e., parallel to the log $\{Cl^-\}$ axis), indicating that the dominant species of those pairs depends on pH but not on $\{Cl^-\}$. For instance, Fe^{2+} is dominant over $FeOH^+$ at pH < 9.50 regardless of the dissolved chloride activity. Similarly,

Table 8.6   Equilibrium relationships used to develop a predominance area diagram for the Fe^{2+}-Cl^--OH^- system

Line	Species Compared	Conditions When Species Have Equal Activities	Equation of Line
1	$Fe^{2+}/FeOH^+$	$\{H^+\} = {}^*K_{OH,1}$	$pH = 9.50$
2	$Fe^{2+}/Fe(OH)_3^-$	$\{H^+\} = {}^*\beta_{OH,3}{}^{1/3}$	$pH = 10.33$
3	$FeOH^+/Fe(OH)_3^-$	$\{H^+\} = ({}^*\beta_{OH,3}/{}^*K_1)^{1/2}$	$pH = 10.75$
4	$Fe^{2+}/FeCl^+$	$\{Cl^-\} = (K_{Cl,1})^{-1}$	$\log\{Cl^-\} = -0.90$
5	$Fe^{2+}/FeCl_2^\circ$	$\{Cl^-\} = (K_{Cl,1}K_{Cl,2})^{-1/2}$	$\log\{Cl^-\} = -0.47$
6	$FeCl^+/FeCl_2^\circ$	$\{Cl^-\} = (K_{Cl,2})^{-1}$	$\log\{Cl^-\} = -0.04$
7	$FeCl^+/FeOH^+$	$\{H^+\}\{Cl^-\} = {}^*K_{OH,1}(K_{Cl,1})^{-1}$	$\log\{Cl^-\} = -10.40 + pH$
8	$FeCl^+/Fe(OH)_3^-$	$\{H^+\}^3\{Cl^-\} = {}^*\beta_{OH,3}(K_{Cl,1})^{-1}$	$\log\{Cl^-\} = -31.90 + 3pH$
9	$FeCl_2^\circ/FeOH^+$	$\{H^+\}\{Cl^-\}^2 = {}^*K_{OH,1}(K_{Cl,1}K_{Cl,2})^{-1}$	$\log\{Cl^-\} = 0.5(-10.44 + pH)$
10	$FeCl_2^\circ/Fe(OH)_3^-$	$\{H^+\}^3\{Cl^-\}^2 = {}^*\beta_3(K_{Cl,1}K_{Cl,2})^{-1}$	$\log\{Cl^-\} = 0.5(-31.94 + 3pH)$

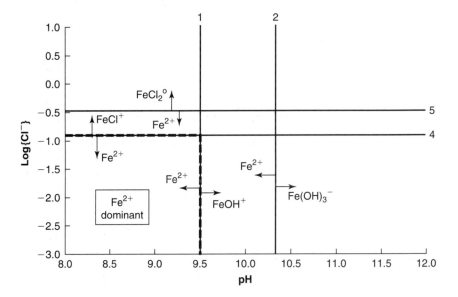

Figure 8.8   Lines showing the four pairwise comparisons between Fe^{2+} and other species in the example system. The arrows indicate the dominant species of the pair on each side of the line. The numbers correspond to the line number column in Table 8.6.

lines 4 and 5 are horizontal, indicating that the identity of the dominant species of those pairs depends on $\{Cl^-\}$ but not on pH. Consideration of all the lines makes it clear that the region of Fe^{2+} dominance covers the area where pH < 9.5 *and* $\log\{Cl^-\} < -0.9$. The borders of this area are outlined on the figure.

If we now wished to identify the region where $FeOH^+$ dominates, we would only need to consider the three pairwise comparisons corresponding to lines 3, 7,

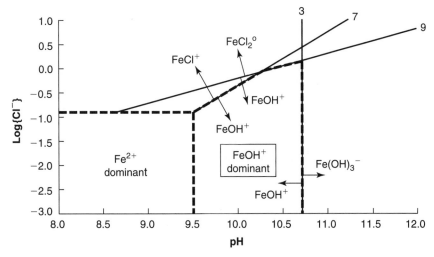

Figure 8.9   Lines showing the pairwise comparisons between $FeOH^+$ and $Fe(OH)_2^\circ$, $FeCl^+$, or $FeCl_2^\circ$, and the area of dominance of $FeOH^+$ inferred from those lines.

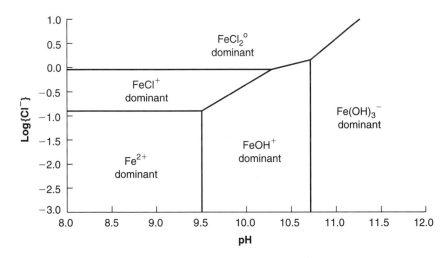

Figure 8.10   Predominance area diagram for the system $Fe^{2+}/Cl^-/OH^-$.

and 9 in Table 8.6, because the comparison with Fe^{2+} has already been made. Lines 7 and 9 are neither horizontal nor vertical, because the equations of those lines include both $\{OH^-\}$ and $\{Cl^-\}$. The region bounded by the FeOH side of these lines is highlighted in Figure 8.9, establishing the area of dominance of $FeOH^+$. Continuation of this process allows completion of the predominance area diagram, which is shown in Figure 8.10.

A review of the patterns and equations of the lines in the final diagram can provide some hints about how this and similar diagrams can be prepared more easily. First, note that it is very easy to draw lines that separate the regions of dominance between a pair of species that differ only in the number of hydroxide or chloride ions they contain. Such lines are vertical (if the species differ in the number of hydroxide ions) or horizontal (if they differ in the number of chloride ions).

The equations relating species that differ in the number of both hydroxide and chloride ligands are slightly more complicated, but they also follow a consistent pattern. The reactions and equilibrium constants relating such species can be written in a generalized form as follows:

$$\text{Fe(OH)}_x^{2-x} + x\text{H}^+ + y\text{Cl}^- \leftrightarrow \text{FeCl}_y + x\text{H}_2\text{O} \qquad \textbf{(8.41)}$$

$$K = \frac{\{\text{FeCl}_y^{2-y}\}}{\{\text{Fe(OH)}_x^{2-x}\}\{\text{H}^+\}^x\{\text{Cl}^-\}^y} \qquad \textbf{(8.42)}$$

The line along which the species being compared are present at equal activity is given by substituting the relationship $\{\text{FeCl}_y^{2-y}\} = \{\text{Fe(OH)}_x^{2-x}\}$ into Equation (8.42), yielding

$$K = \frac{1}{\{\text{H}^+\}^x\{\text{Cl}^-\}^y}$$

$$\log K = -\log\{\text{H}^+\}^x - \log\{\text{Cl}^-\}^y = x\text{pH} - y\log\{\text{Cl}^-\}$$

$$\text{pH} = \frac{1}{x}\log K + \frac{y}{x}\log\{\text{Cl}^-\}$$

$$\log\{\text{Cl}^-\} = \frac{x}{y}\text{pH} - \frac{1}{y}\log K \qquad \textbf{(8.43)}$$

Since $\log\{\text{Cl}^-\}$ and pH are the ordinate and abscissa, respectively, of the predominance area diagram, Equation (8.43) is the equation of a straight line on the diagram with slope equal to x/y; i.e., the slope is the ratio of the number of protons exchanged to the number of chloride ions exchanged in the reaction between the two species being compared.

We now return to the predominance area diagram for the $\text{Fe}^{2+}/\text{Cl}^-/\text{OH}^-$ system to see how this information might have been helpful. As noted, the easiest lines to draw are those that are vertical or horizontal; in the current case, this group includes lines 1 through 6 in Table 8.6, which are shown in Figure 8.11.

If these six lines are drawn on the diagram, the remaining three lines can be drawn immediately by applying Equation (8.43). For instance, Figure 8.11 includes a vertical line along which $\{\text{Fe}^{2+}\} = \{\text{FeOH}^+\}$ (line 1) and a horizontal line along which $\{\text{Fe}^{2+}\} = \{\text{FeCl}^+\}$ (line 4). Therefore, it must be the case that $\{\text{FeOH}^+\} = \{\text{FeCl}^+\}$ at the intersection of these two lines (indicated by a circle in Figure 8.11). Furthermore, according to Equation (8.43), the slope of the line defining the boundary between FeOH^+ and FeCl^+ on the diagram must be 1 [since for these species, $x = y = 1$ in reaction (8.41)]. Since we have now identified the slope of the $\text{FeOH}^+/\text{FeCl}^+$ line and one point that the line passes through, we can draw the line directly.

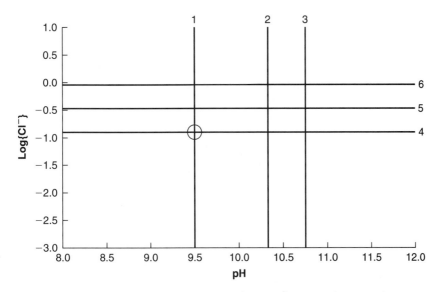

Figure 8.11 Predominance area diagram for the $Fe^{2+}/OH^-/Cl^-$ system, showing only the vertical and horizontal lines that separate regions of dominance of different species. As discussed in the text, the line separating regions of dominance of $FeOH^+$ and $FeCl^+$ must pass through the intersection point that is circled, and it must have a slope of 1.0.

Similarly, the $FeCl^+/Fe(OH)_3^-$ line must pass through the intersection of the $FeCl^+/Fe^{2+}$ and $Fe(OH)_3^-/Fe^{2+}$ lines (lines 4 and 2, respectively) and have a slope of 3 ($x = 3$, $y = 1$); the $FeCl_2^0/Fe(OH)_3^-$ line must pass through the intersection of the $FeCl_2^0/Fe^{2+}$ and $Fe(OH)_3^-/Fe^{2+}$ lines (lines 5 and 2) and have a slope of 1.5 ($x = 3$, $y = 2$); and so on.

The above discussion points out a simple way to draw certain lines on the predominance area diagram and also leads to the conclusion that whenever two lines intersect on a predominance area diagram (say, one for the species A/species B comparison and the other for the species B/species C comparison), a third line must pass through the same intersection point (the line for the species A/species C comparison). Furthermore, once all the relevant comparisons are made, three line segments will emanate from any such intersection point. These line segments divide the space immediately surrounding the intersection point into three regions, in each of which a different species (A, B, or C) is dominant over the other two. Equation (8.43) indicates that the three line segments must all have different slopes. Review of Figure 8.10 confirms that these rules apply to all the intersection points shown.

8.6 SUMMARY OF SOLUBLE ME-L CHEMISTRY

This concludes our exploration of metal chemistry in systems that contain only dissolved species. The key points of the discussion can be summarized as follows.

The cationic charge on dissolved metal ions generates an attraction for the unshared electrons in the orbitals of the oxygen atom of water molecules. This attraction constitutes an Me—H_2O bond that can range from very weak to very strong. The formation of the bond causes the electrons that are shared between the oxygen and the hydrogens in the water molecule to shift toward the oxygen, weakening the O—H bonds, so that the acidity of these water molecules is enhanced. In some cases, the Me—O bond is so strong that both hydrogen ions are released from the water molecule. Although hydrolysis reactions occurring adjacent to a metal ion are just that—the splitting of H_2O—they are often represented as the bonding of a free metal ion (Me^{n+}) to a free hydroxide ion, obscuring the fact that OH^- ions reside in the hydration sphere with other unhydrolyzed water molecules.

Other ligands can replace water molecules or hydroxide ions in the coordination sphere of a metal. Because the coordination sphere is always fully occupied, these reactions always involve ligand exchange. Again, however, the reactions are typically represented as the binding of the ligand to the free metal. Almost all ligands that are environmentally significant bind to metal ions through atoms that have unshared electrons, primarily oxygen, nitrogen, and sulfur. Ligands that bind metals especially strongly and do so via several binding groups are called chelators. Chelators play an important role in the bioenvironmental chemistry of metals, since evidence suggests that chelated metals are often unavailable to biota.

Software for modeling equilibrium chemical speciation in a solution can be easily adapted to the study of metal speciation. This is so because the reactions relating metals and ligands to complexes are direct analogs of those relating protons and bases to conjugate acids. The analogy between metal/ligand and acid/base systems can be extended to the idea of metal ion buffering as well.

Often, we are interested in metal speciation as a function of both pH and the concentration of another ligand in a system. Ideally, the speciation could be represented on a three-dimensional log C–pH–log $\{L\}$ diagram. However, since such diagrams can become very complex, we sometimes choose to display only a horizontal "slice" through the diagram, indicating the species that is dominant in any particular region of pH–log $\{L\}$ space. Diagrams that represent this space are referred to as predominance area diagrams.

We next turn our attention to the characteristic ability of metals to form solids in addition to soluble complexes. As with the analogies between soluble metal/ligand chemistry and acid/base chemistry, analogies between solid/solution equilibrium and gas/solution equilibrium will help us understand the formation of solids in these systems. The major differences that we need to consider arise not from conceptual issues, but from the multiplicity of reactions in which metals can participate to form both soluble and particulate products, adding complexity to the analysis.

8.7   INTRODUCTION TO PRECIPITATION REACTIONS: METAL OXIDES AND HYDROXIDES

As the concentration of metal and hydroxyl ions in a solution increases, some of the metals start forming larger complexes. For instance, in the case of Fe(III), well-defined molecules containing two Fe^{3+} ions (*dimers*) may form as shown in reaction (8.44a), which is commonly written in the short-hand form shown in reaction (8.44b):

$$2Fe(H_2O)_5OH^{2+} \leftrightarrow (H_2O)_5Fe\!-\!O\!-\!Fe(H_2O)_5^{4+} + H_2O \quad \textbf{(8.44a)}$$

$$2FeOH^{2+} \leftrightarrow Fe_2(OH)_2^{4+} \quad\quad\quad\quad\quad\quad\quad \textbf{(8.44b)}$$

Trimers, tetramers, and higher-order complexes may form as well. In each step, an additional Fe-containing sub-unit (a *monomer*) is added. For some metals, the existence of some quite large complexes, such as $Al_8(OH)_{20}^{4+}$, has been well established. Molecules made of several monomers are called polymers. In complexes containing a single metal ion, the metal ion can be thought of as the core or nucleus of the complex. Such complexes are sometimes referred to as **mononuclear complexes.** Correspondingly, complexes containing more than one metal ion are sometimes called **polynuclear complexes.**

If the polymerization process continues, eventually a three-dimensional molecular network builds up, and many of the ions on the interior lose contact with the bulk aqueous solution. In other words, the polymer becomes a separate phase from the water. When the new phase first forms, it may not be visible or easily separated from the bulk solution, complicating any effort to identify the exact conditions when precipitation begins. However, many large metal-containing polymers are stable over only a very limited range of conditions. In such cases, the system changes abruptly from one containing mostly small, soluble monomers and polymers to one containing a distinct particulate metal phase, and the initiation of precipitation is relatively unambiguous.

Although the properties of very large polymers and/or very small particles are extremely interesting, they are not the subject matter of this text. We concern ourselves here only with small molecules (monomers and dimers) which are clearly dissolved and large particles which are clearly not in solution.

As is the case for forming soluble complexes, metals can form solids with a variety of ligands. In fact, the ligands that can form soluble complexes are often the same ones that can combine with the metals to form solids. For instance, many metals of interest can form soluble complexes and/or solids with hydroxide, carbonate, phosphate, and sulfide ions. We now turn to the task of identifying and solving the equations (mass balances, equilibrium equations, etc.) that characterize systems in which solids might precipitate. As always, analysis of the system composition involves finding a set of concentrations that meets the constraints imposed by these equations.

We begin by noting that any equations that describe relationships among soluble species in the system apply regardless of whether or not a solid is present. That is, in any solution containing a metal ion and a ligand (Me, L), if soluble complexes between Me and L can form, they will do so, and the activities of the complexes will be related to the activities of free (uncomplexed) metal and ligand species by the relevant stability constants. The complexes might be present at very low concentrations in some cases, but they will always be present at some non-zero concentration.

By contrast, the fact that a solid *could* form between Me and L (as indicated by the availability of an equilibrium constant for the precipitation reaction) does not ensure that such a solid *will* form and be present at equilibrium. Solids sometimes precipitate in systems that initially contain only dissolved species, while in other systems solids that are present initially dissolve completely before equilibrium is attained. Obviously, if a certain solid is not present, the equilibrium expressions involving it are irrelevant (just as we do not concern ourselves with gas/liquid equilibrium if no gas phase is present).

The uncertainty about whether a particular solid will be present at equilibrium distinguishes these systems from those that we have considered previously. In light of this uncertainty, the only way to conduct the analysis is to make an assumption about whether the solid will be present or absent once the system equilibrates and to solve the system of equations based on this assumption. The computed equilibrium condition can then be checked for consistency with the assumption; if the assumption turns out to be wrong, the alternative assumption must be made, and the process must be repeated. In the following section, we consider how the set of equations characterizing the system differs depending on which assumption is made. First, though, we review material from earlier chapters on how the chemical activity of the solid is defined, and we define some terms applicable to systems where precipitation must be considered.

8.7.1 THE CHEMICAL ACTIVITY OF SOLIDS: DEFINITION AND EXPLANATION

Recall from Chapter 1 that the activity of any species is defined as the product of two terms, one expressing the concentration of the constituent relative to its concentration in the standard state and the other characterizing the chemical environment relative to the environment in the reference state. Recall also that concentrations of constituents in solids are conventionally expressed in terms of mole fractions, and that the standard state concentration is a mole fraction of 1.0, i.e., pure solid.

The reference state environment for molecules in a solid phase is also conventionally defined as the environment in the pure solid. Therefore, if we are actually dealing with a pure solid of, say, $Cd(OH)_2(s)$, both the concentration of $Cd(OH)_2$ molecules in the solid and the environment of those molecules correspond to standard state conditions, so the activity of the solid is 1.0. We will consider almost exclusively solids that are in their standard state, but it is important to recognize that if a solid phase contains a mixture of solid species [e.g., a "solid solution" of $Cd(OH)_2(s)$ and $CdCO_3(s)$], the activity of each constituent is likely to be quite different from 1.0, both because of the dilution of one constituent by the other and because, in such a solid, the molecular environment near each molecule is different from that in the reference state.

Another point that was emphasized in Chapter 1 was the importance of distinguishing between the activity of a pure solid (1.0, dimensionless) that is dispersed in a solution and the concentration of that solid in the dispersion (mass of solid per unit volume of the solution or of the whole suspension, say, milligrams per liter). A brief quantitative justification, based on kinetic molecular theory, as to why the activity of a solid dispersed in an aqueous suspension is independent of the solid's concentration is provided in Appendix 8A.

8.7.2 DEFINITION OF THE SOLUBILITY PRODUCT

Consider the equilibrium expression for dissolution of a metal hydroxide solid and the corresponding equilibrium constant expression, as follows:

$$\text{Me(OH)}_n(s) \leftrightarrow \text{Me}^{n+} + n\text{OH}^- \qquad \textbf{(8.45)}$$

$$K_{s0} = \frac{\{\text{Me}^{n+}\}\{\text{OH}^-\}^n}{\{\text{Me(OH)}_n(s)\}} \qquad \textbf{(8.46)}$$

Conventionally, stability constants involving solids are written with the solid as a reactant and the dissolved metal as a product. Such stability constants are designated by a subscript s, followed by the number of ligands attached to the soluble metal species. That is, the symbol K_{si} is used to represent the equilibrium constant for a reaction with a solid as a reactant and a dissolved complex containing i ligands as the product. Additional species such as uncomplexed ligands might appear on either side of the reaction, as required by stoichiometry. (An alternate designation for the equilibrium constant K_{s0} is K_{sp}, where sp stands for solubility product.) For example, K_{si} values for equilibrium with amorphous $\text{Cd(OH)}_2(s)$ are shown below.[1]

$$\text{Cd(OH)}_2(s) \leftrightarrow \text{Cd}^{2+} + 2\text{OH}^- \qquad \log K_{s0} = -14.27 \quad \textbf{(8.47)}$$

$$\text{Cd(OH)}_2(s) \leftrightarrow \text{CdOH}^+ + \text{OH}^- \qquad \log K_{s1} = -10.35 \quad \textbf{(8.48)}$$

$$\text{Cd(OH)}_2(s) \leftrightarrow \text{Cd(OH)}_2^\circ(aq) \qquad \log K_{s2} = -6.62 \quad \textbf{(8.49)}$$

$$\text{Cd(OH)}_2(s) + \text{OH}^- \leftrightarrow \text{Cd(OH)}_3^- \qquad \log K_{s3} = -5.573 \quad \textbf{(8.50)}$$

$$\text{Cd(OH)}_2(s) + 2\text{OH}^- \leftrightarrow \text{Cd(OH)}_4^- \qquad \log K_{s4} = -5.62 \quad \textbf{(8.51)}$$

Assigning a value of 1.0 to the activity of the solid, the equilibrium constant for the dissolution of a pure metal hydroxide $\text{Me(OH)}_n(s)$ [reaction (8.45), Equation (8.46)] can be simplified as follows:

$$K_{s0} = \frac{\{\text{Me}^{n+}\}\{\text{OH}^-\}^n}{\{\text{Me(OH)}_n(s)\}}\bigg|_{eq} = \{\text{Me}^{n+}\}\{\text{OH}^-\}^n\big|_{eq} \qquad \textbf{(8.46a)}$$

Recall that the activity quotient Q for any reaction is defined as the ratio of the activities of the reaction products divided by that of the reactants; i.e., it has the same form as the equilibrium constant. However, whereas K is the ratio that applies if the system is at equilibrium, Q describes the corresponding ratio for any system, regardless of whether the system is equilibrated. Thus, for a system

---

[1]Note that although $\text{Cd(OH)}_2(s)$ has the same chemical formula as the dissolved dihydroxo complex Cd(OH)_2°, they are very different species. The former is a solid whose activity is 1.0 regardless of its concentration in a solution (assuming it is a pure solid), while the latter is a dissolved species whose activity can vary widely, depending on its concentration and the ionic composition of the solution.

Table 8.7   The K_{s0} values of some solids of interest

Metal	Mineral Name	Formula	Log K_{s0}	Metal	Mineral Name	Formula	Log K_{s0}
Ag^+		$AgOH(s)$	-7.70	Cu^+	Nantokite	$CuCl(s)$	-6.76
		$Ag_2CO_3(s)$	-11.07	Fe^{2+}		$Fe(OH)_2(s)$	-15.90
		$Ag_3PO_4(s)$	-17.55		Siderite	$FeCO_3(s)$	-10.55
		$Ag_2S(s)$	-48.97		Vivianite	$Fe_3(PO_4)_2(s)$	-36.00
		$AgCl(s)$	-9.75			$FeS(s)$	-16.84
Al^{3+}		$Al(OH)_3(s)$	-31.62	Fe^{3+}	Ferrihydrite	$Fe(OH)_3(s)$	-37.11
	Gibbsite	$Al(OH)_3(s)$	-33.23		Goethite	$\alpha\text{-}FeOOH(s)$	-41.50
		$AlPO_4(s)$	-22.50		Lepidocrocite	$\gamma\text{-}FeOOH(s)$	-46.00
Ca^{2+}	Calcite	$CaCO_3(s)$	-8.48		Hematite	$\alpha\text{-}Fe_2O_3(s)$	-40.63
	Aragonite	$CaCO_3(s)$	-8.36	Hg^{2+}		$Hg(OH)_2(s)$	-25.40
	Portlandite	$Ca(OH)_2(s)$	-5.32			$HgO(s)$	-25.55
	Lime	$CaO(s)$	4.80			$Hg(CN)_2(s)$	-39.28
	Gypsum	$CaSO_4(s)$	-4.85			$HgCO_3(s)$	-22.52
	Hydroxylapatite	$Ca_5(OH)(PO_4)_3(s)$	-44.2		Cinnabar	$HgS(s)$	-52.01
Cd^{2+}		$Cd(OH)_2(s)$	-14.27	Ni^{2+}		$Ni(OH)_2(s)$	-17.20
	Otavite	$CdCO_3(s)$	-13.74			$NiCO_3(s)$	-6.84
	Greenockite	$CdS(s)$	-28.85			$Ni_3(PO_4)_2(s)$	-31.30
		$Cd_3(PO_4)_2(s)$	-32.60	Pb^{2+}	Massicot	$PbO(s)$	-15.09
Co^{2+}		$Co(OH)_2(s)$	-15.90		Hydrocerrusite	$Pb_3(CO_3)_2(OH)_2(s)$	-45.46
		$CoCO_3(s)$	-12.80		Cerrusite	$PbCO_3(s)$	-13.13
Cr^{3+}		$Cr(OH)_3(s)$	-33.13		Galena	$PbS(s)$	-28.05
Cu^{2+}		$Cu(OH)_2(s)$	-19.36			$Pb_3(PO_4)_2(s)$	-44.50
	Tenorite	$CuO(s)$	-20.38	Zn^{2+}		$\alpha\text{-}Zn(OH)_2(s)$	-15.55
	Malachite	$Cu_2(OH)_2CO_3(s)$	-33.18			$ZnCO_3 \cdot H_2O(s)$	-10.26
		$CuCO_3(s)$	-9.63			$Zn_3(PO_4)_2(s)$	-36.70
		$Cu_3(PO_4)_2(H_2O)_3(s)$	-35.12			$ZnS(s)$	-21.97
	Covallite	$CuS(s)$	-35.96				

containing a pure solid $Me(OH)_n(s)$, Q_{s0} equals the product $\{Me^{n+}\}\{OH^-\}^n$ under any (equilibrium or non-equilibrium) conditions. Solutions in which $Q_{s0} < K_{s0}$ are said to be **undersaturated** with respect to the solid; in such solutions, if the solid is present, it will dissolve as the system equilibrates, and if no solid is present, none will form.[2] Correspondingly, solutions in which $Q_{s0} > K_{s0}$ are said to be **supersaturated** with respect to the solid, and in such solutions the solid will tend to precipitate. Values of K_{s0} for solids formed by various combinations of metal ions and ligands are listed in Table 8.7.

Example 8.3	In Example 7.8, the process of softening water by inducing the precipitation of calcite, $CaCO_3(s)$, was described. Typically, softening involves addition of lime, $Ca(OH)_2$, to increase the solution pH and, in many cases, sodium carbonate (often referred to by its common name of *soda ash*) to increase $TOTCO_3$. (It might seem paradoxical that lime is added

[2] Formally speaking, if no solid is present, Q_{s0} is meaningless, since computation of Q_{s0} requires a value for the activity of the solid. However, it is common to refer to the product $\{Me^{n+}\}\{OH^-\}^n$ as Q_{s0} even if the system contains no solid.

in a process intended to remove Ca from solution, but the precipitation reaction actually removes enough of both the Ca originally present and the Ca added to achieve the treatment goal, and often this approach is cheaper than using a different base to raise the pH.)

The lime and soda ash doses are often chosen based on a target of *stoichiometric softening,* meaning that $TOTCa = TOTCO_3$. Consider a solution that has been subjected to stoichiometric softening and has a final pH of 10.8. If the solution has reached equilibrium with calcite, what is the concentration of dissolved Ca^{2+}? If the sum of the Ca^{2+} in the original solution and that added with the lime is 180 mg/L Ca^{2+}, how much $CaCO_3(s)$ sludge must be disposed of per liter of water treated?

Solution

The composition of the softened solution can be determined by noting that the solution is in equilibrium with $CaCO_3(s)$ and that $TOTCa = TOTCO_3$. Calculating that α_2 for H_2CO_3 at pH 10.8 is 0.74, we can write

$$K_{s0} = \{Ca^{2+}\}\{CO_3^{2-}\}$$

$$10^{-8.48} = (TOTCa)(0.74\ TOTCO_3) = 0.74\ (TOTCO_3)^2$$

$$TOTCa = TOTCO_3 = 6.7 \times 10^{-5}$$

The total concentration of Ca^{2+} in the system is given as 180 mg/L, which equals $4.5 \times 10^{-3}\ M\ Ca^{2+}$. Since the calculations above indicate that only $6.7 \times 10^{-5}\ M\ Ca^{2+}$ remains in solution, ~98.5% of the Ca^{2+} in the initial solution is converted to $CaCO_3(s)$. The molecular weight of $CaCO_3(s)$ is 100, so the mass of solid generated is $0.985(4.5 \times 10^{-3}\ mol/L)(100,000\ mg/mol)$, or approximately 443 mg/L.

8.7.3   THE LOG C–pH DIAGRAM FOR A SOLUTION IN WHICH A SOLID IS KNOWN OR PRESUMED TO BE PRESENT AT EQUILIBRIUM

As with other equilibria that involve H^+ or OH^- ions and are therefore sensitive to pH, a great deal of information about the solubility of $Me(OH)_n(s)$ solids can be displayed on log C–pH diagrams. We can develop such diagrams by deriving the equations for the activity of each species as a function of pH, subject to the criterion that all the equations be consistent with the equilibrium expression for solubility. For instance, we can derive the equation for the line representing $\{Me^{n+}\}$ on such a diagram by adding n times the reverse of the water dissociation reaction to reaction (8.45), as follows:

$$Me(OH)_n(s) \leftrightarrow Me^{n+} + nOH^- \qquad K_{s0} \qquad \textbf{(8.45)}$$

$$\underline{nH^+ + nOH^- \leftrightarrow nH_2O \qquad\qquad K_w^{-n} \qquad\qquad\quad}$$

$$Me(OH)_n(s) + nH^+ \leftrightarrow Me^{n+} + nH_2O \qquad {}^*K_{s0} \qquad \textbf{(8.52)}$$

$${}^*K_{s0} = \frac{K_{s0}}{K_w^{\ n}} = \frac{\{Me^{n+}\}}{\{H^+\}^n}$$

$$\log {}^*K_{s0} = \log \{Me^{n+}\} + npH$$

$$\log \{Me^{n+}\} = \log {}^*K_{s0} - npH \qquad \textbf{(8.53)}$$

Equation (8.53) describes the equilibrium activity of Me^{n+} in a solution that has equilibrated with $Me(OH)_2(s)$. Although the activity of $Me(OH)_2(s)$ does not appear explicitly in the equation, the presence of the solid is implicit, because the activity of $Me(OH)_n(s)$ was assumed to be 1.0 as part of the derivation.

Equation (8.53) allows us to compute the activity of Me^{n+} as a function of pH in a system equilibrated with $Me(OH)_2(s)$. However, Me^{n+} is only one of the Me species in such a system, because in any system that contains Me^{n+} and OH^- or other ligands, Me—OH and Me—L complexes are bound to form in accordance with their respective equilibrium expressions. We have already analyzed these types of equilibria in systems containing the metal and ligands but no solid, and the fundamental relationships among those dissolved species do not change simply because a solid is present in the system. Therefore, to determine the activities of the complexes in a system equilibrated with the solid, we need only combine the solubility expression that characterizes $\{Me^{n+}\}$ with the expressions for formation of the various complexes.

To do so, we note that the equilibrium reaction between an oxide or hydroxide solid of any metal and a soluble hydroxo complex of the metal can be written as follows:

$$Me(OH)_n(s) + (i - n)OH^- \leftrightarrow Me(OH)_i^{n-i}(aq)$$

Manipulating the equilibrium constant expression for the above reaction, we obtain the desired equations for the activities of Me—OH complexes in equilibrium with the solid:

$$K_{si} = \frac{\{Me(OH)_i^{n-i}\}}{\{Me(OH)_n(s)\}\{OH^-\}^{i-n}}$$

$$\log K_{si} = \log \{Me(OH)_i^{n-i}\} - (i - n) \log \frac{K_w}{\{H^+\}}$$

$$= \log \{Me(OH)_i^{n-i}\} - (i - n) \log K_w - (i - n)pH$$

$$\log \{Me(OH)_i^{n-i}\} = \log K_{si} - 14(i - n) + (i - n)pH \qquad \textbf{(8.54)}$$

According to Equation (8.54), a line representing the dissolved species $Me(OH)_i^{n-i}$ in a system in equilibrium with $Me(OH)_n(s)$ has a slope equal to $i - n$ on a log C–pH diagram.

Applying Equation (8.54) to a system in equilibrium with $Cd(OH)_2(s)$, we derive the following equations for the lines of the various dissolved Cd species:

$$\log \{Cd^{2+}\} = 13.73 - 2pH \qquad \textbf{(8.55)}$$

$$\log \{CdOH^+\} = 3.65 - pH \qquad \textbf{(8.56)}$$

$$\log \{Cd(OH)_2^{\circ}(aq)\} = -6.62 \qquad \textbf{(8.57)}$$

$$\log \{Cd(OH)_3^-\} = -19.57 + pH \qquad \textbf{(8.58)}$$

$$\log \{Cd(OH)_4^{2-}\} = -33.62 + 2pH \qquad \textbf{(8.59)}$$

The log C–pH diagram for the system in equilibrium with $Cd(OH)_2(s)$ is shown in Figure 8.12. The sum of all Cd species gives the total soluble Cd as a

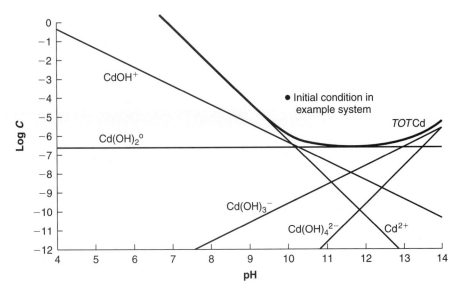

Figure 8.12 Log C–pH diagram showing Cd species in equilibrium with Cd(OH)$_2$(s). The point identified as the initial condition is for an example presented later in this chapter.

function of pH. Because the concentrations of Cd(OH)$_3^-$ and Cd(OH)$_4^{2-}$ increase with pH, the curve representing $TOTCd_{diss}$ is bowl-shaped, and there is a pH where the total solubility of Cd(OH)$_2$(s) passes through a minimum. Such a minimum is a common feature of the solubility diagrams for many metals.

Note that the region of dominance of each Cd(OH)$_i^{2-i}$ species is identical to that in a system containing no solid (Figure 8.2). This result must be obtained, of course, because the presence of the solid cannot alter the equilibrium relationship between two dissolved species. Consequently, the relationship between the total amount of *dissolved* cadmium and the activity of free Cd^{2+} ion at any pH when Cd(OH)$_2$(s) is present is exactly the same as in the case developed earlier for a completely soluble system (Figure 8.2):

$$TOTCd_{diss} = \{Cd^{2+}\}\left(1 + \sum_{i=1}^{4} \beta_{OH,i}\{OH^-\}^i\right) \qquad \textbf{(8.18)}$$

However, for a system in equilibrium with solid Cd(OH)$_2$(s), the K_{s0} expression must also be satisfied (and the equation establishing a fixed value of $TOTCd_{diss}$ is no longer applicable). Substituting for $\{Cd^{2+}\}$ from the K_{s0} relationship into Equation (8.18), we obtain, for a solution in equilibrium with Cd(OH)$_2$(s),

$$TOTCd_{diss} = \frac{K_{s0}}{\{OH^-\}^2}\left(1 + \sum_{i=1}^{4} \beta_i\{OH^-\}^i\right)$$

$$= *K_{s0}\{H^+\}^2\left(1 + \sum_{i=1}^{4} \frac{*\beta_i}{\{H^+\}^i}\right) \qquad \textbf{(8.60)}$$

Equation (8.60) indicates the total dissolved Cd concentration that must be present at any given OH$^-$ concentration in a system in equilibrium with solid

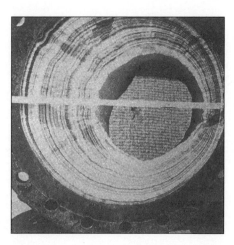

Buildup of solids on the insides of water supply pipes can interfere with their ability to convey water. This photograph shows the $CaCO_3(s)$ that has accumulated in the interior of a pipe over a 20-year period. Conditions favoring the precipitation reaction were maintained in order to minimize pipe corrosion. The addition of complexing agents to the water was recommended in order to gradually dissolve the $CaCO_3(s)$ and improve the pipe's carrying capacity.

| Courtesy of Carus Chemicals.

$Cd(OH)_2(s)$ and that contains hydroxide ions as the only complex-forming ligands. Note that once pH is fixed, the concentrations of all dissolved Cd species, both free Cd^{2+} and Cd—OH complexes, are established.

The lines representing the various $Cd(OH)_i^{2-i}$ species on the log C–pH diagram do not "bend" at all, and in this way they are similar to the lines for the various dissolved species in systems equilibrated with a gas phase. This similarity is not coincidental. Recall that in a system equilibrated with a gas phase, the activity of the dissolved gas is independent of pH and therefore is represented by a horizontal line on a log C–pH diagram. If the dissolved gas is an acid (e.g., H_2CO_3), it equilibrates with its conjugate base (e.g., HCO_3^-). In such a system, the line for the base must intersect the line for the acid at pH = pK_a and have a slope of +1. The lines for species generated by further deprotonation (e.g., CO_3^{2-}) have correspondingly larger slopes. The net result is that the concentration of total dissolved gas (e.g., $TOTCO_3$) increases continuously with increasing pH. If the gaseous species is a base (e.g., NH_3), the same reasoning applies, but the concentration of the conjugate acid, and therefore of the total dissolved gas, increases with decreasing pH.

In systems equilibrated with a solid, the activity of the dissolved species that has the same chemical formula as the solid itself [$Cd(OH)_2^{\,0}$ in the example system] is independent of pH, because that species is formed by dissolution of the solid without any consumption or release of H^+ ions. The $Cd(OH)_2^{\,0}$ can act either as an acid [by combining with additional OH^- ions to form $Cd(OH)_3^-$ and $Cd(OH)_4^{2-}$] or as a base (by releasing OH^- ions to form $CdOH^+$ or Cd^{2+}). As a result, a log C–pH diagram for $Cd(OH)_2(s)$ combines features that are characteristic of the diagrams for both $CO_2(g)$ and $NH_3(g)$. That is, the concentrations of some $Cd(OH)_i^{2-i}$ species increase with increasing pH, while those of others increase with decreasing pH, leading to the characteristic bowl-shaped curve for the total dissolved metal concentration as a function of pH shown in Figure 8.12.

8.7.4   SYSTEMS CONTAINING LIGANDS OTHER THAN OH⁻

If ligands other than OH^- are present in a solution containing $Cd(OH)_2(s)$, then Cd^{2+} will form complexes with those ligands as well, just as it does in the absence of the solid. The extension of Equations (8.18d) and (8.60) to such a situation is straightforward, since the presence of the solid has no effect on the stability constant relationships between Cd^{2+} and dissolved ligands. For instance, in the presence of Cl^-, regardless of whether or not a solid is present, Equation (8.39) applies:

$$TOTCd_{diss} = \{Cd^{2+}\} \left(1 + \sum_{i=1}^{4} \beta_{OH,i}\{OH^-\}^i + \sum_{i=1}^{3} \beta_{Cl,i}\{Cl^-\}^i\right) \quad \textbf{(8.39)}$$

The only difference in the case where solid $Cd(OH)_2(s)$ is present is that, as in Equation (8.60), the activity of the free, uncomplexed Cd^{2+} ion is related to pH via the solubility constant. In such a case, the total soluble cadmium concentration is

$$TOTCd_{diss} = \frac{K_{s0}}{\{OH^-\}^2} \left(1 + \sum_{i=1}^{4} \beta_{OH,i}\{OH^-\}^i + \sum_{i=1}^{3} \beta_{Cl,i}\{Cl^-\}^i\right) \quad \textbf{(8.61)}$$

A log C–pH diagram for a system in equilibrium with $Cd(OH)_2(s)$ and at a fixed $\{Cl^-\}$ of 0.5 M is shown in Figure 8.13. Note that, as in Figure 8.12, the

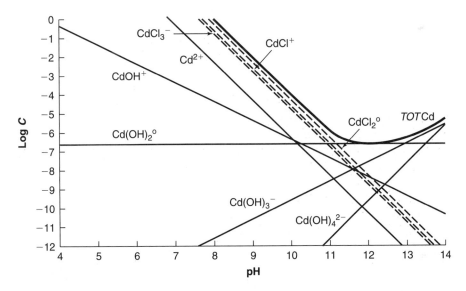

Figure 8.13   Log C–pH diagram for a system in equilibrium with $Cd(OH)_2(s)$ for a fixed Cl^- activity of 0.5.

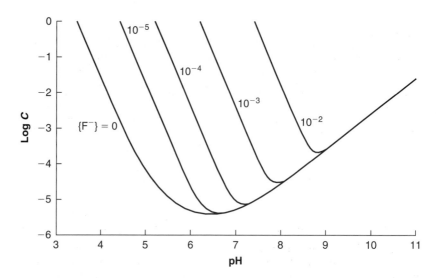

Figure 8.14   Total soluble Al in equilibrium with amorphous $Al(OH)_3(s)$ as a function of the activity of F^- in the solution.

lines for $\{CdCl_x^{2-x}\}$ are all parallel to that for $\{Cd^{2+}\}$. The extension to a case with even more ligands present is straightforward.

The dramatic effect that ligand concentration can have on the total amount of soluble metal at equilibrium is demonstrated for another system in Figure 8.14.

8.8   PRECIPITATION OF SOLIDS OTHER THAN HYDROXIDES

We have now considered several issues of interest in aqueous systems that contain metals, including the effects of ligands on speciation of soluble compounds and the composition of systems in which a hydroxide solid is known to be present. We next extend these considerations to solutions that are in equilibrium with a metal-bearing solid other than the metal hydroxide.

Metal carbonate solids are extremely important natural minerals, and they often precipitate in water and wastewater treatment operations. For instance, $CaCO_3(s)$ is the mineral limestone, and precipitation of $CaCO_3(s)$ is one of the most common approaches for removing hardness from drinking waters. Other metal carbonates are also important both naturally and in engineered processes. Because we will ultimately wish to compare the solubilities of metal carbonate and metal hydroxide solids, and we have already analyzed the composition of systems in equilibrium with $Cd(OH)_2(s)$, the following discussion of carbonate-containing solids focuses on precipitation of $CdCO_3(s)$ as an example system.

We begin by deriving the relevant equations and drawing $\log C$–pH diagrams for two solutions in equilibrium with $CdCO_3(s)$: one containing a fixed concentration of total dissolved carbonate species ($TOTCO_{3,\text{diss}}$) and the other in equilibrium with atmospheric $CO_2(g)$.

8.8.1 SPECIATION IN A SYSTEM WITH FIXED $TOTCO_{3,\text{diss}}$

Cadmium carbonate is a relatively insoluble solid with $K_{s0} = 10^{-13.74}$:

$$CdCO_3(s) \leftrightarrow Cd^{2+} + CO_3^{2-} \qquad K_{s0} = 10^{-13.74} \qquad \textbf{(8.62)}$$

Consider the solubility of $CdCO_3(s)$ in a solution containing $TOTCO_{3,\text{diss}}$ fixed at 10^{-3} M.[3] By manipulation of K_{s0}, we obtain the following expression for $\{Cd^{2+}\}$ in the system:

$$\{Cd^{2+}\}\{CO_3^{2-}\} = 10^{-13.74}$$

$$\log\{Cd^{2+}\} + \log\{CO_3^{2-}\} = -13.74 \qquad \textbf{(8.63)}$$

Equation (8.63) points out that, in a solution in equilibrium with $CdCO_3(s)$, the value of $\{Cd^{2+}\}$ can be determined by knowing $\{CO_3^{2-}\}$, regardless of the value of $\{HCO_3^-\}$ or $\{H_2CO_3\}$. That is, if $\{CO_3^{2-}\}$ is $10^{-5.0}$ and the system is in equilibrium with $CdCO_3(s)$, then $\{Cd^{2+}\}$ will be $10^{-8.74}$, regardless of whether the CO_3^{2-} activity is established by having $\sim 10^{-5.0}$ M $TOTCO_3$ in a solution at pH 12, or $\sim 10^{-3}$ M $TOTCO_3$ in a solution at pH 8.3, or ~ 0.2 M $TOTCO_3$ in a solution at pH 6.3.

Furthermore, any change in $\log\{Cd^{2+}\}$ must be mirrored exactly by a change in $\log\{CO_3^{2-}\}$ that is equivalent in magnitude but opposite in direction. Thus, for a fixed $TOTCO_{3,\text{diss}}$, the shape of the $\{Cd^{2+}\}$ line on a $\log C$–pH diagram is easy to deduce. At pH ≥ 11, the $\{CO_3^{2-}\}$ line is essentially horizontal, so the $\{Cd^{2+}\}$ line must be horizontal also. Only in this way will $\log\{Cd^{2+}\} + \log\{CO_3^{2-}\}$ be constant. In the region between the two carbonate pK_a's, the slope of the $\{CO_3^{2-}\}$ line is $+1$, so the slope of the $\{Cd^{2+}\}$ line must be -1. Similarly, at pH < 6.3, the slope of the $\{CO_3^{2-}\}$ line is $+2$, so that of the $\{Cd^{2+}\}$ line will be -2. Log$\{Cd^{2+}\}$ is shown over a wide pH range for this system in Figure 8.15.

---

[3] It might be more realistic to consider a system in which $TOTCO_3$ is fixed, considering both soluble and particulate species. In that case, precipitation of $CdCO_3(s)$ would reduce the total concentration of dissolved carbonate species. However, the change in $TOTCO_3$ as a function of the amount of solid precipitated in such a system complicates the mathematics considerably and obscures some of the key points of the analysis. Therefore, the system that is modeled is one that, if it were to be set up physically, would require additions of carbonate species to solution whenever $CdCO_3(s)$ precipitated, and removal of carbonate whenever $CdCO_3(s)$ dissolved. Although awkward, such a system could be designed. Systems in which the total concentration of the ligand is fixed are handled well by chemical equilibrium software, and these types of systems are considered later in the chapter.

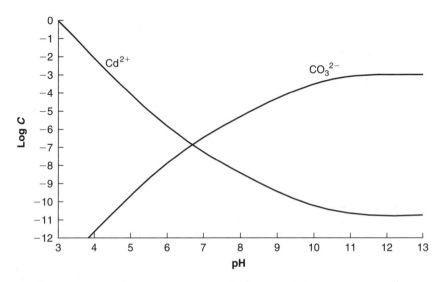

Figure 8.15   The Cd^{2+} and CO_3^{2-} concentrations in a solution containing 10^{-3} M $TOTCO_{3,diss}$ and equilibrated with $CdCO_3(s)$.

Although the solid of interest in this system is $CdCO_3(s)$, the solution contains both Cd^{2+} and OH^-, so we expect dissolved $Cd(OH)_i^{2-i}$ complexes to form, and we should include them on the log C–pH diagram for the system. There are a number of ways to evaluate what the concentrations of these complexes will be once the system equilibrates. Perhaps the easiest approach is to use the curve for the Cd^{2+} activity in conjunction with the equilibrium constant expressions relating the activities of the various complexes to that of Cd^{2+}. For example, because $*K_1$ is the acid dissociation constant relating $\{Cd^{2+}\}$ to $\{CdOH^+\}$, we know that the following equation applies in any solution containing Cd^{2+}, regardless of which, if any, solid is present:

$$\log \{CdOH^+\} = \log \{Cd^{2+}\} + \log *K_1 + pH$$

The above equation is the basis for knowing that the $\{CdOH^+\}$ line always has a slope that is 1 greater than the $\{Cd^{2+}\}$ line on a log C–pH diagram, and that $\{Cd^{2+}\} = \{CdOH^+\}$ at $pH = p*K_1 = 10.08$. Using this information, we can draw the curve for $CdOH^+$ directly; the lines for the other hydroxo complexes can be drawn by following the same general procedures. The concentrations of all the soluble Cd species in equilibrium with $CdCO_3(s)$ are shown in Figure 8.16.

The $TOTCd_{diss}$ curve in the log C–pH diagram for this system has a bowl-like shape similar to that for equilibrium with $Cd(OH)_2(s)$, and for essentially the same reason: at low pH, the activities of soluble species that are more acidic than the solid become very large, and at high pH, the same is true of soluble species

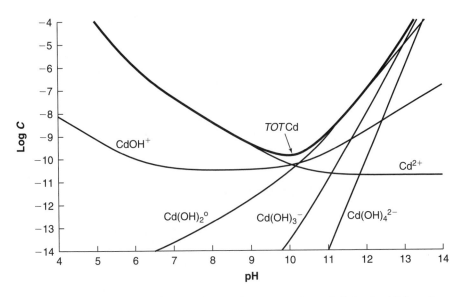

Figure 8.16 Log C–pH diagram showing the concentrations of all dissolved Cd species in the system shown in Figure 8.15.

that are more basic than the solid. The exact value of $TOTCd_{diss}$ at any pH and the pH of minimum solubility differ for different solids, but the general pattern observed here is very consistently observed.

The equation describing $TOTCd_{diss}$ as a function of pH can be written concisely as the sum of the concentrations of free Cd^{2+} and the various complexes. Since the only complexes being considered are hydroxo complexes, the summation is given by Equation (8.18), which is reproduced below:

$$TOTCd_{diss} = \{Cd^{2+}\}\left(1 + \sum_{i=1}^{4} \beta_i\{OH^-\}^i\right) \qquad \textbf{(8.18)}$$

Parallel to the case where the solution is in equilibrium with $Cd(OH)_2(s)$, $\{Cd^{2+}\}$ can be determined as $K_{s0}/\{CO_3^{2-}\}$. Once this value of $\{Cd^{2+}\}$ is computed, Equation (8.18) can be used to compute the total dissolved Cd concentration. The equation for total dissolved Cd in such a case is as follows, which is exactly analogous to Equation (8.60) for systems in equilibrium with $Cd(OH)_2(s)$:

$$TOTCd_{diss} = \frac{K_{s0,CdCO_3(s)}}{\{CO_3^{2-}\}}\left(1 + \sum_{i=1}^{4} \beta_i\{OH^-\}^i\right)$$

$$TOTCd_{diss} = \frac{K_{s0,CdCO_3(s)}}{\alpha_2(TOTCO_{3,diss})}\left(1 + \sum_{i=1}^{4} \beta_i\{OH^-\}^i\right) \qquad \textbf{(8.64)}$$

Massive amounts of $CaCO_3(s)$ can be precipitated from hard water when the water is softened. This photo shows the precipitate forming in a water treatment plant in Austin TX. The disposal of this solid, along with $Al(OH)_3(s)$ and $Fe(OH)_3(s)$ that form in other water treatment operations, is often a major component of the cost of water treatment.

| Bob Daemmrich Photography.

Other complexes such as $CdHCO_3^+$ and $CdCO_3^\circ$ might also form in the solution, and if other ligands such as Cl^- were present, complexes with those ligands would be present as well. Such species could be represented by additional lines on the diagram; although this may take some time and algebra, the procedures are straightforward. As in the cases where we considered complexation of Cd^{2+} by Cl^- or of Al^{3+} by F^- (Figures 8.13 and 8.14, respectively), the formation of additional complexes has no bearing on the other calculations, so these other complexes can only increase the total soluble metal concentration.

Next, consider the case of a solution containing a fixed $TOTCO_{3,\text{diss}}$ of $10^{-1}\,M$ in equilibrium with $CdCO_3(s)$. In such a system, the curve representing the CO_3^{2-} concentration on the log C–pH diagram will be two units higher than the corresponding line in Figure 8.15 (because $TOTCO_3$ is two orders of magnitude larger

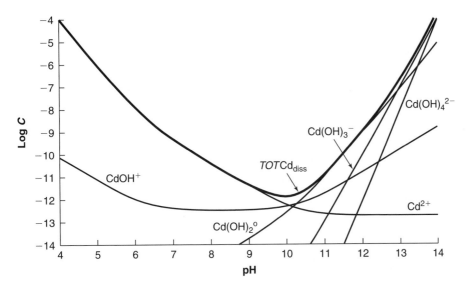

Figure 8.17 Concentrations of Cd^{2+} and Cd—OH complexes in a solution containing 0.1 M $TOTCO_{3,diss}$ and in equilibrium with $CdCO_3(s)$.

in this system than in the one characterized in Figure 8.15), but will otherwise be identical. As a result, the curves representing the equilibrium Cd^{2+} concentration and that of each of the Cd—OH complexes will be two units lower. These lines are shown on Figure 8.17.

8.8.2 SPECIATION IN A SYSTEM WITH FIXED P_{CO_2}

The next case of interest is one in which the solution is open to the atmosphere. The analysis of this system is actually quite easy, because Henry's law establishes a fixed value of $\{H_2CO_3\}$ in such systems, independent of pH. As a result, the CO_3^{2-} concentration is represented by a straight line of slope $+2$ at all pH values. For such a solution to be in equilibrium with $CdCO_3(s)$, the value of $\{Cd^{2+}\}$ must decrease by two log units for each unit increase in pH, because the K_{s0} expression still requires that the sum of $\log\{Cd^{2+}\}$ and $\log\{CO_3^{2-}\}$ be a constant $(= \log K_{s0})$. As a result, $\{Cd^{2+}\}$ is represented on the diagram by a straight line of slope -2. The line representing $\{CdOH^+\}$ is also straight, with a slope of -1, intersecting the $\{Cd^{2+}\}$ line at $pH = p^*K_{OH,1} = 10.08$ (Figure 8.18a). You should be able to derive the equations necessary to compute the concentrations of the other complexes in the system. The log C–pH diagram for a system with fixed P_{CO_2} of $10^{-3.46}$ bar (the partial pressure of CO_2 in the atmosphere) is shown in Figure 8.18.

By analogy with the equations presented above for $TOTCd_{diss}$ in equilibrium with solids in other systems, the total dissolved Cd in an open system equilibrated

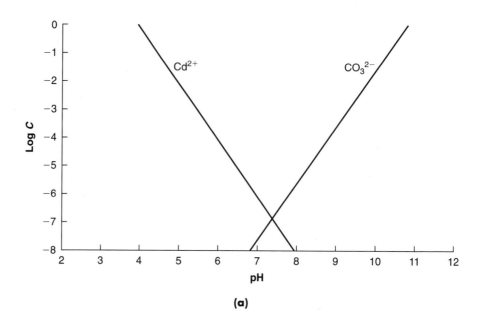

(a)

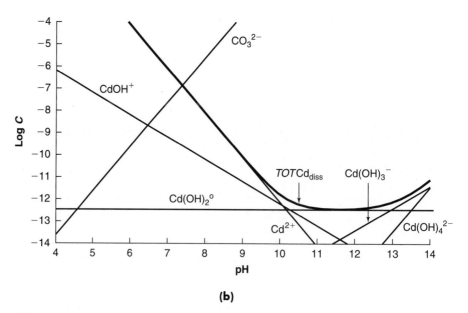

(b)

Figure 8.18 Concentrations of Cd^{2+} and Cd—OH complexes in a solution equilibrated with atmospheric $CO_2(g)$ and with $CdCO_3(s)$. **(a)** Activities of Cd^{2+} and CO_3^{2-}. **(b)** Activities of all Cd species.

with $CdCO_3(s)$ is given by the following equations:

$$TOTCd_{diss} = \frac{K_{s0,CdCO_3(s)}}{\{CO_3^{2-}\}} \left(1 + \sum_{i=1}^{4} \beta_i \{OH^-\}^i\right)$$

$$TOTCd_{diss} = \frac{K_{s0,CdCO_3(s)}}{(P_{CO_2}/H_{CO_2})(K_{a1}K_{a2}/\{H^+\}^2)} \left(1 + \sum_{i=1}^{4} \beta_i \{OH^-\}^i\right) \quad \textbf{(8.65)}$$

8.9 PRECIPITATION AND COMPLEXATION WITH THE SAME LIGAND

As noted above, many ligands can bind with a metal to form either soluble complexes or a precipitate. Indeed, OH^- is the prototypical ligand in this case, capable of forming either hydroxo complexes or hydroxide solids with most metal ions. However, many other ligands share that property and hence affect metal solubility in the same qualitative way that hydroxide ion does. This point is illustrated in the following example.

In the past, wastes generated by industrial processes such as electroplating were a major source of metals to wastewater treatment plants and receiving waters. Steady improvements over the past several decades in process control and in on-site industrial wastewater treatment have dramatically reduced the metal load discharged to the environment from these sources.

| Richard Pasley/Stock Boston.

Example 8.4 | Oxalic acid is a simple dicarboxylic acid, consisting essentially of two carboxyl groups bound to one another: HOOC—COOH. It is an important biomolecule and is also produced if drinking water is ozonated (as a disinfection step). Plot a $\log C$–$\log\{Ox^{2-}\}$ diagram for a solution in equilibrium with $ZnOx(s)$ at pH 7.0. The K_{s0} value for $ZnOx(s)$ and equilibrium constants for formation of the four most significant Zn complexes in the system at pH 7.0 are shown below.

$$\frac{\{ZnOx^{\circ}\}}{\{Zn^{2+}\}\{Ox^{2-}\}} = 10^{4.2} \qquad \frac{\{ZnOH^{+}\}\{H^{+}\}}{\{Zn^{2+}\}} = 10^{-8.96}$$

$$\frac{\{ZnOx_2{}^{2-}\}}{\{Zn^{2+}\}\{Ox^{2-}\}^2} = 10^{6.9} \qquad \frac{\{Zn(OH)_2{}^{\circ}\}\{H^{+}\}^2}{\{Zn^{2+}\}} = 10^{-16.90}$$

$$K_{s0,ZnOx(s)} = 10^{-8.20}$$

Solution

The activity of Zn^{2+} as a function of the Ox^{2-} activity can be derived by direct substitution into the K_{s0} expression. Since the product $\{Zn^{2+}\}\{Ox^{2-}\}$ is constant, the curve for $\{Zn^{2+}\}$ on the diagram is simply a straight line with slope -1. Similarly, plugging into the stability constant expressions, we conclude that the $ZnOx^{\circ}$ and $Zn(Ox)_2{}^{2-}$ activities are represented by straight lines with slopes of 0 and $+1$, respectively. Finally, since the pH is fixed, the $ZnOH^{+}$ and $Zn(OH)_2{}^{\circ}$ lines are parallel to the Zn^{2+} line. The diagram is shown below.

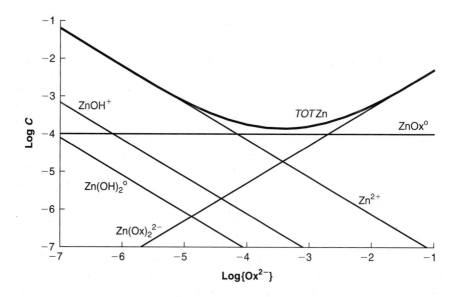

The variation of $TOTZn_{diss}$ with $\log\{Ox\}$ follows the same trend as does $TOTZn_{diss}$ versus pH in a system in equilibrium with $Zn(OH)_2(s)$. The reason, of course, is that in both systems, the activity of free Zn^{2+} increases steadily with decreasing concentration of the ligand (Ox^{2-} or OH^{-}), and the activities of the anionic complexes [$Zn(Ox)_2{}^{2-}$ in one case, $Zn(OH)_3{}^{-}$ and $Zn(OH)_4{}^{2-}$ in the other] increase steadily with increasing ligand concentration. Therefore, in both cases, the total soluble Zn passes through a minimum at some intermediate concentration of the ligand.

The preceding example emphasizes that not only OH^-, but virtually any ligand, can either increase or decrease the solubility of metals, depending on the exact solution conditions and, especially, the ligand concentration. As a result, by a judicious choice of which ligand(s) to add, and their concentrations, we can induce precipitation to remove the metal from solution or complex the metal so that it does not precipitate under conditions where it otherwise would. This duality of functions provides enormous flexibility in the industrial or commercial use of ligands.

8.10   USE OF CHEMICAL EQUILIBRIUM SOFTWARE TO MODEL SYSTEMS IN WHICH SOLIDS ARE KNOWN TO BE PRESENT

If a solid is known to be present in a system, then it represents a separate phase whose activity is known and fixed. The system can therefore be modeled by using chemical equilibrium software exactly as in cases where the solution is equilibrated with a gas of known composition. Recall that, in such cases, the constituent in the non-aqueous phase is input as a Type 3b *species,* and the $\log K$ value for that *species* equals the $\log K$ for forming the *species* from the *components* minus the logarithm of its fixed activity. In the case of a pure solid, the fixed activity of the *species* is 1.0, so the logarithm of the fixed activity is 0.0, and the input $\log K$ value is simply the $\log K$ value for formation of the Type 3b *species* from the *components*.

For instance, consider a solution equilibrated with the mineral hydroxyapatite, with chemical formula $Ca_5(OH)(PO_4)_3(s)$ and $pK_{s0} = 58.20$. This compound is an economically important resource for phosphate, a major component of bones and teeth, and can sometimes be induced to form in advanced wastewater treatment operations as a way of reducing the soluble phosphate concentration prior to discharge of the water. If we wished to model the chemistry of a solution that was equilibrated with the mineral, we would specify it as a Type 3b *species* present at an activity of 1.0. Assuming that Ca^{2+}, H^+, and PO_4^{3-} were used as *components,* we would write the stoichiometry for formation of hydroxyapatite and determine the corresponding equilibrium constant for its formation as follows:

$$5Ca^{2+} + 3PO_4^{3-} + OH^- \rightarrow Ca_5(OH)(PO_4)_3(s) \qquad \log K = -\log K_{s0} = 58.20$$

$$\underline{\phantom{5Ca^{2+} + 3PO_4^{3-} + } H_2O \rightarrow H^+ + OH^- } \qquad \underline{\log K = \log K_w = -14.00}$$

$$\mathbf{5Ca^{2+} + 3PO_4^{3-} - 1H^+ + 1H_2O \rightarrow Ca_5(OH)(PO_4)_3(s)} \qquad \log K = +44.20$$

The input would therefore specify $\overline{Ca_5(OH)(PO_4)_3(s)}$ as a Type 3b *species* with a log K that would be formally computed as log $K = 44.20 - $ log $1.0 = 44.20$.

As pointed out in previous chapters, the program output can be used to determine how much of a Type 3 *species* enters or leaves the solution phase during the equilibration process. In the case of Type 3b *species* that are solids, entry into solution corresponds to dissolution of the solid, and removal from solution corresponds to precipitation. This calculation is demonstrated in the following example.

Example 8.5

Uranium can exist in a number of oxidation states, taking on a charge of +3 to +6. When in the +6 oxidation state, it is covalently bound to two oxygen ions to form the uranyl ion, UO_2^{2+}. This cation can hydrolyze to $UO_2(OH)^+$, with $*K_1 = 10^{-5.09}$. Uranyl ions can also link together and hydrolyze further to form a dimer and a trimer with composition $(UO_2)_2(OH)_2^{2+}$ and $(UO_2)_3(OH)_5^+$, respectively. The stability constants of these complexes, written for reactions in which UO_2^{2+} and H^+ are the reactants, are $10^{-5.64}$ and $10^{-15.59}$, respectively.

You wish to explore the conditions under which precipitation of the uranyl hydroxide solid Schoepite, $UO_2(OH)_2 \cdot H_2O(s)$, can maintain $TOTUO_{2,\text{diss}}$ at less than $10^{-8.0}$ M. Develop the tableau that you would use to prepare a log C–pH diagram for a solution in equilibrium with Schoepite, and determine whether a pH range exists in which the goal can be met. The log K_{s0} for Schoepite is -22.60. Use UO_2^{2+} as a *component*, along with H^+ and H_2O.

Solution

To develop the log C–pH diagram, we can follow the procedure outlined at the end of Chapter 6. This procedure involves inputting a Type 3a *species* $\overline{H^+}$ to fix the activity of the real *species* H^+ (a Type 1 *species*) at each desired value. In addition, we need to input a Type 3b *species* with the composition of Schoepite, made from the *components*.

$$UO_2^{2+} + 2OH^- + H_2O \rightarrow UO_2(OH)_2 \cdot H_2O(s) \qquad -\log K_{s0} = 22.60$$

$$\underline{2H_2O \rightarrow 2H^+ + 2OH^- \qquad \log K_w^2 = -28.00}$$

$$1UO_2^{2+} + 3H_2O - 2H^+ \rightarrow UO_2(OH)_2 \cdot H_2O(s) \qquad \log K = -5.40$$

The logarithm of the equilibrium constant for the above reaction is -5.40, as shown. When we define Schoepite as a Type 3b *species,* we need to input an equilibrium constant that equals -5.40 minus the logarithm of the activity of Schoepite; however, because the solid has an activity of 1.0, the input log K value is simply -5.40.

The input tableau is shown on the following page. The inputs to the program for the total concentrations of $\mathbf{H^+}$ and $\mathbf{UO_2^{2+}}$ are zero. However, the program will report the corresponding total concentrations present at equilibrium at each pH. Defining x_1 as the concentration of H^+ (or, if it is negative, OH^-) that must be added to reach the specified pH, and x_2 as the concentration of solid that dissolves as the solution equilibrates at that pH, the values of $TOTH$ and $TOTUO_2$ in the equilibrium solution will be

$$TOTH = x_1 - 2x_2$$

$$TOTUO_2 = x_2$$

Species	Type	Component H$^+$	Component UO$_2$$^{2+}$	Log K
H$^+$	1	1	0	0.00
UO$_2$$^{2+}$	1	0	1	0.00
OH$^-$	2	−1	0	−14.00
UO$_2$OH$^+$	2	−1	1	−5.09
(UO$_2$)$_2$(OH)$_2$$^{2+}$	2	−2	2	−5.64
(UO$_2$)$_3$(OH)$_5$$^+$	2	−5	3	−15.59
$\overline{H^+}$	3a	1	0	Various*
$\overline{UO_2(OH)_2H_2O}$	3b	−2	1	−5.40
TOTComp$_{eq}$				

Inputs			Input Concentration
$\overline{H^+}$	1	0	x_1
$\overline{UO_2(OH)_2H_2O}$	−2	1	x_2
TOTComp$_{in}$	$(x_1 - 2x_2)$	(x_2)	

| *Value depends on the fixed pH for the given iteration.

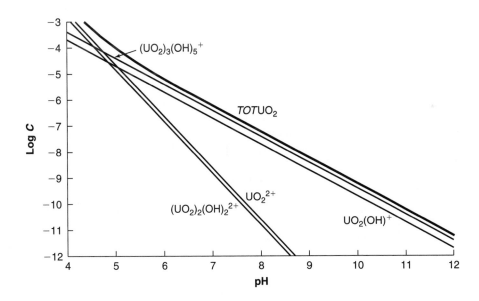

The log C–pH diagram that is generated when the program is run is shown above, with the value of TOTUO$_2$ shown as the boldface line. The TOTUO$_2$ curve does not have the characteristic bowl shape that is associated with most oxide or hydroxide solids. Upon reviewing the inputs, we see why this is so: the right-hand portion of such bowls is formed by species whose concentrations increase with increasing pH. Such species must be more basic (have more OH^- or, equivalently, less H^+) than the solid. In the current case, the only complexes that are being considered have more H^+ than the solid does (as we can see from the $\mathbf{H^+}$ column in the input tableau), so only the left-hand portion of the bowl is relevant. The conclusion is that the goal of $TOTUO_2 < 10^{-8}$ can be met at any pH > 8.8, if the system reaches equilibrium.

Example 8.6 | A supply source for potable water is at pH 8.1 and contains 80 mg/L TOTCa. Its alkalinity, which can be attributed entirely to carbonate the system, is 120 mg/L *as CaCO₃*. The pH of the water is to be increased to 11.0 in a water softening process, causing calcite to precipitate. The precipitate can then be settled and/or filtered out of the water, and the softened water can be distributed. Determine the concentration of $CaCO_3(s)$ that precipitates when the pH is increased if the solution equilibrates with the calcite. Assume that the precipitation is rapid, so that the system can be treated as closed with respect to $CO_2(g)$ exchange with the atmosphere.

Solution

The required inputs for this problem include TOTCa, $TOTCO_3$, and instructions to the program to force the pH of the solution to be 11.0 and for it to be in equilibrium with $CaCO_3(s)$. Appropriate values and techniques for these inputs are as follows. The atomic weight of Ca is 40, so the TOTCa concentration in the untreated water is 2.0×10^{-3} M.

The way that the alkalinity is expressed indicates that the solution contains the same amount of alkalinity as would 120 mg/L of $CaCO_3$ (the solution does not really contain this amount, or any known amount, of $CaCO_3$). As discussed in Chapter 5, the equivalent weight of $CaCO_3$ is 50, so 120 mg/L *as CaCO₃* corresponds to 2.4 meq/L. At the initial pH of the solution (8.1), bicarbonate is the dominant contributor to ALK, so we can make the approximation $\{HCO_3^-\} \approx$ ALK $= 2.4 \times 10^{-3}$. In addition, at that pH, $TOTCO_3 \approx \{HCO_3^-\}$, so $TOTCO_3$ is approximately 2.4×10^{-3} M.

The information needed by the program is TOTCa$_{in}$ and $TOTCO_{3,in}$. It is not necessary to specify how these components are added; any set of inputs that is consistent with the total concentrations will be acceptable. However, in the bottom of our tableau, we normally include information about how the various *components* were added. In this case, for simplicity, we can set up the tableau assuming that 2.0×10^{-3} M Ca^{2+} and 2.4×10^{-3} M CO_3^{2-} were added to prepare the solution, without specifying what other salts were added with those ions. The input value of TOTH is $2.4 \times 10^{-3} + 10^{-8.1} - 10^{-5.9}$, or essentially 2.4×10^{-3}.

To instruct the program to add or remove H^+ and $CaCO_3(s)$ as needed to ensure that the solution has the proper pH and is equilibrated with the solid, we need to include $\overline{H^+}$ and $\overline{CaCO_3(s)}$ as Type 3a and 3b *species*, respectively. At this point, it should no longer be necessary to go through the detailed calculations to demonstrate the calculation of the input log K values for these Type 3 *species*. You should verify that these values are 11.0 and 8.48, respectively.

The initial pH is not needed for this problem, since we are only concerned with the solution composition after the pH has been increased to 11.0.

The input data and the computer-generated equilibrium composition of the solution are shown in the following table. Note that when the program scanned its database, it identified soluble complexes of Ca^{2+} with OH^-, HCO_3^-, and CO_3^{2-} which are included in the table.

| Species | Type | Component | | | Log K | Equil. Concentration |
		H^+	Ca^{2+}	CO_3^{2-}		
H^+	1	1	0	0	0.00	1.00×10^{-11}
Ca^{2+}	1	0	1	0	0.00	9.91×10^{-6}
CO_3^{2-}	1	0	0	1	0.00	3.38×10^{-4}
OH^-	2	-1	0	0	-14.00	1.00×10^{-3}
HCO_3^-	2	1	0	1	-10.33	7.22×10^{-5}
H_2CO_3	2	2	0	1	-16.68	1.62×10^{-9}
$CaOH^+$	2	-1	1	0	12.60	2.50×10^{-7}
$CaHCO_3^-$	2	1	1	1	11.33	7.16×10^{-9}
$CaCO_3^\circ$	2	0	1	1	3.15	4.73×10^{-6}
$\overline{H^+}$	3a	1	0	0	11.00	
$\overline{CaCO_3(s)}$	3b	0	1	1	8.48	
$TOTComp_{eq}$		-9.33×10^{-4}	1.49×10^{-5}	4.15×10^{-4}		

Inputs		H^+	Ca^{2+}	CO_3^{2-}	Input Concentration
H^+		1	0	0	2.4×10^{-3}
Ca^{2+}		0	1	0	2.0×10^{-3}
CO_3^{2-}		0	0	1	2.4×10^{-3}
$\overline{H^+}$		1	0	0	x_1
$\overline{CaCO_3(s)}$		0	1	1	x_2
$TOTComp_{in}$		2.4×10^{-3}	2.0×10^{-3}	2.4×10^{-3}	
		$(+x_1)$	$(+x_2)$	$(+x_2)$	

The total input of Ca into the solution includes 2.0×10^{-3} M in the initial solution plus the Ca that was "added" by the program (x_2) to force the solution to reach equilibrium with $CaCO_3(s)$ at pH 11. Once equilibrium is reached, the computed concentration of dissolved Ca is only 1.49×10^{-5} M. Equating these two terms ($2 \times 10^{-3} + x_2 = 1.49 \times 10^{-5}$ M), we find $x_2 = -1.985 \times 10^{-3}$ M, meaning that >99% of the Ca initially present was removed by the precipitation reaction. The consistency of the calculation is confirmed by the fact that $TOTCO_{3,diss}$ decreases by the same amount, leaving 4.15×10^{-4} M $TOTCO_{3,diss}$ in the equilibrium solution.

8.11 DETERMINING WHETHER A SOLID WILL PRECIPITATE UNDER GIVEN CONDITIONS

The discussion thus far has established an approach for determining the composition of solutions that are known to be in equilibrium with a solid. We next consider systems that are, initially, not in equilibrium with a solid, and we ask these questions: Will a solid precipitate and/or dissolve as the system equilibrates, and if so, how much solid enters or leaves the solution? If more than one solid can form, which one(s) will be present at equilibrium? How will the equilibration process affect other water quality variables, such as pH and $TOTCO_{3,diss}$?

8.11.1 SYSTEMS IN WHICH ONLY ONE SOLID NEED BE CONSIDERED

We begin the analysis by consideration of a system that is well buffered at pH 10.0 and that contains $10^{-4} M\ TOTCd$, and in which we believe $Cd(OH)_2(s)$ is the only solid likely to form. The Cd might be present initially as soluble species, as solid $Cd(OH)_2(s)$, or as a combination of soluble and particulate species. For simplicity, assume that hydroxo complexes are the only ones likely to form in this system. We wish to know if solid $Cd(OH)_2(s)$ will be present when the system equilibrates.

The approach for determining the composition of the equilibrium solution involves making an initial assumption that the solid will or will not be present at equilibrium. If we assume that the solid is absent at equilibrium, then the analysis proceeds by computation of the Cd speciation in a solution containing $10^{-4} M$ total dissolved Cd. The calculations are carried out by substituting 10^{-4} for both $TOTCd_{diss}$ and $\{OH^-\}$ (because pH = 10) in Equation (8.18) and solving for $\{Cd^{2+}\}$. This value of $\{Cd^{2+}\}$ (call it $\{Cd^{2+}\}'$) can then be used to compute $\{Cd^{2+}\}'\{OH^-\}^2$ and thereby determine whether the computed solution composition would be supersaturated with respect to $Cd(OH)_2(s)$. That is, given the specified conditions and assuming that all the Cd is soluble, we can carry out the following calculations:

$$TOTCd_{diss} = \{Cd^{2+}\}'\,(1 + K_1\{OH^-\} + \beta_2\{OH^-\}^2 + \beta_3\{OH^-\}^3 + \beta_4\{OH^-\}^4)$$

$$10^{-4} = \{Cd^{2+}\}'\,[1 + (10^{3.92})(10^{-4.0}) + (10^{7.65})(10^{-4.0})^2$$
$$+ (10^{8.70})(10^{-4.0})^3 + (10^{8.65})(10^{-4.0})^4]$$

$$10^{-4} = \{Cd^{2+}\}'(2.28)$$

$$\{Cd^{2+}\}' = 4.39 \times 10^{-5} = 10^{-4.36}$$

The preceding calculation indicates that if all the Cd in the system is soluble, $\{Cd^{2+}\}$ will be $10^{-4.36}$. At pH 10, the product $\{Cd^{2+}\}'\{OH^-\}^2$ would then be $(10^{-4.36})(10^{-4.0})^2 = 10^{-12.36}$. Since this product is greater than $K_{s0}\,(= 10^{-13.73})$, we conclude that in the absence of the solid, the solution will be supersaturated with respect to $Cd(OH)_2(s)$. As a result, our assumption that no solid would be present at equilibrium must be in error. We therefore abandon the initial assumption and conclude that the solid will indeed be present. Given that the solution is well buffered, we can assume that the final pH will be 10.0, and the equilibrium value of $\{Cd^{2+}\}$ will be the value in equilibrium with $Cd(OH)_2(s)$, i.e., $\{Cd^{2+}\} = K_{s0}/\{OH^-\}^2 = 10^{-5.73}$. Correspondingly, the total soluble Cd will be:

$$TOTCd_{diss} = \{Cd^{2+}\}(1 + K_1\{OH^-\} + \beta_2\{OH^-\}^2 + \beta_3\{OH^-\}^3 + \beta_4\{OH^-\}^4)$$

$$= (10^{-5.73})(2.28) = 4.25 \times 10^{-6}$$

As expected, the total dissolved Cd concentration in equilibrium with the solid is less than the total initially present. The difference between these two values is the amount of Cd that precipitates as the system approaches equilibrium. The concentration of Cd present as $Cd(OH)_2(s)$ is thus $10^{-4.0} - 4.25 \times 10^{-6}$, or 9.57×10^{-5} M; i.e., approximately 96% of the Cd in the system precipitates as the system equilibrates.

The total concentration of the solid in the system is typically reported as its mass concentration (e.g., milligrams per liter), which includes the contribution of hydroxide ions. In the example, this concentration would be

$$[Cd(OH)_2(s)] = (9.57 \times 10^{-5} \, M)(146{,}000 \text{ mg/mol}) = 14.0 \text{ mg/L}$$

The mass of solid that actually forms, as determined by weighing a dried sample of the solid, is often quite a bit more than indicated by calculations such as this one, because significant amounts of water can be incorporated into and remain in the solid during drying.

If $\{Cd^{2+}\}'$ had been less than $K_{s0}/\{OH^-\}^2$, then we would have concluded that the solution containing 10^{-4} M $TOTCd_{diss}$ was undersaturated; i.e., the assumption that all the Cd was soluble would be confirmed. The concentrations of the other dissolved species could then be found based on $\{Cd^{2+}\}'$ and the stability constants. If this had been the case, then we would infer that the equilibrated solution could not contain the solid, and if the solution did come into contact with the solid, some of or all the solid would dissolve as the system approached equilibrium.

The calculations that we would carry out if the initial assumption had been that the solid *was* present at equilibrium are essentially identical to those shown above, but they would be carried out in a different sequence. Specifically, we would have first computed $TOTCd_{diss}$ by assuming that $\{Cd^{2+}\}$ equals $K_{s0}/\{OH^-\}^2$, i.e., using Equation (8.60). We call this value $TOTCd'_{diss}$. Then $TOTCd'_{diss}$ would be compared with the actual total amount of Cd in the system, $TOTCd$. If $TOTCd'_{diss} < TOTCd$, i.e., if the soluble Cd concentration based on the assumption that solid is present was less than the total amount of Cd in the system, then we would conclude that the assumption that some of the Cd was present as a solid was correct. If, on the other hand, $TOTCd'_{diss} > TOTCd$, then we would conclude that the assumption was in error and that all the Cd in the system was soluble. In that case, the speciation would be recomputed by setting $TOTCd_{diss}$ equal to $TOTCd$ and solving for the concentrations of the various complexes from the respective stability constants. In the current example, of course, we would find that the assumption that the solid is present was correct.

Thus, the determination of whether a solid is present at equilibrium can be carried out algebraically, starting with either the assumption that the solid is present or that it is absent; the only penalty associated with making the wrong assumption is that a few additional calculations must be carried out.

If a $\log C$–pH graph describing the system is available, the process is even simpler. On such a graph, the curve showing $TOTCd_{diss}$ as a function of pH represents the term defined above as $TOTCd'_{diss}$; i.e., it is the cumulative concentration of dissolved Cd species in a solution that is in equilibrium with the solid. The point representing the pH and total concentration of Cd in the actual system of interest (which might or might not be an equilibrium system) can also be identified on the graph. If this point is above the $TOTCd'_{diss}$ curve, then $TOTCd_{initial} > TOTCd'_{diss}$ and the solid will be present when the system reaches equilibrium; if it is below the curve, $TOTCd_{initial} < TOTCd'_{diss}$ and the solid will not be present. The point representing the initial conditions in the example system analyzed above is shown in Figure 8.10. The fact that it is above the curve for total dissolved Cd indicates that the solid will be present when the system equilibrates, consistent with our conclusion from the numerical analysis.

A plot showing the total dissolved Cd concentration as a function of pH in a system containing 10^{-4} M $TOTCd$ is shown in Figure 8.19a, and the concentration of $Cd(OH)_2(s)$ in the system is shown in Figure 8.19b. At pH values less than approximately 9.2, total dissolved Cd is 10^{-4} M, no solid precipitates, and the activity of the solid is therefore zero. Under these conditions, the speciation of dissolved Cd is identical to that shown in Figure 8.2, which was developed without ever considering possible precipitation of solids. However, at higher pH values (at least up to pH 14), the solid is present at the concentration given in Figure 8.19b, its activity is 1.0, and the speciation of dissolved Cd is identical to that in Figure 8.12, which was developed by assuming that $Cd(OH)_2(s)$ was always present.

Almost all the Cd in the system precipitates at pH values between approximately 10 and 13, although Figure 8.19a makes it clear that as pH is increased above about 11.3, the solid starts dissolving (since total dissolved Cd starts increasing). Because of this, adjusting solution pH to between 11 and 12 appears to be a good approach for removing Cd from solution, for example, in an industrial water treatment process. The solid could then be filtered out and reused or disposed of, and the water could be discharged. Note that increasing amounts of the solid dissolve as the pH is increased from 13 to 14; if the pH were raised above 14, all the Cd would eventually redissolve. Therefore, in a treatment process, it would be important to neither undershoot nor overshoot the target pH range.

We can summarize the possible scenarios in a system like the example system described above as follows:

1. If the initial solution is undersaturated with respect to $Cd(OH)_2(s)$, i.e., $\{Cd^{2+}\}\{OH^-\}^2 < K_{s0}$, and if no solid is initially present, no solid will precipitate.

2. If the initial solution is undersaturated with respect to $Cd(OH)_2(s)$ and if solid is initially present, the solid will dissolve until the solubility product is satisfied; if the solid dissolves completely before the solubility product is satisfied, then scenario 1 applies.

3. If the initial solution is supersaturated, then $Cd(OH)_2(s)$ will precipitate until the solubility product is satisfied.

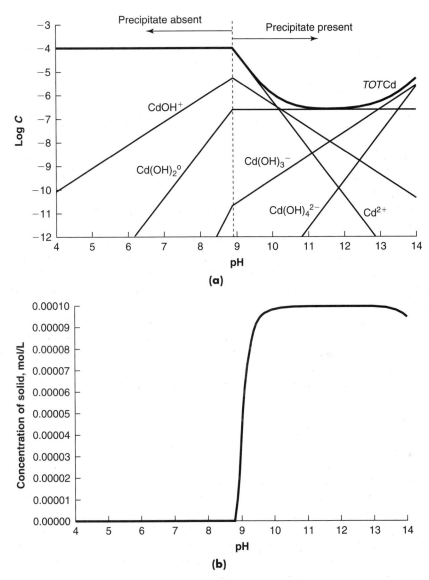

Figure 8.19   Cd speciation in a system containing 10^{-4} M TOTCd. **(a)** Dissolved species. **(b)** Solid $Cd(OH)_2(s)$.

8.11.2   SYSTEMS IN WHICH MORE THAN ONE SOLID MUST BE CONSIDERED

At this point, we have computed the concentration of Cd^{2+} in equilibrium with $Cd(OH)_2(s)$ and $CdCO_3(s)$ for a number of circumstances. However, in each case we assumed that only one solid could be present and that we knew which one it was. Consider next a solution containing both CO_3^{2-} and OH^-, to which Cd^{2+} is steadily added. Will $CdCO_3(s)$ precipitate? Will $Cd(OH)_2(s)$? Which solid will precipitate first? Is it possible for both solids to precipitate? We now turn our attention to answering these questions.

In this system, for any value of TOTCd, the final equilibrium conditions can be described by one of four possible situations: neither solid is present at equilibrium; $Cd(OH)_2(s)$ is present and $CdCO_3(s)$ is not; $CdCO_3(s)$ is present and $Cd(OH)_2(s)$ is not; or both $CdCO_3(s)$ and $Cd(OH)_2(s)$ are present. If we made an initial assumption that neither solid was present at equilibrium, we could compute the equilibrium speciation using the techniques described earlier and synthesized in Equation (8.18) [or Equation (8.40) if ligands other than OH^- are present]. The validity of the assumption could then be checked by computing the products $\{Cd^{2+}\}\{OH^-\}^2$ and $\{Cd^{2+}\}\{CO_3^{2-}\}$. If both solids were undersaturated based on this calculation, then the assumption would be confirmed and the problem would be solved. If, on the other hand, one or both solubility products were exceeded, then the assumption would have been proved incorrect and the problem would have to be re-solved using an alternative assumption.

Let's assume that when the problem is analyzed as described above and the assumption that no solids precipitate is tested, one of the two solubility products is exceeded but the other is not. We would conclude that the former solid will be the only one present at equilibrium. The problem can then be solved by following the procedure described previously for a situation where it is known that a particular solid is present at equilibrium. That analysis culminated in Equation (8.60) for the case where the hydroxide solid was present, and in Equations (8.64) and (8.65) for systems in equilibrium with $CdCO_3(s)$ and with fixed TOTCO$_{3,\text{diss}}$ and fixed $CO_2(g)$ partial pressure, respectively.

The only other possible outcome from the initial computation is that, based on the computed speciation, both solids are supersaturated in the initial solution. Although our first guess might be that both solids would be present when such a solution reached equilibrium, this assumption is not necessarily correct. To understand why, consider a system that is well buffered at pH 10, is in equilibrium with atmospheric CO_2, and is initially supersaturated with both $Cd(OH)_2(s)$ and $CdCO_3(s)$. In such a case, precipitation of either solid reduces TOTCd$_{\text{diss}}$, but $\{OH^-\}$ and $\{CO_3^{2-}\}$ remain constant: $\{OH^-\}$ because the pH is assumed to be well buffered and $\{CO_3^{2-}\}$ because of the combined effects of a constant pH and the equilibration of the solution with atmospheric CO_2.

The Cd^{2+} activity in equilibrium with pure $CdCO_3(s)$ under these conditions is $10^{-12.12}$, and that in equilibrium with pure $Cd(OH)_2(s)$ is $10^{-6.27}$. Clearly, the solution cannot simultaneously have both values of $\{Cd^{2+}\}$. What, then, will the equilibrium condition be? To answer this question, consider the sequence of events that might ensue as the system moves from its initial condition toward equilibrium.

Since both solids are initially supersaturated, it is reasonable to assume that both will begin to precipitate. As they do so, Cd^{2+} is removed from solution, so the solubility quotients of both solids—$Q_{s0,\text{Cd(OH)}_2(s)}$ and $Q_{s0,\text{CdCO}_3(s)}$—begin to decrease. This process is shown schematically in Figure 8.20a.

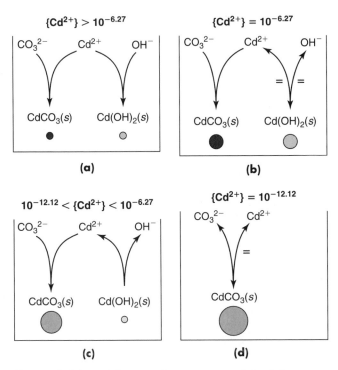

Figure 8.20   Schematic of precipitation and dissolution processes in example system. **(a)** Both $Cd(OH)_2(s)$ and $CdCO_3(s)$ are supersaturated in the initial solution, so both start to precipitate. **(b)** After ~95% of the Cd^{2+} has precipitated, the solution is in equilibrium with $Cd(OH)_2(s)$, so that solid stops precipitating. $CdCO_3(s)$ is still supersaturated. **(c)** $CdCO_3(s)$ continues to precipitate, removing Cd^{2+} from solution. This causes $Cd(OH)_2(s)$ to become undersaturated, so that solid begins to dissolve. **(d)** Eventually, enough Cd^{2+} has been removed from solution so that the solution is equilibrated with $CdCO_3(s)$. At this Cd^{2+} concentration, $Cd(OH)_2(s)$ is undersaturated, so it is completely dissolved.

Eventually, the solubility quotient of one of the solids is reduced to the point where $Q_{s0} = K_{s0}$ for that solid. In the example system, this point is reached first for $Cd(OH)_2(s)$, when $\{Cd^{2+}\}$ has declined to $10^{-6.27}$, and $TOTCd_{diss}$ is about $5 \times 10^{-6}\ M$ (i.e., approximately 95% of the dissolved Cd has precipitated). $CdCO_3(s)$ is still supersaturated under these conditions, so it continues to precipitate (Figure 8.20b). [Note that the form of the precipitated Cd, i.e., whether most of the Cd has been removed from solution as $CdCO_3(s)$ or $Cd(OH)_2(s)$, is unimportant for this portion of the analysis.]

Because $CdCO_3(s)$ continues to precipitate, $\{Cd^{2+}\}$ and the product $\{Cd^{2+}\}\{OH^-\}^2$ both decrease. As a result, $Cd(OH)_2(s)$ becomes undersaturated, and some of the $Cd(OH)_2(s)$ that precipitated previously begins to dissolve. That is, under these conditions ($\{Cd^{2+}\}$ slightly less than $10^{-6.27}$), $CdCO_3(s)$ is precipitating to reduce its degree of supersaturation, and $Cd(OH)_2(s)$ is simultaneously dissolving in an attempt to bring the solution back to saturation with respect to it (Figure 8.20c).

The $CdCO_3(s)$ is bound to win this "battle" for control of the solution composition, because as the reactions progress, $CdCO_3(s)$ accumulates and $Cd(OH)_2(s)$ is depleted. Eventually, the last bit of $Cd(OH)_2(s)$ dissolves, and $CdCO_3(s)$ continues to precipitate until $\{Cd^{2+}\}$ is $10^{-12.12}$. At that point, the system contains and is equilibrated with $CdCO_3(s)$, and it is undersaturated with respect to $Cd(OH)_2(s)$, which is, accordingly, not present. Thus, for the given system conditions, even though both solids were supersaturated initially, only $CdCO_3(s)$ can be present at equilibrium (Figure 8.20d).

Generalizing the above result, we conclude that at any pH in this system [equilibrated with the given $CO_2(g)$ partial pressure], the only solid that can be present at equilibrium is the one that equilibrates with the lower Cd^{2+} activity. Furthermore, because the ratio $\{Cd^{2+}\}/TOTCd_{diss}$ is the same at a given pH regardless of which solid is present, the lower Cd^{2+} activity always corresponds to a lower value of $TOTCd_{diss}$. The same can be said of the ratio of Cd^{2+} to any of the Cd—OH complexes. To demonstrate this point, the activities of all Cd species that would be present in the example system if it were equilibrated with each solid are shown in Table 8.8. Because all the values in equilibrium with $CdCO_3(s)$ are lower than the corresponding values in equilibrium with $Cd(OH)_2(s)$, the criterion for determining which solid can be present at equilibrium can be based on any of these species; i.e., the only solid that can be present at equilibrium is the one that equilibrates with the lower activity of any selected Cd-containing species.

Based on the preceding discussion, the only way that both $Cd(OH)_2(s)$ and $CdCO_3(s)$ could be present simultaneously in an equilibrium solution is if both solids are in equilibrium with the same solution. In such a solution, the solids must also be in equilibrium with one another. We can therefore assess the characteristics of solutions in equilibrium with both solids by considering the equilibrium constant for the reaction relating the two solids. The corresponding equilibrium

Table 8.8   Comparison of activities of various species in equilibrium with $CdCO_3(s)$ and $Cd(OH)_2(s)$ at pH 10, in a system equilibrated with atmospheric $CO_2(g)$

Species (1)	Equilibrium with $CdCO_3(s)$ (2)	Equilibrium with $Cd(OH)_2(s)$ (3)	Ratio of Column 3 to Column 2
Cd^{2+}	7.59×10^{-13}	5.37×10^{-7}	7.08×10^{5}
$CdOH^{+}$	6.31×10^{-13}	4.47×10^{-7}	7.08×10^{5}
$Cd(OH)_2^{\circ}$	3.39×10^{-13}	2.40×10^{-7}	7.08×10^{5}
$Cd(OH)_3^{-}$	3.80×10^{-16}	2.69×10^{-10}	7.08×10^{5}
$Cd(OH)_4^{2-}$	3.39×10^{-20}	2.40×10^{-14}	7.08×10^{5}
$TOTCd_{diss}$	1.73×10^{-12}	1.22×10^{-6}	7.08×10^{5}

reaction can be derived by adding the reaction for dissolution of one solid to the reaction for precipitation of the other, as shown below:

$$CdCO_3(s) \leftrightarrow Cd^{2+} + CO_3^{2-} \qquad\qquad K_{s0,CdCO_3(s)}$$

$$Cd^{2+} + 2OH^- \leftrightarrow Cd(OH)_2(s) \qquad\qquad K_{s0,Cd(OH)_2(s)}^{-1}$$

$$CdCO_3(s) + 2OH^- \leftrightarrow Cd(OH)_2(s) + CO_3^{2-} \qquad K = \dfrac{K_{s0,CdCO_3(s)}}{K_{s0,Cd(OH)_2(s)}}$$

If both pure solids are present at equilibrium, the activity of each is 1.0.[4] The mathematical form and the value of the equilibrium constant for the overall reaction are then

$$K = \frac{\{Cd(OH)_2(s)\}\{CO_3^{2-}\}}{\{CdCO_3(s)\}\{OH^-\}^2} = \frac{\{CO_3^{2-}\}}{\{OH^-\}^2} = \frac{10^{-13.73}}{10^{-14.27}} = 10^{+0.54} \qquad \textbf{(8.66)}$$

Equation (8.66) indicates that in any system equilibrated with both solids, the ratio of carbonate ion activity to the square of the hydroxide ion activity must be $10^{+0.54}$. If, in a given solution, that activity quotient is greater than $10^{+0.54}$, the reaction will tend to proceed to the left [$Cd(OH)_2(s)$ dissolving and $CdCO_3(s)$ precipitating], and if it is less than $10^{+0.54}$, the reaction will proceed to the right. As with any equilibrium process, these changes will tend to reduce the degree of disequilibrium.

The reaction proceeds until $Q = K$, at which point the two solids are equilibrated with each other. However, if all the $Cd(OH)_2(s)$ dissolves before equilibrium is achieved, $CdCO_3(s)$ becomes the only solid in the system and the equilibrium reaction involving the two solids becomes irrelevant. In the example system, the combination of fixed pH and fixed P_{CO_2} maintains $\{OH^-\}$ at $10^{-4.0}$ and $\{CO_3^{2-}\}$ at $10^{-1.3}$, no matter how much of each solid precipitates or dissolves. As a result, $\{CO_3^{2-}\}/\{OH^-\}^2$ is maintained at $10^{+6.7}$, equilibrium between the two solids can never be achieved, and $Cd(OH)_2(s)$ cannot be present at equilibrium.

An even simpler criterion for equilibrium between $Cd(OH)_2(s)$ and $CdCO_3(s)$ can be derived by writing the reaction between these solids in terms of the H_2CO_3 activity rather than the activities of OH^- and CO_3^{2-}, as follows:

$$CdCO_3(s) + 2OH^- \leftrightarrow Cd(OH)_2(s) + CO_3^{2-} \qquad K = \frac{K_{s0,CdCO_3(s)}}{K_{s0,Cd(OH)_2(s)}} = 10^{+0.54}$$

$$CO_3^{2-} + 2H_2O \leftrightarrow H_2CO_3 + 2OH^- \qquad K = \frac{K_w^2}{K_{a1}K_{a2}} = 10^{-11.32}$$

$$CdCO_3(s) + 2H_2O \leftrightarrow Cd(OH)_2(s) + H_2CO_3 \qquad K = 10^{-10.78}$$

---

[4] The assignment of a value of 1.0 for the activity of each solid presumes that the solids are present as distinct, pure phases. That is, if a pure $CdCO_3(s)$ solid phase is present, its mole fraction in that phase and therefore its activity are 1.0. If a pure $Cd(OH)_2(s)$ is in the same system, its activity will also be 1.0. This situation differs from that in which the two solids are present as a *solid solution*, in which case they are intimately mixed in a single solid phase, each with a mole fraction <1.0.

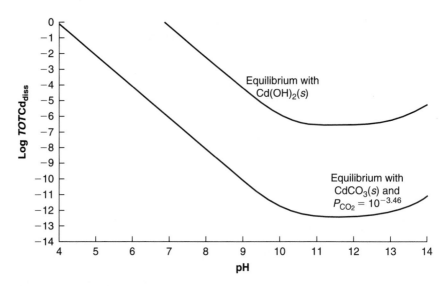

Figure 8.21   The total dissolved Cd concentration in equilibrium with $Cd(OH)_2(s)$ or $CdCO_3(s)$ in a system equilibrated with atmospheric $CO_2(g)$.

The form of the equilibrium expression for the reaction written in this way is simply $\{H_2CO_3\} = 10^{-10.78}$. In other words, the two solids can be in equilibrium with one another only if the dissolved H_2CO_3 activity is $10^{-10.78}$. Again, for the example system, we conclude that the two solids cannot be equilibrated with one another, because the fixed P_{CO_2} maintains the H_2CO_3 activity at $10^{-4.94}$. Thus, the impossibility of having $Cd(OH)_2(s)$ and $CdCO_3(s)$ present together in an equilibrium solution in contact with atmospheric $CO_2(g)$ applies not only at pH 10, but at *any* pH. If a solution initially contained both solids, the $Cd(OH)_2(s)$ would always dissolve completely as equilibrium was approached.

This result can also be inferred by preparing a graph showing the total dissolved Cd concentration in equilibrium with $Cd(OH)_2(s)$ or $CdCO_3(s)$ for a system in equilibrium with the atmosphere (Figure 8.21). The curve for equilibrium with $CdCO_3(s)$ is always below that for equilibrium with $Cd(OH)_2(s)$, consistent with the result that a solution in equilibrium with $Cd(OH)_2(s)$ will always be supersaturated with respect to $CdCO_3(s)$ in a system with $P_{CO_2(g)} = 10^{-3.46}$ bar.

Consider now the situation that would apply in a *closed* system containing 10^{-3} *M* $TOTCO_{3,\text{diss}}$. If we plot the total dissolved Cd concentration in equilibrium with each solid on a single log C–pH diagram (Figure 8.22), regions where zero, one, or both solids are supersaturated can be identified.

The point where the two curves intersect is the only point where the two solids are in equilibrium with the same amount of $TOTCd_{\text{diss}}$; i.e., it is the only point at which both solids can be present at equilibrium, for the given total

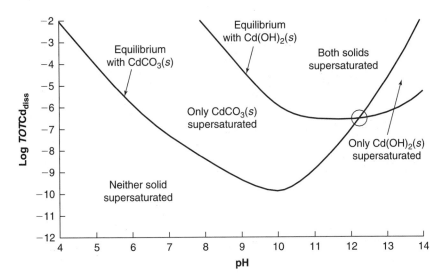

Figure 8.22   Curves showing $TOTCd_{diss}$ in equilibrium with $Cd(OH)_2(s)$ or $CdCO_3(s)$, for a system with 10^{-3} M $TOTCO_{3,diss}$. Each curve divides the space into a region where the solid is undersaturated and a region where it is supersaturated. The two curves together therefore define four regions where zero, one, or both solids are supersaturated. The circled point indicates the only condition where the same $TOTCd_{diss}$ is in equilibrium with both solids.

dissolved carbonate concentration. For the conditions shown (10^{-3} M $TOTCO_{3,diss}$), that point is at pH 12.23. At that pH, $\{CO_3^{2-}\} = 0.988 \times 10^{-3}$, the ratio $\{CO_3^{2-}\}/\{OH^-\}^2$ equals $10^{+0.54}$, and $\{H_2CO_3\} = 10^{-10.78}$, consistent with the calculations above. Evaluation of the Cd speciation confirms that at pH 12.23 the concentration of each Cd species is the same for equilibrium with either solid.

As an overall summary, Figure 8.23 shows the concentrations of all Cd species (both soluble species and solids) in an equilibrium system containing 10^{-4} M $TOTCd$ and 10^{-3} M $TOTCO_{3,diss}$. All of the Cd is soluble at pH < 5, with Cd^{2+} being the dominant species. Almost all the Cd in the system is present in solids at pH values above ~5.5, but the identity of the solid containing the Cd changes abruptly at pH 12.23: at pH < 12.23, only $CdCO_3(s)$ is present, and at pH > 12.23, only $Cd(OH)_2(s)$ is present. Only if the pH is exactly 12.23 can both solids be present.

Example 8.7

A sample of seawater is filtered and dosed with $CdCO_3(s)$ particles at a total concentration of 10^{-3} mol/L. Assume (1) the system is closed, (2) the seawater has been bubbled with CO_2-free air so it initially contains no dissolved carbonate species, and (3) seawater is well buffered at pH = 8.0 and contains 0.5 M Cl^-. What is the equilibrium speciation of Cd? Does any solid Cd compound exist at equilibrium?

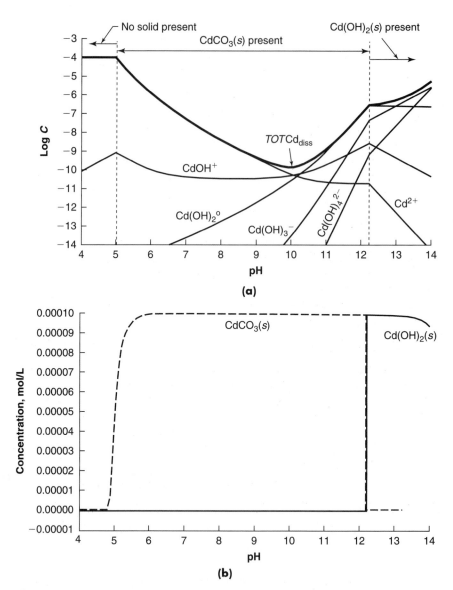

Figure 8.23   The Cd speciation in a system containing 10^{-4} M TOTCd and 10^{-3} M $TOTCO_3$, considering possible precipitation of $Cd(OH)_2(s)$ or $CdCO_3(s)$. **(a)** Soluble Cd species; **(b)** Cd-containing solids. Values on the y axis extend to negative numbers so that the conditions where the solids' concentrations are zero can be seen more clearly.

Solution

To solve this problem, we need to make an initial assumption about which solid(s) will be present at equilibrium. Arbitrarily, the first assumption we make is that all the $CdCO_3(s)$ dissolves. We can then solve for the concentrations of all Cd-containing species and test the assumption by determining whether the solubility products of any Cd-containing solids are exceeded. Assuming that the only soluble complexes that need to be considered are the chloro and hydroxo complexes, we have

$$TOTCd_{diss} = \{Cd^{2+}\} + \{CdOH^+\} + \{Cd(OH)_2^{\,\circ}\} + \{Cd(OH)_3^-\} + \{Cd(OH)_4^{2-}\}$$
$$+ \{CdCl^+\} + \{Cd(Cl)_2^{\,\circ}\} + \{CdCl_3^-\} + \{CdCl_4^{2-}\}$$

$$= \{Cd^{2+}\} \left(1 + \sum_{i=1}^{4} \beta_{OH,i}\{OH^-\}^i + \sum_{i=1}^{3} \beta_{Cl,i}\{Cl^-\}^i\right)$$

$$= \{Cd^{2+}\}[1 + (10^{3.92})(10^{-6.0}) + (10^{7.65})(10^{-6.0})^2 + (10^{8.70})(10^{-6.0})^3$$
$$+ (10^{8.65})(10^{-6.0})^4 + (10^{1.98})(0.5) + (10^{2.60})(0.5)^2 + (10^{2.40})(0.5)^3]$$

$$= \{Cd^{2+}\}(1 + 10^{-2.08} + 10^{-4.35} + 10^{-9.30} + 10^{-15.35} + 10^{+1.68}$$
$$+ 10^{+2.00} + 10^{+1.50})$$

$$= \{Cd^{2+}\}(180)$$

$$\{Cd^{2+}\} = \frac{TOTCd}{180} = \frac{10^{-3.0}}{180} = 10^{-5.26}$$

Note that the concentration of free Cd^{2+} ions is much greater than that of any of the hydroxo complexes, but it still represents a small fraction of the total dissolved Cd because the chloro complexes (represented by the final three terms in the summation) are present at much higher concentrations.

The above calculations indicate that if all the cadmium dissolves, $\{Cd^{2+}\}$ is $10^{-5.26}$. Because we are assuming that all the $CdCO_3(s)$ that was added goes into solution, and dissolution of the solid is the only source of carbonate species to the solution, $TOTCO_3 = 10^{-3}$. Combining this information with the pK_a values for the carbonate system and the known pH of 8.0, we find $\{CO_3^{2-}\} = 10^{-5.33}$. With that value, we can check to see whether the assumption of total dissolution is valid:

$$\{Cd^{2+}\}\{OH^-\}^2 = (10^{-5.26})(10^{-6.0})^2 = 10^{-17.26} < K_{s0,Cd(OH)_2(s)}$$

$$\{Cd^{2+}\}\{CO_3^{2-}\} = (10^{-5.26})(10^{-5.33}) = 10^{-10.59} > K_{s0,CdCO_3(s)}$$

$Cd(OH)_2(s)$ is undersaturated, so the assumption that it is not present at equilibrium is valid, but $CdCO_3(s)$ is supersaturated and will be present at equilibrium (i.e., the solid will not dissolve completely). We must therefore recompute the equilibria with the constraint that the solubility product for $CdCO_3(s)$ is satisfied:

$$\{Cd^{2+}\} = \frac{K_{s0,CdCO_3(s)}}{\{CO_3^{2-}\}} = \frac{K_{s0,CdCO_3(s)}}{\alpha_2\{TOTCO_{3,diss}\}}$$

Based on the stoichiometry of the solid, the total concentrations of Cd and CO_3 dissolving must be equal, so $TOTCd_{diss} = TOTCO_{3,diss}$. Furthermore, the calculation above indicates that uncomplexed Cd^{2+} ion represents only $\frac{1}{180}$ of the total dissolved Cd. That calculation is valid regardless of which solids, if any, are present at equilibrium, so $TOTCO_{3,diss} = 180\{Cd^{2+}\}$. Combining this information with the above equation gives

$$K_{s0,CdCO_3(s)} = \{Cd^{2+}\}[\alpha_2 (TOTCO_{3,diss})] = \{Cd^{2+}\}[10^{-2.33}(180\{Cd^{2+}\})] = 0.84\,\{Cd^{2+}\}^2$$

$$\{Cd^{2+}\} = \left(\frac{10^{-13.73}}{0.84}\right)^{0.5} = 1.49 \times 10^{-7} = 10^{-6.83}$$

The concentrations of all other Cd species in the system can be calculated based on this result, so the problem is completely solved. In particular, $TOTCd_{diss}$ is

$$TOTCd_{diss} = 180(10^{-6.83}) = 2.68 \times 10^{-5} = 10^{-4.57}$$

The vast majority of the dissolved cadmium is present as chloro complexes, with $\{CdCl_2^{\,0}\} > \{CdCl^+\} > \{CdCl_3^-\} > \{Cd^{2+}\}$. The exact concentration of these species and of the hydroxo complexes can be calculated by using the known values of $\{Cd^{2+}\}$, $\{Cl^-\}$, $\{OH^-\}$, and the stability constants. At equilibrium, some $CdCO_3(s)$ is present. The concentration of the solid can be calculated from a mass balance:

$$[CdCO_3(s)] = TOTCd - TOTCd_{diss} = 10^{-3.0}\,M - 10^{-4.57}\,M = 9.73 \times 10^{-4}\,M$$

Thus, only about 2.7% of the solid dissolves, and the remainder is present as $CdCO_3(s)$.

8.11.3 THE RELATIVE SOLUBILITIES OF SOLID POLYMORPHS

It is fairly common for several solids to have the same chemical formula and to differ from one another only in their crystal structure. Such solids are called *polymorphs* of one another. In other cases, the compositions of two solids differ only in the number of water molecules incorporated into their structures. In such cases, the form of the solubility product for the solids is essentially identical. For instance, the dissolution reactions and the solubility product expressions for four Fe(III) oxide or hydroxide solids are shown in Table 8.9.

The final two columns in the table indicate that for any of these solids to be in equilibrium with a solution, the product $\{Fe^{3+}\}\{OH^-\}^3$ must be a specific value. Obviously, since this product can have only one value in a given solution, no two of the solids can be equilibrated with the same solution. Furthermore, the

Table 8.9   Equilibrium constants for dissolution of four forms of iron oxyhydroxide

Name	Reaction	K_{s0} Expression	Log K_{s0}
Ferrihydrite (amorphous iron hydroxide)	$Fe(OH)_3(s) \leftrightarrow Fe^{3+} + 3OH^-$	$K_{s0} = \dfrac{\{Fe^{3+}\}\{OH^-\}^3}{\{Fe(OH)_3(s)\}} = \{Fe^{3+}\}\{OH^-\}^3$	-37.11
Goethite α-FeOOH	$\alpha\text{-FeOOH}(s) + H_2O \leftrightarrow Fe^{3+} + 3OH^-$	$K_{s0} = \dfrac{\{Fe^{3+}\}\{OH^-\}^3}{\{\alpha\text{-FeOOH}(s)\}} = \{Fe^{3+}\}\{OH^-\}^3$	-41.50
Lepidocrocite γ-FeOOH	$\gamma\text{-FeOOH}(s) + H_2O \leftrightarrow Fe^{3+} + 3OH^-$	$K_{s0} = \dfrac{\{Fe^{3+}\}\{OH^-\}^3}{\{\gamma\text{-FeOOH}(s)\}} = \{Fe^{3+}\}\{OH^-\}^3$	-40.63
Hematite α-Fe$_2$O$_3$	$0.5Fe_2O_3(s) + 1.5H_2O \leftrightarrow Fe^{3+} + 3OH^-$	$K_{s0} = \dfrac{\{Fe^{3+}\}\{OH^-\}^3}{\{Fe_2O_3(s)\}^{0.5}\{H_2O\}^{1.5}} = \{Fe^{3+}\}\{OH^-\}^3$	-44.00

relative values of K_{s0} for the different solids indicate that the solubility decreases in the order ferrihydrite > lepidocrocite > goethite > hematite. As a result, any solution in equilibrium with one of the first three solids will be supersaturated with respect to hematite. The conclusion is that no solution that contains any of the first three solids can, in truth, be an equilibrium solution; conversion of the solid to hematite will always be favored.

Nevertheless, the solids listed in the upper rows of the table are significant in the natural environment, and they can be prepared and used in the laboratory. These solids are commonly referred to as being meta-stable, meaning that they do not normally convert to more stable solids in the time frame of interest. In such cases, we can treat any of the solids as though it were the most stable one that could form in the system, ignoring the possibility of precipitation of more stable polymorphs. This situation arises commonly when we wish to model the initial formation of solids, such as in cases where metals are precipitated as part of a water or wastewater treatment process. The multiplicity of potential solids that might form in such situations makes clear the importance of specifying which solids are being considered in any given analysis.

8.12   USE OF CHEMICAL EQUILIBRIUM SOFTWARE TO MODEL SYSTEMS IN WHICH PRECIPITATION AND DISSOLUTION ARE POSSIBLE, BUT NOT ENSURED

Use of chemical equilibrium software to determine the equilibrium composition of solutions in which solids might or might not be present mimics the steps that are followed in analyzing such systems manually. That is, the program must make an assumption about whether each solid will be present in the equilibrated system, solve the equations using that assumption, then check whether the assumption is valid and, if not, make an alternate one.

As noted earlier in the chapter, if a solid is known to be present at equilibrium (or if we want to compute the solution composition for a system that is assumed to be in equilibrium with the solid), the solid is input as a Type 3b *species*. On the other hand, if we want the program to consider the possibility that a solid is present, but we do not want to insist that equilibrium with the solid be incorporated into the reported solution composition, we must identify the solid differently. Solids that fall into this category have historically been labeled as Type 4 or Type 5 *species*. Type 4 *species* are usually referred to as *precipitated solids* and Type 5 *species* as *dissolved solids*. (For a reminder of what each *species* Type represents and a slightly more detailed discussion of Type 4 and Type 5 *species,* refer to the discussion of software nomenclature in Appendix 6A.)

The rules for assigning stoichiometric coefficients and equilibrium constants to these types of *species* are identical to those for Type 3b *species,* but the program handles the information differently. If a solid is input as a Type 4 *species,* the program carries out the initial calculations assuming that the solid is present. Then, after each iteration, it checks to see whether the specified inputs provide enough of each *component* to generate the computed solution composition. If the inputs are large enough to meet the requirements of equilibrium with the solid, then the assumption is retained; if not, the next iteration is based on the assumption that the solid is not present.

Conversely, if the solid is specified as a Type 5 *species,* the initial calculations are carried out assuming that the solid is not present. After each iteration, the program checks to see if the computed solution composition exceeds the solubility product of the solid. If it does not, then the assumption is retained; but if it does, then the next iteration is performed assuming that the solid is present. In essence, then, the program shuttles solids back and forth between Type 4 and Type 5 *species* until it performs a calculation in which the assumptions about all possible solids are determined to be correct.

Just a few years ago, when the software was being used with much less powerful computers than are available now, the benefits of making correct assumptions about the presence or absence of individual solids were sometimes dramatic. However, that is no longer the case. Currently, many programs make the default choice to treat all solids that might be present at equilibrium as Type 5 *species* (i.e., the programs make the assumption that the solids do not precipitate), and it is probably not worth the effort to override these default options. If the program determines that the solids are present at equilibrium, those solids are often reported as Type 4 *species* in the output, so it is important to understand what this group of *species* represents.

Example 8.8 | Example 1.10 used an algebraic approach to determine if any solid would remain at equilibrium (and if so, how much) if we added 1 g/L AgCl(s) (7.0×10^{-3} M) to a solution in which the activity of Cl^- is 0.1 and the ionic strength is 0.12 M. K_{s0} for AgCl(s) is $10^{-9.75}$. Prepare the input tableau to evaluate this system using chemical equilibrium software. (Although Ag^+ can form a range of soluble complexes with OH^- and Cl^-, we will ignore these reactions in this example, so that the result can be compared to that obtained in Chapter 1.)

Solution

In this problem, we do not know in advance whether AgCl(s) will be present at equilibrium, so we input the solid as a Type 5 *species*. The stoichiometry and log K values for the solid are the same as if it were a solid that we knew would be present (a Type 3b *species*). Also, since the AgCl(s) that was added is part of the initial composition of the system, we include it in our calculations of the total input concentrations of the *components*. This step can be thought of as treating the AgCl(s) as though it all dissolves when it is added to the solution, and then running the program to determine whether any precipitate forms.

The input tableau and the computed equilibrium composition of the solution are shown below. The program was instructed to account for ionic strength effects using the Davies equation via a toggle switch in one of the menus.

Species	Type	Component Ag$^+$	Cl$^-$	Log K	Equil. Concentration
Ag$^+$	1	1	0	0.00	3.05×10^{-9}
Cl$^-$	1	0	1	0.00	1.00×10^{-1}
$\overline{\text{AgCl}(s)}$	5	1	1	9.75	
*TOT*Comp$_{eq}$		3.05×10^{-9}	1.00×10^{-1}		

Inputs				Input Concentration
Cl$^-$	0	1		0.100
AgCl(s)	1	1		7.0×10^{-3}
*TOT*Comp$_{in}$	$7.0 \times 10^{-3}\ (-x)$	$0.107\ (-x)$		

In the table, x represents the concentration of AgCl(s) that is present once the system reaches equilibrium, so that $7 \times 10^{-3} - x$ and $0.107 - x$ represent the concentrations of dissolved Ag$^+$ and Cl$^-$, respectively. When the calculations are actually carried out, the program reports that *TOT*Ag$_{diss} = 3.05 \times 10^{-9}$. We can use this information to compute that x is very close to 7.0×10^{-3}, i.e., a negligible portion of the solid dissolves, consistent with the result obtained in Chapter 1.

8.13   THE GIBBS PHASE RULE

From the very beginning of the discussion of manual approaches for solving equilibrium problems in Chapter 3, we have formulated the problems as comprising a group of unknowns and a set of simultaneous equations that can be used to solve for those unknowns. In each case, the number of equations had to match the number of unknowns, and a group of other parameter values had to be known in advance (things like equilibrium constants; the total amounts of certain chemicals in a system; the known, fixed activity of some species, etc.). The relationship among the total number of variables in a chemical equilibrium problem, the number of variables that can and/or must be specified, and the number of variables that can then be determined by solving the available equations is formalized in a statement known as the *Gibbs phase rule* (or, simply, the *phase rule*). This rule states that the number of degrees of freedom F in a chemical system equals the number of components C minus the number of phases P plus two:

$$F = C - P + 2 \qquad \textbf{(8.67)}$$

The number of degrees of freedom indicates how many independent constraints can be (and need to be) specified in order to completely define the system composition at equilibrium. These constraints can be system parameters that we specify, such as temperature or pressure, the activity of a particular chemical species, or the total concentration of a group of species (e.g., $TOTCO_3$), or they can be other requirements, such as the requirement that the total charge on the dissolved cations equal that on the dissolved anions.

The term "components" as used in the phase rule is defined in different ways by different people, but these differences are primarily semantic. The differences relate to issues such as whether the activity of H_2O is considered as an unknown and the equation $\{H_2O\} = 1.0$ is included as a system constraint, or $\{H_2O\}$ is treated as not being an unknown in the first place. Those who make different choices about this issue view the problem as having different numbers of unknowns, but they end up solving the same set of equations. For our purposes, the most convenient definition of components for use in conjunction with the Gibbs phase rule is the same one we use when analyzing a problem using chemical equilibrium software: it is the minimum number of chemicals that can be used to generate all the chemical species present in the equilibrium system. In fact, the phase rule was the basis for choosing the term *components* for use in the software jargon.

The phases in a chemical system might include a gas, one or more liquids, and a number of different solids. Phases are defined by the possibility of mixing at the molecular level and by the presence of an interface with any other phase. Thus, each distinct solid in a system (each solid with a different composition or crystal structure) represents a separate phase. For instance, a system containing metallic iron, an iron oxyhydroxide solid (e.g., goethite), and iron carbonate solid would have three solid phases. On the other hand, a solid that contains a more or less random distribution of different elements [e.g., a solid consisting of a $CaCO_3(s)$ crystal structure in which 10% of the Ca^{2+} ions are randomly replaced by Cd^{2+} ions] would be a single phase, often referred to as a solid solution.

Similarly, a well-ordered crystal containing repeating units or layers of SiO_2 and Al_2O_3 (e.g., aluminosilicate clays) would be a single phase. Although multiple liquid phases exist in some environmental systems of interest (e.g., organic solvents in some contaminated groundwaters), the focus of this text is strictly on aqueous solutions, so we will only consider systems with one liquid phase. Because gases always mix at the molecular level and never form an interface with one another, there can be only one gas phase in a given system at equilibrium.

One of the simplest examples of the application of the phase rule is for a system containing only $H_2O(l)$, H^+, and OH^-. Any two of these three species can serve as a component set for the system for the purposes of the Gibbs phase rule; the remaining species cannot be included as a component because it could be made by a combination of the other two. Therefore, for this system, $C = 2$. The system consists of only the one phase ($P = 1$), so according to Equation (8.67), $F = 3$. One of the degrees of freedom is associated with the requirement for electroneutrality, and the other two are normally used in specifying the system's temperature and pressure. Once these parameters are specified, the state of the system is completely defined. (The temperature and pressure establish the value of K_w, and that value plus the electroneutrality condition establishes the concentrations of H^+ and OH^-.)

If the same system has two phases (e.g., ice and liquid water), the number of degrees of freedom decreases to one. In such a case, we could arbitrarily specify the pressure, for instance, as being 1.0 bar, but at that point the temperature would be established as 0°C. We cannot arbitrarily specify both the pressure and the temperature of a system containing two H_2O phases, because once one of those parameters is set, the other is established; for instance, it is simply not possible to have an equilibrium system at 1.0 bar total pressure that contains both ice and water (and no other components) and that is at any temperature other than 0°C. Similarly, if we specified that all three phases of H_2O were present ($P = 3$) in a system containing only H_2O, H^+, and OH^-, the system would have to be at the triple point of water; i.e., both temperature and pressure would be constrained to known values.

In almost all systems of interest in this text, the temperature and total pressure of the system are assumed to be known (typically 25°C and 1.0 bar). Even in cases where values of these parameters are not specified explicitly, they are implicitly established when we state the equilibrium constants for the reactions that take place in the system. Furthermore, since we are always interested in systems that contain dissolved ions, the charge balance places another constraint on the system. Therefore, the systems of interest always have at least three pre-specified constraints, so the number of remaining degrees of freedom in these systems is $C - P - 1$.

Although we have not considered the phase rule explicitly in previous chapters, the analyses we have conducted have always been consistent with it. For instance, if a solution is made by adding HAc to water, the system has three components. These components might be chosen to be H_2O, H^+, and Ac^-; they might also be chosen to be H_2O, H^+, and HAc, or H^+, OH^-, and HAc, or any of a variety of other combinations (including combinations of imaginary chemicals). However, no matter what component set is chosen for the system, it must comprise three chemicals. Therefore, for such a system, $C = 3$, $P = 1$, and $C - P - 1 = 1$; i.e., the system has only one degree of freedom (other than temperature and pressure, and assuming that we wish to impose the electroneutrality constraint). This degree of freedom is utilized when we specify that the total acetate concentration is 10^{-3} M or that the pH is 5.0, or make any other chemical assignment. Once this degree of freedom is utilized, the equilibrium composition of the solution is fully established.[5]

It is not possible to determine a unique equilibrium composition for a system in which the number of parameters that have been specified is less than the number of degrees of freedom. An example would be to ask the question, What is the pH of a solution made by adding HAc to pure water? Obviously, any pH could be

---

[5]The phase rule is, in effect, a statement of mathematical possibilities, not all of which are chemically realistic. Mathematically, one could meet all the constraints of the equilibrium constants and the mass balance and charge balance equations by specifying a negative total concentration of a component. In this way, one could even get the pH to an alkaline value by adding "negative" amounts of HAc. Thus, mathematically, it is possible to choose any value one wants for pH in the example system (i.e., the choice really is unconstrained), even though achieving that pH might not be possible in a realistic system.

obtained, depending on the concentration added. Similarly, it is usually not possible to find any equilibrium composition that is consistent with more constraints than specified by the number of degrees of freedom. An example of this situation would be to ask the question, What is the concentration of Ac^- in a system made by adding 10^{-3} M HAc to pure water if the pH is 6.5? The problem with this question is that the pH of a solution of 10^{-3} M HAc is fixed by the given addition of HAc, and one cannot cause such a solution to have a pH of 6.5 (except by adding a base, like NaOH, which then changes the number of components in the system). Although this particular violation of the phase rule is obvious and would be unlikely to go unnoticed, one might inadvertently describe a system that violates the phase rule when dealing with complex systems having lots of phases.

Unfortunately, Gibbs phase rule violations are sometimes reported incorrectly by chemical equilibrium software. Conformance with the Gibbs phase rule is checked by most software packages before each iteration, and if a violation occurs, the program terminates. The user is then alerted to the problem and must correct the error. If the violation occurs on the first iteration, chances are that too many species were input with known, fixed activities, in which case the problem can be solved by reducing the number of such specifications. However, phase rule violations can sometimes be generated by the program in the midst of solving the problem, if it "guesses" that more solids are present than can truly be present at equilibrium. This type of phase rule violation is software-related and can be overcome by instructing the software to ignore some of the possible precipitates that might form. The user must then check the computed final equilibrium condition to see if the ignored solids might in fact precipitate (by comparing Q_{s0} with K_{s0}). Some strides have been made to overcome this problem, but it still occurs in many programs.

Example 8.9 | Apply the Gibbs phase rule to determine the number of degrees of freedom in the following solutions. Assume that the temperature and total pressure of each system are fixed.

    a. A mixture of sodium carbonate and sodium bicarbonate that has been adjusted to pH 7.0 by addition of acetic acid.

    b. The solution in part (*a*) if it continuously equilibrates with atmospheric $CO_2(g)$ as the other chemicals are added.

    c. A solution that contains Na^+, Ca^{2+}, Cl^-, and the four species of the H_xPO_4 acid/base group and that is in equilibrium with hydroxyapatite, $Ca_5(OH)(PO_4)_3(s)$.

    d. A solution in equilibrium with both $CdCO_3(s)$ and $Cd(OH)_2(s)$, in which $TOTCO_{3,\text{diss}}$ is 10^{-3} M. The solution also contains some Na^+, added along with some of the carbonate.

    e. The same solution as in part (*d*) but prepared by adding only H_2CO_3, $CdCO_3(s)$, and $Cd(OH)_2(s)$ to water, so that the solution contains no Na^+.

Solution

Each of the solutions of interest contains ions, so the electroneutrality condition in solution is a constraint in addition to the temperature and pressure. Therefore, the number of degrees of freedom is given by $F = C - P - 1$. This value of F in each system is then determined as follows.

a. The solution can be made by a combination of five components (e.g., H_2O, H^+, Na^+, H_2CO_3, and Ac^-), and it contains only one phase (the solution). Thus, there are three degrees of freedom available. One degree of freedom is associated with the constraint that the pH is 7.0. We could use the other two degrees of freedom to specify, for example, the concentrations of sodium carbonate and sodium bicarbonate added, or the total concentrations of carbonate and acetate in the system, or any other two parameters that characterize the system.

b. The number of components in the system is the same as in part (a), because $CO_2(g)$ can be made from the existing components. However, we do have an additional phase to consider (the gas phase), so the number of degrees of freedom remaining is reduced to one. Thus, if we specified TOTAc, for example, the system would be completely defined.

c. Although it is not mentioned explicitly, the solution must, of course, contain H^+ and OH^-, which cannot be made from the ions listed. Therefore, the system contains six components (an acceptable component set would comprise H_2O, H^+, Ca^{2+}, Na^+, Cl^-, and PO_4) and two phases (solution and hydroxyapatite), so $F = C - P - 1 = 3$. If, for instance, we specified the concentrations of the three salt ions (Ca^{2+}, Na^+, and Cl^-), the system's chemistry would be completely defined. (Pick arbitrary values for the activities of these three ions, and convince yourself that the phosphate activity is established by these choices.)

d. This system contains five components (e.g., H_2O, H^+, Na^+, Cd^{2+}, and CO_3^{2-}) and three phases. It is also subject to the constraint imposed that TOTCO$_3$ is 10^{-3} M. It therefore has no remaining degrees of freedom ($F = 5 - 3 - 1 = 1$, with the one degree of freedom used by imposing the constraint on TOTCO$_3$). This result is consistent with the conclusion we reached earlier, that a solution equilibrated with 10^{-3} M TOTCO$_{3,diss}$ can be in equilibrium with both $CdCO_3(s)$ and $Cd(OH)_2(s)$ only at pH 12.23. Given the pH and known TOTCO$_3$ concentration, the concentrations of all the H_xCO_3 species are also established under these conditions.

e. The system contains only four components, but it still has three phases, and we are trying to impose the constraint that TOTCO$_3$ is 10^{-3} M. This system is overly constrained ($F = -1$) and violates the phase rule; no system can meet the specified conditions and be at equilibrium.

8.14 INTEGRATED ANALYSIS OF A MODEL SYSTEM: TITRATION OF A CARBONATE-CONTAINING SOLUTION WITH $Zn(NO_3)_2$

To illustrate the integrated application of many of the concepts presented in this chapter, we next carry out a fairly detailed analysis of an aqueous solution containing a variable concentration of Zn and a known initial concentration of total carbonate species. Specifically, we analyze the composition of a solution that initially contains 10^{-2} M Na_2CO_3 when it is titrated with $Zn(NO_3)_2$ up to a maximum concentration of 0.02 M TOTZn. We might be interested in such a simulation, for instance, if we worked for an industry that generated a small flow of a concentrated Zn-containing waste and also a much larger flow of a weakly alkaline waste solution that could be reasonably represented as $10^{-2} M$ Na_2CO_3. The question we wish to explore is: In what ratio should we mix the two solutions to maximize the amount of Zn that precipitates? We assume, as a first approximation, that the volume ratio of the two solutions is such that the alkaline solution is not significantly diluted when the Zn-containing solution is mixed into it.

In the analysis, we consider the interactions of Zn with both hydroxide and carbonate ligands, the effects of these interactions on solution pH, and the possible precipitation of both $Zn(OH)_2(s)$ and $ZnCO_3(s)$. The pH of the solution and $TOTCO_{3,diss}$ are allowed to change in response to the removal of OH^- and CO_3^{2-} from solution by precipitation reactions.

Solubility and stability constants for the relevant Zn-containing species are shown below. No Zn—NO_3 complexes or solids are expected to form. The solubility constants listed are for the most soluble form of each solid listed in the database used by many chemical equilibrium programs, specifically amorphous $Zn(OH)_2(s)$ and the mineral smithsonite, $ZnCO_3(s)$.

Ligand	Reaction Designation	Log K	Ligand	Reaction Designation	Log K
OH^-	$K_{OH,1}$	5.04	CO_3^{2-}	$K_{CO_3,1}$	5.30
	$\beta_{OH,2}$	11.10	CO_3^{2-}	$\beta_{CO_3,2}$	9.63
	$\beta_{OH,3}$	13.60	HCO_3^-	$K_{HCO_3,1}$	2.07
	$\beta_{OH,4}$	14.80	CO_3^{2-}	$K_{s0,ZnCO_3(s)}$	-10.00
	$K_{s0,Zn(OH)_2(s)}$	-15.55			

We start by determining the pH of the $10^{-2} M$ Na_2CO_3 solution. This determination is quite trivial and can be accomplished either by drawing a log C–pH diagram and finding the point where the proton condition is satisfied, or by using chemical equilibrium software. The result is that the solution pH is 11.13.

The changes in system composition for various additions of the Zn-containing solution into the Na_2CO_3 solution can be represented as a titration using chemical equilibrium software. The inputs specified for the initial solution are 0.01 M $TOTCO_3$, and 0 M each of $TOTH$, $TOTZn$, and $TOTNO_3$. The titration is then simulated by solving for the equilibrium speciation numerous times, each time using slightly larger input values for $TOTZn$ and $TOTNO_3$. For the calculations shown here, $TOTZn$ and $TOTNO_3$ were increased from 0 to 0.02 M and 0.04 M, respectively, in 100 steps, so that each step represents an addition of $2 \times 10^{-4} M$ $Zn(NO_3)_2$. Solution pH was calculated by the program, along with the concentrations of all dissolved *species*. Amorphous $Zn(OH)_2(s)$ and smithsonite were included as Type 5 *species;* i.e., they were considered as possible precipitates, but

Table 8.10 Input tableau for the initial $Zn(NO_3)_2$ addition in the example system

Species	Type	Component					Log K
		H^+	Na^+	Zn^{2+}	CO_3^{2-}	NO_3^-	
H^+	1	1	0	0	0	0	0.00
Na^+	1	0	1	0	0	0	0.00
Zn^{2+}	1	0	0	1	0	0	0.00
CO_3^{2-}	1	0	0	0	1	0	0.00
NO_3^-	1	0	0	0	0	1	0.00
OH^-	2	−1	0	0	0	0	−14.00
H_2CO_3	2	2	0	0	1	0	16.68
HCO_3^-	2	1	0	0	1	0	10.33
$ZnOH^+$	2	−1	0	1	0	0	−8.96
$Zn(OH)_2^\circ$	2	−2	0	1	0	0	−16.90
$Zn(OH)_3^-$	2	−3	0	1	0	0	−28.40
$Zn(OH)_4^{2-}$	2	−4	0	1	0	0	−41.20
$ZnCO_3^\circ$	2	0	0	1	1	0	5.30
$Zn(CO)_2^{2-}$	2	0	0	1	2	0	9.63
$ZnHCO_3^+$	2	1	0	1	1	0	12.40
$Zn(OH)_2(s)$	5	−2	0	1	0	0	−12.45
$ZnCO_3(s)$	5	0	0	1	1	0	10.00

Inputs							Concentration
Na_2CO_3		0	2	0	1	0	10^{-2}
$Zn(NO_3)_2$		0	0	1	0	2	2×10^{-4}

the solution was not forced to equilibrate with either solid. The input tableau for the solution after the first addition of 10^{-4} M $Zn(NO_3)_2$ is shown in Table 8.10.

Before discussing the results of the simulation, we can carry out a few calculations to anticipate what we might find. At the initial solution pH (11.13), approximately 86% of the total dissolved carbonate is present as CO_3^{2-}. The activity of free Zn^{2+} that would be in equilibrium with $ZnCO_3(s)$ in such a solution is $K_{s0,ZnCO_3(s)}/\{CO_3^{2-}\}$, or $10^{-7.94}$. A similar calculation for $Zn(OH)_2(s)$ indicates that the activity of Zn^{2+} in equilibrium with that solid would be $10^{-9.81}$. Of course, the total amount of Zn that must be added to increase the activity of free Zn^{2+} to either of these values is much larger, because much of the added Zn will form complexes with OH^-, CO_3^{2-}, or HCO_3^-. Nevertheless, the calculation does tell us that, at least under the initial conditions, $Zn(OH)_2(s)$ is less soluble and therefore more likely to precipitate than $ZnCO_3(s)$.

We can also determine the conditions that would have to be met for both solids to be equilibrated with a single solution, and therefore with each other.

This calculation is based on the reaction

$$ZnCO_3(s) + 2OH^- \leftrightarrow Zn(OH)_2(s) + CO_3^{2-} \tag{8.68}$$

The equilibrium constant for this reaction is $K_{s0,ZnCO_3(s)}/K_{s0,Zn(OH)_2(s)}$, or $10^{+5.55}$, indicating that for the two solids to be equilibrated with each other, the ratio $\{CO_3^{2-}\}/\{OH^-\}^2$ must equal $10^{+5.55}$. By combining reaction (8.68) with the water dissociation reaction and the acid dissociation reactions for H_2CO_3, we can conclude that the equilibrium constant for the following reaction is $10^{-5.77}$, meaning that whenever these two solids are equilibrated, the H_2CO_3 activity must be $10^{-5.77}$.

$$ZnCO_3(s) + 2H_2O \leftrightarrow Zn(OH)_2(s) + H_2CO_3 \tag{8.69}$$

We now turn to the results of the simulations. These results can be summarized in a series of graphs, the first two of which are shown in Figures 8.24 and 8.25.

Figure 8.24 shows the pH change as the titration proceeds. As expected, since Zn^{2+} behaves as an acid in virtually all its reactions in this system, the pH declines as Zn^{2+} is added to the solution. The reactions responsible for this change include not only hydrolysis (formation of Zn—OH complexes), but also formation of both solids and of Zn—CO$_3$ complexes, all of which consume bases that are originally in solution. Although we might expect the high concentration of carbonate species in the original solution to buffer the pH effectively, the pH actually drops fairly steadily to slightly <10 before it stabilizes. We will return to interpret this result after reviewing more of the output.

Figure 8.25 shows the distribution of added Zn among the three potential phases in the system: the solution, $Zn(OH)_2(s)$, and $ZnCO_3(s)$. With even a small addition of Zn to the system, $Zn(OH)_2(s)$ starts precipitating. The amount of this

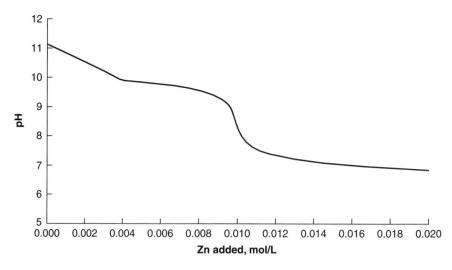

Figure 8.24   The pH change as a function of $Zn(NO_3)_2$ addition in the example system.

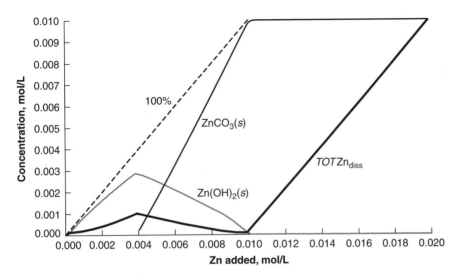

Figure 8.25 Comparison of Zn speciation among the two solid forms and the dissolved phase in the example system. The line labeled 100% indicates a concentration equal to 100% of the Zn added.

solid that precipitates steadily increases as more Zn is added, as does the concentration of total soluble Zn, until about 0.004 mol/L of Zn has been added. At that point, $ZnCO_3(s)$ starts precipitating, and that solid is present throughout the remainder of the titration. The $Zn(OH)_2(s)$ that precipitated early in the titration begins dissolving at this point and continues to do so as the titration proceeds, until it is completely dissolved when approximately 0.01 M Zn has been added. Figure 8.25 also indicates that as the $Zn(OH)_2(s)$ dissolves (from TOTZn values of approximately 0.004 to 0.010 mol/L), the total dissolved Zn concentration actually declines as more Zn is added to the system.

Consistent with the observations about when each solid is present, the solubility quotient for $Zn(OH)_2(s)$ ($Q_{s0,Zn(OH)_2(s)}$) equals the corresponding solubility product $K_{s0,Zn(OH)_2(s)}$ when $0.0001 < TOT$Zn < 0.01 M and is less than $K_{s0,Zn(OH)_2(s)}$ thereafter, whereas the solubility quotient for $ZnCO_3(s)$ is less than its solubility product when TOTZn is less than 0.004 M and equals its solubility product for all TOTZn > 0.004 M (Figure 8.26).

The speciation of dissolved Zn throughout the titration is shown in Figure 8.27. Until $ZnCO_3(s)$ begins to precipitate, the concentration of the soluble complex $Zn(CO_3)_2{}^{2-}$ increases steadily, and this species accounts for the majority of the dissolved Zn. When $ZnCO_3(s)$ starts precipitating, the concentration of $Zn(CO_3)_2{}^{2-}$ starts declining, but it remains the major dissolved Zn species until TOTZn is approximately 0.009 M. Thereafter, the hydroxo complexes and especially free Zn^{2+} dominate the dissolved speciation.

Finally, the distribution of carbonate in the system is shown in Figure 8.28. All the carbonate added initially remains in solution, primarily as $HCO_3{}^-$ and

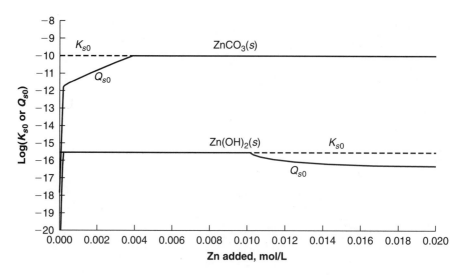

Figure 8.26   Comparison of the solubility quotient and the solubility product for $Zn(OH)_2(s)$ and $ZnCO_3(s)$ for the example system.

$CO_3{}^{2-}$, until $ZnCO_3(s)$ starts precipitating. However, by the time 0.01 M Zn has been added, almost all the carbonate has been removed from solution as $ZnCO_3(s)$. Note that during the portion of the titration when both $ZnCO_3(s)$ and $Zn(OH)_2(s)$ are present (*TOT*Zn between 0.004 and 0.010 M), the H_2CO_3 concentration is constant and equal to $10^{-5.77}$, consistent with the calculations shown above.

Another point worth noting is that whenever $Zn(OH)_2(s)$ is present, the concentration of the dissolved, dihydroxo complex $\{Zn(OH)_2{}^\circ\}$ is steady at $10^{-4.45}$, and whenever $ZnCO_3(s)$ is present, the concentration of the complex $ZnCO_3{}^\circ$ is steady at $10^{-4.70}$. These results stem from the fact that we can write reactions forming these two species directly from the corresponding solids, without production or consumption of any other species:

$$ZnCO_3(s) \leftrightarrow ZnCO_3{}^\circ \tag{8.70}$$

$$Zn(OH)_2(s) \leftrightarrow Zn(OH)_2{}^\circ \tag{8.71}$$

The equilibrium constants for reactions (8.70) and (8.71) are

$$K_{s1,ZnCO_3(s)} = \frac{\{ZnCO_3{}^\circ\}}{\{ZnCO_3(s)\}} = \{ZnCO_3{}^\circ\} = 10^{-4.70} \tag{8.72}$$

$$K_{s2,Zn(OH)_2(s)} = \frac{\{Zn(OH)_2{}^\circ\}}{\{Zn(OH)_2(s)\}} = \{Zn(OH)_2{}^\circ\} = 10^{-4.45} \tag{8.73}$$

Equations (8.72) and (8.73) indicate that the activities of $ZnCO_3{}^\circ$ and $Zn(OH)_2{}^\circ$ are constants, independent of pH and of the concentration of any other dissolved

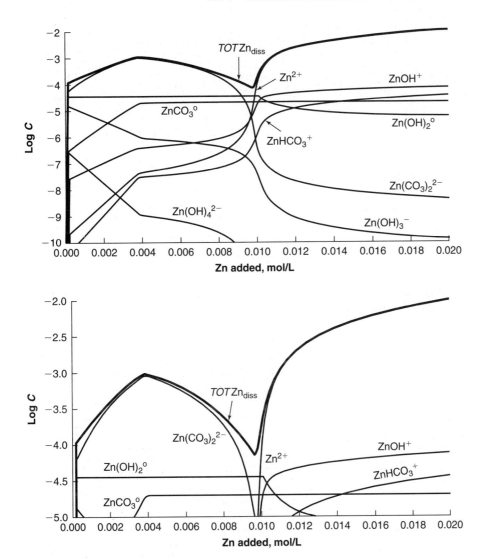

Figure 8.27 Speciation of dissolved Zn in the example system. The lower graph shows the same data as the upper one, but with an expanded y axis.

species, when the corresponding solid is present. Extending this observation to other systems, we can conclude that the activity of any species that has the same chemical composition of a solid is constant when that solid is present.

Now that we have reviewed all the changes in system speciation during the titration, we can interpret the results in an integrated fashion and begin to understand some of the details that seemed a bit perplexing in isolation. One interesting way to display some of the changes in the system is to plot the total soluble Zn concentration as a function of system pH, as shown in Figure 8.29.

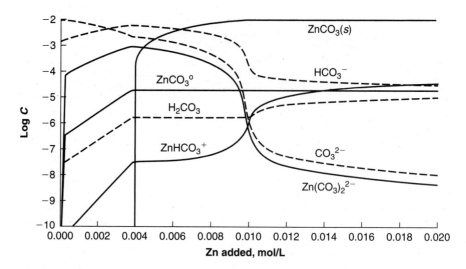

Figure 8.28   Distribution of carbonate-containing species during the simulated titration of the example system.

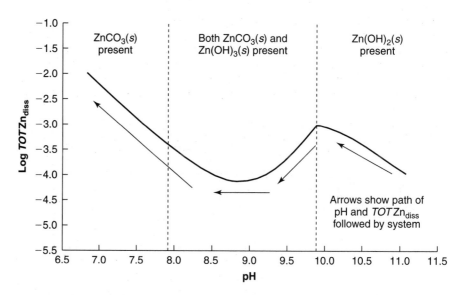

Figure 8.29   Progress of the example titration in terms of changes in pH and total soluble Zn.

Initially, and until the pH of the solution declines to about 8, the dominant soluble form of Zn is $Zn(CO_3)_2^{2-}$. Also, $Zn(OH)_2(s)$ becomes supersaturated with the first addition of 2×10^{-4} M Zn, and during the first part of the titration, each addition of $Zn(NO_3)_2$ leads to precipitation of more solid. Therefore, at the beginning of the titration, the overall changes in the system can be characterized primarily by a combination of the following two reactions:

$$Zn(NO_3)_2 + 2OH^- \leftrightarrow Zn(OH)_2(s) + 2NO_3^-$$

$$Zn(NO_3)_2 + 2CO_3^{2-} \leftrightarrow Zn(CO_3)_2^{2-} + 2NO_3^-$$

Since the formation of $Zn(CO_3)_2^{2-}$ by the second reaction shown above does not release or consume H^+, only the first reaction has a significant effect on solution pH. So at the start of the titration, each increment of 2×10^{-4} M Zn consumes approximately 4×10^{-4} M OH^-. Although the carbonate acid/base system does buffer the pH during this process, the total carbonate concentration is only 10^{-2} M, and almost all of the CO_3^{2-} initially present is converted to HCO_3^- or Zn-CO$_3$ complexes; i.e., the buffering capacity is consumed by the point where TOTZn reaches $\sim$0.003 M. The pH is actually more strongly buffered after most of the CO_3^{2-} has been converted to HCO_3^- for reasons that are described later.

Because $Zn(OH)_2(s)$ precipitates with the first increment of Zn added, the system can be characterized by a point on the $Zn(OH)_2(s)$ solubility curve. The minimum solubility of $Zn(OH)_2(s)$ is near pH 11.4, and the initial pH of the solution is 11.15, so the relevant point on the solubility curve is on the acid side of the point of minimum solubility. In this region of the curve, Zn solubility increases with decreasing pH.

Subsequent additions of Zn lower the pH and cause the total dissolved Zn to increase. Because of the increase in total soluble Zn with each addition of $Zn(NO_3)_2$ and because $\alpha_{Zn^{2+}}$ increases with decreasing pH, the activity of Zn^{2+} increases quite dramatically with further additions of titrant, as shown in Figure 8.27. The declining pH causes the concentration of CO_3^{2-} to decrease, but not as rapidly as the concentration of Zn^{2+} increases. Therefore, the product of these two concentrations increases steadily, eventually reaching $K_{s0,ZnCO_3(s)}$. The pH at this point in the titration is 9.89 (corresponding to TOTZn = 0.004 M, as shown in Figure 8.26).

In contrast to the earlier situation, this pH is greater than the pH of minimum solubility for $ZnCO_3(s)$, for the given TOTCO$_3$. Therefore, as the titration proceeds from that point, each addition of $Zn(NO_3)_2$ causes the pH to decline very slightly, forcing some $Zn(OH)_2(s)$ to dissolve and some $ZnCO_3(s)$ to precipitate. These two reactions, which dominate the next part of the titration, can be combined and represented as follows:

$$Zn(OH)_2(s) + Zn(NO_3)_2 + 2HCO_3^- \rightarrow 2ZnCO_3(s) + 2NO_3^- + 2H_2O$$

The extremely low solubility of $ZnCO_3(s)$ under the given pH and TOTCO$_3$ conditions causes essentially all the Zn that dissolves from $Zn(OH)_2(s)$ and essentially all the Zn added as titrant to precipitate as $ZnCO_3(s)$. If the reaction proceeded in isolation, $Zn(OH)_2(s)$ would be converted to $ZnCO_3(s)$, the Zn added as titrant would precipitate as $ZnCO_3(s)$, and the pH would remain stable (because the reaction shown above does not involve any release or consumption of H^+). It is primarily for this reason that the solution pH is so well buffered between *TOT*Zn values of 0.004 and 0.009 M. However, of course, the reaction does not proceed in complete isolation. The removal of HCO_3^- perturbs the equilibrium of the carbonate system, causing a group of other interrelated adjustments in speciation, the net effect of which is to cause the pH to decline slightly.

Because the solubility of Zn in equilibrium with $ZnCO_3(s)$ decreases with decreasing pH under the given conditions, the decline in pH forces precipitation of additional $ZnCO_3(s)$ and causes the odd result noted above: addition of $Zn(NO_3)_2$ as a titrant leads to a decline in the concentration of total dissolved Zn. Still, these latter reactions are quantitatively much less significant than the one listed above, so the pH change is very slow throughout the range where $Zn(OH)_2(s)$ is dissolving. Note that as the system moves through this range, the changes in solution conditions are always such that the ratio $\{OH^-\}^2/\{CO_3^{2-}\}$ remains constant and equal to $K_{s0,Zn(OH)_2(s)}/K_{s0,ZnCO_3(s)}$, i.e., $10^{-5.55}$.

The process described above continues until all the $Zn(OH)_2(s)$ has dissolved. This point is reached when slightly more than 0.01 M $Zn(NO_3)_2$ has been added, and the solution pH is approximately 7.9. At this point, enough Zn has been added to precipitate virtually all the carbonate, so *TOT*CO_3 declines precipitously, and the majority of any subsequent Zn addition yields an almost equivalent increase in dissolved *TOT*Zn. Furthermore, since the pH at this point has declined to a value where Zn^{2+} is the dominant dissolved species (p*K_1 = pK_{a1} = 8.96), a relatively small fraction of the Zn that enters solution hydrolyzes to release H^+. The solution is poorly buffered at pH 7.9, and some of the added Zn does hydrolyze, so the pH drops slowly. However, with each addition of Zn the pH declines less, and the system pH eventually becomes very stable near 6.8.

Based on Figure 8.29, if we wished to maximize Zn precipitation, the optimum mixing ratio of the two streams would be such that the mixed solution contained 0.0097 M total Zn. With this ratio, the equilibrium pH would be 8.87, and the total dissolved Zn would be 7.3×10^{-5} M, corresponding to >99% precipitation. If the goal were simply to remove the majority of the Zn, then a wide range of mixing ratios would be acceptable. For instance, at least 95% of the added Zn would precipitate if *TOT*Zn were anywhere from 0.0073 to 0.0103 M. However, if the goal was to minimize the dissolved Zn concentration, it would be important to achieve the desired mixing ratio with a high degree of accuracy. (Keep in mind that these calculations assume the system reaches equilibrium within the time frame available for treatment, which might not be a good assumption.)

The above analysis demonstrates the wealth of information and understanding that can be acquired when the simple concepts of acid/base and metal/ligand chemistry are combined with the power of computer-based numerical solution techniques. The numerical output from the analysis could be generated in a matter of minutes, and the results include some aspects that would have been very difficult to foresee by a strictly intuitive or conceptual analysis. The same results could have been obtained using a numerical or graphical analysis, but either of these approaches would have required an immense amount of time and effort.

In particular, a manual or graphical analysis would have been complicated by the fact that the precipitation reaction causes the concentrations of total dissolved Zn and CO_3 and the solution pH to decline simultaneously. If we wished to represent this process using graphs like those we have used to solve simpler problems, we would have had to draw a separate three-dimensional $\log C$–pH–$\log\{CO_3^{2-}\}$ diagram (corresponding to the current value of TOTZn) for each addition of $Zn(NO_3)_2$. By contrast, analysis of the problem using chemical equilibrium software automatically includes correct procedures for considering the simultaneous changes in TOTZn, TOTCO$_3$, and TOTH caused by all the reactions taking place. As a result, a single set of input data leads directly to the complete speciation and pH profile for any mixing ratio. On the other hand, generating the numerical output without having a good grasp of the underlying chemistry would leave us with a theoretical solution, but no useful insights into *why* the system behaves the way it does. The analysis thus serves as an example of the synergism that can be achieved between computational and conceptual approaches to these types of problems.

8.15 PREDOMINANCE AREA DIAGRAMS CONSIDERING POSSIBLE PRECIPITATION OF SOLIDS

The final section of this chapter returns to the preparation of predominance area diagrams, extending the prior analysis to include the possibility that a solid will precipitate in the system. Often, this aspect of predominance area diagrams is what makes them valuable, because they can tell us the broad conditions under which a given species is expected to be present primarily in soluble versus particulate form. This distinction is often of greater interest than the determination of regions where particular dissolved species are dominant over other dissolved species, e.g., for a preliminary analysis of treatability.

The inclusion of solids in predominance area diagrams raises a few issues that need not be considered if only soluble species are being considered. First, if only soluble species are being considered, the diagrams are independent of the total concentration of metal in the system. That is, the ratio of activities of, for example, $FeOH^+$ and $FeCl^+$ depends on $\{OH^-\}$ and $\{Cl^-\}$ but not on TOTFe, just as the ratio of $\{Ac^-\}$ to $\{HAc\}$ depends on pH but not on TOTAc. However, when one adds the possibility that a solid might form, the total concentration of Fe in the system becomes an important variable. This is because, at a given pH, the solid is expected to be present at high total metal concentration and absent at low total metal concentration, even if all other system conditions remain the same.

Second, because the activity of the solid is 1.0 regardless of how much solid is present, we need to be clear as to whether *predominance* refers to activity or concentration. That is, if a solid is present at a concentration that is less than that of some dissolved species, but at an activity (1.0) that is greater than that of any dissolved species, should the solid be considered the dominant metal-containing species in the system? The convention that has been adopted is to define a solid to be dominant whenever it is present, and soluble species to be dominant only if no solid is present.

The predominance area diagram including consideration of solids is prepared by following essentially the same steps as described previously. That is, for each species in the system, we make sequential, pairwise comparisons with all the other species in the system to identify the boundaries of the region (if one exists) where that species is dominant over all the others. If the species is a solid, we assign it an activity of 1.0, and we assign the soluble species with which it is being compared an activity equal to TOTMe.

The reason that the comparison is with TOTMe is that the goal of the comparison is simply to decide whether the solid is expected to be present. Consider, for example, the pairwise comparison of the activities of $Fe(OH)_2(s)$ and $FeCl^+$ in the Fe/OH/Cl system. In a region of the diagram where $FeCl^+$ is the dominant soluble species, its concentration will be close to $TOT Fe_{diss}$; and if it is dominant over the solid (i.e., if the solid is not present), then $TOT Fe_{diss}$ will equal TOTFe. Thus, in this region of the diagram, it makes sense to test for possible supersaturation of the solid using the assumption that $\{FeCl^+\} = TOT$Fe. The same logic applies to other dissolved species in regions where they are dominant. Therefore, the assessment of whether the solid is dominant is always based on computations in which the concentration (and if the solution is assumed to be ideal, the activity) of the dissolved species is assumed to equal TOTFe.

For instance, the comparison between $FeCl^+$ and $Fe(OH)_2(s)$ in a system containing 10^{-6} M TOTFe is carried out by writing the reaction relating these two species and determining the corresponding equilibrium constant:

$$Fe(OH)_2(s) + 2H^+ \leftrightarrow Fe^{2+} + 2H_2O \qquad *K_{s0}$$

$$Fe^{2+} + Cl^- \leftrightarrow FeCl^+ \qquad K_{Cl,1}$$

$$Fe(OH)_2(s) + Cl^- + 2H^+ \leftrightarrow FeCl^+ + 2H_2O \qquad K_{(8.74)} \qquad \textbf{(8.74)}$$

$$K_{(8.74)} = \frac{\{FeCl^+\}}{\{Cl^-\}\{H^+\}^2} = *K_{s0}K_{Cl,1} = 10^{13.66} \qquad \textbf{(8.75)}$$

Equation (8.75) can be converted to a logarithmic form that represents the relationship as a straight line in $\log\{Cl^-\}$–pH space as follows:

$$\frac{\{FeCl^+\}}{\{Cl^-\}\{H^+\}^2} = 10^{13.66}$$

$$\log\{FeCl^+\} - \log\{Cl^-\} + 2pH = 13.66$$

$$\log\{Cl^-\} = 2pH + \log\{FeCl^+\} - 13.66 \qquad \textbf{(8.76)}$$

Equation (8.76) indicates the pH at which any activity of $FeCl^+$ and Cl^- can be in equilibrium with $Fe(OH)_2(s)$. Substituting $TOTFe$ $(= 10^{-6})$ for $\{FeCl^+\}$, Equation (8.76) simplifies to

$$\log\{Cl^-\} = 2pH - 19.66$$

Generalizing from the above comparison, we see that the reaction relating the solid to any complex can be written as follows:

$$Fe(OH)_2(s) + xH^+ + yCl^- \leftrightarrow Fe(OH)_{2-x}Cl_y^{x-y} + xH_2O \qquad \textbf{(8.77)}$$

The corresponding line separating the region of precipitation from the region where no precipitation occurs is then

$$\log\{Cl^-\} = \frac{x}{y}pH + \frac{1}{y}\log\{Fe(OH)_{2-x}Cl_y^{x-y}\} - \frac{1}{y}\log K_{(8.77)}$$

or, if $y = 0$

$$pH = \frac{1}{x}\log K_{(8.77)} - \frac{1}{x}\log\{Fe(OH)_{2-x}^{x}\}$$

The equations of the lines separating the region of dominance of the solid from that of each soluble complex in the example system are summarized in Table 8.11.

Each line divides the predominance area diagram into two regions: one where the solid would form if the soluble species were present at a concentration of $TOTFe$ and one where it would not. Put another way, on one side of the line, the solid is sufficiently insoluble that it prevents the soluble species from being present at a concentration equal to $TOTFe$. On the other side of the line, the concentration of the soluble species would have to be larger than $TOTFe$ for the solid to precipitate, so all the metal remains in solution.

Table 8.11   Equilibrium relationships between dissolved species and $Fe(OH)_2(s)$ used to develop a predominance area diagram for the $Fe^{2+}/Cl^-/OH^-$ system, including consideration of $Fe(OH)_2(s)$

Species Being Compared	Conditions When Species Have Equal Activities	Equation of Line
$Fe(OH)_2(s)$, Fe^{2+}	$\dfrac{\{Fe^{2+}\}}{\{H^+\}^2} = \dfrac{K_{s0}}{K_w^{\ 2}} = 10^{12.10}$	$pH = -\dfrac{1}{2}\log\{Fe^{2+}\} + 6.05$
$Fe(OH)_2(s)$, $FeOH^+$	$\dfrac{\{FeOH^+\}}{\{H^+\}} = \dfrac{K_{s0}*K_1}{K_w^{\ 2}} = 10^{2.60}$	$pH = -\log\{FeOH^+\} + 2.60$
$Fe(OH)_2(s)$, $Fe(OH)_3^-$	$\{Fe(OH)_3^-\}\{H^+\} = \dfrac{K_{s0}*\beta_3}{K_w^{\ 2}} = 10^{-18.90}$	$pH = \log\{Fe(OH)_3^-\} + 18.90$
$Fe(OH)_2(s)$, $FeCl^+$	$\dfrac{\{FeCl^+\}}{\{Cl^-\}\{H^+\}^2} = \dfrac{K_{s0}K_{1,Cl}}{K_w^{\ 2}} = 10^{13.00}$	$\log\{Cl^-\} = 2pH + \log\{FeCl^+\} - 13.00$
$Fe(OH)_2(s)$, $FeCl_2^{\,\circ}$	$\dfrac{\{FeCl_2^{\,\circ}\}}{\{Cl^-\}^2\{H^+\}^2} = \dfrac{K_{s0}K_{1,Cl}K_{2,Cl}}{K_w^{\ 2}} = 10^{13.04}$	$\log\{Cl^-\} = pH + \dfrac{1}{2}\log\{FeCl_2^{\,\circ}\} - 6.52$

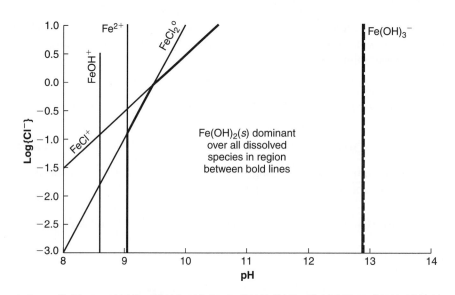

Figure 8.30   Predominance area diagram showing regions of dominance of $Fe(OH)_2(s)$ in comparison with each of five soluble species, for $TOTFe = 10^{-6}$ M. The solubility of the dissolved species is $>10^{-6}$ M on the side of the line where its name is shown and $<10^{-6}$ M on the other side.

The five lines identified in Table 8.11 are plotted in Figure 8.30. For four of the lines [those that contain zero or one OH group, i.e., fewer than $Fe(OH)_2(s)$], the solid is dominant over the soluble species to the right of the line, and the soluble species is dominant to the left. By contrast, for the $Fe(OH)_2(s)/Fe(OH)_3^-$ couple, the solid is dominant to the left.

When the region of dominance of $Fe(OH)_2(s)$ is overlain on the predominance area diagram developed earlier considering only the soluble species, the solid occupies a portion of the regions where $FeOH^+$, $Fe(OH)_3^-$, and $FeCl_2^\circ$ had been identified as dominant previously (Figure 8.31). None of the relationships between pairs of soluble species are altered by this analysis, so the regions of the diagram outside the area where the solid is dominant are identical to those determined earlier.

Although the predominance area diagram was developed here in two stages (considering soluble species first, then the solid), in general there is no need for such an approach. One simply identifies all possible species that might be dominant, chooses one, and starts deriving the equations needed to make pairwise comparisons to determine the conditions under which that species is dominant over all the others. When that task is completed, another species is chosen and the process is repeated, as many times as necessary until all possible species have been considered. In more complex systems, other solids (e.g., carbonates or sulfides) might be considered in addition to hydroxides. These systems have the added

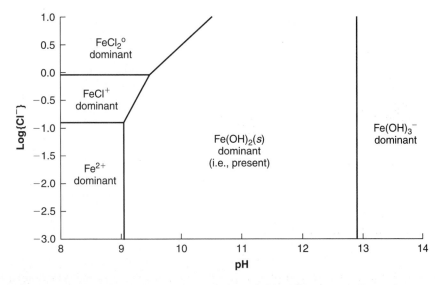

Figure 8.31 The predominance area diagram for the $Fe^{2+}/Cl^-/OH^-$ system, for 10^{-6} M TOTFe, and considering possible precipitation of $Fe(OH)_2(s)$.

complexity that the speciation of the precipitating anion also depends on pH, so the number of calculations required to develop the entire diagram can become substantial. However, the approach is the same as that outlined here.

SUMMARY

This chapter has covered the multitude of reactions of metal ions with other constituents of aqueous solutions. The defining characteristic of these interactions is the interaction between the metal ion and unshared electron pairs that are typically associated with oxygen, nitrogen, or sulfur atoms in other chemicals (including water). The virtually unlimited number of chemicals in environmental samples that contain O, N, or S donor atoms leads to the possibility that metals can exist in a correspondingly large number of species in an equilibrium solution. Because the speciation of metals has an enormous effect on their mobility and bioavailability, understanding speciation is critical to an understanding of metal behavior in the environment.

The interactions of metals with donor atoms can yield both soluble and solid reaction products. When the product is soluble, the molecule with the donor atom is referred to as a ligand, and the product is referred to as a complex. The strength of a metal-ligand complex is characterized by an equilibrium constant referred to as a stability constant. With increasing stability constant and increasing concentration of the ligand in solution, the portion of TOTMe that is present as the complex increases, and in many solutions complexes account for the over-whelming majority of dissolved metal species.

When the concentrations of certain metals and ligands increase, they start polymerizing rather than forming increasing concentrations of small complexes, eventually growing into a distinct solid phase. A solid has an activity that is in-dependent of its concentration in a suspension and, according to the almost uni-versal choice of the standard state for solids, pure solids have an activity of 1.0. Many of the ligands that can form solids with metals are the same ones that form soluble complexes, so those ligands might impose limits on metal solubility under some circumstances and provide a means for increasing metal solubility in others.

In many cases, if the ligand can form anionic complexes with the metal, the shape of the overall metal solubility curve as a function of pH is bowl-like. That is, the metal is highly soluble at low pH due to the large equilibrium concentra-tions of the uncomplexed metal ion and cationic complexes, and highly soluble at high pH due to the high equilibrium concentrations of anionic complexes. In between these extremes (at slightly alkaline pH for many of the most common metals used commercially), metal solubility passes through a minimum that is often very low (<0.1 μM). By taking advantage of these dramatic changes in solubility, metals can sometimes be removed from solution very efficiently. However, in such processes, it is important to recognize that missing the target pH in either direction can lead to dramatic increases in the equilibrium solubil-ity. It is also important to remember that metals often do not reach equilibrium with solid phases in the time frame of minutes to hours that is typical of treat-ment processes.

If multiple solids containing a given type of metal ion are supersaturated in a solution, at least one of those solids will be present when the system reaches equilibrium. However the others might or might not be present, depending on how the precipitation of the most stable solid alters the solution chemistry. De-termination of the equilibrium composition in such cases requires that we make assumptions about which solids will be present, solve the equations that charac-terize the system based on those assumptions, and then modify the assumptions as needed. Software for solving chemical equilibrium problems is extremely handy in such cases, since it can rapidly carry out numerous iterations based on the different assumptions. When the software is used, it is important to distin-guish between solids that are known to be present at equilibrium and those that might or might not be present, depending on the solution chemistry.

The total number of phases that can be present at equilibrium is related to the number of independent chemical components that can be used to describe a sys-tem's composition and the number of independent parameters that can be speci-fied to characterize the system via the Gibbs phase rule.

Because of the multiplicity of species that can form in systems containing metals and ligands, simplified diagrams that identify the dominant metal species but do not display the concentrations or activities of non-dominant species are sometimes prepared. Such diagrams can incorporate consideration of solids as well as soluble species and can therefore display critical information about metal speciation over a very wide range of conditions.

Problems

Note: In the following problems, ignore possible precipitation of solids unless the problem indicates that you should consider that possibility.

1. A solution at pH 7.0 contains Cu^{2+} and NH_3.

   a. What is the ratio $\{CuOH^+\}/\{Cu^{2+}\}$?

   b. What is the ratio $\{NH_4^+\}/\{NH_3\}$?

   c. Are there any hydroxo complexes of Cu present at concentrations greater than that of $CuOH^+$?

   d. If $\{NH_4^+\} = 10^{-3.0}$, which $Cu(NH_3)_x^{2+}$ complex will be present at the greatest concentration?

   e. For a system with $TOTCu = 10^{-3} M$ and at pH 7.0, what are $TOTNH_3$ and $\{Cu^{2+}\}$ when $Cu(NH_3)_4^{2+}$ first becomes the dominant Cu species?

2. A solution is made by adding $0.1 M\ HgSO_4$ to $0.15 M$ HCl plus $0.03 M$ NaCl.

   a. Write mass balances for total Hg, SO_4, and Cl.

   b. Write out any additional equations you would need to consider, aside from the stability constants for the various Hg complexes, to solve for solution pH and speciation. Use the Davies equation to estimate activity coefficients.

   c. Solve the set of equations generated in parts (a) and (b), ignoring adjustments for non-ideality.

   d. Re-solve part (c), including the effects of non-ideality. (Note: If you use chemical equilibrium software to solve the problem, you might need to relax the mass balance and/or the error tolerance to avoid a problem with exceeding the maximum allowable number of iterations.)

3. Draw a log C–log $\{S_2O_3^{2-}\}$ diagram to show silver speciation in a system containing various $S_2O_3^{2-}$ concentrations at pH 9.0 and pH 13.0. ($S_2O_3^{2-}$ is the thiosulfate ion, a complexing agent used extensively in industry, especially in photoprocessing.) Use a total silver concentration of $10^{-5} M$, and consider the range $-12 < \log\{S_2O_3^{2-}\} < -3$, assuming $\gamma = 1.0$ for all solutes. Show all species that are present at concentrations $>10^{-12} M$. Your graph should follow usual conventions (i.e., increasing concentrations from left to right on the abscissa and from bottom to top on the ordinate).

   a. What is the value of $\{S_2O_3^{2-}\}$ at the points where $\{Ag^+\} = \{AgS_2O_3^-\}$ and where $\{Ag^+\} = Ag(S_2O_3)_2^{3-}$, for the system at pH 9? What are the corresponding values at pH 13? Explain this result briefly.

   b. The species $AgS_2O_3^-$ should be dominant over a range of values of $\{S_2O_3^{2-}\}$ in the graph at each pH. Explain briefly why the region of $AgS_2O_3^-$ dominance covers a smaller range of $\{S_2O_3^{2-}\}$ at pH 13 than at pH 9.

c. A solution buffered at pH 9 and containing 10^{-5} M *TOT*Ag is titrated with S_2O_3. Plot the concentration of complexed Ag (i.e., the sum of the concentrations of all Ag complexes) as $TOTS_2O_3$ is increased from 0 to 10^{-5} M. What are the concentrations of all species at the end of the titration, i.e., when $TOTAg = TOTS_2O_3 = 10^{-5}$ M? Based on this result, is thiosulfate a relatively strong or weak complexing agent for silver? Explain your reasoning.

4. A solution contains 3×10^{-4} M $CuCl_2$, 10^{-3} M NH_4Cl, and 5×10^{-4} M $Ca(OH)_2$. The lime dissociates completely, and the only complexes of significance are those of Cu with OH^-, Cl^-, or NH_3. A $\log C$–pH diagram, showing all Cu-containing species present at concentrations greater than 10^{-6} M, is shown below.

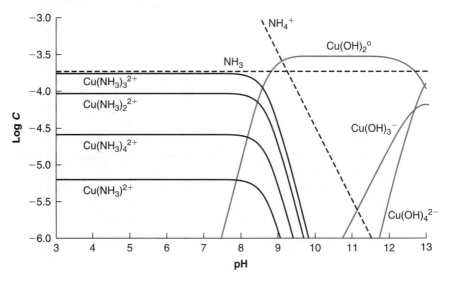

a. Find the pH of the solution either from the $\log C$–pH diagram in conjunction with a proton condition or *TOT*H equation, or by using a software package. (Note: Not all the commonly available problems include data for Cu—NH_3 complexes in their default database. If the software you are using does not include those complexes, add them to the database manually for this analysis.)

b. The Henry's law constant for ammonia is 0.017 bar/(mol/L). What partial pressure of ammonia is in equilibrium with the solution in part (*a*)?

c. What is the pH of the system if it equilibrates with a gas phase in which the partial pressure of ammonia is $10^{-5.5}$ bar? Develop a $\log C$–pH diagram for the system in contact with this gas phase.

5. What is the equilibrium partial pressure of ammonia over a solution at pH 8.0 in which the concentrations of $Hg(NH_3)^{2+}$ and $Hg(NH_3)_3^{2+}$ equal one another?

6. NaOH is added to a solution containing 10^{-4} M $Hg(NO_3)_2$ and 10^{-2} M NaCl to adjust the pH to 8.5, while maintaining the solution in equilibrium with a gas

with P_{NH_3} of 10^{-6} bar. List the species you expect to be present in the equilibrated solution, and write out the equations that could be used to determine the final composition of the solution. Choose the dihydroxo mercury complex, $Hg(OH)_2^\circ$, as the reference species for the proton condition or *TOT*H equation. Insert numerical values in the equations whenever those values are known.

7. The input table for a chemical equilibrium computer program for a system of interest is shown below. The table includes the input data for forming various *species* from the chosen *components*. The output indicates that, at equilibrium, the following *component* concentrations apply: $\{Hg^{2+}\} = 10^{-9.54}$, $\{CO_3^{2-}\} = 10^{-5.69}$, $\{Cl^-\} = 10^{-2.51}$, $\{NH_4^+\} = 10^{-4.73}$, $\{H^+\} = 10^{-8.85}$.

a. What is the chemical formula of *species* I?
b. What is the equilibrium concentration of *species* F?

			Component			
Species	H^+	Hg^{2+}	CO_3^{2-}	Cl^-	NH_4^+	Log K
A	−1	0	0	0	0	−14.00
B	1	0	1	0	1	6.33
C	2	0	1	0	0	16.68
D	−1	0	0	0	1	−9.25
E	−1	1	0	0	1	−0.4
F	−2	1	0	0	2	−0.9
G	0	1	0	1	0	6.74
H	0	1	0	2	0	13.22
I	−1	1	0	1	0	3.67

8. Draw a pH–log$\{S_2O_3\}$ predominance area diagram for a system containing 10^{-5} *M* *TOT*Ag. Consider thiosulfate concentrations in the range $-12 < \log\{S_2O_3^{2-}\} < -3$ and pH in the range of 2 to 14.5. Consider possible precipitation of $AgOH(s)$.

9. A water at pH 9.8 contains $10^{-3.0}$ *M* HCO_3^- and is in equilibrium with $CaCO_3(s)$. Find the total alkalinity and dissolved Ca concentration in the water.

10. A wastewater is characterized as follows:

Fe^{2+}	2.0 mg/L
Alkalinity	2.5 meq/L
Calcium	65 mg/L
PO_4-P	10 mg/L
NH_4-N	140 mg/L
pH	7.5
Ionic strength	$10^{-1.5}$

The following solids might form in the water:

	pK_{s0}
$CaHPO_4(s)$	6.9
$Ca_3(PO_4)_2(s)$	28.92
$Ca_5(PO_4)_3OH(s)$	58.2
$FeCO_3(s)$	10.55
$CaCO_3(s)$	8.48
$Fe_2(NH_4)(PO_4)(s)$	15.2

The alkalinity was determined by titration to an endpoint of 4.7. Assume that any alkalinity not attributable to the H_xPO_4 and NH_3/NH_4^+ groups is provided by the carbonate system. Formulate and compute ion activity products for each of the solids. Use the Davies equation to estimate activity coefficients. Which solids are supersaturated?

11. The solubility of a solid is sometimes determined by putting a weighed sample of the solid in water of known initial composition, letting it equilibrate, then removing and reweighing the solid. During equilibration the solution is closed to the atmosphere. Using the Davies equation to compute activity coefficients, determine the solubility of calcite, $CaCO_3(s)$, in grams per liter, in pH 12 solutions of ionic strength 0.01 and 0.1 M. What would the solubility be in a solution with $I = 0.1\ M$ at pH 8.0?

12. A solution is prepared by adding $10^{-5}\ M\ AgNO_3$ and $5 \times 10^{-6}\ M\ Na_2S_2O_3$ to water. It is then titrated with NaCl. Compute the speciation of the system during the titration. Consider *TOT*Cl concentrations from $10^{-6}\ M$ to the Cl concentration of seawater (0.5 M). Plot the data as $\log C$ versus $\log(TOTCl)$. (*Hint:* To characterize such a range using chemical equilibrium software, it is easiest to evaluate the speciation at fixed Cl^- activities at, for example, 0.1 log unit intervals and then to convert the output data so that the speciation can be plotted as a function of *TOT*Cl rather than $\{Cl^-\}$. Consider the presence of Ag complexes with OH^-, $S_2O_3^{2-}$, and Cl^-, and also the possible precipitation of $AgOH(s)$ and $AgCl(s)$.)

13. Consider a treated domestic wastewater containing 20 mg/L PO_4-P. You wish to reduce the P concentration to 1 mg/L by precipitation of $AlPO_4(s)$.

    a. Prepare a $\log C$–pH diagram showing curves for the activity of PO_4^{3-} in the untreated and treated solutions. Also show the Al^{3+} activity that would cause the PO_4^{3-} concentrations in each solution to be in equilibrium with $AlPO_4(s)$.

    b. Alum (hydrated aluminum sulfate, with a typical composition of $Al_2(SO_4)_3 \cdot 18H_2O$) is added to the initial solution to provide $10^{-3}\ M\ TOTAl$. Add a line to the diagram prepared in part (*a*) representing $\{Al^{3+}\}$ as a function of pH before any solid forms. (Hydrolysis reactions are typically much faster than precipitation reactions.) In which pH range(s) is $AlPO_4(s)$ supersaturated?

c. If the pH is well buffered at 7.5, how much alum (in mg/L) must be added to the original solution to achieve the treatment goal? Keep in mind that one Al ion precipitates for each PO_4 ion removed. For now, ignore possible precipitation of $Al(OH)_3(s)$.

d. Add a line to the diagram showing the Al^{3+} activity in equilibrium with gibbsite [a crystalline form of $Al(OH)_3(s)$], which has log $K_{s0} = -33.23$. Discuss the consequences of this equilibrium for your answer to parts (b) and (c).

14. Sodium cyanide is slowly added to a solution containing a relatively small amount of mercuric ion (Hg^{2+}). Analysis of the solution composition at various points in the titration indicates that all the mercury remains in solution when small amounts of the cyanide are added, some mercury precipitates when intermediate amounts are added, and the precipitate then redissolves when large amounts of cyanide are added. Explain these changes.

15. Sunda and Guillard (*J. Marine Research* **34**, 511–529 (1976)) reported the results shown in the figure below for the growth of the alga *Thalassiosira* in seawater that had been dosed with Cu^{2+} and/or irradiated with ultraviolet (UV) light. The UV irradiation destroys some of the natural organic chelating agents in the water. When the water was irradiated, algal growth was inhibited. (The irradiation occurred before the algae were added, so the irradiation did not affect them directly.) When either 3×10^{-8} or 1×10^{-7} M Cu^{2+} was added to the irradiated seawater, the inhibition of algal growth was much greater. However, when 10^{-5} M EDTA was added to the irradiated seawater, algal growth was at least as great as or perhaps a bit greater than, that in the control (untreated) sample. Explain these results.

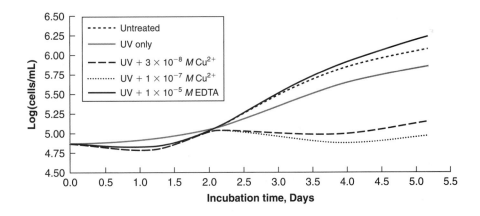

16. a. Determine the total molar concentration of dissolved Zn in equilibrium with $ZnCO_3(s)$ in a system with 10^{-2} M $TOTCl$ and 5×10^{-3} M $TOTCO_3$, at pH 8.0.

    b. Draw a log *C*–pH diagram showing all the dissolved Zn species in a system in equilibrium with $ZnCO_3(s)$ and always containing 10^{-2} *M* TOTCl and 5×10^{-3} *M* TOTCO_3. Ignore possible precipitation of $ZnO(s)$.

    c. Draw a log *C*–pH diagram showing all the dissolved Zn species in a system in equilibrium with $ZnO(s)$ and always containing 10^{-2} *M* TOTCl and 5×10^{-3} *M* TOTCO_3. Ignore possible precipitation of $ZnCO_3(s)$.

    d. Redraw the lines for $\{Zn^{2+}\}$ from parts (*b*) and (*c*) on a separate graph. Discuss the significance of the intersection point and the implications of the graph in pH regions on the acid and base side of the intersection point. Also redraw the lines for $\{ZnCO_3^\circ\}$ from parts (*b*) and (*c*) on the graph. Compare the intersection point for these two lines with the intersection point for the $\{Zn^{2+}\}$ lines, and explain the resulting observation.

17. One gram of solid $Cd(OH)_2(s)$ is dispersed in 1.0 L of a solution at pH 9.0 containing 1.0 mg/L Cd^{2+}. Assuming the pH is well buffered and that Cd—OH complexes are negligible, do you expect some solid to dissolve, some dissolved Cd to precipitate, or no change in the solution composition?

18. A system containing 10^{-2} *M* TOTCl and 10^{-3} *M* TOTCO_3, initially at pH 9.0, is titrated with $AgNO_3$ until 1.0 *M* Ag has been added. The carbonate species are the only weak bases present in the system in significant concentrations. Answer the following questions about the system at various stages of the titration. (*Note:* You can answer this question by simulating the titration and determining the sequence of precipitation reactions, but it is also possible to reach the correct conclusion simply by manipulating the various solubility constants and the information about the initial solution composition, without actually determining the Ag^+ concentration at the points of interest. Either approach is acceptable.)

    a. Ignoring the formation of any soluble complexes (i.e., assuming all the dissolved Ag is Ag^+), determine whether $AgOH(s)$, $AgCl(s)$, or $Ag_2CO_3(s)$ precipitates first.

    b. The titration continues until a second solid precipitates. Which solid is it, and what are the solution conditions?

    c. The titration continues. Will the third solid eventually precipitate? If so, what are the conditions when this occurs? If not, explain (words or equations) why you reached that conclusion.

19. (Thanks to D. Spyridakis.) A monument made of $CaCO_3$ rock (calcite) has been exposed to the weathering action of the CO_2 of the atmosphere over a period of 2550 years. Starting around 1950, the weathering of the monument drastically increased, primarily as a result of the elevated SO_2 content of the atmosphere, from the burning of fossil fuels.

    Assume that rainfall and P_{CO_2} have been constant over the years at 40 cm/yr and $10^{-3.5}$ bar, respectively, and that the SO_2 content of the atmosphere has been 1 ppm (by volume) since 1950. Henry's constant for $SO_2(g)$ is $10^{-0.10}$ bar/(mol/L). Calculate the maximum possible extent of weathering (dissolution of $CaCO_3$) for the 2500 years preceding 1950 and for the period from 1950 to 2000. Express the result as grams of $CaCO_3(s)$ dissolved per cm^2 of cross-sectional area of the monument. Carry out the calculations based on two sets of assumptions. First, assume that the rain is in equilibrium with the atmosphere before striking the monument, but that no more gases dissolve as it drips down the monument's surface. Second, assume that the rain is in continuous equilibrium with the air as it drips.

20. "pHistory of water" (thanks to J. Ferguson).

   a. A raindrop is formed in a cloud and comes to equilibrium with $P_{CO_2} = 10^{-3.5}$ bar. The water droplet then passes over a smokestack and absorbs 10^{-5} M H_2SO_4. Draw a log C–pH graph for the system and find the pH.

   b. The droplet next passes over a feedlot and comes to equilibrium with $P_{NH_3} = 10^{-10.2}$ bar. The droplet remains in equilibrium with atmospheric CO_2 as it equilibrates with the NH_3. Modify the graph to include NH_3, and find the pH and total dissolved N.

   c. The droplet falls to earth where it hits a rock containing $SiO_2(s)$. The rock starts dissolving according to the reaction $SiO_2(s) + 2H_2O \leftrightarrow H_4SiO_4°$. The reaction takes place rapidly enough that re-equilibration with the atmosphere does not occur. Redraw the graph to include silicic acid, and find the pH one last time.

21. Often lime is added to wastewater in order to remove phosphorus from solution by forming a Ca—PO_4 solid. Since wastewaters typically contain $TOTCO_3$ concentrations of a few millimolar, $CaCO_3(s)$ may also form. Consider a waste containing 2×10^{-3} equiv/L alkalinity. The initial pH is 8.1, and $TOTPO_4$ is 2×10^{-4} M. Lime is to be added until at least 95% of the initial phosphorus precipitates. Ignore non-ideal behavior of solutes.

   a. As the lime is added, the Ca^{2+} activity and the pH both increase. Draw a log C–pH diagram showing $\{Ca^{2+}\}$ in equilibrium with each of the two possible solid phases: calcite, $CaCO_3(s)$, and hydroxyapatite, $Ca_5(OH)(PO_4)_3(s)$. Which solid will be the first to precipitate? Assume the alkalinity is comprised entirely of carbonate and phosphate species and that the initial Ca^{2+} concentration is negligible.

   b. Determine the pH at which the first solid precipitates.

   c. Add a line to your diagram showing $\{Ca^{2+}\}$ in equilibrium with hydroxyapatite once the treatment goal has been met. Is it likely that $CaCO_3(s)$ will be present in this system?

   d. As more lime is added, the pH continues to increase and eventually both solids are precipitated. Determine how much Ca has been added and the solution composition at that point.

22. Because cyanide is a strong complexing agent, it is used to prevent metals from precipitating in some electroplating operations.

   a. Consider a solution at pH 10 in equilibrium with mercuric oxide, $HgO(s)$. Compute the amount of $TOTCN$ that would be in solution if the overlying air had $P_{HCN} = 10^{-7}$ bar, assuming that $Hg(CN)_2(s)$ does not form. What can you conclude from the result?

b. Now consider the possibility that $Hg(CN)_2(s)$ does precipitate. What is the concentration of $TOTCN$ that would be in a pH 10 solution in equilibrium with this solid and with $P_{HCN} = 10^{-7.0}$ bar? Ignore the possible formation of $HgO(s)$ for this part of the problem.

c. A solution is initially in contact with both $HgO(s)$ and $Hg(CN)_2(s)$ at pH 10 and under a fixed pressure of $P_{HCN} = 10^{-7.0}$ bar. The total Hg contained in the two solids is 0.75 M. There is initially no dissolved Hg in the solution. What is the value of $\{Hg(CN)_3{}^-\}$ at equilibrium? Is there any solid remaining in the system once equilibrium is attained? Assume the solution is well buffered, and ignore activity corrections.

23. Polyphosphates, made of several phosphate groups linked together in a chain, are strong metal-complexing agents. Sometimes these ligands are added to drinking water in order to form complexes with iron (Fe^{3+}) that enters the water via corrosion of the pipes. If the ligands are not added, the iron precipitates and causes the tap water to appear red. By adding the polyphosphates, the iron that enters solution remains dissolved and does not cause the water to become colored. For a solution at pH 7.0 containing 1 mg/L $TOTFe(III)$:

a. What fraction of the Fe precipitates if no complexing agents are added?

b. Compute the total amount of tripolyphosphate ($P_3O_{10}{}^{5-}$) that must be added to cause all the Fe to remain in solution.

For $H_5P_3O_{10}$: $pK_{a1} < 2$, $pK_{a2} = 2.2$, $pK_{a3} = 2.6$, $pK_{a4} = 5.6$, $pK_{a5} = 7.9$. For Fe^{3+}—tripolyphosphate complexes:

$$Fe^{3+} + H_2P_3O_{10}{}^{3-} \leftrightarrow FeH_2P_3O_{10}{}^{0} \qquad \log K = 5.10$$

$$Fe^{3+} + H_3P_3O_{10}{}^{2-} \leftrightarrow FeH_3P_3O_{10}{}^{+} \qquad \log K = 5.04$$

24. Although Mg^{2+} is soluble under most conditions encountered in natural waters and wastewaters, it can form a solid named *struvite* [$MgNH_4PO_4(s)$] in solutions that have relatively high concentrations of ammonia and phosphate species. These conditions might be found in tanks where sewage sludge is anaerobically digested. In such systems, microorganisms that grew under aerobic conditions die due to the absence of oxygen and are preyed upon by others that are better adapted to the anaerobic conditions. In the process, large amounts of ammonia and phosphate are released from the breakdown of proteins, nucleic acids, ATP (adenosine triphosphate, the chemical fuel that microorganisms produce and use internally), and other biomolecules. In some waste treatment systems, struvite has precipitated in the pipes exiting the anaerobic digester, forming a thick, solid layer adjacent to the pipe wall and, over time, dramatically reducing the cross-sectional area available for flow.

Consider a solution that contains 4×10^{-3} M $TOTMg$, 7×10^{-3} M $TOTNH_3$, and 9×10^{-4} M $TOTPO_4$ and that is contacting a struvite layer on a pipe as it exits a digester. Use chemical equilibrium software to determine the pH values in the range of 6 to 10 where additional struvite will precipitate and the values where some of the struvite on the pipe will dissolve. K_{s0} for struvite is $10^{-12.4}$.

APPENDIX 8A: AN EXPLANATION FOR THE CONSTANCY OF THE ACTIVITY OF PURE SOLIDS, BASED ON MASS ACTION CONSIDERATIONS

Consider a solid $X(s)$ that is suspended in three different solutions, each of volume 1.0 L. In solutions I and II, the concentration of $X(s)$ is 1.0 g/L, but in solution I the solid is present as a single spherical particle, and in solution II it is present as 1000 identical, spherical particles, each with mass 1.0 mg. Solution III contains only a single, 1.0-mg particle. Because of the different sizes and amounts of solid in the three systems, the relative concentrations of surface area are $1:10^6:10^3$ in solutions I, II, and III, respectively.

If $X(s)$ can dissolve by reaction with water molecules to form $X(aq)$, we might reasonably expect the rate of reaction to be proportional to the amount of solid surface in contact with the solution. That is, we might represent the dissolution reaction as an elementary reaction between the solid surface and water, as shown in reaction (8.78):

$$X(s)_{surf} + H_2O \rightarrow X(aq) \tag{8.78}$$

where $X(s)_{surf}$ represents a molecule of X on the surface of the solid, i.e., in contact with the solution.

The corresponding reaction rate can be written as the product of the concentration of X on the solid surface ($\sigma_{X,surf}$, with units such as moles of X per m^2 of surface) and the concentration of surface area in the system. Note that $\sigma_{X,surf}$ is the same in all three systems, because the amount of X per unit of surface area has nothing to do with the total amount of X in the system (most of which is on the interior of the solids). Thus, the rates of dissolution in the three solutions are expected to be as follows:

$$r_{X(s)\rightarrow X(aq),I} = k_f \sigma_{X,surf} A_I \tag{8.79a}$$

$$r_{X(s)\rightarrow X(aq),II} = k_f \sigma_{X,surf} A_{II} \tag{8.79b}$$

$$r_{X(s)\rightarrow X(aq),III} = k_f \sigma_{X,surf} A_{III} \tag{8.79c}$$

where k_f is the forward reaction rate constant and A_i is the surface area concentration in system i, in units such as m^2/L.

The kinetic view of equilibrium, applied to this example, postulates that at equilibrium the precipitation reaction proceeds at the same rate as the dissolution reaction. Following the same reasoning as above and considering that the precipitation reaction requires a collision between a dissolved $X(aq)$ molecule and the surface of a particle,[6] we expect the rate of precipitation to be proportional to the concentrations of both $X(aq)$ and surface molecules of $X(s)$. That is, the precipitation reaction might be represented as an elementary reaction with the following stoichiometry:

$$X(aq) + X(s)_{surf} \rightarrow X\!-\!X(s)_{surf} \tag{8.80}$$

where $X\!-\!X(s)_{surf}$ represents the solid that has acquired an additional molecule of X by the precipitation reaction.

Thus, the rates of precipitation in the three systems are expected to be given by equations of the following form:

$$r_{X(aq)\rightarrow X(s),I} = k_r\{X(aq)\}A_I \tag{8.81a}$$

$$r_{X(aq)\rightarrow X(s),II} = k_r\{X(aq)\}A_{II} \tag{8.81b}$$

$$r_{X(aq)\rightarrow X(s),III} = k_r\{X(aq)\}A_{III} \tag{8.81c}$$

At equilibrium, the forward and reverse reaction rates can be equated, yielding the following equations:

$$k_f\sigma_{X,surf}A_I = k_r\{X(aq)\}A_I \tag{8.82a}$$

$$k_f\sigma_{X,surf}A_{II} = k_r\{X(aq)\}A_{II} \tag{8.82b}$$

$$k_f\sigma_{X,surf}A_{III} = k_r\{X(aq)\}A_{III} \tag{8.82c}$$

Any of the forms of Equation (8.82) can be manipulated to yield the equilibrium constant for the dissolution reaction:

$$K_{eq} = \frac{k_f}{k_r} = \frac{\{X(aq)\}}{\sigma_{X,surf}} \tag{8.83}$$

Of course, we would expect the equilibrium constant for a reaction between a solid and its dissolved counterpart to be, in fact, constant, so it is not surprising that analysis of all three systems yields the same value of K_{eq}. Furthermore, if we consider $\sigma_{X,surf}$ to be a measure of the activity of the solid phase, we see that the equilibrium constant is the ratio of the activity of the reaction product [dissolved $X(aq)$] to the activity of the reactant [solid $X(s)$], which is entirely consistent with our previous experience. However, upon closer consideration, we might also note an important difference between this equilibrium constant and those that we have encountered previously. Specifically, the activity of solid X is invariant, regardless of the concentration of $X(s)$ in the system. For instance, for this example, we expect the concentration of X

---

[6]Precipitation of a completely new particle is possible, but is much less likely than the growth of existing particles.

atoms at the surface of the solid (i.e., molecules per unit area of surface) to be the same in all three systems, independent of the number or size of particles in the system. We therefore conclude that the activity of dissolved X, i.e., $\{X(aq)\}$, that is present at equilibrium is identical in all three systems and that the concentration and state of dispersion of the solid in the solution do not affect the equilibrium solubility.[7]

---

[7]If particles are extremely small (in the colloidal size range), the activity of the solid does differ from that of larger particles, because the small radius of curvature of the surface places additional stress on the bonds between molecules at the surface. Therefore, even though the surface density of atoms might be similar in large and small particles, the smaller particles tend to be slightly more soluble. This effect is generally negligible for particles larger than approximately 1 μm in diameter.

REDOX CHEMISTRY

9.1   INTRODUCTION: DEFINITIONS OF OXIDATION, REDUCTION, REDOX REACTIONS, AND OXIDATION STATE

This chapter describes oxidation/reduction, or more simply, **redox** reactions—chemical reactions in which electrons are transferred from one atom to another. The loss of electrons is called **oxidation,** and an atom that has lost electrons is said to be **oxidized.** Correspondingly, the acquisition of electrons is called **reduction,** and an atom that has gained electrons is said to be **reduced.** Redox reactions involve simultaneous oxidation of one atom and reduction of another.

Since any reduced element is capable of releasing electrons that can combine with (i.e., reduce) another element, reduced compounds are sometimes referred to as **reducing agents** or **reductants.** Similarly, oxidized elements are capable of acquiring electrons from other elements; in the process, the (initially) oxidized element becomes reduced and acts as an **oxidizing agent** or **oxidant.** Reductants are also commonly referred to as **electron donors,** and oxidants as **electron acceptors,** especially in the microbiological and biochemical literature.

The charge or **oxidation number** (also called **oxidation state**) of any atom is the difference between the number of protons in the atom's nucleus and the number of electrons assigned to that atom. In a redox reaction, the oxidation number of the atom being oxidized increases (i.e., it becomes either more positive or less negative), and that of the atom being reduced decreases.

Redox reactions are of overwhelming importance in environmental chemistry. Their significance stems primarily from two characteristics. First, changes in the behavior of elements when they participate in redox reactions can be enormous. To cite just a few examples, Cr(III)[1] species are very insoluble cations under normal conditions and pose little risk to health, whereas Cr(VI) species are very soluble anions and are potent carcinogens. Similarly, S(VI) species (e.g., sulfate ion) are highly soluble, non-volatile, and relatively innocuous, while S(−II) species (H_2S and S^{2-}, for instance) form very insoluble metal precipitates and are quite toxic. Finally, we are all familiar with the very different properties of metallic iron and rust [Fe(0) in the former case, compounds of Fe(II) or Fe(III) in the latter]. The cost associated with preventing and/or repairing the effects of this redox reaction runs into billions of dollars annually.

A second factor contributing to the importance of redox reactions is that often they are strongly inhibited kinetically, so that systems that are severely out of equilibrium (Q/K values different from 1.0 by many orders of magnitude) can be stable for long periods of time. This characteristic represents an important distinction between many redox reactions and most other reactions that we have considered up to this point (e.g., acid/base, gas/liquid, metal complexation, and metal precipitation reactions). Because the Gibbs energy that is released when a reaction proceeds is directly related to how far that reaction is from equilibrium [recall from Chapter 2 that $\Delta \overline{G}_r = RT \ln(Q/K)$], redox reactions often release much more energy, when they do proceed, than other reactions.

The ability to capture the energy released in redox reactions is at the core of virtually all metabolic activity. Specifically, organisms produce catalysts (enzymes) that reduce the kinetic inhibition of redox reactions and allow the reactions to approach equilibrium more rapidly. The energy released as a result of those reactions is captured and used for life processes; if the reactions proceeded at any significant rate in the absence of the catalyst, the disequilibrium would be reduced or eliminated outside the cell, and insufficient energy would

---

[1]The representation X(n) is commonly used to designate the oxidation state of an element X, where n is the oxidation number assigned to the element, often given as a Roman numeral. The representation is meant to include all the species in the system in which X is present with that oxidation number.

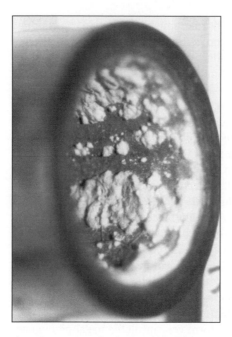
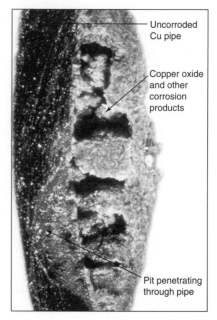

The redox reactions leading to internal corrosion of water supply pipes can cause multiple problems. Aside from outright failure of the pipe, the products of corrosion can clog the pipe or enter the water supply, increasing the consumer's intake of metals (Fe, Cu, Pb) and leading to complaints of "red water." These photographs show the formation of a Zn-containing precipitate in a galvanized pipe section and the penetration of a pit almost all the way through a copper pipe section.

Left: HDR Engineering, Inc.; Right: Reprinted from *Internal Corrosion of Water Distribution Systems,* 2nd edition, by permission. Copyright © 1996, American Water Works Association.

remain for the cell to survive. (Indeed, all aerobic organisms are themselves thermodynamically unstable; they would burn up spontaneously if the approach to equilibrium between cellular material and molecular oxygen were fast.)

In this chapter, redox reactions of environmental importance are introduced, with a major focus on the energy changes associated with those reactions. The chapter relates to and synthesizes information from virtually all the preceding chapters of the text, while simultaneously pointing out the unique features of redox reactions. In addition to a good deal of new terminology, which is defined at various stages throughout the chapter, the topics covered include the quantification of the Gibbs energy of electrons, the analysis of redox speciation and its display in $\log C$–pe diagrams, and an introduction to electrochemistry and the functioning of electrodes.

Example 9.1	Determine the Q/K ratio and the molar Gibbs energy of reaction for the following redox reaction, in which the sugar glucose ($C_6H_{12}O_6$) and dissolved oxygen are converted to carbonic acid.

$$C_6H_{12}O_6 + 6O_2(aq) \leftrightarrow 6H_2CO_3$$

This reaction is representative of reactions in which organisms oxidize sugar to power their metabolism. Assume that the solution initially contains 4.5 mg/L glucose (25 μmol/L), 8 mg/L $O_2(aq)$ (250 μmol/L), and 10^{-5} mol/L H_2CO_3. Compare the Gibbs energy for reaction of one mole of oxygen in the above reaction with the corresponding value for dissolution of atmospheric oxygen into a solution that contains 2 mg/L $O_2(aq)$ if the concentration of $O_2(aq)$ in equilibrium with the atmosphere is 10 mg/L. The standard Gibbs energies of formation of the chemicals of interest are -908.01, 16.40, and -623.04 kJ/mol for dissolved glucose, $O_2(aq)$, and H_2CO_3, respectively.

Solution

The equilibrium constant for the glucose reaction can be computed from the following relationships, where the second relationship applies at 25°C:

$$\ln K = -\frac{\Delta \bar{G}_r{}^\circ}{RT} \qquad \log K = -\frac{\Delta \bar{G}_r{}^\circ}{5.71 \text{ kJ/mol}}$$

The standard Gibbs energy of the reaction is

$$\Delta \bar{G}_r{}^\circ = 6\bar{G}_{H_2CO_3}^\circ - \bar{G}_{C_6H_{12}O_6}^\circ - 6\bar{G}_{O_2(aq)}^\circ = -2928.6 \text{ kJ/mol of reaction}$$

Plugging the value of $\Delta \bar{G}_r{}^\circ$ into the expression for the equilibrium constant, we find $\log K = 512.9$, i.e., $K = 10^{512.9}$.

The reaction quotient for the reaction under the given conditions is

$$Q = \frac{\{H_2CO_3\}^6}{\{C_6H_{12}O_6\}\{O_2(aq)\}^6} = \frac{(10^{-5})^6}{(25 \times 10^{-6})(250 \times 10^{-6})^6} = 1.64 \times 10^{-4}$$

Thus, under reasonable conditions, the reaction is out of equilibrium by >500 orders of magnitude! Clearly, if the redox reaction proceeded to reach equilibrium, essentially all the glucose would be consumed. However, because the reaction has a very large activation energy, it proceeds at a negligible rate in the absence of a catalyst. Organisms provide such catalysts (in the form of enzymes), allowing the reaction to proceed and allowing them to capture and utilize the energy stored in the chemical bonds of the sugar and oxygen molecules.

The molar Gibbs energy of the reaction under the given conditions can be computed as

$$\Delta \bar{G}_r = \left(5.71 \frac{\text{kJ}}{\text{mol}}\right) \log \frac{Q}{K} = -2950 \frac{\text{kJ}}{\text{mol}}$$

Because the molar energy of reaction is defined for one mole of stoichiometric reaction, the above calculation corresponds to reaction of 1 mole of glucose with 6 moles of oxygen. The Gibbs energy change per mole of oxygen reacting is therefore one-sixth of the value computed above, or -491.7 kJ/mol.

The oxygen dissolution reaction is simply

$$O_2(g) \leftrightarrow O_2(aq)$$

Concentrations of 2 and 10 mg/L $O_2(aq)$ correspond to 6.25×10^{-5} and 3.125×10^{-4} mol/L, respectively. Therefore, the values of Q, K, and $\bar{G}_r$ for O_2 dissolution in the system of interest are as follows:

$$Q = \frac{\{O_2(aq)\}_{actual}}{\{O_2(g)\}} = \frac{6.25 \times 10^{-5}}{0.21} = 2.98 \times 10^{-4}$$

$$K = \frac{\{O_2(aq)\}_{eq}}{\{O_2(g)\}} = \frac{3.125 \times 10^{-4}}{0.21} = 1.49 \times 10^{-3}$$

$$\frac{Q}{K} = \frac{2.98 \times 10^{-4}}{1.49 \times 10^{-3}} = 0.20$$

$$\Delta \bar{G}_r = \left(5.71 \frac{kJ}{mol}\right) \log \frac{Q}{K} = -3.99 \frac{kJ}{mol}$$

The Gibbs energy change of both the redox reaction and the oxygen dissolution reaction is negative, meaning that both reactions will proceed to the right spontaneously. However, the Gibbs energy per mole of oxygen for the redox reaction is larger by a factor of approximately 123, reinforcing the point made above about the relatively large amounts of energy available from such reactions.

9.1.1   DETERMINING OXIDATION NUMBERS

Although determining the oxidation number of an atom is simple in concept, it is not always easy in practice, because ambiguity arises as to which atom a particular electron "belongs" to. Indeed, since the formation of chemical bonds involves the sharing of electrons between or among atoms, all electrons that are involved in bond formation should logically be assigned only partially to any single atom. While such a practice is common in some fields of chemistry, the convention for defining oxidation numbers in redox reactions is to assign each electron in a chemical compound entirely to a specific atom. One rule and several important conventions for assigning oxidation numbers to elements of environmental significance are summarized in Table 9.1.

Table 9.1   Conventions for assigning charge to various atoms

---

Rule (can never be violated): Charge must be conserved; i.e., the charge on a molecule must equal the sum of the charges on the constituent atoms.

Conventions:

1. H has an oxidation number of +1.

2. O has an oxidation number of −2.

3. N has an oxidation number of −3 when bonded only to H or C, as it is in many organic compounds.

4. S has an oxidation number of −2 when bonded only to H or C, as it is in many organic compounds.

Application: The above conventions are applied in the order given, but can be violated if necessary to avoid violating the charge conservation rule.

| Determine the average oxidation number of the following elements: | **Example 9.2** |

a. C in (i) CO_2, (ii) CO_3^{2-}, (iii) HCO_3^-, and (iv) CO_3^{2-}.
b. C in acetic acid, $C_2H_4O_2$.
c. C in the amino acid glycine, $C_2H_5O_2N$.
d. Cr in $Cr(OH)_3^0$, $Cr_2O_7^{2-}$, and $HCrO_4^-$.
e. H and O in hydrogen peroxide, H_2O_2.
f. H in $H_2(g)$.

Solution

a. i. By convention 2, in carbon dioxide, each oxygen atom contributes a charge of -2. Since the molecule is neutral, the carbon must contribute a charge of $+4$. Therefore, its oxidation state or oxidation number is $+4$.

ii. In a carbonate ion, each oxygen atom has a charge of -2, so the three oxygen atoms contribute a total charge of -6 to the ion. Since the overall charge on the ion is -2, the carbon must once again have a charge of $+4$.

iii, iv. A similar calculation to that in part (ii) indicates that the carbon atom has a charge of $+4$ in both bicarbonate and carbonate ion. Thus, neither hydration/dehydration nor any acid/base reaction among carbonate species is a redox reaction. This result can be extended to all hydration/dehydration and acid/base reactions.

b. In the acetate ion, the three H atoms contribute a charge of $+1$ each, and each of the two O atoms contributes a charge of -2. Together, the H and O therefore contribute a net charge of -1, which is the charge on the molecule. Thus, the average oxidation state of the C atoms is zero.

c. In glycine, the charge contributions of the five H, two O, and one N atoms are $+5$, -4, and -3, respectively, for a total of -2. The molecule is neutral, so the total charge contributed by the two C atoms is $+2$, and the average charge on the C atoms is $+1$.

d. Charge balances analogous to those in parts (a) through (c) indicate that the oxidation number of Cr is $+3$ in $Cr(OH)_3^0$ and $+6$ in both $Cr_2O_7^{2-}$ and $HCrO_4^-$. Note that when $Cr(OH)_3^0$ is oxidized to either $Cr_2O_7^{2-}$ or $HCrO_4^-$, the charge on the molecule decreases even though the charge on the central metal ion increases.[2]

e. The first two conventions cannot both be applied while following the rule of charge conservation, so convention 1 is followed and convention 2 is violated. The average oxidation number of H in H_2O_2 is $+1$, and that of O is -1.

f. Convention 1 cannot be applied without violating the rule that charge must be conserved, so the convention is ignored and the rule is followed. The oxidation number of each H atom in $H_2(g)$ is zero.

In Example 9.2b above, we found that the average charge on C atoms in acetic acid is zero. However, the molecular structure of acetic acid suggests that neither carbon atom in the molecule is, in fact, uncharged. Acetic acid can be

---

[2] It is common to refer to the entire molecule as being oxidized or reduced, since in reality it is the molecule that is losing or gaining electrons; the assignment of those electrons to specific atoms is a convention that is not rigorously defensible.

reasonably represented as a combination of a methyl (methane-like) group and a carboxyl (carbonate-like) group, as shown below.

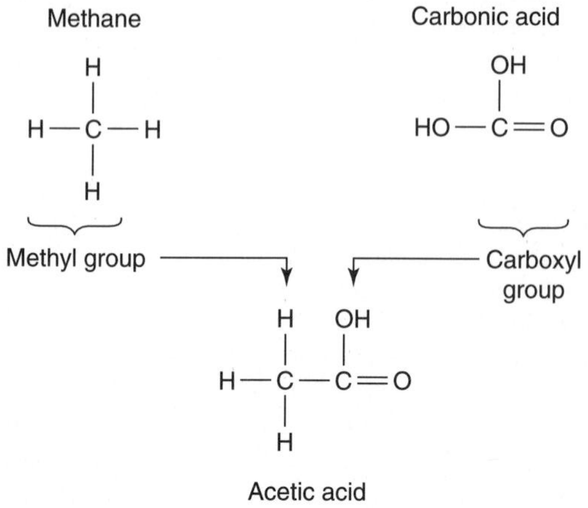

The oxidation number of C is $+4$ in carbonic acid and -4 in methane. Thus, it is most reasonable to think of the carbon in acetic acid as having an average oxidation number of zero due to the balancing of two non-zero charges on the individual carbon atoms. The same type of situation can cause the average oxidation number of an atom in a molecule to be non-integral, even though charges (i.e., electrons) can exist only in whole-number units. At times, analysis of the charge on individual atoms in a compound can provide insight into its behavior; in other cases, the average oxidation state of all the atoms of a given type in a molecule is all that interests us. In any case, it is important to remember that the procedure described above yields a value for the average charge on a group of atoms and does not necessarily indicate the charge on any particular atom in a molecule.

9.1.2   BALANCING REDOX REACTIONS

Free electrons, like free protons, are highly unstable in aqueous solution. As a result, the concentration of free electrons in any solution is always vanishingly small, and all the electrons released by one element must combine with another; i.e., no reaction releasing electrons to solution can take place in isolation, nor can electrons be acquired from bulk solution. At times, we know the reactants and products of a redox reaction, but the stoichiometry is not obvious. In such a case, we can use the information that free aquo electrons cannot be a significant reactant or product in the reaction to help balance the reaction, following the algorithm summarized in Table 9.2.

Table 9.2   Algorithm for balancing redox reactions

1. Determine oxidation numbers of all atoms in the reaction. Identify those atoms that are oxidized and those that are reduced in the reaction.

2. Choose one of the oxidized or reduced species to have a stoichiometric coefficient of 1.

3. Assign other stoichiometric coefficients that are established unambiguously by the choice made in step 2.

4. Given that the total number of electrons lost by atoms that are oxidized must equal the total number of electrons gained by those that are reduced (electrons must be conserved) and given the formula of each compound, determine stoichiometric coefficients for the other oxidized and reduced species in the reaction.

5. Add H_2O to either side of the reaction to balance the oxygen atoms.

6. Add H^+ to either side of the reaction to balance charge.

7. Hydrogen should automatically balance at this point. Check the H balance to verify.

Balance a redox reaction in which ferrous ion, Fe^{2+}, is converted to ferric hydroxide solid, $Fe(OH)_3(s)$, and dichromate ion, $Cr_2O_7^{2-}$, is converted to the chromium dihydroxo complex $Cr(OH)_2^+$. [*Note*: This type of reaction is often used to convert Cr from its more toxic to its less objectionable oxidation state ($+6$ to $+3$).]

Example 9.3

Solution

The unbalanced reaction can be written as follows:

$$\_Fe^{2+} + \_Cr_2O_7^{2-} \leftrightarrow \_Fe(OH)_3(s) + \_Cr(OH)_2^+$$

In the reaction, Fe(II) is oxidized to Fe(III), and Cr(VI) is reduced to Cr(III). Choosing the stoichiometric coefficient of $Cr_2O_7^{2-}$ to be 1, it is clear that the coefficient of $Cr(OH)_2^+$ must be 2, since this is the only way that the Cr atoms can balance.

Each Cr atom undergoing the conversion from Cr(VI) to Cr(III) gains three electrons, so for the given coefficients, six electrons are transferred in the overall reaction. Each Fe(II) atom releases only one electron when it is converted to Fe(III), so six Fe(II) ions must be oxidized to provide enough electrons to reduce one $Cr_2O_7^{2-}$ ion. The coefficients on the four redox-active species are therefore as follows:

$$6Fe^{2+} + 1Cr_2O_7^{2-} \leftrightarrow 6Fe(OH)_3(s) + 2Cr(OH)_2^+$$

The above reaction has 7 oxygen atoms on the reactant side and 22 on the product side. To balance the oxygen, we add 15 water molecules to the reactant side of the reaction:

$$6Fe^{2+} + 1Cr_2O_7^{2-} + 15H_2O \leftrightarrow 6Fe(OH)_3(s) + 2Cr(OH)_2^+$$

Finally, we note that the total charge on the reactant side is $+10$, while that on the product side is $+2$. Adding eight H^+ ions to the product side balances the charge and, as it must, also balances the H atoms (30 on each side). The final, balanced reaction is as follows:

$$6Fe^{2+} + 1Cr_2O_7^{2-} + 15H_2O \leftrightarrow 6Fe(OH)_3(s) + 2Cr(OH)_2^+ + 8H^+$$

Note that H^+ and H_2O appear in the reaction, but neither hydrogen nor oxygen is oxidized or reduced; their oxidation states are $+1$ and -2, respectively, in all compounds on both the reactant and product sides of the reaction.

9.1.3 REDOX HALF-CELL REACTIONS

It is sometimes convenient to think of electron transfers as taking place in two steps—one in which electrons are released by the species being oxidized and another in which they are consumed by the species being reduced. The individual reactions releasing or consuming electrons are often called *half-reactions*. However, even though unattached electrons are extremely unstable, reactions showing them as reactants or products are legitimate "whole" reactions; in this text, the term *half-cell reaction* is used instead of *half-reaction*. Redox half-cell reactions that can be combined to generate the overall reaction in the preceding example [oxidation of Fe^{2+} to $Fe(OH)_3(s)$ and reduction of $Cr_2O_7^{2-}$ to $Cr(OH)_2^+$] are shown below.

$$Fe^{2+} + 3H_2O \leftrightarrow Fe(OH)_3(s) + 3H^+ + e^- \tag{9.1}$$

$$Cr_2O_7^{2-}(aq) + 10H^+ + 6e^- \leftrightarrow 2Cr(OH)_2^+ + 3H_2O \tag{9.2}$$

Redox half-cell reactions representing oxidation of any compound and reduction of any other can easily be combined (on paper) to form a virtually limitless variety of overall redox reactions.

One half-cell reaction that is of overwhelming importance in environmental systems is that for the reduction of dissolved oxygen, because of the large amount of energy released by that reaction and the requirement for oxygen by many bacteria and all animals. The maximum amount of oxygen that can potentially be consumed by oxidation of the organic and inorganic constituents of a solution is therefore of interest. However, direct measurement of oxygen consumption is usually not a convenient approach for assessing the total oxygen-consuming capacity of a solution, both because the reaction can be slow and because the potential O_2 consumption might be far greater than the amount of O_2 available. Analysis of individual chemicals that react with oxygen is also impractical, because the solutions of interest typically contain a multitude of discrete chemicals, each present at a low concentration.

To address this problem, the chemical oxygen demand (COD) test has been developed. In this test, a very strong, very acidic oxidizing agent (chromic acid, which dimerizes in concentrated solutions to form dichromate ion, $Cr_2O_7^{2-}$) is added to the test solution, along with some catalysts and other reactants, and the mixture is heated to boiling. Under these conditions, most of the oxidizable chemicals in the original solution are oxidized. The difference between the chromic acid dose and the amount of chromic acid remaining at the end of the test indicates how much Cr(VI) was reduced to Cr(III), and therefore how many electrons were released by the substances that were oxidized. Because, if oxygen were the oxidant, it would have to acquire the same number of electrons as did the Cr(VI), one can use the results of the COD test [i.e., the concentration of Cr(VI) consumed] to compute how much oxygen would be required to carry out the same reactions.

Since oxygen is a weaker oxidant than Cr(VI), and since the environmental conditions would not be as favorable for the oxidation reactions as they are in the COD test, the actual oxygen consumption in the environment is virtually always less than the value computed based on the COD test. Nevertheless, the test provides a relatively quick, consistent estimate of the maximum O_2 consumption that could be caused by constituents in the water, and it is therefore a useful indicator of one aspect of water quality. The COD test is also often used to assess the efficiency of a treatment process intended to remove oxidizable compounds from a water (based on a comparison of the COD values of the influent and effluent).

Assume that a wastewater contains phenol (C_6H_6O) as a major constituent. | **Example 9.4**

 a. Write the redox half-cell reaction for oxidation of phenol to carbon dioxide, and combine this reaction with the half-cell reaction for $Cr_2O_7^{2-}$ reduction to $Cr(OH)_3(s)$ to develop the overall reaction for oxidation of phenol by dichromate.
 b. A solution contains 15 mg/L phenol (1.6×10^{-4} M). How much Cr(VI) is reduced by the phenol in a COD test? How much COD does the phenol contribute to the solution, expressed as mg O_2/L?

Solution

 a. The half-cell reactions for phenol oxidation and dichromate reduction are

$$C_6H_6O + 11H_2O \leftrightarrow 6CO_2 + 28H^+ + 28e^-$$
$$Cr_2O_7^{2-}(aq) + 8H^+ + 6e^- \leftrightarrow 2Cr(OH)_3(s) + H_2O$$

Because the net reaction between phenol and dichromate cannot include free electrons, $28/6 = 4.67$ mol of dichromate must be reduced for each mole of phenol that is oxidized. Adding the reactions in this ratio (14 times the dichromate reaction plus 3 times the phenol reaction), we can write the overall reaction is as follows:

$$14Cr_2O_7^{2-}(aq) + 3C_6H_6O + 19H_2O + 28H^+ \leftrightarrow 28Cr(OH)_3(s) + 18CO_2$$

 b. According to the balanced redox reaction, 14 moles of $Cr_2O_7^{2-}$ react with 3 moles of phenol. Therefore, the amount of $Cr_2O_7^{2-}$ reacting with phenol in the given solution is $(14/3)(1.6 \times 10^{-4}$ M), or 7.45×10^{-4} M. Each mole of $Cr_2O_7^{2-}$ contains 2 moles of Cr(VI), both of which are reduced in the reaction, so 1.49×10^{-3} M Cr(VI) is reduced in the test.

The half-cell reaction for reduction of $O_2(aq)$ indicates that 4 moles of electrons are consumed per mole of oxygen that gets reduced:

$$O_2(aq) + 4H^+ + 4e^- \leftrightarrow 2H_2O$$

Since only 3 moles of electrons are consumed per mole of Cr(VI) that gets reduced in the COD test, any chemical that is oxidized by reaction with x moles of Cr(VI) could in theory be oxidized by $0.75x$ moles of O_2. (Note that this conversion is based on stoichiometry only. Both the driving force and the kinetics of the reaction would be different for the two oxidants, so the reaction with O_2 might not be spontaneous or sufficiently rapid to observe.) In the current case, this conversion ratio means that the phenol could potentially react with $(0.75)(1.49 \times 10^{-3}\ M)$, or $1.12 \times 10^{-3}\ M\ O_2$. This concentration of O_2 corresponds to 35.7 mg/L O_2, which is how the COD would be reported conventionally.

Note that the calculation of the Cr(VI):O_2 ratio in part (b) of the preceding example is independent of the identity of the compound being oxidized. That is, for *any* compound, reduction of x mol/L Cr(VI) in the COD test corresponds to potential reduction of $0.75x$ mol/L O_2 in the environment. In conventional units, the conversion is 0.42 mg of O_2 per milligram of Cr(VI), i.e., 0.42 mg COD/mg Cr(VI) consumed.

Redox reactions between metallic (zero-valent) iron and some chlorinated organic compounds can dechlorinate the organics. The recent discovery of this reaction has spawned a number of field experiments in which iron is placed in a ditch that intercepts contaminated groundwater, in an attempt to dechlorinate the trace organics in the water. The photograph shows an experimental site where such a process is being tested.

| Photo courtesy of U.S. Geological Survey.

9.1.4 THE ACTIVITY OF FREE ELECTRONS AND EQUILIBRIUM CONSTANTS FOR REDOX HALF-CELL REACTIONS

Consider the following generic redox half-cell reaction, in which "Ox" represents an oxidized species and "Red" represents the conjugate reduced species, i.e., the species that forms when Ox is reduced by gaining n_e electrons:

$$Ox + n_e e^- \leftrightarrow Red \tag{9.3}$$

$$K_{(9.3)} = \frac{\{Red\}}{\{Ox\}\{e^-\}^{n_e}} \tag{9.4}$$

If there are multiple species on either side of the reaction, the same equations apply, with the understanding that {Red} represents the product of the activities of all species on the right-hand side of the reaction, and {Ox} represents the corresponding product of the activities of all species on the left-hand side, excluding the electrons.

To evaluate $K_{(9.3)}$ by plugging values into Equation (9.4), we would have to evaluate $\{e^-\}$. Although defining the activity of free electrons based on the convention used for most other dissolved species (i.e., a standard state concentration of 1.0 mol/L and a reference state of infinite dilution) presents no conceptual problem, it does present a practical one: as noted above, electrons are tremendously unstable as a free dissolved species, so the concentration of e^- in almost any real solution is immeasurably small.

This difficulty is overcome by defining the standard state for dissolved electrons (i.e., the conditions under which the activity of electrons is 1.0 by definition) differently from that for most other dissolved species. To understand the justification for this approach, recall that the chemical activity of a species quantifies the *relative* reactivity of that species in different systems. Therefore, we can define the activity of e^- in some arbitrarily chosen standard system (standard state) to be 1.0, and measure the activity of e^- in other systems relative to this standard. Note that the concentration of free dissolved (hydrated) electrons is presumed to be finite, albeit extremely small, in all these systems.

By convention, the standard state for dissolved electrons is defined to be a solution in equilibrium with H^+ and $H_2(g)$, both present at an activity of 1.0. An activity of 1.0 for H^+ ions corresponds to pH = 0 and, assuming that $H_2(g)$ behaves according to the ideal gas law, an activity of 1.0 for $H_2(g)$ corresponds to $P_{H_2} = 1.0$ bar. The species of interest are related by the following redox half-cell reaction:

$$H^+ + e^- \leftrightarrow \frac{1}{2}H_2(g) \tag{9.5}$$

Because, by definition, $\{e^-\} \equiv 1.0$ if reaction (9.5) is at equilibrium and if both $\{H_2(g)\}$ and $\{H^+\}$ are 1.0, the equilibrium constant for reaction (9.5) must equal 1.0:

$$K_{(9.5)} = \frac{\{H_2(g)\}^{1/2}}{\{H^+\}\{e^-\}} = 1.0 \tag{9.6}$$

Knowing the value of $K_{(9.5)}$, we can use Equation (9.6) to compute the activity of e^- in any other equilibrium system containing H^+ and $H_2(g)$. For instance, in a pH 7.0 solution in equilibrium with a hydrogen partial pressure of 10^{-5} bar, the value of $\{e^-\}$ would be

$$\{e^-\} = \frac{\{H_2(g)\}^{1/2}}{K_{(9.5)}\{H^+\}} = \frac{(10^{-5})^{0.5}}{(1.0)(10^{-7.0})} = 10^{+4.5}$$

Using the "p" shorthand, we can represent the activity of electrons in this system as pe $= -4.5$ (i.e., $-\log\{e^-\} = -4.5$).

Once the value of $\{e^-\}$ in a solution is established, the equilibrium constant of any redox half-cell reaction that is at equilibrium in the solution can be determined, in theory, by analysis of the activities of the species participating in that reaction. For instance, assume that the solution described above (with pe $= -4.5$) also contains Fe^{2+} and Fe^{3+} in equilibrium with one another. These species can be interconverted via reaction (9.7):

$$Fe^{3+} + e^- \leftrightarrow Fe^{2+} \tag{9.7}$$

By evaluating the activities of Fe^{2+} and Fe^{3+} in the solution, we could determine K_{eq} for the Fe^{3+}/Fe^{2+} redox couple by

$$K_{(9.7)} = \frac{\{Fe^{2+}\}}{\{Fe^{3+}\}\{e^-\}} = \frac{\{Fe^{2+}\}}{\{Fe^{3+}\}(10^{4.5})} \tag{9.8}$$

For this particular reaction, the equilibrium constant $K_{(9.7)}$ equals $10^{+13.03}$. The type of analysis described above has been conducted for many redox couples, and the equilibrium constants for those couples have been tabulated in numerous collections. A list of some half-cell reactions of importance in environmental systems is provided in Table 9.3.

The equilibrium constants given in Table 9.3 can be combined with others (e.g., K_a, K_H, or K_i values) to yield $\log K$ values for redox half-cell reactions between other species. For instance, if we wished to know $\log K$ for the half-cell reaction in which SO_4^{2-} is reduced to $H_2S(aq)$, we could simply combine the given reaction for reduction of SO_4^{2-} to HS^- with K_{a1} for $H_2S(aq)$ to obtain

Reaction	Log K
$SO_4^{2-} + 8e^- + 9H^+ \leftrightarrow HS^- + 4H_2O$	33.66
$HS^- + H^+ \leftrightarrow H_2S(aq)$	6.99
$SO_4^{2-} + 8e^- + 10H^+ \leftrightarrow H_2S(aq) + 4H_2O$	40.65

It should be apparent that the definition of electron activity bears many similarities to that for H^+ activity. For instance, the concentration of free dissolved protons in solution is always negligible. Nevertheless, we choose to define the *activity* of H^+ in a way that makes its value finite. Specifically, we assign $\{H^+\}$ a value of 1.0 in an ideal solution in which $\{H_2O\} = 1.0$ and in which the sum of the concentrations of all species of the type $H_{2n+1}O_n{}^+$ is 1.0 M. Based on this (arbitrary) definition, we can compute unambiguous values of $\{H^+\}$ from experimental data in any other system.

Table 9.3 Equilibrium constants for some environmentally important redox half-cell reactions*

Reaction	Log K	pe$^\circ$	pe$^\circ$(W)	E_H°, mV
$NO_3^- + 2e^- + 2H^+ \leftrightarrow NO_2^- + H_2O$	28.57	14.29	7.28	843
$NO_3^- + 8e^- + 10H^+ \leftrightarrow NH_4^+ + 3H_2O$	119.08	14.89	6.14	878
$NO_3^- + 8e^- + 9H^+ \leftrightarrow NH_3(aq) + 3H_2O$	109.83	13.73	5.85	809
$NO_3^- + 3e^- + 4H^+ \leftrightarrow NO(g) + 2H_2O$	48.40	16.13	6.80	952
$2NO_3^- + 10e^- + 12H^+ \leftrightarrow N_2(g) + 6H_2O$	210.34	21.03	12.63	1241
$NO_2(g) + 2e^- + 2H^+ \leftrightarrow NO(g) + H_2O$	53.60	26.80	19.80	1581
$N_2O(g) + 2e^- + 2H^+ \leftrightarrow N_2(g) + H_2O$	59.79	29.89	22.89	1764
$SO_4^{2-} + 8e^- + 9H^+ \leftrightarrow HS^- + 4H_2O$	33.68	4.21	−3.67	248
$SO_4^{2-} + 8e^- + 10H^+ \leftrightarrow H_2S(aq) + 4H_2O$	40.67	5.08	−3.67	299
$SO_4^{2-} + 2e^- + 2H^+ \leftrightarrow SO_3^{2-} + H_2O$	27.16	13.58	6.58	801
$SeO_4^{2-} + 2e^- + 4H^+ \leftrightarrow H_2SeO_3 + H_2O$	36.32	18.16	4.16	1071
$H_3PO_4 + 2e^- + 2H^+ \leftrightarrow H_3PO_3 + H_2O$	−10.10	−5.05	−12.05	−298
$AsO_4^{3-} + 2e^- + 2H^+ \leftrightarrow AsO_3^{3-} + H_2O$	5.29	2.64	−4.36	156
$CrO_4^{2-} + 3e^- + 8H^+ \leftrightarrow Cr^{3+} + 4H_2O$	77.00	25.66	7.00	1514
$OCN^- + 2e^- + 2H^+ \leftrightarrow CN^- + H_2O$	−4.88	−2.44	−9.44	−144
$2H^+ + 2e^- \leftrightarrow H_2(g)$	0.00	0.00	−7.00	0
$2H^+ + 2e^- \leftrightarrow H_2(aq)$	3.10	1.55	−5.45	92
$O_2(g) + 4H^+ + 4e^- \leftrightarrow 2H_2O$	83.12	20.78	13.78	1226
$O_2(aq) + 4H^+ + 4e^- \leftrightarrow 2H_2O$	86.00	21.50	14.50	1268
$O_2(aq) + 2e^- + 2H^+ \leftrightarrow H_2O_2(aq)$	26.34	13.17	6.17	777
$H_2O_2(aq) + 2e^- + 2H^+ \leftrightarrow 2H_2O$	59.59	29.80	22.80	1758
$O_3(g) + 2e^- + 2H^+ \leftrightarrow O_2(g) + H_2O$	70.12	35.06	28.06	2069
$Cl_2(aq) + 2e^- \leftrightarrow 2Cl^-$	47.20	23.60	23.60	1392
$ClO_3^- + 6e^- + 6H^+ \leftrightarrow Cl^- + 3H_2O$	147.02	24.50	17.50	1446
$HOCl + 2e^- + H^+ \leftrightarrow Cl^- + H_2O$	50.20	25.10	21.60	1481
$ClO_2 + 5e^- + 4H^+ \leftrightarrow Cl^- + 2H_2O$	126.67	25.33	19.73	1495
$ClO_2^- + 4e^- + 4H^+ \leftrightarrow Cl^- + 2H_2O$	109.06	27.27	20.26	1609
$HOBr + 2e^- + H^+ \leftrightarrow Br^- + H_2O$	45.36	22.68	19.18	1338
$2HOBr + 2e^- + 2H^+ \leftrightarrow Br_2(aq) + 2H_2O$	53.60	26.80	20.27	1581
$BrO_3^- + 6H^+ + 6e^- \leftrightarrow Br^- + 3H_2O$	146.1	24.35	17.35	1437
$Al^{3+} + 3e^- \leftrightarrow Al(s)$	−85.71	−28.57	−28.57	−1686
$Zn^{2+} + 2e^- \leftrightarrow Zn(s)$	−25.76	−12.88	−12.88	−760
$Ni^{2+} + 2e^- \leftrightarrow Ni(s)$	−7.98	−3.99	−3.99	−236
$Pb^{2+} + 2e^- \leftrightarrow Pb(s)$	−4.27	−2.13	−2.13	−126
$Cu^{2+} + e^- \leftrightarrow Cu^+$	2.72	2.72	2.72	160
$Cu^{2+} + 2e^- \leftrightarrow Cu(s)$	11.48	5.74	5.74	339
$Fe^{3+} + e^- \leftrightarrow Fe^{2+}$	13.03	13.03	13.03	769
$Hg_2^{2+} + 2e^- \leftrightarrow 2Hg(l)$	26.91	13.46	13.46	794
$Ag^+ + e^- \leftrightarrow Ag(s)$	13.51	13.51	13.51	797
$Pb^{4+} + 2e^- \leftrightarrow Pb^{2+}$	28.64	14.32	14.32	845
$2Hg^{2+} + 2e^- \leftrightarrow Hg_2^{2+}$	30.79	15.40	15.40	908
$MnO_2(s) + 2e^- + 4H^+ \leftrightarrow Mn^{2+} + 2H_2O$	41.60	20.80	6.80	1227
$Mn^{3+} + e^- \leftrightarrow Mn^{2+}$	25.51	25.51	25.51	1505
$MnO_4^- + 5e^- + 8H^+ \leftrightarrow Mn^{2+} + 4H_2O$	127.82	25.56	14.36	1508
$Co^{3+} + e^- \leftrightarrow Co^{2+}$	33.10	33.10	33.10	1953

*For reactions included in the Mineql+ database, values are from that database. Other values are computed from the Gibbs energy of reaction or from various other reference sources. The meaning of the terms in the last three columns is discussed later in the chapter.

The situation is similar for free electrons, with the defining reaction for electron activity being a redox reaction [reaction (9.5)], rather than an acid/base reaction. An important qualitative difference between H^+ and e^- is that H^+ is relatively stable when bound to one or more water molecules, whereas e^- is not (i.e., H_3O^+ is a weakly stable species, while H_2O^- is much less stable). However, this distinction is not important in terms of how $\{H^+\}$ and $\{e^-\}$ are defined.

The electron activity is an indicator of how likely elements are to be in a reduced versus an oxidized state at equilibrium, just as the activity of H^+ is an indicator of how likely molecules are to be in a protonated (acidic) versus a deprotonated (basic) state. This point can be illustrated by manipulating Equation (9.8) to demonstrate that $\{e^-\}$ is a direct measure of the relative activities of Fe^{2+} and Fe^{3+} at equilibrium:

$$K_{(9.7)} = \frac{\{Fe^{2+}\}}{\{Fe^{3+}\}\{e^-\}} \tag{9.8}$$

$$\{e^-\} = \frac{1}{K_{(9.7)}} \frac{\{Fe^{2+}\}}{\{Fe^{3+}\}} \tag{9.9}$$

Equations (9.8) and (9.9) indicate that any ratio of oxidized to reduced iron may be present in an equilibrium solution, and that each such ratio is associated with a specific value of $\{e^-\}$; the larger the value of $\{e^-\}$, the larger the equilibrium ratio of $\{Fe^{2+}\}$ to $\{Fe^{3+}\}$. Similarly, any activity ratio of an acid and its conjugate base may exist in an equilibrium solution, and a given ratio corresponds to a specific value of $\{H^+\}$.

9.2   THE MOLAR GIBBS ENERGY OF ELECTRONS

As shown in Chapter 2, the standard Gibbs energy change of any reaction is related to the equilibrium constant for that reaction by

$$\Delta \bar{G}_r{}^\circ = -RT \ln K = -2.303RT \log K \tag{9.10}$$

When applied to the $H_2(g)/H^+$ half-cell reaction [reaction (9.5), Equation (9.6)], Equation (9.10) can be used to establish the standard Gibbs energy of electrons as follows:

$$\frac{1}{2}H_2(g) \leftrightarrow H^+ + e^- \qquad K \equiv 1.0 \tag{9.5}$$

$$\Delta \bar{G}_r{}^\circ = -RT \ln K = \bar{G}_{H^+}{}^\circ + \bar{G}_{e^-}{}^\circ - \frac{1}{2}\bar{G}_{H_2(g)}{}^\circ \tag{9.11a}$$

$$-RT \ln 1.0 = 0 + \bar{G}_{e^-}{}^\circ - \frac{1}{2}(0) \tag{9.11b}$$

$$\bar{G}_{e^-}{}^\circ = 0 \tag{9.11c}$$

The molar Gibbs energy of e^- in any solution can then be determined as

$$\bar{G}_{e^-} = \bar{G}_{e^-}^{\circ} + RT \ln\{e^-\} \tag{9.12}$$

Substituting -2.303pe for $\ln\{e^-\}$, we obtain the following simple expression for the molar Gibbs energy of electrons at 25°C.[3]

$$\bar{G}_{e^-} = -5.71\text{pe} \ \text{kJ/mol} \tag{9.13}$$

9.3 An Alternative Way of Expressing Electron Activity: The E_H Scale

The molar energy of electrons, like any value of energy, can be expressed in a variety of ways. In Chapter 2, the following equation was derived for the electrical potential energy of a solute i:

$$\overline{\overline{\text{PE}}}_{i,\text{elec}} = z_i F \Psi \tag{9.14}$$

where $\overline{\overline{\text{PE}}}_{i,\text{elec}}$ is in kilojoules per mole of i, Ψ is in volts, and F equals 96.48 (kJ/equiv of i)/V. In words, the equation indicates that the product $z_i F \Psi$ is the electrical energy, in kilojoules per mole, that a substance with molar charge z_i has when exposed to an electrical potential of Ψ volts. Although the expression is clearly applicable most directly to electrical systems, it also can be used to define a "virtual" electrical potential in cases where no such potential exists. That is, one way of thinking about Equation (9.14) is that it characterizes what the electrical potential Ψ *would have to be* at a location for a species with a charge z_i at that location to have the specified amount of electrical energy per mole.

Because of the link between electrons and electrical energy, it is common to express the molar *electrochemical* energy of electrons, $\overline{\overline{\text{PE}}}_{e^-,\text{ec}}$, in terms of equivalent electrical potentials. That is, we can express $\overline{\overline{\text{PE}}}_{e^-,\text{ec}}$ as

$$\overline{\overline{\text{PE}}}_{e^-,\text{ec}} = z_{e^-} F E_H = -F E_H \tag{9.15}$$

where E_H is analogous to Ψ in Equation (9.14); i.e., it is the voltage to which electrons would have to be subjected to have an amount of electrical energy per mole that equals the electrochemical energy per mole that the electrons really do have in the system of interest. The designation E_H has been used historically for this purpose, with the E indicating that the value is an electrical potential and the H indicating that the baseline for measuring this potential is the assignment

---

[3]According to the approach used here, the activity of free electrons in solution is defined based on reaction (9.5) and Equation (9.6), and $\bar{G}_{e^-}^{\circ} = 0$ is derived as a result of the relationship between the standard free energy of reaction and the equilibrium constant. These results are often presented in the reverse order. That is, the standard free energy of formation of electrons in solution is defined to be zero, and the equilibrium constant of 1.0 for reaction (9.5) is derived. The approaches are effectively identical, since the assignment of a value to one of the terms ($\bar{G}_{e^-}^{\circ}$ or K) leads directly to the correct value for the other.

of $E_H = 0$ in a system containing H^+ and $H_2(g)$ in their standard states. [This assignment is made in defining the scale for $\bar{G}_{e^-}$ so, according to Equation (9.16) below, it must apply to the scale for E_H as well.] E_H is referred to as the **oxidation potential** or **redox potential** of the solution.

Recall that $\overline{\overline{PE}}_{i,ec} \equiv \bar{G}_i + z_i F \Psi_{ext}$ and that, in the vast majority of systems of interest, $\Psi_{ext} = 0$. Therefore, for those systems, we can substitute $\bar{G}_i$ for $\overline{\overline{PE}}_{i,ec}$ in Equation (9.15) to obtain

$$\bar{G}_{e^-} = -FE_H \tag{9.16}$$

Equation (9.16) is frequently presented as a fundamental relationship, and throughout this chapter, we will assume that it applies (i.e., that $\Psi_{ext} = 0$) unless it is explicitly stated otherwise. However, it is important to remember that E_H is a measure of the electrochemical potential energy of electrons in a system, not just their Gibbs energy, and that in cases where a solution is subjected to an external electrical field, Equation (9.15) applies, and Equation (9.16) does not.

An additional caveat is that the use of voltage as the units for E_H can easily lead to a misperception that the parameter is referring to the actual electrical potential in the solution. It is important to recognize that this is not the case (remember that E_H is a *virtual* potential). The electrical potential is zero in any bulk solution unless an electrical field is imposed from outside the solution phase.

Combining Equation (9.16) with Equations (9.12) and (9.13), we find

$$E_H = \frac{2.303\,RT}{F}\,\text{pe} \tag{9.17}$$

and, for a system at 25°C,

$$E_H = (0.059\,\text{V})\text{pe} \tag{9.18}$$

Equations (9.13) and (9.18) indicate that when the pe of a solution at 25°C increases by 1 unit (i.e., when the electron activity decreases by a factor of 10), the Gibbs energy of the electrons in the solution decreases by 5.71 kJ/mol, and the redox potential of the solution increases by 59 mV. These simple relationships among the three parameters emphasize the fact that they are really just different ways of describing the same quantity—the chemical potential energy of exchangeable electrons in the system.

9.4   DEFINITION OF e^o AND pe^o

Just as it is convenient to define a generic term K_a to describe the equilibrium constant of certain acid/base reactions, some conventional terminology has developed for certain types of redox reactions. Specifically, e^o **is defined as the equilibrium constant for a redox half-cell reaction in which the reduced species releases one electron to form the corresponding oxidized species** [reaction (9.19)]:

$$\frac{1}{n_e}\text{Red} \leftrightarrow \frac{1}{n_e}\text{Ox} + e^- \qquad \textbf{(9.19)}$$

The generic redox reaction defined previously [reaction (9.3)] can be converted into reaction (9.19) by dividing the stoichiometric coefficients by n_e, so the value of e^o can be represented as follows:[4]

$$e^o \equiv K_{(9.19)} = K_{(9.3)}^{1/n_e} = \frac{\{\text{Ox}\}^{1/n_e}}{\{\text{Red}\}^{1/n_e}}\{e^-\} \qquad \textbf{(9.20)}$$

An equivalent definition that is sometimes used and whose validity is obvious from inspection of Equation (9.20) is that e^o **is the equilibrium value of** $\{e^-\}$ **in a system in which the activities of all other species in the reaction are 1.0,** i.e., a system in which all species other than the electron are in their standard states. For this reason, e^o is sometimes called the **standard electron activity** for the reaction.

By definition, pe^o is the negative logarithm of e^o:

$$pe^o = -\log\{e^o\} = -\log\left(\frac{\{\text{Ox}\}^{1/n_e}}{\{\text{Red}\}^{1/n_e}}\{e^-\}\right) \qquad \textbf{(9.21a)}$$

$$pe^o = \log\left(\frac{\{\text{Red}\}^{1/n_e}}{\{\text{Ox}\}^{1/n_e}}\frac{1}{\{e^-\}}\right) \qquad \textbf{(9.21b)}$$

The term in parentheses in Equation (9.21b) is the equilibrium constant for the following reduction half-cell reaction:

$$\frac{1}{n_e}\text{Ox} + e^- \leftrightarrow \frac{1}{n_e}\text{Red} \qquad \textbf{(9.22)}$$

Thus, if we wish to describe the equilibrium constant for the redox reaction in terms of pe^o rather than e^o, we can state that **pe^o, i.e., $-\log\{e^o\}$, is the base-10 logarithm of the equilibrium constant for the redox half-cell reaction written as a reduction, normalized to a one-electron transfer** [reaction (9.22)].

By analogy with the designation of e^o as the standard electron activity, pe^o is sometimes called the **standard pe** of the reaction. When expressed in terms of E_H, pe^o is designated $E_H{}^o$ and is called the **standard oxidation potential, standard redox potential,** or simply the **standard potential** of the reaction.

The above definitions of e^o and pe^o are universally accepted and are completely consistent with one another; in essence, they are two different ways of describing the same relationship. However, some confusion can arise based on the fact that e^o is most easily defined by reference to the reaction written as an oxidation, whereas pe^o is most easily defined by reference to the reaction written

---

[4]Keep in mind that the reactant and product sides of the reaction might include other species that do not undergo changes in redox state. For instance, H_2O and/or H^+ is often required to balance the reaction. In such a case, the numerator of the fraction in Equation (9.20) includes the activities of all the reaction products, and the denominator includes those of all the reactants.

as a reduction. The second definition, establishing that $pe°$ is $\log K$ for the one-electron reduction reaction, is the one that is most widely used in environmental engineering and science, and it is the one that we will rely on here. Correspondingly, although the information in a redox reaction can be shown equally well whether the reaction is written as an oxidation or a reduction, the convention is to write the reactions as reductions, just as the convention for acid/base reactions is to write them as dissociations, i.e., with the proton as a product.

According to the above definitions, the larger the value of $e°$ (and the lower the value of $pe°$), the more the oxidation reaction is favored, or equivalently, the more stable the oxidized species is, compared to the reduced species. The analogy to acid/base systems is that the lower the pK_a value, the more the dissociation reaction is favored and the more stable the dissociated species is. Put another way, just as a strong acid (characterized by a high K_a and low pK_a) has a strong tendency to give up an H^+ ion, a strong reductant (characterized by a high $e°$ and low $pe°$) has a strong tendency to give up an e^-.

The equilibrium speciation of a given redox couple in a given solution depends on the relative values of the pe (which characterizes the solution) and the $pe°$ value (which characterizes the particular redox couple). The higher the pe of the solution and the lower the $pe°$ of the couple, the more favored the oxidized species is over the reduced species. Solutions with high pe values are referred to as being strongly oxidized or oxidizing (analogous to being strongly alkaline), and solutions characterized by a low pe are referred to as being reduced or reducing. If we wish to alter the pe of a solution, we can do so by adding an oxidized or reduced species (which will increase or lower the solution pe, respectively), just as we add a base or an acid to raise or lower solution pH.

The analogies discussed above and several others between acid/base and redox equilibria are summarized in Table 9.4. The major difference between these two groups of reactions is simply in the conventions used for writing the reactions: as noted above, the convention for writing redox reactions is to show them as reductions, with the electron as a reactant, whereas the convention for acid/base reactions is to write them as dissociations, with the proton as a product.

Example 9.5

Three redox half-cell reactions are shown below, along with their $pe°$ values.

a. What are the equilibrium constants for the three reactions?
b. Of the six metal species shown, which is the strongest oxidant, and which is the strongest reductant?
c. If a solution that has attained redox equilibrium at pH 0 contains all the species shown, and if Pb^{2+} and Pb^{4+} are present at equal activities, which form of dissolved copper and which form of dissolved cobalt would you expect to be dominant?

$$Pb^{4+} + 2e^- \leftrightarrow Pb^{2+} \qquad pe° = 28.64$$
$$Cu^{2+} + e^- \leftrightarrow Cu^+ \qquad pe° = 2.72$$
$$Co^{3+} + e^- \leftrightarrow Co^{2+} \qquad pe° = 33.1$$

Table 9.4   Analogies between acid/base and redox half-cell reactions

	Redox Reactions	Acid/Base Reactions
Species being transferred	e^-	H^+
Molecules lacking e^- or H^+	Oxidized species	Base
Molecules having e^- or H^+	Reduced species	Acid
General reaction, written in conventional form	$Ox + n_e e^- \leftrightarrow Red$	$Acid \leftrightarrow base + n_H H^+$
Equilibrium constant for reaction written in conventional form	$K_{reduction} = \dfrac{\{Red\}}{\{Ox\}\{e^-\}^{n_e}}$	$K_{dissoc'n} = \dfrac{\{base\}\{H^+\}^{n_H}}{\{acid\}}$
$\log K$	$\log K = \log \dfrac{\{Red\}}{\{Ox\}} + n_e pe$	$\log K = \log \dfrac{\{base\}}{\{acid\}} - n_H pH$
Ways of defining e^o and K_a (Two equivalent ways of expressing each definition are shown; the most common way of expressing it is shown in **boldface**)	$e^o \equiv K$ for the reaction: $$\dfrac{1}{n_e} Red \leftrightarrow \dfrac{1}{n_e} Ox + e^-;$$ **pe^o = log K for the reaction:** $$\dfrac{1}{n_e} Ox + e^- \leftrightarrow \dfrac{1}{n_e} Red$$	**$K_a \equiv K$ for the reaction:** $$\dfrac{1}{n_H}(Acid) \leftrightarrow \dfrac{1}{n_H}(base) + H^+;$$ $pK_a = \log K$ for the reaction: $$\dfrac{1}{n_H}(base) + H^+ \leftrightarrow \dfrac{1}{n_H}(acid)$$
Relationship between pe^o or pK_a and $\log K$	$pe^o = \dfrac{1}{n_e} \log K_{reduction}$	$pK_a = \dfrac{1}{n_H} \log K_{deprotonation}$
Conditions at "crossover" point*	$\{Ox\} = \{Red\}$ when $pe = pe^o$	$\{Acid\} = \{base\}$ when $pH = pK_a$
Condition away from crossover point*	At $pe > pe^o$: $\{Ox\} > \{Red\}$ At $pe < pe^o$: $\{Red\} > \{Ox\}$	At $pH > pK_a$: $\{Base\} > \{acid\}$ At $pH < pK_a$: $\{Acid\} > \{base\}$
Solution condition	High pe: oxidizing Low pe: reducing	High pH: alkaline Low pH: acid

*Relationships shown apply only if the stoichiometric coefficients on Ox and Red (or acid and base) are equal (see below).

Solution

a. As indicated in Table 9.4, the value of $\log K$ for a reduction half-cell reaction is $n(pe^o)$, where n is the number of electrons transferred. Therefore, the $\log K$ values of the three reactions above are 57.28, 2.72, and 33.1, respectively.

b. The oxidants in the above reactions are the more highly charged species of each pair (Pb^{4+}, Cu^{2+}, and Co^{3+}). The strongest oxidant is the species that has the strongest tendency to acquire electrons from other species (thereby oxidizing those species). The tendency to acquire electrons increases as pe^o increases, so the strongest oxidant listed is Co^{3+}. Similarly, the strongest reductant is the reduced species in the reaction with the lowest pe^o value, i.e., Cu^+.

c. If Pb^{2+} and Pb^{4+} are present with equal activities in an equilibrium solution, then the pe of the solution must equal pe^o for the corresponding half-cell reaction, i.e., $pe = 28.64$. This value of pe is more oxidizing than pe^o for the Cu^{2+}/Cu^+ couple, so the oxidized form of copper (Cu^{2+}) is dominant. However, it is more reducing than the pe^o value for the Co^{3+}/Co^{2+} couple, so the reduced form of that couple (Co^{2+}) dominates.

A visual representation of the pe scale is shown along with the analogous pH scale in Figure 9.1. At various values along the pe scale, dominance switches from the oxidized form of the redox couple to the reduced form. For reasons that are explained later in the chapter, the crossover pe value for many conjugate redox pairs is at pe° only if the solution pH is 0.0. The pe at which dominance switches at pH 7 [designated pe°(W)] is often of interest, and this value is also shown in the diagram; the formal calculation of these values is demonstrated later in the chapter.

The information on the pH scale in Figure 9.1a can be interpreted as showing that HAc and OCl⁻ cannot both be dominant in the same solution, because HAc is dominant only at pH < 4.75 while OCl⁻ is dominant only at pH > 7.60. If we make the approximation that the interconversion of acid and base is essentially complete at pH = pK_a, then we could say further than HAc and OCl⁻ cannot coexist in any equilibrium solution.

By the same token, H_2S and HOCl cannot both be dominant redox species in an equilibrium solution at pH 0 (Figure 9.1b) or pH 7 (Figure 9.1c). Making the same approximation as above, then, we might say that these species cannot coexist at equilibrium at either pH 0 or pH 7; if these two species are added to a solution, the H_2S will be oxidized, releasing electrons to (and thereby reducing) HOCl.

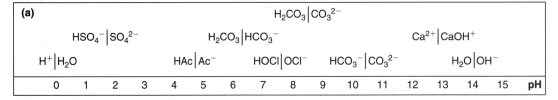

Figure 9.1   Location of various conjugate pairs on acid/base and redox scales. **(a)** pK_a values on a pH scale; **(b)** and **(c)**, pe° and pe°(W) values on a pe scale.

To generalize the above result, at pH 0.0, any oxidized species will (eventually) oxidize any reduced species that is part of a redox couple with a lower pe^o. On the other hand, oxidized species are unreactive with any reduced species that is part of a couple with a higher pe^o. An interesting conclusion from this analysis is that HOCl and OCl$^-$ should never be stable in an aqueous solution at pH 0 or 7; if such a solution is prepared, we predict that essentially all the HOCl and OCl$^-$ would oxidize water to $O_2(g)$ and would be converted to Cl$^-$ as equilibrium was approached.

The range of typical pe and pH values in a few natural environments is shown schematically in Figure 9.2. In theory, using this figure as a guideline, we could make reasonable predictions about the dominant form of any element in a given environment. In practice, the impact of pe on speciation is not always as dramatic as one would predict because of the tendency for many redox reactions to remain far from equilibrium. In fact, the speciation of many redox pairs does not change at all in response to the addition of an oxidant or reductant, even if the computed

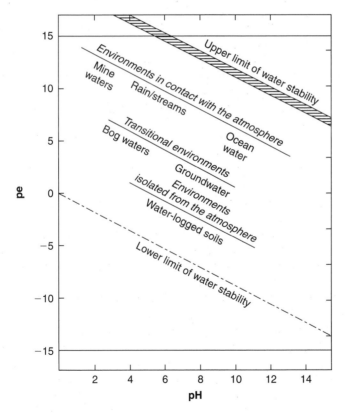

Figure 9.2   Typical pe and pH values in natural aquatic systems.

Adapted from R. Garrels and C. Christ. SOLUTIONS, MINERALS, AND EQUILIBRIA. Harper-Row (1965).

change in equilibrium speciation is large. For this reason, one must be very cautious when interpreting or applying equilibrium calculations for such systems.

Redox speciation can be represented on a log C–pe diagrams that, not surprisingly, bear many similarities to log C–pH diagrams. For instance, for a system containing $10^{-4}\,M\,TOTFe$ at a pH value that is low enough that formation of hydroxo complexes can be ignored and in which $\gamma = 1.0$ for both Fe^{2+} and Fe^{3+}, we can determine the fraction of the total Fe in the system in each oxidation state as follows:

$$\alpha_{Fe^{2+}} = \frac{\{Fe^{2+}\}}{\{Fe^{2+}\} + \{Fe^{3+}\}} = \frac{1}{1 + \{Fe^{3+}\}/\{Fe^{2+}\}} \qquad \textbf{(9.23a)}$$

$$\alpha_{Fe^{2+}} = \frac{1}{1 + e^\circ/\{e^-\}} = \frac{\{e^-\}}{\{e^-\} + e^\circ} \qquad \textbf{(9.23b)}$$

$$\alpha_{Fe^{3+}} = 1 - \alpha_{Fe^{2+}} \qquad \textbf{(9.24a)}$$

$$\alpha_{Fe^{3+}} = \frac{e^\circ}{\{e^-\} + e^\circ} = \frac{1}{1 + \{e^-\}/e^\circ} \qquad \textbf{(9.24b)}$$

Equations (9.23b) and (9.24b) are essentially identical to the equations for α_0 and α_1 for a monoprotic acid, with $\{e^-\}$ in the Fe^{3+}/Fe^{2+} system taking the place of $\{H^+\}$ in an HA/A$^-$ system and e° corresponding to K_a. By combining this result with the value for $TOTFe$, we can prepare the log C–pe diagram, which is shown in Figure 9.3. As in a log C–pH diagram, the lines for the two species have slopes that differ by 1 (the number of electrons exchanged when one species is converted to the other); extrapolation of the linear parts of the curves shows that they would intersect at the point pe $= pK_{(9.7)}$ and log $C = $ log $\{TOTFe\}$; and the curvature near this point causes the curves to intersect at pe $= pK_{(9.7)}$ and log $C = $ log $\{TOTFe\} - 0.3$.

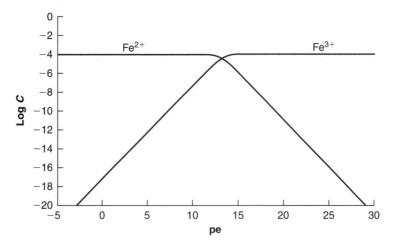

Figure 9.3   Log C–pe diagram for a solution containing $10^{-4}\,M\,TOTFe$, considering only free metal ions as possible dissolved species.

9.4.1 COMPUTING PE FROM SPECIES ACTIVITIES: THE NERNST EQUATION

Manipulation of Equation (9.21b) provides a way to calculate pe in any equilibrium solution of known composition if the pe^o values of redox-active couples in the system are available. Specifically, we can rearrange the equation as follows:

$$pe^o = \log\left(\frac{\{Red\}^{1/n_e}}{\{Ox\}^{1/n_e}} \frac{1}{\{e^-\}}\right) \qquad (9.21b)$$

$$pe^o = \log\frac{\{Red\}^{1/n_e}}{\{Ox\}^{1/n_e}} + pe \qquad (9.25)$$

$$pe = pe^o - \frac{1}{n_e}\log\frac{\{Red\}}{\{Ox\}} \qquad (9.26a)$$

$$E_H = E_H^{\,o} - \frac{2.303RT}{n_e F}\log\frac{\{Red\}}{\{Ox\}} \qquad (9.26b)$$

Equation (9.26), in either of the forms shown, is known as the **Nernst equation** and is one of the fundamental equations of redox analysis. The Nernst equation can be used to determine the equilibrium pe (or E_H) of a redox couple under any conditions, whether the constituents are in their standard states or not. The relationship expressed by the Nernst equation between the electron activity and the equilibrium redox speciation is shown schematically in Figure 9.4 for three redox couples [HOCl/Cl$^-$, SO$_4^{2-}$/H$_2$S(aq), and H$^+$/H$_2$(g)].

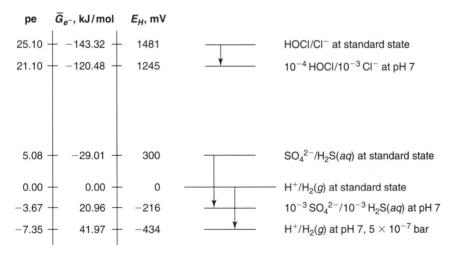

pe	$\bar{G}_{e^-}$, kJ/mol	E_H, mV		
25.10	−143.32	1481		HOCl/Cl$^-$ at standard state
21.10	−120.48	1245		10^{-4} HOCl/10^{-3} Cl$^-$ at pH 7
5.08	−29.01	300		SO$_4^{2-}$/H$_2$S(aq) at standard state
0.00	0.00	0		H$^+$/H$_2$(g) at standard state
−3.67	20.96	−216		10^{-3} SO$_4^{2-}$/10^{-3} H$_2$S(aq) at pH 7
−7.35	41.97	−434		H$^+$/H$_2$(g) at pH 7, 5×10^{-7} bar

Figure 9.4   Schematic of the relationship between pe and solution composition expressed by the Nernst equation. The arrows indicate the effect of changing solution composition from the standard state to the non-standard state conditions shown for each redox couple.

| **Example 9.6** | A solution at pH 7.0 contains Mn^{2+} at an activity of 10^{-5}, as well as some $MnO_2(s)$. The redox half-cell reaction for the $Mn^{2+}/MnO_2(s)$ reaction is shown below. What is the pe of the solution, assuming the system is at equilibrium? |

$$MnO_2(s) + 4H^+ + 2e^- \leftrightarrow Mn^{2+} + 2H_2O \qquad pe^\circ = 20.8$$

Solution

Using the Nernst equation, and noting that the activity of solid MnO_2 is 1.0, we compute the pe as follows:

$$pe = pe^\circ - \frac{1}{2} \log \frac{\{Mn^{2+}\}\{H_2O\}^2}{\{MnO_2(s)\}\{H^+\}^4}$$

$$pe = 20.8 - \frac{1}{2} \log \frac{(10^{-5.0})(1.0)^2}{(1.0)(10^{-7.0})^4} = +9.3$$

| **Example 9.7** | The potentially fatal effects of massive arsenic ingestion are well known, but in recent years the effects of long-term, low-level exposures have been hotly debated. As a conservative response to the heightened concern, the U.S. Environmental Protection Agency (EPA) is expected to lower the maximum contaminant level (MCL) for arsenic in drinking water below its current (year 2000) level of 50 $\mu g/L$, by approximately an order of magnitude. Arsenic occurs naturally primarily in two oxidation states: As(III) and As(V). It is acidic in both oxidation states, but much more so as As(V), in which case it is a chemical analog of phosphoric acid [i.e., P(V)]. |

The pK_a values of arsenous acid, H_3AsO_3, and arsenic acid, H_3AsO_4, are shown below and indicate that most H_3AsO_3 remains fully protonated at pH 7, whereas H_3AsO_4 is extensively deprotonated. Because of this difference in ionic charge on the dominant species, it is much easier to remove As(V) than As(III) from solution by adsorption and ion-exchange processes that might be employed at a water treatment plant.

	pK_{a1}	pK_{a2}	pK_{a3}
H_3AsO_3	9.23	12.10	13.41
H_3AsO_4	2.24	6.76	11.60

Workers at a water treatment plant want to ensure that >99% of the *TOT*As in the water is in the oxidized form prior to feeding it to an ion-exchange process. The pH of the water is 6.7. Assuming the arsenic equilibrates rapidly in response to solution pe, what is the minimum target pe that they should shoot for?

$$AsO_4{}^{3-} + 2H^+ + 2e^- \leftrightarrow AsO_3{}^{3-} + H_2O \qquad pe^\circ = 2.64$$

Solution

We could use the Nernst equation in conjunction with the above reaction to determine the pe at which the activity of $AsO_4{}^{3-}$ is at least 99 times as large as that of $AsO_3{}^{3-}$. However, that information would do us little good, since neither of the fully deprotonated species represents a significant fraction of *TOT*As. Since the pH of the water is well below pK_{a1}

for H_3AsO_3, we can assume that essentially all the As(III) is present as fully protonated arsenous acid. On the other hand, the pH is near pK_{a2} for H_3AsO_4, so the As(V) will be distributed approximately equally as $H_2AsO_4^-$ and $HAsO_4^{2-}$. As a result, $TOTAs(V)$ can be approximated as either $2\{H_2AsO_4^-\}$ or $2\{HAsO_4^{2-}\}$, and the target pe will be such that $\{H_2AsO_4^-\} \approx \{HAsO_4^{2-}\} \approx 50\{H_3AsO_4\}$.

To find the target pe, it is easiest to convert the redox reaction given in the problem statement into one that relates dominant species. Since $\{HAsO_4^{2-}\} \approx \{H_2AsO_4^-\}$, we can treat either of these species as dominant. If we choose $HAsO_4^{2-}$, the conversion into the desired reaction is as follows:

$$AsO_4^{3-} + 2H^+ + 2e^- \leftrightarrow AsO_3^{3-} + H_2O \qquad \log K = 2pe^o = 5.28$$

$$HAsO_4^{2-} \leftrightarrow AsO_4^{3-} + H^+ \qquad \log K = \log K_{a3} = -11.60$$

$$\underline{AsO_3^{3-} + 3H^+ \leftrightarrow H_3AsO_3 \qquad \log K = \log (K_{a1}K_{a2}K_{a3})^{-1} = 34.74}$$

$$HAsO_4^{2-} + 4H^+ + 2e^- \leftrightarrow H_3AsO_3 + H_2O \qquad \log K = 28.42 \qquad pe^o = \frac{\log K}{2} = 14.22$$

We can now apply the Nernst equation to the reaction between the dominant species to determine the target pe:

$$pe = pe^o - \frac{1}{n_{e^-}} \log \frac{\{Red\}}{\{Ox\}} = 14.22 - \frac{1}{2} \log \left(\frac{\{H_3AsO_3\}\ \{H_2O\}}{\{HAsO_4^{2-}\}\ \{H^+\}^4} \right)$$

$$= 14.22 - \frac{1}{2} \log \left[\left(\frac{1}{50}\right) \frac{1.0}{(10^{-6.7})^4} \right] = 1.67$$

9.4.2 USING PE^o VALUES TO DETERMINE OVERALL REDOX EQUILIBRIUM CONSTANTS

The definition of pe^o is particularly useful for quickly computing the value of the equilibrium constant for an overall redox reaction composed of any two half-cell reactions. For instance, say we wanted to write the balanced overall reaction and determine the equilibrium constant for oxidation of H_2S to SO_4^{2-} by reduction of HOCl to Cl$^-$. This reaction is sometimes used to convert S($-$II) species to S(VI) species as sewage enters a wastewater treatment plant, both to reduce odors and to prevent toxic S($-$II) species from entering the biologically active reactor.

Since the half-cell reaction corresponding to pe^o is always written for a one-electron reduction, adding the "pe^o reaction" involving the species being reduced (HOCl) to the opposite of the pe^o reaction for the species being oxidized yields a balanced overall reaction from which the free electron species is eliminated. Furthermore, since pe^o equals $\log K$ for a one-electron reduction, $\log K$ for the overall reaction is the sum of pe^o for the redox couple being reduced and $-pe^o$ for the couple being oxidized. For example, for the reaction of HOCl with H_2S, summation of the pe^o reaction for SO_4^{2-}/H_2S and the reverse of the pe^o

reaction for HOCl/Cl⁻ yields the following overall reaction and equilibrium constant:

$$\frac{1}{8}H_2S + \frac{1}{2}H_2O \leftrightarrow \frac{1}{8}SO_4^{2-} + \frac{5}{4}H^+ + e^- \qquad \log K = -pe_{SO_4^{2-}/H_2S}^o = -5.08$$

$$\frac{1}{2}HOCl + \frac{1}{2}H^+ + e^- \leftrightarrow \frac{1}{2}Cl^- + \frac{1}{2}H_2O \qquad \log K = pe_{HOCl/Cl^-}^o = 25.1$$

$$\frac{1}{8}H_2S + \frac{1}{2}HOCl \leftrightarrow \frac{1}{2}Cl^- + \frac{1}{8}SO_4^{2-} + \frac{3}{4}H^+ \qquad \log K = pe_{HOCl/Cl^-}^o - pe_{SO_4^{2-}/H_2S}^o = 20.02$$

Although the electron does not appear explicitly as either a reactant or a product, the overall redox reaction obtained from this procedure always characterizes a one-electron transfer. Because of the way the reaction is generated, the value of log K for the overall reaction is often designated pe°; to minimize confusion with pe° values for redox half-cell reactions, in this text, the pe° value for overall reactions will be referred to as $pe_{Ox/Red}^o$, where Ox and Red are the oxidant and reductant, respectively.[5]

Example 9.8 | Using the half-cell reactions shown in Table 9.3, write balanced reactions and determine pe° values for the following overall reactions:

    a.   i. Oxidation of $NH_3(aq)$ to NO_3^- by reduction of $O_2(g)$ to H_2O.
           ii. Oxidation of cyanide (CN^-) to cyanate (OCN^-) by reduction of $Cl_2(aq)$ to Cl^-.
          iii. Oxidation of Fe^{2+} to Fe^{3+} by reduction of H^+ to $H_2(g)$.
    b. What is the equilibrium constant for the following reaction?

$$NH_3(aq) + 2O_2(aq) \leftrightarrow NO_3^- + H_2O + H^+$$

Solution

    a. Each overall reaction can be obtained by adding the half-cell reaction of the substance being reduced to the reverse of the half-cell reaction of the substance being oxidized. The results are shown below.

       i. $\frac{1}{4}O_2(g) + H^+ + e^- \leftrightarrow \frac{1}{2}H_2O$ $\log K = pe^o = 20.78$

           $\frac{1}{8}NH_3(aq) + \frac{3}{8}H_2O \leftrightarrow \frac{1}{8}NO_3^- + \frac{9}{8}H^+ + e^-$ $\log K = -pe^o = -13.73$

           $\frac{1}{8}NH_3(aq) + \frac{1}{4}O_2(g) \leftrightarrow \frac{1}{8}NO_3^- + \frac{1}{8}H_2O + \frac{1}{8}H^+$ $\log K = pe_{O_2/NH_3}^o = 7.05$

---

[5]Although use of the same symbol (pe°) to represent the logarithm of the equilibrium constant for both a one-electron reduction half-cell reaction and a balanced reaction from which free electrons have been eliminated could be confusing, the meaning of the term is usually obvious from context.

ii.

$$\frac{1}{2}Cl_2\,(aq) + e^- \leftrightarrow Cl^- \qquad\qquad \log K = pe^o = 23.60$$

$$\frac{1}{2}CN^- + OH^- \leftrightarrow \frac{1}{2}CNO^- + \frac{1}{2}H_2O + e^- \qquad \log K = -pe^o = 16.44$$

$$\frac{1}{2}Cl_2(g) + \frac{1}{2}CN^- + OH^- \leftrightarrow Cl^- + \frac{1}{2}OCN^- + \frac{1}{2}H_2O \qquad \log K = pe^o_{Cl_2/CN^-} = 40.04$$

iii. $\quad H^+ + e^- \leftrightarrow \frac{1}{2}H_2(g) \qquad\qquad\qquad\qquad \log K = pe^o = 0.00$

$$Fe^{2+} \leftrightarrow Fe^{3+} + e^- \qquad\qquad\qquad\qquad \log K = -pe^o = -13.07$$

$$H^+ + Fe^{2+} \leftrightarrow \frac{1}{2}H_2(g) + Fe^{3+} \qquad\qquad \log K = pe^o_{H^+/Fe^{2+}} = -13.03$$

b. The reaction of interest is the same as that in part $(a)(i)$, except that in this part of the question all the stoichiometric coefficients in $(a)(i)$ are multiplied by 8; i.e., we want to determine $\log K$ for an eight-electron transfer. By multiplying the reaction in part $(a)(i)$ by 8, we obtain

$$NH_3(aq) + 2O_2(g) \leftrightarrow NO_3^- + 2H_2O + H^+ \qquad \log K = 8pe^o_{O_2/NH_3} = 56.40$$

Note that the reaction in part $(a)(ii)$ of the above example can be modified to replace OH^- by H^+ by adding the water dissociation reaction to the overall redox reaction, yielding

$$\frac{1}{2}Cl_2(aq) + \frac{1}{2}CN^- + \frac{1}{2}H_2O \leftrightarrow Cl^- + \frac{1}{2}OCN^- + H^+$$

$$\log K = pe^o_{Cl_2/CN^-} = 26.04$$

This modification changes the overall reaction and therefore changes the values of pe^o and $\log K$, even though it does not change the stoichiometry of any of the molecules that are undergoing oxidation or reduction. Because of situations like this, it is important to fully describe any redox reaction under discussion; identification of only the species undergoing oxidation or reduction can sometimes be insufficient to describe the reaction unambiguously.

9.4.3 REDOX REACTIONS INVOLVING EXCHANGE OF BOTH ELECTRONS AND PROTONS; DEFINITION OF $PE_{PH}{}^o$

Often, the acidity of an oxidant differs from that of its conjugate reductant; the most common situation is for oxidation to increase the acidity of the molecule, causing protons to be released along with electrons in the oxidation half-cell reaction. In such cases, the speciation in the system depends not only on solution pe, but also on pH.

For instance, consider a generic half-cell reaction in which n_H protons and n_e electrons combine with the oxidized species to form the reduced species

$$Ox + n_H H^+ + n_e e^- \leftrightarrow Red \qquad\qquad \textbf{(9.27)}$$

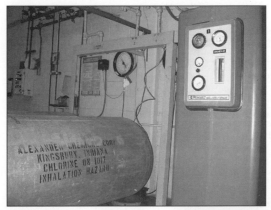

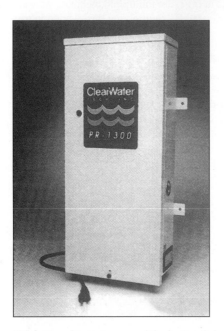

Disinfection of water is accomplished primarily by addition of oxidants to the water that react with critical components of microorganisms to kill them. Chlorine-based disinfection is most common at large treatment plants in the United States, but ozone is also used in those situations and for smaller on-site disinfection operations. Iodine is a convenient disinfectant for use by individuals in remote areas.

| Top: Paul R. Anderson; Bottom left: Clearwater Tech; Bottom right: Richard Thom/Visuals Unlimited.

For reaction (9.27), pe° equals $1/n_e$ times the logarithm of the equilibrium constant for the reaction. Writing out that equation and then rearranging it slightly, we obtain

$$\text{pe}^\circ = \frac{1}{n_e} \log K_{(9.27)} = \frac{1}{n_e} \log \frac{\{\text{Red}\}}{\{\text{Ox}\}} + \frac{n_H}{n_e} \text{pH} + \frac{n_e}{n_e} \text{pe} \quad \textbf{(9.28a)}$$

$$\frac{1}{n_e} \log \frac{\{\text{Red}\}}{\{\text{Ox}\}} = \text{pe}^\circ - \text{pe} - \frac{n_H}{n_e} \text{pH} \quad \textbf{(9.28b)}$$

Equation (9.28b) indicates that the larger the value of pe, the more the oxidized species is favored over the reduced species, consistent with the trends described in Table 9.4. However, we cannot conclude that the activities of the oxidized and reduced species are equal at pe = pe^o, because the right-hand side of the equation depends on solution pH as well as pe. In fact, the equation shows that dominance changes from the reduced to the oxidized species at pe = pe^o only if pH = 0; in solutions at other pH values, pe^o has no special significance with respect to species dominance.

The relationships shown earlier in the chapter for redox couples that are not pH-dependent (i.e., that {Ox} = {Red} at pe = pe^o, and that the oxidized species is dominant when pe > pe^o and the reduced species is dominant when pe < pe^o) provide simple and convenient guidelines for organizing redox reactions mentally and for ranking the relative tendencies for different redox reactions to proceed. Unfortunately, the complexity injected into these relationships when the reactions are pH-dependent is unavoidable. However, some of that complexity can be reduced by defining a new parameter, applicable when the pH of a system is known. This parameter, which we will designate pe$_{pH}^o$, is defined mathematically as follows:

$$pe_{pH}^o \equiv pe^o - \frac{n_H}{n_e} pH \qquad (9.29)$$

Substituting this expression into Equation (9.28b), we obtain

$$\frac{1}{n_e} \log \frac{\{Red\}}{\{Ox\}} = pe_{pH}^o - pe \qquad (9.30)$$

pe$_{pH}^o$ can be viewed as a conditional pe^o that is specific to the given pH. The usefulness of pe$_{pH}^o$ is that it tells us directly the pe at which dominance changes from the oxidized to the reduced species if the system is at the specified pH. By far the most common pH value used when reporting pe$_{pH}^o$ values is 7.0; pe$_{7.0}^o$ is frequently represented as pe^o(W), with the W signifying (neutral) water. Thus, for the SO_4^{2-}/SO_3^{2-} couple (pe^o = 13.58), pe^o(W) is 13.58 − (2/2)(7), or 6.58. From this value, we know that in a solution at pH 7, dominance switches between {SO_3^-} and {SO_4^-} at pe = 6.58.

By comparing pe^o(W) values of various redox couples, we can infer the relative stabilities of the oxidized and reduced species in those couples under conditions that are typical of natural systems and many engineered water or wastewater treatment systems. Specifically, the higher the value of pe^o(W), the more strongly oxidizing the redox couple is at pH 7.0 and the more stable the reduced species is compared to the oxidized species. For this reason, equilibrium constants for redox reactions are often reported in environmental engineering and science as pe^o(W) values, rather than pe^o values. When pe^o(W) values are given, they are usually for reactions written involving the dominant form of the redox-active species, so that pe^o(W) corresponds to the pe where the majority of the redox-active element changes from its oxidized to its reduced state. This point is illustrated in the following example.

Example 9.9

The log C–pe diagram for the Fe(III)/Fe(II) system at a pH where hydrolysis is negligible was shown in Figure 9.3, based on the pe° value of $+13.03$ for the Fe^{3+}/Fe^{2+} couple. Compute $pe^\circ(W)$ for this couple and also for the redox couple comprised of the dominant Fe(III) and Fe(II) species at pH 7, which are $Fe(OH)_2^+$ and Fe^{2+}, respectively. Also, develop the log C–pe diagram for a pH 7 solution containing 10^{-4} M *TOT*Fe and relate it to the two $pe^\circ(W)$ values that you determined. Hydrolysis constants for Fe^{2+} and Fe^{3+} are as follows:

	p*$K_{OH,1}$	p*$K_{OH,2}$	p*$K_{OH,3}$	p*$K_{OH,4}$
Fe^{2+}	9.50	11.07	10.43	
Fe^{3+}	2.19	3.48	7.93	8.00

Solution

The redox reaction for the Fe^{3+}/Fe^{2+} couple does not involve transfer of protons ($n_H = 0$), so $pe^\circ(W)$ is the same as pe°; i.e., it is 13.03. On the other hand, the redox reaction for the $Fe(OH)_2^+/Fe^{2+}$ reaction does involve release of H^+ when the Fe^{2+} is oxidized. The reaction and corresponding equilibrium constant are shown below, based on summation of the Fe^{3+}/Fe^{2+} reaction and the reactions for hydrolysis of Fe^{3+}:

$$Fe^{3+} + e^- \leftrightarrow Fe^{2+} \qquad\qquad \log K = pe^\circ_{Fe^{3+}/Fe^{2+}} = 13.03$$

$$\underline{Fe(OH)_2^+ + 2H^+ \leftrightarrow Fe^{3+} + 2H_2O \qquad\qquad \log {}^*\beta_{OH,2}^{-1} = 5.67}$$

$$Fe(OH)_2^+ + 2H^+ + e^- \leftrightarrow Fe^{2+} + 2H_2O \qquad \log K = pe^\circ_{Fe(OH)_2^+/Fe^{2+}} = 18.70$$

Applying Equation (9.29) to this result, we find

$$pe^\circ_{Fe(OH)_2^+/Fe^{2+}}(W) = 18.70 - \frac{2}{1}(7.0) = 4.70$$

The log C–pe diagram for pH $= 7.0$ is shown below. Consistent with the above calculations and considerations, $\{Fe^{3+}\} = \{Fe^{2+}\}$ at pe $= 13.03$, just as in Figure 9.3. However, that is not the pe at which dominance switches between Fe(II) and Fe(III)

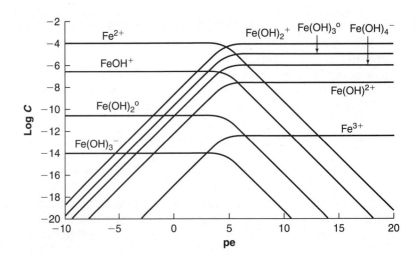

species. Rather, that pe is 4.70, i.e., pe^o(W) for the reaction between the dominant species in each oxidation state. Note how grossly wrong we would have been about the concentrations of TOT[Fe(II)] and TOT[Fe(III)] at, say, pe = 10 if we had considered only the reactions between the unhydrolyzed species: at that pe, $\{Fe^{2+}\} = 10^{3.03}\{Fe^{3+}\}$, but TOTFe(II) = $10^{-5.30}$ TOTFe(III).

When a system is at pH 7 and the standard pe is expressed as pe^o(W), the system pe is sometimes represented as pe(W), and the Nernst equation can be written as pe(W) = pe^o(W) − $(1/n_e)$log($\{Red\}/\{Ox\}$). In that case, it is important to recognize that $\{H^+\}$ is not included in the logarithmic argument even if H^+ appears in the reaction, because its effect on pe has already been accounted for in the conversion of pe^o to pe^o(W). The value of pe(W), i.e., the pe value of a real solution at pH 7.0, is always the same as that of pe; the inclusion of the W is simply a reminder that the solution is at pH 7.0.

In some cases, especially when we are interested in the redox behavior of a metal that is very strongly complexed or chelated, it is convenient to work with the pe^o values of the reaction between the dominant species in solution, rather than the pe^o of the uncomplexed metal ions. This approach has the same advantage as illustrated in the preceding example; i.e., it allows us to assess directly the pe value at which most of the metal shifts between its oxidized and reduced states in the given system. This point is illustrated in the following example.

Example 9.10

Both Fe(II) and Fe(III) form extremely strong complexes with cyanide ion (CN$^-$). As a result, the dominant Fe species in solutions containing cyanide are often $Fe(CN)_6^{4-}$ (ferrocyanide) and/or $Fe(CN)_6^{3-}$ (ferricyanide), respectively. The stability constants for formation of these complexes (β_6 values) are $10^{45.61}$ for $Fe(CN)_6^{4-}$ and $10^{52.63}$ for $Fe(CN)_6^{3-}$. In a solution in which $Fe(CN)_6^{4-}$ and $Fe(CN)_6^{3-}$ are expected to be the dominant soluble Fe species, at what pe would most of the Fe(III) in solution be reduced to Fe(II)?

Solution

An alternative way of asking the question is: What is the pe^o value for the $Fe(CN)_6^{3-}/Fe(CN)_6^{4-}$ redox couple? To determine this value, we simply combine the Fe^{3+}/Fe^{2+} half-cell reaction with the two complexation reactions, as follows:

Reaction	Log K
$Fe^{3+} + e^- \leftrightarrow Fe^{2+}$	13.03
$Fe^{2+} + 6CN^- \leftrightarrow Fe(CN)_6^{4-}$	45.61
$Fe(CN)_6^{3-} \leftrightarrow Fe^{3+} + 6CN^-$	−52.63
$Fe(CN)_6^{3-} + e^- \leftrightarrow Fe(CN)_6^{4-}$	6.01

Because the final reaction shown is a one-electron reduction, the log K value is pe^o. Therefore, in the given solution, most of the $Fe(CN)_6^{3-}$ would be converted to $Fe(CN)_6^{4-}$, and the dominant oxidation state of Fe would shift from III to II, when the pe dropped below 6.01.

9.5 ENERGY CHANGES ACCOMPANYING REDOX REACTIONS

The fact that redox reactions might proceed at imperceptible rates even under conditions that are extremely far from equilibrium has been noted a few times in this chapter. Nevertheless, it is often possible to induce the reactions to proceed rapidly by changing the conditions slightly. Thus, for instance, coal that is stable indefinitely at room temperature burns readily when ignited. Also, by changing the redox environment, we can often induce a change in the opposite direction from the one that would otherwise occur. For instance, the natural tendency of most zero-valent metals is to oxidize (corrode) in the normal atmosphere. However, by appropriate manipulation of the pe of the system (either chemically or by perturbing the system electrically), we can reduce oxidized metals to their zero-valent state, thereby preparing materials of much greater value. In this section, we explore more thoroughly the thermodynamics of such reactions.

9.5.1 THE GIBBS ENERGY CHANGE ACCOMPANYING REDOX REACTIONS

The standard Gibbs energy change of any reaction is related to the equilibrium constant for that reaction by

$$\Delta \overline{G}_r^{\,\circ} = -RT \ln K = -2.303 RT \log K \qquad (9.31)$$

As noted above, the $\log K$ value for a balanced redox reaction involving transfer of one electron can be obtained as the difference between the pe° values of the reduction and oxidation half-cell reactions. Correspondingly, $\log K$ for an n_e-electron transfer is just n_e times this difference

$$\log K = n_e \, \Delta pe^\circ \qquad (9.32)$$

where $\Delta pe^\circ = pe^\circ_{red} - pe^\circ_{ox}$, and pe°_{red} and pe°_{ox} are the pe° values of the reduction and oxidation half-cell reactions, respectively.

Combining Equations (9.31) and (9.32), we can write the standard molar Gibbs energy of a redox reaction as

$$\Delta \overline{G}_r^{\,\circ} = -2.303 n_e RT \, \Delta pe^\circ \qquad (9.33)$$

and at 25°C

$$\Delta \overline{G}_r^{\,\circ} = -(5.71 n_e \, \Delta pe^\circ) \qquad \text{kJ/mol} \qquad (9.34)$$

Equations (9.33) and (9.34) can be converted to an expression that applies under any conditions (i.e., not just standard state conditions) by substituting $\Delta \overline{G}_r - RT \ln Q$ for $\Delta \overline{G}_r^{\,\circ}$ on the left and using the Nernst equation to express Δpe° as $\Delta pe - RT \ln Q$ on the right. The result is a simple relationship between the molar Gibbs energy of reaction and the difference in pe values of the half-cell reactions:

$$\Delta \overline{G}_r = -2.303 n_e RT \, \Delta pe \qquad\qquad (9.35)$$

and, at 25°C,

$$\Delta \overline{G}_r = -(5.71 n_e \, \Delta pe) \qquad \text{kJ/mol} \qquad (9.36)$$

where $\Delta pe = pe_{red} - pe_{ox}$.

Equations (9.35) and (9.36) indicate that the molar Gibbs energy of a reaction is proportional to the difference in the pe values associated with the two half-cell reactions. On one level, this result makes intuitive sense: as we saw earlier, pe is a direct measure of the molar Gibbs energy of electrons in a system, so the difference in pe between two systems indicates the difference in electron energy levels between them. If a path is available for electron transfer between the systems, we expect electrons to transfer from the lower to the higher pe (higher to lower $\overline{G}_{e^-}$) system, releasing a corresponding amount of Gibbs energy as they do. And, if the two "systems" are different redox couples in a given solution, the transfer of electrons will oxidize the couple at lower pe and reduce the couple at higher pe. The process will continue until the pe values of the two couples are equalized, at which point the driving force for the transfer will have dissipated. Though these reactions might not proceed rapidly, they certainly will occur as the system approaches equilibrium.

The scenario described above is, in essence, just a restatement of the conclusion we reached earlier when developing the concept of the pe scale. In that analysis, the relative electron affinities of different redox pairs were compared based on pe^o values, with pe^o interpreted as an equilibrium constant. Then the dominant species of each redox pair at a given pe was determined, and the direction of electron transfer between redox couples was inferred. Here, the same idea is derived by considering electron energy levels rather than the electron affinity implied by the equilibrium constant.

The only difficulty with the preceding description is the idea that different redox couples, all in the same solution, can have different pe values. Certainly a well-mixed solution has only one pH, one temperature, one Cl^- activity, etc.; so how can it have more than one pe value? The answer is that systems can be characterized by a uniform value of some variable only if they are in equilibrium with respect to that variable. For instance, a solution at thermal equilibrium has a single temperature, but not all solutions are at thermal equilibrium. If temperature varies from one location to another in a solution, then the question "What is the temperature of the solution?" has no answer. One might compute the theoretical temperature of the solution once it equilibrates or give a long answer describing the temperature as a function of location, but in either case the original question would be meaningless.

The same situation applies to a redox system that is not at equilibrium, except that the variation is from one redox couple to the next, rather than from one location to another. The pe of a given redox couple can be defined as the pe that would cause the reaction to be at equilibrium, for the extant activities of the oxidized species, reduced species, H^+, and any other chemicals in the reaction. For instance, if the activities of HOCl and Cl^- were 10^{-4} and 10^{-3}, respectively, in a pH 7 solution, we could compute the pe of the HOCl/Cl^- redox couple from the Nernst equation:

$$pe = pe^\circ - \frac{1}{2}\log\frac{\{Cl^-\}}{\{HOCl\}\{H^+\}}$$

$$pe = 25.1 - \frac{1}{2}\log\frac{10^{-3}}{(10^{-4})(10^{-7})} = 21.1$$

If the activities of both $H_2S(aq)$ and SO_4^{2-} were 10^{-3} in the same solution, the pe of that couple could be computed as well. Since the pH of the solution is 7.0, we can use the value of $pe^\circ(W)$ (equal to -3.67) for this calculation:

$$pe^\circ(W) = -3.67 = \frac{1}{8}\log\frac{\{H_2S(aq)\}}{\{SO_4^{2-}\}} + pe$$

$$pe = -3.67 - \frac{1}{8}\log\frac{10^{-3}}{10^{-3}} = -3.67$$

These calculations indicate that the pe of the solution would have to be 21.10 for the HOCl/Cl^- couple to be in equilibrium and -3.67 for the SO_4^{2-}/H_2S couple to be in equilibrium. (Note that these are the values shown in Figure 9.4.) Clearly, these activities of S-containing and Cl-containing species could not coexist in a solution at equilibrium, any more than different temperatures could be present in a solution at thermal equilibrium. Nevertheless, the specified chemical composition could characterize a non-equilibrated system. In such a case, we might be interested in predicting the equilibrium composition of the system (i.e., what reaction will occur, and to what extent, as the system approaches equilibrium) and in calculating the energy change that would accompany the predicted reactions.

Assuming that the system is at 25°C, we can apply the relationship $\bar{G}_{e^-} = -(5.71pe)$ kJ/mol to compute the Gibbs energy of electrons in the HOCl/Cl^- and SO_4^{2-}/H_2S couples to be -120.48 and 20.96 kJ/mol, respectively. Thus, the electrons associated with the SO_4^{2-}/H_2S couple are at substantially higher energy than those associated with the HOCl/Cl^- couple, and Gibbs energy (that is to say, chemical potential energy) could be released if they transferred from the former to the latter. This, then, is the direction in which we expect the reaction to proceed. We can even infer the quantitative result from this analysis: for each mole of electrons transferred, $20.96 - (-120.48) = 141.44$ kJ of Gibbs energy will be released. The overall process is shown schematically in Figure 9.5. Note that, in all these calculations, the energy is quantified *per mole of e^- transferred*; if we were interested in the Gibbs energy change per mole of stoichiometric reaction, we would need to multiply the value -141.44 kJ/mol by n_e.

Assuming that the acid/base reactions in the system (HOCl/OCl^- and H_2S/HS^-) are in equilibrium, each acid is at the same Gibbs energy level as a

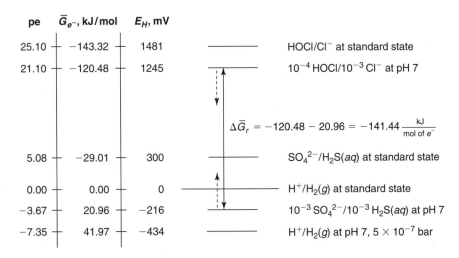

Figure 9.5   Revision of Figure 9.4 showing the molar Gibbs energy of reaction for oxidation of $H_2S(aq)$ by HOCl under the conditions shown. The dashed arrows indicate the direction of change as the reaction proceeds (see text).

combination of its conjugate base and H^+. As a result, $\Delta \overline{G}_r$ for the redox reaction is the same regardless of which of the acid/base species participates. To convince yourself of this point, try computing Δpe for the same solution, but base the analysis on the OCl^-/Cl^- and HS^-/SO_4^{2-} redox couples instead of the $HOCl/Cl^-$ and H_2S/SO_4^{2-} couples.

The Gibbs energy change of the reaction described above could also be derived by combining the half-cell reactions and applying the relationship $\Delta \overline{G}_r = RT \ln(Q/K)$ to the overall reaction. The steps involved in that process are shown below. However, dealing with the two half-cell reactions independently and applying Equation (9.35) or (9.36) to solve for $\Delta \overline{G}_r$ is often easier and more convenient.

$$\frac{1}{8}H_2S(aq) + \frac{1}{2}H_2O \leftrightarrow \frac{1}{8}SO_4^{2-} + \frac{5}{4}H^+ + e^- \qquad \log K = -5.08 \quad \textbf{(9.37)}$$

$$\frac{1}{2}HOCl + \frac{1}{2}H^+ + e^- \leftrightarrow \frac{1}{2}Cl^- + \frac{1}{2}H_2O \qquad \log K = 25.10 \quad \textbf{(9.38)}$$

$$\frac{1}{8}H_2S(aq) + \frac{1}{2}HOCl \leftrightarrow \frac{1}{2}Cl^- + \frac{1}{8}SO_4^{2-} + \frac{3}{4}H^+ \qquad \log K = 20.02 \quad \textbf{(9.39)}$$

$$\Delta \overline{G}_r = RT \ln \frac{Q}{K} = \left(5.71 \frac{kJ}{mol}\right) \log \frac{Q}{K}$$

$$\Delta \overline{G}_r = \left(5.71 \frac{kJ}{mol}\right) \log \frac{\{SO_4^{2-}\}^{1/8}\{Cl^-\}^{1/2}\{H^+\}^{3/4}/(\{H_2S(aq)\}^{1/8}\{HOCl\}^{1/2})}{10^{20.02}}$$

$$= -141.44 \frac{kJ}{mol} \qquad \qquad \textbf{(9.40)}$$

9.5.2 RELATING ΔG, PE, AND E_H TO REDOX EQUILIBRIUM CONSTANTS AND THE EXTENT OF DISEQUILIBRIUM

The preceding sections have introduced a number of relationships among $\overline{G}_{e^-}$, pe, and E_H. When these relationships are combined with expressions for K and Q, a truly dizzying array of equations can be developed, each of them useful under slightly different conditions. Several of these relationships are collected in Table 9.5.

In the table, the first row describes the energy level of electrons in the system. This energy can also be interpreted via the Nernst equation as the energy level of

Table 9.5 Conversion equations among parameters that describe electron energy levels for a half-cell reaction and energy changes accompanying an n-electron reduction reaction*

1	Energy of exchangeable electrons in a system	$\overline{G}_{e^-} = \overline{G}_{e^-}^\circ + RT \ln\{e^-\} = -2.303RT\,\text{pe} = -FE_H$
2	Nernst equation: energy of exchangeable electrons associated with a given half-cell reaction	$\text{pe} = \text{pe}^\circ - \dfrac{1}{n_e} \log \dfrac{\{Red\}}{\{Ox\}}$ $E_H = E_H^\circ - \dfrac{2.303RT}{n_e F} \log \dfrac{\{Red\}}{\{Ox\}}$
3	Gibbs energy of reaction in terms of energy of electrons associated with the two half-cell reactions†	$\Delta\overline{G}_r = n_e \Delta\overline{G}_{e^-}$ $\Delta\overline{G}_r = -2.303RT\,\Delta\text{pe}$ $\Delta\overline{G}_r = -n_e F\,\Delta E_H$
4	Gibbs energy of reaction in terms of extent of disequilibrium	$\Delta\overline{G}_r = \Delta\overline{G}_r^\circ + 2.303RT \log Q = 2.303RT \log \dfrac{Q}{K}$
5	Change in electron energy in a reaction in terms of extent of disequilibrium†	$\Delta\overline{G}_r = \Delta\overline{G}_r^\circ + 2.303RT \log Q$ $\Delta\text{pe} = \Delta\text{pe}^\circ - \dfrac{RT}{n_e} \log Q$ $\Delta E_H = \Delta E_H^\circ - \dfrac{2.303RT}{n_e F} \log Q$
6	Equilibrium constant in terms of energy change at standard state†	$\log K = \dfrac{\Delta\overline{G}_r^\circ}{2.303RT}$ $\log K = n_e\,\Delta\text{pe}^\circ$ $\log K = \dfrac{n_e F}{2.303RT} \Delta E_H^\circ$
7	Conditions for conversion of reactants to products†	$\Delta\overline{G}_r < 0$ $\Delta\text{pe} > 0$ $\Delta E_H > 0$
8	Conditions at equilibrium†	$\Delta\overline{G}_r = \Delta\text{pe} = \Delta E_H = 0$

*At 25°C, $2.303RT = 5.71$ kJ/mol and $2.303RT/F = 59$ mV.
†$\Delta\text{pe} = \text{pe}_{red} - \text{pe}_{ox}$ where pe_{red} and pe_{ox} are the pe values of the reduction and oxidation half-cell reactions, respectively. Similarly, $\Delta E_H = E_{H,red} - E_{H,ox}$.

electrons associated with a redox couple in the solution; expressions relating to this interpretation are shown in row 2. Keep in mind that no absolute value of $\overline{G}_{e^-}$, pe, or E_H can be defined without reference to some arbitrary baseline; the baseline for all three expressions in the first two rows of the table is the $H_2(g)/H^+$ half-cell reaction at standard state.

The equations in the rows labeled 3, 4, and 5 describe the change in energy that accompanies transfer of electrons from one redox couple to another. The important distinction here is that whereas rows 1 and 2 describe the energy levels of electrons in the system as is (i.e., with each electron residing on its current host), rows 3 through 5 describe the energy change that occurs when electrons shift from one host to another.

Rows 6 through 8 of the table relate to the likelihood that such shifts will occur. Row 6 expresses the equilibrium constant for a redox reaction in terms of the change in electron energy under standard state conditions; row 7 indicates the condition that corresponds to a downhill (favorable) gradient in electron energy; and row 8 indicates that when this gradient disappears, the system is at equilibrium and will react no more.

9.6 SPECIATION IN REDOX SYSTEMS: LOG C–PE DIAGRAMS

Having established the fundamental concepts of redox chemistry, we can now proceed to explore in greater detail how pe controls, and is controlled by, redox speciation in a system. We use as our example system a solution in which $5 \times 10^{-4} \, M \, TOT$Cl is distributed among Cl^-, HOCl, and OCl^-. pe° for the $HOCl/Cl^-$ couple is 25.1, and pK_a for HOCl is 7.60. Because the speciation is pH-dependent, we can anticipate that the log C–pe diagrams for the system will look different at different pH values. Therefore, our approach will be to derive the equations needed to draw the diagrams leaving pH as an adjustable parameter and then insert different pH values into the equations to explore the effect of pH.

The example system contains three Cl-containing species whose activities are related by three equations: a redox equilibrium constant (e.g., pe° for the $HOCl/Cl^-$ couple), an acid/base equilibrium reaction (K_a for $HOCl/OCl^-$), and the mass balance on Cl. The easiest way to proceed is to write the mass balance and then eliminate two of the three quantities in it by substitutions based on the equilibrium equations, as follows.

The mass balance on Cl (assuming, as usual, ideal behavior of the solutes) is

$$TOTCl = \{HOCl\} + \{OCl^-\} + \{Cl^-\} \qquad \textbf{(9.41)}$$

We can use the acidity constant expression to write $\{OCl^-\}$ in terms of $\{HOCl\}$, as $\{OCl^-\} = 10^{pH-pK_a}\{HOCl\}$. Similarly, we can use the redox equilibrium expression to write $\{Cl^-\}$ in terms of $\{HOCl\}$. Because log K for a one-electron reduction reaction is pe°, $K = 10^{pe°}$. The relevant reaction and equilibrium constant are therefore

$$\frac{1}{2}HOCl + \frac{1}{2}H^+ + e^- \leftrightarrow \frac{1}{2}Cl^- + \frac{1}{2}H_2O \qquad (9.42)$$

$$K_{(9.42)} = \frac{\{Cl^-\}^{1/2}}{\{HOCl\}^{1/2}\{H^+\}^{1/2}\{e^-\}} \qquad (9.43a)$$

$$\{Cl^-\} = 10^{2pe^\circ}\{HOCl\}\{H^+\}\{e^-\}^2 \qquad (9.43b)$$

Substituting 10^{-pH} and 10^{-2pe} for $\{H^+\}$ and $\{e^-\}^2$, respectively, Equation (9.43b) becomes

$$\{Cl^-\} = 10^{2pe^\circ - pH - 2pe}\{HOCl\} \qquad (9.43c)$$

Finally, the expressions for $\{OCl^-\}$ and $\{Cl^-\}$ can be substituted into the mass balance on Cl to yield an expression for $\{HOCl\}$ as a function of pe for any given $TOTCl$ and pH:

$$TOTCl = \{HOCl\} + 10^{pH-pK_a}\{HOCl\} + 10^{2pe^\circ - pH - 2pe}\{HOCl\} \qquad (9.44a)$$

$$\{HOCl\} = \frac{TOTCl}{1 + 10^{pH-pK_a} + 10^{2pe^\circ - pH - 2pe}} \qquad (9.44b)$$

Equation (9.44b) provides the information needed to draw the curve for HOCl on the $\log C$–pe diagram. Once the value of $\{HOCl\}$ is determined, the corresponding values of $\{OCl^-\}$ and $\{Cl^-\}$ can be computed from the equilibrium expressions, so that lines for all three species can be drawn, completing the diagram. The resulting diagrams for three pH values are shown in Figure 9.6.

Several general observations apply to all three diagrams in Figure 9.6. First, as we would expect, in all cases the two oxidized species (HOCl and OCl^-) dominate at high pe, and the reduced species (Cl^-) dominates at low pe. However, the crossover pe differs depending on the solution pH. This observation is a direct result of the fact that interconversion of HOCl and Cl^- involves transfer not only of electrons, but also of a proton. Second, the lines for HOCl and OCl^- are always parallel, because interconversion of HOCl and OCl^- involves transfer of a proton, but not of electrons, and each diagram characterizes systems over a range of pe values but all at the same pH. Specifically, the line for HOCl is always 3.6 log units above that for OCl^- at pH 4.0, 0.6 log unit above it at pH 7.0, and 2.4 log units below it at pH 10.0. As a result, OCl^- is never dominant at pH 4.0 or 7.0, and HOCl is never dominant at pH 10.0. Third, the slope of the Cl^- line is always 2 less than that of the HOCl and OCl^- lines, consistent with the fact that two electrons are exchanged in the reaction between one molecule of Cl^- and one of HOCl or OCl^-. Finally, the redox reaction of interest [reaction (9.42)] has the form of reaction (9.27), so pe°_{pH} for the HOCl/Cl^- couple can be computed from Equation (9.29), using $n_e = 1$ and $n_H = \frac{1}{2}$. The computed values of pe°_{pH} are 23.1, 21.6, and 20.1 at pH 4.0, 7.0, and 10.0, respectively; as the diagrams show, $\{HOCl\} = \{Cl^-\}$ at pe $= pe^\circ_{pH}$ in each system.

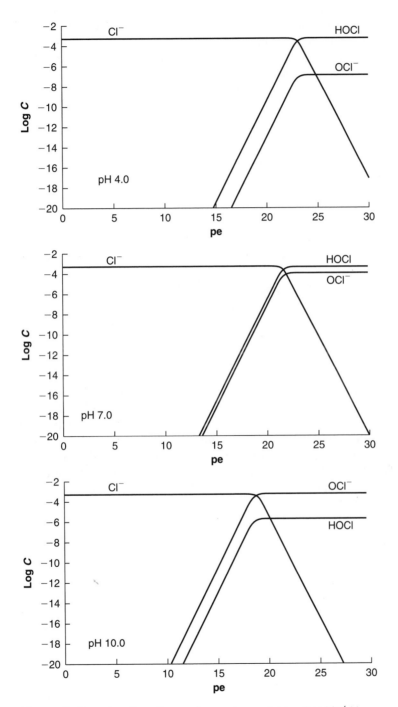

Figure 9.6 Log C–pe diagram for a system containing 5×10^{-4} M $TOTCl$ at pH 4, 7, and 10.

Although we can determine the crossover pe at which OCl^- and Cl^- have equal activities from the above analysis, that value is computed somewhat more easily by writing a reaction relating these two species (OCl^- and Cl^-) directly. This can be accomplished by combining the pe^o and pK_a expressions as follows:

$$\frac{1}{2}HOCl + \frac{1}{2}H^+ + e^- \leftrightarrow \frac{1}{2}Cl^- + \frac{1}{2}H_2O \qquad K = 10^{25.1}$$

$$\frac{\frac{1}{2}H^+ + \frac{1}{2}OCl^- \leftrightarrow \frac{1}{2}HOCl \qquad\qquad K_a^{-0.5} = 10^{+3.8}}{H^+ + \frac{1}{2}OCl^- + e^- \leftrightarrow \frac{1}{2}Cl^- + \frac{1}{2}H_2O \qquad K_{(9.45)} = 10^{28.9} \qquad (pe^o = 28.9)}$$

$$\textbf{(9.38)}$$

Knowing pe^o, we can find the pe where $\{OCl^-\} = \{Cl^-\}$ by computing pe^o_{pH} for the corresponding reaction:

$$pe^o_{pH} = pe^o - \frac{1}{1}pH = pe^o - pH$$

$$\{OCl^-\} = \{Cl^-\} \qquad \text{when} \qquad pe = 28.9 - pH$$

This result is consistent with those shown in Figure 9.6.

As is the case with $\log C$–pH diagrams, the relationships shown in the $\log C$–pe diagrams can be thought of as a summary of the results obtained by solving the equilibrium and mass balance equations simultaneously. As such, the lines on the diagram are applicable regardless of what other redox couples are present in the system, and the lines for several redox couples can be combined on a single diagram to characterize a complex solution. An example demonstrating this point is provided next, after which we consider how chemical equilibrium software can be used to develop $\log C$–pe diagrams and to determine the equilibrium pe of a solution.

9.6.1 DETERMINING THE EQUILIBRIUM SPECIATION IN A REDOX SYSTEM WITH KNOWN INPUTS

The diagrams in Figure 9.6 describe the equilibrium speciation of Cl as a function of pe at various pH values. If we knew the pH and pe of a solution, we could use these types of diagrams to determine the equilibrium concentration of each Cl-containing species in the system. A second type of analysis we might be interested in is the prediction of pe in a system with known inputs. In this section, we develop the $\log C$–pe diagram for the S(VI)/S(−II) system at pH 7.0 and then combine the results with those for the Cl(+I)/Cl(−I) system to predict the equilibrium speciation in a system containing both of these redox couples.

For this analysis, we consider an example system containing 10^{-4} M $TOTS$, in a solution fixed at pH 7.0. The pK_a's for the $H_2S/HS^-/S^{2-}$ species are 6.99 and

12.92, respectively, so at pH 7.0, $\{H_2S\} \approx \{HS^-\}$, and both these species are present at concentrations far greater than that of S^{2-}. The redox reaction between HS^- and SO_4^{2-}, which has a pe° of 4.21, is as follows:

$$SO_4^{2-} + 9H^+ + 8e^- \leftrightarrow HS^- + 4H_2O \tag{9.45}$$

By definition, pe° is the logarithm of the equilibrium constant for a one-electron reduction reaction. Thus, pe° is the logarithm of the equilibrium constant for the above reaction with all the coefficients divided by 8:

$$pe° = \log \frac{\{HS^-\}^{1/8}}{\{SO_4^{2-}\}^{1/8}\{H^+\}^{9/8}\{e^-\}} \tag{9.46a}$$

$$pe° = \frac{1}{8}\log \frac{\{HS^-\}}{\{SO_4^{2-}\}} + \frac{9}{8}pH + pe \tag{9.46b}$$

To draw the log C–pe diagram for a system at pH 7.0, it is convenient to compute $pe_{7.0}°$, i.e., $pe°(W)$:

$$pe°(W) = pe° - \frac{9}{8}(7.0) = -3.67 \tag{9.47}$$

Thus, in this system, $\{HS^-\} = \{SO_4^{2-}\}$ at $pe = pe°(W) = -3.67$. Rewriting Equation (9.46b) in terms of $pe°(W)$, we obtain

$$pe°(W) = \frac{1}{8}\log \frac{\{HS^-\}}{\{SO_4^{2-}\}} + pe$$

$$8pe° = -29.34 = \log \frac{\{HS^-\}}{\{SO_4^{2-}\}} + 8pe$$

Since the SO_4^{2-}/HS^- conversion is an eight-electron transfer and each of the species of interest contains one S atom, the slopes of the lines for HS^- and SO_4^{2-} on the log C–pe diagram are different by 8. Also, since the activities of H_2S and HS^- are equal (because the pH of the system equals pK_{a1}), the lines for H_2S and HS^- overlap. As noted above, the activity of S^{2-} is negligible at pH 7.0 regardless of the pe. We therefore conclude that at $pe < -3.67$, the dominant species are HS^- and H_2S, each of which is present at a concentration of approximately 0.5 $TOTS$, and at $pe > -3.67$, SO_4^{2-} is dominant and present at a concentration of approximately $TOTS$. The log C–pe diagram, showing all the sulfur species and also H^+ and OH^-, is presented in Figure 9.7.

Next, consider a solution prepared by adding 5×10^{-4} M HOCl and 1×10^{-4} M NaHS to water, with the pH fixed at 7.0. What will the speciation and the value of pe at equilibrium be? The relevant data from Figure 9.6b (for the HOCl/OCl$^-$ system at pH 7.0) and Figure 9.7 (for the S(VI)/S(−II) system) are shown together in Figure 9.8, which provides an overview of the equilibrium speciation as a function of pe. The Na$^+$ concentration is characterized by a horizontal line at log $C = -4$, which has been left out to minimize clutter.

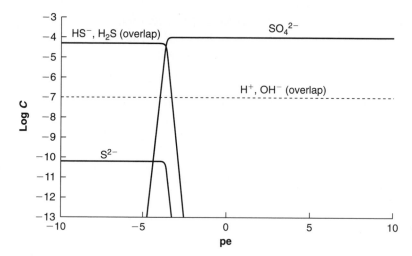

Figure 9.7   Log C–pe diagram for a system containing 10^{-4} M TOTS at pH 7.0.

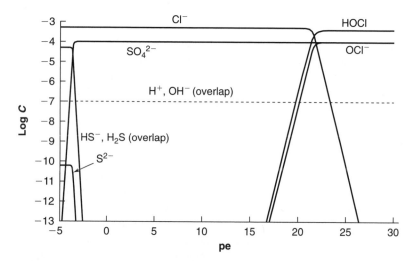

Figure 9.8   Log C–pe diagram for a pH 7 solution containing 10^{-4} M TOTS and 5×10^{-4} M TOTCl.

As with acid/base problems, if we knew exactly how the solution was prepared (all chemicals that were added and their input concentrations), the information in the diagram could be combined with a charge balance to determine the equilibrium pe and speciation. However, the solution pH is fixed at 7.0, and we do not know which acids and bases, or how much of each, have been added to establish that pH. As a result, we are unable to write a complete charge balance for the system. Nevertheless, we can determine the equilibrium redox speciation by writing an *electron condition* (EC) or *TOTe* equation that operates identically to a proton condition or *TOT*H equation. That is, it allows us to keep track of electron transfers that occur in the system, even if we do not have complete knowledge about the redox-inactive species in solution.

For each element in the system, we can choose any oxidation state to define the electron reference levels. Arbitrarily choosing $H(+I)$, $O(-II)$, $S(-II)$, and $Cl(+I)$ as the reference levels, we see that H^+, OH^-, H_2O, $HOCl$, OCl^-, H_2S, HS^-, and S^{2-} are all at the reference level, since these species can be formed from reference level species without adding or removing electrons. The only non-reference level species in the system are Cl^- and SO_4^{2-}, which are 2-e^- excess and 8-e^- deficient species, respectively, based on the reactions by which they form from reference level species. The EC table is therefore as follows:

	e^- **Excess**				
	−8	**0**	**+1**	**+2**	**Concentration**
Species present at equilibrium	SO_4^{2-}	H^+, OH^-, H_2O H_2S, HS^-, S^{2-} $HOCl$, OCl^- Na^+	e^-	Cl^-	
Input species		$HOCl$ $NaHS$			5×10^{-4} 1×10^{-4}

The initial electron excess and deficiency are both zero, since only reference level species ($HOCl$ and HS^-) were added to prepare the solution, so the *TOTe* and EC equations are

$$TOTe_{in} = TOTe_{eq}$$

TOTe equation: $\qquad 0 = 2[Cl^-] - 8[SO_4^{2-}] + [e^-]$ **(9.48a)**

EC: $\qquad\qquad 8[SO_4^{2-}] = 2[Cl^-] + [e^-]$ **(9.48b)**

The determination of the equilibrium pe is then simply a matter of identifying the pe at which the EC is satisfied.

Like the proton condition, *TOT*H, and charge balance equations, the *TOTe* equation is a relationship among concentrations, not activities. That is, when writing a charge balance, we compute the charge associated with carbonate ions as −2 times the CO_3^{2-} concentration (not −2 times its activity), because each carbonate ion carries a −2 charge regardless of the carbonate activity coefficient. Similarly, if HAc is a reference species, each Ac^- ion in solution represents a deficit of one proton, regardless of the activity coefficient for Ac^-.

If the standard state concentration of every species in the equation is 1.0 mol/L and if the solution is assumed to be ideal, then the distinction between activity and concentration becomes irrelevant as a practical matter. We have generally made this assumption when writing proton condition or *TOT*H equations. However, in the current situation, we need to recognize that even in an ideal solution, the electron activity is very different from its concentration.

Because the concentration of dissolved free electrons is negligible in all solutions of interest, the term for free electrons in EC and *TOTe* equations is always negligible; e.g., Equation (9.48b) can be approximated as

$$8[SO_4^{2-}] = 2[Cl^-] + [\cancel{e^-}] \qquad \textbf{(9.48c)}$$

Indeed, recognizing that electrons are present at a negligible concentration and that the only species that we need to include in the upper portion of the electron balance table are those that are present in significant concentration at equilibrium, we could have left e^- out of the table in the first place. We will use that approach in preparing all electron balance tables in the future.

Assuming the solutes behave ideally, the concentrations in Equation (9.48c) can be replaced by activities, yielding the following equation for use in conjunction with the log C–pe diagram:

$$8\{SO_4^{2-}\} = 2\{Cl^-\} \qquad \textbf{(9.48d)}$$

The portion of the log C–pe diagram where the EC is satisfied for the example system is shown with expanded scales in Figure 9.9. The equilibrium pe is around 21.25, where virtually all the sulfide has been oxidized to sulfate [total $S(-II)$ at equilibrium is $<10^{-144}$!], and 80% of the hypochlorite has been reduced to chloride. As shown in the figure, the ratio of {HOCl} to {OCl$^-$} remains the same as before the reaction, because the pH remains at 7.0. The same is true of the relative values of {H$_2$S}, {HS$^-$}, and {S^{2-}}, although the activities of all three of these species are negligibly small. The equilibrium pe of the solution is quite high, because even after essentially all the sulfide is oxidized to sulfate, strongly oxidizing Cl$(+I)$ species (HOCl and OCl$^-$) are present at a total concentration of 10^{-4} M. These species are capable of oxidizing other reduced species that might enter the solution subsequently.

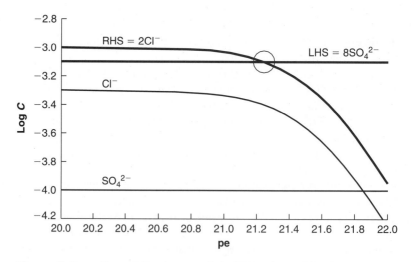

Figure 9.9   Figure 9.8 redrawn, with the RHS and LHS of the electron condition equation shown in bold, and the point where RHS = LHS is circled.

It was noted above that many elements can exist in oxidation states that differ by several electrons (an eight-electron difference between the most oxidized and most reduced forms of an element is common) and that often more than one electron is added or released from an element in a single step. Frequently, electrons are transferred in pairs or multiples of two, as is the case for both of the redox couples in the above example. These tendencies differentiate redox reactions from most acid/base reactions: while acids can be multiprotic, it is generally the case that the protons are released one at a time, and it is rare for the most protonated and the most deprotonated species of environmental significance to differ by more than four protons.

The practical implication of these differences is apparent in Figure 9.8. Specifically, the transfer of multiple electrons increases the absolute values of the slopes of the lines for the non-dominant species, thereby dramatically shrinking the pe region over which the oxidized and reduced species are both present in significant concentrations. For instance, in the case of the sulfur species, the non-dominant species is present at an insignificant concentration relative to the dominant one at a pe value just a few tenths of a pe unit away from the crossover point.

Some data for complexation and redox equilibria of copper are given below. A chloride-free, acidic solution is prepared, and enough copper metal, $Cu(s)$, and solid cuprous chloride, $CuCl(s)$, are added to it that some of each solid remains in the equilibrated solution. Assume that copper species are the only ones that undergo redox reactions in the system, and that the initial acid addition is sufficient to prevent significant hydrolysis of either Cu^{2+} or Cu^+. | **Example 9.11**

a. Show the concentrations of all soluble species in this system on a log C–pe diagram.
b. Use a charge balance to find the equilibrium pe of the system. (*Note:* Without knowing the initial composition of the solution, we cannot write a complete charge balance. However, we can write a balance that expresses the constraint that the net charge that enters solution as a result of the reactions with the solids must be zero.)
c. Draw curves on the diagram from part (*a*) to represent $TOTCu_{diss}$ and $TOTCl_{diss}$. What do the values of these two terms at the equilibrium pe of the system tell us about the relative amounts of each solid that dissolves?

1. $Cu^{2+} + 2e^- \leftrightarrow Cu(s)$ $pe^\circ = 5.74$ $\log K = 11.48$
2. $Cu^{2+} + e^- \leftrightarrow Cu^+$ $pe^\circ = \log K = 2.72$
3. $CuCl(s) \leftrightarrow Cu^+ + Cl^-$ $\log K_{s0} = -6.76$
4. $Cu^+ + 2Cl^- \leftrightarrow CuCl_2^-$ $\log \beta_2 = 5.5$
5. $Cu^+ + 3Cl^- \leftrightarrow CuCl_3^{2-}$ $\log \beta_3 = 5.7$
6. $Cu^{2+} + Cl^- \leftrightarrow CuCl^+$ $\log K_1 = 0.43$
7. $Cu^{2+} + 2Cl^- \leftrightarrow CuCl_2^\circ$ $\log \beta_2 = 0.16$

Solution

a. Since the two solids $Cu(s)$ and $CuCl(s)$ are known to be present at equilibrium, we can assign activities of 1.0 to those species and use this information to determine the equations of the lines on the log C–pe diagram for the soluble species. We begin by considering species whose concentrations do not depend on the chloride concentration. For instance, reaction 1 indicates that

$$11.48 = \log\frac{\{Cu(s)\}}{\{Cu^{2+}\}\{e^-\}^2}$$

Since the activity of $Cu(s)$ is 1.0, the above equation can be rearranged to show the activity of Cu^{2+} as a function of pe and nothing else:

$$\log\{Cu^{2+}\} = -11.48 + 2pe$$

To determine the dependence of $\{Cu^+\}$ on pe, we can use the above equation in conjunction with the redox reaction between Cu^{2+} and Cu^+, or we can derive the equilibrium constant for the redox reaction between Cu^+ and $Cu(s)$. Taking the latter approach, we subtract reaction 2 from reaction 1 to obtain

$$Cu^+ + e^- \leftrightarrow Cu(s) \qquad \log K = pe^\circ = 8.76$$

$$8.76 = \log\frac{\{Cu(s)\}}{\{Cu^+\}\{e^-\}}$$

$$\log\{Cu^+\} = -8.76 + pe$$

Since the system is specified to be in equilibrium with $CuCl(s)$ and since $\{Cu^+\}$ is known as a function of pe from the above equation, the dependence of $\{Cl^-\}$ on pe can be determined as well:

$$\log K_{s0,CuCl(s)} = -6.76 = \log\{Cu^+\} + \log\{Cl^-\}$$
$$\log\{Cl^-\} = -6.76 - \log\{Cu^+\}$$
$$= -6.76 + 8.76 - pe = 2.00 - pe$$

Finally, knowing the activities of Cu^+, Cu^{2+}, and Cl^- as a function of pe, we can compute the activities of the various complexes from the equilibrium constant expressions for reactions 4 through 7:

$$\log\{CuCl_2^-\} = 5.7 + (-8.76 + pe) + 2(2.00 - pe) = -0.94 - pe$$
$$\log\{CuCl_3^{2-}\} = 5.5 + (-8.76 + pe) + 3(2.00 - pe) = 2.74 - 2pe$$
$$\log\{CuCl^+\} = 0.43 + (-11.48 + 2pe) + (2.00 - pe) = -9.05 + pe$$
$$\log\{CuCl_2^\circ\} = -0.16 + (-9.05 + pe) + (2.00 - pe) = -7.21$$

The diagram is shown on the following page, displaying the activities of all seven soluble species and also $TOTCu_{diss}$ and $TOTCl_{diss}$ as a function of pe.

b. A charge balance on the species that enter solution due to dissolution of the solids, assuming that the pH is not altered by the redox reactions (a reasonable assumption, since the solution was acidified to prevent hydrolysis), is

$$2\{Cu^{2+}\} + \{Cu^+\} + \{CuCl^+\} = \{CuCl_2^-\} + 2\{CuCl_3^{2-}\} + \{Cl^-\}$$

The charge balance is satisfied at pe ≈ 4.4, where $2\{Cu^{2+}\} \approx \{Cl^-\}$. The concentrations of all soluble species at this pe value are as follows:

$\{Cu^{2+}\} = 10^{-2.68}$	$\{Cu^+\} = 10^{-4.36}$	$\{CuCl_2^\circ\} = 10^{-7.32}$
$\{Cl^-\} = 10^{-2.40}$	$\{CuCl_2^-\} = 10^{-3.66}$	
$\{CuCl^+\} = 10^{-4.65}$	$\{CuCl_3^{2-}\} = 10^{-5.86}$	

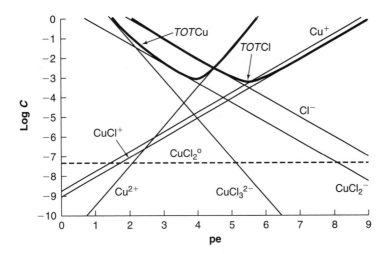

c. At the equilibrium pe, *TOT*Cl is approximately twice *TOT*Cu. Because the only source of Cl to the solution is dissolution of CuCl(s), *TOT*Cl indicates the mass of this solid that has dissolved. The fact that *TOT*Cu is only one-half of this value indicates that one-half of the Cu that dissolved from CuCl(s) must have been removed from solution, and the only way this could have occurred (without removing Cl simultaneously) is if metallic Cu(s) precipitated.

   Thus, the overall equilibration process involves dissolution of CuCl(s) to release Cu^+ and Cl^- and reduction of some Cu(I) to Cu(0). The electrons for this reduction are abstracted from other Cu(I) atoms, generating Cu(II) species. Reactions in which an element in a given oxidation state is simultaneously oxidized and reduced to produce two other species are sometimes called *disproportionation* reactions.

9.7 MODELING REDOX SPECIATION WITH CHEMICAL EQUILIBRIUM SOFTWARE

In principle, the approaches described in previous chapters for modeling equilibrium speciation with available software can be applied directly to redox-active systems. Intuitively, the simplest approach would be to define the hydrated electron as a *component* and to input redox reactions just like acid/base or metal complexation reactions; e.g., we could choose Fe^{3+} and e^- as *components* and then define Fe^{2+} as a Type 2 *species* formed by combining one unit each of Fe^{3+} and e^-:

$$1\,Fe^{3+} + 1\,e^- \leftrightarrow 1\,Fe^{2+}$$

However, if we actually try to implement that approach, we run into some awkward situations that programmers have chosen to circumvent by clever, but not necessarily transparent, logic.

   The first issue that arises relates to the "disconnect" between electron concentration and activity. Fortunately, in the numerical analysis of redox systems, this issue poses no serious problem: we can include e^- as a *component* whose activity must be determined, and not include e^- as a *species* at all. (This is the programming analog of leaving e^- out of the top portion of the electron balance table.) That way, when the program computes the mass balance on e^-, no term will be included for the concentration of free e^- in the equilibrium solution. Most programs carry out this step automatically whenever e^- is included as a *component*.[6]

However, another issue arises if we try to follow the approach used to input information about acid/base systems as a model for inputting information about redox reactions. Assume, for example, that we wanted to model the Fe(III)/Fe(II) redox system, and we chose $\mathbf{Fe^{3+}}$ and e^- as *components*. We would then have to rewrite all the equilibrium reactions for formation of Fe^{2+} complexes as combinations of $\mathbf{Fe^{3+}}$, e^-, and ligands, because all *species* must be defined in terms of *components* and Fe^{2+} would not be a *component* in the given system. While this procedure could be carried out, it is very inconvenient, because all available compilations of Fe(II)—L stability constants are written with Fe^{2+} as a reactant. Therefore, we would much prefer to input information about reactions between Fe^{2+} and ligands directly, as combinations of those molecules.

One approach for dealing with this issue might be to choose $\mathbf{Fe^{2+}}$ and $\mathbf{Fe^{3+}}$ as *components* and to form the Type 2 *species* e^- by a reaction involving subtraction of an $\mathbf{Fe^{3+}}$ ion from an $\mathbf{Fe^{2+}}$ ion. However, the same type of problem then arises if we wish to consider another redox reaction in the same system. For instance, if we wanted to consider the redox reaction between SO_4^{2-} and HS^-, and if we chose $\mathbf{SO_4^{2-}}$ as a *component,* we could not write the reaction for formation of HS^- as a simple combination of $\mathbf{SO_4^{2-}}$ and e^-, because e^- would not be a *component*. Rather, HS^- would have to be identified as a *species* that was created by the following reaction:

$$1SO_4^{2-} + 8Fe^{2+} - 8Fe^{3+} + 9H^+ - 4H_2O \rightarrow 1HS^- \qquad \textbf{(9.49)}$$

Once again, although the above reaction is formally correct and could be accommodated by the software, it is a very inconvenient way to represent the formation of HS^-.

The approach that has been developed to deal with this situation involves defining a new Type 3 dummy *species* for each redox reaction to be considered. In this text, such *species* are referred to as Type 3c *species*. For instance, for the iron system, a Type 3c dummy *species* named $\overline{Fe^{3+}/Fe^{2+}}$ can be defined, with the following stoichiometry and equilibrium constant relationship:

$$1Fe^{3+} + 1e^- - 1Fe^{2+} \leftrightarrow 1\overline{Fe^{3+}/Fe^{2+}} \qquad \textbf{(9.50)}$$

$$K_{(9.50)} = \frac{\{\overline{Fe^{3+}/Fe^{2+}}\}\{Fe^{2+}\}}{\{Fe^{3+}\}\{e^-\}} \qquad \textbf{(9.51)}$$

---

[6]As a practical matter, many software packages require that every *component* be included somewhere in the list of *species*, so e^- is usually listed as a Type 6 *species* (species not considered). However, the effect is the same as if the *species* e^- were totally excluded from the input.

If $\overline{Fe^{3+}/Fe^{2+}}$ is assigned a fixed activity of 1.0, Equation (9.51) becomes[7]

$$K_{(9.50)} = \frac{\{\overline{Fe^{3+}/Fe^{2+}}\}\{Fe^{2+}\}}{\{Fe^{3+}\}\{e^-\}} = \frac{\{Fe^{2+}\}}{\{Fe^{3+}\}\{e^-\}} \qquad \textbf{[9.52]}$$

Thus, assigning the dummy *species* $\overline{Fe^{3+}/Fe^{2+}}$ a fixed activity of 1.0 causes the program to satisfy the equality $K_{(9.50)} = \{Fe^{2+}\}/(\{Fe^{3+}\}\{e^-\})$. The right-hand side of Equation (9.52) is the inverse of the expression for $e^o_{Fe^{3+}/Fe^{2+}}$. Therefore, by inputting $\{e^o_{Fe^{3+}/Fe^{2+}}\}^{-1}$ as the value of $K_{(9.50)}$, we can force the program to satisfy the redox equilibrium expression. That is, to cause the program to account properly for the Fe^{3+}/Fe^{2+} redox equilibrium, we can input a Type 3c *species* with the stoichiometry shown in reaction (9.50) and an equilibrium constant of

$$K_{(9.50)} \equiv \frac{1}{e^o_{Fe^{3+}/Fe^{2+}}} \qquad \log K_{(9.50)} = pe^o_{Fe^{3+}/Fe^{2+}}$$

Note that, in the formal logic followed by the program, the *species* Fe^{2+} cannot be formed strictly by a linear combination of Fe^{3+} and e^- because, based on reaction (9.50), the combination of one Fe^{3+} and one e^- generates both an Fe^{2+} ion and the *species* $\overline{Fe^{2+}/Fe^{3+}}$. As a result, Fe^{2+}, Fe^{3+}, and e^- are mutually independent, so it is allowable to choose all three as *components*. This, in turn, allows us to write reactions for the formation of Fe^{2+} and Fe^{3+} complexes in the conventional way. A similar approach can be used to include any other redox reaction in the program's analysis.

The reactions defining Type 3c *species* can be written based on any valid version of the redox reaction, i.e., for the reaction written as an oxidation or reduction, and for any number of electrons being transferred. To make life simple, we can arbitrarily establish a convention of writing the reaction as an n-electron reduction reaction, in which case the appropriate log K value is $n(pe^o)$.

Determine the stoichiometry and equilibrium constant for a Type 3c *species* that could be used to include the redox reaction between HS^- and SO_4^{2-} in a chemical equilibrium program, using H_2O, H^+, e^-, HS^-, and SO_4^{2-} as *components*. | **Example 9.12**

Solution

The pe^o value for the SO_4^{2-}/HS^- couple is 4.21. We can write the reaction for formation of the Type 3c *species* $\overline{SO_4^{2-}/HS^-}$ and determine its log K value as follows:

$$1SO_4^{2-} + 9H^+ + 8e^- \rightarrow 1HS^- + 4H_2O + 1\overline{SO_4^{2-}/HS^-} \qquad \textbf{(9.53)}$$

$$K_{(9.53)} = \frac{\{\overline{SO_4^{2-}/HS^-}\}\{HS^-\}\{H_2O\}^4}{\{SO_4^{2-}\}\{H^+\}^9\{e^-\}^8} = (10^{4.21})^8 = 10^{33.68}$$

$$\log K_{(9.53)} = 8pe^o_{SO_4^{2-}/HS^-} = 33.68 \qquad \textbf{(9.54)}$$

[7]Although it has not been emphasized in the previous discussion of Type 3 *species*, the programs automatically assign a fixed activity of 1.0 to all such species. Thus, nothing need be done by the user to make the desired assignment in this case.

Including $\overline{SO_4^{2-}/HS^-}$ as a Type 3c *species* with the stoichiometry shown in reaction (9.53) and a $\log K$ equal to 33.68 will cause the program output to satisfy the $SO_4^{2-}/H_2S(aq)$ redox equilibrium relationship, while allowing HS^-, SO_4^{2-}, and e^- to all be used as *components*.

In addition to allowing us to choose a convenient set of *components,* the approach described above makes it easier to include a selective subset of redox reactions in the calculations. For instance, if we had reason to believe that one redox couple (e.g., Fe^{3+}/Fe^{2+}) would reach equilibrium but that another (e.g., SO_4^{2-}/HS^-) would not, we could include a Type 3c *species in the* input for the first couple but not the second; without inclusion of $\overline{SO_4^{2-}/HS^-}$, the program would not recognize any reaction that converts SO_4^{2-} to HS^-.

Many existing programs include Type 3c *species* for various redox couples in their databases, circumventing the need for the user to define the *species* or input equilibrium constants for their formation. However, there are also many redox reactions that are not included in the default database, and in those cases the user must follow the algorithm provided above to cause the program to consider the redox reactions in the calculations.

Once the general approach for inputting redox reactions is understood, $\log C$–pe diagrams can be generated using the programs by following essentially the same steps as are used to develop $\log C$–pH diagrams. That is, redox reactions are included using dummy *species* as described above, and a Type 3a *species* made of one e^- and nothing else (i.e., a *species* $\overline{e^-}$) is also defined. This assignment serves the same purpose as the Type 3a *species* $\overline{H^+}$ that we define when investigating speciation in a system at fixed pH.[8] Then the $\log K$ value for forming $\overline{e^-}$ is systematically altered, and the solution composition is computed at each value of $\log K$. The output can then be plotted on a $\log C$–pe diagram.

Example 9.13 | Develop the input table needed to prepare the $\log C$–pe diagram shown in Figure 9.8, for a system at pH 7.0 that contains 5×10^{-4} *TOT*Cl and 1×10^{-4} *TOT*S.

Solution

Choosing Cl^-, $HOCl$, HS^-, SO_4^{2-}, H^+, and e^- as *components,* we can input the acid/base reactions forming OCl^-, $H_2S(aq)$, S^{2-}, and OH^- as in previous chapters. Also, we can use the result of Example 9.12 to define a Type 3c *species* $\overline{SO_4^{2-}/HS^-}$ that will account for the redox reactions of sulfur, so the only additional reaction we need to define

---

[8]The main difference between $\overline{H^+}$ and $\overline{e^-}$ is that $\overline{H^+}$ has an analog Type 1 *species* that contributes to the mass balance on **H**$^+$, whereas the analog to $\overline{e^-}$ is a Type 6 *species* that does not contribute to the mass balance on **e**$^-$. This difference has no effect on how the Type 3 *species* are used to fix pH or pe.

is one for the Type 3c *species* $\overline{HOCl/Cl^-}$. The reaction is

$$1\textbf{HOCl} + 1\textbf{H}^+ + 2e^- \rightarrow 1\overline{\textbf{HOCl/Cl}^-} + 1\textbf{Cl}^- + 1\textbf{H}_2\textbf{O}$$

Log K for the above reaction is $2pe^\circ_{HOCl/Cl^-}$, or 50.2.

Since pH is fixed at 7.0, we define a Type 3a *species* $\overline{H^+}$ with a log K of +7.0. Then to determine the speciation at various pe values so that we can draw the diagram, we also define a dummy Type 3a *species* $\overline{e^-}$, to which we will assign a range of log K values. Each log K for this *species* corresponds to a different, fixed pe, at which the speciation of the solution will be determined before moving to the next value.

Since we are not solving for the equilibrium pe in a particular system, we are free to assume that the Cl and S were added in any form; at each pe, the program will allocate TOTCl and TOTS as dictated by the relevant equilibria. For the current example, it was assumed that the Cl and S were input as HOCl and SO_4^{2-}, respectively.

The complete input table is shown below. (Note that e^- is included in the table as a Type 6 *species*, although it could just as well be left out, since it has no effect on the calculations.)

Species	Type	Cl⁻	HOCl	HS⁻	SO_4^{2-}	H⁺	e^-	Log K
Cl⁻	1	1	0	0	0	0	0	0.00
HOCl	1	0	1	0	0	0	0	0.00
HS⁻	1	0	0	1	0	0	0	0.00
SO_4^{2-}	1	0	0	0	1	0	0	0.00
H⁺	1	0	0	0	0	1	0	0.00
OH⁻	2	0	0	0	0	−1	0	−14.00
H_2S	2	0	0	1	0	1	0	6.99
S^{2-}	2	0	0	1	0	−1	0	−12.99
OCl⁻	2	0	1	0	0	−1	0	−7.60
$\overline{e^-}$	3a	0	0	0	0	0	1	Various
$\overline{H^+}$	3a	0	0	0	0	1	0	7.00
$\overline{HOCl/Cl^-}$	3c	−1	1	0	0	1	2	50.20
$\overline{SO_4^{2-}/HS^-}$	3c	0	0	−1	1	9	8	33.68
e^-	6	0	0	0	0	0	1	0.00

Inputs								Input Concentration
SO_4^{2-}		0	0	0	1	0	0	1×10^{-4}
HOCl		0	1	0	0	0	0	5×10^{-4}
$\overline{e^-}$		0	0	0	0	0	1	x_1
$\overline{H^+}$		0	0	0	0	1	0	x_2
$\overline{HOCl/Cl^-}$		−1	1	0	0	1	2	x_3
$\overline{SO_4^{2-}/HS^-}$		0	0	−1	1	9	8	x_4
TOTComp$_{in}$		$-x_3$	$5 \times 10^{-4} + x_3$	$-x_4$	$10^{-4} + x_4$	$x_2 + x_3 + 9x_4$	$x_1 + 2x_3 + 8x_4$	

9.8   OXIDATION AND REDUCTION OF WATER

9.8.1   THE LOG C–PE DIAGRAM FOR THE $H_2/H_2O/O_2$ REDOX SYSTEM

Redox-active species such as HOCl and HS⁻ may or may not be present in any given water sample. However, one redox-active species that is always present is water itself, which can undergo either oxidation or reduction according to the following reactions:

Oxidation: $2H_2O \leftrightarrow O_2(aq) + 4H^+ + 4e^-$ **(9.55)**

Reduction: $2H_2O + 2e^- \leftrightarrow H_2(aq) + 2OH^-$ **(9.56)**

Reactions (9.55) and (9.56) are often combined with Henry's law constants for O_2 and H_2, respectively, so that the reactions characterize equilibrium between the aqueous phase and gaseous H_2 or O_2, rather than dissolved H_2 or O_2. The various forms of these reactions and the corresponding equilibrium constants and pe° values are shown in Table 9.6.

The equilibrium constant expressions for the above reactions yield the following equations for the activities of O_2 and H_2 in redox equilibrium with water:

$$\log\{O_2(aq)\} = -86.00 + 4pH + 4pe \qquad \text{(9.57)}$$

$$\log\{O_2(g)\} = -83.12 + 4pH + 4pe \qquad \text{(9.58)}$$

$$\log\{H_2(aq)\} = -3.10 - 2pH - 2pe \qquad \text{(9.59)}$$

$$\log\{H_2(g)\} = -2pH - 2pe \qquad \text{(9.60)}$$

These reactions and equilibrium constants indicate that any aqueous solution that is at equilibrium with respect to the redox reactions of water contains some dissolved oxygen and hydrogen, the activities of which are controlled jointly by

Table 9.6   Redox reactions of water

Two Representations of the Water Oxidation Reaction			
$2H_2O \leftrightarrow O_2(aq) + 4H^+ + 4e^-$	$K_{(9.55)} = 10^{-86.0}$	pe° = 21.50	**(9.55)**
$O_2(aq) \leftrightarrow O_2(g)$	$H_{O_2} = 10^{2.88}$		
$2H_2O \leftrightarrow O_2(g) + 4H^+ + 4e^-$	$K_{(9.63)} = 10^{-83.12}$	pe° = 20.78	**(9.63)**

Two Representations of the Water Reduction Reaction			
$2H_2O + 2e^- \leftrightarrow H_2(aq) + 2OH^-$	$K_{(9.56)} = 10^{-31.1}$	pe° = −15.55	**(9.56)**
$2H^+ + 2OH^- \leftrightarrow 2H_2O$	$K = K_w^{-2} = 10^{28.0}$		
$H_2(aq) \leftrightarrow H_2(g)$	$H_{H_2} = 10^{3.1}$		
$2H^+ + 2e^- \leftrightarrow H_2(g)$	$K_{(9.64)} = 10^{0.0} = 1.0$	pe° = 0	**(9.64)**

solution pH and pe. A $\log a_i$–pe diagram for the $H_2(g)/H_2O/O_2(g)$ redox system at pH values of 4, 7, and 10 is shown in Figure 9.10. The activities of both dissolved and gaseous H_2 and O_2 in equilibrium with water at pH 7.0 are shown in Figure 9.11.

The lines representing the activities of H_2 and O_2 in Figures 9.10 and 9.11 have slopes of -2 and $+4$, respectively, and they have no curvature. This latter feature is reminiscent of the H^+ and OH^- lines on $\log C$–pH diagrams; in both

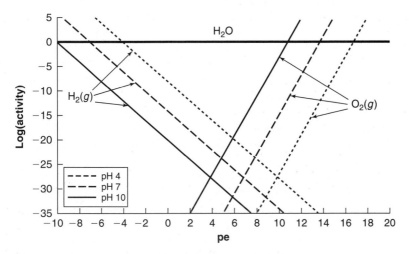

Figure 9.10 Log a–pe diagram for the $H_2(g)/H_2O/O_2(g)$ system at various pH values. Note that the y axis is the logarithm of the activity of the species shown, not the logarithm of its concentration. Thus, for the gases, the value on the ordinate is the logarithm of the partial pressure.

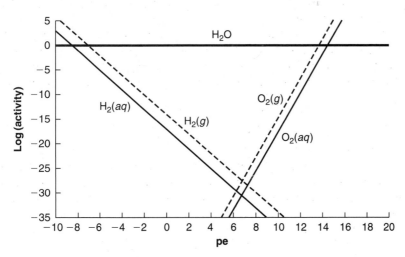

Figure 9.11 Activities of dissolved and gaseous H_2 and O_2 in equilibrium with water for a system at pH 7.0.

cases, this feature is a direct consequence of the assumption that $\{H_2O\} = 1.0$ under all conditions. This assumption is tantamount to assuming that the water provides an infinite source of H^+, OH^-, H_2, or O_2, so that, at least in theory, the activities of these species can grow without bound.

In reality, upper limits are placed on the equilibrium values of $\{H_2(aq)\}$ and $\{O_2(aq)\}$ in most solutions by Henry's law in combination with the atmospheric concentrations of these two gases. In the normal atmosphere, the partial pressures of gaseous hydrogen and oxygen are 5×10^{-7} and 0.21 bar, respectively. Figure 9.10 indicates that, for a solution at pH 7.0, the normal atmospheric partial pressures of $H_2(g)$ and $O_2(g)$ are in redox equilibrium with pure water at pe = -3.8 and pe = 13.6, respectively. Put another way, if a solution is in redox equilibrium with dissolved $H_2(aq)$ and has a pe < -3.8, it is supersaturated with respect to atmospheric $H_2(g)$ according to Henry's law, and hydrogen gas will leave solution. Similarly, if pe > 3.8, the solution is undersaturated according to Henry's law, and atmospheric $H_2(g)$ will enter solution; only if pe = 3.8 are the H_2 /H_2O redox reaction and the $H_2(g)$/$H_2(aq)$ Henry's law relationship both satisfied, for the partial of $H_2(g)$ in the normal atmosphere. Corresponding statements apply to oxygen, except in that case the critical pe is 13.6.

Transfer of $H_2(g)$ or $O_2(g)$ between the atmosphere and solution in response to the types of disequilibrium described above is not usually noticeable for two reasons. First, equilibration of the $H_2(aq)$/H_2O/$O_2(aq)$ redox system is very slow in the absence of a catalyst, so under normal circumstances neither hydrogen nor oxygen is consumed or generated at a detectable rate by redox reactions involving water. Second, any gas transfer that does occur proceeds by diffusion across the air/water interface, with no impact on the system that is visible to the naked eye. For these reactions to be easily observed, the pe of solution must be driven to extreme values where the equilibrium partial pressure of hydrogen or oxygen is greater than the ambient total pressure. In such cases, a pure $H_2(g)$ or $O_2(g)$ phase can form in the solution, and visible bubbles of these gases can be generated in the water, especially if a catalyst is available that accelerates the equilibration process. Demonstrations of this phenomenon, in which $H_2(g)$ is generated at the surface of a piece of metal immersed in an acidic solution, are common in introductory chemistry classes. In these demonstrations, the metal serves as both a source of electrons and a catalyst.

Example 9.14

In what pe range will gas bubbles of pure hydrogen be generated in a solution at pH 2.0 if $P_{tot} = 1.0$ bar?

Solution

For hydrogen bubbles to form, the partial pressure of $H_2(g)$ must be >1 bar. According to Equation (9.60), the pe value in a solution at pH 2.0 and in which the activity of $H_2(g)$ is 1.0 is given by

$$\text{pe} = -\text{pH} - \frac{1}{2} \log \{H_2(g)\} = -2.0 - \frac{1}{2}\log 1.0 = -2.0$$

Thus, $H_2(g)$ bubbles should appear if the pe of the solution is less than or equal to -2.0.

9.8.2 THE IMPOSSIBILITY OF REDOX EQUILIBRIUM BETWEEN H_2O AND THE ATMOSPHERE

Based on Equation (9.60), for a solution to be in equilibrium with the atmosphere, which contains H_2 at a partial pressure of 5×10^{-7} bar, the pe and pH must be related by

$$pe = -\frac{1}{2} \log (5 \times 10^{-7}) - pH \qquad \textbf{(9.61a)}$$

$$pe = 3.15 - pH \qquad \textbf{(9.61b)}$$

Similarly, according to Equation (9.58), for a solution to be in equilibrium with atmospheric O_2 (i.e., with an oxygen partial pressure of 0.21 bar), the pe and pH must be related by

$$pe = \frac{1}{4}(83.12 + \log 0.21 - 4pH) \qquad \textbf{(9.62a)}$$

$$pe = 20.6 - pH \qquad \textbf{(9.62b)}$$

Clearly, Equations (9.61b) and (9.62b) cannot be satisfied simultaneously. We therefore conclude that, at their normal atmospheric concentrations, oxygen and hydrogen cannot simultaneously be in redox equilibrium with H_2O in any aqueous solution. To determine the extent of disequilibrium of the $H_2/H_2O/O_2$ system under typical environmental conditions, we can generate the following overall reaction for redox dissociation of water into $H_2(g)$ and $O_2(g)$ (analogous to acid dissociation of H_2O into H^+ and OH^-):

$$2H_2O \leftrightarrow O_2(g) + 4H^+ + 4e^- \qquad K = 10^{-83.12} \qquad \textbf{(9.63)}$$

$$\underline{4H^+ + 4e^- \leftrightarrow 2H_2(g) \qquad\qquad K = 10^{0.0} \qquad\quad \textbf{(9.64)}}$$

$$2H_2O \leftrightarrow O_2(g) + 2H_2(g) \qquad K = 10^{-83.12} \qquad \textbf{(9.65)}$$

Reaction (9.65) indicates that for liquid water to be in equilibrium with gaseous oxygen and hydrogen, the product $\{O_2(g)\}\{H_2(g)\}^2$ must equal $10^{-83.12}$. In the atmosphere, the actual value of that product is $(0.21)(5 \times 10^{-7})^2$, or 5×10^{-14}. Remarkably, this value is more than *69 orders of magnitude larger than the equilibrium value*. The same result would be obtained if we carried out the calculations using dissolved $H_2(aq)$ and $O_2(aq)$ concentrations, if we assumed that these dissolved species were in equilibrium with the atmosphere. Thus, under ambient environmental conditions, there is an enormous driving force favoring the reaction of H_2 and O_2 to form H_2O, regardless of whether the H_2 and O_2 are gases in the atmosphere or dissolved in an aqueous solution.

According to the above calculations, if a solution containing $H_2(aq)$ and $O_2(aq)$ were isolated from the atmosphere and the redox reaction proceeded to equilibrium, essentially all the H_2 would be consumed (the concentration of O_2 is much larger than that of H_2, so the solution would run out of H_2 before a significant fraction of the O_2 was consumed). However, despite the large driving force for such a reaction, the two gases coexist in the non-equilibrated condition because, normally, the rate of the redox reaction is extremely slow and because inputs of $H_2(g)$ and $O_2(g)$ from the atmosphere replenish the dissolved gases as fast as they are consumed by the reaction. As a result of these competing processes, the reaction remains in a continuous steady-state of extreme disequilibrium.

9.8.3 CONSIDERING REDOX ACTIVITY OF WATER WHEN ANALYZING PE OF A SOLUTION

To illustrate the potential effect of the $H_2/H_2O/O_2$ redox system on other redox couples in solution, we next reanalyze the example system considered above (containing sulfide and hypochlorite species) with the assumption that the $H_2/H_2O/O_2$ redox group reaches equilibrium. For the purposes of the example, we assume that the solution is initially in equilibrium with the atmosphere with respect to the gas dissolution reactions; i.e., the concentrations of dissolved O_2 and H_2 are consistent with Henry's law, with $P_{O_2(g)} = 0.21$ bar and $P_{H_2(g)} = 5 \times 10^{-7}$ bar. However, we also assume that the system is closed, i.e., that no gas exchange occurs between the atmosphere and the solution after the initial solution is prepared.

The log C–pe diagram for the system is identical to Figure 9.8, except that we must add lines representing the activities of H_2O, $H_2(aq)$, and $O_2(aq)$ at pH 7. These lines can be drawn based on reactions (9.55) and (9.56) and are the same as those shown in Figure 9.11. The modified diagram is shown in Figure 9.12.

The electron condition (EC) and *TOTe* equation for the system ignoring $H_2/H_2O/O_2$ redox reactions were developed earlier. Using that EC as a template, we can retain H(+I), O(−II), S(−II), and Cl(+I) as the electron reference levels. Then, since four electrons must be removed from reference level species to form one molecule of $O_2(aq)$ [reaction (9.55)], O_2 must be a four-electron-deficient species. Similarly, according to reaction (9.56), $H_2(aq)$ is a two-electron-excess species. The electron balance table for the system is therefore as follows:

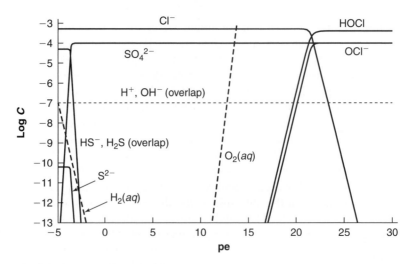

Figure 9.12   Figure 9.8 modified to include equilibria among H_2O, $O_2(aq)$, and $H_2(aq)$.

		e^- Excess			
	-8	**-4**	**0**	**+2**	**Concentration**
Species present at equilibrium	SO_4^{2-}	$O_2(aq)$	H^+, OH^-, H_2O H_2S, HS^-, S^{2-} $HOCl, OCl^-$ Na^+	$H_2(aq)$ Cl^-	
Input species			$HOCl$ $NaHS$		5×10^{-4} 1×10^{-4}
				$H_2(aq)$	Equilibrium w/atmosphere
		$O_2(aq)$			Equilibrium w/atmosphere

Since we are assuming that the dissolved O_2 and H_2 concentrations in the initial solution are in equilibrium with the atmosphere, we can use Henry's law to determine those values.

$$\{O_2(aq)\}_{in} = \frac{\{O_2(g)\}}{H_{O_2}} = \frac{0.21 \text{ bar}}{10^{2.88} \text{ bar/(mol/L)}} = 2.77 \times 10^{-4} \frac{\text{mol}}{\text{L}}$$

$$\{H_2(aq)\}_{in} = \frac{\{H_2(g)\}}{H_{H_2}} = \frac{5 \times 10^{-7} \text{ bar}}{10^{3.1} \text{ bar/(mol/L)}} = 3.97 \times 10^{-10} \frac{\text{mol}}{\text{L}}$$

Because $O_2(aq)$ is a four-electron-deficient species, the initial electron excess contributed by dissolved O_2 is $-4\{O_2(aq)\}$ or $-1.107 \times 10^{-3} M$. Similarly, H_2 is a two-electron-excess species, so it contributes an initial excess of $8 \times 10^{-10} M$ electrons. The terms in the *TOTe* equation are therefore

$$TOTe_{in} = 8 \times 10^{-10} - 1.107 \times 10^{-3}$$

$$TOTe_{eq} = 2\{H_2(aq)\}_{eq} + 2\{Cl^-\}_{eq} - 4\{O_2(aq)\}_{eq} - 8\{SO_4^{2-}\}_{eq}$$

Then, equating $TOTe_{in}$ with $TOTe_{eq}$,

$$-1.107 \times 10^{-3} = 2\{H_2(aq)\} + 2\{Cl^-\} - 4\{O_2(aq)\} - 8\{SO_4^{2-}\}$$

$$4\{O_2(aq)\} + 8\{SO_4^{2-}\} = 2\{H_2(aq)\} + 2\{Cl^-\} + 1.107 \times 10^{-3}$$

The portion of Figure 9.12 containing the point where the *TOTe* is satisfied is shown with expanded axes in Figure 9.13. The equilibrium pe is 13.62. At this pe value, virtually all the Cl is in the reduced form; i.e., it is present as Cl^-; in fact, the computed value of $TOTCl(+I) < 10^{-19}$. On the other hand, virtually all the S is oxidized (present as SO_4^{2-}). While the sulfur speciation is roughly the same as in the prior analysis (in which the $H_2/H_2O/O_2$ equilibria were not considered), the speciation of Cl is different: the Cl is much more completely reduced in the system where the $H_2/H_2O/O_2$ equilibria are considered, because in the current analysis, $Cl(+I)$ oxidizes not only the sulfide species (to SO_4^{2-}) but also H_2O, to form $O_2(aq)$. As a result, oxygen is generated in the solution; and, at equilibrium, the activity of $O_2(aq)$ is greater than that in the initial solution. If this solution were

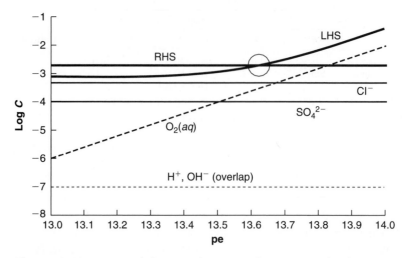

Figure 9.13   Expanded version of a portion of Figure 9.12, also showing the RHS and LHS of the electron condition.

allowed to equilibrate with atmospheric O_2, an amount of $O_2(aq)$ equal to the amount that was generated by the redox reaction would transfer into the gas phase, thereby returning the $O_2(aq)$ concentration to its original value.

The equilibrium speciation in the systems with and without consideration of the $H_2/H_2O/O_2$ redox reactions is summarized in the following table.

Equilibrium speciation in a system prepared by adding 10^{-4} S($-$II) and 5×10^{-4} Cl($+$I) to water at a fixed pH of 7.0

	$H_2/H_2O/O_2$ System Does Not Participate	$H_2/H_2O/O_2$ Equilibrates; $\{O_2\}_{init} = 2.64 \times 10^{-4}$, $\{H_2\}_{init} = 4.0 \times 10^{-10}$
pe	21.24	13.62
HOCl	8.00×10^{-5}	5.48×10^{-20}
OCl$^-$	2.00×10^{-5}	1.38×10^{-20}
Cl$^-$	4.00×10^{-4}	5.00×10^{-4}
H_2S	1.03×10^{-203}	1.00×10^{-142}
HS$^-$	1.03×10^{-203}	1.00×10^{-142}
S^{2-}	1.03×10^{-210}	1.00×10^{-149}
SO_4^{2-}	1.0×10^{-4}	1.00×10^{-4}
$O_2(aq)$	N/A	3.02×10^{-4}
$H_2(aq)$	N/A	4.57×10^{-45}

By comparison of the pe°(W) values [-3.67 for SO_4^{2-}/H_2S and $+14.50$ for $O_2(aq)/H_2O$], we see that sulfide is a stronger reducing agent than H_2O at pH 7.0. Therefore, in the example system, sulfide has a greater affinity for the electrons released by $Cl(+I)$ than does H_2O. As a first approximation, we can model the reaction as though all the sulfide is oxidized to sulfate before any H_2O is oxidized to O_2, much as we sometimes make the approximation that stronger bases become fully protonated before weaker ones are protonated at all. Thus, the electrons released by $Cl(+I)$ species can be thought of as first reacting with sulfide, consuming 8×10^{-4} M electrons, after which the remaining 1×10^{-4} M $Cl(+I)$ species oxidize 0.5×10^{-5} M H_2O to $O_2(aq)$.

---

a. Construct a log C–pe diagram showing the speciation of nitrogen in a pH 8.0 solution that is in equilibrium with atmospheric $N_2(g)$ ($P_{N_2} = 0.79$ bar). Consider the possibility that dissolved N can be present as $N(0)$ [in the form of $N_2(aq)$], $N(-III)$ (as NH_3 and NH_4^+), or $N(+V)$ (as NO_3^-). The following equilibrium data apply:	**Example 9.15**

$$H_{N_2} = 1513 \text{ bar/(mol/L)}$$

$$K_{a,NH_4^+} = 9.25$$

$$pe^\circ_{NO_3^-/N_2(g)} = 21.03 \quad \text{for the reaction } 2NO_3^- + 12H^+ + 10e^- \leftrightarrow N_2(g) + 6H_2O$$

$$pe^\circ_{N_2(g)/NH_4^+} = 4.64 \quad \text{for the reaction } N_2(g) + 8H^+ + 6e^- \leftrightarrow 2NH_4^+$$

b. If the solution in part (*a*) is in redox equilibrium with atmospheric $O_2(g)$, what is the nitrogen speciation?

Solution

a. The line representing the activity of $N_2(aq)$ is easy to draw on a log C–pe diagram for this system because the solution is in equilibrium with atmospheric $N_2(g)$. The activity and concentration of $N_2(aq)$ are therefore fixed by Henry's law at a value given by

$$\{N_2(aq)\} = \frac{\{N_2(g)\}}{H_{N_2}} = \frac{0.79}{1513} = 5.2 \times 10^{-4}$$

The activities of NO_3^- and NH_4^+ in the system at any pe can be determined by the pe° values for redox reactions between those species and $N_2(g)$ and the known solution pH, as follows. The reactions for the $NO_3^-/N_2(g)$ and $N_2(g)/NH_4^+$ redox couples are

$$2NO_3^- + 12H^+ + 10e^- \leftrightarrow N_2(g) + 6H_2O \quad pe^\circ = 21.03 \quad \log K = 10(pe^\circ) = 210.30$$

$$N_2(g) + 8H^+ + 6e^- \leftrightarrow 2NH_4^+ \quad pe^\circ = 4.64 \quad \log K = 6(pe^\circ) = 27.84$$

Substituting the known values of pH and the activity of $N_2(g)$ into the above equations yields expressions for the activities of nitrate and ammonium ions:

$$210.30 = \log \frac{\{N_2(g)\}}{\{NO_3^-\}^2\{H^+\}^{12}\{e^-\}^{10}} = \log\{N_2(g)\} - 2\log\{NO_3^-\} + 12pH + 10pe$$

$$\log\{NO_3^-\} = -57.20 + 5pe$$

$$27.84 = \log \frac{\{NH_4^+\}^2}{\{N_2(g)\}\{H^+\}^8\{e^-\}^6} = 2\log\{NH_4^+\} - \log\{N_2(g)\} + 8pH + 6pe$$

$$\log\{NH_4^+\} = -18.13 - 3pe$$

Finally, the activity of $NH_3(aq)$ can be determined, since at pH 8.0 it is always 1.25 log units less than the activity of NH_4^+. Thus

$$\log \{NH_3\} = \log \{NH_4^+\} - 1.25 = -18.13 - 3pe - 1.25 = -19.38 - 3pe$$

Alternatively, we can develop the diagram with the aid of a software package; this process turns out to be much simpler. For this analysis, we choose NO_3^-, $N_2(aq)$, NO_3^-, H^+, and e^- as *components*. The key inputs, in addition to those characterizing the acid/base reactions, are one Type 3b *species*, $\overline{N_2(g)}$, to account for equilibrium with the gas phase and two Type 3c *species*, $\overline{NO_3^-/N_2(aq)}$ and $\overline{N_2(aq)/NH_4^+}$, to account for the redox reactions.

Log $K_{\overline{N_2(g)}}$ can be computed using the rule for Type 3b gases in Chapter 7. Because the dissolved *species* with the same stoichiometry as the gas is $N_2(aq)$, which is a *component*, log K for forming it is 0.0, and the calculation of log $K_{\overline{N_2(g)}}$ yields

$$\log K_{\text{Type 3b}} = \log K_{\substack{\text{corresponding} \\ \text{species}}} \underset{\text{dissolved species}}{} + \log H_i - \log P_i$$

$$\log K_{\overline{N_2(g)}} = 0.0 + 3.18 - (-0.10) = 3.28$$

The log K values for the two Type 3c *species* can be determined by combining the given redox reactions with the Henry's law reaction. The results are that log K for $\overline{NO_3^-/N_2(aq)}$ and $\overline{N_2(aq)/NH_4^+}$ are 207.12 and 31.02, respectively. The calculation is left as an exercise. The complete input table and the log C–pe diagram are shown below.

| Species | Type | Component | | | | | Log K |
		NO_3^-	$N_2(aq)$	NH_4^+	H^+	e^-	
NO_3^-	1	1	0	0	0	0	0.00
$N_2(aq)$	1	0	1	0	0	0	0.00
NH_4^+	1	0	0	1	0	0	0.00
H^+	1	0	0	0	1	0	0.00
OH^-	2	0	0	0	-1	0	-14.00
$NH_3(aq)$	2	0	0	1	-1	0	-9.25
$\overline{e^-}$	3a	0	0	0	0	1	Various
$\overline{H^+}$	3a	0	0	0	1	0	8.0
$\overline{N_2(g)}$	3b	0	1	0	0	0	3.28
$\overline{NO_3^-/N_2(aq)}$	3c	2	-1	0	12	10	207.12
$\overline{N_2(aq)/NH_4^+}$	3c	0	1	-2	8	6	31.02
e^-	6	0	0	0	0	1	0.00

As in the case of log C–pH diagrams for systems in which one of the acid/base species is a dissolved gas that is present at fixed activity, the activities of all the species are represented by straight lines. Note that the slope of each line equals the number of electrons necessary to form one molecule of that species from the species present at fixed activity [in this case $N_2(aq)$].

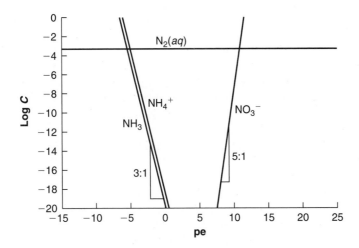

b. Equation (9.62b) indicates that a solution in redox equilibrium with atmospheric oxygen must have a pe value given by

$$pe = 20.6 - pH$$

Thus, a pH 8.0 solution in equilibrium with atmospheric $O_2(g)$ must have pe = 12.6. Plugging this value into the equations derived in part (*a*) for the various nitrogen species, we find

$$\{N_2(aq)\} = 5.2 \times 10^{-4} \qquad \log\{NO_3^-\} = -57.20 + 5pe = 5.80$$
$$\log\{NH_4^+\} = -18.13 - 3pe = -55.97 \qquad \log\{NH_3\} = \log\{NH_4^+\} - 1.25 = -57.23$$

The result indicates that if redox equilibrium with atmospheric O_2 were reached, essentially all the dissolved nitrogen in the system would be present in the most oxidized form, i.e., as NO_3^-. Furthermore, the theoretical equilibrium value of $\{NO_3^-\}$ is unattainably large ($10^{6.1}$). This result means that, as a practical matter, the system would never reach equilibrium with atmospheric N_2 and O_2. Rather, one of two scenarios would apply. Either the system would reach some non-equilibrium, steady-state condition in which N_2 and O_2 continuously dissolve from the gas phase and are slowly but continuously removed from solution by the redox reaction; or, if the gas-phase volume were limited, so much oxygen and nitrogen would dissolve and react that the gas-phase concentrations of these species would be reduced substantially. In most natural systems, the former situation applies, with the redox reaction proceeding so slowly that dissolved N_2 and O_2 are close to equilibrium with the atmosphere and far from redox equilibrium with one another.

9.9 REDOX TITRATIONS, REDOX BUFFERING, AND THE GEOCHEMICAL REDOX SEQUENCE

Just as the pH of a solution responds to additions of acids or bases, the pe responds to additions of reducing or oxidizing agents. And, like the changes in pH, the changes in pe depend strongly on the initial conditions: pe changes only modestly when redox-active species enter solution under some conditions, and it changes dramatically when the same species enter under other conditions. Put another way, the pe of a solution can be well or poorly buffered.

If a solution is titrated with a strong reductant, electrons transfer from that titrant to the oxidized species that has the greatest affinity for them under the extant conditions (assuming the solution equilibrates after each addition of titrant). To a good approximation, each oxidant in the system is converted completely from its oxidized to its reduced form before any of the next oxidant reacts.

For instance, consider a solution that is buffered at pH 8.0 and that contains 10^{-3} *M TOT*Cl distributed among HOCl, OCl$^-$, and Cl$^-$, and 5×10^{-4} *M TOT*N, distributed among NO$_3{}^-$, NO$_2{}^-$, NH$_3$(*aq*), and NH$_4{}^+$. Assume also that the initial pe of the solution is 25 (highly oxidizing conditions), that the redox reactions equilibrate rapidly, and that the H$_2$/H$_2$O/O$_2$ system is kinetically inactive. A log *C*–pe diagram for the system (Figure 9.14) indicates that HOCl, OCl$^-$, and NO$_3{}^-$ are the only species present at significant concentrations in such a solution.

Now consider what happens as a strong reductant, say, NaHS, is gradually added to the system. Since pe°_{pH} for the SO$_4{}^{2-}$/HS$^-$ couple under the given conditions is -4.79, the stable S species in the initial solution is SO$_4{}^{2-}$. Therefore, there will be a strong driving force for transfer of electrons from the sulfide species to hypochlorite species, which are the strongest oxidants in the system. As occurred in the HOCl/HS$^-$ example system explored earlier in the chapter, the added sulfide is expected to react quantitatively with Cl($+$I) species, maintaining the equilibrium concentration of sulfide at undetectable levels until essentially all the hypochlorite has been reduced to Cl$^-$.

When the titration starts, the pe of the solution would be expected to decline rapidly to near 21 ($pe^\circ_{8.0}$ for the OCl$^-$/Cl$^-$ couple), where it would hover until essentially all the Cl($+$I) had reacted. The dose of NaHS required to consume all

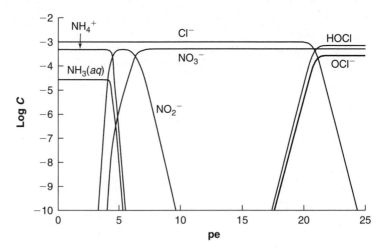

Figure 9.14   Log *C*–pe diagram for a system containing 10^{-3} *M TOT*Cl and 5×10^{-4} *M TOT*N, assuming that N can exist in the -3, $+3$, and $+5$ oxidation states.

the Cl(+I) is easy to compute by an electron balance: each Cl(+I) molecule undergoing reduction consumes two electrons, and each S(−II) molecule undergoing oxidation provides eight electrons, so the NaHS requirement to reduce all the Cl(+I) is TOTCl/4, or 2.5×10^{-4} M. Further additions of NaHS will drive the pe rapidly to a value close to the next-lower pe^o_{pH} of a significant redox couple in the system (in this case, the NO_3^-/NO_2^- couple, for which $pe^o_{8.0}$ is ~6.3), where it will be buffered until essentially all the N(+V) is converted to N(III).

If the titration were continued, the pe would be poorly buffered between 6.3 and 4.4, but then the NO_2^-/NO_4^+ couple would provide substantial buffering near pe 4.4. Once almost all the nitrogen had been reduced to the −III state, the pe would drop quickly to near pe^o_{pH} for the SO_4^{2-}/HS^- couple (−4.79), where it would remain.

Like the pH, the pe of a solution can also be buffered by equilibration of the solution with other phases. We encountered this situation already when considering the $H_2/H_2O/O_2$ system. Analysis of that system indicated that a solution that is buffered at pH 7.0 and is equilibrated with atmospheric O_2 must have a pe of 13.6. If an oxidant were added to the solution that could oxidize water to O_2, some $O_2(aq)$ would volatilize in order to maintain equilibrium with the O_2 in the atmosphere. The net effect would be that the pe would be maintained (buffered) exactly at the original value. An analogous process would buffer the pe if a reducing agent were added, except in this case the dissolved O_2 would oxidize the reducing agent and O_2 would then enter from the gas phase to replenish the $O_2(aq)$ concentration.

Solids can also buffer the pe of solutions, acting in the same way as the atmosphere in the example above. That is, they provide both a source and a sink of oxidizing/reducing power for the solution by dissolving or precipitating in response to changes in solution composition. One important example of such a system involves the simultaneous equilibration of the solid calomel, $Hg_2Cl_2(s)$, and pure liquid mercury, Hg(l), in a solution containing a high concentration of Cl^-. The highly buffered pe of this mixture is the key feature of calomel reference electrodes, which are discussed more fully later in the chapter.

We can compute the pe of such a mixture as a function of the chloride activity in the solution by noting that if the two pure condensed phases are both in redox equilibrium with the solution, they must be in equilibrium with each other as well. The corresponding reaction, which has a pe^o value of 4.53, is

$$\frac{1}{2} Hg_2Cl_2(s) + e^- \leftrightarrow Hg(l) + Cl^- \qquad \textbf{(9.66)}$$

Because the activity of each pure phase is 1.0, the Nernst equation for this system yields the following simple relationship between pe and $\{Cl^-\}$:

$$pe = pe^o - \frac{1}{1} \log \frac{\{Hg(l)\}\{Cl^-\}}{\{Hg_2Cl_2(s)\}^{1/2}} = pe^o - \log\{Cl^-\} \qquad \textbf{(9.67)}$$

$$pe = 4.53 - \log\{Cl^-\} \qquad \textbf{(9.68)}$$

Equation (9.68) indicates that the pe of a solution in equilibrium with both liquid mercury and calomel depends only on the dissolved chloride activity.

Thus, by placing both of those phases in a solution in which the activity of Cl^- is approximately constant, we can fix (buffer) the pe of the solution. If a reducing agent enters the solution, some calomel is reduced to $Hg(l)$; but as long as some calomel remains in the system, its activity is unchanged by the reaction. Similarly, even though liquid mercury is generated when the calomel is reduced, its activity remains fixed at 1.0. The only dissolved species whose concentration changes is Cl^-; but if that concentration is somehow held almost constant, the pe of the solution remains essentially constant as well.

When the $Hg_2Cl_2(s)/Hg(l)/Cl^-$ system described above is used in reference electrodes, the Cl^- activity is maintained at a very nearly constant value by adding much more Cl^- to the solution than the amount that is likely to be generated or consumed by inter-conversion of $Hg_2Cl_2(s)$ and $Cl(l)$ when the electrode is used. Typically, the solution is made by saturating pure water with KCl salt, which yields a solution with a Cl^- activity of approximately 2.5. Such electrodes are called *standard calomel electrodes* (SCEs).

Example 9.16

   a. What is the pe of the solution in an SCE if the electrode is filled with saturated KCl, with $\{Cl^-\} = 2.5$?

   b. Although the oxidation number of mercury in calomel is $+1$, dissolved Hg(I) is unusual in that it occurs primarily as a dimer, Hg_2^{2+}. Because of this, the solubility product (K_{s0}) of calomel is defined by the following reaction:

$$Hg_2Cl_2(s) \leftrightarrow Hg_2^{2+} + 2Cl^- \qquad \log K_{s0} = -17.84 \qquad \textbf{(9.69)}$$

Use the solubility product and the fact that pe° for the $Hg^{2+}/Hg(l)$ couple is 14.42 to compute the activities of Hg^{2+} and Hg_2^{2+} in the electrode solution. Are these activities well buffered?

Solution

   a. The pe can be computed by direct substitution of $\{Cl^-\} = 2.5$ into Equation (9.68). The result is that pe $= 4.13$.

   b. Based on K_{s0} for calomel and the known Cl^- activity, the activity of Hg_2^{2+} in a solution equilibrated with calomel is

$$\{Hg_2^{2+}\} = \frac{K_{s0,Hg_2Cl_2(s)}}{\{Cl^-\}^2} = \frac{10^{-17.84}}{(2.5)^2} = 10^{-18.64} \qquad \textbf{(9.70)}$$

Similarly, the activity of Hg^{2+} can be computed from pe° for the $Hg^{2+}/Hg(l)$ couple and the pe of the solution, obtained in part (*a*):

$$Hg^{2+} + 2e^- \rightarrow Hg(l)$$

$$pe = pe^\circ - \frac{1}{2}\log\frac{\{Hg(l)\}}{\{Hg^{2+}\}}$$

$$4.13 = 14.42 - \frac{1}{2}\log\frac{1.0}{\{Hg^{2+}\}}$$

$$\log\{Hg^{2+}\} = -20.59$$

The pe and Cl^- activity of the solution are both highly buffered, and the above analysis indicates that the activities of Hg^{2+} and Hg_2^{2+} depend solely on pe and $\{Cl^-\}$; so the activities of the dissolved mercury species are highly buffered as well, even though the values of those activities are exceedingly small.

Natural redox titrations are carried out continuously near the bottom of water bodies with intense bioactivity. The titrations are mediated by organisms that oxidize organic matter and reduce the strongest oxidant available to them. In this case, the preference for the strongest oxidant can be viewed as a purely chemical phenomenon or as a result of natural selection: the strongest oxidant is the one that is farthest from equilibrium with the organic matter, so the reaction between it and organic matter releases the most Gibbs energy. Organisms that can mediate these reactions and capture the energy are at a competitive advantage over others, so they proliferate until essentially all of that oxidizing agent is consumed. Thereafter, either those organisms or, more often, a different microbial community starts utilizing the next most powerful oxidant to gain energy from the remaining organic matter. When this sequence occurs in lake sediments, the major oxidants, in order of preference, are typically $O_2(aq)$, NO_3^-, $Mn(+IV)$, $Fe(+III)$, SO_4^{2-}, and, finally, organic molecules that are more oxidized than the organic matter being used as food.

The sequence described above might occur over time in a batch system, but in a natural system, it is more often manifested as a steady-state gradient in space, especially in sediments. That is, organic matter diffuses downward from the water-sediment interface, along with the dissolved oxidants [$O_2(aq)$, NO_3^-, and SO_4^{2-}]. In the upper sediment layers, the oxygen is consumed, while the NO_3^- and SO_4^{2-} remain in solution. At some depth, the oxygen is almost completely consumed, and the local organisms use NO_3^- as the electron acceptor for their energy-producing metabolic reactions.

Once all the NO_3^- is depleted (at a lower depth), other groups of organisms use the available Fe and Mn, which are often present as solids in the sediments. When these metal atoms are reduced to their $+II$ state, they typically dissolve and diffuse away from the solid; those that diffuse upward (to more oxidizing regions) might reprecipitate. At still lower depths, the Fe and Mn are depleted, and organisms use SO_4^{2-} as the oxidant. Finally, when even that source of oxidizing power is consumed, some organisms can obtain relatively small amounts of energy by mediating redox reactions in which complex organic molecules are converted to carbonate species [$C(+IV)$] and methane [$C(-IV)$]. In these reactions, referred to as *fermentation* reactions, electrons are simply transferred among different carbon atoms in the organic molecules, without the direct involvement of any other inorganic chemicals.

9.10 PE–PH PREDOMINANCE AREA DIAGRAMS

Because pe can vary by many orders of magnitude over time and space, and because non-dominant redox species are often present in insignificant concentrations, pe–pH predominance area diagrams offer a convenient way to describe the major changes expected in speciation in redox-active systems. In such diagrams, the dominant form of each element (considering both acid/base and redox equilibria) is identified for all possible combinations of pH and pe, but the concentrations of non-dominant species are not shown.

The procedure for preparing such a diagram is identical to the procedure described previously for other types of predominance area diagrams. Specifically, we choose one species arbitrarily and use equilibrium constant expressions relating that species to each of the other species in the system to draw a line in pe–pH space, separating the region where the species of interest is dominant over the other species in that reaction. The ultimate result, considering the lines comparing the species of interest with all its conjugate partners, is either the identification of a region where that species is dominant or a conclusion that the species is never dominant. We then choose a second species and repeat the procedure, continuing until the dominance regions of all possible species have been identified. Although the procedure can start with any species and proceed in any order, it is often convenient to proceed systematically, e.g., from the most oxidized and most acidic species to progressively more reduced and alkaline species.

As an example, the steps involved in preparing a pe–pH predominance area diagram for nitrogen are described next, considering the species NH_3, NH_4^+, NO_2^-, and NO_3^-. The oxidation number of N is -3 in NH_3 and NH_4^+, $+3$ in NO_2^-, and $+5$ in NO_3^-.

In this analysis, we arbitrarily choose NH_4^+ as the first species whose area of dominance will be determined. The region in which NH_4^+ dominates over NH_3 is trivial to identify, because these two species have the same oxidation number and form an acid/base conjugate pair. As a result, NH_4^+ dominates over NH_3 at $pH < pK_a$, regardless of pe.

The calculations to determine the other boundaries of the region where NH_4^+ predominates can be simplified by considering the generic equation for the line along which the reduced and oxidized species of any given redox reaction have equal activity. Recall that, earlier in the chapter, we applied the Nernst equation to the generic redox reaction shown below to derive Equation (9.28b).[9]

$$Ox + n_H H^+ + n_e e^- \leftrightarrow Red \qquad \textbf{(9.71)}$$

$$pe = pe^\circ - \frac{n_H}{n_e} pH + \frac{1}{n_e} \log \frac{\{Ox\}}{\{Red\}} \qquad \textbf{(9.28b)}$$

---

[9]Note that the manipulation of the Nernst equation leading to Equation (9.28b) assumes that the oxidized and reduced species have equal stoichiometric coefficients. If they do not [e.g., for comparison of $N_2(aq)$ and $NH_3(aq)$], then the line along which the activities are equal depends on the total concentration of the constituent in the system.

Therefore, if the activities of the oxidized and reduced species equal one another, the following relationship applies:

$$\text{pe} = \text{pe}^{\circ} - \frac{n_H}{n_e}\text{pH} \qquad \text{when} \qquad \{\text{Red}\} = \{\text{Ox}\} \qquad \textbf{(9.72)}$$

Applying Equation (9.72) to the reactions of NH_4^+ with NO_2^- and NO_3^-, we obtain the following:

Red/Ox	pe°	n_H	n_e	Equation for Line Along Which {Red} = {Ox}
NH_4^+/NO_2^-	15.08	8	6	$\text{pe} = 15.08 - \frac{4}{3}\text{pH}$
NH_4^+/NO_3^-	14.88	10	8	$\text{pe} = 14.88 - \frac{5}{4}\text{pH}$

Based on either the equilibrium constant expression or the reaction stoichiometry, it should be clear that in each case NH_4^+ dominates at lower pe or lower pH than the values on the line, and NO_2^- or NO_3^- dominates at higher pe or pH.

Having compared the concentration of NH_4^+ with each of the other N-containing species in the system, we can identify its region of dominance, as shown in Figure 9.15.

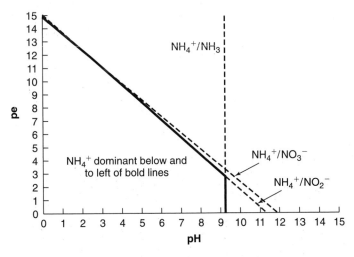

Figure 9.15 The NH_4^+ region of dominance in a system also containing NH_3, NO_2^-, and NO_3^-. Along each dashed line, the activity of NH_4^+ equals that of another N-containing species in pe–pH space. The bold line follows the NO_2^-/NH_4^+ line from pH 9.25 to 2.46, and the NO_3^-/NH_4^+ line at pH < 2.46.

The same procedure can be used to determine the equations of the lines separating the regions of dominance for the pairs NH_3/NO_2^-, NH_3/NO_3^- and NO_2^-/NO_3^-, yielding the following results:

Red/Ox	pe°	n_H	n_e	Equation for Line Along Which {Red} = {Ox}
NH_3/NO_2^-	13.54	7	6	$pe = 13.54 - \dfrac{7}{6}pH$
NH_3/NO_3^-	13.73	9	8	$pe = 13.73 - \dfrac{9}{8}pH$
NO_2^-/NO_3^-	14.28	2	2	$pe = 14.28 - pH$

Using the information in the first two rows of this table and the K_a expression for NH_4^+, the area of dominance of NH_3 can be circumscribed. Similarly, the information in the final row can be combined with that already shown to define the areas of dominance of NO_2^- and NO_3^-. The final diagram is shown in Figure 9.16.

It is worth noting that, like predominance area diagrams covering ligand–pH space, those covering pe–pH space are convenient for identifying regions where the system is well buffered. That is, at any boundary where the dominant species changes, if the reaction between the species on opposite sides of the line involves H^+ exchange, then the pH of the solution will be buffered, and if it involves e^- exchange, the pe will be buffered. The larger the number of protons or electrons exchanged (and, of course, the larger the total concentration of the chemical in the system), the greater the buffering capacity will be.

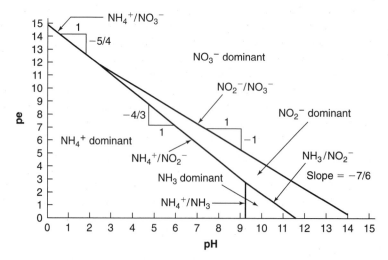

Figure 9.16   Completed pe–pH predominance area diagram considering NH_4^+, NH_3, NO_2^-, and NO_3^-.

9.11 REDOX REACTIONS AND ELECTROCHEMISTRY

9.11.1 ANODES, CATHODES, AND PATHWAYS OF ELECTROCHEMICAL REACTIONS

Throughout this chapter, the analogies between electron transfer in redox reactions and proton transfer in acid/base reactions have been emphasized. We now consider the most important *difference* between these reactions, and the characteristic that makes redox reactions unique among chemical reactions: the ability of electrons to travel easily through wires or other electrical conductors. This characteristic, in conjunction with the slowness of many redox reactions via direct molecular collisions, allows a degree of external control over redox reactivity that is not available when one is dealing with most other types of reactions.

Reactions that proceed very slowly are understood to have high activation energies. In simplistic terms, a high activation energy in a redox reaction implies that there is no "convenient" path for the electrons to take as they move from the species being oxidized to the one being reduced, and a great deal of energy must be provided either to free the electrons from the reductant or to get the electrons past some other barrier interfering with their attachment to the oxidant. The kinetic inhibition can be surmounted either by providing energy to the system to overcome the activation energy barrier along the existing path or by providing an alternative reaction pathway with a less substantial activation energy barrier. Examples of these two alternatives include providing a spark to start coal burning and using enzymes to catalyze metabolic reactions, respectively.

The surfaces of pure metals are good catalysts for many redox reactions occurring in solution. For instance, dissolved molecular hydrogen and oxygen react with one another to form water at a negligible rate if the only pathway for electron transfer is direct collision of the molecules, but the reaction proceeds immensely more quickly in the presence of metallic platinum. The catalysis is attributed to the tendency of hydrogen and oxygen molecules to bind to the metal's surface (a process referred to as *adsorption*) and to exchange electrons readily with the metal. This fact, in conjunction with the availability of delocalized, highly mobile electrons in the metal, allows electron transfer to proceed more rapidly via a hydrogen-to-metal-to-oxygen pathway than a direct hydrogen-to-oxygen pathway (Figure 9.17).

When the reaction is catalyzed by the platinum, the H_2 oxidation and O_2 reduction half-cell reactions are still linked in the sense that they can only proceed in tandem, with electrons being released at the same rate that they are consumed.[10] However, unlike in the direct-collision pathway, the two electron transfer steps can take place at two different locations. That is, because the electrons

---

[10]In theory, electrons could simply accumulate in the metal, allowing the oxidation reaction to proceed independently of the reduction reaction. However, in reality, the number of electrons that can accumulate in a metal beyond those that are part of the metallic structure is extremely small, so the rates of the two half-cell reactions are in fact closely linked.

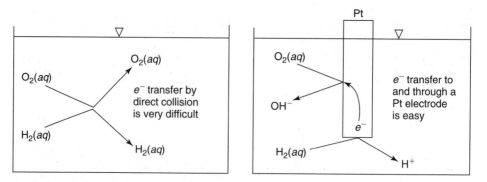

Figure 9.17   Schematic contrasting the rates of electron transfer from $H_2(aq)$ to $O_2(aq)$ in the absence and presence of metallic Pt as a catalyst.

are so mobile in the metal, it is not necessary for the H_2 and O_2 to bind to the platinum adjacent to one another. The location on the metal where the oxidation occurs is called the **anode,** the location where the reduction occurs is called the **cathode,** and both the anode and cathode are referred to as **electrodes.** Anodes and cathodes might be distributed more or less randomly around the surface of the metal, or they might be localized.

9.11.2   QUANTIFYING THE DRIVING FORCE AND ENERGY CHANGE IN ELECTROCHEMICAL REACTIONS

Next, consider a slightly different scenario, in which the water is separated into two portions and a Pt electrode is immersed in each. The electrodes can be connected electrically via a switch, but initially the switch is in the open (disconnected) position. The two solutions are then equilibrated with different gases: one gas contains H_2 at its atmospheric partial pressure, but no O_2, and the other contains O_2 at *its* atmospheric partial pressure, but no H_2. Also, assume that both solutions contain identical concentrations of salts that are not redox-active (e.g., each might contain some NaCl), that both are well buffered at pH 7.0 and 25°C, and that the solutions are connected through a semipermeable barrier that allows small amounts of ions to diffuse from one solution to the other, but does not allow extensive mixing of the two solutions. The scenario is depicted schematically in Figure 9.18.

When the gases enter the solution, some H_2 or O_2 adsorbs to each electrode, and each redox half-cell reaction approaches equilibrium. As the reactions proceed, electrons leave the cathode to reduce oxygen molecules in one half-cell, and electrons enter the anode as hydrogen molecules are oxidized in the other half-cell. As a result of the reactions, OH^- accumulates in the solution where $O_2(aq)$ is reduced, and H^+ accumulates in the solution where $H_2(aq)$ is oxidized. The generation of these reaction products causes a temporary charge imbalance between the two solutions that is rapidly eliminated by diffusion of cations

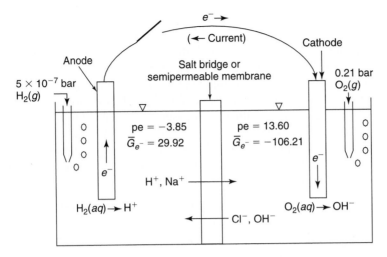

Figure 9.18 Schematic of an electrochemical cell. The anodic half-cell is in equilibrium with atmospheric $H_2(g)$ but no $O_2(g)$, and the cathodic cell is in equilibrium with atmospheric $O_2(g)$ but no $H_2(g)$. The pe values shown apply when the circuit is open, and the flows of electrons and salt ions apply when the circuit is closed. The molar Gibbs energy of the electrons is shown in kilojoules per mole.

(Na^+ and H^+) from the anode chamber into the cathode chamber, and of anions (Cl^- and OH^-) in the opposite direction.

Applying the Nernst equation to each half-cell, we can compute that once equilibrium is attained, the pe will be $+13.6$ and -3.85 in the solutions in contact with the oxygen- and hydrogen-containing gases, respectively. Although the pe of each solution changes substantially as the solution equilibrates with the electrode, the amount of reaction that occurs (in terms of the mass of oxygen or hydrogen consumed, or the number of electrons transferred into or out of the electrodes) is exceedingly small. Correspondingly, the charge imbalance that is generated and the amount of diffusion needed to counter it are also very small.

Consider now the thermodynamic status of electrons in the electrodes. As in any phase, the activity of the electrons (or any chemical species) can be characterized based on their concentration and their environment. The concentration of free electrons in each metal is huge compared to the number that transfer in or out, so this concentration is effectively constant as the redox reactions proceed. However, the environment in the electrodes, and specifically their electrical potential, changes substantially when even a small number of electrons enter or leave. Therefore, this is the factor that has the greatest effect on the electrons' potential energy and on which we will focus.

The electrical potential in each electrode is generated by the excess or deficiency of electrons resulting from the reactions with dissolved hydrogen or oxygen. This potential is manifested nearly uniformly throughout each electrode. Thus, because electrons transfer into the anode, it will be characterized by an almost uniform (negative) potential, Ψ_{anode}; correspondingly, the cathode will develop a positive potential, $\Psi_{cathode}$.

As each half-cell reaction proceeds, the magnitude of the corresponding Ψ increases. The reaction continues until the electrons in the electrode have the same electrochemical potential as those in the solution, at which point no potential energy would be released by transfer of additional electrons, and the reaction has reached equilibrium. Recalling that E_H can be defined as the electrical potential to which electrons would have to be exposed in order to have the same electrochemical potential energy as the electrons in a solution of interest, we see that the scenario in the two half-cells is, in essence, the experimental realization of that conceptual definition. That is,

$$\Psi_{cathode} = E_{H,\text{oxygenated} \atop \text{solution}} = (0.059 \text{ V}) \text{ pe}_{\text{oxygenated} \atop \text{solution}} = +0.802 \text{ V}$$

$$\Psi_{anode} = E_{H,\text{hydrogenated} \atop \text{solution}} = (0.059 \text{ V}) \text{ pe}_{\text{hydrogenated} \atop \text{solution}} = -0.227 \text{ V}$$

The electrical potentials of the electrodes propagate into solution a short distance, orienting the water molecules near the electrode surfaces and attracting or repelling ions in much the same way that ions orient the water molecules that surround them.

Now consider what happens if the switch between the two electrodes is closed. Because the electrons in the anode are at a higher electrochemical potential than those in the cathode, they will travel through the wire from the former to the latter. This process serves as a sink for electrons that would otherwise accumulate in the anode, while replenishing the electron supply in the cathode. Although the E_H values of the solutions remain at $+0.802$ and -0.227 V (being the hypothetical electrical potential to which electrons *would* have to be exposed to be in equilibrium with the solutions), $\Psi_{cathode}$ and Ψ_{anode} become lower than $+0.802$ V and higher than -0.227 V, respectively. Thus, the electrochemical potential in the entire system decreases continuously in the sequence of hydrogenated solution > anode > cathode > oxygenated solution, driving electrons in the corresponding direction. The electron flow and reactions continue until the electron activity is identical throughout the system; in theory, if the gases are continuously supplied and the pH is maintained in both cells, the reaction will continue indefinitely.

The system described above comprises a complete electrical circuit, with the driving force for electron flow through the wire being provided by the chemical (redox) disequilibrium between the two parts of the system. Note that this scenario is essentially identical to one in which the H_2 and O_2 and a Pt electrode are all present in a single solution; the only difference is that, in the current scenario, all the micro anodes are in one part of the system (the one containing H_2) and all the micro cathodes are in the other, so the electrons all flow in the same direction. In essence, we have rectified the electron flow.

If, in the system with rectified electron flow, an electrical appliance is placed in the circuit, the Gibbs energy of the overall redox reaction will be expended "pushing" the electrons through the appliance. Thus, the chemical reaction could be used to drive an electrical system. This is, of course, the principle of operation of a chemical battery. In such systems, a solution containing a large salt concentration called a *salt bridge* is sometimes used to link the two solutions to ensure that diffusion of salt ions does not become the bottleneck limiting the reaction rate.

9.11.3   REDOX REACTIONS IN SYSTEMS WITH EXTERNALLY IMPOSED ELECTRICAL POTENTIALS

The preceding discussion describes a situation in which electrodes are inserted into solutions to catalyze reactions that are favorable but are kinetically inhibited. We next explore how electrical potentials applied to electrodes and controlled externally can facilitate or impede redox reactions.

Continuing with the same example system, imagine that an external circuit is used to force the potential of the electrode in the hydrogenated solution to a value that is 1.5 V more positive than of the other electrode. The exact values of the potentials are not critical, but to make the example concrete, we will assume that the electrodes in the hydrogenated and oxygenated solutions have (Ψ_{ext}) potentials of $+0.9$ and -0.6 V, respectively.

Near each electrode, the electrical potential propagates into solution some distance, eventually decaying to zero. Although the profile of potential versus distance from each electrode is important for a variety of practical reasons, all that matters for now is that immediately adjacent to each electrode, the electrical potential is very nearly equal to that in the electrode. Then, if a packet of solution with the bulk composition is brought near the electrode, its E_H value will change from the E_H of the bulk solution to $E_{H,bulk} + \Psi_{ext}$, i.e.,

$$E_H|_{\substack{\text{with imposed} \\ \text{potential}}} = E_H|_{\substack{\text{no imposed} \\ \text{potential}}} + \Psi_{ext} \tag{9.73}$$

Thus, in the example system, the E_H values in the two solutions near the electrodes are

Hydrogenated solution: $E_{H,\text{hydrogenated}} = -0.227 \text{ V} + 0.900 \text{ V} = +0.673 \text{ V}$

Oxygenated solution: $E_{H,\text{oxygenated}} = +0.802 \text{ V} - 0.600 \text{ V} = +0.202 \text{ V}$

The key result is that the relative E_H values have been reversed, indicating that now, near the electrodes, the electrons in the hydrogenated solution are more stable (have lower total electrochemical potential energy) than those in the oxygenated solution. Therefore, electrons transfer out of the oxygenated solution (via oxidation of H_2O to O_2), through the electrodes and wire, and into the hydrogenated solution (via reduction of H_2 to H_2O). Thus, the imposition of the electrical potential has reversed the direction of the reaction. The thermal analog of this situation is air conditioning, in which an input of electrical energy drives a process that would otherwise be thermodynamically unfavorable (extracting heat from a cool environment and transferring it to a warmer one).

It should be clear that this result (reversal of the direction in which the reaction proceeds) would have been obtained regardless of the absolute values of the imposed potentials; the only criterion is that the imposed potential difference be larger than (and in the opposite direction from) the natural potential difference that develops because of the different chemical compositions of the two solutions. Where the absolute potentials *do* come into play is in controlling the kinetics of the electrode reactions, a topic which is beyond the scope of the current discussion. This topic is addressed in many texts on electrochemistry and electrode dynamics.[11]

In industrial settings, external electrical potentials are applied to drive all sorts of reactions in a direction that they would not proceed spontaneously, with the most obvious examples being electroplating and metal refining. The approach is sometimes also used to prevent an undesirable reaction from occurring, such as by imposing a positive potential on pipes to prevent them from corroding.

Example 9.17

According to the Nernst equation, the E_H of a solution at pH 7 that contains dissolved oxygen in equilibrium with the atmosphere is 0.802 V. What reaction occurs if an electrode is inserted into a solution that has been equilibrated with air if the electrode potential is controlled at 1.0 V? Consider two scenarios: one in which the solution is not in contact with any gas phase and another in which the solution is continuously bubbled with air.

Solution

The electrode surface is at a higher electrical potential than the adjacent solution, so electrons transfer from solution to the electrode. The electrons are withdrawn from water molecules, converting them to dissolved oxygen and H^+ via the following reaction:

$$2H_2O \rightarrow O_2(aq) + 4H^+ + 4e^-$$

If the solution were not in contact with a gas phase, the accumulation of $O_2(aq)$ in solution would cause the electrochemical potential of the solution to increase. The process would continue until $E_H = 1.0$ V, at which point the solution and electrode would be in equilibrium and no further reaction would occur.

On the other hand, if the solution were continuously bubbled with air to maintain the $O_2(aq)$ activity in equilibrium with the atmosphere, the reaction would proceed indefinitely. The system is not a perpetual motion machine; energy is being expended to remove the electrons that transfer into the electrode and maintain its potential at the fixed value, and it is this input of energy that induces a reaction (H_2O oxidation) that would not otherwise occur. In theory, this process could provide an approach for generating O_2-enriched air.

9.11.4   REDOX ELECTRODES, REFERENCE ELECTRODES, AND THE MEASUREMENT OF E_H

If, in the system shown in Figure 9.18, the switch is closed and the gas supply is halted, the pe values of the two solutions will approach one another. As they do

---

[11]An excellent introduction to this topic is provided in *Modern Electrochemistry* by Bockris and Reddy (Plenum, 1973).

so, the driving force for both half-cell reactions decreases, and once the pe values in the solutions are equal, the driving force disappears. If the electrode surface is a good catalyst for the two half-cell reactions and the wire provides negligible resistance, the reaction will proceed to equilibrium quickly. On the other hand, if a large electrical resistance is placed in the wire, the driving force for electron transfer will exist, but the reaction will proceed very slowly, because the transfer of electrons will be impeded. This arrangement describes, in simplest terms, the workings of a potentiometer or voltmeter. A known, extremely high resistance R is placed between two points of different electrical potential, and the (very small) current i that flows across the resistance is measured. The magnitude of the current is then used to determine the voltage difference V between the two points by Ohm's law: $V = iR$.

The potential difference that the voltmeter actually measures is $\Delta\Psi$, i.e., the difference in electrical potential between the two electrodes. However, if the electrodes are good catalysts, the pe of each electrode will be almost identical to that in the corresponding solution, so the reading can be equated (approximately) to the difference in electrochemical potential between the two solutions, i.e., to ΔE_H. To the extent that an electrical potential difference exists between either solution and the electrode in it, the measured value of $\Delta\Psi$ will differ from ΔE_H between the solutions. The difference in potential between the electrode and the solution adjacent to it is called the *liquid junction potential*.

Recall that the $H^+/H_2(g)$ system under standard conditions has pe $= E_H = 0$ by definition. Therefore, if a solution whose E_H is to be determined is placed in one half-cell and the $H^+/H_2(g)$ system under standard conditions is placed in the other, and if the electrodes equilibrate with both solutions rapidly, then the measured $\Delta\Psi$ (that is, $\Psi_{sol'n} - \Psi_{std.H^+/H_2}$) will indicate the value of E_H in the unknown solution directly.

A system similar to that shown in Figure 9.18 is set up, but with different solutions bathing the electrodes. The solution on the left is at pH 0 and is bubbled with $H_2(g)$ at $P_{H_2} = 1.0$ bar [i.e., the left-hand portion of the system is the standard $H^+/H_2(g)$ half-cell], and the solution on the right contains 10^{-4} *M TOTO*Cl and 5×10^{-3} Cl$^-$ at pH 7.6. What is the value of E_H of the solution on the right?

A voltmeter is placed in the circuit between the two electrodes. What is the reading on the voltmeter? (*Note:* In schematics such as Figure 9.18, the convention is to assume that the voltmeter is set up such that its output equals E_H of the solution on the right minus E_H of the solution on the left.)

Example 9.18

Solution

The redox reaction of interest occurring in the right-hand cell is

$$HOCl + H^+ + 2e^- \leftrightarrow Cl^- + H_2O \qquad pe^\circ = 25.1$$

Because the pH equals pK_a for HOCl, we know that {HOCl} $= 0.5${*TOTO*Cl}, or 5×10^{-5}. Substituting this value into the Nernst equation, we can determine the energy level of the electrons associated with the half-cell reaction and express it in terms of E_H as follows:

$$pe = pe^\circ - \frac{1}{2}\log\frac{\{Cl^-\}\{H_2O\}}{\{HOCl\}}$$

$$pe = 25.1 - \frac{1}{2}\log\frac{(10^{-3.0})(1.0)}{(5 \times 10^{-5})} = 24.45$$

$$E_{H,HOCl/Cl^-} = (0.059 \text{ V})pe = 1.443 \text{ V}$$

The voltmeter measures the difference in potentials between the two sides of the system. In the current case, the potential of the right-hand side is 1.443 V, and that of the left-hand side is zero, so the meter will read +1.443 V.

Example 9.19

Chloride ion is added to the solution described in the previous example until the meter reads +1.405 V. How much Cl^- has been added?

Solution

Again using the Nernst equation, this time in terms of E_H, and keeping $\{HOCl\} = 5 \times 10^{-5}$, we can find $\{Cl^-\}$ in the solution:

$$E^\circ_{H,HOCl/Cl^-} = (0.059 \text{ V})pe^\circ = 1.481 \text{ V}$$

$$E_H = E_H^\circ - \frac{0.059 \text{ V}}{2}\log\frac{\{Cl^-\}\{H_2O\}}{\{HOCl\}}$$

$$1.405 \text{ V} = 1.481 \text{ V} - (0.0295 \text{ V})[\log\{Cl^-\} - \log(5 \times 10^{-5})]$$

$$\{Cl^-\} = 10^{-1.72} = 0.019$$

The Cl^- concentration in the solution is 0.019 M. Since the original solution contained 0.001 M Cl^-, 0.018 M Cl^- was added.

A system comprising the pH 0.0 solution equilibrated with $H_2(g)$ at 1.0 bar plus the electrode immersed in it is referred to as a *standard hydrogen cell* or a *standard hydrogen electrode* (SHE). The SHE is a convenient conceptual construct because, by maintaining the pH and partial pressure of $H_2(g)$ constant, the electrode potential can be held steadily at 0 V. In reality, however, the need for the specialized gas phase makes the use of such an electrode inconvenient. It should be apparent, though, that *any* electrode whose potential can be maintained at a constant, known value would serve the same purpose. For instance, if a solution (call it solution A) had a fixed potential of −100 mV when measured against the standard hydrogen electrode, it could be used in a standard electrode. If the potential of a solution B were determined to be 40 mV greater than that of solution A in a setup like that shown in Figure 9.18, we would know that the potential of solution B relative to an SHE would be −60 mV, i.e., $E_{H,B} = -60$ mV.

Any electrode whose potential is fixed (or so strongly buffered that it can be treated as if it were fixed) is called a *reference electrode*. The buffering can be accomplished by fixing the activities of all the reactants that participate in the redox reaction in the cell (e.g., by adding them to solution in great excess over the amounts that are likely to be generated or consumed) or by equilibrating the solution with a different phase. For instance, the standard hydrogen electrode relies on the presence of H^+ in solution at an activity of 1.0 (pH 0.0) and equilibration of the solution with $H_2(g)$ at an activity of 1.0. Similarly, the standard calomel electrode (SCE) described earlier in the chapter relies on equilibration of a saturated KCl solution with calomel, $Hg_2Cl_2(s)$, and pure liquid mercury, $Hg(l)$, to establish a fixed pe of 4.13 ($E_{H,SCE} = 244$ mV) (see Example 9.16).

Another common reference electrode is the silver/silver chloride electrode, in which both $Ag(s)$ and $AgCl(s)$ are equilibrated with a saturated KCl solution. In this case, the potential is fixed by the following equilibrium:

$$AgCl(s) + e^- \leftrightarrow Ag(s) + Cl^- \qquad pe^\circ = 3.76 \qquad E_H^\circ = 222 \text{ mV}$$

The pe and E_H of this type of electrode [when the $AgCl(s)$ and $Ag(s)$ are immersed in saturated KCl solution with $\{Cl^-\} = 2.5$] are 3.36 and 199 mV, respectively.

What potential difference would be recorded by a voltmeter between the solution in Example 9.19 and an SCE?

Example 9.20

Solution

The potential of the solution measured against the standard hydrogen electrode was 1405 mV, and the SCE has a fixed potential of $+244$ mV, so the potential difference between the solution and the SCE is $(1405 - 244)$ mV, or 1161 mV.

In practice, reference electrodes are enclosed in a glass or plastic tube, which is then inserted into the solution to be analyzed. A small salt bridge is built into the side of the electrode to establish good electrical contact with the solution. A similar arrangement is often used to enclose the other electrode (the *working* electrode), whose potential is intended to change in response to changing conditions in the sample being analyzed. Thus, all the components shown in Figure 9.18 are present in the combination of reference electrode and working electrode in a conventional electrode setup. Sometimes, the reference electrode and working electrode are combined in a single housing, in which case the unit is referred to as a *combination* electrode. A schematic of a setup for analyzing the E_H of a solution using a Pt working electrode and a calomel reference electrode is shown in Figure 9.19.

PRACTICAL CONSIDERATIONS AND LIMITATIONS IN THE USE OF ELECTRODES When the E_H of a solution is measured using a set of electrodes, some current must flow, so some conversion of oxidized species to reduced species, or vice versa, must occur in the reference electrode. For instance, if the solution being analyzed has an E_H that is lower (more reducing) than a silver/silver chloride reference

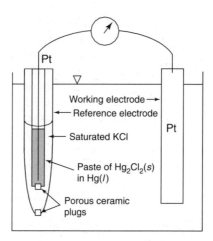

Figure 9.19   A calomel reference electrode (on left) connected to a platinum working electrode for the measurement of solution E_H. If the Pt electrode is selectively responsive to the $H^+/H_2(g)$ couple, and if solution is bubbled with $H_2(g)$ at a known partial pressure, the reading on the voltmeter can be related to the pH of the solution.

electrode immersed in it, some of the silver ions in solution will be reduced to silver metal. The depletion of silver ions lowers the product $\{Ag^+\}\{Cl^-\}$ in solution, causing some solid silver chloride to dissolve. However, assuming that the reservoir of $AgCl(s)$ is large enough that some of it remains present, the activities of $Ag(s)$ and $AgCl(s)$ remain at 1.0. Furthermore, the mass of $AgCl(s)$ that dissolves is very small (because of the high resistance in the voltmeter), so the change in Ag^+ and Cl^- concentrations is negligible, and the E_H of the electrode remains effectively fixed at 199 mV.

Although the conceptual basis for measuring E_H (or, equivalently, pe) is relatively straightforward, in practice the measurements are not as simple as theory predicts. Some of the reasons for this result have already been alluded to, such as the possibility that the liquid junction potential is significant. More commonly, the problem that arises is that the solutions being analyzed contain many chemical species other than the redox couple of interest, including other redox couples. Any of these species might bind to (adsorb to) the electrode surface and change its catalytic properties in ways that are specific to the system being analyzed, i.e., in ways that cannot be predicted or corrected for by any simple procedure.

Furthermore, the electrode might respond to any of the redox couples in the system. In such a case, the meaning of the measured value of E_H is not obvious. This situation might be considered analogous to trying to analyze a system in which $\{Ac^-\} = \{HAc\}$ and $\{OCl^-\} = \{HOCl\}$. The first equality indicates that the solution is at pH 4.7, and the second that it is 7.6. In fact, if neither of these acid/base reactions were equilibrated, it might be that the actual value of $-\log\{H_{2x+1}O_x\}$ is neither 4.7 nor 7.6, but some other value altogether. We generally do not concern ourselves with such situations when analyzing acid/base systems, because most such systems equilibrate rapidly; i.e., the hypothetical situation described above is never a realistic possibility. On the other hand, the coexistence in solution of several redox couples, all disequilibrated to different extents, is the norm. As a result, even if the electrode is operating in accordance with theory, the significance of the measured value is often unclear.

The ideal situation would be to develop electrodes that respond only to a particular redox couple of interest, and then to use those electrodes to determine the E_H or pe characterizing that reaction. Electrodes that approach such behavior have been developed for a number of redox couples. If the electrode is set up in such a way that one of the redox-active species is present at a fixed activity, the measured E_H can be used to determine the activity of the other species of that couple.

For instance, if the Pt (working) electrode in Figure 9.19 were sensitive to the E_H of the $H^+/H_2(g)$ redox couple, and if the sample were bubbled with $H_2(g)$ at $P_{H_2(g)} = 1.0$ bar, then the E_H of the electrode would depend only on the H^+ activity in solution. As a result, the setup could be used to determine solution pH. As with the SHE, the actual measurement of pH using such a setup is impractical; however, the same principle is applied in many commercial electrodes. Such electrodes are commonly referred to as *ion-selective electrodes,* although they are used for analyzing both ions and neutral species.

In truth, the electrical potential developed in many modern ion-selective electrodes does not rely directly on a redox reaction. Rather, the working electrode consists of a thin glass membrane across which a potential develops if the solutions on the two sides of the membrane have different compositions. For instance, glass membranes can be prepared that develop a potential difference in response to different H^+ activities on the two sides. If the H^+ activity on one side is fixed (typically by placing the membrane at the tip of an electrode housing and filling the housing with $1.0\ M$ acid), then the potential across the membrane reflects the H^+ activity in the solution in which it is immersed. Thus, the $1\ M$ acid serves the same role as bubbling the solution with hydrogen gas in the idealized (but impractical) pH electrode described above. By modifying the composition of the glass, its potential can be made to respond selectively to ions other than H^+, so that glass electrodes are useful as selective ion electrodes for a number of dissolved species. In other cases, metallic electrodes that respond selectively to the activity of specific ions and that closely resemble the setup shown in Figure 9.19 are useful.

SUMMARY

This chapter describes the qualitative importance and quantitative interpretation of redox reactions — reactions in which electrons are transferred from one molecule to another. Because free, hydrated electrons are extremely unstable, all the electrons released are presumed to bind immediately to other molecules. By convention, the electrons are assigned to specific atoms in each molecule, so that certain atoms are viewed as losing electrons (becoming oxidized) and others as gaining them (becoming reduced).

Even though free electrons are so unstable, it is convenient to write reactions in which they are reactants or products in order to compare their relative affinity for different molecules. By convention, the equilibrium constant for a reaction in which one electron is released from a substance being oxidized is designated e°. The value of e° characterizes a redox couple in a manner that is analogous to how the value of K_a characterizes an acid/base couple. Specifically, increasing e° (corresponding to decreasing pe°) indicates increasing stability of the oxidized species relative to the reduced species, in much the same way that increasing K_a indicates increasing stability of the base relative to its conjugate acid. However, because redox conjugate pairs often differ in their acidities, the exact pe value where dominance switches between the oxidized and reduced species can depend on pH as well as pe°.

The historical development of electrochemical concepts in different fields has led to disparate terminology to describe similar concepts. As a result, electron energy is expressed in terms as varied as the molar Gibbs energy of the electrons (kJ/mol), the pe of the system (dimensionless), and the E_H of the system (volts). However, in all cases, the essential idea is that, given the opportunity, electrons transfer from higher- to lower-energy environments, and that the energy they release in doing so can be captured and used productively.

The equilibrium speciation of a redox couple under a specific set of conditions can be determined by using the Nernst equation. Redox speciation can also be represented over a wide range of conditions on $\log C$–pe diagrams, which serve essentially the same role as $\log C$–pH diagrams do in acid/base systems. Because redox speciation often depends on pH as well as pe, another convenient way to display redox speciation is on a pe–pH predominance area diagram. Whereas $\log C$–pe diagrams provide information about the concentration of non-dominant species that cannot be shown on a predominance area diagram, they do not allow one to display speciation as a function of both pH and pe on a single sheet, as is possible on a predominance area diagram.

Because electrons are often exchanged in multiples of two, the pe range over which the oxidized and reduced species are both significant contributors to the mass balance is often exceedingly small. This fact is reflected in the slopes of the lines representing non-dominant species on $\log C$–pe diagrams, which are often steeper than those on $\log C$–pH diagrams. Over the narrow regions of pe where the equilibrium redox speciation is changing dramatically, the solution pe is well buffered, just as solution pH is buffered in regions of the pH scale where the acid/base speciation is changing most rapidly.

Although comparisons between acid/base and redox systems are extensive and useful, they must be tempered by the understanding that whereas most acid/base reactions reach equilibrium rapidly, many redox reactions do not. The extremely high activation energies that impede the progress of redox reactions mean that although equilibrium calculations can correctly describe the thermodynamically stable state, the system might reside in a grossly different state for indefinitely long periods. The most important example of a thermodynamically unstable, but kinetically stable system is pure water in contact with the atmosphere. If $H_2(aq)$ and $O_2(aq)$ are equilibrated with the gas phase via Henry's law, an enormous driving force favors their conversion to liquid water, but that reaction proceeds at a negligible rate under normal environmental conditions. Similarly, oxidation of organic matter by O_2 is strongly favored thermodynamically, but that reaction does not proceed at a significant rate in the absence of catalysts such as enzymes.

Because the Gibbs energy released when a reaction proceeds is directly related to the extent of disequilibrium, redox reactions tend to be more energetic than most other environmentally significant reactions. The energy release accompanying these reactions can be thought of as reflecting the different energies of the electrons in the various species. By mediating redox reactions, catalyzing their progress, and capturing the energy released as they proceed, organisms obtain the energy needed to power virtually all metabolic activity.

The ability of electrons to flow freely through electrical conductors distinguishes them from all other reactants in chemical processes and connects redox chemistry to electrochemistry. This property of electrons allows us to power machinery via redox reactions, measure the electrochemical potential difference between two solutions and thereby explore solution composition, and drive redox reactions in directions opposite from their natural tendency.

Problems

1. Bacteria oxidize organic matter to gain energy to drive their metabolic reactions. Consider a pH 8.0 solution containing 10^{-2} M total acetate species in continuous equilibrium with atmospheric oxygen and carbon dioxide. Using only the information provided below and the K_a values for the carbonate and acetate systems, determine how much energy is released by the bacterial oxidation of 10^{-4} mole of acetate ion via the reaction shown. Assume that the system is at 25°C and well buffered and that the acid/base reactions equilibrate instantaneously. $P_{O_2} = 0.21$ bar; $P_{CO_2} = 10^{-3.46}$ bar; $H_{CO_2} = 10^{-1.48}$.

$$CH_3COO + 2O_2(aq) \leftrightarrow 2HCO_3^- + H^+$$

Species	$\bar{G}_i^\circ$, kJ/mol
CH_3COO^-	−372.3
$O_2(g)$	0.0
H^+	0.0
H_2O	−237.18
$H_2CO_3(aq)$	−623.20

2. (a) Plot a log C–pe diagram for bromine species in a solution containing 0.25 mg/L TOTBr at pH 7.5. Consider Br^-, $HOBr$, OBr^-, BrO_3^-, and $Br_2(aq)$.
(b) What would the equilibrium speciation of Br be in a pH 7.0 solution initially containing 0.25 mg/L Br^- and 25 mg/L Cl^- if HOCl were added at a

dose of 2 mg/L *as* Cl_2. (*Note*: See Example 1.4 for a reminder of how to interpret these units for describing the dose of HOCl.)

3. The addition of iron metal to an acidic solution containing mercuric ion can lead to the reduction of Hg^{2+} to Hg° or the reduction of H^+ to $H_2(g)$. The hydrogen gas that is generated forms bubbles and migrates out of solution.

    You wish to use iron filings to treat a pH 1.5 solution containing 180 μg/L *TOT*Hg, 10 mg/L *TOT*Fe(II), and 10^{-3} M Cl^-. What is the Gibbs energy change for oxidation of metallic iron to Fe(II) in this system, in kilojoules per mole of Fe, for reaction with each possible oxidant (H^+ or Hg^{2+})? Use the stability constants for complexation of Fe^{2+} and Hg^{2+} by Cl^- given in Table 8.3.

4. The attached graph indicates the speciation of the As(V)/As(III) and the Fe(III)/Fe(II) redox couples at pH 7, for a system containing 2×10^{-4} M *TOT*Fe and 1×10^{-4} M *TOT*As and in which no solids precipitate. To minimize clutter, only the two most dominant species of each element in each oxidation state are shown for the given pH. All the lines continue without changing slope as they extend beyond the pe region shown.

    a. What are the strongest oxidizing agent and strongest reducing agent of the species shown?

    b. If a solution that is prepared by mixing 2×10^{-4} M $FeCl_2$, 5×10^{-5} M Na_3AsO_3, and 5×10^{-5} M Na_3AsO_4 is buffered at pH 7, what will the equilibrium pe and the solution composition be? Assume that redox reactions of water are kinetically inhibited.

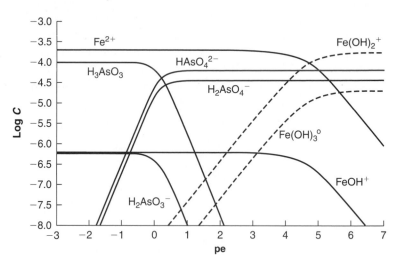

5. An aluminum pot is filled with tap water at pH 7.0 that contains 25 mg/L Cl^- and 1 mg/L HOCl *as* Cl_2. What is $TOTAl_{diss}$ in the solution after it equilibrates? You may assume that the reaction proceeds until all the HOCl is consumed.

6. A solution at pH 7.0 contains 10 mg/L *TOT*Cr(VI). Ferrous sulfate ($FeSO_4$) is added to the solution to provide 10 mg/L of *TOT*Fe(II). Write a redox reaction

between the Fe^{2+} and $HCrO_4^-$ that shows the generation of $Fe(OH)_3(s)$ and $Cr(OH)_3(s)$ as products.

a. After 1 mg/L of Fe(II) has reacted, both $Fe(OH)_3(s)$ and $Cr(OH)_3(s)$ are present in the solution. What is the molar Gibbs energy of reaction for the redox reaction under these conditions?

b. Compute E_H of each half-cell reaction and ΔE_H of the overall redox reaction under the conditions specified in part (a).

c. What is the concentration of TOTFe(II) at equilibrium?

d. Repeat part (b) for the equilibrated solution.

7. Review Example 9.11. If 1.0 g CuCl(s) and 1.0 g Cu(s) are added to 1 L of the acidified solution water, will zero, one, or both solids be present at equilibrium? If one or both solids are present at equilibrium, compute the concentration of each solid that is present; if neither solid is present, compute the solution composition.

8. During turnover of a lake, reduced bottom water is mixed with oxidized surface water. Assume that bottom and surface waters with the following compositions mix in a 1:1 ratio.

Bottom water: $\quad TOT[Fe(II)] = 1.5 \times 10^{-3} \qquad TOT[S(-II)] = 3 \times 10^{-4}$
$$\{SO_4^{2-}\} = 1.0 \times 10^{-3}$$

Surface water: $\qquad \{O_2(aq)\} = 3 \times 10^{-4} \qquad \{SO_4^{2-}\} = 1.3 \times 10^{-3}$
$$\{NO_3^-\} = 1.2 \times 10^{-4}$$

Use chemical equilibrium software to prepare a log C–pe diagram for the mixture for the range $-20 < $ pe $ < 20$, and compute the expected equilibrium condition, assuming that the solution pH is 7.5. Assume that no exchange of oxygen between the atmosphere and the solution occurs in the time frame of interest. The analysis should include consideration of elements in the following oxidation states: Fe(II), Fe(III), S(II), S(VI), N(−III), N(0), and N(V). For the Fe species, S(II), and N(−III), also consider acid/base reactions. Write out the electron condition for the system, using Fe^{2+}, SO_4^{2-}, $N_2(aq)$ and H_2O as reference species, and confirm that it is satisfied at the equilibrium condition reported in the program output.

9. Hexavalent chromium Cr(VI), can exist as either a monomer ($H_xCrO_4^{x-2}$) or a dimer ($Cr_2O_7^{2-}$). H_2CrO_4 is chromic acid, and $Cr_2O_7^{2-}$ is dichromate ion. Cr(VI) is used extensively in several industries, including metal processing and leather tanning; it also used to be used as an oxidant to destroy organic matter in a number of applications (e.g., COD tests, cleaning laboratory glassware), although those uses are diminishing. Because Cr(VI) is a carcinogen, its discharge into receiving waters and sewage systems is tightly regulated, and efforts are being made to reduce or eliminate its use in many of these applications. In those situations where Cr(VI) is still used, the most common approach for removing it from wastewater is to reduce it to Cr(III), a much less toxic and less soluble form of Cr. The reduction is sometimes accomplished by addition of Fe(II) to the water.

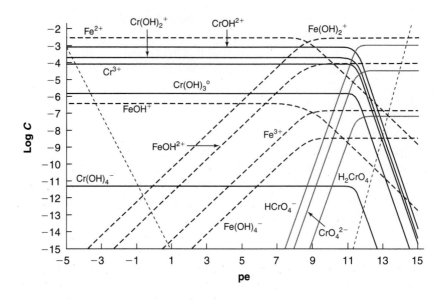

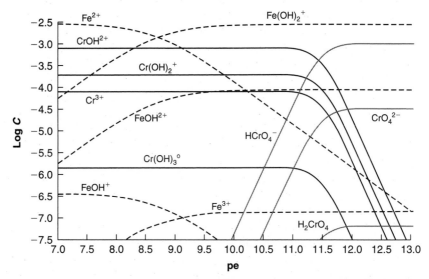

Log C–pe diagrams for a pH 5 system containing $1.1 \times 10^{-3} M$ *TOT*Cr and $3 \times 10^{-3} M$ *TOT*Fe are shown above. Based on the information in the diagram, answer the following questions.

a. What is pe° for the H_2CrO_4/Cr^{3+} redox couple?

b. What is pK_{a2} for chromic acid, H_2CrO_4?

c. Write an electron condition (EC) for a system made by adding $3 \times 10^{-3} M$ $FeCl_2$ to a solution initially containing $1.1 \times 10^{-3} M$ Cr(VI) and no Cr(III) at pH 5. Ignore redox reactions of Cl^-, but include all other species that appear in the EC, even if you believe they are negligible. Assuming the system is well buffered at this pH, what are the approximate values of pe and TOTCr(VI) at equilibrium?

d. Assuming that Cr(VI) species constitute the only weak acid/base system in the initial solution, how would the addition of the $FeCl_2$ and the subsequent reactions affect the alkalinity and pH of the solution (increase, decrease, no change)? Explain briefly.

10. Once H_2S is formed in an anaerobic sediment or bottom water, it may react with dissolved O_2 in the overlying aerobic water. The contact between the two reactants might result from simultaneous diffusion of sulfide up and oxygen down, creating a sharp interface where they meet, or from the mixing of oxygenated water with sulfide-bearing sediment as a result of a storm or burrowing benthic organisms. The reaction is often mediated by bacteria, although it also occurs in the absence of microorganisms. A lake water at pH 6.5 contains 1×10^{-3} M SO_4^{2-} and 10^{-5} M TOTS($-$II). Assuming the solution is at equilibrium with respect to redox reactions, determine the pe of the water. What is the equilibrium value of P_{O_2}?

11. Recall from Chapter 8 that chelating agents are sometimes added to process solutions to prevent the precipitation of metals. Although such chelating agents might be advantageous in a production process, their presence in the waste stream produced by such processes is often problematic. You have been asked to recommend a treatment process to remove copper from a solution at pH 1.0 that contains 5×10^{-3} M each of Cu^{2+} and EDTA. Determine the minimum concentration of TOTCu$_{diss}$ that could be attained by increasing the solution pH to any desired value between 1 and 13 to precipitate $Cu(OH)_2(s)$. Compare that concentration with the concentration that could be achieved by adding metallic aluminum to the solution to reduce the Cu^{2+} and precipitate it as $Cu(s)$ if the solution remains at pH 1.0. Consider the acid/base chemistry of Cu^{2+}, Al^{3+}, and EDTA in the first part of the problem, and the $Cu(II)/Cu(0)$ and $Al(III)/Al(0)$ redox couples in the second part. [*Note*: This question is based on a waste treatment process developed and used in the 1990s by Boeing Co., in which scrap aluminum was added to a waste solution to precipitate strongly chelated $Cu(II)$. The process worked well, but had to be abandoned when it was determined that, under some circumstances, HNO_3 in the waste was being reduced to the potentially toxic gases $NO(g)$ and $NO_2(g)$.]

chapter

10

ADSORPTION REACTIONS

CHAPTER OUTLINE

10.1 INTRODUCTION

10.1.1 CHAPTER OVERVIEW

Adsorption, the accumulation of molecules at the interface between two phases, plays a critical role in the transport, bioavailability, and fate of contaminants and naturally occurring trace compounds in both natural and engineered aquatic systems. For instance, adsorption can cause both nutrients and contaminants to accumulate in a soil column, in the sludge of a water treatment plant, or in the silt that settles in a river delta, rather than remaining in the aqueous solution that passes through such systems. These reactions help determine whether the adsorbable species are available to organisms (and if so, which ones)

Many contaminants in aquatic systems are adsorbed on the surfaces of particles suspended in the water. This photograph shows the formation of the Ohio River by the confluence of the relatively clear Allegheny River and the much more turbid Monongahela River in Pittsburgh, PA. The fate of the particles after the waters mix will dictate the bioavailability and effects of any contaminants associated with them.

| Mark E. Gibson/Visuals Unlimited.

and what chemical reactions they might undergo. Furthermore, while the fate of adsorbed molecules is determined, in some ways, by the fate of the particles to which they are bound, the interaction is not one-way: by altering the surface charge on particles, adsorbed species alter the tendency for particles to collide and coagulate or to bind to larger nearby particles (e.g., soil particles).

Adsorption also plays an important role in many phase transfer reactions. For instance, accumulation of certain molecules at the air/water interface can have a significant effect on the rate at which oxygen and carbon dioxide transfer across the interface, and adsorption is a critical step in solid dissolution and precipitation reactions. Finally, as noted in Chapter 9, many oxidation/reduction reactions take place on surfaces, in which case adsorption of the reactants is a necessary preliminary step.

In engineered water treatment systems, adsorption often provides the most cost-effective means of removing contaminants from solution to extremely low levels. One important application is the removal of trace organic contaminants by adsorption onto activated carbon, a manufactured solid with an extremely large specific surface area (surface area per unit mass of solid). Ion exchange, in which certain types of ions are removed from solution and are replaced by others that are less objectionable, has wide application for the removal of inorganic

Granular activated carbon is a very effective adsorbent for hydrophobic contaminants. In this photograph, water is being sprayed out of a rotating arm onto the top of a column of granular activated carbon (GAC). As the water falls through the column, many trace organic compounds will adsorb to the GAC.

| E. R. Degginger/Color-Pic Inc.

chemicals from water, ranging from home water treatment systems for softening to industrial process water production and wastewater treatment. In other water treatment applications, precipitation of metal hydroxides is induced in order to provide adsorbent surfaces to which undesirable constituents such as metals or natural organic matter can adsorb, after which the solid and associated contaminants can be settled or filtered out of the water.

In this chapter, after a brief introduction to adsorption terminology, we explore the equilibrium relationship between dissolved and adsorbed solutes, paying special attention to the role of electrostatic interactions when ions adsorb to charged surfaces. The chapter also includes a discussion of the application of chemical equilibrium software to adsorption modeling.

Interpretation of adsorption phenomena in natural waters requires consideration of acid/base chemistry, metal-ligand chemistry, and thermodynamic and electrochemical principles, and it has come to rely heavily on the application of chemical equilibrium software.[1] Because of the inherent importance of adsorption processes, and because the analysis of those processes synthesizes so many of the topics covered in previous chapters, this chapter and the text conclude with

[1] An excellent and very accessible presentation covering the fundamentals of adsorption reactions and their significance in natural systems is available in *Chemistry of the Solid-Water Interface* by Stumm (Wiley, 1992).

a relatively extensive section describing adsorption phenomena in natural and engineered aquatic systems and linking those phenomena to basic water chemistry principles.

10.1.2 TERMINOLOGY

Given the wide variety of applications in which adsorption is important, it is not surprising that several conceptual models have been developed to describe the process. Unfortunately, no single terminology for describing adsorptive systems has been adopted. Some terms that have a well-accepted meaning are defined below, while others that are associated with a specific model or application are described later in the chapter.

Adsorption is the accumulation of a substance at or near an interface relative to its concentration in the bulk solution. **Desorption** is the reverse of adsorption; i.e., it is the release of an adsorbed substance to the bulk solution. The substance that adsorbs is called the **adsorbate,** and the solid to which it binds is called the **adsorbent.** In principle, these definitions differentiate adsorption from *absorption,* which is the accumulation of a substance in the interior of a non-aqueous phase. However, since it is often not possible to distinguish between adsorption and absorption, the two processes are often treated jointly, in which case the above terms are sometimes written as **sorption, sorbate,** and **sorbent,** respectively.

The amount of material sorbed per unit amount of adsorbent is called the **adsorption density** and is commonly represented in equations by the letter q or the Greek letter Γ. The adsorption density can be quantified as the sorbed mass per unit surface area (with units such as mg/m^2) or per unit mass of sorbent (e.g., mg sorbate/g sorbent). The amount sorbed can be normalized to the volume of solution (e.g., mg sorbed per liter) by multiplying the adsorption density by the concentration of surface area or adsorbent in the system (m^2/L or g/L, respectively).

For a given system composition, there exists a unique equilibrium distribution of adsorbate between the dissolved and adsorbed states. The equilibrium relationship between the adsorption density and the soluble adsorbate concentration at a given temperature is called an **adsorption isotherm.**

The surfaces of suspended solids can sometimes acquire an electrical charge, much as dissolved species do. This surface charge can enhance or impede sorption of ions from solution. If the tendency for a species to adsorb depends strongly on its identity and not simply on the surface charge, the primary driving force for the reaction is attributed to specific chemical interactions between it and the surface, and the adsorbates are referred to as being *specifically* or *chemically adsorbed* (or *chemisorbed*). Chemically adsorbing molecules can bind to the surface even under conditions where electrostatic interactions oppose adsorption (much as Cl^- might bind to the negatively charged species $CuCl_3^-$ to form $CuCl_4^{2-}$). By contrast, *physically adsorbed* or *non-specifically adsorbed* species are attracted to the surface primarily or exclusively via electrostatic interactions.

10.2   TWO VIEWS OF THE INTERFACE AND ADSORPTION EQUILIBRIUM

Near the boundary between any two phases, a (usually microscopic) region exists wherein the physicochemical environment is different from that in either bulk phase, causing the molecules to behave differently from those far from the interface. Adsorbed molecules reside in this region and, colloquially, can be considered to be "half in solution and half out." Two broad conceptual models have been developed to describe equilibrium between adsorbed species and those in bulk solution. In simple terms, the key difference between these two models is which "half" of the adsorbed species they focus on. That is, one model treats adsorbed molecules as dissolved species and makes adjustments to account for the fact that these species happen to be attached to solids, while the other model treats the adsorbed species as though they have been removed from solution, forming a separate phase as different from the dissolved phase as a solid or gas. Mathematically, these two models differ in how they define the concentration, activity, and activity coefficient of adsorbed species, as described below.

10.2.1   ADSORPTION AS A SURFACE COMPLEXATION REACTION

In adsorption models that focus on the similarity between adsorbed species and other species that are truly in solution, binding of the adsorbate to the adsorbent is viewed as similar to the binding of a proton to a base, a metal to a ligand, or an electron to an oxidant. The metal/ligand analogy is the one that is most often made, with the adsorbent sometimes described as a collection of *surface ligands* and the adsorption reaction referred to as a **surface complexation** reaction. This conceptual view is shown schematically in Figure 10.1. The surface ligands are presumed to correspond to specific sites on the adsorbent surface, so the model is sometimes referred to as a *site-binding* model. In some of the more complex versions of such models, different adsorbed molecules are envisioned to reside at different locations (i.e., different distances from the surface) in the interfacial zone.

An extreme example of an adsorbent that fits this model is an ion exchange resin. Such resins typically consist of a crosslinked, polymeric skeleton containing a high concentration of identifiable, fixed, charged sites. When the resin is exposed to an aqueous solution, adsorbable ions can bind to these sites. If the solution composition changes, these adsorbed ions can be replaced by other ions from solution, hence the name ion exchange.

To model adsorbents that have binding sites in fixed locations on the surface, the number of sites per unit mass of solid is estimated and multiplied by the concentration of solid in the system, yielding the concentration of sites in units comparable to those of dissolved species (moles per liter of solution). This quantity can be represented as $TOT \equiv S$, expressed in moles per liter, where S designates a surface site and the symbol $\equiv$ designates the bond to the underlying solid. The total site concentration is distributed among sites occupied by water ($\equiv SH_2O$) and by various adsorbates ($\equiv SA$, $\equiv SB$, etc.). Because the amount of

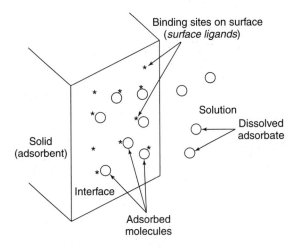

Figure 10.1 Schematic representation of an adsorptive solid/solution interface according to the site-binding model. The adsorbed molecules are located in the interfacial plane. The asterisks represent binding sites on the surface.

water adsorbed is never of interest in aquatic systems, sites occupied by water are often referred to as unoccupied sites and represented as $\equiv S$.

A surface complexation reaction and the corresponding equilibrium constant can be written for occupation of surface sites by each adsorbate in the system, i.e.,

$$A(aq) + \equiv SH_2O \leftrightarrow \equiv SA + H_2O \tag{10.1}$$

Soluble adsorbate + Hydrated surface site $\leftrightarrow$ adsorbed A + water

$$K_{ads} = \frac{\{\equiv SA\}\{H_2O\}}{\{\equiv SH_2O\}\{A\}} \tag{10.2}$$

To evaluate K_{ads} in Equation (10.2), we need to establish conventions for quantifying the activity of surface species; i.e., we need to choose a standard state for these species. In this model, the logical choice is to define the standard state in the same way as for most species that are in true solution. That is, we define the standard concentration for adsorbed species to be 1.0 mol/L and define the reference state as infinite dilution.[2] If we assume that the solution is ideal, then the activity of $\equiv SA$ can be quantified simply as its molar concentration in the overall solution, ignoring the fact that the ligands are at the surface of a solid.[3]

---

[2] Recall from Chapter 1 that use of the infinite dilution reference state means that $\gamma_{\equiv SA} = 1.0$ if each $\equiv SA$ species in the real system has the same reactivity as it would in an infinitely dilute solution. Also, in Chapter 2, we saw that the phrase "the same reactivity" can be interpreted as "the same molar Gibbs energy". We will see later in the chapter that for ionic adsorbates we need to add the criterion that the surface is uncharged to completely define the reference state for adsorbed species.

[3] The question of whether activity coefficients based on the bulk solution composition (i.e., the ionic strength) should be applied to adsorbed species has been debated for years and is still unresolved. Most current models do not apply activity coefficients, effectively assuming that adsorbed species behave ideally regardless of the solution composition.

Pairs of adsorption reactions are sometimes combined to emphasize the fact that when ions adsorb, other ions must be brought into or expelled from the interfacial region to maintain the electroneutrality of the region. Since this approach is used most often to describe ion exchange equilibria, the assumption is that the reaction involves release of ions with the same total charge as the charge of the ions that are adsorbing, rather than simultaneous adsorption of oppositely charged ions. If the adsorbing and desorbing ions have the same molar charge, the reaction and corresponding equilibrium expression are

$$\equiv SB + A^{n+}(aq) \leftrightarrow \equiv SA + B^{n+}(aq) \tag{10.3}$$

$$K_{\text{exch,A–B}} = \frac{\{\equiv SA\}\{B\}}{\{\equiv SB\}\{A\}} = \frac{\{\equiv SA\}/\{A\}}{\{\equiv SB\}/\{B\}} \tag{10.4}$$

The ratio of adsorbed i to dissolved i is a measure of the affinity of the surface for i under the given conditions. Therefore, the ratio

$$\frac{\{\equiv SA\}/\{A\}}{\{\equiv SB\}/\{B\}}$$

can be considered to characterize the preference of the adsorbent for species A over species B for a particular set of system conditions. This latter ratio is sometimes referred to as a selectivity coefficient ($K_{\text{sel,A–B}}$) or separation factor. For reaction (10.3), the selectivity coefficient equals the equilibrium constant for the reaction.

Ion exchange reactions can also involve ions with different charges; for instance, one doubly charged ion might bind to the solid and cause two singly charged ions to be released. In such a reaction, electroneutrality is maintained, and the exchange is equivalent on a charge basis, but not on a molar basis. The appropriate form for the equilibrium constant for such a reaction is not always obvious and can depend on the details of the binding (e.g., whether the divalent ion actually binds to both sites vacated by the monovalent ions, or whether it binds to one of those sites and the other site remains unoccupied). This topic is considered in greater detail later in the chapter.

10.2.2   ADSORPTION AS A PHASE TRANSFER REACTION

As noted above, an alternative to the site-binding model for adsorption treats the interfacial region as a phase that is distinct from the solution (sometimes called the *interphase*) and treats adsorption as a phase transfer reaction. One version of this model envisions the adsorption reaction to proceed identically as described above, i.e., by binding of adsorbate molecules to identifiable, fixed surface sites, so it is also a site-binding model of adsorption. However, rather than quantify the activity of adsorbed species using a concentration scale based on the number of moles of adsorbed molecules per liter of solution, this model uses a concentration scale based on the mole fraction of adsorbate in the surface phase. That is, the concentration of $\equiv SA$ is defined as the number of moles of $\equiv SA$ on the surface as a fraction of the total number of moles of surface sites ($TOT \equiv S$). When this convention is used, the reference state for adsorbed species is a surface mole fraction of 1.0.

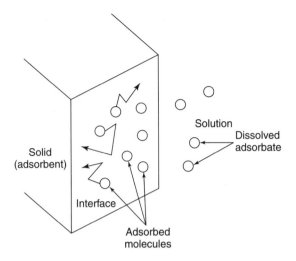

Figure 10.2   Schematic representation of adsorbed species according to the phase transfer model. Adsorbed molecules are envisioned to move freely in the two-dimensional *interphase* region near the surface.

In theory, activity coefficients could be applied to account for changes in the surface environment based on the composition of the surface phase; however, in practice, the assumption is always made in these cases that the surface phase is ideal, so that the activity of $\equiv$SA is just equated with its surface mole fraction.

In a second version of the phase transfer model for adsorption, the adsorbed molecules are thought of not as bonding to specific surface sites, but rather as moving around the surface freely in much the same way that molecules move about in a gas phase. This model is depicted in Figure 10.2. In this model, even if the adsorbed concentration is small, adsorbate is considered to occupy the entire interphase, in the same sense that a bulk gas phase expands to occupy all the space available to it. Correspondingly, this model has no explicit limit on the amount of adsorbate that can bind to the surface (as opposed to the models described above, in which case the maximum value of $\equiv$SA is $TOT\equiv$S).

Quantitative modeling using this paradigm is based on the phase transfer reaction

$$A(aq) \leftrightarrow \equiv A(\text{interphase}) \tag{10.5}$$

$$K_p = \frac{\{\equiv A(\text{interphase})\}}{\{A(aq)\}} \tag{10.6}$$

for which the equilibrium constant K_p is sometimes referred to as a *partition coefficient*. As is apparent from Equation (10.6), K_p bears greater similarity to a Henry's law constant than to a metal-ligand complexation constant. And, in fact, the activity of adsorbed A in this model is quantified by reference to something called the *surface pressure*, just as the activity of a three-dimensional gas is quantified by reference to its (volumetric) pressure. Briefly, the idea is that if one could encircle some of the adsorbed molecules, their motion would exert a force along the perimeter of the circle, and that force could be quantified as the surface pressure, in units of force per length. Logically enough, the surface pressure increases with increasing adsorption density. As in the site-binding phase transfer model, the concentration scale used in the gas transfer model of adsorption is based on mole fractions in the surface phase.

10.2.3   WHICH MODEL AND WHICH EQUILIBRIUM EXPRESSION ARE BEST?

All three models of adsorption described above (the site-binding, surface complexation model; the site-binding, phase transfer model; and the gas transfer model) represent idealized cases; the real situation is not likely to conform to any of these models completely or to be identical in all adsorptive systems. Furthermore, each model has both a great deal of flexibility and some unavoidable limitations, so that none of them can be declared universally better than the others. However, strong preferences have developed for modeling certain adsorptive systems by one model or another. When hydrophilic adsorbates bind to mineral adsorbents, strong bonds are envisioned to form at specific surface sites, so the two site-binding models are preferred. Conversely, the driving force for sorption of hydrophobic adsorbates to non-mineral adsorbents (e.g., organic solids, activated carbon) is thought to be dominated by unfavorable adsorbate-water interactions; i.e., the driving force is viewed more as water pushing the adsorbate molecules out of solution than as the adsorbent attracting them. As a result, adsorption can be highly favorable in these systems even if direct binding of adsorbate to surface sites is weak, and the gas transfer model is preferred.

Because the focus of this text is primarily on hydrophilic solutes and because the site-binding models are more easily integrated into the chemical equilibrium software discussed in previous chapters, those models are emphasized here. However, the gas transfer model is also widely used in environmental engineering, especially as a tool for modeling adsorption onto activated carbon in systems containing multiple hydrophobic adsorbates. The mathematical formulation of that model is elaborated in a number of publications.[4]

10.3   QUANTITATIVE REPRESENTATIONS OF ADSORPTION EQUILIBRIUM: THE ADSORPTION ISOTHERM

As noted above, any equation that relates the amount of adsorbate at the surface to that in solution in systems that have reached equilibrium is called an adsorption isotherm. Isotherm equations can summarize large amounts of data in a concise way and can also be used to predict adsorbate behavior under conditions that have not been investigated experimentally.

---

[4]See, for example, J. C. Crittenden et al., "Prediction of Multicomponent Adsorption Equilibria in Background Mixtures of Unknown Composition," *Water Research, 19:* 1537–1548 (1985).

Adsorption isotherms are typically derived empirically by gathering data for the adsorption density q as a function of the dissolved concentration of the adsorbate and then attempting to fit the data to simple equations. Alternatively, isotherm equations can be derived from theory, based on models for the interactions among the surface, dissolved adsorbate, and adsorbed molecules.

Once the isotherm has been developed, it can be used to predict the distribution of adsorbate between the surface and solution for other, hypothetical conditions. For instance, assuming the system reaches equilibrium, the adsorbent dose c_{solid} needed to reduce the dissolved adsorbate concentration from some initial value c_{init} to a lower target value c_{fii} can be computed as follows:

$$q \equiv \frac{c_{init} - c_{fin}}{c_{solid}}$$

$$c_{solid} = \frac{c_{init} - c_{fin}}{q} \qquad \text{(10.7)}$$

The isotherm can be used to relate q to c_{fii}. Using this relationship to substitute for q in the denominator of Equation (10.7) allows calculation of c_{solid} for any given values of c_{init} and c_{fii}.

Adsorption of CrO_4^{2-} onto three different minerals is studied, yielding good fits to the isotherms shown in the graph below. You wish to reduce the concentration of CrO_4^{2-} in a wastewater from 0.2 to 0.02 mmol/L (roughly 10 to 1 mg Cr/L) by sorption onto the best of these three adsorbents. In this case, "best" means the one that can achieve the treatment goal using the least amount of solid.   **Example 10.1**

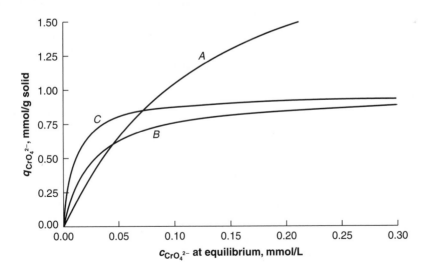

a. Which adsorbent would be preferred, and why?
b. What adsorbent dose (g/L) is required?

Solution

  a. When the solution and solid have equilibrated, c_{eq} will be 0.02 mmol/L, so the solid that has the largest adsorption density in equilibrium with this concentration (i.e., solid C) will be the preferred one.
  b. The adsorption density of adsorbent C in equilibrium with a dissolved concentration of 0.02 mmol/L is approximately 0.65 mmol CrO_4^{2-}/g solid. The required dose of adsorbent can be computed from Equation (10.7) as

$$c_{solid} = \frac{(0.20 - 0.02) \text{ mmol } CrO_4^{2-}/L}{0.65 \text{ mmol } CrO_4^{2-}/g \text{ solid}} = 0.277 \text{ g solid/L}$$

To accurately represent the behavior of an adsorbate over a wide range of conditions, an adsorption isotherm must take into account characteristics of the adsorbent, the solution, and their interactions. One convenient approach for characterizing the surface and its affinity for adsorbates is by a function that we will call the *cumulative surface site distribution function* $\Phi_{cum}(K)$, defined as the surface density of sites (i.e., sites/m^2) having adsorption equilibrium constants less than or equal to K for a given adsorbate. We will refer to the derivative of Φ_{cum} with respect to K, i.e., $d\Phi_{cum}/dK$, as the *differential site distribution function,* $\Phi_{diff}(K)$.

Plots of Φ_{cum} and Φ_{diff} versus K for the simplest imaginable case—an adsorbent on which all the sites are equivalent, and for a system in which the affinity of the sites for the adsorbate is independent of solution conditions—are shown in Figure 10.3. Consider first the plot of Φ_{cum} for adsorbate A. The equilibrium constant for sorption of A onto any surface site, under any conditions, is K_1. Because the adsorbent has no sites with binding constants for A less than K_1, $\Phi_{cum,A}$ is zero at all $K < K_1$. The adsorbent has a large number of sites with a binding constant for A equal to K_1, so $\Phi_{cum,A}$ increases as a step at this K value. In fact, since there are no surface sites with binding constant $>K_1$, this step increase is from zero to the total surface site density, which we designate $\Phi_{cum,A}^{max}$. The slope of $\Phi_{cum,A}$ versus K is zero at all values other than K_1 and infinite at $K = K_1$, so the plot of $\Phi_{diff,A}$ versus K consists of only an infinitely thin and infinitely tall spike at $K = K_1$.

Figure 10.3 also shows plots of Φ_{cum} and Φ_{diff} versus K for a second adsorbate B. The fact that the step increase in Φ_{cum} (and, correspondingly, the spike in Φ_{diff}) occurs at a value of K that is larger for adsorbate B than adsorbate A indicates that the affinity of the surface sites is greater for adsorbate B. Other than that, the curves are identical for the two adsorbates; in particular, they have identical values of Φ_{cum}^{max}, because the total population of sites is the same regardless of which adsorbate is considered.

Plots of Φ_{cum} and Φ_{diff} versus K characterize the distribution of sites on the surface in terms of their affinity for a given adsorbate. The two functions provide identical information, and the choice of how to display the data is solely a matter of convenience and personal preference. Depending on the solution conditions, anywhere from a negligible fraction to almost all of the sites will be occupied,

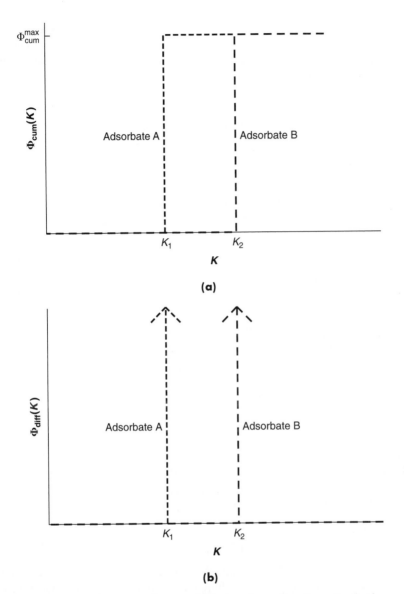

Figure 10.3 **(a)** Φ_{cum} as a function of K for a surface with uniform sites that have an affinity for each adsorbate that is independent of solution conditions. **(b)** Φ_{diff} for the adsorbent shown in part (a).

and it is this relationship between the solution conditions and site occupation that is characterized by the adsorption isotherm. Thus, isotherms can be viewed as the synthesis of a site distribution function for an adsorbent (independent of the solution conditions) with information about the solution composition (independent of the adsorbent).

In the following section, we derive the isotherm equation for simple systems such as those characterized in Figure 10.3. We then derive other isotherm equations that have been found to apply other systems, in each case interpreting the new isotherm as a response to a change in either the solution composition or the site distribution function.

10.3.1 THE LANGMUIR ISOTHERM

ONE-ADSORBATE SYSTEMS   Each curve in Figure 10.3 characterizes a system with a single adsorbate whose binding to the surface can be described by a single value (e.g., K_1 for adsorbate A) under all conditions. This situation is, in essence, the one described by reaction (10.1) and Equation (10.2), with K_{ads} in Equation (10.2) corresponding to K_1 in Figure 10.3. That equation is repeated below in slightly rearranged form, representing sites that are occupied by water as $\equiv S$:

$$K_{ads}\{\equiv S\}\{A\} = \{\equiv SA\}\{H_2O\} \qquad (10.8)$$

The activity of H_2O equals 1.0, and using the site-binding surface complexation model, we can equate the activities of $\equiv S$ and $\equiv SA$ with their concentrations in moles per liter. Then, using a balance on binding sites to substitute for $\{\equiv S\}$ in Equation (10.8), we obtain the following:

$$K_{ads}(TOT\equiv S - \{\equiv SA\})\{A\} = \{\equiv SA\}$$

$$K_{ads}(TOT\equiv S)\{A\} = \{\equiv SA\}(1 + K_{ads}\{A\})$$

$$\{\equiv SA\} = \frac{K_{ads}\{A\}}{1 + K_{ads}\{A\}} TOT\equiv S \qquad (10.9)$$

Dividing both sides of Equation (10.9) by the adsorbent concentration in the system (g/L), the left-hand side becomes the adsorption density q_A (moles adsorbed per gram of adsorbent). The ratio of $TOT\equiv S$ to the adsorbent concentration is the maximum possible adsorption density of A, commonly represented as q_{max}, so Equation (10.9) can be written in the form of an adsorption isotherm as

$$q_A = \frac{K_{ads}\{A\}}{1 + K_{ads}\{A\}} q_{max} \qquad (10.10)$$

Equation (10.10) is called the **Langmuir isotherm.** The isotherm incorporates two constants, one (q_{max}) establishing the maximum adsorption density and the other (K_{ads}) establishing the affinity of the adsorbent for the adsorbate. The isotherm equation is plotted in Figure 10.4. Under conditions where $K_{ads}\{A\} \ll 1$, the denominator in Equation (10.10) is approximately equal to 1, and the isotherm becomes linear: $q_A \approx K_{ads}\{A\}q_{max}$. On the other hand, if $K_{ads}\{A\} \gg 1$, the denominator is approximately equal to $K_{ads}\{A\}$, and $q_A \approx q_{max}$. Other factors being equal, an increase in the adsorption equilibrium constant increases the slope of the isotherm curve at low concentrations of dissolved adsorbate, and an increase in the value of q_{max} increases both the slope of the isotherm at low $\{A\}$ and the adsorption density at high $\{A\}$.

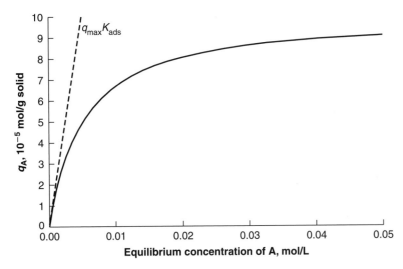

Figure 10.4   Graphical representation of the Langmuir isotherm. For the isotherm shown, $K_{ads} = 200$ L/mol and $q_{max} = 10^{-4}$ mol/g.

To test whether experimental adsorption data fit the Langmuir isotherm, the data are often manipulated so that they can be plotted according to either of the following linearized versions of Equation (10.10):

$$\frac{1}{q_A} = \frac{1}{q_{max}K_{ads}}\frac{1}{\{A\}} + \frac{1}{q_{max}} \qquad \textbf{(10.11)}$$

$$\frac{\{A\}}{q_A} = \frac{1}{q_{max}}\{A\} + \frac{1}{q_{max}K_{ads}} \qquad \textbf{(10.12)}$$

Based on the above equations, if a plot of q_A^{-1} versus $\{A\}^{-1}$ or a plot of $\{A\}/q_A$ versus $\{A\}$ is linear, the data are consistent with the Langmuir isotherm. However, a plot based on Equation (10.11) tends to compress the data at relatively high values of $\{A\}$ into a small region near the y axis, while one based on Equation (10.12) does the same to data at relatively low values of $\{A\}$. Given the ease with which non-linear regression analysis can be done, it seems wiser to use that approach to test the data for their fit to the unmodified isotherm equation [Equation (10.10)] directly, rather than to test the fit to either linearized equation.

Although K_{ads} has the form of an equilibrium constant and is widely referred to as such, the values used to compute it are frequently concentrations, rather than activities. As a result, K_{ads} is commonly reported as a dimensional quantity, with units that depend on those that are used to quantify the concentrations of adsorbate and adsorbent. In that case, according to the terminology adopted in this text, K_{ads} is really a partitioning coefficient, not an equilibrium constant.

The fractional occupation of surface sites is often of interest in adsorptive systems, and this parameter can be evaluated for systems that obey the Langmuir

Table 10.1   Comparison of fractional occupation of various types of binding sites in acid/base, metal/ligand, redox, and adsorption systems[1]

Fraction of weak base sites protonated as a function of $\{H^+\}$	$HA \xleftrightarrow{K_a} H^+ + A^-$	$\dfrac{\{HA\}}{\{TOTA\}} = \dfrac{K_a^{-1}\{H^+\}}{1 + K_a^{-1}\{H^+\}}$
Fraction of ligand sites occupied as a function of $\{Me^{n+}\}$	$Me + L \xleftrightarrow{K_1} MeL$	$\dfrac{\{MeL\}}{\{TOTL\}} = \dfrac{K_1\{L\}}{1 + K_1\{L\}}$
Fraction of oxidant sites occupied as a function of $\{e^-\}$	$Red \xleftrightarrow{pe^\circ} Ox + e^-$	$\dfrac{\{Red\}}{\{TOTC\}} = \dfrac{\{e^\circ\}^{-1}\{e^-\}}{1 + \{e^\circ\}^{-1}\{e^-\}}$
Fraction of surface sites occupied as a function of $\{A\}$	$\equiv S + A \xleftrightarrow{K_{ads}} \equiv SA$	$\dfrac{\{\equiv SA\}}{TOT \equiv S} = \dfrac{K_{ads}\{A\}}{1 + K_{ads}\{A\}}$

[1] $TOTA = A^- + HA$; $TOTL = L + MeL$; $TOTC = O_x + Red$; $TOT \equiv S = \equiv S + \equiv SA$

isotherm simply by dividing both sides of Equation (10.10) by q_{max}:

$$\text{Fractional occupation of surface sites} = \frac{\{\equiv SA\}}{TOT \equiv S} = \frac{q_A}{q_{max}} = \frac{K_{ads}\{A\}}{1 + K_{ads}\{A\}}$$

(10.13)

This result is compared with the fraction combined for various other types of reactions we have studied in Table 10.1. The only difference among the equations characterizing acid/base, metal/ligand, redox, and adsorptive systems relates to conventions for writing the reactions: K_a and e° are defined based on dissociation reactions, whereas K_1 and K_{ads} are defined based on association reactions. Comparisons based on other indicators of speciation would, of course, exhibit strong similarities as well.

Thus, at least for adsorptive systems that follow the Langmuir isotherm, treating surface sites as ordinary, dissolved ligands allows adsorption to fit comfortably into our existing paradigm for analyzing speciation in any solution. The equations and some of the terminology used to describe adsorption are a bit different from the corresponding equations and terminology applied to soluble equilibria, but this is more a reflection of historical differences in the way the systems were studied than of any fundamental differences among them.

COMPETITIVE LANGMUIR ADSORPTION   As noted above, the Langmuir adsorption isotherm for a single adsorbate represents the simplest of all possible cases of interest. The first modification to that scenario that we will consider is one in which the solution composition is changed so that *both* curves in Figure 10.3 apply, i.e., we will consider a change in solution composition such that two adsorbates are present and can compete for a single population of surface sites.

Historically, a good deal of interest has been devoted to the study and modeling of such *competitive adsorption*. For the given scenario, an expression like Equation (10.8) applies to each adsorbate. Writing the equilibrium expression for each one and then taking their ratio, we obtain

$$\{\equiv SA\} = \frac{K_{ads,A}\{\equiv S\}\{A\}}{\{H_2O\}} \tag{10.14a}$$

$$\{\equiv SB\} = \frac{K_{ads,B}\{\equiv S\}\{B\}}{\{H_2O\}} \tag{10.14b}$$

$$\frac{\{\equiv SB\}}{\{\equiv SA\}} = \frac{K_{ads,B}\{B\}}{K_{ads,A}\{A\}} \tag{10.15}$$

The site balance for this case is

$$TOT\equiv S = \{\equiv S\} + \{\equiv SA\} + \{\equiv SB\} \tag{10.16}$$

Using Equation (10.15) to substitute for $\{\equiv SB\}$ in the site balance, then using the resulting equation to substituting for $\{\equiv S\}$ in the equilibrium constant expression for $\{\equiv SA\}$ [Equation (10.14a)], and setting the activity of H_2O to 1.0, we obtain

$$\{\equiv SA\} = K_{ads,A}\{A\}\left(TOT\equiv S - \{\equiv SA\} - \frac{K_{ads,B}\{B\}}{K_{ads,A}\{A\}}\{\equiv SA\}\right)$$

$$\{\equiv SA\} = \frac{K_{ads,A}\{A\}}{1 + K_{ads,A}\{A\} + K_{ads,B}\{B\}} TOT\equiv S \tag{10.17}$$

Finally, dividing through by the concentration of adsorbent yields the isotherm equation for adsorbate A in a system where it competes with species B for surface sites:

$$q_A = \frac{K_{ads,A}\{A\}}{1 + K_{ads,A}\{A\} + K_{ads,B}\{B\}} q_{max} \tag{10.18}$$

The analogous expressions for $\{\equiv SB\}$ are

$$\{\equiv SB\} = \frac{K_{ads,B}\{B\}}{1 + K_{ads,A}\{A\} + K_{ads,B}\{B\}} TOT\equiv S \tag{10.19}$$

$$q_B = \frac{K_{ads,A}\{B\}}{1 + K_{ads,A}\{A\} + K_{ads,B}\{B\}} q_{max} \tag{10.20}$$

Generalizing the above result to the case where j adsorbates compete for the sites leads to the following adsorption isotherm for any species i:

$$q_i = \frac{K_{ads,i}a_i}{1 + \displaystyle\sum_{all\, j} K_{ads,j}a_j} q_{max} \tag{10.21}$$

The numerator of Equation (10.21) is the same as in the equation for non-competitive adsorption [Equation (10.10)], and the denominator differs only in that it contains a term of the form $K_{ads,j}a_j$ for each adsorbate in the system instead of just for the target adsorbate. Thus, each additional adsorbate in the system increases the denominator of Equation (10.21) and decreases the amount of i adsorbed. As would be expected, the larger the activity of the competing adsorbates (a_j) and the greater their tendency to sorb ($K_{ads,j}$), the larger is their effect on sorption of i.

While we have not encountered equations that look much like Equations (10.17) through (10.21) in previous chapters, it is important to recognize that this is strictly because of the ways in which we have chosen to represent the data. In fact, if we considered a solution containing Cd and Zn along with a single ligand (say, EDTA) and carried out calculations similar to those above, we would obtain for the concentrations of Cd−EDTA and Zn−EDTA:

$$\{\text{Cd}-\text{EDTA}\} = \frac{K_{1,Cd}\{\text{Cd}^{2+}\}}{1 + K_{1,Cd}\{\text{Cd}^{2+}\} + K_{1,Zn}\{\text{Zn}^{2+}\}}\,TOT\text{EDTA} \quad \textbf{(10.22a)}$$

$$\{\text{Zn}-\text{EDTA}\} = \frac{K_{1,Zn}\{\text{Zn}^{2+}\}}{1 + K_{1,Cd}\{\text{Cd}^{2+}\} + K_{1,Zn}\{\text{Zn}^{2+}\}}\,TOT\text{EDTA} \quad \textbf{(10.22b)}$$

The similarity of Equations (10.22a) and (10.22b) to Equations (10.17) and (10.19) is apparent, indicating that, as in the single-adsorbate systems described above, competitive Langmuir adsorption bears a strong similarity to the reactions involving only soluble species.

Although *competitive complexation* is accurately represented by Equations (10.22a) and (10.22b), those equations are virtually never derived or shown explicitly. Ironically, this is so because competition among metals for whatever ligands are available is assumed to always be relevant and therefore not particularly noteworthy. By contrast, adsorption processes are often viewed with a specific target adsorbate in mind, and experiments are often conducted to compare adsorption of that species in systems where it is the sole adsorbate with that in more complex systems containing multiple adsorbates. Because of the focus on a particular adsorbate in such cases, equations describing competitive adsorption are of primary interest and hence have received substantial attention.

Example 10.2 | The Langmuir isotherm for sorption of chromate onto adsorbent C in Example 10.1 has $q_{max} = 0.96$ mmol CrO_4^{2-}/g solid and $K_{ads} = 104$ L/mmol CrO_4^{2-}. The value of K_{ads} for sorption of sulfate onto the solid is 20 L/mmol SO_4^{2-}. q_{max} is the same for the two adsorbates when expressed as moles of adsorbate per gram of solid. If the solution to be treated contains 2.0 mmol/L sulfate in addition to the chromate, how large an adsorbent dose is needed to meet the CrO_4^{2-} removal target?

Solution

The isotherms for the two adsorbates are

$$q_{CrO_4^{2-}} = \frac{(0.96 \text{ mmol/g})(104 \text{ L/mmol})\{CrO_4^{2-}\}}{1 + (104 \text{ L/mmol})\{CrO_4^{2-}\} + (20 \text{ L/mmol})\{SO_4^{2-}\}}$$

$$q_{SO_4^{2-}} = \frac{(0.96 \text{ mmol/g})(20 \text{ L/mmol})\{SO_4^{2-}\}}{1 + (104 \text{ L/mmol})\{CrO_4^{2-}\} + (20 \text{ L/mmol})\{SO_4^{2-}\}}$$

In addition to the adsorption isotherms, we know that the mass balance equations on CrO_4^{2-} and SO_4^{2-} must be satisfied:

$$TOTCrO_4 = \{CrO_4^{2-}\} + q_{\{CrO_4^{2-}\}}c_{solid}$$

$$TOTSO_4 = \{SO_4^{2-}\} + q_{\{SO_4^{2-}\}}c_{solid}$$

Since the equilibrium concentration of CrO_4^{2-} is specified, we have four equations in four unknowns (the concentration of dissolved SO_4^{2-}, the adsorption densities of CrO_4^{2-} and SO_4^{2-}, and the adsorbent dose). When these equations are solved simultaneously, we obtain

$$c_{solid} = 1.59 \text{ g/L} \qquad q_{CrO_4^{2-}} = 0.11 \text{ mmol/g}$$

$$\{SO_4^{2-}\} = 0.74 \text{ mmol/L} \qquad q_{SO_4^{2-}} = 0.79 \text{ mmol/g}$$

Approximately two-thirds of the SO_4^{2-} adsorbs, occupying a significant fraction of the surface sites. As a result, almost six times as much adsorbent is required to achieve the treatment goal as in the non-competitive system.

10.3.2 ADSORPTION IN SYSTEMS WITH NON-UNIFORM SURFACE SITES

Although the Langmuir isotherm is adequate for describing many systems, it fails in many others. To understand the reasons for these deviations from idealized behavior, we need to shift our focus from the similarities between adsorption and complexation by dissolved ligands to the ways in which these reactions might differ.

In developing the Langmuir isotherm, we assumed a site distribution function that assigned a single value to K_{ads} for the reaction between a given adsorbate and all surface sites. The analogous situation in solution is to assign a single value of K_1 to the complexation reaction of Cd^{2+} with Cl^- or with EDTA, which seems eminently reasonable. However, adsorptive surfaces are neither uniformly flat nor infinite in extent, so some sites are in the middle of a crystal face of the solid, and others are on edges, in corners, etc. Even on a perfectly flat surface, different surface atoms might be attached to the underlying bulk solid via different numbers or types of bonds. Any of these factors might cause different sites to have different affinities for adsorbate molecules; i.e., a single surface might contain both Cl^--strength and EDTA-strength binding sites for Cd^{2+}, and perhaps many other sites of intermediate strength. Thus, it might be appropriate to treat the surface as a collection of many different ligands, rather than as a group of identical ligands.

Second, in the Langmuir analysis, we implicitly assumed that binding of adsorbates to the surface had no effect on the value of K_{ads} at other sites. Again, the analogous assumption applied to dissolved ligands—that formation of a Cd−EDTA complex in one part of the solution has no effect on K_1 for this reaction elsewhere in the system—seems intuitive. However, adsorbent sites are not independent, and a reaction at one point on the surface *can* affect the chemistry of other parts of the surface, in at least two ways. First, if an ion adsorbs, it can redistribute electrical charge on the surface, causing other ions to be attracted to the surface either more or less strongly than they would have been if the ion had not adsorbed. And, second, because adsorbates all share the same surface, the likelihood that they will interact directly with one another is greater than if each complex were an independent molecule in solution.

Thus, the Langmuir model represents the surface as a collection of identical and non-interacting ligands that can bind to adsorbates to form non-interacting complexes, and we anticipate that all these assumptions might be violated at times. In the following sections, we consider the surface site distribution functions that are obtained if we make different (and perhaps more realistic) assumptions about the system composition and behavior, and we assess how the changes in those functions affect the adsorption isotherm.

THE MULTISITE LANGMUIR ISOTHERM   Evidence for the presence of more than one type of surface site on an adsorbent often derives from competitive adsorption experiments. The mass balance on sites in the preceding analysis of competitive adsorption [Equation (10.16)] assumed that all adsorbates compete for a single population of uniform sites. This population corresponds to Φ_{cum}^{max} in the site distribution function and q_{max} in the adsorption isotherm. If two adsorbates have different values of q_{max} evaluated in isolation (i.e., in non-competitive systems), then at least some of the sites must be available to one adsorbate but not the other. In such cases, the system is sometimes represented as having a few distinct populations of sites that are not all accessible to all adsorbates, e.g., one group available only to adsorbate A, one available only to adsorbate B, and one available to both A and B.

If a surface contains groups of distinct sites, it might be modeled simply by extending the Langmuir model to represent the surface as several ligands, rather than a single ligand, similar to a solution containing a mix of metal-binding ligands. A concentration and an equilibrium binding constant could be established for each type of site, and the system could be modeled following the same steps as are used above to develop the one-site model. The cumulative and differential surface site distributions for a hypothetical adsorbent containing three site-types is shown in Figure 10.5.

As was true for the adsorbent considered earlier (Figure 10.3), the adsorbent characterized in Figure 10.5 has no sites with binding constants for A less than K_1, so Φ_{cum} is zero at all $K < K_1$, and the function undergoes a step increase at $K = K_1$. However, in this case, the sites with binding constant K_1 represent only a fraction of the total surface sites, so Φ_{cum} does not increase all the way to Φ_{cum}^{max}. Rather, it increases to a value corresponding to the surface concentration of these "type-1"

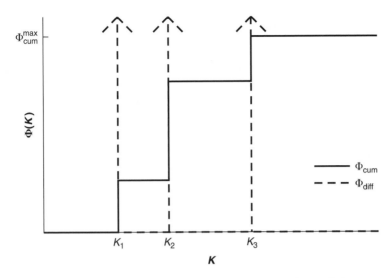

Figure 10.5 Φ_{cum} and Φ_{diff} as a function of K for a surface with three types of site, each of which has an affinity for the adsorbate that is independent of solution conditions.

sites. It then remains constant between K_1 and K_2, because the surface has no sites with binding constants in that range. At K_2, the function has another step increase (reflecting the inclusion of the type-2 sites into Φ_{cum}), and then another at K_3, at which point Φ_{cum} increases to Φ_{cum}^{max}. As before, at each value where Φ_{cum} increases as a step, Φ_{diff} has, in theory, an infinitely thin and infinitely tall spike.

Note that, although the hypothetical surface we are considering has three different types of sites, each of those sites is assumed to have a single binding constant for the adsorbate, independent of how much adsorbate is in solution or is bound to the other sites. In other words, each type of site is assumed to have the characteristics of a single-site system and to behave as it would if the other sites did not exist. In such a system, the various site-types would be occupied such that the Langmuir isotherm was satisfied for each site independently. The only linkage among the sites is indirect, in that the solution to the Langmuir isotherm equation for each site must account for the fact that all the sites are equilibrated with the same aqueous solution (and, in particular, with the same value of {A}). Complex natural adsorbent mixtures such as soils are often represented in this way, and in some cases attempts are made to associate the different sites with different mineral phases in the mixtures.

The most common multisite Langmuir models represent the surface as having two groups of sites: a small number of sites that bind the adsorbate strongly and a much larger number of sites where the strength of the adsorptive bond is weaker. The overall isotherm for a single adsorbate in such a system is a linear addition of the isotherms for the two sites:

$$q_{A,tot} = q_{A,1} + q_{A,2} = \frac{q_{max,1}K_{ads,A,1}\{A\}}{1 + K_{ads,A,1}\{A\}} + \frac{q_{max,2}K_{ads,A,2}\{A\}}{1 + K_{ads,A,2}\{A\}} \qquad \textbf{(10.23)}$$

Equation (10.23) is easily generalized to any arbitrary number N of surface sites as follows:

$$q_{\text{A,tot}} = \sum_{i=1}^{N} q_{\text{A},i} = \sum_{i=1}^{N} \frac{q_{\max,i} K_{\text{ads,A},i}\{A\}}{1 + K_{\text{ads,A},i}\{A\}} \tag{10.24}$$

The adsorption density on each type of site and the overall adsorption density on a hypothetical, two-site surface are shown in Figure 10.6.

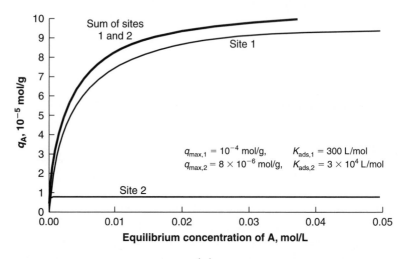

(a)

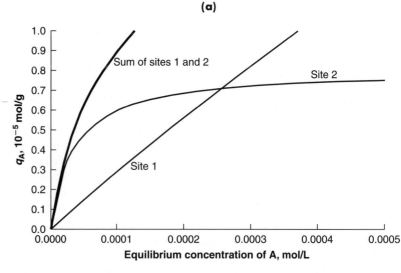

(b)

Figure 10.6   Adsorption of one adsorbate on a surface with two types of sites, each of which is occupied in accord with the Langmuir isotherm. Part (b) of the figure shows the data at very low q and $\{A\}$.

An example system in which adsorption of Zn^{2+} onto quartz was modeled using a two-site Langmuir isotherm is shown in Figure 10.7 below, along with the reactions and equilibrium constants used to model the data. For reasons that are explained later in the chapter, the surface sites in this system are considered to be oxide groups and hence are represented as $\equiv SO$, rather than simply as $\equiv S$. The surface concentrations of strong (St) and weak (Wk) sites were reported as 0.033 and 3.84 $\mu mol/m^2$, respectively. Derive conditional Langmuir isotherms for sorption to the two sites at pH 5 (the isotherms are conditional in the sense that they are applicable only at the given pH). According to the model, what fraction of the total adsorbed zinc is strongly adsorbed at equilibrium Zn^{2+} activities of 10^{-7}, 10^{-6}, 10^{-5}, and 10^{-4}?

Example 10.3

$$\equiv S_{St}OH + Zn^{2+} \leftrightarrow \equiv S_{St}OZn^+ + H^+ \qquad \log K_{St} = 0.85$$

$$\equiv S_{Wk}OH + Zn^{2+} \leftrightarrow \equiv S_{Wk}OZn^+ + H^+ \qquad \log K_{Wk} = -2.40$$

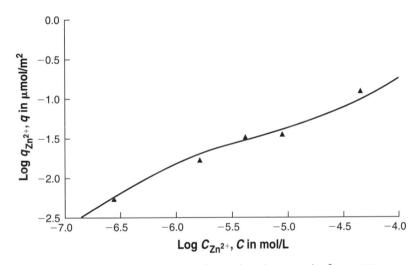

Figure 10.7   Data and model simulations for adsorption of Zn^{2+} at pH 5 onto a surface with both strong and weak binding sites. In this simulation, the stronger binding sites represented 0.86% of the total sites and had an adsorptive binding constant ~1800 times that of the weak sites. The curvature in the model simulation occurs because the strong binding sites are nearing complete coverage.

Adapted from J. A. Davis, J. A. Coston, D. B. Kent, and C. P. Fuller, *Environmental Science and Technology*, Vol. 32, 2820–2828 (1998).

Solution

Substituting the value $\{H^+\} = 10^{-5}$ into the equilibrium constant expressions yields the following conditional adsorption equilibrium constants:

$$K_{ads,St}^{pH\,5} = \frac{\{\equiv S_{St}OZn^+\}}{\{\equiv S_{St}OH\}\{Zn^{2+}\}} = \frac{K_{St}}{\{H^+\}} = \frac{10^{0.85}}{10^{-5.0}} = 10^{5.85}$$

$$K_{ads,Wk}^{pH\,5} = \frac{\{\equiv S_{Wk}OZn^+\}}{\{\equiv S_{Wk}OH\}\{Zn^{2+}\}} = \frac{K_{Wk}}{\{H^+\}} = \frac{10^{-2.40}}{10^{-5.0}} = 10^{2.60}$$

The total concentrations of strong and weak sites correspond to the q_{max} values for the sites, so the conditional Langmuir isotherms can be written as follows:

$$q_{St} = \frac{10^{5.85}\{Zn^{2+}\}}{1 + 10^{5.85}\{Zn^{2+}\}}\left(0.033\ \frac{\mu mol}{m^2}\right)$$

$$q_{Wk} = \frac{10^{2.60}\{Zn^{2+}\}}{1 + 10^{2.60}\{Zn^{2+}\}}\left(3.84\ \frac{\mu mol}{m^2}\right)$$

By substituting $\{Zn^{2+}\}$ values of 10^{-7}, 10^{-6}, 10^{-5}, and 10^{-4}, the adsorption densities on the two types of sites, and the fraction of the adsorbed zinc that is bound to strong sites, are computed as follows:

$\{Zn^{2+}\}$	q_{St}, $\mu mol/m^2$	q_{Wk}, $\mu mol/m^2$	q_{St}/q_{tot}
10^{-7}	0.0022	0.0002	0.93
10^{-6}	0.0137	0.0015	0.90
10^{-5}	0.0289	0.0152	0.66
10^{-4}	0.0325	0.1470	0.18

The fraction of the adsorbed Zn that is bound to the strong sites is shown in the final column. As we would expect, the surface speciation shifts from being dominated by strongly bound to weakly bound Zn^{2+} as the equilibrium dissolved Zn^{2+} activity and the overall adsorption density increase.

Although the number of different types of sites which could be present on a given solid surface is, in theory, unlimited, adsorption can often be modeled by considering just two or three types of sites, implying that just a few site types dominate adsorption in these systems.

The approximation is sometimes made that if a surface contains several different types of sites, the sites are occupied sequentially from strongest to weakest, similar to the assumption that is often made regarding protonation of bases. However, in the case of adsorption, we are often interested in sites that are present at vastly different concentrations as well as having different binding constants. For instance, in the above example, the binding constant for Zn^{2+} adsorption to the strong sites was 1800 times as large as that to the weak sites. However, the total concentration of strong sites was <1% as large as the concentration of weak sites. As a consequence, both groups of sites made significant contributions to the total adsorption density over a very wide range in $\{Zn^{2+}\}$. Under the circumstances, the assumption of sequential occupation is as likely to be misleading as it is helpful, so it is generally better to recognize that the different types of sites are actually occupied simultaneously than to rely on the simplifying assumption that they are occupied sequentially.

CHARACTERIZING SURFACES WITH A SEMICONTINUOUS DISTRIBUTION OF SITE TYPES
The remarks above notwithstanding, many systems exist in which it is not possible to represent experimental data satisfactorily by modeling the surface as a collection of just a few site types. Often, such surfaces behave as though they

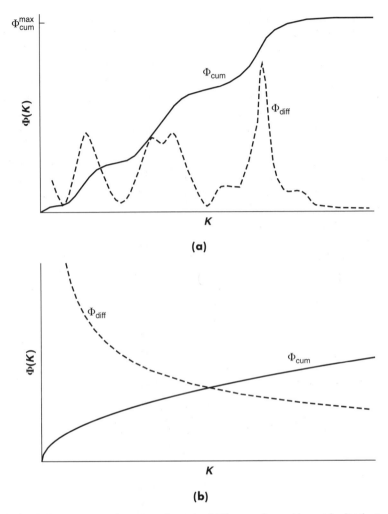

Figure 10.8 Φ_{cum} and Φ_{diff} as a function of K for a surface with a wide distribution of site types. **(a)** A surface with sites whose affinity for the adsorbate follows no obvious pattern. **(b)** A surface with a large concentration of sites with low affinity for the adsorbate and steadily smaller concentrations of sites with steadily higher K values.

have sites with a virtually continuous distribution of binding energies. Plots of Φ_{cum} and Φ_{diff} versus K are shown for two such surfaces in Figure 10.8. The adsorbent characterized in Figure 10.8a has a highly irregular distribution of site-types covering a wide range of K values. The range of K values for the adsorbent characterized in Figure 10.8b is also large and continuous, but by contrast with that shown in part (a) of the figure, in this case Φ_{cum} and Φ_{diff} might be described fairly accurately by some simple function of K. For both surfaces, Φ_{diff} is large in regions where Φ_{cum} increases rapidly, and Φ_{diff} is small or zero in regions where Φ_{cum} increases slowly or not at all, respectively.

Table 10.2   Various subsets of surface binding sites, as expressed by the surface site distribution functions

Concentration of sites with $K < K_1$	$$\Phi_{cum}(K_1) = \int_0^{K_1} \Phi_{diff}\, dK$$	**(10.25)**
Concentration of sites with $K > K_1$	$$TOT \equiv S - \Phi_{cum}(K_1) = \int_{K_1}^{\infty} \Phi_{diff}\, dK$$	**(10.26)**
Concentration of sites with $K_1 < K < K_2$	$$\Phi_{cum}(K_2) - \Phi_{cum}(K_1) = \int_{K_1}^{K_2} \Phi_{diff}\, dK$$	**(10.27)**
Total concentration of all sites	$$TOT \equiv S = \int_0^{\infty} \Phi_{diff}\, dK$$	**(10.28)**

Some useful relationships that describe specific portions of the Φ_{cum} and Φ_{diff} functions for these solids are summarized in Table 10.2.

A heterogeneous surface might have sites with any distribution of adsorptive equilibrium constants, so there is no theoretical or fundamental reason to expect the surface site distribution function to have a particular form. However, empirically, many systems behave more or less as shown in Figure 10.8b. That is, they appear to comprise a very small number of sites with very strong affinity for the adsorbate (large equilibrium constant), and steadily increasing numbers of sites with steadily decreasing affinities. We will develop the isotherm corresponding to one particular such distribution shortly. First, though, we consider a general approach for developing the isotherm for a complex surface such as that characterized in Figure 10.8a.

As noted earlier, it is neither practical nor particularly useful to describe the surfaces of adsorbents by a distribution function that consists of numerous individual steps. The purpose of an isotherm is to simplify the analysis, and characterizing dozens of hypothetically distinct sites on a surface would be more work than could be justified. Nevertheless, we can use the analysis developed above for adsorbents containing a few discrete sites to guide our analysis of the more complex surfaces.

Consider how we proposed to develop the isotherm for a system that contained three site-types. The approach was to consider each site-type independently, write the equilibrium expression for adsorption to that site (the Langmuir isotherm), and then solve the three equilibrium expressions simultaneously with the mass balance on A, subject to the constraint that the Langmuir expression for each of the three site-types had to use the same value for {A}. The total adsorption density on the solid could then be computed as the sum of the adsorption densities on the three site-types.

The analysis for a system with a continuous site distribution is essentially identical, except that instead of considering three macroscopic steps in the Φ_{cum} function, we divide Φ_{cum} (conceptually) into a large number of differential steps. In that case, the operation analogous to summing the adsorption densities on the three site-types is integration of the adsorption densities over all the site-types. The analysis can be developed quantitatively as follows.

Consider a system containing an adsorbent with a semi-continuous range of site-types that has equilibrated with an adsorbate A, present in solution at an activity {A}. We can represent the adsorption density of A on a group of sites that have K_{ads} values for A that fall within a given small range by defining functions analogous to Φ_{cum} and Φ_{diff}, except that they refer only to the sites that are occupied, rather than all the sites in that group. That is, we define the *cumulative site occupation function* $q'_{cum}(K, \{A\})$ as the adsorption density of A on sites that have adsorption equilibrium constants less than or equal to K, when the activity of dissolved A is {A}. Correspondingly, the *differential site occupation function* $q'_{diff}(K, \{A\})$ is the derivative of $q'_{cum}(K, \{A\})$ with respect to K. Note that, unlike Φ_{cum} and Φ_{diff}, q'_{cum} and q'_{diff} depend on {A}. That is, logically enough, the adsorption density on a specified subset of surface sites depends not only on the affinity of those sites for the adsorbate but also on the adsorbate activity in the solution.

Conceptually, the site distribution function, the site occupation function, and the equilibrium activity of dissolved A are related to one another by the following word equation.

$$\begin{pmatrix} \text{Adsorption density} \\ \text{on sites with binding} \\ \text{constants between } K_1 \\ \text{and } K_1+\Delta K, \text{ for the} \\ \text{given } \{A\} \end{pmatrix} = \begin{pmatrix} \text{Total concentration} \\ \text{of sites with} \\ \text{binding constants} \\ \text{between } K \text{ and} \\ K+\Delta K \end{pmatrix} \begin{pmatrix} \text{Fractional coverage of} \\ \text{sites based on a } K \text{ value} \\ \text{in the middle of the} \\ K \text{ to } K+\Delta K \text{ range,} \\ \text{for the given } \{A\} \end{pmatrix}$$

$$(10.29)$$

We can convert the word expressions in Equation (10.29) into mathematical terms as follows. The term on the left is the increment in total adsorption density attributable to the group of sites being considered and can be represented as $q'_{cum}(K+\Delta K, \{A\}) - q'_{cum}(K, \{A\})$. If we multiply and divide this expression by ΔK, and let ΔK become differentially small, the expression simplifies to

$$\frac{q'_{cum}(K+\Delta K, \{A\}) - q'_{cum}(K, \{A\})}{\Delta K}\Delta K = \frac{dq'_{cum}}{dK}\, dK = q'_{diff}\, dK \quad (10.30)$$

By analogous reasoning, the first term on the right in Equation (10.29) can be expressed as $\Phi_{cum}(K+\Delta K) - \Phi_{cum}(K)$ and then written in terms of Φ_{diff} as follows:

$$\frac{\Phi_{cum}(K+\Delta K) - \Phi_{cum}(K)}{\Delta K}\Delta K = \frac{d\Phi_{cum}}{dK}\, dK = \Phi_{diff}\, dK \quad (10.31)$$

Finally, assuming that adsorption onto this group of sites can be characterized by the Langmuir isotherm, we can use Equation (10.13) to express the second term on the right in Equation (10.29) as $K\{A\}/(K\{A\} + 1)$. Substituting the three mathematical expressions for the corresponding word statements in

Equation (10.29) and rearranging slightly, we obtain

$$q'_{diff}dK = \Phi_{diff}\frac{K_1\{A\}}{K_1\{A\} + 1}dK \tag{10.32}$$

The overall adsorption density for the given value of $\{A\}$ is the summation of the adsorption densities on all subsets of surface sites, i.e., on all sites with K values between 0 and ∞. If the summation is conducted over differential increments in K, it becomes an integration, which we can represent as follows:

$$q_{tot} = \int_0^\infty q'_{diff}dK = \int_0^\infty \Phi_{diff}\frac{K\{A\}}{K\{A\} + 1}dK \tag{10.33}$$

Equation (10.33) expresses the adsorption density of A in terms of its activity in solution and Φ_{diff}. Thus, if Φ_{diff} is known as a function of K, the expression on the right can be integrated to yield an equation for q_{tot} as a function of $\{A\}$, i.e., the adsorption isotherm equation.

THE FREUNDLICH ISOTHERM   As noted above, many adsorbents have a site distribution function like that shown in Figure 10.8b. One mathematical function that fits this pattern and that has found widespread application represents the concentration of sites as a function that decreases geometrically as the value of the equilibrium constant increases, i.e., $\Phi_{diff} = \alpha K^{-n}$, where α and n are constants and $0 \leq n \leq 1$.[5] The function is assumed to apply from $K = 0$ to some finite, maximum value K_{max}. Substitution of this function into Equation (10.33) yields the following expression for the adsorption density for any given value of $\{A\}$:

$$q_{tot} = \int_0^{K_{max}} \alpha K^{-n}\frac{K\{A\}}{1 + K\{A\}}dK = \int_0^{K_{max}} \frac{\alpha K^{1-n}\{A\}}{1 + K\{A\}}dK \tag{10.34}$$

Although the upper limit of integration of Equation (10.33) is formally $K = \infty$, the assumption is that $\Phi_{diff} = \alpha K^{-n}$ only up to K_{max} and that $\Phi_{diff} = 0$ at higher values of K, so the integration in Equation (10.34) is only up to K_{max}. For any given $\{A\}$, the integral on the right in Equation (10.34) has a closed-form solution, yielding the following simple isotherm equation:

$$q_{tot} = k_f\{A\}^n \tag{10.35}$$

where $k_f = \alpha RT\pi/\sin(n\pi)$, and R and T are the universal gas constant and the absolute temperature, respectively.[6] The relationship in Equation (10.35), known

---

[5] Since the adsorption equilibrium constant can be expressed as $K_{ads} = \exp[-\Delta\bar{G}_{ads}°/(RT)]$, where $\Delta\bar{G}_{ads}°$ is the standard molar Gibbs energy of the adsorption reaction, another way to describe the given distribution of site types is

$$\Phi_{diff} = \alpha K_{ads}^{-n} = \alpha\left[\exp\left(\frac{-\Delta\bar{G}_{ads}°}{RT}\right)\right]^{-n} = \alpha\exp\left[-\frac{n(-\Delta\bar{G}_{ads}°)}{RT}\right]$$

The equation shows that the assumption about the site distribution is equivalent to an assumption that the site concentrations decrease exponentially with increasingly negative standard Gibbs energy of adsorption, i.e., increasingly favorable adsorption energy.

[6] In much of the literature, n is defined as the inverse of its definition here; in those cases, the exponent in Equation (10.35) is written as $1/n$.

as the **Freundlich isotherm,** has been shown to fit many experimental data sets quite well. Like the single-site Langmuir isotherm, a Freundlich isotherm is fully defined by two fitting parameters: k_f, which describes the adsorption density under standard conditions ($q = k_f$ when $\{A\} = 1$), and n, which indicates how dramatically the binding strength changes as the adsorption density changes. Graphical representations of the Freundlich isotherm for a few values of k_f and n are shown in Figure 10.9.

Figure 10.9 shows the cumulative adsorption onto all surface sites in a system that follows the Freundlich isotherm. Based on the development of the isotherm

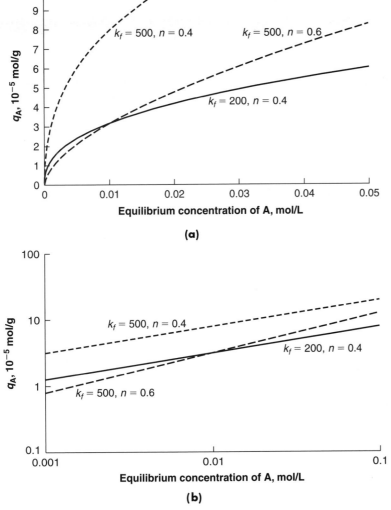

Figure 10.9 Graphical representation of the Freundlich isotherm on **(a)** linear and **(b)** logarithmic coordinates. The values of k_f shown have units of $\mu mol/g/(mol/L)^n$.

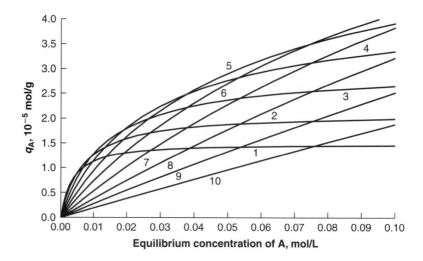

Figure 10.10   The distribution of binding sites and their occupation in a system characterized by a Freundlich adsorption isotherm. The different curves represent different groups of sites with steadily increasing surface concentration and decreasing K_{ads}. For this simulation, the curves correspond to the relationship $\Phi_{diff}(K) = 26K^{-0.5}$. Sorption onto each type of site is in accord with the Langmuir isotherm. In the limit envisioned by the Freundlich isotherm, the distribution of K_{ads} values is continuous rather than discrete.

equation, the total adsorption density is thought to represent the summation of the adsorption densities onto a wide variety of sites, with the Langmuir isotherm applying to each group of sites individually. A plot showing the adsorption density on different groups of surface sites in a system that follows the Freundlich isotherm is provided in Figure 10.10. Consistent with the idea that each group of sites fills in accord with the Langmuir isotherm, for any given value of {A} some sites of all types are occupied, with the high-energy (i.e., high-affinity) sites being closer to saturation (i.e., complete coverage) than the low-energy sites.

Example 10.4 | For the system characterized in Figure 10.10, estimate the total adsorption density at {A} values of 0.01 and 0.10 M, based on the amount of A adsorbed on the 10 site types shown. How does the distribution of the adsorbed material among the different site types compare under these two conditions?

Solution

The total adsorption density is simply the summation of the adsorption densities on the 10 site types. For the given simulation, the values of q when c_{eq} is 0.01 and 0.1 M are 7.9×10^{-5} and 28.9×10^{-5} mol/g, respectively (corresponding to $n = 0.56$).

At the lower {A}, sorption is preferentially on the low-density, high-affinity sites, with site type 3 having the highest adsorption density. At the higher {A}, the low-density, high-affinity sites are almost fully occupied, and more of the adsorption occurs on higher-density, lower-affinity sites. Under these conditions, site type 6 has the highest adsorption density.

The result shown in Equation (10.35) is valid only for values of n between 0 and 1. A value of $n = 1$ indicates that the binding strength is the same on all sites, so the equation reduces to a linear isotherm ($q_A = k\{A\}$). If $n < 1$, the average binding strength decreases with increasing surface coverage; the lower the value of n, the more dramatic the decrease.

Sometimes data conform to Equation (10.35) with a value of $n > 1$ over a limited range of $\{A\}$ values, implying that as the adsorption density increases, the affinity of the surface for the adsorbate also increases. Such a result is sometimes used as evidence of multilayer adsorption or conversion of adsorbed species into a precipitated, separate solid phase at the surface. However, this trend cannot apply indefinitely. There is an upper limit of $\{A\}$ at which the trend must change, because otherwise it would predict that the affinity of the surface for the adsorbate eventually becomes infinitely strong, which is thermodynamically impossible.

Example 10.5

Freundlich isotherms are commonly reported for sorption of metals onto mineral adsorbents, as shown below for sorption of Cd^{2+} onto $Fe(OH)_3(s)$ [from Benjamin and Leckie, *Journal of Colloid and Interface Science, 79:* 209 (1981)]. A system containing a fixed concentration of 10^{-4} M $Fe(OH)_3(s)$ at pH 7.2 is titrated with $Cd(NO_3)_2$, over a range of *TOT*Cd from 10^{-7} to 10^{-5} M. Plot the concentrations of dissolved and adsorbed Cd^{2+} and the fraction of the Cd^{2+} that is adsorbed as a function of *TOT*Cd.

Solution

The Freundlich isotherm characterizing the system at pH 7.2 can be determined from the graph to be $q = 27.5\{Cd^{2+}\}^{0.66}$, where q is in moles of Cd per mole $Fe(OH)_3(s)$. The distribution of Cd for any value of *TOT*Cd can be determined by solving the mass balance

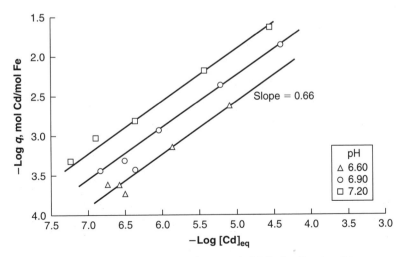

Adapted from M. M. Benjamin and J. O. Leckie, *Journal of Colloid and Interface Science,* Volume 79, 209–221 (1981).

equation in conjunction with the isotherm equation:

$$TOTCd = \{Cd^{2+}\} + q_{Cd}c_{solid}$$

$$TOTCd = \{Cd^{2+}\} + 27.5\{Cd^{2+}\}^{0.66}(10^{-4})$$

Inserting various values of $TOTCd$ and solving the above equation for $\{Cd^{2+}\}$ by trial and error yields the desired information, which is plotted below. Both adsorbed and dissolved Cd^{2+} increase as Cd is added to the system, but the dissolved concentration increases more rapidly than the adsorbed portion, so the fraction sorbed decreases.

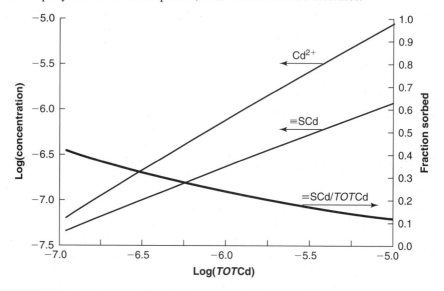

COMPETITIVE ADSORPTION IN SYSTEMS CHARACTERIZED BY THE FREUNDLICH ISOTHERM
In systems where adsorption is characterized by the Freundlich isotherm, competitive effects can be significant if even a small fraction of the surface is occupied, since the sites that are preferentially occupied by the competing adsorbate might be the ones that bind the target adsorbate most strongly. That is, if there are only a small number of strong binding sites on a solid, a relatively small concentration of a strongly competing adsorbate might leave all these sites occupied. In that case, even though there are plenty of sites still available on the surface, the competitive effect would be significant, since the target adsorbate would have to bind to relatively weak sites.

To model such a situation, a logical conceptual approach would be to apply the competitive Langmuir model to each group of sites. Use of such an approach is reasonably simple if the adsorbates are characterized by Freundlich isotherms with the same n value. In such a case, the k_f values provide a direct measure of the relative affinity of the surface for the two adsorbates. That is, if $n_A = n_B$ and $k_{f,A}/k_{f,B} = 2$, we can infer that every site on the surface has twice the attraction for species A as for species B. If, however, the adsorbates are characterized by different values of n, then A might outcompete B for sorption on some sites, and

B might outcompete A for sorption on others. In such a case, additional assumptions or measurements relating the two isotherms are required in order to model competitive interactions.

The additional information required can be represented as a so-called interaction parameter a_{AB}; in a system with m competing adsorbates, the values of $m - 1$ interaction parameters are needed. The resulting isotherm equation for the competitive system is

$$q_A = k_{f,A} c_A \left(\sum_{j = A, B, C...} a_{Aj} c_A \right)^{n_A - 1} \tag{10.36}$$

where A, B, C, ... are the various adsorbates competing for the surface. The value of the interaction parameter for an adsorbate with itself (e.g., a_{AA}) is 1.0, so Equation (10.36) reduces to the conventional Freundlich isotherm for a system containing only one adsorbate.[7]

COMPARISON OF MULTISITE AND FREUNDLICH ISOTHERMS The preceding discussion describes the two most common approaches for modeling surfaces that are thought to have groups of sites with different affinities for an adsorbate. The multisite Langmuir model allows one to model any distribution of surface site concentrations and adsorption binding constants, but it suffers from the fact that each postulated type of site increases the number of adjustable parameters in the model by two. It is rarely possible to resolve experimental data well enough to justify proposing more than two types of sites on a given surface, and even then the evaluation of the four constants must be viewed somewhat skeptically.

On the other hand, the Freundlich model presupposes a great deal of variability in adsorption binding constants, but postulates a specific form for the site distribution function, i.e., a specific relationship between the binding constants and the site concentrations. Because the Freundlich isotherm is a two-parameter model, it does not possess quite as much flexibility as multisite Langmuir models. Nevertheless, it often provides a remarkably good fit to experimental data over a range of several orders of magnitude in adsorbate concentration, and it is therefore frequently used to describe equilibrium adsorption relationships.

It should be emphasized that the isotherms presented above have found popular acceptance because they are mathematically simple and because they seem to fit experimental data, not because there is some fundamental physical or chemical reason to expect them to apply. Unfortunately, a false distinction between these isotherms has become widely accepted, whereby the Langmuir isotherm is viewed as theoretically based and the Freundlich isotherm as strictly empirical. Such a distinction is incorrect on both counts: as shown above, both the Langmuir and Freundlich isotherms can be derived by starting from a conceptual (theoretical) view of the surface and the adsorption process, but in both cases arbitrary

---

[7]For additional details on this approach, see C. Sheindorf, M. Rebhun, and M. Sheintuch, "A Freundlich-Type Multicomponent Isotherm," *J. Colloid Interface Sci.* 79, 136–142 (1981).

assumptions about the surface site distribution function are needed. Thus, in both cases, empirical data are needed to justify that the assumptions are applicable to a given situation.

Other functional relationships could be (and have been) postulated between surface site densities and binding constants, and any such relationship would be perfectly acceptable as the basis of an isotherm if the experimental data supported it. Correspondingly, the fact that a data set fits a given equation should not be taken as proof that the conceptual model for that isotherm is correct.

10.4   USING CHEMICAL EQUILIBRIUM COMPUTER MODELS TO SOLVE FOR SPECIATION IN ADSORPTIVE SYSTEMS

10.4.1   MODELING SINGLE- AND MULTIPLE-SITE LANGMUIR ADSORPTION

Incorporating adsorption reactions into software that calculates chemical equilibrium speciation is straightforward if the reactions can be represented by the Langmuir equation. In that case, we can just treat the surface as a conventional ligand and the reactions as complexation reactions. To model the surface as having multiple types of sites, with adsorption obeying the Langmuir isotherm on each of them, we simply represent the surface as a group of several different ligands and proceed as usual. No modifications are required to account for competitive adsorption; if two or more adsorbates compete for the same group of sites, we just input the corresponding reactions, and the competition is considered implicitly.

Example 10.6

In a study of Pb^{2+} and H^+ adsorption onto a volcanic soil [Papini et al., *Environmental Science and Technology, 33:* 4457 (1999)], the soil was modeled as a mixture of three site types, designated $\equiv MO$, $\equiv SO$, and $\equiv TO$ (this representation was chosen because the authors considered all the surface binding sites to be oxide ions that were bound to the solid in different ways). Both Pb^{2+} and H^+ were assumed to compete for $\equiv MO$ and $\equiv SO$ sites, while only H^+ could bind to $\equiv TO$ sites. The adsorption data were modeled based on the following seven reactions:

Reaction	Log K
$\equiv MO^- + H^+ \leftrightarrow \equiv MOH$	4.95
$\equiv MO^- + Pb^{2+} \leftrightarrow \equiv MOPb^+$	4.70
$\equiv SO^- + H^+ \leftrightarrow \equiv SOH$	9.65
$\equiv SOH + H^+ \leftrightarrow \equiv SOH_2^+$	7.58
$\equiv SO^- + Pb^{2+} \leftrightarrow \equiv SOPb^+$	9.91
$\equiv TO^- + H^+ \leftrightarrow \equiv TOH$	11.71
$\equiv TOH + H^+ \leftrightarrow \equiv TOH_2^+$	4.70

The surface concentrations of $\equiv MO$, $\equiv SO$, and $\equiv TO$ sites were estimated as 2.33×10^{-2}, 3.24×10^{-2}, and 6.74×10^{-2} mol/g, respectively.

a. Prepare the input tableau to determine the speciation in a suspension containing 10 g/L of the soil and 10^{-4} M TOTPb at pH 5.0.

b. Characterize the protonation/deprotonation of the $\equiv SO$ sites as a function of pH in the absence of any Pb^{2+}.

Solution

a. The input tableau is prepared exactly as it would be if the surface sites were three different types of dissolved ligands and the adsorbed species were complexes.

				Component			
Species	**Type**	**H^+**	**Pb^{2+}**	**$\equiv MOH$**	**$\equiv SOH$**	**$\equiv TOH$**	**Log K**
H^+	1	1	0	0	0	0	0.00
Pb^{2+}	1	0	1	0	0	0	0.00
$\equiv MOH$	1	0	0	1	0	0	0.00
$\equiv SOH$	1	0	0	0	1	0	0.00
$\equiv TOH$	1	0	0	0	0	1	0.00
OH^-	2	−1	0	0	0	0	−14.00
$\equiv MO^-$	2	−1	0	1	0	0	−4.95
$\equiv MOPb^+$	2	−1	1	1	0	0	−0.25
$\equiv SO^-$	2	−1	0	0	1	0	−9.65
$\equiv SOH_2^+$	2	1	0	0	1	0	7.58
$\equiv SOPb^+$	2	−1	1	0	1	0	0.26
$\equiv TO^-$	2	−1	0	0	0	1	−11.71
$\equiv TOH_2^+$	2	1	0	0	0	1	4.70
$\overline{H^+}$	3a	1	0	0	0	0	−5.00

Inputs							**Input Concentration**
$\overline{H^+}$		1	0	0	0	0	x
Pb^{2+}		0	1	0	0	0	10^{-4}
Solid		0	0	2.33×10^{-2}	3.24×10^{-2}	6.74×10^{-2}	10
TOTComp$_{in}$		x	10^{-4}	0.233	0.324	0.674	

Conventionally, singly protonated sites ($\equiv MOH$, $\equiv SOH$, and $\equiv TOH$) are chosen as *components*. The input tableau for the system is shown above. Note that the input values of the equilibrium constants are not exactly those given in the problem statement, since the input values must correspond to formation of the *species* from a combination of *components*, and not all the given reactions meet that criterion.

The line labeled Solid in the Inputs portion of the tableau indicates the concentration of adsorbent in grams per liter and the "stoichiometry" associated with the addition of the adsorbent to the system: each gram of solid added contributes 2.33×10^{-2} mol of $\equiv MOH$, 3.24×10^{-2} mol of $\equiv SOH$, and 6.74×10^{-2} mol of $\equiv TOH$.

b. By changing the value of log K for the species $\overline{H^+}$, we can change the fixed pH at which the calculations are carried out, and thereby develop the speciation of the surface as a function of pH.

In the absence of Pb, the $\equiv$SO sites are distributed among doubly protonated, singly protonated, and unprotonated forms. Dominance shifts from $\equiv$SOH$_2{}^+$ to $\equiv$SOH to $\equiv$SO$^-$ as pH increases, just as it would for a soluble, diprotic acid. The modeling results obtained in the study are shown below.

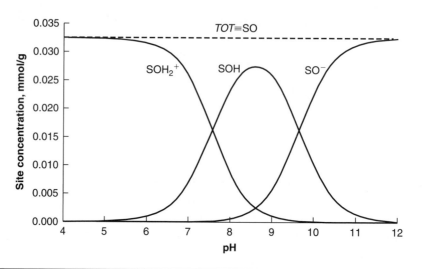

10.4.2   USING CHEMICAL EQUILIBRIUM SOFTWARE TO MODEL THE FREUNDLICH ISOTHERM

Chemical equilibrium software can also be used to compute the distribution of adsorbate in systems characterized by Freundlich isotherms, but in this case the appropriate data inputs are not as intuitive, since there is no direct analog to the Freundlich isotherm in solution phase equilibrium. The approach used by the program to solve for the equilibrium speciation in these situations is based on the fact that the Freundlich isotherm has a form that is very similar to the equilibrium constant for the following hypothetical reaction:

$$\equiv S + nA \leftrightarrow \equiv SA_n \tag{10.37}$$

$$K_{(10.37)} = \frac{\{\equiv SA_n\}}{\{\equiv S\}\{A\}^n} \tag{10.38}$$

If reaction (10.37) were input to the program just like any other reaction, with $K_{(10.37)}$ as the equilibrium constant, the program would compute the value of $\{\equiv SA_n\}$ as follows:

$$\{\equiv SA_n\} = K_{(10.37)}\{\equiv S\}\{A\}^n \tag{10.39}$$

Equation (10.39) bears a strong resemblance to the equation that we would use to calculate $\{\equiv SA\}$ according to the Freundlich isotherm, based on the isotherm parameters:

$$\{\equiv SA\} = k_f c_{solid}\{A\}^n \qquad \textbf{(10.40)}$$

Comparing Equations (10.39) and (10.40), we see a correspondence between $K_{(10.37)}$ and k_f, and between $\{\equiv S\}$ and c_{solid}. That is, if reaction (10.37) is input to the program with an equilibrium constant equal to k_f, and if the value of $\{\equiv S\}$ is forced to equal c_{solid}, then the value that the program computes for $\{\equiv SA_n\}$ will equal the concentration of adsorbed A that is consistent with the given Freundlich isotherm. Since, in any given system, c_{solid} is fixed, $\{\equiv S\}$ can be forced to equal c_{solid} by declaring $\equiv S$ to be a Type 3a *species*. Thus, by inputting an appropriate combination of reactions, equilibrium constants, and *species,* we can cause the program to carry out calculations that are consistent with the Freundlich isotherm, even though that isotherm is not directly analogous to any solution phase relationship.

One other modification to the program that is required to allow it to model Freundlich adsorption is that, even though the concentration of adsorbed A is computed using the equilibrium constant for reaction (10.37), the mass balance on A must be modified to account for the fact that only one molecule of A (not n molecules) is consumed by the reaction. This modification, as well as the assignment of $\equiv S$ as a Type 3a *species* whose equilibrium concentration is fixed at c_{solid}, is carried out in the background by many available programs. That is, if the user indicates that the data should be modeled using a Freundlich isotherm, he or she is prompted for values of k_f, n, and c_{solid}. The program then carries out the necessary data manipulations and reports the equilibrium concentration of adsorbed A, so the user need not be familiar with the details of how the data were manipulated.

The generic chemical equilibrium programs described in this text are not currently able to model competitive adsorption in systems where adsorption obeys Freundlich isotherms. Separate programs designed specifically to solve those types of problems are available. Most such programs use the gas transfer model of adsorption, rather than surface complexation as the basis for these calculations.[8]

10.5   INTERACTIONS BETWEEN CHARGED SURFACES AND IONIC ADSORBATES: THE EFFECT OF SURFACE POTENTIAL ON ADSORPTION OF IONS

In addition to the non-uniformity of surface sites, adsorption differs from formation of soluble complexes with respect to the electrical potential that the chemicals experience. As discussed in Chapter 2, we can identify two contributions to the electrical potential experienced by a dissolved ion. One contribution, which

---

[8]Competition is usually represented in the gas transfer model of adsorption according to *ideal adsorbed solution theory* (IAST). This approach is described in the article by Crittenden et al. referred to in footnote 4.

we referred to as Ψ_{chem}, arises from the charged groups that are in solution near the ion. This potential gives rise to the conventional activity coefficient applicable to the ion, as estimated by the Debye–Huckel equation or similar expressions. Because the only system-specific parameter in these equations is the ionic strength (I), we can model all the interactions of interest related to Ψ_{chem} based solely on the value of I.

The second contribution to the electrical potential experienced by an ion is generated outside the solution and was designated Ψ_{ext} in Chapter 2.[9] In the current analysis, the external potential of concern originates at the surface of a charged adsorbent particle and applies both to ions that are chemically bound to the surface and to ions in solution near the surface. By contrast with Ψ_{chem}, Ψ_{ext} depends on the properties of the system composition in ways that cannot be reduced to a function of a single variable. Therefore, the effect of Ψ_{ext} on adsorption cannot be computed based on the value of any simple surrogate parameter; it must be computed based on the detailed composition of each system of interest.

We can interpret the effects of Ψ_{ext} on adsorption in the context of the surface site distribution functions $\Phi_{cum,A}$ and $\Phi_{diff,A}$ by considering the same idealized, single-site surface that we considered in our original development of the Langmuir isotherm. However, we now assume that the adsorbate is a cation and that the adsorbent can acquire a surface charge. The surface charge generates a surface potential, which we will assume is uniform at any given distance from the surface, decaying to zero in the bulk solution. Thus, in this system, the magnitude of Ψ_{ext} is viewed as varying from some finite value Ψ_0 at the surface to zero at some distance far (on a molecular scale) from the surface.

Plots of Φ_{cum} versus K for this system under three conditions are shown in Figure 10.11. In the absence of surface charge ($\Psi = 0$), Φ_{cum} jumps from zero to Φ_{cum}^{max} at K_1, as in the previous analysis. On the other hand, when Ψ_0 is non-zero, the figure shows that Φ_{cum} has the same shape, but the step change occurs at a higher K value if $\Psi_0 < 0$ and a lower K value of $\Psi_0 > 0$. This shift can be rationalized as follows.

The tendency of ions to reside at any location where Ψ_{ext} is non-zero is affected by that potential. Thus, in systems with non-zero Ψ_0, the net tendency for an ion to adsorb represents the combined effects of any chemical bonding between the ion and the surface and the electrical attraction or repulsion that the ion experiences. If, for instance, Ψ_0 is positive, then all the adsorptive sites on the hypothetical one-site surface will still be identical (just as they are when the surface is uncharged), but they will have a net affinity for A that is less than that corresponding to K_1. Similarly, if the surface is negatively charged, the net binding constant will be greater than K_1.

If the main mechanism by which the surface acquires charge is by adsorption of A, the result will be that the surface binding constant for A changes (in this case, becoming weaker) as more A adsorbs. In such a case, the first molecule of

---

[9]Conventionally, Ψ_{ext} is written simply as Ψ, with the implicit understanding that it describes only potentials that are imposed externally. However, for clarity, we will continue to make the distinction between Ψ_{ext} and Ψ_{chem} throughout the chapter.

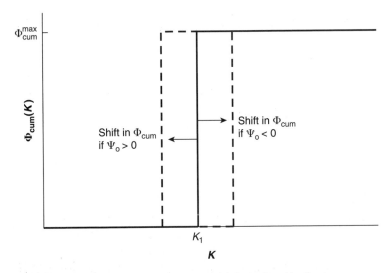

Figure 10.11 Φ_{cum} as a function of K for a surface like that in Figure 10.3, showing the effect of electrical interactions with the surface, assuming that the adsorbate is a cation. The trends would be in the opposite direction if the adsorbate were an anion.

A that sorbs will be properly modeled by applying a binding constant of K_1. Once a substantial amount of A has bound, the equilibrium constant characterizing adsorption will be lower. However, unlike the case of a surface with non-uniform sites on which the binding constant for the original molecule remains at K_1 when subsequent molecules adsorb, in this case the binding constants for *all* adsorbed A molecules decline uniformly. Furthermore, the changes in the electrical potential caused by adsorption of A affect not only those molecules, but all adsorbed ions. As a result, the electrical effects of adsorption of A will alter the net affinity of the surface for all ionic adsorbates in the system.

As is clear from the above discussion, the key distinction between analysis of adsorption in systems with and without electrical interactions is the need to calculate Ψ_{ext} as a function of location in the former systems. These calculations are complicated by the fact that ion adsorption and the electrical potential are implicitly linked to one another: the extent of ion adsorption affects the electrical potential at the interface, and vice versa. An additional complicating factor is that whereas adsorption involves a discrete ion and a discrete site (at least in the site-binding models), the electrical potential near the surface is a continuous or *mean-field* parameter that reflects the integrated effects of all the species in the interfacial region. A schematic showing the discrete and mean-field representations of the interface is presented in Figure 10.12.

To quantify all the interactions in this system, we need to first choose a physical model for the interfacial structure, i.e., a model that describes how ions are distributed on and near the adsorbent particle. The basic model for the interfacial structure that is currently used almost universally was first developed in the 1910s and 1920s. This model represents the interfacial region as comprising

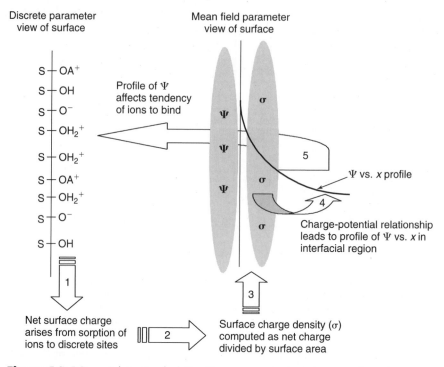

Figure 10.12   Schematic of relationships among adsorbed charge, surface charge density, and the profile of electrical potential near the interface, and the links between discrete and mean field parameters.

one layer of molecules that is part of the solid surface and one or more layers adjacent to the surface where adsorbate molecules might reside.

The general approach for computing the distribution of ions between the interface and solution using this structural model can be summarized as follows. First, assumptions are made about what types of ions can reside in each layer. Then, isotherm equations are used to estimate the adsorption density of each ion of interest. This estimate might, for instance, be based on the amount of adsorption that would be expected if the surface were uncharged. The amount of charge in each layer is next computed based on the number of anions and cations in that layer (arrow 1 in Figure 10.12). Because the isotherm equations and the calculations of adsorbed charge characterize the behavior of individual ions, they reflect a discrete parameter view of the surface. The link between that view and the mean-field view is forged by converting the adsorbed charge in each layer to a charge density σ, equal to the amount of adsorbed charge per unit surface area. This computation, represented by arrow 2 in Figure 10.12, converts the adsorbed charge from a discrete to a continuous variable by treating the charge as though it were spread uniformly on the surface (arrow 3).

The difference in charge densities between the different adsorbing layers can be related to the change in electrical potential (another mean-field parameter)

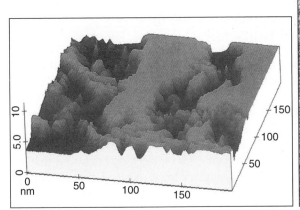

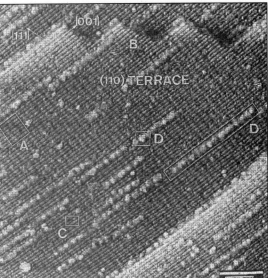

Sophisticated instrumentation is now able to provide fantastically high resolution images of adsorbent surfaces. The photograph on the left shows a graphite surface that has been coated with copper by electrodeposition from solution. Note that the scale in all three dimensions is in nm. The image on the right shows an atomic-scale view of the surface of $TiO_2(s)$, on which areas of ordered structure as well as crystal defects can be identified.

Left: Reprinted from *Surface Science*, Vol. 338, pp. 31–40, 1995, with permission of Elsevier Science. Right: Reprinted from *Surface Science*, Vol. 337, pp. 17–30, 1995, with permission of Elsevier Science.

between the layers via basic principles of physics and electricity. Once this computation is carried out, the computed electrical potential can be used to modify the effective adsorption equilibrium constants for the various ions in each layer, thereby providing a link from the mean-field representation back to the discrete representation of the system. Using these modified equilibrium constants, the extent of adsorption of discrete ions can be recomputed, and the process can be repeated. By iterating through this loop, a set of conditions can eventually be found in which all the equations characterizing the system are satisfied simultaneously.

In principle, the process described above can be used to determine the equilibrium state of any adsorptive system. However, the interfacial structure has still not been fully elucidated, and debate continues about the number of layers that exist and the types of adsorbates that bind in each layer. Because the mathematical models that correspond to the different assumptions all have enough flexibility to fit much of the available data, identification of the most appropriate model for a given system requires either additional data or new analytical tools. Instruments that can probe the interfacial region with fantastic resolution have recently been developed, and as these tools are applied to more systems, there is likely to be some convergence in the modeling approaches.

In the following section, additional details are provided to expand on the discussion above of adsorption of charged adsorbates onto charged surfaces.

The section begins with a discussion of how adsorbent particles acquire charge in the first place, followed by a description of the assumed ion distribution at the surface and in the adjacent solution according to the most popular current models. Finally, some key quantitative features of the models are presented and explained.

10.5.1 THE ORIGINS OF SURFACE CHARGE ON SUSPENDED PARTICLES

Although the charge associated with the billions of cations in any particle is almost exactly balanced by the charge of the anions,[10] there is often a slight imbalance between the two. For naturally occurring particles, the most important ways in which this imbalance is generated are *isomorphic substitution,* acid/base reactions of ions that comprise the solid surface, and adsorption of ions from solution.[11]

Isomorphic substitution is important mostly for clays and describes a process whereby Al^{3+} and Si^{4+} ions in the clay lattice are replaced by other ions that are of similar size (isomorphous), but (typically) of lower positive charge. The net effect is that the particle acquires a negative charge, while the solution acquires a positive charge.

As suggested in Example 10.6, solids can also acquire charge via acid/base reactions at their surfaces. Various iron oxide minerals provide excellent examples of this behavior and have been studied extensively both for that reason and because of their importance in natural systems (as components of soils, sediments, and aquifers) and water treatment systems (where they form due to addition of iron salts as coagulants).

At the surface of an iron oxide particle, Fe(III) and O(–II) ions are exposed to solution. These ions are bonded to oppositely charged ions in the interior of the solid, so they do not express their full ionic charge at the surface. In the models that have been used most extensively in the past 30 years, these surface ions have been assigned net charge values of $+1$ and -1, respectively. (The same is true for other metal oxide and hydroxide solids, regardless of the identity of the metal ion.) More recently, an approach whereby the charge on the surface ions is computed based on the coordination environment of atoms in the solid has been proposed and shown to yield predictions that are in close accord with many experimental observations. According to these calculations, the average charge on a surface atom is not necessarily integral, being in many cases $+1/2$ or $-1/2$. However, for our purposes, simply noting that the surface ions are charged is sufficient, so we will proceed by making the simplifying assumption that their charges are $+1$ and -1, respectively.[12]

---

[10] As an order-of-magnitude estimate, there are about a billion molecules in a 1-μm-diameter particle.

[11] The statement earlier in the chapter that the interfacial region must be electroneutral still applies. However, that electroneutrality condition applies to the combination of the surface layer of the solid and the layers of nearby adsorbed molecules. The non-zero surface charge being referred to here is the charge associated with the ions that comprise the solid surface and a subset of the adsorbed ions. The details of which ions are included in the calculation are described below.

[12] The newer model has been developed and championed by van Riemsdijk and coworkers in numerous publications. It is described in detail by Hiemstra and van Riemsdijk in *J. Colloid Interface Sci. 179,* 488–508 (1996), and is referred to as the CD-MUSIC (charge *distribution-multi*site complexation) model.

The charges on the surface ions can be neutralized by binding of OH^- to the $\equiv Fe$ sites and H^+ to the $\equiv O$ sites, so that a surface consisting entirely of $\equiv FeOH$ and $\equiv OH$ groups would be neutral. Since each $\equiv OH$ group is presumed to be bonded to an Fe atom in the solid, that species can be considered to be an $\equiv FeOH$ as well, so we can view the whole surface as being comprised of $\equiv FeOH$ groups.

Considering the $\equiv FeOH$ groups to be "half in solution and half out," we expect that they can undergo some of the same reactions with solution species that dissolved, hydrated $FeOH^{2+}$ ions do. Thus, for instance, the group can become protonated, converting the OH^- ligand to an H_2O and forming a species $\equiv FeOH_2^+$. Similarly, the group might become deprotonated to form $\equiv FeO^-$. These surface acidity reactions can be written just like acidity reactions in solution, as follows:

$$\equiv FeOH_2^+ \leftrightarrow \equiv FeOH + H^+ \qquad \textbf{(10.41)}$$

$$\equiv FeOH \leftrightarrow \equiv FeO^- + H^+ \qquad \textbf{(10.42)}$$

Like any acidity reaction, reactions (10.41) and (10.42) are expected to depend on solution pH (as suggested by Example 10.6), so that $\equiv FeOH_2^+$ groups dominate at low pH and $\equiv FeO^-$ groups dominate at high pH. These trends have been confirmed in a variety of ways, most directly by titrations of suspensions of adsorbents with acid or base. By comparison of the titration curves in the presence and absence of the adsorbent, the amount of H^+ or OH^- attached to the surface can be inferred via a process exactly analogous to that for determining the amount of H^+ or OH^- consumed by a dissolved weak base or acid. The surface charge per unit mass or area of solid is then computed as $q_{H^+} - q_{OH^-}$.

The surface charge of a variety of particles is shown as a function of pH in Figure 10.13. The pH at which the surface is neutral is a characteristic of the

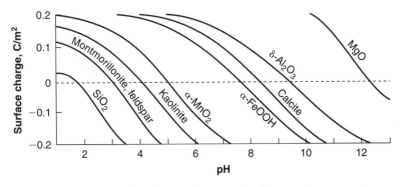

Figure 10.13   Surface charge of various colloidal materials, computed based on the sorption of H^+ and OH^- ions (by comparison of acid/base titration curves in the presence and absence of the colloid).

From W. Stumm and J. Morgan, *Aquatic Chemistry: Chemical Equilibria and Rates in Natural Waters*, 3rd Ed., Copyright © 1996. Reprinted by permission of John Wiley & Sons, Inc.

Table 10.3   PZCs of some oxides, clays, and other solids*

Material	pH$_{PZC}$	Material	pH$_{PZC}$
α-Al$_2$O$_3$	9.1	δ-MnO$_2$	2.8
α-Al(OH)$_3$	5.0	β-MnO$_2$	7.2
γ-AlOOH	8.2	SiO$_2$	2.0
CuO	9.5	TiO$_2$	6.2
Fe$_3$O$_4$	6.5	Feldspars	2–2.4
α-FeOOH	7.8	Kaolinite	4.6
α-Fe$_2$O$_3$	8.5	Montmorillonite	2.5
Fe(OH)$_3$ ferrihydrite	8.5	Albite	2.0
MgO	12.4	Chrysotile	>10

*From W. Stumm, *Chemistry of the Solid-Water Interface* (Wiley, 1992). Note that all values are approximate; different methods for preparing the solids can yield PZC values that are different by several tenths of a pH unit.

solid and is referred to as the **point of zero charge (PZC).**[13] The PZCs of several metal oxides, clays, and other minerals are presented in Table 10.3.

The third important way that adsorbent particles acquire charge is, in essence, identical to the surface acid/base reactions discussed above, except that the idea is extended to reactions with other solutes. That is, species other than H$^+$ and OH$^-$ can also adsorb, by replacing either the H$^+$ ion or the entire $\equiv$OH$_x$ group at the surface. If the charge of the adsorbing group is either greater or less than that of the group(s) released, and the surface charge is defined to include those groups, the reaction will cause the net surface charge to change. Representative such reactions for sorption of SO$_4{}^{2-}$ and Ni^{2+} onto an iron oxide surface are shown below:

$$\equiv\text{FeOH} + \text{SO}_4{}^{2-} \leftrightarrow \equiv\text{FeSO}_4{}^- + \text{OH}^- \qquad \textbf{(10.43)}$$

$$\equiv\text{FeOH} + \text{Ni}^{2+} \leftrightarrow \equiv\text{FeONi}^+ + \text{H}^+ \qquad \textbf{(10.44)}$$

Although reactions (10.43) and (10.44) are written as though the neutral surface species participates, they could just as well be written showing either a fully protonated or deprotonated surface site as the reactant; the difference is just like choosing to write the formation of CaCO$_3$ by a reaction of Ca^{2+} with H$_2$CO$_3$, HCO$_3{}^-$, or CO$_3{}^{2-}$. When a reaction such as one of those shown above proceeds, the surface protonation changes in a way that satisfies all the relevant equilibrium expressions (i.e., those for surface acid/base reactions and other adsorption reactions). The net effect on surface charge reflects a composite of all these reactions.

---

[13] Because other ions can also bind to the surface and alter the overall charge, the point of zero charge (where the overall surface charge is zero) might differ from the point of zero net proton charge (PZNPC, where q$_{H^+}$ equals q$_{OH^-}$). See Sposito, *Environmental Science and Technology, 32:* 2815 (1998), for additional discussion of this topic.

Because of the almost limitless variety of adsorbent particles and possible adsorbates in natural waters, we might expect that the surfaces of adsorbent particles would have a wide range of net charge, depending on the identity of the adsorbent and the composition of the solution. However, it turns out that the ubiquity of natural organic matter, its moderately strong acidity, and its strong tendency to adsorb cause virtually all particles in natural waters to have a net negative surface charge. Neutralization of the negative surface charge on these particles is a key factor in getting the particles to coagulate or to attach to filter media in water treatment operations.

10.5.2   THE DISTRIBUTION OF IONS NEAR A CHARGED INTERFACE AND THE CORRESPONDING CHARGE-POTENTIAL RELATIONSHIP

As described above, a number of conceptual and mathematical models exist for characterizing the interfacial region, differing primarily in how they envision the distribution of ions near the interface and the corresponding profile of electrical potential from the surface to the bulk solution. The discussion here focuses on the most commonly applied three-layer model of the adsorptive region (the *triple-layer model,* or TLM); the one- and two-layer models that are in common use can be represented as limiting cases of that more general model.[14] To date, the models have been applied primarily to oxide adsorbents, which are assumed to be hydrated and therefore to have a layer of $\equiv OH_x$ groups all along the surface (either from the oxide structure or from bound water molecules) when the solid is suspended in pure water. The surface structure of an iron oxide adsorbent according to the triple-layer model is shown in Figure 10.14.

Some adsorbates are presumed to bind directly to either a surface oxide ion or an underlying Fe ion (effectively replacing a surface $\equiv OH_x$ group). Molecules that adsorb in this manner must lose at least the water of hydration on the side nearest the solid. The surface complexes formed by such reactions are relatively strong and are sometimes referred to as *inner-sphere* complexes. H^+ and OH^- are always presumed to bind in this way. A plane that runs through the centers of the surface oxide ions and any adsorbates that are bound either directly to them or to surface Fe ions is typically referred to as either the *surface* plane or *naught* (o) *plane*.

In some models, other ions also are presumed to bind to the surface via chemical bonds, but to retain all their waters of hydration. They are therefore separated from the surface by a water molecule and form weaker complexes (sometimes called *ion pair* or *outer-sphere* complexes) with the surface. A plane through these ions is designated the *beta* (β) *plane*.

[14]The triple-layer model was first fully elaborated by Davis et al. [*J. Colloid Interface Sci., 63*: 480–499 (1978) and *67*: 90–107 (1978)] and has since been used by that group and others in numerous adsorption modeling efforts.

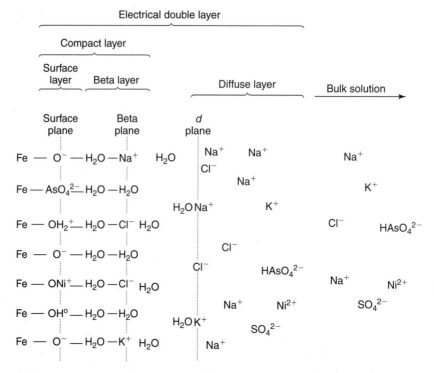

Figure 10.14   Schematic of the interfacial structure as envisioned in the triple-layer model. The charges shown on species in the naught plane represent the sum of the charge from that species plus the Fe atom on the surface; i.e., they represent the total charge expressed at the site. The space between the beta plane and the beginning of the diffuse layer is occupied by hydration waters of the ions in the beta and d planes. Some models do not include a beta layer, in which case the d plane is separated from the surface layer only by the waters of hydration between the ions in the surface plane and the d plane.

At distances farther from the surface than the β plane, ions are assumed to be attracted to or repelled from the surface solely by electrostatic interactions. That is, only non-specific adsorption occurs in this region, increasing the concentration of oppositely charged species (*counterions*) and decreasing that of like-charged species (*co-ions*) compared to their concentrations in bulk solution (the co-ions are considered to be negatively adsorbed). Over a distance of a few molecular diameters, the net charge of these counterions and co-ions completely neutralizes the charge in the naught and beta layers, and the concentrations of the ions approach their values in bulk solution.

The region in which the concentrations of the non-specifically adsorbed ions differ from their values in bulk solution is called the *diffuse layer*. By definition, then, the charge in the diffuse layer is equal and opposite to that in the surface and beta layers:

$$\sigma_d = -(\sigma_o + \sigma_\beta) \qquad\qquad \textbf{(10.45)}$$

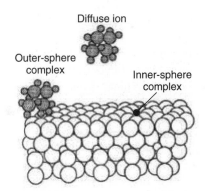

Figure 10.15 A hypothetical interfacial structure comprising a solid (open circles); an adsorbate that has lost the water molecules from its primary hydration sphere and binds directly to the surface, forming an *inner-sphere* complex (black circle); an adsorbate that is bonded to the surface, but retains its waters of hydration, forming an *outer-sphere* complex (gray circles); and an adsorbate that is attracted to the surface via electrostatic interactions, but is not bonded directly to the surface (a *diffuse* ion). The terminology is described in greater detail later in the text.

From THE CHEMISTRY OF SOILS by Garrison Sposito, copyright—1989 by Oxford University Press, Inc. Used by permission of Oxford University Press, Inc.

The diffuse layer starts a short distance farther from the surface than the β layer (or a short distance from the naught layer in models that do not include the β layer). The plane of closest approach of ions in the diffuse layer to the surface is called the *d plane*. The region containing the surface and beta layers (and therefore containing all specifically adsorbed ions) is sometimes referred to as the *compact layer*, and the combination of the compact layer and the diffuse layer is called the *electrical double layer*. Ions adsorbed in each of the three layers are depicted in Figure 10.15.

The charge-potential relationships in the various layers are central to the quantitative analysis of adsorption. The relationship between the charge in the diffuse layer and the potential at its inner boundary (the *d* plane) was determined in 1910 for an idealized system that treated the ions as point charges, and this relationship is used by all current models. Except for geometric considerations, the analysis is just like that for the distribution of ions surrounding a central ion, i.e., the analysis that yields a prediction for the ion activity coefficient.[15] The result for a solution of symmetric electrolyte (one in which the magnitude of the charge on the cations, $|z^+|$, is the same as that on the anions, $|z^-|$) for a system at 25°C is

$$\sigma_d = -0.1174 c_s^{0.5} \sinh \frac{zF\Psi_d}{2RT} \qquad \textbf{(10.46)}$$

where σ_d is the equivalent charge density of the *d* plane, in C/m², c_s is the electrolyte concentration in mol/L, z is the absolute value of the ionic charge number

[15] In fact, the relationship describing the adsorptive system was derived 13 years before the Debye–Huckel equation.

of the electrolyte ions, F is the Faraday constant, Ψ_d is the electrical potential in the d plane, and the product RT has its usual meaning. The parameter σ_d is referred to above as an equivalent charge density, because it does not refer to charges that reside *in* the d plane, but rather to the net charge throughout the diffuse layer, treated *as though* it were all in the d plane. Equation (10.46) is known as the *Gouy-Chapman equation.*

When combined with the model for non-specific adsorption of ions as a function of Ψ (developed below), Equation (10.46) can be converted into an expression for Ψ versus distance from the d plane. For most situations of interest in natural aquatic systems ($\Psi < 25$ mV), Ψ decays approximately exponentially. The characteristic distance for this decay (the distance needed for Ψ to decay by a factor of e, sometimes called the Debye length and usually designated by the Greek letter κ) varies with the inverse square root of the ionic strength, i.e., with $I^{-0.5}$, i.e., $\Psi_x = \Psi_d \exp(-\kappa x)$. The approximate range of κ values in natural waters is from ~ 10 nm for 10^{-3} M NaCl to ~ 0.4 nm for seawater.

The charge-potential relationships in the compact layer are less well agreed upon than that in the diffuse layer. The most common modeling approach is to assume that the potential changes linearly in this region, with one slope between the surface and beta planes and another one between the beta and d planes. The (assumed) fixed ratio of the change in potential between two layers to the charge is referred to as the *capacitance,* making an analogy between the parallel layers of adsorbed charge and a parallel-plate capacitor (or, in models incorporating a beta layer, two capacitors in series). The relationships are

$$C_1 = \frac{\Psi_o - \Psi_\beta}{\sigma_o} \tag{10.47}$$

$$C_2 = \frac{\Psi_\beta - \Psi_d}{-\sigma_d} = \frac{\Psi_d - \Psi_\beta}{\sigma_d} \tag{10.48}$$

where C_1 and C_2 are the capacitances of the inner (o-to-β) and outer (β-to-d) layers, respectively, typically expressed in units of coulombs per unit area per volt (farads per unit area).[16]

The capacitances of the compact layers of adsorbents are not well known, and in any case the idea that all the adsorbates are aligned in a plane and that the charge is distributed uniformly on that plane is clearly an idealization. Therefore, C_1 and C_2 are usually used as fitting parameters to improve the match of model calculations to experimental data, with the constraint that the values be physically reasonable. Values on the order of 1 to 3 F/m^2 for C_1 and 0.2 to 1 F/m^2 for C_2 are often reported, although values outside those ranges have also been found.

---

[16]Parallel-plate capacitors have equal and opposite charges on the two plates. When two capacitors are in series, the middle plate has a charge that is the opposite of the sum of the charges on the outer plates. Thus, in the current case, the model consists of a capacitor near the surface with a charge density of σ_o and one farther from the surface with a charge density of $-\sigma_d$.

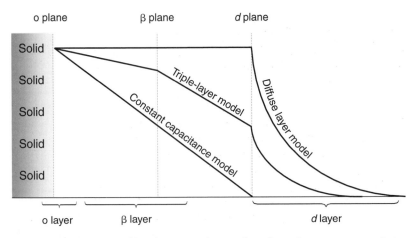

Figure 10.16   Profiles of Ψ versus distance from the surface, as envisioned in three surface complexation adsorption models.

Adapted from J. Westall and H. Hohl, *Advances in Colloid and Interface Science,* Vol. 12, 265–294 (1980).

Two relatively simple and widely used models for adsorption of ions to charged interfaces do not include a beta layer and make opposite, limiting-case assumptions about where the ions that neutralize the charge in the naught layer reside. The *constant capacitance* model assumes that all these ions are in the d plane, so that there is no diffuse layer, whereas the *diffuse layer* model assumes that all the neutralizing charge is in the diffuse layer, so that $\Psi_o = \Psi_d$. A schematic of the potential-distance relationship through the interfacial region according to the three models mentioned above is shown in Figure 10.16, and the key characteristics of the models are compared in Table 10.4.

10.5.3   THE EFFECT OF ELECTRICAL POTENTIAL ON DISTRIBUTION OF IONS IN A SOLUTION

CRITERIA FOR EQUILIBRIUM FOR A REACTION THAT INVOLVES CHANGES IN BOTH CHEMICAL AND ELECTRICAL POTENTIAL   As discussed in Chapter 2, the criterion for equilibrium in any isolated system is that the potential energy of the system (PE_{tot}) be minimized. An equivalent statement is that the change in the total potential energy of the system accompanying a differential shift in the state of the system must be zero. When applied to chemical reactions in the absence of an externally imposed electrical potential, this criterion can be expressed in terms of the molar Gibbs energy of reaction as $\Delta \bar{G}_r|_{eq} = 0$. However, the adsorption of ions to a charged interface is accompanied by changes in both chemical and electrical potential energy. The criterion for equilibrium of such a reaction is that the change in molar *electrochemical* potential energy accompanying a differential amount of reaction be zero, and not that $\Delta \bar{G}_r$ alone (i.e., the *chemical* component of the electrochemical potential) be zero.

Table 10.4 Comparison of the charge distribution and electrical potential profile near the surface in three surface complexation adsorption models*

Model	Charge Distribution
Constant capacitance	• Specifically adsorbed species in o plane • No β layer • Charge equal and opposite to that in the o plane provided by non-specifically sorbed ions, all in d plane
Diffuse layer	• Specifically adsorbed species in o plane • No β layer • Charge equal and opposite to that in the o plane provided by non-specifically sorbed ions, starting in d plane and distributed throughout d layer
Triple-layer	• Dehydrated specifically adsorbed species in o plane • Hydrated specifically adsorbed species in o plane • Charge equal and opposite to that in the combined o and β planes provided by non-specifically sorbed ions, starting in d plane and distributed throughout d layer

| *Profile of Ψ near the interface is shown in Figure 10.16.

The extension of the relationship between Gibbs energy and the equilibrium constant to systems where electrical potential energy must also be considered was introduced briefly in Chapter 2. Specifically, the molar electrochemical potential energy of a species i ($\overline{\overline{PE}}_i$) is defined as the sum of its chemical potential energy and the component of its electrical potential energy that is imposed externally:

$$\overline{\overline{PE}}_i = \overline{G}_i + z_i F \Psi_{ext} \qquad (10.49)$$

As was also shown in Chapter 2, the effect of Ψ_{ext} on the electrochemical activity of i can be represented concisely by an activity coefficient $\gamma_{i,ext}$:

$$a_{i,ec} = a_{i,chem}\gamma_{i,ext} = c_i\gamma_{i,chem}\gamma_{i,ext} \qquad (10.50)$$

where $\gamma_{i,ext} = \exp[z_i F \Psi_{ext}/(RT)]$. Alternatively, the two activity coefficients in Equation (10.50) can be lumped into a single *electrochemical activity coefficient* $\gamma_{i,ec}$, yielding

$$a_{i,ec} = \gamma_{i,ec}c_i \qquad (10.51)$$

where $\gamma_{i,ec} \equiv \gamma_{i,chem}\gamma_{i,ext}$.

In Equations (10.50) and (10.51), $\gamma_{i,chem}$ is the conventional activity coefficient that accounts for solute-water and solute-solute interactions; $\gamma_{i,ext}$, on the other hand, accounts for the effects of the externally imposed electrical potential on the reactivity of i, and $\gamma_{i,ec}$ accounts for all these effects collectively.

For reasons that will become clear shortly, it is important to understand that $z_i F \Psi_{ext}$ is the *increment* in the overall electrochemical potential energy of i that is induced when Ψ_{ext} is imposed, and hence it includes any change in potential energy associated with the rearrangement of the ions surrounding i. That is, we can associate electrical interactions between an ion and its neighbors in the absence of an external electrical field with Ψ_{chem} and $\gamma_{i,chem}$, but any changes in these interactions when an external field is present are associated with Ψ_{ext} and $\gamma_{i,ext}$. As a result, in a given system, $\gamma_{i,chem}$ depends only on the bulk composition of the solution (i.e., the ionic strength) and is independent of Ψ_{ext}.

Essentially all the important thermodynamic relationships developed in Chapter 2 can be rewritten in terms of electrochemical parameters instead of purely chemical parameters, for application in systems where Ψ_{ext} is important. In all cases, the electrochemical forms of the equations reduce to the chemical forms if $\Psi_{ext} = 0$. Some of the key definitions and relationships are collected in Table 10.5. Note that Ψ_{ext}° (the electrical potential in the reference state) is defined to be zero, simplifying many of the equations.

THE DISTRIBUTION OF IONS IN THE DIFFUSE LAYER By using the expressions in Table 10.5, the effects of the non-zero electrical potential near the surface of an adsorbent can be incorporated into the analysis of adsorptive equilibrium relatively easily. To demonstrate this, we first consider the concentration profile of a species A in the diffuse layer adjacent to a charged surface once the system has reached equilibrium.

The movement of A from bulk solution (where $\Psi_{ext} = 0$), to a point x in the diffuse layer (where $\Psi_{ext,x} \neq 0$) can be represented by the following simple

Table 10.5 Comparison of chemical and electrochemical expressions describing species' activity and equilibrium relationships

Chemical Term or Equation (Applicable if $\Psi = 0$)	Electrochemical Term or Equation (Applicable for Any Value of Ψ)
$\gamma_{i,chem}$	$\gamma_{i,ec} = \gamma_{i,chem}\gamma_{i,elec}$
$\bar{G}_i^{\circ}$	$\overline{\overline{PE}}_i^{\circ} = \bar{G}_i^{\circ} + \Psi_{ext}^{\circ} = \bar{G}_i^{\circ}$
$a_{i,chem} = \gamma_{i,chem}c_i$	$a_{i,ec} = \gamma_{i,ec}c_i = \gamma_{i,chem}\gamma_{i,elec}c_i$
$Q_{chem} \equiv \dfrac{\prod\limits_{products} (a_{i,chem})^{\nu_i}}{\prod\limits_{reactants} (a_{i,chem})^{\nu_i}}$	$Q_{ec} \equiv \dfrac{\prod\limits_{products} (a_{i,ec})^{\nu_i}}{\prod\limits_{reactants} (a_{i,ec})^{\nu_i}}$
$\bar{G}_i = \bar{G}_i^{\circ} + RT \ln a_{i,chem}$	$\overline{\overline{PE}}_i = \overline{\overline{PE}}_i^{\circ} + RT \ln a_{i,ec}$ $= \bar{G}_r^{\circ} + RT \ln a_{i,ec}$
$\Delta\bar{G}_r = \Delta\bar{G}_r^{\circ} + RT \ln Q_{chem}$	$\Delta\overline{\overline{PE}}_r = \Delta\overline{\overline{PE}}_r^{\circ} + RT \ln Q_{ec}$ $= \Delta\bar{G}_r^{\circ} + RT \ln Q_{ec}$
$\Delta\bar{G}_r\big|_{\text{at equilibrium}} = 0$	$\Delta\overline{\overline{PE}}_r\big|_{\text{at equilibrium}} = 0$
$K_{eq} \equiv Q_{chem}\big|_{\text{at equilibrium}}$	$K_{eq} \equiv Q_{ec}\big|_{\text{at equilibrium}}$

chemical reaction:

$$A_{\text{bulk},\Psi_{\text{ext}}=0} \leftrightarrow A_{x,\Psi_{\text{ext},x}} \tag{10.52}$$

Applying the criterion that, at equilibrium, $\Delta\overline{\overline{\text{PE}}}_r = 0$ in such a system, we obtain

$$\overline{\overline{\text{PE}}}_{A,\text{bulk}} = \overline{\overline{\text{PE}}}_{A,x} \tag{10.53}$$

$$\overline{G}_A{}^\circ + RT \ln a_{A,\text{chem,bulk}} + z_A F\Psi_{\text{ext,bulk}} = \overline{G}_A{}^\circ + RT \ln a_{A,\text{chem},x} + z_A F\Psi_{\text{ext},x} \tag{10.54}$$

Subtracting $\overline{G}_A{}^\circ$ from both sides, dividing through by RT, and substituting $c_A\gamma_{A,\text{chem}}$ for $a_{A,\text{chem}}$, we obtain

$$\ln a_{A,\text{chem,bulk}} = \ln a_{A,\text{chem},x} + \frac{z_A F\Psi_{\text{ext},x}}{RT}$$

$$\frac{a_{A,\text{chem},x}}{a_{A,\text{chem,bulk}}} = \exp\left(-\frac{z_A F\Psi_{\text{ext},x}}{RT}\right)$$

$$\frac{c_{A,x}\gamma_{A,\text{chem},x}}{c_{A,\text{bulk}}\gamma_{A,\text{chem,bulk}}} = \exp\left(-\frac{z_A F\Psi_{\text{ext},x}}{RT}\right) \tag{10.55}$$

Because $\gamma_{A,\text{chem}}$ is independent of Ψ_{ext}, $\gamma_{A,\text{chem},x} = \gamma_{A,\text{chem,bulk}}$, so Equation (10.55) yields the concentration profile of A as a function of the local (externally imposed) potential:

$$c_{A,x} = c_{A,\text{bulk}} \exp\left(-\frac{z_A F\Psi_{\text{ext},x}}{RT}\right) = c_{A,\text{bulk}}\gamma_{A,\text{ext},x}{}^{-1} \tag{10.56}$$

As noted above, for $\Psi_{\text{ext},x}$ values less than about 25 mV, the decay of $\Psi_{\text{ext},x}$ with distance is approximately exponential, with a characteristic length κ that depends on the ionic strength of the solution. The ionic distribution computed according to Equation (10.56) for an example system containing 0.01 M NaCl ($\kappa \approx 3$ nm) is shown in Figure 10.17.

THE EFFECT OF Ψ_{ext} ON SPECIFICALLY ADSORBING IONS If ions specifically adsorb, they experience the external electrical potential in the plane of adsorption and also bond chemically to the surface. The adsorption reaction can be written just as it is in the absence of an electrical potential, but to analyze the extent of adsorption at equilibrium, the effect of the potential on the electrochemical activities of the reacting molecules must be considered. For instance, consider the following generic adsorption reaction and the corresponding equilibrium constant:

$$\equiv\text{SOH} + A^{2+} \leftrightarrow \equiv\text{SOA}^+ + H^+ \tag{10.57}$$

$$K_{\text{ads,ec}} = \frac{\{\equiv\text{SOA}^+\}_{\text{ec}}\{H^+\}_{\text{ec}}}{\{\equiv\text{SOH}\}_{\text{ec}}\{A^{2+}\}_{\text{ec}}} \tag{10.58}$$

where $\{i\}_{\text{ec}}$ is the electrochemical activity of species i and the ec has been added to the subscript of K_{ads} as a reminder that it refers to a ratio of electrochemical

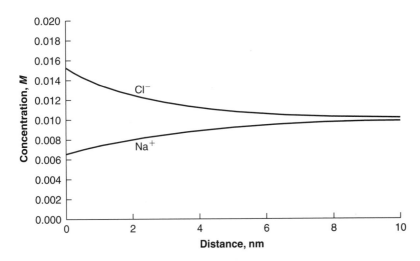

Figure 10.17   Distribution of Na$^+$ and Cl$^-$ ions in the diffuse layer near an interface with $\Psi_d = 25$ mV, in a solution of 0.01 M NaCl.

activities. In the adsorption literature, $K_{ads,ec}$ is called the *intrinsic adsorption equilibrium constant* K_{int}. Because both the chemical and the electrical interactions at the surface are accounted for in the evaluation of the $\{i\}_{ec}$ terms, K_{int} characterizes the distribution of A between the surface and solution in any system, regardless of the location of adsorbed ions or the profile of Ψ_{ext} in the surface region.

Often K_{int} is expressed as the product of a term that includes chemical activities but no electrical factors, and another that accounts for the effect of the external electric field on the species' activities via the appropriate ratio of $\gamma_{i,ext}$ values:

$$K_{int} = \frac{\{\equiv SOA^+\}_{chem}\{H^+\}_{chem}}{\{\equiv SOH\}_{chem}\{A^{2+}\}_{chem}} \frac{\gamma_{\equiv SOA^+,ext}\gamma_{H^+,ext}}{\gamma_{\equiv SOH,ext}\gamma_{A^{2+},ext}} \qquad \textbf{(10.59)}$$

$$K_{int} = K_{app} \frac{\gamma_{\equiv SOA^+,ext}\gamma_{H^+,ext}}{\gamma_{\equiv SOH,ext}\gamma_{A^{2+},ext}} \qquad \textbf{(10.60)}$$

K_{app} is called the *apparent adsorption equilibrium constant,* and, as before, $\gamma_{i,ext}$ equals $\exp[z_i F\Psi_{i,ext}/(RT)]$, where $\Psi_{i,ext}$ is the value of Ψ_{ext} at the location where species i resides. Because Ψ_{ext} is zero by definition in bulk solution, $\gamma_{i,ext}$ equals 1.0 for dissolved ions, i.e., for H$^+$ and A^{2+} in this example. (This statement has no bearing on $\gamma_{i,chem}$ for these species, which can have any value.) On the other hand, $\gamma_{\equiv SOH,ext}$ and $\gamma_{\equiv SOA^+,ext}$, would typically not equal 1.0, but would have values that depend on the system composition and on the locations where H$^+$ and A^{2+} bind.

K_{app} can be thought of as a conditional equilibrium constant, in that it describes a ratio of chemical activities for the electrical potentials extant in the system. That is, K_{app} would have a different value if some change were made to the system that altered the Ψ_{ext} profile near the surface. Note how different this situation is from reactions that take place entirely in solution or from adsorption reactions that do not involve charged interfaces: in those cases, a change in conditions might change the values of individual terms in the chemical activity quotient, but they do not change the equilibrium constant itself.

K_{int} is a more fundamental characterization of the equilibrium relationship than K_{app}, and it is much more useful in adsorption modeling efforts, because it is a true constant; it can therefore be tabulated and used in calculations applicable to any system containing the given adsorbate and adsorbent. On the other hand, K_{app} is usually of greater interest for practical applications, because in those cases the question being addressed is usually, How is the adsorbate distributed between the surface and solution, for the given electrical potential profile? Also K_{app} can be somewhat easier to understand intuitively than K_{int}, since it relates the same types of parameters (chemical activities) as do equilibrium constants for other reactions of interest in the system.

In some literature, the ratio of $\gamma_{i,ext}$ terms in Equation (10.60) is represented as K_{elec}^{-1}, in which case the apparent equilibrium constant can be expressed as $K_{app} = K_{int}K_{elec}$. While this approach has gained some currency, designating the net effect of the $\gamma_{i,ext}$ values as K_{elec} can be a bit misleading, because the letter K is usually used for constants, and the whole reason that the term containing $\gamma_{i,ext}$ values is separated from K_{int} is that it is *not* constant. In the context of the discussion here, representing the net electrical activity coefficient as $\gamma_{ext,overall}$, and therefore expressing K_{app} as $K_{int}/\gamma_{ext,overall}$, seems preferable.

To understand the relationship between K_{int} and K_{app} more clearly, consider the reaction shown in (10.57), and assume that both A^{2+} and H^+ bind in the naught plane. Writing $\gamma_{i,ext}$ for each species in terms of the value of Ψ_{ext} that it experiences, we can relate K_{int} to K_{app} as follows:

$$K_{int} = K_{app} \exp\left(\frac{z_{\equiv SOA^+}F\Psi_o}{RT}\right)\exp\left(\frac{z_{H^+}F\Psi_{bulk}}{RT}\right)\exp\left(-\frac{z_{\equiv SOH}F\Psi_o}{RT}\right)$$
$$\times \exp\left(-\frac{z_{A^{2+}}F\Psi_{bulk}}{RT}\right) \tag{10.61}$$

$$K_{int} = K_{app} \exp\frac{(z_{\equiv SOA^+} - z_{\equiv SOH})F\Psi_o}{RT} = K_{app} \exp\frac{(+1)F\Psi_o}{RT} \tag{10.62}$$

Generalizing the steps that led to Equation (10.62), we see that $\gamma_{i,ext}$ for any dissolved species will always be 1.0, and that $\gamma_{ext,overall}$, considering all adsorbing and desorbing species, will always be $\exp[\Delta z\, F\Psi_{ext}/(RT)]$, where Δz is the net change in the adsorbed charge, and Ψ_{ext} refers to the location where the adsorbed species reside (i.e., it could be Ψ_o, Ψ_β, or some other value, depending on the interfacial model being employed). For instance, in the current example, Δz is $+1$ (because one H^+ ion is replaced in the naught plane by one A^{2+} ion, so the net change in adsorbed charge is $+1$), and Ψ_{ext} is Ψ_o (because the change in charge is in the naught layer).

If the reaction causes the adsorbed charge to change in both the naught and beta layers, then one term like the one shown above is needed for each layer. That is, representing the change in charge in the naught and beta planes as Δz_o and

Δz_β, respectively, a generalized form of Equation (10.62) can be written as

$$K_{int} = K_{app} \exp \frac{[(\Delta z_o)\Psi_o + (\Delta z_\beta)\Psi_\beta]F}{RT} \qquad \textbf{(10.63)}$$

In some models (particularly the one referred to in footnote 11, in which surface sites are allowed to have fractional charge), the charge on an adsorbing ion is distributed between two or more layers, rather than all of it being assigned to a single layer. For instance, when a $PO_4{}^{3-}$ ion adsorbs, two of the oxygen atoms in the ion are assumed to bind directly to the surface. The ionic charge is assumed to be distributed equally over the four oxygen atoms in the $PO_4{}^{3-}$ ion, so the charge on each one is -0.75. As a result, according to this model, the charge entering the naught layer when a phosphate ion adsorbs is -1.5 esu. The remaining -1.5 esu on the ion is then assigned to a layer farther from the surface, where the potential is different from that in the naught layer.

In many cases, models with the charge of an adsorbed species distributed between two layers can reproduce experimental data among multiple layers better than models that assign all the ionic charge to a single layer. However, regardless of how the charge is allocated, the general approach for considering electrostatic effects when modeling ion adsorption is as described above. In the interest of simplicity, the following discussion uses the assumption that all the ionic charge resides in a single plane when an ion adsorbs.

10.5.4   PUTTING IT ALL TOGETHER: THE EQUATIONS DESCRIBING IONIC ADSORPTION

We now have all the pieces that we need to describe ionic adsorption quantitatively in the context of our conceptual model of the interfacial region. Consider, for instance, a system in which an adsorbate A^{2+} can bind in both the naught and beta layers of an oxide adsorbent [reactions (10.64) and (10.65), respectively):

$$\equiv SOH + A^{2+} \leftrightarrow \equiv SOA^+ + H^+ \qquad \textbf{(10.64)}$$

$$\equiv SOH + A^{2+} + H_2O \leftrightarrow \equiv SO{-}H_2O{-}A^+ + H^+ \qquad \textbf{(10.65)}$$

Note that, in reaction (10.65), adsorbed A^{2+} resides in the beta layer, but the H^+ that is released as part of the reaction comes from the naught layer. Therefore, Δz_o is -1, and Δz_β is $+2$.

The naught layer can also undergo protonation or deprotonation, which can be represented by the following K_a-type reactions:

$$\equiv SOH_2{}^+ \leftrightarrow \equiv SOH + H^+ \qquad \textbf{(10.66)}$$

$$\equiv SOH \leftrightarrow \equiv SO^- + H^+ \qquad \textbf{(10.67)}$$

Assume that, as in many of the problems solved in previous chapters, the (intrinsic) equilibrium constants for all the reactions are known, along with the total concentrations of A and $\equiv$S ($TOTA$, $TOT\equiv$S). To characterize the surface environment, the values of C_1 and C_2 also must be known (which is somewhat analogous to the need to know the ionic strength if activity coefficients in solution are to be considered). Finally, for this exercise, we will assume that the pH is known and fixed, although in principle one could compute the pH if all the inputs were known. What, then, would be the speciation of the system at equilibrium?

For the scenario described, the system has 12 unknowns, which are listed in Table 10.6, along with the 12 equations available to solve for them. The standard

Table 10.6   Unknowns and equations available to solve for speciation in a system where species A can sorb in either the o or the β layer[†]

Unknowns (12)

Activity of soluble species:	$\{A^{2+}\}$
Activity of surface species:	$\{\equiv SO^-\}, \{\equiv SOH\}, \{\equiv SOH_2^+\}, \{\equiv SOA^+\}, \{\equiv SO^- - H_2O - A^{2+}\}$
Potentials:	$\Psi_o, \Psi_\beta, \Psi_d$
Charge densities:	$\sigma_o, \sigma_\beta, \sigma_d$

Equations (12)

Mass balances (2):

MB on A:
$$TOTA = \{A^{2+}\} + \{\equiv SOA^+\} + \{\equiv SO^- - H_2O - A^{2+}\}$$

MB on $\equiv \overline{S}$:
$$\overline{TOT\equiv S} = \{\equiv SO^-\} + \{\equiv SOH\} + \{\equiv SOH_2^+\} + \{\equiv SOA^+\} + \{\equiv SO^- - H_2O - A^{2+}\}$$

Mass action expressions for species that bind in the o layer (3):

$$K_{a_1,int} = \frac{\{\equiv SOH\}_{ec}\{H^+\}_{ec}}{\{\equiv SOH_2^+\}_{ec}} = \frac{\{\equiv SOH\}\{H^+\}}{\{\equiv SOH_2^+\}} \exp\left(-\frac{F\Psi_o}{RT}\right)$$

$$K_{a_2,int} = \frac{\{\equiv SO^-\}_{ec}\{H^+\}}{\{\equiv SOH\}_{ec}} = \frac{\{\equiv SO^-\}\{H^+\}}{\{\equiv SOH\}} \exp\left(-\frac{F\Psi_o}{RT}\right)$$

$$K_{A,int,o} = \frac{\{\equiv SOA^+\}_{ec}\{H^+\}_{ec}}{\{\equiv SOH\}_{ec}\{A^{2+}\}_{ec}} = \frac{\{\equiv SOA^+\}\{H^+\}}{\{\equiv SOH\}\{A^{2+}\}} \exp\left(-\frac{F\Psi_o}{RT}\right)$$

Mass action expression for binding A in the β layer (1):

$$K_{A,int,\beta} = \frac{\{\equiv SO^- - H_2O - A^{2+}\}_{ec}\{H^+\}}{\{\equiv SOH\}_{ec}\{A^{2+}\}\{H_2O\}} = \frac{\{\equiv SO^- - H_2O - A^{2+}\}\{H^+\}}{\{\equiv SOH\}\{A^{2+}\}\{H_2O\}} \exp\frac{F(-\Psi_o + 2\Psi_\beta)}{RT}$$

Charge balance in each layer (3):

$$\sigma_o = \{\equiv SOH_2^+\} + \{\equiv SOA^+\} - \{\equiv SO^-\} - \{\equiv SO^- - H_2O - A^{2+}\}$$
$$\sigma_\beta = 2\{\equiv SO^- - H_2O - A^{2+}\}$$
$$\sigma_d = (\sigma_o + \sigma_\beta)$$

Charge-potential relationships (3):

$$C_1 = \frac{\Psi_o - \Psi_\beta}{\sigma_o}$$

$$C_2 = \frac{\Psi_\beta - \Psi_d}{-\sigma_d} = \frac{\Psi_d - \Psi_\beta}{\sigma_d}$$

$$\sigma_d = 0.1174I^{0.5} \sinh \frac{zF\Psi_d}{RT}$$

[†]Solution pH is assumed to be known. Values shown as $\{i\}$ with no subscript are chemical (as opposed to electrochemical) activities. Chemical activity coefficients of all species are assumed to be 1.0.

state for surface species is defined as a concentration of 1.0 M and an environment corresponding to that on the unoccupied and uncharged solid (i.e., with all the sites in the singly protonated, neutral $\equiv$SOH form). The chemical activities of surface species can therefore be equated with their molar concentrations, since any deviation from the reference state environment is accounted for separately, as part of the $\gamma_{ext,overall}$ term. Note that, in accord with the charge exchange accompanying reaction (10.65) (adsorption of A in the β layer), $\gamma_{ext,overall}$ for that reaction includes the terms $\Delta z_o = -1$ and $\Delta z_\beta = 2$.

Since the number of equations matches the number of unknowns, a unique solution to the problem is accessible. The algebra is most easily carried out with the aid of chemical equilibrium software, as described shortly.

If one were trying to determine the K_{int} values from experimental data, the same set of equations would be used, but the known information would include the equilibrium speciation instead of the K_{int} values. The K_{int} values would then be calculated by determining K_{app} under a range of conditions and using the above equations to find the K_{int} values that best fit the data. A limited data bank of K_{int} values is available for adsorption of simple species (mostly uncomplexed metals) onto a few common adsorbents, and more such compilations are likely to appear in the future.[17]

10.5.5   SUMMARY OF ADSORPTION UNDER CONDITIONS WHERE ELECTROSTATIC INTERACTIONS ARE SIGNIFICANT

To summarize, electrical interactions between charged surfaces and ionic adsorbates can modify the apparent adsorption equilibrium constant for ions at all surface sites. The magnitude of these effects depends not only on the identity of the adsorbate and the surface, but also on the composition of the system as a whole. The apparent adsorption equilibrium constant, which describes the relationship among chemical activities at the surface and in solution, can be viewed as the combination of a constant intrinsic value and a variable electrostatic contribution, with the latter term being, effectively, a combination of electrical activity coefficients. Ions in the diffuse layer can be treated using the same model construct, but without any chemical binding component (i.e., with $K_{int} = 1.0$).

At equilibrium, the overall charge around an adsorbent particle, including the fixed charge, the charge of specifically adsorbed ions, and the charge in the diffuse layer, must be zero. Therefore, if an ion adsorbs, other processes must combine to expel some like-charged species from the interfacial region and/or bring oppositely charged species into the interfacial region. These processes need not occur at the point where the adsorbing ion resides, but they must take place in the interfacial region. Therefore, adsorption of an ion is *always* associated with either adsorption or desorption of another ion; it never involves exchange exclusively with water molecules.

---

[17] The most extensive such compilation is for sorption of metal ions and oxyanions onto ferrihydrate, in Dzombak and Morel, *Surface Complex Modeling* (Wiley, 1990).

Sorption reactions that lead to a net change of surface charge distribution affect and are affected by the electrical potential near the surface; sorption reactions that leave the charge in each interfacial layer unaltered are not affected by that potential, even if the individual species moving to and from the surface as part of the reaction are charged.

Although a number of models exist for the structure and properties of charged interfaces and the binding of ions to surfaces, the details of these models have not been verified independently. Modern analytical tools are beginning to provide a great deal more insight into these issues. Over time, this new information as well as a continually expanding data bank of intrinsic adsorption constants will enhance the capability to predict adsorption of ions in such systems a priori.

10.5.6 Accounting for Surface-Adsorbate Electrical Interactions in Chemical Equilibrium Computer Models

Available chemical equilibrium software can be used to solve for speciation in systems where electrostatics are significant by defining the electrical interaction terms as pseudo-*components*. To understand how this process works, it might be easiest to think about how we would go about the process manually (if we had a *lot* of time to devote to trial and error calculations!). Recall that to determine the equilibrium speciation in a system the program guesses the activities of the *components* and solves for the corresponding concentrations of the *species* (via the equilibrium constants and estimates of the γ_{chem} values). It then checks the guesses by seeing if the mass balances on the *components* are satisfied.

Looking at the 12 equations that describe the example system above, we see that if we chose $\equiv SOH$, H^+, and A^{2+} as *components,* and if we made initial guesses for Ψ_o and Ψ_β, we could calculate the corresponding concentrations of all the *species* in the equilibrium system. Based on those concentrations, we could compute the charge density in the naught and beta layers. We could then check the mass balances on the *components* to see if they were satisfied, and also see whether the charge-potential relationships in the two inner layers were satisfied; if so, we would know that our guesses were correct, and if not, we would have to make other guesses for the activities of the *components* and the Ψ values and begin again. Thus, the process is similar to that used in previous systems we have studied, except that guesses must be made for two additional variables (Ψ_o and Ψ_β), and the test for whether these guesses are correct involves the charge-potential relationships rather than mass balances. [Note that σ_d and Ψ_d are not part of the trial-and-error process; once the process is completed, σ_d can be computed from the electroneutrality condition specified by Equation (10.45), and Ψ_d can be computed from Equation (10.46).]

The software carries out essentially the process described above. That is, it makes guesses for the potentials in the two inner layers and treats those potentials as *component*-like units. Special flags alert the program that, rather than carry out a mass balance test on these pseudo-*components,* it should test the accuracy of the guesses by comparing the surface charge density in each layer as computed by the speciation in that layer with the surface charge density as computed by the corresponding charge-potential relationship. Most programs provide simple ways of declaring which structural model of the interface (triple-layer, constant capacitance, or diffuse layer) the user wishes to apply.

Because of the form in which the Ψ values appear in the mass action expressions, it is more convenient for the program to guess values of $\exp[-F\Psi/(RT)]$ than to guess values of Ψ itself, but obviously such a guess is fundamentally equivalent to guessing Ψ directly. The benefit of using $\exp[-F\Psi/(RT)]$ as the pseudo-*component* is apparent if we rewrite the mass action expressions for the example system discussed above in which the reactions of interest are reactions (10.64) through (10.67). Defining $\exp[-F\Psi/(RT)]$ as $f(\Psi)$ the mass action equations describing the four reactions are

$$\{\equiv SOH_2^+\} = K_{a_1,\text{int}}^{-1}\{\equiv SOH\}\{H^+\}[f(\Psi_o)]^1[f(\Psi_\beta)]^0$$

$$\{\equiv SO^-\} = K_{a_2,\text{int}}\{\equiv SOH\}\{H^+\}^{-1}[f(\Psi_o)]^{-1}[f(\Psi_\beta)]^0$$

$$\{\equiv SOA^+\} = K_{A,\text{int,o}}\{\equiv SOH\}\{A^{2+}\}[f(\Psi_o)]^1[f(\Psi_\beta)]^0$$

$$\{\equiv SO^--H_2O-A^{2+}\} = K_{A,\text{int},\beta}\{\equiv SOH\}\{A^{2+}\}[f(\Psi_o)]^{-1}[f(\Psi_\beta)]^{+2}$$

or, in general,

{Adsorbed *species i*}

$$= K_{\text{int},i}\prod(components\text{ needed to form }i)^{\nu_i}[f(\Psi_o)]^{\Delta z_{i,o}}[f(\Psi_\beta)]^{\Delta z_{i,\beta}} \quad \textbf{(10.68)}$$

where ν_i is the stoichiometric coefficient of *species i* in the adsorption reaction and $\Delta z_{i,j}$ is the corresponding change in charge in layer j. The input tableau for the example system is shown below. Note that the $f(\Psi)$ terms are listed as Type 6 *species,* indicating that they should not be considered in mass balance calculations.

Species	**Type**	**$\equiv SOH$**	**A^{2+}**	**H^+**	**$f(\Psi_o)$**	**$f(\Psi_\beta)$**	**Log K**
$\equiv SOH$	1	1	0	0	0	0	0.0
A^{2+}	1	0	1	0	0	0	0.0
H^+	1	0	0	1	0	0	0.0
OH^-	2	0	0	-1	0	0	-14.0
$\equiv SOH_2^+$	2	1	0	1	1	0	$\log K_{a_1,\text{int}}^{-1}$
$\equiv SO^-$	2	1	0	-1	-1	0	$\log K_{a_2,\text{int}}$
$\equiv SOA^+$	2	1	1	-1	1	0	$\log K_{A,\text{int,o}}$
$\equiv SO^--H_2O-A^{2+}$	2	1	1	-1	-1	2	$\log K_{A,\text{int},\beta}$
$\overline{H^+}$	3a	0	0	1	0	0	pH
$f(\Psi_o)$	6	0	0	0	1	0	N/A
$f(\Psi_\beta)$	6	0	0	0	0	1	N/A

The header row spans: *Component* over $\equiv SOH$, A^{2+}, H^+, $f(\Psi_o)$, $f(\Psi_\beta)$.

Example 10.7 | When they bind to $Fe(OH)_3(s)$, H^+, HCO_3^-, and CO_3^{2-} are thought to form inner-sphere (i.e., o layer) surface complexes according to the following reactions.

Reaction	Log K
$\equiv FeOH + H^+ \leftrightarrow \equiv FeOH_2^+$	6.51
$\equiv FeOH \leftrightarrow \equiv FeO^- + H^+$	−9.13
$\equiv FeOH + H_2CO_3 \leftrightarrow \equiv FeCO_3H^0 + H_2O$	2.90
$\equiv FeOH + H_2CO_3 \leftrightarrow \equiv FeCO_3^- + H_2O + H^+$	−5.09

What would the inputs to a chemical equilibrium speciation program be for formation of these four surface species, using $\equiv FeOH$, H^+, and CO_3^{2-} as *components*?

Solution

The translation of the input data from the form given as conventional reactions to the form needed for the computer is fairly straightforward. The two reactions for sorption of carbonate species must be converted to reactions using only *components* to form the surface species, as follows:

$$\equiv FeOH + 2H^+ + CO_3^{2-} \leftrightarrow \equiv FeCO_3H^0 + H_2O \qquad \log K = 19.58$$

$$\equiv FeOH + H^+ + CO_3^{2-} \leftrightarrow \equiv FeCO_3^- + H_2O \qquad \log K = 11.59$$

In this problem, no species adsorb into the β layer, so we need not include $f(\Psi_\beta)$ as a pseudo-*component*. Furthermore, because the surface species that is used as the *component* for forming all surface complexes has no net charge in the o layer, the net change in charge in that layer (Δz_o) is simply the amount of charge that resides there when the adsorbate binds. Taking the above considerations into account, the input data for forming the surface species are shown below.

| Species | Type | Component | | | | Log K |
		H^+	CO_3^{2-}	$\equiv FeOH$	$f(\Psi_o)$	
$\equiv FeOH_2^+$	2	1	0	1	1	6.51
$\equiv FeO^-$	2	−1	0	1	−1	−9.13
$\equiv FeCO_3H^0$	2	2	1	1	0	19.58
$\equiv FeCO_3^-$	2	1	1	1	−1	11.59

10.6   ADSORPTION ISOTHERMS IN TWO SPECIAL CASES: BIDENTATE ADSORPTION AND SURFACE PRECIPITATION

10.6.1   INTRODUCTION

Before completing our consideration of adsorption isotherms, we consider two special cases that might be very important in certain systems, but for which the form of the isotherm has not been fully elucidated. These cases involve scenarios in which individual adsorbate molecules bind to more than one surface site or in which the accumulation of adsorbate at the surface is sufficient for it to form a surface precipitate.

In the derivation of the Langmuir isotherm above, we assumed that every adsorbed molecule occupies a single surface site. While that conceptual model is often adequate, there are cases where more than one surface site is thought to be involved in binding a molecule at the surface. Such bonding arrangements were hypothesized a number of years ago based on structural models of the surface and on analogies with reactions in solution between metals and multidentate ligands. Also, as noted at the beginning of the chapter, this type of bonding is often assumed to be involved when divalent ions bind to ion exchange resins. More recently, surface spectroscopic studies have provided more direct support for the existence of such bidentate surface species.

Extending the analogy between adsorption and formation of metal/ligand complexes, the binding of an adsorbate to multiple surface sites is commonly referred to as *multidentate* adsorption. Although many important molecules, including NOM molecules, have the potential to bind to several surface sites, the only case that has been considered in any detail to date is bidentate adsorption, and that is the case that is discussed below.

In the second special case analyzed below, we consider the possibility that the adsorption density of ions (typically, metal ions) becomes large enough that a new phase (e.g., a metal oxide) precipitates on the adsorbent surface. This scenario is depicted schematically in Figure 10.18. In such a case, if the precipitate covers enough of the original surface, we would expect the system to behave more or less like a suspension of the pure precipitated adsorbate.

Empirically, the manifestation of this process is that the adsorption density gradually increases in response to increasing dissolved adsorbate concentration, but then, rather than approaching some maximum value, the apparent adsorption density increases without bound. In truth, the latter observation does not reflect

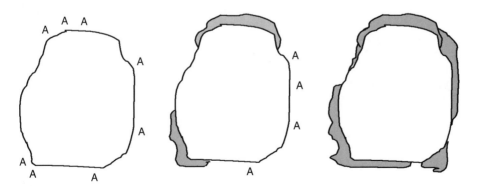

Figure 10.18   Schematic showing an adsorbate A that is first bound to discrete sites on the adsorbent, but that gradually forms a surface precipitate covering much of the solid, so that the surface of the final product behaves almost like that of a particle of pure A.

continuous adsorption onto the original adsorbent, but rather the formation of increasing amounts of the new precipitate. If the precipitate is a metal hydroxide and solution pH is fixed, the solubility product imposes a limit on the amount of the adsorbate that can be dissolved. As a result, addition of an increment of metal to a system where a surface precipitate has formed leads to precipitation of virtually all the added metal. If adsorption is quantified simply by analyzing the amount of metal that is removed from solution, the precipitation reaction can easily be mistaken for adsorption, causing the apparent adsorption density to increase indefinitely.

Experimental investigations of adsorption over a range of c_{eq} that is intended to span both adsorption and surface precipitation have been fairly rare. In those studies, although the endpoints of this process (adsorption at very low dissolved adsorbate concentrations, and precipitation at high concentrations) are reasonably well defined, the transition between the two states has not been easily distinguishable. Recently, use of advanced surface spectroscopic techniques has begun to shed greater light on this transition, although as yet, few generalizations are possible.

Attempts to model the transition between adsorption and surface precipitation have been even rarer than experimental studies of the process. One approach, proposed by Farley et al.,[18] is based on the formation of an ideal *solid solution* or *mixed solid* at the surface. The solid solution is postulated to form at very low adsorption densities and to become gradually more enriched with adsorbate as the adsorption density increases. This model forms the basis of the discussion below.

10.6.2 BIDENTATE SURFACE COMPLEXATION

Conceptually, the modeling of bidentate binding can be carried out almost identically to that for monodentate binding. For instance, the adsorption reaction and equilibrium constant can be represented as follows:

$$\equiv S_2 + A \leftrightarrow \equiv S_2A \tag{10.69}$$

$$K_{bd} = \frac{\{\equiv S_2A\}}{\{\equiv S_2\}\{A\}} \tag{10.70}$$

The bidentate surface ligand is shown as $\equiv S_2$, as opposed to $2\{\equiv S\}$, to indicate that even though two surface groups are involved, they must be adjacent to one another; i.e., they cannot be on different parts of the surface. Thus, the expression indicates that the concentration of $\equiv S_2A$ is related to the concentration of pairs of adjacent free surface sites.

The site balance in such a system must reflect the fact that two sites are occupied for each molecule of A that sorbs, i.e.,

$$TOT \equiv S = \{\equiv S\} + 2\{\equiv S_2A\} \tag{10.71}$$

Equation (10.71) indicates correctly that the maximum concentration of adsorbed A in the system equals one-half of $TOT \equiv S$. Note that the unoccupied

---

| [18]K. J. Farley, D. A. Dzombak, and F. M. M. Morel, *Journal of Colloid and Interface Science*, 106:226 (1985).

surface species $\equiv S_2$ does not appear in the site balance, because all the sites that can participate in bidentate bonding are assumed to also be capable of participating in monodentate bonding; hence, those sites are already counted in the mass balance in the $\{\equiv S\}$ term.

Equations (10.70) and (10.71) and the mass balance on $TOTA$ provide three independent equations characterizing the system, but those three equations contain four unknowns (the activities of A, $\equiv S_2A$, $\equiv S_2$, and $\equiv S$), so we need a fourth equation to solve for the equilibrium speciation. In particular, we need to know how the concentration (activity) of available bidentate sites ($\equiv S_2$) is related to that of available monodentate sites ($\equiv S$).

The relationship between these two parameters depends on the surface geometry and the exact pattern in which the surface sites get occupied. To understand why, consider the small portion of a hypothetical surface shown schematically in Figure 10.19. The schematic shows a 3×3 grid of $\equiv S$ sites. This grid provides 12 possible binding locations for a single bidentate adsorbate: 8 around the edges and 4 involving the center site. Even though it would be impossible for more than 4 of those sites to be occupied simultaneously, theoretical considerations indicate that the proper way to analyze this system is to treat the unoccupied surface as having 12 $\equiv S_2$ sites.

Now consider the situation once a single molecule has adsorbed via bidentate bonding. Assuming that any given $\equiv S$ site is capable of binding to only one molecule, the number of $\equiv S_2$ sites remaining is either 6 or 8, depending on whether the adsorbed molecule binds to the interior or along the edge of the grid. Thus, the attachment of a single molecule to the surface via a bidentate bond decreases the number of $\equiv S_2$ sites remaining on the surface by much more than just the one site that it occupies, and that number can vary depending on the location where the molecule binds. Both of these results differ from the case

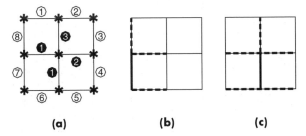

(a)   **(b)**   **(c)**

Figure 10.19   Schematic of a 3×3 surface site array. In **(a)**, asterisks are $\equiv S$ sites, and lines represent potential bidentate binding sites; numbers in open circles identify bidentate sites on the edge of the array, and those in filled circles identify binding sites in the center. In parts **(b)** and **(c)**, the bold lines show sites where a molecule has adsorbed via a bidentate bond to an edge and a center site, respectively. The dashed lines show other locations that are thereby precluded from becoming occupied subsequently. As a result, only 8 and 6 sites remain available in (b) and (c), respectively.

for monodentate binding, and they complicate the analysis of bidentate binding considerably.

To date, no closed-form analytical expression has been found that can model the relationship between $\{\equiv S\}$ and $\{\equiv S_2\}$ over the full range from very low to very high surface coverage. However, for up to approximately 80% surface site occupancy, the relationship can be fit reasonably well ($\pm 10\%$) by the following equation:

$$\{\equiv S_2\} \approx \frac{k\{\equiv S\}}{TOT \equiv S} \tag{10.72}$$

where k depends on the surface site geometry. For instance, for a square grid of $n \times n$ sites, $k = 2 - 2/n$.

Equation (10.72) provides the fourth equation needed to solve for the system speciation. The resulting isotherm equation for bidentate bonding onto a surface with an $n \times n$ site grid is

$$q_A \approx \left(1 + \frac{1 - \sqrt{8 K_{bd}\{A\} + 1}}{4 K_{bd}}\right) q_{max} \tag{10.73}$$

where $q_{A,max}$ corresponds to $TOT \equiv S/2$. Isotherms for other surface site geometries and for competitive adsorption between pairs of bidentate adsorbates or between a mono- and a bidentate adsorbate require numerical solutions.

Given the difficulties of assessing the number of bonds between an adsorbate and the surface and the possibility that that number might depend on system conditions even for a given adsorbent-adsorbate pair, it is not clear that numerical solutions relating $\{\equiv S_2\}$ to $\{\equiv S\}$ for complex systems will ever be developed or be useful. The point of the discussion here is simply to demonstrate qualitatively how and why bidentate adsorption requires a different analysis and leads to a different isotherm from monodentate adsorption.

10.6.3   SURFACE PRECIPITATION

As noted above, the discussion of surface precipitation provided here follows the model of Farley et al. (1985). In the model, a fixed portion of the adsorbent is assumed to participate in formation of the mixed solid. The activity of the adsorbate in the (idealized) mixed solid phase is presumed to be proportional to its molar concentration in that phase, which, in turn, is closely related to the apparent adsorption density. Defining q_{sp} as the apparent adsorption density attributable to surface precipitation [i.e., the mass of adsorbate bound to the solid (in the mixed solid phase) per unit mass of solid in the system], the equation describing q_{sp} as a function of the dissolved adsorbate concentration can be derived by combining the solubility product expression with a mass balance. The result is an equation that is analogous to an adsorption isotherm:

$$q_{sp} = \frac{\phi(c_{eq}/c'_{eq})}{1 - c_{eq}/c'_{eq}} = \frac{\phi K' c_{eq}}{1 - K' c_{eq}} \tag{10.74}$$

where ϕ = mass of adsorbent that contributes to the mixed solid phase, normalized to the adsorbent surface area (e.g., mg/m^2)

c'_{eq} = adsorbate concentration that would be in equilibrium with the pure precipitated solid (e.g., if the precipitating solid is a metal hydroxide, $c'_{eq} = K_{s0,Me(OH)_2(s)}/\{OH^-\}^2$)

$K' = 1/c'_{eq}$

The form of Equation (10.74) is similar to the Langmuir isotherm, except that the sign in the denominator is opposite in the two expressions (plus in the isotherm, minus in the expression for surface precipitation). The equation includes two constants: one that describes the amount of the adsorbent that participates in the reaction (ϕ) and is analogous to q_{max} in the Langmuir isotherm, and one related to the thermodynamic driving force for binding, analogous to K_{ads}.

Equation (10.74) indicates that at low c_{eq}, the amount of adsorbate in the surface precipitate is a linear function of c_{eq}, but that as c_{eq} approaches $1/K'$, the amount at the surface gets very large (Figure 10.20). The equilibrium dissolved adsorbate concentration c_{eq} can never exceed c'_{eq}, since c'_{eq} is the adsorbate concentration in equilibrium with the pure solid; as c_{eq} approaches c'_{eq}, the mixed surface solid approaches becoming a pure precipitate of the adsorbate. Since c_{eq} cannot exceed c'_{eq}, any additional adsorbate that is added to solution precipitates, and the apparent adsorption density grows without bound.

The total apparent adsorption density (q_{app}) can be computed as $q_{sp} + q_L$, where q_L is the true adsorption density on the original solid, which is presumed to follow a Langmuir isotherm. Figure 10.21a shows the contributions of q_{sp} and q_L to q_{app} for two hypothetical systems, one in which adsorption is relatively strong and one in which it is weak. In both cases, adsorption accounts for most of the adsorbate binding at low c_{eq} (c_{eq} less than ~200 and ~500 in the systems with weak and strong adsorption, respectively), and surface precipitation dominates at high c_{eq}.

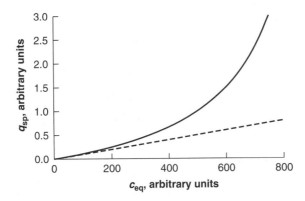

Figure 10.20 The apparent adsorption density attributable to surface precipitation as a function of the equilibrium concentration of dissolved adsorbate, according to the Farley et al. model for surface precipitation. For this plot, $\phi = 1000$, $K' = 0.001$ (both in arbitrary units). As $\phi K'$ approaches a value of 1.0, the surface precipitate approximates a pure precipitate of the adsorbate, and q_{app} increases without bound. The dashed line corresponds to $q_{app} = \phi K' c_{eq}$.

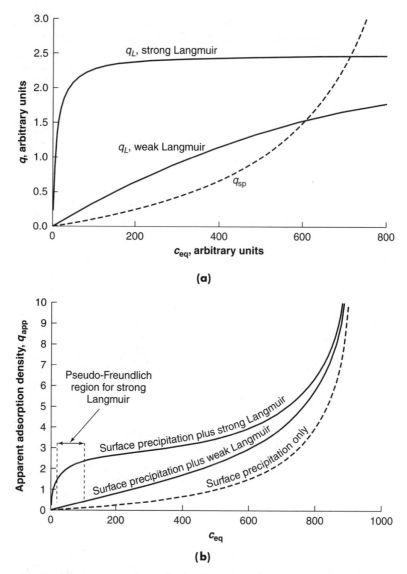

Figure 10.21 Apparent adsorption isotherms for systems in which both Langmuir adsorption and surface precipitation can occur simultaneously. **(a)** Isotherms for adsorption and surface precipitation separately. **(b)** Apparent isotherms for simultaneous adsorption and surface precipitation.

Assuming that the adsorption and surface precipitation processes occur in parallel (e.g., on different parts of the surface), the apparent isotherm that one would derive from experiments is the sum of the Langmuir and surface precipitation isotherms, which is shown for each system in Figure 10.21b. In the system with strong Langmuir binding, the empirical isotherm would look much like a simple Langmuir isotherm at low to moderate c_{eq}, and like the isotherm for surface precipitation at higher c_{eq}. Thus, in this case, the isotherm provides a fairly clear indication of the transition from an adsorption-dominated system to one dominated by surface precipitation. However, in the system with weaker adsorption, the curve is qualitatively similar to one for surface precipitation alone.

Farley et al. suggested that much of the metal ion adsorption data in the literature might be characterized by relatively strong adsorption and might fall in the region of Figure 10.21b that is indicated as the pseudo-Freundlich region. (The absolute values of c_{eq} and q_{app} in this region depend on the particular metal ion and adsorbent under study, but the shape of the curve would be the same for any system in which adsorption is relatively strong.) They also showed that the relationship between q_{app} and c_{eq} in this region can be fit quite well by a Freundlich isotherm. The implication is that a system that obeys a Freundlich isotherm might actually be undergoing a combination of Langmuir-type adsorption and surface precipitation, rather than adsorption onto a surface with a wide range of site affinities (the classical explanation for such behavior, presented earlier in the chapter). This issue is likely to be resolved by direct probing of the surface in the future.

This completes our consideration of adsorption principles and their applications in adsorption modeling. Although the number of physical and chemical factors that affect sorption is large, especially when electrical interactions between the adsorbate and adsorbent are significant, each of these interactions can be dealt with in the context of reactions presented earlier in the text. Furthermore, although solving the resulting equations manually would require an overwhelming effort for any but the simplest of systems, those equations can be solved using the same chemical equilibrium software packages that are used to solve for speciation in purely soluble systems, once some relatively slight modifications are made to the inputs to those programs.

In the final section of the chapter, we turn our attention to the practical applications of the material presented above.

10.7 Sorption onto Inorganic Solids in Natural and Engineered Aquatic Systems

10.7.1 Sorption of Natural Organic Matter During Coagulation of Drinking Water

Sorption onto inorganic solids plays a central role in controlling the behavior of natural organic matter (NOM) as well as many trace inorganic elements in natural systems, and the sorbed species, in turn, affect the behavior of the particles. In this section, we consider the interaction of NOM with oxide minerals, since those interactions determine the fate of NOM in conventional water treatment operations.

As noted in previous chapters, NOM consists, in large part, of polymeric breakdown products from the decay of plant material. These molecules have complex structures, characterized by a skeleton of linked aliphatic chains and aromatic rings. A variety of functional groups are attached to the NOM skeleton, of which carboxylic acid groups are dominant. Typically, titration of a sample of NOM from pH 2.0 to 12.0 indicates that one proton is released for each eight to ten carbon atoms in the sample, mostly at pH < 6. Many NOM molecules are thought to contain at least a few dozen carbon atoms, and hence they are likely to be polyanionic at neutral pH. For the range of NOM and particle concentrations typically present in natural water bodies, the affinity of many of the organic molecules for the surfaces leads to a situation in which the particles are relatively highly loaded with NOM. As noted previously, these reactions dramatically reduce the variation in behavior among particles of different chemical composition, causing virtually all particles to behave as though they have acidic, organic surfaces.

If small amounts of an adsorbent are added to a batch of water containing dissolved NOM, the fresh adsorbent rapidly acquires a negative surface charge due to sorption reactions, no matter what its surface charge would be in pure (organic-free) water. However, if relatively large amounts of the adsorbent are added, the pool of adsorbable molecules remaining in solution is depleted, the ratio of adsorbed NOM to bare (hydrated) surface decreases, and the underlying solid has progressively more influence on the surface properties.

In water treatment and many wastewater treatment systems, the most important inorganic adsorbents are hydrous oxides of iron and aluminum, which form when salts of Fe^{3+} and Al^{3+} are added to the water as coagulants. The reactions of these metals with the water and dissolved constituents are varied and complex. Both Fe^{3+} and Al^{3+} form very insoluble hydroxides. The $Fe(OH)_3(s)$ and $Al(OH)_3(s)$ that form when the metal salts are mixed rapidly into a solution at near-neutral pH are usually poorly crystallized, gelatinous structures with very large specific surface areas. These solids are good adsorbents for many polar, weakly ionic species, including many NOM molecules. Al^{3+} and Fe^{3+} ions also form complexes and solids with NOM, so it is often difficult to distinguish between the formation of metal-NOM precipitates and sorption of NOM onto hydrous metal oxide precipitates. In most cases, all the reactions mentioned above proceed simultaneously, yielding a mix of products that depends on the pH, the concentrations of other ions in solution, hydraulic conditions, etc.

The sorption of NOM to oxide surfaces is commonly represented as a reaction between the carboxylic functional groups of the organic molecule and a surface site:

$$\equiv SOH + R{-}COO^-(aq) \leftrightarrow \equiv S{-}OCO{-}R + OH^- \qquad \textbf{(10.75)}$$

Near neutral pH, one effect of the reaction is to release OH^- from the surface into solution, a result that is confirmed by the observation that, in most cases, the solution pH and alkalinity both increase as a result of the reaction. A corollary of this observation is that the tendency for the sorption reaction to proceed can be enhanced by lowering the solution pH, and such a trend is indeed observed. The overall effect of adding the metal salt might be to lower pH and consume alkalinity, since the hydrolysis of the metal and precipitation of the hydrous oxides consume OH^- ions. Thus, in many cases, the sorption reactions simply reduce the magnitude of the decrease in pH, rather than actually cause the pH to increase.

At sufficiently low pH, both the surface sites and the acid groups become protonated, and the dominant reaction becomes

$$\equiv SOH_2^+ + R-COOH(aq) \leftrightarrow \ \equiv S-OCO-R + H_3O^+ \quad \textbf{(10.76)}$$

That is, at low pH, the net reaction can release hydronium ions to solution, in which case the reaction becomes less favorable as the solution becomes more acidic. The combination of acid/base reactions affecting the surface and the NOM molecules leads to a characteristic pattern in which sorption goes through a maximum at some pH. The exact pH of the maximum depends on the solids and NOM molecules involved, but for typical water treatment conditions, in which the adsorbent is a freshly precipitated hydrous oxide of either aluminum or iron, the pH of maximum adsorption is usually around 4.5 to 6.0. This trend is shown for both NOM and some model carboxylic acids in Figure 10.22.

The model reactions shown above are written to emphasize the interaction between the surface and a single functional group on an NOM molecule, and they do not consider explicitly the fact that most NOM molecules are polyanionic. This property of the NOM can affect the reaction in two important ways. First, different parts of a single molecule might bind to the surface at two or more different sites, increasing the overall binding strength and making it quite unlikely that the molecule will desorb unless solution conditions are changed (individual bonds might break and reform, but the chances of all the bonds breaking at once are small). Second, the part of the molecule that does not bind directly to the surface typically carries a negative charge, and the sorption reaction adds negative charge to the surface. Since electroneutrality must be maintained near the surface, the reaction must be accompanied by addition of cations and/or release of anions from the diffuse layer. The ions participating in this part of the reaction are likely to be the major anions and cations in solution (e.g., Na^+, Ca^{2+}, HCO_3^-, Cl^-), and the changes in their bulk concentration as a result of the reaction are negligible in most cases. Nevertheless, a complete description of the process should at least acknowledge their role.

The maximum sorption density of NOM on hydrous oxides is probably controlled by a combination of limited site availability, buildup of negative surface charge, and factors that control the size of NOM molecules. The importance of surface charge is suggested by the fact that increasing the concentration of dissolved divalent cations (particularly Ca^{2+}) usually increases the tendency of NOM to sorb. This effect is probably manifested by complexation of Ca^{2+} with NOM molecules, lowering the negative charge on the surface and thereby decreasing the repulsive force acting on additional NOM molecules approaching

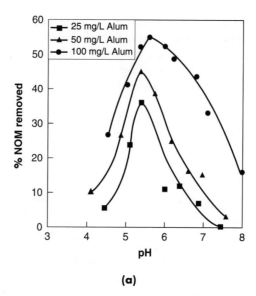

(a)

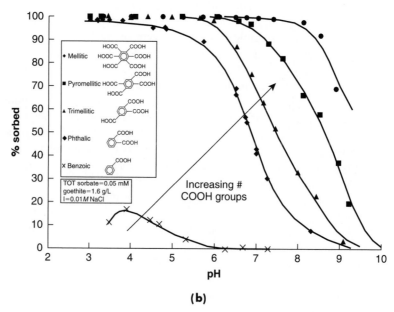

(b)

Figure 10.22   **(a)** Removal of natural organic matter (quantified as TOC) by addition of aluminum sulfate (alum) as a function of pH.

Adapted from M. J. Semmens and T. K. Field, *Journal AWWA*, Vol. 72, No. 8, 476–483 (1980).

(b) Sorption of some well-characterized carboxylic acids as a function of pH.

Reprinted with permission from C. R. Evanko and D. A. Dzombak, *Environmental Science and Technology*, Vol. 32, 2846–2855. Copyright 1998 American Chemical Society.

the surface.[19] The Ca^{2+} might also decrease the space occupied by individual molecules by neutralizing part of the negative charge on those molecules and allowing parts of the molecule to coil up.

10.7.2 SORPTION OF METALS

Ferric and aluminum salts are also used in many industrial waste treatment processes, particularly those designed to remove metals from solution, in much the same way as they are used in drinking water treatment. For instance, in systems where pH adjustment is used to precipitate metal hydroxides, ferric chloride ($FeCl_3$) or alum might be added to coagulate the colloidal metal hydroxide particles that form, generating a suspension of larger particles that can subsequently be settled out of solution. The historical understanding of the usefulness of these chemicals focused on their ability to coagulate colloidal matter and facilitate its removal from the aqueous phase. However, it is now apparent that metal sorption onto the precipitated $Fe(OH)_3$ or $Al(OH)_3$ is frequently an important component of the overall metal removal that occurs in such systems. Iron, aluminum, and other oxide minerals are also important adsorbents for metals in river systems, the oceans, sediments, and soils.

The characteristic dependence of metal sorption on solution pH is shown in Figure 10.23. Similar diagrams apply for most metals that form cationic dissolved species (Mn^{2+}, Co^{2+}, Co^{3+}, Cr^{3+}, Ni^{2+}, Zn^{2+}, Hg^{2+}, Pb^{2+}, etc.). Typically, for a given metal concentration, sorption increases dramatically with increasing pH at some pH that depends on the details of the system. This pattern is often referred to as a *pH-adsorption edge*.

Removal of metal ions from solution via precipitation is characterized by a similar pattern to that shown in Figure 10.23. However, in many systems, metal removal in the presence of adsorbents begins at a pH value one to several units below the pH at which precipitation first becomes apparent in systems without the adsorbent (Figure 10.24), indicating clearly that the removal mechanism is adsorption.

The tendency of metal cations to sorb to oxide surfaces is highly correlated with their tendency to undergo hydrolysis reactions in solution. Furthermore, although the tendency for different oxides to bind metals varies widely, depending on the acidity of the surface (PZC) and the specific surface area, the relative tendency for different metals to bind to them is quite consistent. This trend is summarized in the following sequence:

Tendency to sorb to oxides:

$$Ca^{2+} < Cd^{2+}, Ni^{2+} < Zn^{2+}, Co^{2+}, Cu^{+} < Cu^{2+} < Pb^{2+} < Cr^{3+}, Fe^{3+}, Hg^{2+}$$

---

[19] Note that, at equilibrium, it is not possible to distinguish whether NOM—Ca^{2+} complexes first form in solution and then adsorb, or whether the NOM adsorbs and then binds to a dissolved Ca^{2+}. In all probability, both processes occur. In any case, they lead to the same equilibrium condition.

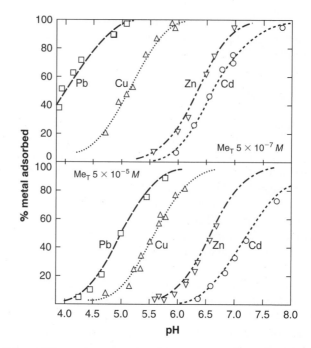

Figure 10.23   The pH-adsorption edges for sorption of Pb^{2+}, Cu^{2+}, Zn^{2+}, and Cd^{2+} onto 10^{-3} M $Fe(OH)_3(s)$. Ionic strength was 0.1 M in all solutions.

Figure from "Multi-site adsorption of Cd, Cu, Zn, and Pb on amorphous iron oxyhydroxide" by M. M. Benjamin and J. O. Leckie in *Journal of Colloid and Interface Science*, Volume 79, 209–221, copyright © 1981 by Academic Press, reproduced by permission of the publisher. All rights of reproduction in any form reserved.

Sorption of metals onto hydrous oxides is often characterized by Freundlich isotherms with values of *n* between about 0.4 and 0.7. Since this observation applies even at quite low adsorption densities, the oxides apparently have substantial variations among surface sites, including small numbers of sites with binding strengths much greater than an average site. It is often difficult to identify a maximum adsorption density in such systems, since metal sorption merges into metal surface precipitation as regions of the surface become densely occupied.

10.7.3   SORPTION OF METALLIC OXYANIONS

Many elements that are classified as metals when in their zero-valent oxidation state are more stable as oxyanions than as cations when they dissolve in water; examples include SeO_3^{2-} and SeO_4^{2-}, AsO_3^{3-} and AsO_4^{3-}, CrO_4^{2-}, MoO_4^{2-}, PO_4^{3-}, etc. These weakly basic species and/or their protonated conjugate acids can react with oxide surfaces in much the same way as NOM anions do, with minor differences attributable to the fact that they are smaller and bind to only one or two surface sites (Figure 10.25). As is the case with NOM sorption, the

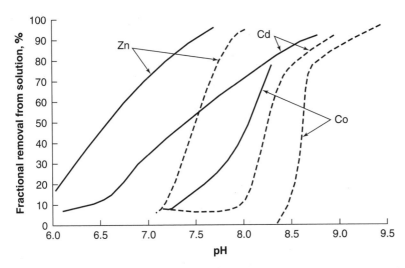

Figure 10.24 Removal of Zn^{2+}, Cd^{2+}, and Co^{2+} from solution in the presence (solid lines) and absence (dashed lines) of $Fe(OH)_3(s)$. In the presence of $Fe(OH)_3(s)$, the metals are removed by adsorption at pH values under which precipitation from bulk solution does not occur. All systems contain 10^{-3} M Me^{2+} and either 0 or 10^{-3} M $Fe(OH)_3(s)$.

| Adapted from M. M. Benjamin, *Environmental Science and Technology*, Vol. 17, 686–692 (1983).

reactions release OH^- to solution and cause the surface to acquire negative charge, and sorption often goes through a maximum value under mildly acidic conditions; the exact pH of the maximum depends on the pK_a of the adsorbate as well as the charge of the adsorbent.

Sorption of the oxyanionic metals also bears some similarities to sorption of cationic metals, with the inverse dependence on pH. Whereas sorption of the cations can be compared with hydrolysis reactions, sorption of the anions can be compared to complexation with a dissolved metal ion (Figure 10.26). For some anions, a transition from an adsorbed state to formation of a surface precipitate is possible (e.g., $AlPO_4$); however, the sorption density of many of these ions seems to reach an upper limit imposed by availability of sites and/or buildup of negative surface charge, reducing the likelihood that the ions will form surface precipitates with the metal ion of the adsorbent phase. In the case of freshly precipitated hydrous iron oxide, the maximum adsorption density is often in the range of 0.1 mole adsorbed per mole of Fe.

10.7.4 EFFECTS OF COMPLEXING AGENTS ON METAL SORPTION

Many industrial wastewaters contain complexing agents that bind to dissolved metal ions. Indeed, the complexing agents are often added to the industrial process specifically to prevent precipitation of the metals. In the case of a few ligands (e.g., NH_3), the ligand is sufficiently similar to a water molecule that the

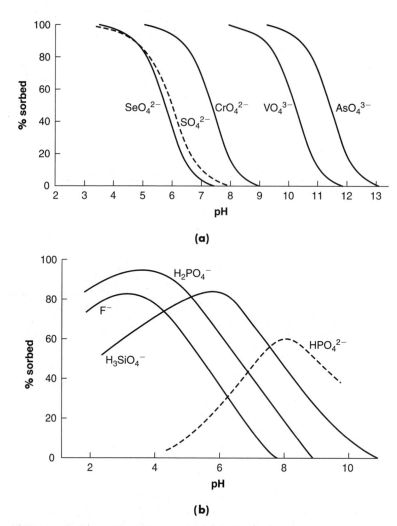

Figure 10.25 The pH-adsorption edges for various anions.

Adapted from W. Stumm and J. Morgan, AQUATIC CHEMISTRY: CHEMICAL EQUILIBRIA AND RATES IN NATURAL WATERS, 3rd Ed. John Wiley & Sons, Inc., New York (1996).

impact on metal sorption is negligible. However, in most cases, the complexing agents reduce the likelihood that the metal will bind directly to an adsorbent surface. The simplest explanation for this observation is that the surface ligand and dissolved ligand compete for the free metal ion, so the more metal that is bound by the dissolved ligand, and the more strongly that metal is bound, the less that is available to bind to the surface (Figure 10.27). In the limit, if the complexed metal does not sorb at all, complexation reduces overall metal sorption by the same percentage as it reduces the concentration of uncomplexed (aquo) metal ions in solution.

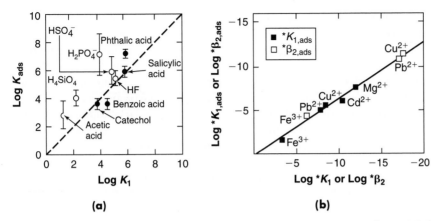

Figure 10.26 Correlations of sorption with complexation reactions in solution. **(a)** The K_{ads} for binding of various acids to $Al(OH)_3(s)$ (filled circles) or $Fe(OH)_3(s)$ (open circles) versus the corresponding constants for forming soluble complexes with Fe^{3+} or Al^{3+}.

| Adapted from L. Sigg and W. Stumm, *Colloids and Surfaces*, Vol. 2, 101–117 (1981).

(b) Equilibrium constants for metals binding to one $(*K_{1,ads})$ or two $(*\beta_{2,ads})$ OH groups on the surface of $SiO_2(s)$ versus the corresponding constant for the metals reacting with one or two H_2O molecules to form $MeOH^{n-1}$ or $Me(OH)_2^{n-2}$, respectively. The asterisk indicates that the equilibrium constants are for the reaction with protonated ligands (H_2O for OH complexes, SOH for SOMe complexes).

| Adapted from P. Schindler, *Osterreichischie Chemie-Zeitschrift*, Vol. 86, 141–146 (1985).

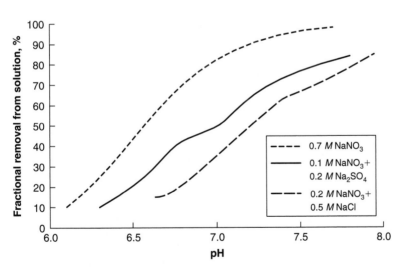

Figure 10.27 Effect of complexing ligands on adsorption of Cd onto $Fe(OH)_3(s)$. All solutions contain 5×10^{-7} M TOTCd and 10^{-3} M $Fe(OH)_3(s)$ and have an ionic strength of 0.7 M. Both Cl^- and SO_4^{2-} can form soluble complexes with Cd^{2+}, whereas NO_3^{2-} cannot.

| Adapted from M. M. Benjamin and J. O. Leckie, *Environmental Science and Technology*, Vol. 16, 162–170 (1982).

If the metal—ligand complex can adsorb, the overall sorption of metal reflects the combined sorption of free metal ions and complexes. This situation is particularly interesting in cases where the ligand contains two or more functional groups that can bind independently to the surface and the metal ion. Typically, ligands bond to the surface through oxygen or other electron-donating atoms, so their adsorption pattern (particularly the pH dependence of adsorption) is similar to that of NOM or oxyanions. Therefore, they tend to adsorb under conditions where free metal ions do not, and if they bind to metal ions and the surface simultaneously, they can enhance rather than interfere with metal sorption.

A system containing silver (Ag^+) ions and thiosulfate ($S_2O_3^{2-}$) provides a good example of such behavior. Thiosulfate ions can be thought of as resulting from replacement of one of the oxide ions in sulfate by an S^{2-} ion ($^-S{-}SO_3^-$). Complexation of thiosulfate with Ag^+ ions occurs almost exclusively through the sulfide, whereas thiosulfate sorption to an oxide surface occurs through the oxygen ions. Therefore, although the complexation reaction effectively eliminates the possibility that the silver ion will bind directly to the surface, Ag^+ can nevertheless be removed from solution if it is complexed with the thiosulfate ion and the complex binds to the surface via the oxygen ions. Because of this reaction, addition of the ligand can dramatically enhance Ag sorption under some conditions.

Figure 10.28 characterizes sorption of Ag and S_2O_3 onto $Fe(OH)_3(s)$ as a function of pH in systems where each ion is present alone, and also when they are

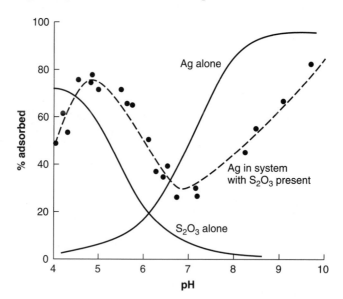

Figure 10.28   Sorption of 4×10^{-7} M Ag and/or S_2O_3 onto 10^{-3} M $Fe(OH)_3(s)$ as a function of pH in systems where each adsorbate is present alone (solid lines) and in a system where they are both present (data points). Ionic strength is 0.1 M in all cases.

Adapted from J. A. Davis and J. O. Leckie, *Environmental Science and Technology*, Vol. 12, 1309–1315 (1978).

present together. In the former case, S_2O_3 sorption decreases steadily from pH 4 to 7, and Ag sorption increases steadily from pH 5 to 8 (solid lines). However, when both are present, the Ag sorbs in parallel with the S_2O_3 at pH 5 to 7, suggesting that while one end of the S_2O_3 binds to the surface, the other end remains bound to Ag. At higher pH in the combined system, the S_2O_3 does not bind to the surface, and it competes with the surface for the Ag, so Ag sorption is diminished at pH > 7 compared to the case where S_2O_3 is absent.

A similar situation has been shown to apply to Fe, Al, and Cu in the presence of NOM: complexation with the NOM prevents direct bonding of the metals with inorganic surfaces, but if the NOM itself adsorbs, the metals are carried along with it. In such systems, metal removal from solution can be more extensive than if the metals were present in solution as free aquo ions.

As is clear from the Ag–S_2O_3 system, the effects of complexing ligands on metal sorption can be complicated, since both the free metals and the ligands undergo acid/base reactions that alter their speciation and the tendency for the complexes to form. When the acid/base reactions of the surface are also considered, along with the possibilities for surface precipitation and/or dissolution of the solid, it becomes apparent why predicting the detailed effects of complexation on adsorption is very problematic. Nevertheless, some general trends can be identified, and experimental results can usually be interpreted by assuming that one or two of the possible reactions dominate.

10.7.5 EMPIRICAL EVIDENCE FOR SURFACE PRECIPITATION

Fairly clear evidence of the formation of a surface precipitate is sometimes available from its effect on the properties of the adsorbent. For instance, Figure 10.29a shows the *electrophoretic mobility* (the velocity of colloidal particles when exposed to an electrical field, divided by the magnitude of the electric field) of pure $SiO_2(s)$ and pure $Co(OH)_2(s)$ as a function of pH. In a system containing $SiO_2(s)$ and dissolved Co^{2+}, the mobility of the particles is similar to that of pure $SiO_2(s)$ at low pH, but it gradually deviates from that pattern between pH 4 and 8. Presumably, this pattern is observed because some Co^{2+} binds to the $SiO_2(s)$ surface, reducing the negative charge on the particle's surface. At pH 8.0, however, a dramatic change occurs, and the electrophoretic mobility approaches that of pure $Co(OH)_2(s)$. The most logical explanation for this change is that $Co(OH)_2(s)$ has precipitated on the $SiO_2(s)$ surface, causing the particle to behave almost like one of pure $Co(OH)_2(s)$.

Figure 10.29b demonstrates another effect of surface precipitation. This figure indicates that, in the absence of Co^{2+}, chromate sorption onto $Fe(OH)_3(s)$ decreases as pH increases from ~6 to ~9 (line Cr/C). When Co^{2+} is present, however, it binds to the $Fe(OH)_3(s)$ in the range of pH 7 to 8 (line Co/A), and the CrO_4 readsorbs (line Cr/A and triangles). The sorption of the CrO_4 at higher pH in the presence of $Fe(OH)_3(s)$ and Co^{2+} is quite similar to that in a system containing $Co(OH)_2(s)$ and no $Fe(OH)_3(s)$ (line Cr/B), suggesting that the

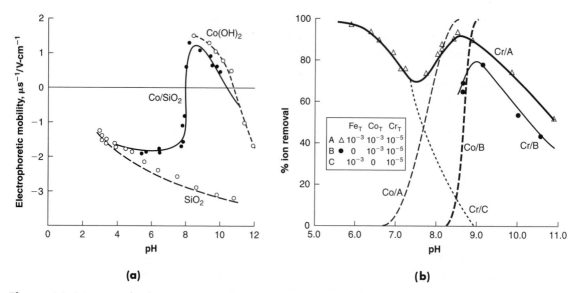

(a)   **(b)**

Figure 10.29   **(a)** The electrophoretic mobility of particles in a system containing 0.002 g/L $SiO_2(s)$ alone, 10^{-4} M *TOT*Co alone, or both $SiO_2(s)$ and Co. In the presence of the $SiO_2(s)$ particles, the Co^{2+} can adsorb or precipitate onto the $SiO_2(s)$ surface.

| Adapted from R. O. James and T. Healy, *Journal of Colloid and Interface Science*, Volume 40, 53–64 (1972).

(b) Removal of CrO_4 and Co^{2+} from solutions containing various combinations of CrO_4, Fe, and Co. When Fe is present, it is essentially all precipitated as $Fe(OH)_3(s)$ and can adsorb either the Co or CrO_4. The labels indicate the species and the system represented by the data; e.g., Co/B indicates that the line represents removal of Co^{2+} from solution B. When Co is present, it can adsorb onto $Fe(OH)_3(s)$ or precipitate as $Co(OH)_2(s)$, in which case it can adsorb CrO_4.

| Adapted from M. M. Benjamin, *Environmental Science and Technology*, Vol. 17, 686–692 (1983).

readsorption of CrO_4 in the $Fe(OH)_3(s) + Co^{2+}$ system is caused by precipitation of a surface layer of $Co(OH)_2(s)$. Note that, in this case, the data suggest that a $Co(OH)_2(s)$ surface precipitate forms under conditions where $Co(OH)_2(s)$ would not precipitate from bulk solution.

SUMMARY

Adsorption, or the binding of molecules at interfaces, is a critical process in natural water systems and engineered water treatment systems. At the most basic level, adsorption determines whether a chemical species follows the path of solids or solution in a system. Adsorption also often has dramatic effects on the rates of redox and mineral precipitation and dissolution reactions, the availability of the adsorbing species to organisms, and the interactions among colloidal particles suspended in a solution. For all these reasons, adsorption reactions have important geological and ecological implications.

Adsorption of hydrophilic species (most inorganic chemicals, natural organic matter, and other ionizable organic species) is commonly modeled as a surface complexation reaction, with surface sites behaving essentially as complexing ligands. Often, both the adsorbates and the surface ligands can participate in acid/base reactions, causing the net amount of sorption to depend strongly on solution pH. Competition for surface sites among adsorbates, competition for adsorbate between the surface and dissolved ligands, and electrostatic interactions at the surface also all play important roles in determining the distribution of adsorbate molecules between the solution and the adsorbent particles.

In some cases, the net result of all the factors affecting adsorption can be synthesized into a relatively simple equation relating the adsorption density to the concentrations of adsorbate and surface sites in the system. Such equations are called adsorption isotherms, and they are often useful for characterizing the sorption process over a limited range of conditions. However, there are fundamental reasons (e.g., the heterogeneity of all surfaces, the effects of adsorbates on the surface electrical potential, interactions among adsorbed species) why simple isotherms are expected to fail to describe such systems completely.

By incorporating more of the complexities of adsorption reactions into models, the range of applicability of those models can be extended. Acid/base chemistry, both in solution and at the surface, is a central component of the sorption of ionizable adsorbates onto inorganic surfaces, and its incorporation into adsorption models is now fairly well established. Attempts to account for the relationships among electrical potential near the surface and the adsorption of charged species in adsorption modeling are also reasonably well developed, but are still being modified to take into account new information about surface speciation and structure. Models accounting for surface heterogeneity and interactions among adsorbed species (including formation of surface precipitates) are less developed.

Because of the central role of adsorption in so many environmental processes, it remains an area of intense research interest. The many topics being addressed currently include adsorption onto mixed assemblages of solids, which are much more common in nature than are the pure solids that have been studied most extensively in the past; the adsorption of complex polyelectrolytes like NOM and its effect on sorption of other ions; the factors that control whether and to what extent adsorbed species catalyze redox reactions; and the microscopic structure of the interfacial region. These efforts are likely to dramatically enhance our understanding of and ability to model adsorption reactions in the future.

Problems

1. Stillings et al. (*Environmental Science and Technology, 32:* 2856–2864) report the following data for adsorption of oxalate species onto a natural aluminosilicate mineral. The data were collected in solutions at pH 3 to 5, and the authors found that the total adsorption of oxalate was independent of pH in this range.

   a. Estimate best-fit parameters to model the data according to the Langmuir and Freundlich isotherms.

   b. Use the isotherms derived in part (*a*) to estimate the adsorption density at an equilibrium dissolved oxalate concentration of 20 mmol/L. Which isotherm would you use if you needed to predict q_{Ox} for those conditions?

$\{Ox\}_{diss,eq}$, mmol/L	q_{Ox}, $\mu mol/m^2$	$\{Ox\}_{diss,eq}$, mmol/L	q_{Ox}, $\mu mol/m^2$
0.00	0.00	3.45	16.84
0.65	8.03	3.49	18.17
0.94	6.50	3.54	18.07
1.00	2.60	6.50	23.69
1.54	16.74	6.53	23.38
1.65	11.52	6.81	18.20
1.93	10.93		

2. In water treatment systems, adsorption processes are often designed as packed columns: the columns are filled with adsorbent particles, and contaminants are removed from a solution as it passes through the column. Conceptually, this process is very similar to what occurs when contaminants adsorb from slowly moving groundwater onto mineral and soil particles that it contacts. The fate of contaminant in a single batch of fluid in such a system can be simulated, to a first approximation, by treating the system as a series of batch reactors. That is, movement of water downstream through the packed column can be represented as movement from one batch reactor to the next in a series.

The compound methyl-isoborneol (MIB) can give water an earthy musty odor, even when the compound is present in solution at concentrations of just a few nanograms per liter. The MIB can be removed from solution by adsorption onto activated carbon (AC). Gillogly et al. (*Journal American Water Works Association, 90*(1): 98–108) report that MIB sorption from a solution of pure water (no competing adsorbates) can be characterized by the Freundlich isotherm $q = 9.6c^{0.5}$, where c is in nanograms per liter and q is in nanograms of MIB per milligram of AC. Compare the final soluble concentration of MIB in two systems each receiving an influent containing 100 μg/L MIB, and in each of which a total dose of 5 mg AC/L is applied. In one system, the activated carbon is all added to a single batch reactor, and in the other, the water flows through four sequential reactors, each receiving a dose of 1.25 mg AC/L. Assume that no AC transfers from one reactor to the next.

3. Sorption of arsenate onto an ion exchange resin that is initially loaded with Cl^- has been found to obey a Langmuir isotherm with $q_{max} = 150\ \mu mol/g$ and $K_{ads} = 0.6\ L/\mu mol$. In a comparable experiment, the adsorption equilibrium constant for sulfate is found to be twice as large as that for arsenate, whereas q_{max} is the same for both adsorbates, when all concentrations are expressed on a molar basis.

Compare the dose of resin required to reduce the arsenate concentration in a raw water from 1.0 to 0.07 μmol/L in a solution containing negligible sulfate with that in a solution containing 150 mg/L sulfate.

4. Equation (10.24) suggests that if multiple adsorbents are present in a system, each one behaves independently, and they are linked only through the fact that

they must all equilibrate with the same solution. However, there are cases where the assumption of independence among adsorbents seems to fail. For instance, the two figures below show cases where adsorption actually *decreases* when a second adsorbent is added to the system. Suggest some possible explanations for these observations.

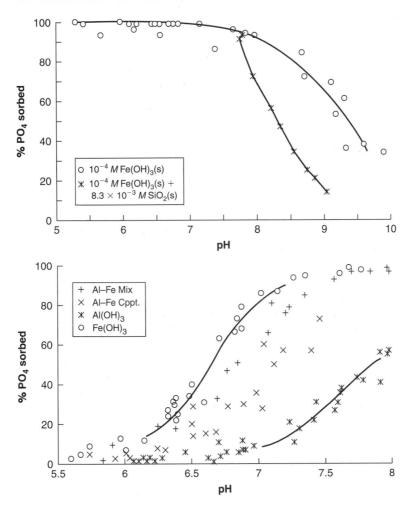

Upper figure: fractional removal of 10^{-6} M phosphate from two solutions containing 10^{-4} M $Fe(OH)_3(s)$. Phosphate does not sorb onto $SiO_2(s)$ under these conditions, and the addition of $SiO_2(s)$ to the system containing $Fe(OH)_3(s)$ causes sorption of phosphate to decrease. Lower figure: Both $Fe(OH)_3(s)$ and $Al(OH)_3(s)$ can adsorb Cd from solution, as shown by the solid lines. However, when the two solids are both present, the sorption of Cd is intermediate between the single-adsorbate cases, rather than additive. The concentration of each solid is 10^{-3} M in all cases. In the mixed system, the two solids were formed separately and then combined; in the coprecipitated system, the two solids were precipitated simultaneously in the same solution.

Adapted from P. R. Anderson and M. M. Benjamin, *Environmental Science and Technology*, Vol. 24, 692 (1990).

5. Data for sorption of Cd onto ferrihydrite, $Fe(OH)_3(s)$, is shown in Figure 10.27 for solutions with and without two complexing ligands. One hypothesis to explain the data is that free Cd^{2+} can adsorb to the solid, but complexed Cd species cannot.

   a. Based on the data point where 50% of TOTCd is adsorbed in the absence of the ligands, estimate the apparent equilibrium constant (i.e., ignore surface electrostatics) for the following reaction. Use the Davies equation to estimate $\gamma_{Cd^{2+}}$, and assume that $\gamma_{\equiv FeOH} = \gamma_{\equiv FeOCd^+} = 1.0$.

   $$\equiv FeOH + Cd^{2+} \leftrightarrow \equiv FeOCd^+ + H^+$$

   Assume $TOT \equiv FeO = 8.7 \times 10^{-4}$. Use the estimated value of K_{ads} to model the entire pH-adsorption edge for Cd^{2+}, and compare the predictions with the experimental results. Assume that, in the pH range of interest, essentially all the surface sites are singly protonated (i.e., ignore the species $\equiv FeO^-$ and $\equiv FeOH_2^+$). Ignore hydrolysis of Cd^{2+}.

   b. Compute the distribution of dissolved Cd species (i.e., determine the α values for the various dissolved Cd species) in the solution containing $0.5\ M\ Cl^-$. Use the Davies equation to estimate activity coefficients. Predict the pH-adsorption edge for this system by assuming that $\equiv FeO^-$ and Cl^- compete for the Cd^{2+} ions. How well does your prediction match the experimental data? Suggest an approach that might improve the fit. (*Hint:* Think about what species might adsorb.)

6. As noted in this chapter and in Chapter 8, complexing or chelating agents can enhance the rate of mineral dissolution. In these so-called ligand-promoted dissolution processes, dissolution is thought to proceed via a sequence

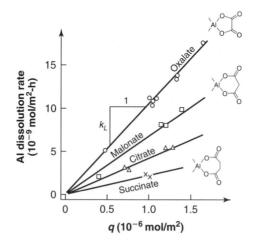

Enhancement in the rate of dissolution of $Al_2O_3(s)$ attributable to the presence of four different carboxylic acids.

| Adapted from G. Furrer and W. Stumm, *Geochimica et Cosmochimica Acta*, Vol. 50, 1847–1860 (1986).

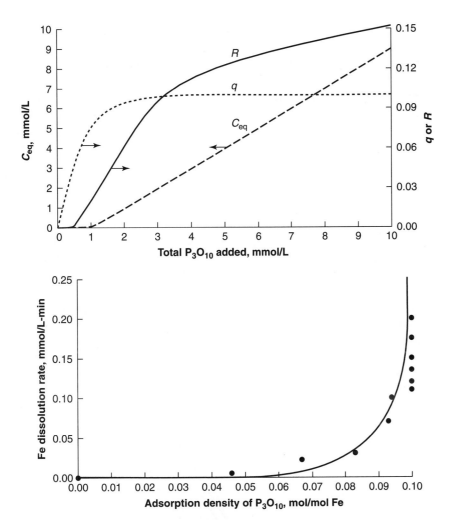

Distribution of P_3O_{10} between solution and the solid, and the rate R of Fe dissolution, for various additions of total P_3O_{10} to a system containing 10^{-2} M $Fe(OH)_3(s)$ (upper figure), and Fe dissolution rate as a function of the P_3O_{10} adsorption density in these systems (lower figure). At adsorption densities less than q_{max} (equal to 0.1 mol P_3O_{10}/mol Fe), dissolution is slow and is not proportional to q. When $q \approx q_{max}$, the dissolution rate increases with increasing amounts of P_3O_{10} in excess of the adsorption capacity (i.e., P_3O_{10} in solution).

Adapted from C.-F. Lin and M. M. Benjamin, *Environmental Science and Technology*, Vol. 24, 126–134 (1990).

involving adsorption of the ligand, complexation of a metal ion in the solid's surface layer, and release of the complex into solution. Often, the rate of dissolution is found to be proportional to the adsorption density of the ligand, as shown in the figure on the preceding page. However, in other cases, almost no dissolution is evident until enough complexing agent has been added to saturate the available surface sites, as shown in the two figures shown above.

By writing out all the reactions that you expect to take place in such systems and considering the different ways that these reactions might compete with one another, suggest an explanation for the different types of behavior. (*Hint:* Tripolyphosphate species ($H_xP_3O_{10}^{5-x}$) are much stronger complexing agents and also much stronger adsorbates than the other ligands shown.)

7. Ferrihydrite is to be used to treat a photoprocessing waste solution that contains Ag^+, $S_2O_3^{2-}$, Na^+, and NO_3^- ions. Any of these ions can adsorb to the ferrihydrite surface, with the first two binding much more strongly than the last two. Write out the reactions that you would have to consider to compute the equilibrium composition of the solution, and prepare the input table to model the system using chemical equilibrium software, if the triple-layer model is to be used to characterize adsorption. Consider acid/base reactions of the surface, sorption of the four free ions, complexation of Ag^+ with $S_2O_3^{2-}$, and sorption of $AgS_2O_3^{\circ}$ complexes. Assume that free Ag^+ and $S_2O_3^{2-}$ ions adsorb in the o layer, that Na^+ and NO_3^- adsorb in the β layer, and that when $AgS_2O_3^{\circ}$ sorbs, the S_2O_3 portion of the molecule resides in the naught layer, but the Ag portion resides in the beta layer.

8. The following figure shows linearized fits to the Langmuir isotherm according to Equation (10.12) for sorption of Cd onto goethite (α-FeOOH) at various temperatures. Estimate the molar enthalpy of the adsorption reaction, and predict the adsorption density of Cd in equilibrium with a dissolved concentration of 5×10^{-5} M at 2°C. Assume $\gamma_{Cd^{2+}} = 1.0$.

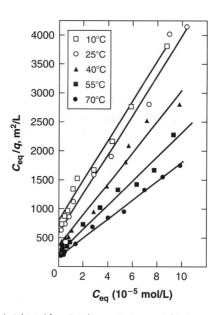

Adapted from B. Johnson, *Environmental Science and Technology*, Vol. 24, 112–118 (1990).

THERMODYNAMIC DATA AND EQUILIBRIUM CONSTANTS

A.1

Thermodynamic Properties; Table of G_f^0, H_f^0 and $\bar{S}^0$ Values for Common Chemical Species in Aquatic Systems[1] Valid at 25°C, 1 atm Pressure and Standard States[2]

| Species | Formation from the Elements | | Entropy | Reference[3] |
	G_f^0 (kJ mol^{-1})	H_f^0 (kJ mol^{-1})	$\bar{S}^0$ J mol^{-1} K^{-1}	
Ag (Silver)				
Ag (metal)	0	0	42.6	NBS
Ag$^+$ (aq)	77.12	105.6	73.4	NBS
AgBr	−96.9	−100.6	107	NBS
AgCl	−109.8	−127.1	96	NBS
AgI	−66.2	−61.84	115	NBS
Ag$_2$S(α)	−40.7	−29.4	14	NBS
AgOH(aq)	−92	—	—	NBS
Ag(OH)$_2^-$(aq)	−260.2	—	—	NBS
AgCl(aq)	−72.8	−72.8	154	NBS
AgCl$_2^-$(aq)	−215.5	−245.2	231	NBS
Al (Aluminum)				
Al	0	0	28.3	R
Al^{3+}(aq)	−489.4	−531.0	−308	R
AlOH^{2+}(aq)	−698	—	—	S
Al(OH)$_2^+$(aq)	−911	—	—	S
Al(OH)$_3$(aq)	−1115	—	—	S
Al(OH)$_4^-$(aq)	−1325	—	—	S
Al(OH)$_3$ (amorph)	−1139	—	—	R
Al$_2$O$_3$ (corundum)	−1582	−1676	50.9	R
AlOOH (boehmite)	−922	−1000	17.8	R
Al(OH)$_3$ (gibbsite)	−1155	−1293	68.4	R

(*continued*)

Species	Formation from the Elements		Entropy	Reference[3]
	G_f^0 (kJ mol^{-1})	H_f^0 (kJ mol^{-1})	$\overline{S}^0$ J mol^{-1} K^{-1}	
Al$_2$Si$_2$(OH)$_4$ (kaolinite)	−3799	−4120	203	R
KAl$_3$Si$_3$O$_{10}$(OH)$_2$ (muscovite)	−1341	—	—	G
Mg$_5$Al$_2$Si$_3$O$_{10}$(OH)$_8$ (chlorite)	−1962	—	—	R
CaAl$_2$Si$_2$O$_8$ (anorthite)	−4017.3	−4243.0	199	R
NaAlSiO$_3$O$_8$ (albite)	−3711.7	−3935.1	—	R
As (Arsenic)				
As (α metal)	0	0	35.1	NBS
H$_3$AsO$_4$(aq)	−766.0	−898.7	206	NBS
H$_2$AsO$_4{}^-$(aq)	−748.5	−904.5	117	NBS
HAsO$_4{}^{2-}$(aq)	−707.1	−898.7	3.8	NBS
AsO$_4{}^{3-}$(aq)	−636.0	−870.3	−145	NBS
H$_2$AsO$_3{}^-$(aq)	−587.4			NBS
Ba (Barium)				
Ba^{2+}(aq)	−560.7	−537.6	9.6	R
BaSO$_4$ (barite)	−1362	−1473	132	R
BaCO$_3$ (witherite)	−1132	−1211	112	R
Be (Beryllium)				
Be^{2+}(aq)	−380	−382	−130	NBS
Be(OH)$_2$(α)	−815.0	−902	51.9	NBS
Be$_3$(OH)$_3{}^{3+}$	−1802	—	—	NBS
B (Boron)				
H$_3$BO$_3$(aq)	−968.7	−1072	162	NBS
B(OH)$_4{}^-$(aq)	−1153.3	−1344	102	NBS
Br (Bromide)				
Br$_2$(l)	0	0	152	NBS
Br$_2$(aq)	3.93	−2.59	130.5	NBS
Br$^-$(aq)	−104.0	−121.5	82.4	NBS
HBrO(aq)	−82.2	−113.0	147	NBS
BrO$^-$(aq)	−33.5	−94.1	42	NBS
C (Carbon)				
C (graphite)	0	0	152	NBS
C (diamond)	3.93	−2.59	130.5	NBS
CO$_2$(g)	−394.37	−393.5	213.6	NBS
H$_2$CO$_3$*(aq)[4]	−623.2	−699.7	187.0	R
H$_2$CO$_3$(aq) ("true")	~ −607.1	—	—	S

Species	Formation from the Elements		Entropy	Reference[3]
	G_f^0 (kJ mol^{-1})	H_f^0 (kJ mol^{-1})	$\overline{S}^0$ J mol^{-1} K^{-1}	
$HCO_3^-(aq)$	−586.8	−692.0	91.2	S
$CO_3^{2-}(aq)$	−527.9	−677.1	−56.9	NBS
$CH_4(g)$	−50.75	−74.80	186	NBS
$CH_4(aq)$	−34.39	−89.04	83.7	NBS
$CH_3OH(aq)$	−175.4	−245.9	133	NBS
$HCOOH(aq)$	−372.3	−425.4	163	NBS
$HCOO^-(aq)$	−351.0	−425.6	92	NBS
$HCN(aq)$	119.7	107.1	124.6	NBS
$CN^-(aq)$	172.4	150.6	94.1	NBS
$CH_3COOH(aq)$	−396.6	−485.8	179	NBS
$CH_3COO^-(aq)$	−369.4	−486.0	86.6	NBS
$C_2H_5OH(aq)$	−181.8	−288.3	149	NBS
$NH_2CH_2COOH(aq)$	−370.8	−514.0	158	NBS
$NH_2CH_2COO^-(aq)$	−315.0	−469.8	119	NBS
Ca (Calcium)				
$Ca^{2+}(aq)$	−553.54	−542.83	−53	R
$CaOH^+(aq)$	−718.4	—	—	NBS
$Ca(OH)_2(aq)$	−868.1	−1003	−74.5	NBS
$Ca(OH)_2$ (portlandite)	−898.4	−986.0	83	R
$CaCO_3$ (calcite)	−1128.8	−1207.4	91.7	R
$CaCO_3$ (aragonite)	−1127.8	−1207.4	88.0	R
$CaMg(CO_3)_2$ (dolomite)	−2161.7	−2324.5	155.2	R
$CaSiO_3$ (wollastonite)	−1549.9	−1635.2	82.0	R
$CaSO_4$ (anhydrite)	−1321.7	−1434.1	106.7	R
$CaSO_4 \cdot 2H_2O$ (gypsum)	−1797.2	−2022.6	194.1	R
$Ca_5(PO_4)_3OH$ (hydroxyapatite)	−6338.4	−6721.6	390.4	R
Cd (Cadmium)				
Cd (γ metal)				
Cd^{2+} (aq)	−77.58	−75.90	−73.2	R
$CdOH^+(aq)$	−284.5			R
$Cd(OH)_3^-(aq)$	−600.8			R
$Cd(OH)_4^{2-}(aq)$	−758.5			R
$Cd(OH)_2(aq)$	−392.2			R
$CdO(s)$	−228.4	−258.1	54.8	
$Cd(OH)_2$ (precip.)	−473.6	−560.6	96.2	R
$CdCl^+(aq)$	−224.4	−240.6	43.5	R

(*continued*)

Species	Formation from the Elements		Entropy	Reference[3]
	G_f^0 (kJ mol^{-1})	H_f^0 (kJ mol^{-1})	$\overline{S}^0$ J mol^{-1} K^{-1}	
CdCl$_2$(aq)	−340.1	−410.2	39.8	R
CdCl$_3^-$(aq)	−487.0	−561.0	203	R
CdCO$_3$(s)	−669.4	−750.6	92.5	R
Cl (Chlorine)				
Cl$^-$(aq)	−131.3	−167.2	56.5	NBS
Cl$_2$(g)	0	0	223.0	NBS
Cl$_2$(aq)	6.90	−23.4	121	NBS
HClO(aq)	−79.9	−120.9	142	NBS
ClO$^-$(aq)	−36.8	−107.1	42	NBS
ClO$_2$(aq)	117.6	74.9	173	NBS
ClO$_2^-$(aq)	17.1	−66.5	101	NBS
ClO$_3^-$(aq)	−3.35	−99.2	162	NBS
ClO$_4$(aq)	−8.62	−129.3	182	NBS
Co (Cobalt)				
Co (metal)	0	0	30.04	R
Co^{2+}(aq)	−54.4	−58.2	−113	R
Co^{3+}	−134	−92	−305	R
HCoO$_2^-$(aq)	−407.5	—	—	NBS
Co(OH)$_2$(aq)	−369	−518	134	NBS
Co(OH)$_2$ (blue precip.)	−450			NBS
CoO	−214.2	−237.9	53.0	R
Co$_3$O$_4$ (cobalt spinel)	−725.5	−891.2	102.5	R
Cr (Chromium)				
Cr (metal)	0	0	23.8	NBS
Cr^{2+}(aq)	—	−143.5	—	NBS
Cr^{3+}(aq)	−215.5	−256.0	308	NBS
Cr$_2$O$_3$ (eskolaite)	−1053	−1153	81	R
HCrO$_4^-$(aq)	−764.8	−878.2	184	R
CrO$_4^{2-}$(aq)	−727.9	−881.1	50	R
Cr$_2$O$_7^{2-}$(aq)	−1301	−1490	262	R
Cu (Copper)				
Cu(metal)	0	0	33.1	NBS
Cu$^+$(aq)	50.0	71.7	40.6	NBS
Cu^{2+}(aq)	65.5	64.8	−99.6	NBS
Cu(OH)$_2$(aq)	−249.1	−395.2	−121	NBS

Species	Formation from the Elements		Entropy	Reference[3]
	G_f^0 (kJ mol^{-1})	H_f^0 (kJ mol^{-1})	$\overline{S}^0$ J mol^{-1} K^{-1}	
HCuO$_2^-$(aq)	−258	—	—	
CuS (covellite)	−53.6	−53.1	66.5	NBS
Cu$_2$S(α)	−86.2	−79.5	121	NBS
CuO (tenorite)	−129.7	−157.3	43	NBS
CuCO$_3$ · Cu(OH)$_2$ (malachite)	−893.7	−1051.4	186	NBS
2CuCO$_3$ · Cu(OH)$_2$ (azurite)		−1632		NBS
F (Fluorine)				
F$_2$(g)	0	0	202	NBS
F$^-$(aq)	−278.8	−332.6	−13.8	NBS
HF(aq)	−296.8	320.0	88.7	NBS
HF$_2^-$ (aq)	−578.1	−650	92.5	NBS
Fe (Iron)				
Fe (metal)	0	0	27.3	NBS
Fe^{2+}(aq)	−78.87	−89.10	−138	NBS
FeOH$^+$(aq)	−277.3	—	—	NBS
Fe^{3+}(aq)	−4.60	−48.5	−316	NBS
FeOH^{2+}(aq)	−229.4	−324.7	−29.2	NBS
Fe(OH)$_2^+$(aq)	−438	—	—	NBS
Fe(OH)$_2^-$(aq)	−659	—	—	NBS
Fe$_2$(OH)$_2^{4+}$(aq)	−467.3	—	—	NBS
FeS$_2$ (pyrite)	−160.2	−171.5	52.9	R
FeS$_2$ (marcasite)	−158.4	−169.4	53.9	R
FeO(s)	−251.1	−272.0	59.8	R
Fe(OH)$_2$ (precip.)	−486.6	−569	87.9	NBS
α-Fe$_2$O$_3$ (hematite)[5]	−742.7	−824.6	87.4	R
Fe$_3$O$_4$ (magnetite)	−1012.6	−1115.7	146	R
α-FeOOH (goethite)[5]	−488.6	−559.3	60.5	R
FeOOH (amorph)[5]	−462	—	—	S
Fe(OH)$_3$ (amorph)[5]	−699(−712)			S
FeCO$_3$ (siderite)	−666.7	−737.0	105	R
Fe$_2$SiO$_4$ (fayalite)	−1379.4	−1479.3	148	R
H (Hydrogen)				
H$_2$(g)	0	0	130.6	NBS
H$_2$(aq)	17.57	−4.18	57.7	NBS

(continued)

Species	Formation from the Elements		Entropy	Reference[3]
	G_f^0 (kJ mol^{-1})	H_f^0 (kJ mol^{-1})	$\overline{S}^0$ J mol^{-1} K^{-1}	
H$^+$(aq)	0	0	0	NBS
H$_2$O(l)	−237.18	−285.83	69.91	NBS
H$_2$O$_2$(aq)	−134.1	−191.1	144	NBS
HO$_2^-$(aq)	−67.4	−160.3	23.8	NBS
H$_2$O(g)	−228.57	−241.8	188.72	R
Hg (Mercury)				
Hg(l)	0	0	76.0	NBS
Hg$_2^{2+}$(aq)	153.6	172.4	84.5	NBS
Hg^{2+}(aq)	164.4	171.0	−32.2	NBS
Hg$_2$Cl$_2$ (calomel)	−210.8	265.2	192.4	NBS
HgO (red)	−58.5	−90.8	70.3	NBS
HgS (metacinnabar)	−43.3	−46.7	96.2	NBS
HgI$_2$ (red)	−101.7	−105.4	180	NBS
HgCl$^+$(aq)	−5.44	−18.8	75.3	NBS
HgCl$_2$(aq)	−173.2	−216.3	155	NBS
HgCl$_3^-$(aq)	−309.2	−388.7	209	NBS
HgCl$_4^{2-}$(aq)	−446.8	−554.0	293	NBS
HgOH$^+$(aq)	−52.3	−84.5	71	NBS
Hg(OH)$_2$(aq)	−274.9	−355.2	142	NBS
HgO$_2^-$(aq)	−190.3	—	—	NBS
I (Iodine)				
I$_2$ (crystal)	0	0	116	NBS
I$_2$(aq)	16.4	22.6	137	NBS
I$^-$(aq)	−51.59	−55.19	111	NBS
I$_3^-$(aq)	−51.5	−51.5	239	NBS
HIO(aq)	−99.2	−138	95.4	NBS
IO$^-$(aq)	−38.5	−107.5	−5.4	NBS
HIO$_3$(aq)	−132.6	−211.3	167	NBS
IO$_3^-$	−128.0	−221.3	118	NBS
Mg (Magnesium)				
Mg (metal)	0	0	32.7	R
Mg^{2+}(aq)	−454.8	−466.8	−138	R
MgOH$^+$(aq)	−626.8	—	—	S
Mg(OH)$_2$(aq)	−769.4	−926.8	−149	NBS
Mg(OH)$_2$ (brucite)	−833.5	−924.5	63.2	R

Species	Formation from the Elements		Entropy	Reference[3]
	G_f^0 (kJ mol^{-1})	H_f^0 (kJ mol^{-1})	$\overline{S}^0$ J mol^{-1} K^{-1}	
Mn (Manganese)				
Mn (meta)	0	0	32.0	R
Mn^{2+}(aq)	−228.0	−220.7	−73.6	R
Mn(OH)$_2$ (precip.)	−616			S
Mn$_3$O$_4$ (hausmannite)	−1281			S
MnOOH (α manganite)	−557.7			S
MnO$_2$ (manganate) (IV) (MnO$_{1.7}$ − MnO$_2$)	−453.1			S
MnO$_2$ (pyrolusite)	−465.1	−520.0	53	R
MnCO$_3$ (rhodochrosite)	−816.0	−889.3	100	R
MnS (albandite)	−218.1	−213.8	87	R
MnSiO$_3$ (rhodonite)	−1243	−1319	131	R
N (Nitrogen)				
N$_2$(g)	0	0	191.5	NBS
N$_2$O(g)	104.2	82.0	220	NBS
NH$_3$(g)	−16.48	−46.1	192	NBS
NH$_3$(aq)	−26.57	−80.29	111	NBS
NH$_4^+$(aq)	−79.37	−132.5	113.4	NBS
HNO$_2$(aq)	−42.97	−119.2	153	NBS
NO$_2^-$(aq)	−37.2	−104.6	140	NBS
HNO$_3$(aq)	−111.3	−207.3	146.	NBS
NO$_3^-$(aq)	−111.3	−207.3	146.4	NBS
Ni (Nickel)				
Ni^{2+}(aq)	−45.6	−54.0	−129	R
NiO (bunsenite)	−211.6	−239.7	38	R
NiS (millerite)	−86.2	−84.9	66	R
O (Oxygen)				
O$_2$(g)	0	0	205	NBS
O$_2$(aq)	16.32	−11.71	111	NBS
O$_3$(g)	163.2	142.7	239	NBS
OH$^-$(aq)	−157.3	−230.0	−10.75	NBS
P (Phosphorus)				
P (α, white)	0	0	41.1	
PO$_4^{3-}$(aq)	−1018.8	−1277.4	−222	NBS

(continued)

Species	Formation from the Elements		Entropy	Reference[3]
	G_f^0 (kJ mol^{-1})	H_f^0 (kJ mol^{-1})	$\overline{S}^0$ J mol^{-1} K^{-1}	
$HPO_4^{2-}(aq)$	−1089.3	−1292.1	−33.4	NBS
$H_2PO_4^-(aq)$	−1130.4	−1296.3	90.4	NBS
$H_3PO_4(aq)$	−1142.6	−1288.3	158	NBS
Pb (Lead)				
Pb (metal)	0	0	64.8	NBS
$Pb^{2+}(aq)$	−24.39	−1.67	10.5	NBS
$PbOH^+(aq)$	−226.3	—	—	NBS
$Pb(OH)_3^-(aq)$	−575.7			NBS
$Pb(OH)_2$ (precip.)	−452.2			NBS
PbO (yellow)	−187.9	−217.3	68.7	NBS
PbO_2	−217.4	−277.4	68.6	NBS
Pb_3O_4	−601.2	−718.4	211	NBS
PbS	−98.7	−100.4	91.2	NBS
$PbSO_4$	−813.2	−920.0	149	NBS
$PbCO_3$ (cerussite)	−625.5	−699.1	131	NBS
S (Sulfur)				
S (rhombic)	0	0	31.8	NBS
$SO_2(g)$	−300.2	−296.8	248	NBS
$SO_3(g)$	−371.1	−395.7	257	NBS
$H_2S(g)$	−33.56	−20.63	205.7	NBS
$H_2S(aq)$	−27.87	−39.75	121.3	NBS
$S^{2-}(aq)$	85.8	33.0	−14.6	NBS
$HS^-(aq)$	12.05	−17.6	62.8	NBS
$SO_3^{2-}(aq)$	−486.6	−635.5	−29	NBS
$HSO_3^-(aq)$	−527.8	−626.2	140	NBS
$H_2SO_3^*(aq)$	−537.9	−608.8	232	NBS[6]
$H_2SO_3(aq)$ ("true")[6]	∼−534.5			S
$SO_4^{2-}(aq)$	−744.6	−909.2	20.1	NBS
$HSO_4^-(aq)$	−756.0	−887.3	132	NBS
Se (Selenium)				
Se (black)	0	0	42.4	NBS
$SeO_3^{2-}(aq)$	−369.9	−509.2	12.6	NBS
$HSeO_3^-(aq)$	−431.5	−514.5	135	NBS
$H_2SeO_3(aq)$	−426.2	−507.5	208	NBS
$SeO_4^{2-}(aq)$	−441.4	−599.1	54.0	NBS
$HSeO_4^-(aq)$	−452.3	−581.6	149	NBS

Species	Formation from the Elements		Entropy	Reference[3]
	G_f^0 (kJ mol^{-1})	H_f^0 (kJ mol^{-1})	$\bar{S}^0$ J mol^{-1} K^{-1}	
Si (Silicon)				
Si (metal)	0	0	18.8	NBS
SiO$_2$ (α, quartz)	-856.67	-910.94	41.8	NBS
SiO$_2$ (α, cristobalite)	-855.88	-909.48	42.7	NBS
SiO$_2$ (α, tridymite)	-855.29	-909.06	43.5	NBS
SiO$_2$ (amorph)	-850.73	-903.49	46.9	NBS
H$_4$SiO$_4$(aq)	-1316.7	-1468.6	180	NBS
Sr (Strontium)				
Sr^{2+}(aq)	-559.4	-545.8	-33	R
SrOH$^+$(aq)	-721	—	—	NBS
SrCO$_3$ (strontianite)	-1137.6	-1218.7	97	R
SrSO$_4$ (celestite)	-1341.0	-1453.2	118	R
Zn (Zinc)				
Zn, metal	0	0	29.3	NBS
Zn^{2+}(aq)	-147.0	-153.9	112	NBS
ZnOH$^+$(aq)	-330.1			NBS
Zn(OH)$_2$(aq)	-522.3			NBS
Zn(OH)$_3^-$(aq)	-694.3			NBS
Zn(OH)$_4^{2-}$(aq)	-858.7			NBS
Zn(OH)$_2$ (solid β)	-553.2	-641.9	81.2	R
ZnCl$^+$(aq)	-275.3			NBS
ZnCl$_2$(aq)	-403.8			NBS
ZnCl$_3^-$(aq)	-540.6			NBS
ZnCl$_4^{2-}$(aq)	-666.1			S
ZnCO$_3$ (smithsonite)	-731.6	-812.8	82.4	NBS

[1]Table is reproduced from *Aquatic Chemistry: Chemical Equilibria and Rates in Natural Waters*, 3rd Ed., by W. Stumm and J. M. M. Morgan. Copyright © 1996, John Wiley and Sons. Reprinted by permission of John Wiley & Sons, Inc. The quality of the data is highly variable; the authors do not claim to have critically selected the "best" data. For information on precision of the data and for a more complete compendium which includes less common substances, the reader is referred to the references. For research work, the original literature should be consulted.

[2]Thermodynamic properties taken from Robie, Hemingway, and Fisher are based on a reference state of the elements in their standard states at 1 bar. This change in reference pressure has a negligible effect upon the tabulated values for the condensed phases. [For gas phases only data from NBS (reference state = 1 atm) are given].

[3]NBS: D. D. Wagman et al., Selected Values of Chemical Thermodynamic Properties, U.S. National Bureau of Standards, Technical Notes 270-3 (1968), 270-4 (1969), 270-5 (1971). R: R. A. Robie, B. S. Hemingway, and J. R. Fisher, *Thermodynamic Properties of Minerals and Related Substances at 298.15 K and 1 Bar (10^5 Pascals) Pressure and at Higher Temperatures*, Geological Survey Bulletin No. 1452, Washington D.C., 1978. S: Other sources (e.g., computed from data in *Stability Constants*).

[4]H$_2$CO$_3^*$ = CO$_2$(aq) + "true" H$_2$CO$_3$.

[5]The thermodynamic stability of oxides, hydroxides, or oxyhydroxides of Fe(III) depends on mode of preparation, age, and molar surface. Reported solubility products (K_{s0} = {Fe^{3+}}{OH$^-$}^{3}) range from 10$^{-37.3}$ to 10$^{-43.7}$. Correspondingly FeOOH may have G_f^0 values between -452 J mol^{-1} (freshly precipitated amorphous FeOOH) and -489 J mol^{-1} (aged goethite). If the precipitate is written as Fe(OH)$_3$, its G_f^0 values vary from -692 to -729 J mol^{-1}.

[6]H$_2$SO$_3^*$ = SO$_2$(aq) + "true" H$_2$SO$_3$.

A.2

Table 3.2   Chemical formulas and acidity constants of some important acids[1]

Name	Formula	pK_{a1}	pK_{a2}	pK_{a3}	pK_{a4}
Nitric acid	HNO_3	−1.30			
Trichloroacetic acid	CCl_3COOH	−0.5			
Hydrochloric acid	HCl	<0			
Sulfuric acid	H_2SO_4	<0	1.99		
Hydronium ion	H_3O^+	0.00	14.00		
Chromic acid	H_2CrO_4	0.86	6.51		
Oxalic acid	$(COOH)_2$	0.90	4.20		
Dichloroacetic acid	$CHCl_2COOH$	1.1			
Sulfurous acid	H_2SO_3	1.86	7.30		
Phosphoric acid	H_3PO_4	2.16	7.20	12.35	
Arsenic acid	H_3AsO_4	2.24	6.76		
Monochloroacetic acid	$CH_2ClCOOH$	2.86			
Salicylic acid	$C_6H_4OHCOOH$	2.97	13.70		
Citric acid	$C_3H_4OH(COOH)_3$	3.13	4.72	6.33	
Hydrofluoric acid	HF	3.17			
Benzoic acid	C_6H_5COOH	4.20			
Pentachlorophenol	C_6Cl_5OH	4.7			
Acetic acid	CH_3COOH	4.76			
Carbonic acid	H_2CO_3	6.35	10.33		
Hydrogen sulfide	H_2S	6.99	12.92		
Hypochlorous acid	$HOCl$	7.60			
Cupric ion	Cu^{2+}	8.00	5.68		
2-Chloro-phenol	C_6H_4ClOH	8.53			
Hypobromous acid	$HOBr$	8.63			
Zinc ion	Zn^{2+}	8.96	8.94		
Arsenous acid	H_3AsO_3	9.23	12.10		
Hydrocyanic acid	HCN	9.24			
Boric acid	H_4BO_4	9.24			
Ammonium ion	NH_4^+	9.25			
2,4-Dichloro-phenol	$C_6H_3Cl_2OH$	9.43			
Silicic acid	H_4SiO_4	9.84	13.20		
Phenol	C_6H_5OH	9.98			
Cadmium ion	Cd^{2+}	10.08	10.27	12.95	14.05
Calcium ion	Ca^{2+}	12.60			

[1] The equilibrium constants given are consistent with the database in the chemical equilibrium computer program Mineql[+]. This choice has been made to facilitate correspondence of computations carried out manually with the output from that program. Equilibrium constants for reactions of interest that are not in the program's database have been selected from a range of other sources.

A.3

Table 7.2   Henry's constants of some environmentally important gases

Compound	H^1	Compound	H^1
Nitrogen	1560	Hydrogen sulfide	9.8
Hydrogen	1260	Chloroform	4.0
Carbon monoxide	1050	Sulfur dioxide	0.81
Oxygen	790	Bromoform	0.70
Methane	776	Benzene	0.22
Ozone	107	Hydrogen cyanide	0.040
Carbon dioxide	28.8	Ammonia	0.017
Tetrachloroethylene	20	Acetic acid	0.0013
Trichloroethylene	11		

[1]Based on standard state concentrations given in units of bars and moles per liter at 20°C.

Values from Stumm and Morgan (*Aquatic Chemistry*, Wiley, 1996), Langmuir (*Aqueous Environmental Geochemistry*, Prentice-Hall, 1997), Staudinger et al. [*Journal of the American Water Works Association 82*, 1, 73–79 (1990)], and references therein.

A.4

Table 8.2   Stability constants for complexation of metals by OH^-

	i	Log K_i	Log $*K_i$	Log β_i	Log $*\beta_i$
Ag^+	1	2.00	−12.00	2.00	−12.00
	2	2.00	−12.00	4.00	−24.00
Al^{3+}	1	9.01	−4.99	9.01	−4.99
	2	8.89	−5.11	17.90	−10.10
	3	8.10	−5.90	26.00	−16.00
	4	7.00	−7.00	33.00	−23.00
Ca^{2+}	1	1.40	−12.60	1.40	−12.60
Cd^{2+}	1	3.92	−10.08	3.92	−10.08
	2	3.73	−10.27	7.65	−20.35
	3	1.05	−12.95	8.70	−33.30
	4	−0.05	−14.05	8.65	−47.35
Co^{2+}	1	4.80	−9.20	4.80	−9.20
	2	4.90	−9.10	9.70	−18.30
	3	1.10	−12.90	10.80	−31.20
Cr^{3+}	1	10.00	−4.00	10.00	−4.00
	2	8.38	−5.62	18.38	−9.62
	3	6.87	−7.13	25.25	−16.75
	4	2.98	−11.02	28.23	−27.77
Cu^{2+}	1	6.00	−8.00	6.00	−8.00
	2	8.32	−5.68	14.32	−13.68
	3	0.78	−13.22	15.10	−26.90
	4	1.30	−12.70	16.40	−39.60
Fe^{2+}	1	4.50	−9.50	4.50	−9.50
	2	2.93	−11.07	7.43	−20.57
	3	3.57	−10.43	11.00	−31.00
Fe^{3+}	1	11.81	−2.19	11.81	−2.19
	2	10.52	−3.48	22.33	−5.67
	3	6.07	−7.93	28.40	−13.60
	4	6.00	−8.00	34.40	−21.60
Hg^{2+}	1	10.60	−3.40	10.60	−3.40
	2	11.30	−2.70	21.90	−6.10
	3	10.26	−3.74	20.86	−21.14
Mg^{2+}	1	2.21	−11.79	2.21	−11.79
Ni^{2+}	1	4.14	−9.86	4.14	−9.86
	2	4.86	−9.14	9.00	−19.00
	3	3.00	−11.00	12.00	−30.00
Pb^{2+}	1	6.29	−7.71	6.29	−7.71
	2	4.59	−9.41	10.88	−17.12
	3	3.06	−10.94	13.94	−28.06
	4	2.36	−11.64	16.30	−39.70
Zn^{2+}	1	5.04	−8.96	5.04	−8.96
	2	6.06	−7.94	11.10	−16.90
	3	2.50	−11.50	13.60	−28.40
	4	1.20	−12.80	14.80	−41.20

A.4

Table 8.3 Stability constants for some metal-ligand complexes

	CO_3^{2-}	SO_4^{2-}	Cl^-	F^-	NH_3	PO_4^{3-}	EDTA	CN^-	HS^-
Ag^+		AgL 1.29	AgL 3.27 AgL$_2$ 5.27 AgL$_3$ 5.29 AgL$_4$ 5.51	AgL 0.36			AgL 7.36 AgHL	AgL$_2$ 20.38 AgL$_3$ 21.4	AgL 14.05 AgL$_2$ 18.45
Al^{3+}		AlL 4.92		AlL 7.01 AlL$_2$ 12.75 AlL$_3$ 17.02 AlL$_4$ 19.72 AlL$_5$ 20.8 AlL$_6$ 20.5			AlL 19.8 AlHL 22.5		
Ca^{2+}		CaL 2.91		CaL 0.94		CaHL 15.08			
Cd^{2+}	CdL 5.4 CdL$_3$ 6.22 CdHL 12.4	CdL 2.46 CdL$_2$ 3.5	CdL 1.98 CdL$_2$ 2.6 CdL$_3$ 2.4	CdL 1.1 CdL$_2$ 1.5		CdL 3.9	CdL 16.28 CdHL 2.9	CdL 5.32 CdL$_2$ 10.37 CdL$_3$ 14.83 CdL$_4$ 18.29	CdL 10.17 CdL$_2$ 16.53 CdL$_3$ 18.71 CdL$_4$ 20.9
Co^{2+}		CoL 2.5	CoL 0.5				CoL 18.6 CoHL 21.6		
Cr^{3+}		CrL 1.34	CrL -0.25 CrL$_2$ -0.96	CrL 4.92		CrH$_2$L 22.29			
Cu^{2+}	CuL 6.73 CuL$_2$ 9.83 CuHL 13.6	CuL 2.31	CuL 0.43 CuL$_2$ 0.16 CuL$_3$ -2.29 CuL$_4$ -4.59	CuL 1.26	CuL 5.8 CuL$_2$ 10.7 CuL$_3$ 14.7 CuL$_4$ 17.6	CuHL 16.6	CuL 18.78 CuHL 11.2		CuL$_3$ 25.9

(continued)

Table 8.3 (Continued)

	CO$_3^{2-}$	SO$_4^{2-}$	Cl$^-$	F$^-$	NH$_3$	PO$_4^{3-}$	EDTA	CN$^-$	HS$^-$
Fe^{2+}		FeL 2.25	FeL 0.9			FeH$_2$L 22.25	FeL 16.7 FeHL 20.1	FeL$_6$ 52.44 FeHL$_6$ 50 FeH$_2$L$_6$ 45.61	
Fe^{3+}		FeL 3.92 FeL$_2$ 5.42	FeL 1.48 FeL$_2$ 2.13 FeL$_3$ 1.13	FeL 6.2 FeL$_2$ 10.8 FeL$_3$ 14		FeHL 17.78	FeL 27.8 FeHL 29.4	FeL$_6$ 52.63	FeL$_2$ 8.95 FeL$_3$ 10.99
Hg^{2+}		HgL 1.39	HgL 6.75 HgL$_2$ 13.12 HgL$_3$ 14.02 HgL$_4$ 14.43	HgL 1.98	HgL 8.76 HgL$_2$ 17.43 HgL$_3$ 18.4 HgL$_4$ 19.17			HgL 18.07 HgL$_2$ 34.55 HgL$_3$ 38.3 HgL$_4$ 41.31	HgL$_2$ 37.72
Mg^{2+}									
Ni^{2+}	NiL 6.87 NiL$_2$ 10.11 NiHL 12.47	NiL 2.29 NiL$_2$ 1.02	NiL 0.4 NiL$_2$ 0.96	NiL 1.3			NiL 20.33 NiHL 11.56	NiL$_2$ 14.59 NiL$_3$ 22.64 NiH$_3$L$_3$ 43.95	
Pb^{2+}	PbL 7.24 PbL$_2$ 10.64 PbHL 13.2	PbL 2.75 PbL$_2$ 3.47	PbL 1.6 PbL$_2$ 1.8 PbL$_3$ 1.7 PbL$_4$ 1.38	PbL 1.25 PbL$_2$ 2.56 PbL$_3$ 3.42 PbL$_4$ 3.1			PbL 17.86 PbHL 9.68 PbH$_2$L 6.22	PbL$_4$ 10.6	PbL 15.27 PbL$_2$ 16.57
Zn^{2+}	ZnL 5.3 ZnL$_2$ 9.6 ZnHL 12.4	ZnL 2.37 ZnL$_2$ 3.28	ZnL 0.43 ZnL$_2$ 0.45 ZnL$_3$ 0.5 ZnL$_4$ 0.2	ZnL 1.15	ZnL$_4$ 44.54	ZnHL 15.7	ZnL 16.44 ZnHL 9	ZnL$_2$ 11.07 ZnL$_3$ 16.05 ZnL$_4$ 16.72	ZnL 14.94 ZnL$_2$ 16.1

A.5

Table 8.7   The K_{s0} values of some solids of interest

Metal	Mineral Name	Formula	Log K_{s0}	Metal	Mineral Name	Formula	Log K_{s0}
Ag^+		$AgOH(s)$	-7.70	Cu^+	Nantokite	$CuCl(s)$	-6.76
		$Ag_2CO_3(s)$	-11.07	Fe^{2+}		$Fe(OH)_2(s)$	-15.90
		$Ag_3PO_4(s)$	-17.55		Siderite	$FeCO_3(s)$	-10.55
		$Ag_2S(s)$	-48.97		Vivianite	$Fe_3(PO_4)_2(s)$	-36.00
		$AgCl(s)$	-9.75			$FeS(s)$	-16.84
Al^{3+}		$Al(OH)_3(s)$	-31.62	Fe^{3+}	Ferrihydrite	$Fe(OH)_3(s)$	-37.11
	Gibbsite	$Al(OH)_3(s)$	-33.23		Goethite	$\alpha\text{-FeOOH}(s)$	-41.50
		$AlPO_4(s)$	-22.50		Lepidocrocite	$\gamma\text{-FeOOH}(s)$	-46.00
Ca^{2+}	Calcite	$CaCO_3(s)$	-8.48		Hematite	$Fe_2O_3(s)$	-40.63
	Aragonite	$CaCO_3(s)$	-8.36	Hg^{2+}		$Hg(OH)_2(s)$	-25.40
	Portlandite	$Ca(OH)_2(s)$	-5.32			$HgCO_3(s)$	-22.52
	Lime	$CaO(s)$	4.80		Cinnabar	$HgS(s)$	-52.01
	Gypsum	$CaSO_4$	-4.85				
	Hydroxylapatite	$Ca_5(OH)(PO_4)_3(s)$	-44.2	Ni^{2+}		$Ni(OH)_2(s)$	-17.20
Cd^{2+}		$Cd(OH)_2(s)$	-14.27			$NiCO_3(s)$	-6.84
	Otavite	$CdCO_3(s)$	-13.74			$Ni_3(PO_4)_2(s)$	-31.30
	Greenockite	$CdS(s)$	-28.85	Pb^{2+}	Massicot	$PbO(s)$	-15.09
		$Cd_3(PO_4)_2(s)$	-32.60		Hydrocerrusite	$Pb_3(CO_3)_2(OH)_2(s)$	-45.46
Co^{2+}		$Co(OH)_2(s)$	-15.90		Cerrusite	$PbCO_3(s)$	-13.13
		$CoCO_3(s)$	-12.80		Galena	$PbS(s)$	-28.05
Cr^{3+}		$Cr(OH)_3(s)$	-33.13			$Pb_3(PO_4)_2(s)$	-44.50
Cu^{2+}		$Cu(OH)_2(s)$	-19.36	Zn^{2+}		$\alpha\text{-Zn(OH)}_2(s)$	-15.55
	Tenorite	$CuO(s)$	-20.38			$ZnCO_3(H_2O)(s)$	-10.26
	Malachite	$Cu_2(OH)_2CO_3(s)$	-33.18			$Zn_3(PO_4)_2(s)$	-36.70
		$CuCO_3(s)$	-9.63			$ZnS(s)$	-21.97
		$Cu_3(PO_4)_2(H_2O)_3(s)$	-35.12				
	Covallite	$CuS(s)$	-35.96				

Table 9.3   Equilibrium constants for some environmentally important redox half-cell reactions*

Reaction	Log K	pe°	pe°(W)	$E_H°$, mV
$NO_3^- + 2e^- + 2H^+ \leftrightarrow NO_2^- + H_2O$	28.57	14.29	7.28	843
$NO_3^- + 8e^- + 10H^+ \leftrightarrow NH_4^+ + 3H_2O$	119.08	14.89	6.14	878
$NO_3^- + 3e^- + 4H^+ \leftrightarrow NO(g) + 2H_2O$	48.40	16.13	6.80	952
$2NO_3^- + 10e^- + 12H^+ \leftrightarrow N_2(g) + 6H_2O$	210.34	21.03	12.63	1241
$NO_2(g) + 2e^- + 2H^+ \leftrightarrow NO(g) + H_2O$	53.60	26.80	19.80	1581
$N_2O(g) + 2e^- + 2H^+ \leftrightarrow N_2(g) + H_2O$	59.79	29.89	22.89	1764
$SO_4^{2-} + 8e^- + 9H^+ \leftrightarrow HS^- + 4H_2O$	33.66	4.21	−3.67	248
$2H_2SO_3 + 4e^- + 2H^+ \leftrightarrow S_2O_3^{2-} + 3H_2O$	27.00	6.75	3.25	398
$SO_4^{2-} + 2e^- + 2H^+ \leftrightarrow SO_3^{2-} + H_2O$	27.16	13.58	6.58	801
$SeO_4^{2-} + 2e^- + 4H^+ \leftrightarrow H_2SeO_3 + H_2O$	36.32	18.16	4.16	1071
$H_3PO_4 + 2e^- + 2H^+ \leftrightarrow H_3PO_3 + H_2O$	−10.10	−5.05	−12.05	−298
$AsO_4^{3-} + 2e^- + 2H^+ \leftrightarrow AsO_3^{3-} + H_2O$	5.29	2.64	−4.36	156
$CrO_4^{2-} + 3e^- + 8H^+ \leftrightarrow Cr^{3+} + 4H_2O$	77.00	25.66	7.00	1514
$OCN^- + 2e^- + 2H^+ \leftrightarrow CN^- + H_2O$	−4.88	−2.44	−9.44	−144
$2H^+ + 2e^- \leftrightarrow H_2(g)$	0.00	0.00	−7.00	0
$2H^+ + 2e^- \leftrightarrow H_2(aq)$	3.10	1.55	−5.45	92
$O_2(g) + 4H^+ + 4e^- \leftrightarrow 2H_2O$	83.12	20.78	13.78	1226
$O_2(aq) + 4H^+ + 4e^- \leftrightarrow 2H_2O$	86.00	21.50	14.50	1268
$O_2(aq) + 2e^- + 2H^+ \leftrightarrow H_2O_2(aq)$	26.34	13.17	6.17	777
$H_2O_2(aq) + 2e^- + 2H^+ \leftrightarrow 2H_2O$	59.59	29.80	22.80	1758
$O_3(g) + 2e^- + 2H^+ \leftrightarrow O_2(g) + H_2O$	70.12	35.06	28.06	2069
$Cl_2(aq) + 2e^- \leftrightarrow 2Cl^-$	47.20	23.60	23.60	1392
$ClO_3^- + 6e^- + 6H^+ \leftrightarrow Cl^- + 3H_2O$	147.02	24.50	17.50	1446
$HOCl + 2e^- + H^+ \leftrightarrow Cl^- + H_2O$	50.20	25.10	21.60	1481
$ClO_2 + 5e^- + 4H^+ \leftrightarrow Cl^- + 2H_2O$	126.67	25.33	19.73	1495
$ClO_2^- + 4e^- + 4H^+ \leftrightarrow Cl^- + 2H_2O$	109.06	27.27	20.26	1609
$HOBr + 2e^- + H^+ \leftrightarrow Br^- + H_2O$	45.36	22.68	19.18	1338
$2HOBr + 2e^- + 2H^+ \leftrightarrow Br_2(aq) + 2H_2O$	53.60	26.80	20.27	1581
$Al^{3+} + 3e^- \leftrightarrow Al(s)$	−85.71	−28.57	−28.57	−1686
$Zn^{2+} + 2e^- \leftrightarrow Zn(s)$	−25.76	−12.88	−12.88	−760
$Ni^{2+} + 2e^- \leftrightarrow Ni(s)$	−7.98	−3.99	−3.99	−236
$Pb^{2+} + 2e^- \leftrightarrow Pb(s)$	−4.27	−2.13	−2.13	−126
$Cu^{2+} + e^- \leftrightarrow Cu^+$	2.72	2.72	2.72	160
$Cu^{2+} + 2e^- \leftrightarrow Cu(s)$	11.48	5.74	5.74	339
$Fe^{3+} + e^- \leftrightarrow Fe^{2+}$	13.03	13.03	13.03	769
$Hg_2^{2+} + 2e^- \leftrightarrow 2Hg(l)$	26.91	13.46	13.46	794
$Ag^+ + e^- \leftrightarrow Ag(s)$	13.51	13.51	13.51	797
$Pb^{4+} + 2e^- \leftrightarrow Pb^{2+}$	28.64	14.32	14.32	845
$2Hg^{2+} + 2e^- \leftrightarrow Hg_2^{2+}$	30.79	15.40	15.40	908
$MnO_2(s) + 2e^- + 4H^+ \leftrightarrow Mn^{2+} + 2H_2O$	41.60	20.80	6.80	1227
$Mn^{3+} + e^- \leftrightarrow Mn^{2+}$	25.51	25.51	25.51	1505
$MnO_4^- + 5e^- + 8H^+ \leftrightarrow Mn^{2+} + 4H_2O$	127.82	25.56	14.36	1508
$Co^{3+} + e^- \leftrightarrow Co^{2+}$	33.10	33.10	33.10	1953

*For reactions included in the Mineql+ database, values are from that database. Other values are computed from the Gibbs energy of reaction, based on the data in Appendix, or from various other reference sources.

B

LIST OF IMPORTANT EQUATIONS

The following list of useful equations and expressions is a compilation culled from the entire text. It is provided to help you locate a particular equation quickly or to check the exact form of an equation if you are uncertain of a detail. Correct interpretation and usage require careful attention to the context in which the equation is presented in the main body of the text.

$$P_i = \frac{n_i}{V_{tot}} RT = c_{g,i} RT \qquad (1.1)$$

$$V_i = \frac{n_i}{P_{tot}} RT \qquad (1.2)$$

$$I = \frac{1}{2} \sum_{\text{all ions}} c_i z_i^2 \qquad (1.6)$$

Table 1.4a   Various equations for predicting activity coefficients in aqueous solutions

Name and Equation	Notes and Approximate Range of Applicability
Debye–Huckel $$\log_{10}(\gamma_{\text{D-H}}) = -Az^2 I^{1/2}$$	$A = 1.82 \times 10^6 (\varepsilon T)^{-3/2}$, where ε is the dielectric constant of the medium. For water at 25°C, $A = 0.51$; $z =$ ionic charge; Applicable at $I < 0.005\ M$
Extended Debye–Huckel $$\log_{10}(\gamma_{\text{Ext.D-H}}) = -Az^2 \frac{I^{1/2}}{1 + BaI^{1/2}}$$	$a \equiv$ ion size parameter (see Table 1.4b) $B = 50.3(\varepsilon T)^{1/2}$; for water at 25°C, $B = 0.33$; Appropriate in solutions where one salt dominates ionic strength; Applicable at $I < 0.1\ M$
Davies $$\log_{10}(\gamma_{\text{Davies}}) = -Az^2 \left\{ \frac{I^{1/2}}{1 + I^{1/2}} - 0.2I \right\}$$	Applicable at $I < 0.5\ M$
Specific interaction models $$\log_{10}(\gamma_{\text{Pitzer}}) = \log_{10}(\gamma_{\text{Ext.D-H}}) + \sum_j B_{ij} I m_j$$	B_{ij} is specific interaction term between ions i and j; m_j is molality (mole/kg solution) of j; Applicable at $I < 1\ M$; Additional terms can extend range to higher ionic strengths[a]

[a]See Pitzer (*J. Solution Chem.* **4**, 249–265, 1975), Millero (*Geochim. Cosmochim. Acta* **47**, 2121–2129, 1983).

$$\gamma = \frac{a_A}{\dfrac{c_A|_{\text{real}}}{c_A|_{\text{std. state}}}} \tag{1.7d}$$

$$\ln \frac{K_{eq}|_{T_2}}{K_{eq}|_{T_1}} = \frac{\Delta \bar{H}_r}{R}\left(\frac{1}{T_1} - \frac{1}{T_2}\right) \tag{1.18}$$

$$S_{\text{tot}} = k_B \ln W \tag{2.7}$$

$$\bar{S}_{i,\text{therm}} = R \ln Q_i \tag{2.14a}$$

$$\bar{S}_{i,\text{config}} = R \ln \frac{1}{X_i} \tag{2.14b}$$

$$\overline{PE}_{\text{elec}} \equiv \Psi = \frac{\partial PE_{\text{elec}}}{\partial \sigma} \tag{2.22}$$

$$\Delta G \equiv \Delta H - T\,\Delta S \tag{2.38}$$

$$\mu_i \equiv \frac{\partial G_{\text{tot}}}{\partial n_i}\bigg|_{\text{constant } n_{j \neq i}, P, T} \tag{2.43}$$

$$\overline{\overline{PE}}_{i,\text{elec}}\left(\frac{\text{kJ}}{\text{mol of } i}\right) = \left\{z_i e N_A\left(\frac{C}{\text{mol of } i}\right)\right\}\left\{10^{-3}\frac{\text{kJ}}{\text{C-V}}\right\}\{\Psi_{\text{chem}}\,(V)\} \tag{2.49}$$

$$\bar{G}_i = \bar{G}_i^\circ + RT \ln a_i \tag{2.54b}$$

$$\overline{\overline{PE}}_{i,\text{ec}} = \bar{G}_i + z_i F \Psi_{\text{ext}} \tag{2.67}$$

$$\bar{G}_i = \bar{G}_i^\circ + \left(5.71\frac{\text{kJ}}{\text{mol}}\right) \log a_i \tag{2.73}$$

$$\Delta \bar{G}_r = \Delta \bar{G}_r^\circ + RT \ln Q \tag{2.80}$$

$$\Delta \bar{G}_r^\circ = -RT \ln K_{eq} \tag{2.86a}$$

$$K_{eq} = \exp\left(-\frac{\Delta \bar{G}_r^\circ}{RT}\right) \tag{2.86b}$$

$$\Delta \bar{G}_r = -\left(5.71\frac{\text{kJ}}{\text{mol}}\right) \log \frac{Q}{K_{eq}} \tag{2.87}$$

$\Delta \bar{G}_r < 0$: Reaction proceeds to form products **(Table 2.2)**

$\Delta \bar{G}_r > 0$: Reaction proceeds to form reactants **(Table 2.2)**

$\Delta \bar{G}_r = 0$: Reaction is at equilibrium **(Table 2.2)**

$$\frac{d}{dT} \ln K_{eq} \approx \frac{\Delta \bar{H}_{r,25°C}^\circ}{R}\frac{1}{T^2} \tag{2.106}$$

$$\ln \frac{K_{eq}|_{T_2}}{K_{eq}|_{T_1}} = \frac{\Delta \bar{H}_r^\circ}{R}\left(\frac{1}{T_1} - \frac{1}{T_2}\right) \qquad \textbf{(2.109)}$$

$$\ln \frac{K_{eq}|_{P_2}}{K_{eq}|_{P_1}} = -\frac{\Delta \bar{V}_r^\circ (P_2 - P_1)}{RT} \qquad \textbf{(2.110)}$$

$$K_a = \frac{\{H^+\}\{A^-\}}{\{HA\}} \qquad \textbf{(3.10)}$$

$$K_b = K_a^{-1} K_w \qquad \textbf{(3.15a)}$$

$$\alpha_0 = \frac{\{H^+\}}{K_a + \{H^+\}} = \frac{1}{\dfrac{K_a}{\{H^+\}} + 1} \qquad \textbf{(3.20)}$$

$$\alpha_1 = \frac{K_a}{K_a + \{H^+\}} = \frac{1}{1 + \dfrac{\{H^+\}}{K_a}} \qquad \textbf{(3.21)}$$

$$\alpha_i \equiv \frac{\{H^+\}^{-i} \displaystyle\prod_{j=0}^{i} K_{aj}}{\displaystyle\sum_{k=0}^{n}\left(\{H^+\}^{-k} \displaystyle\prod_{j=0}^{k} K_{aj}\right)} \quad (\text{where } K_{a0} \equiv 1.0) \qquad \textbf{(3.35)}$$

$$\beta = \frac{dc_{OH^- \text{ added}}}{d\text{pH}} = -\frac{dc_{H^+ \text{added}}}{d\text{pH}} \qquad \textbf{(5.12)}$$

$$\beta_w = 2.3(\{H^+\} + \{OH^-\}) \qquad \textbf{(5.20a)}$$

$$\beta_{HA} = 2.3(TOTA)\alpha_0\alpha_1 \qquad \textbf{(5.20b)}$$

$$\beta = 2.3(\{H^+\} + \{OH^-\} + (TOTA)\alpha_0\alpha_1) \qquad \textbf{(5.20c)}$$

$$\beta_{H_nA} = 2.3(TOTA) \sum_{j>i}^{n} \sum_{i=0}^{n} (j - i)^2 \alpha_i \alpha_j \qquad \textbf{(5.23)}$$

$$c_{G,i} = \frac{n_i}{V} = \frac{P_i}{RT} \qquad \textbf{(7.1)}$$

$$H = \frac{\{A(g)\}}{\{A(aq)\}}\bigg|_{eq} \qquad \textbf{(7.12)}$$

$$pe^\circ = \log\left[\frac{\{Red\}^{1/n_e}}{\{Ox\}^{1/n_e}} \frac{1}{\{e^-\}}\right] \qquad \textbf{(9.21b)}$$

$$E_H = E_H^\circ - \frac{2.303RT}{n_e F} \log \frac{\{Red\}}{\{Ox\}} \qquad \textbf{(9.26b)}$$

$$\frac{1}{n_e} \log \frac{\{Red\}}{\{Ox\}} = pe^{\circ} - pe - \frac{n_H}{n_e} pH \qquad \textbf{(9.28b)}$$

$$\frac{1}{n_e} \log \frac{\{Red\}}{\{Ox\}} = pe^{\circ}_{pH} - pe \qquad \textbf{(9.30)}$$

$$\Delta \bar{G}^{\circ}_r = -2.303 n_e RT \, \Delta pe^{\circ} \qquad \textbf{(9.33)}$$

$$\Delta \bar{G}^{\circ}_r = -(5.71 n_e \, \Delta pe^{\circ}) \frac{kJ}{mol} \qquad \textbf{(9.34)}$$

Table 9.5   Conversion equations among parameters that describe electron energy levels for a half-cell reaction energy changes accompanying an n-electron reduction reaction[a]

1	Energy of exchangeable electrons in a system	$\bar{G}_{e^-} = \bar{G}_{e^-}^{\circ} + RT \ln\{e^-\} = -2.303RT \, pe = -FE_H$
2	Nernst equation: energy of exchangeable electrons associated with a given half-cell reaction	$pe = pe^{\circ} - \dfrac{1}{n} \log \dfrac{\{Red\}}{\{Ox\}}$ $E_H = E_H^{\circ} - \dfrac{2.303RT}{n_e F} \log \dfrac{\{Red\}}{\{Ox\}}$
3	Gibbs energy of reaction in terms of energy of electrons associated with the two half-cell reactions[a]	$\Delta \bar{G}_r = n_e \Delta \bar{G}_{e^-}$ $\Delta \bar{G}_r = -2.303RT \, \Delta pe$ $\Delta \bar{G}_r = -n_e F \, \Delta E_H$
4	Gibbs energy of reaction in terms of extent of disequilibrium	$\Delta \bar{G}_r = \Delta \bar{G}_r^{\circ} + 2.303RT \log Q = 2.303RT \log \dfrac{Q}{K}$
5	Change in electron energy in a reaction in terms of extent of disequilibrium[a]	$\Delta \bar{G}_r = \Delta \bar{G}_r^{\circ} + 2.303RT \log Q$ $\Delta pe = \Delta pe^{\circ} - \dfrac{RT}{n_e} \log Q$ $\Delta E_H = \Delta E_H^{\circ} - \dfrac{2.303RT}{n_e F} \log Q$
6	Equilibrium constant in terms of energy change at standard state[a]	$\log K = \dfrac{\Delta \bar{G}_r^{\circ}}{2.303RT}$ $\log K = n_e \, \Delta pe^{\circ}$ $\log K = \dfrac{n_e F}{2.303RT} \, \Delta E_H^{\circ}$
7	Conditions for conversion of reactants to products[a]	$\Delta \bar{G}_r < 0$ $\Delta pe > 0$ $\Delta E_H > 0$
8	Conditions at equilibrium[a]	$\Delta \bar{G}_r = \Delta pe = \Delta E_H = 0$

[a] $\Delta pe = pe_{red} - pe_{ox}$, and pe_{red} and pe_{ox} are the pe values of the reduction and oxidation half-cell reactions, respectively. Similarly, $\Delta E_H = E_{H,red} - E_{H,ox}$.

$$q_A = \frac{K_{ads}\{A\}}{1 + K_{ads}\{A\}} q_{max} \tag{10.10}$$

$$q_i = \frac{K_{ads,i}a_i}{1 + \sum_{all\ j}(K_{ads,j}a_j)} q_{max} \tag{10.21}$$

$$q_{tot} = \int_0^\infty q'_{diff}\,dK = \int_0^\infty \Phi_{diff}\frac{K\{A\}}{K\{A\} + 1}\,dK \tag{10.33}$$

$$q_{tot} = k_f\{A\}^n \tag{10.35}$$

$$\sigma_d = -0.1174c_s^{0.5}\sinh\left(\frac{zF\Psi_d}{2RT}\right) \tag{10.46}$$

$$C_1 = \frac{\Psi_o - \Psi_\beta}{\sigma_o} \tag{10.47}$$

$$C_2 = \frac{\Psi_\beta - \Psi_d}{-\sigma_d} = \frac{\Psi_d - \Psi_\beta}{\sigma_d} \tag{10.48}$$

$$K_{int} = K_{app}\exp\left(\frac{\{(\Delta z_o)\Psi_o + (\Delta z_\beta)\Psi_\beta\}F}{RT}\right) \tag{10.63}$$

NOMENCLATURE

{ }	indicates chemical activity of species inside braces, dimensionless
[ ]	indicates concentration of species inside brackets, mol/L
a	ion size parameter for Debye-Huckel Equation, Å
a_i, $\{i\}$	chemical activity of species i, dimensionless
A	parameter in Debye-Huckel Equation; for water at 25°C, A = 0.5
ALK	alkalinity, (H$^+$ equiv)/L
B	parameter in Debye-Huckel Equation; for water at 25°C, B = 0.33
B_{ij}	specific interaction term in specific interaction model for activity coefficient, dimensionless
c_i	concentration of species i, mol/L
$c_{g,i}$	gas-phase concentration of i, mol/(L of gas)
C	in the Gibbs phase rule, the number of components used to prepare the system
C_1, C_2	capacitances of the compact layers of charge in the electrical double layer, F/m^2 (farads per m^2)
CB	charge balance
Comp	a *component,* as the term is used in chemical equilibrium software
e°	equilibrium constant for an oxidation reaction written for release of one electron, dimensionless
eq	value of a term at equilibrium (when used as a subscript)
E	energy, kJ
E*	activation energy in the collision model, kJ/mol
E_{Ar}	energy term in the Arrhenius rate expression, similar to an activation energy, kJ/mol
E_H	electrochemical potential, V
$E_H{}^\circ$	standard electrochemical potential, V
f	OH$^-$ added as a fraction of *TOTA*
F	in the Gibbs phase rule, the number of independent parameter values (degrees of freedom) that can be specified
F	Faraday constant, kJ/eq-V (or C/V)
g	gravitational constant, m/s^2
G	Gibbs energy, J (note: for $\Delta \bar{G}_i$, $\Delta \bar{G}_r$, and $\Delta \bar{G}_r^\circ$, see $\Delta \bar{X}_i$, $\Delta \bar{X}_r$, and $\Delta \bar{X}_r^\circ$, respectively)
h	Planck constant, J-s
h	height above datum level, m
H	enthalpy, J (note: for $\Delta \bar{H}_i$, $\Delta \bar{H}_r$, and $\Delta \bar{H}_r^\circ$, see $\Delta \bar{X}_i$, $\Delta \bar{X}_r$, and $\Delta \bar{X}_r^\circ$, respectively)
H	Henry's constant, dimensionless [although often given as a partitioning constant, with units of bar/(mol/L)]
I	ionic strength, mol/L
J_i	flux of species i, kg/m^2-s
k_{Ar}	frequency factor in the Arrhenius rate expression, dimensions same as those of the rate constant for the reaction under consideration
k_B	Boltzmann constant, J/K

k_f	pre-exponential constant in the Freundlich adsorption isotherm, dimensions depend on value of isotherm exponential constant n		pe_{pH}°	logarithm (base 10) of the conditional equilibrium constant corresponding to pe°, except incorporating the assumption that pH equals the value specified in the subscript, dimensionless
k_f, k_r	forward and reverse rate constants for chemical reactions (dimensions depend on reaction being considered and conventions for writing the rate expression)		$pe^\circ(W)$	logarithm (base 10) of the conditional equilibrium constant corresponding to pe°, except incorporating the assumption that pH = 7.0, dimensionless
^cK	conditional equilibrium constant, dimensionless		pX	negative base-10 logarithm of the value of X
K_{ads}	equilibrium constant for adsorption, dimensionless		P	in the Gibbs phase rule, the number of phases in the system under consideration
K_{ai}	acid dissociation constant, dimensionless		P_i	partial pressure of i, bar
K_{app}	apparent equilibrium constant for an adsorption reaction, dimensionless		PC	proton condition
K_b	basicity constant, dimensionless		PE	potential energy, J
K_{eq}	equilibrium constant, dimensionless		$\overline{PE}_{grav}$	gravitational potential energy per unit mass (gravitational potential), J/kg
K_i	equilibrium constant for formation of MeL_i from MeL_{i-1} plus one molecule of ligand L, dimensionless		$\overline{PE}_{elec}$	electrical potential energy per unit charge (electrical potential), J/C
$*K_i$	equilibrium constant for formation of MeL_i from MeL_{i-1} plus one molecule of protonated ligand, dimensionless		$\overline{PE}_{elec,i}$	electrical potential energy per mole of i, J/mol
K_{int}	intrinsic equilibrium constant for an adsorption reaction, dimensionless		$\overline{\overline{PE}}_{i,ec}$	electrochemical potential energy per mole of i, J/mol
K_w	equilibrium constant for dissociation of water, dimensionless		q	thermal energy transferred into a system of interest, kJ
LHS	value of the left-hand side of an equation		q'_{cum}	the cumulative site occupation function, mol/g or mol/m^2
m	mass, kg			
m_i	molality of species i, mol/(kg solution)		q'_{diff}	the differential site occupation function, mol/g or mol/m^2
MB	mass balance			
M	molarity, mol/L		q_i	adsorption density of species i, mol/g or mol/m^2
n	exponential constant in the Freundlich adsorption isotherm, dimensionless			
n_e, n_H	number of electrons or protons participating in a redox reaction		q_{max}	maximum possible adsorption density on a surface, mol/g or mol/m^2
n_i	number of moles of species i		Q	reaction quotient, dimensionless
N	normality, equiv/L		r_A	radius of molecule A
N_A	Avogadro's number, dimensionless		r_{d-c}	diffusion-controlled reaction rate, mol/L-s
pe°	logarithm (base 10) of the equilibrium constant for a reduction reaction written for consumption of one electron (equal to $-\log\{e^\circ\}$), dimensionless		R	universal gas constant, kJ/mol-K
			RHS	value of the right-hand side of an equation
			$\equiv S$	a surface site on an adsorbent

$\equiv S_2$	a bidentate surface site (consisting of two adjacent unoccupied $\equiv S$ sites)	α_i	fraction of *TOT*A present in the form indicated by i; if i is simply a number, then α_i represents the fraction of an acid H_nA present in the form $H_{n-i}A$, dimensionless
S	entropy, J/mol-K (note: for $\Delta \bar{S}_i$, $\Delta \bar{S}_r$, and $\Delta \bar{S}_r^\circ$, see $\Delta \bar{X}_i$, $\Delta \bar{X}_r$, and $\Delta \bar{X}_r^\circ$, respectively)		
S_{config}	configurational entropy, J/mol-K	β_i	equilibrium constant for formation of MeL_i from free aquo Me plus i molecules of ligand L, dimensionless
Spec	a *species,* as the term is used in chemical equilibrium software		
S_{therm}	thermal entropy, J/mol-K	$*\beta_i$	equilibrium constant for formation of MeL_i from free aquo Me plus i molecules of protonated ligand, dimensionless
SCE	standard calomel electrode		
SHE	standard hydrogen electrode	β_{H_nA}	buffer intensity contributed by acid/base group, $(H^+$ equiv)/(L-pH unit)
$\equiv SO$	a surface site on an oxide adsorbent		
T	absolute temperature, K	β_{tot}	total buffer intensity of a solution, $(H^+$ equiv)/(L-pH unit)
*TOT*A	total concentration of species A in the system, mol/(L of solution)		
$TOT A_{diss}$	total concentration of species A dissolved in the system, mol/L	β_w	buffer intensity contributed by water, $(H^+$ equiv)/(L-pH unit)
		γ_i	activity coefficient of species i, dimensionless
U	internal energy, J (note: for $\Delta \bar{U}_i$, $\Delta \bar{U}_r$, and $\Delta \bar{U}_r^\circ$, see $\Delta \bar{X}_i$, $\Delta \bar{X}_r$, and $\Delta \bar{X}_r^\circ$, respectively)	$\gamma_{i,chem}$	activity coefficient attributable to solvent and other solutes, dimensionless
V_i	partial volume of species i, L	$\gamma_{i,ec}$	activity coefficient considering solvent, other solutes, and electrical potentials arising outside the solution, dimensionless
W	number of microstates of a system yielding a given macrostate, dimensionless		
		ε	dielectric constant of the medium, dimensionless
X	mole fraction, dimensionless		
$\bar{X}_i$	molar value of X, where X can be G, H, S, or U; units correspond to those of X per mole of i	ε	extent of reaction, dimensionless
		$\Delta \eta$	number of moles of stoichiometric reaction occurring, mol
$\Delta \bar{X}_r$	change in total X in the system, per mole of stoichiometric reaction, where X can be G, H, S, or U; units correspond to those of X per mole of stoichiometric reaction	κ	Debye length ("thickness" of the diffuse layer near an adsorbent), nm
		μ	viscosity, kg/m-s
		μ_i	chemical potential of species i
		v_x	velocity of an object x, m/s
$\Delta \bar{X}_r^\circ$	value of $\Delta \bar{X}_r$ when all reactants and products are in their standard states; same units as $\Delta \bar{X}_r$	ν_i	stoichiometric coefficient of i in a reaction of interest, dimensionless
		π	Mathematical constant pi, dimensionless
$\bar{X}$	in the context of chemical equilibrium software, designates a dummy *component* with chemical formula X	σ	electrical charge, C
		σ_o	charge density in the surface (naught plane) of an adsorbent, C/m^2
y_i	mole fraction in the gas phase, dimensionless	σ_β	charge density in the beta plane of an adsorbent, C/m^2
z (z_i)	molar charge (of species i), mol charge/mol i	σ_d	charge density in the diffuse layer of an adsorbent, C/m^2

Φ_{cum} the cumulative surface site distribution function, mol/g or mol/m^2

Φ_{diff} the differential surface site distribution function, mol/g or mol/m^2

Ψ electrical potential energy per unit charge (electrical potential), J/C

Ψ_{chem} contribution to Ψ of a solute i arising from other solutes

Ψ_{ext} contribution to Ψ of a solute i originating outside the solution

Ψ_o electrical potential at the surface (naught plane) of an adsorbent, V

Ψ_{β} electrical potential in the beta plane of an adsorbent, V

Ψ_d electrical potential at the beginning of the diffuse layer of an adsorbent, V

INDEX